Industrial Applications of Microbial Secondary Metabolites

This comprehensive volume presents the fascinating and diverse world of microbial secondary metabolites, exploring their biosynthesis, ecological roles, and wide-ranging applications in biotechnology, pharmacy, agriculture, and various other industries. The initial chapters provide a foundational overview of microbial secondary metabolites and their immense value. The subsequent chapters discuss microbial strain engineering, enzymatic pathways, and metagenomic approaches to unlock hidden biosynthetic potential. Emphasis has also been given to how these natural products contribute to sustainable development, including eco-friendly agricultural practices and stress alleviation in crops through plant–microbe interactions. The book further explores source-specific contributions, with dedicated chapters on bacterial, fungal, algal, and marine microbial metabolites. Innovative applications are also highlighted in various chapters with recent advancements in technologies such as microbial mycoherbicides for weed control, viral-derived metabolites, and the promise of marine microbial communities. Special attention has also been given to sustainable exploitation and the integration of these compounds into functional products and solutions for global challenges.

This book serves as a valuable reference for researchers and students of microbiology, biotechnology, and environmental sciences.

The editors would welcome the fruitful suggestions from the readers to be incorporated into future editions.

Industrial Applications of Microbial Secondary Metabolites

A Modern Approach

Edited by
Neeraj Verma, Deepak Mishra,
Ayodhya Prasad Pandey and Rajbeer Singh Gaur

CRC Press
Taylor & Francis Group
Boca Raton London New York

CRC Press is an imprint of the
Taylor & Francis Group, an **informa** business

Designed cover image: Neeraj Verma

First edition published 2026
by CRC Press
2385 NW Executive Center Drive, Suite 320, Boca Raton FL 33431

and by CRC Press
4 Park Square, Milton Park, Abingdon, Oxon, OX14 4RN

CRC Press is an imprint of Taylor & Francis Group, LLC

ISBN: 978-1-032-51549-6 (hbk)
ISBN: 978-1-032-52869-4 (pbk)
ISBN: 978-1-003-40883-3 (ebk)

DOI: 10.1201/9781003408833

Typeset in Times
by Apex CoVantage, LLC

Contents

Preface

Scientists and businesspeople have long recognized the value of bacterial secondary compounds. These complex, physiologically active molecules are produced by many microorganisms and are revolutionizing medicine, farming, biotechnology, and environmental science. Secondary metabolites are signaling molecules, defense compounds, or microbial ecosystem competitors. Primary metabolites are essential for cell growth and function. Their environmental relevance and application in industry processes, where they enable progress and new ideas, make them important.

Industrial Applications of Microbial Secondary Metabolites: A Modern Approach reviews the theory and practice of microbial secondary metabolism in detail. It teaches researchers, businesspeople, and students how bacteria create high-value secondary chemicals and how they are being employed in novel ways. The book begins with an outline of microbial secondary metabolites and the concept. It does this by discussing history and progress. After that, we'll discuss microbial strains as cellular factories, showing how microorganisms can naturally and artificially manufacture industry-relevant chemicals.

This book focuses on microorganism secondary metabolites and their sources. This theme highlights the biochemical potential of bacteria, fungi, algae, viruses, mycorrhizae, and sea microorganisms. Our detailed analysis of each microbe shows how their particular compounds work and where they dwell in their environment to generate them. Biomolecular microbial metabolism and secondary metabolite production paths are thoroughly covered in the book to assist readers in understanding biosynthesis. It emphasizes complicated genetic and enzymatic regulation. Also covered are enzymes that speed up novel or sophisticated metabolic changes and help make secondary metabolites. The book concludes with chapters on marine, mycorrhizal, rhizosphere, bacterial, fungal, and viral microorganisms. Each microbe's compounds, biosynthetic processes, and commercial uses are examined in these chapters.

Top microbial biotechnology and secondary metabolism experts selected the material. This ensures it is academically sound and understandable to a wide spectrum of scientists. As global issues like antibiotic resistance, environmental harm, and the demand for long-lasting technologies worsen, microbial secondary metabolites have great promise. This book should spur new research, encourage cross-disciplinary collaboration, and lead to more widespread adoption of microbial advances in business and other domains.

The editors will be pleased to receive comments on the style and content of this volume for inclusion in future editions.

About the Editors

Neeraj Verma, Ph.D. (botany), with specialization in plant pathology, has been working as a professor, Department of Agriculture Science, Faculty of Agriculture Science and Technology, AKS University, Satna, India, since 2012. He is currently the Head of the Department of Agriculture Science, Faculty of Agriculture Science and Technology, AKS University, Satna, MP, India. He is also Director, research. Earlier, he was an assistant specialist (plant virology), University of California–Riverside, Riverside, California, USA. He has over 25 years of research and teaching experience and has published more than 60 international and national research papers, several book chapters, and ten books. He has worked on various research projects funded by government agencies and attended several national and international conferences. He is an editor and reviewer of professional journals.

Deepak Mishra is a professor of biotechnology in the Department of Biotechnology, Faculty of Life Science, AKS University, Satna. With a career spanning over 17 years, he qualified for GATE in 2011 and completed his PhD in biotech in 2014. His teaching and research interest are in the field of plant biotechnology, molecular biology, biochemistry, microbiology, and mushroom technology. He has authored 45 research papers in prestigious national, as well as international, journals and 30 book chapters. He is also actively involved in collaborative research initiatives. Six students have been awarded a doctorate in the field of biotechnology under his guidance, and eight students are doing their PhD on various aspects of biotechnology under his supervision. He has attended several international and national conferences and seminars as an invited speaker.

Ayodhya Prasad Pandey is working as an assistant professor in the Department of Genetics and Plant Breeding, Faculty of Agricultural Science and Technology, AKS University, Satna. He has more than 12 years of experience in teaching both undergraduate and postgraduate students and conducting research. Mr. Pandey has written over 50 book chapters and more than 25 research papers published in well-known national and international journals. He has also written two books in the field of genetics and plant breeding. He has taken part in several national and international workshops, conferences, and faculty development programs and has guided more than ten postgraduate students in their research work and successfully led several research projects. He is also a member of various professional societies.

Rajbeer Singh Gaur is an assistant professor, Department of Genetics and Plant Breeding, Faculty of Agriculture Science and Technology, AKS University, in Satna. With more than 14 years of experience, he has taught both undergraduate and postgraduate students. Mr. Gaur has written over 40 book chapters and published more than 20 research papers in reputed national and international journals. He has also written two books on genetics and plant breeding. He has attended national and international conferences, workshops, and training programs and has mentored more than 11 postgraduate students with their research.

1 Microbial World
A Treasure of Secondary Metabolites

*Seema Singh
(Forest Pathology Discipline, Forest Protection Division, Forest Research Institute, Dehradun, India; seema07jan@gmail.com)

Ambika
(Forest Pathology Discipline, Forest Protection Division, Forest Research Institute, Dehradun, India; ambikabudakoti@gmail.com)

Charul Kainthola
(Forest Pathology Discipline, Forest Protection Division, Forest Research Institute, Dehradun, India; tinni0509@gmail.com)

Nidhi Gupta
(Forest Pathology Discipline, Forest Protection Division, Forest Research Institute, Dehradun, India; nidhi88gupta5@gmail.com)

Abhishek Kumar
(Forest Pathology Discipline, Forest Protection Division, Forest Research Institute, Dehradun, India; abhishek259kumar@gmail.com)

*Amit Pandey
(Institute of Forest Productivity Ranchi, Jharkhand, India; amiticfre@gmail.com)

*Corresponding Authors

1.1 INTRODUCTION

The microbial world is a mysterious land of natural products. Microorganisms' incredible capability of producing a myriad of natural products or metabolites reflects their evolutionary adaptation to the changing environment. In addition to producing primary metabolites, microbes also produce secondary metabolites, or idiolites in idiophase, which are required for their interaction with the environment and other living organisms. Secondary metabolites are not essential for the growth of microorganisms but may act as signaling molecules or protective/defensive compounds during communication with other organisms and unfavorable environmental conditions (Pandey et al., 2021). Based on structure, functions, and biosynthesis, secondary metabolites are classified into five main classes: terpenoids and steroids, fatty acid–derived substances and polyketides, alkaloids, non-ribosomal polypeptides, and enzyme cofactors (Thirumurugan et al., 2018). Secondary metabolites are synthesized by seven different metabolic pathways, which include the peptide pathway, polyketide synthase pathway, non-ribosomal polypeptide synthase pathway, hybrid pathway, shikimate pathway, beta-lactam synthetic pathway, and carbohydrate pathway (Ullah et al., 2020). Microbial secondary metabolites are antimicrobial compounds, toxins, competition weapons, symbiosis compounds, communication agents, pigments, nutrient scavengers, *etc*. (Fouillaud and Dufossé, 2022).

Microorganisms are an important source of secondary metabolites. Numerous microorganisms from different habitats have been reported as producers of novel secondary metabolites. Yadav et al. (2019) reviewed microbial biodiversity with secondary metabolites–producing capabilities and inferred that novel secondary metabolites were produced by 41% bacteria, 15% archaea, and 44% eukarya

DOI: 10.1201/9781003408833-1

(1% dinoflagellates, 36% fungi, and 7% algae). These microbes belonged to diverse phyla, *viz.*, Actinobacteria (23%), Ascomycota (31%), Bacillariophyta (1%), Bacteroidetes (1%), Basidiomycota (1%), Chlorophyta (3%), Crenarchaeota (1%), Cyanobacteria (2%), Dinoflagellata (1%), Euryarchaeota (12%), Eustigmatophyceae (1%), Firmicutes (8%), Haptophyta (1%), Mucoromycota (1%), Proteobacteria (9%), and Rhodophyta (1%). The most dominant secondary metabolite–producing microorganisms are *Streptomyces*, *Acremonium*, *Aspergillus*, *Halococcus*, *Fusarium*, *Bacillus*, *Pestalotiopsis*, *Penicillium*, *Pseudomonas*, *Halobacterium*, *Haloarcula*, and *Amycolatopsis*.

Microbes and their metabolites have remarkably contributed to the welfare of human beings since ancient times (Gupta et al., 2022, 2024). They have been an integral part of the agriculture, medical, nutraceutical, pharmaceutical, and industrial sectors. Secondary metabolites have been used as enzyme inhibitors, antimicrobial agents, antioxidants, cholesterol-lowering agents, herbicides, surfactants, antitumor agents, insecticides, immunosuppressants, colorants, antiparasitic agents, plant growth promoters, antiaging agents, *etc.* (Vaishnav and Demain, 2011; Jain et al., 2023, 2024).

The present chapter reviews a brief history of SMs, followed by an account of secondary metabolites produced by various microorganisms, such as bacteria (general and with special reference to actinomycetes), fungi, algae, and archaea.

1.2 HISTORY OF MICROBIAL SECONDARY METABOLITES

Long before the discoveries of advanced sciences, secondary metabolites have been known to mankind. They were being used in different forms, for example, dyes, perfumes, spices, medicines, poisons, pharmacological agents, cosmetics, *etc.*, and were widely exploited. Most of the secondary metabolites that have been used for centuries had been obtained from plants, but microorganisms like fungi, bacteria, actinomycetes, or algae have also been known to secrete certain products that have toxic, hallucinogenic, and aphrodisiacal effects (Bennett and Bentley, 1989).

For ages, microorganisms have been contributing to the production of beneficial products serving various purposes. A long history is known about the manufacturing of gallic acid by use of *Aspergillus* sp. to hydrolyze plant tannins in the 1700s (Thorn and Raper, 1945). Mannitol, a widely utilized carbohydrate, was also reported to have been isolated from higher fungi in the year 1813 (Birkinshaw et al., 1931). Certain complex fungal metabolites (kojic acid, mycophenolic acid, penicillic acid, *etc.*) had already been discovered in the early twentieth century. In 1922, Harold Raistrick, a chemist, and his colleagues studied systematics of fungal metabolites by growing pure fungal cultures on synthetic media at Ardeer Factory of Nobel's Explosives Co. Ltd. at Stevenson (Scotland) and continued their research even after 1929, at the London School of Hygiene and Tropical Medicine, where they isolated more than 200 metabolites from mold fungi (Raistrick, 1949). The importance of metabolite production by mold fungi was less appreciated by microbiologists as fungi were considered contaminants of pure bacterial cultures, but the discovery of penicillin changed the paradigm. Alexander Fleming, in 1929, discovered the antibacterial activity of culture filtrates of *Penicillium notatum* and named the mold juice "penicillin" (Fleming, 2001). The discovery of penicillin in 1929 was historical, and it was used as an effective, inexpensive antibiotic during World War II. It led to the beginning of the antibiotic era, which declared the golden age of antibiotics, and encouraged further discoveries and investigations to be done on microbial metabolites.

In *Biochemistry of Fungi*, Tatum (1944) studied the processes by which metabolic products are formed from mold fungi. He also researched primary and secondary intermediates. He later shared the Nobel Prize for the insight that the inherent capacities for the formation of secondary metabolites are genetically controlled. Foster (1949), in his classic work *Chemical Activities of Fungi*, emphasized that products of primary condensation can participate in "secondary condensations." Tatum and Foster must be credited for stating that natural products have been derived from primary intermediates of normal metabolism; however, it was Bu'Lock (1965) who coined the term "secondary metabolite" and observed that antibiotics are not only the agents acting upon other organisms but are organic products themselves. Gaden (1959), before Bu'Lock, categorized all microbial products taxonomically, based on fermentation processes, into three classes. Class I metabolites are concentrated as a result of primary metabolism (*e.g.*, lactate and ethanol), Class II metabolites are produced by side reactions of direct metabolic processes (*e.g.*, itaconate and citrate), and Class III metabolites are accumulated by cells independently (*e.g.*, penicillin).

Research and commercialization of antibiotics flourished in the 1940s with the commercial production of penicillin. These antibiotics were most commonly brought into use to treat infections of wounded soldiers during World War II. Considering other microbes in the production of secondary metabolites, there were certain misconceptions. Few strains of the bacterium *Serratia marcescens* produced prodigiosin, a red pigment which, for ages, had been mistaken for fresh blood and regarded as a food contaminant. The product was accounted as superstitious by higher Church authorities during the Middle Ages. Another filamentous fungus, *Claviceps purpurea*, which was found to be the causal organism of ergot of rye disease, contaminated rye grains with ergot alkaloids, which, on ingestion, caused gangrene among people. That furious disease, which was associated with convulsions and hallucinations, led to an epidemic throughout medieval Europe and was termed "St. Anthony's fire" (Gaughran, 1969).

After the golden era of antibiotic discovery, in the 1940s and the 1950s, nearly all the major antibacterial antibiotics had been discovered. In the late 1950s and 1960s, antibacterial as well as antifungal compounds were mostly recovered from *Streptomyces* (70–80%). Also, this period was the beginning of the discovery of metabolites with other antitumor, antiviral, and non-antibiotic (enzyme inhibition) activities. The next period of the 1970s and 1980s marked a decrease in research but an increase in the number of new compounds that were mainly analogues of known compounds. Also, along with the dominant actinomycete metabolites (65–70%), the rare actinomycete products increased, and the chemical structure of the discovered compounds was determined. Since the 1990s, an exponential increase in new metabolites has continued, but only a few new chemical types have been seen. The challenges of multidrug resistance and enhanced recurrence of human pathogens (*e.g.*, HIV, mycobacteria, *etc.*) led to the adaptation of new techniques for screening metabolites. There was an increase in compounds with other biological activities, in fungal metabolites (up to 50%), and chemical synthesis of complicated structures (Bérdy, 2005).

1.3 MICROORGANISMS: PRODUCERS OF SECONDARY METABOLITES

Approximately 32,000–34,000 bioactive compounds and 25,000–27,000 antibiotics (anti-infective, antitumor, and antiviral compounds) have been known to be produced by microorganisms (Bérdy, 2012).

1.3.1 Bacteria

Bacteria are a rich source of natural products, especially secondary metabolites. Various studies have reported the production of bioactive secondary metabolites from bacterial species belonging to different bacterial phyla, namely, Actinobacteria, Bacteroidetes, Cyanobacteria, Firmicutes, and Proteobacteria (α, β, and γ) (Yadav et al., 2019). Till 2010, unicellular bacteria produced approximately 7,000 bioactive metabolites and ~5,000 antibiotic compounds, and actinobacteria produced ~12,000 bioactive metabolites and ~10,000 antibiotic compounds (Bérdy, 2012). The presence of bacteria in different habitats, such as in soil, water bodies, and extreme environments; on the surface of living and non-living matter, and inside other living beings, represents their metabolic diversity. Secondary metabolite–producing bacteria present in varied environments and actinomycetes are addressed in the following sections.

1.3.1.1 Rhizospheric Bacteria

These are present near plant roots and referred to as plant growth–promoting rhizobacteria (PGPR). As the name suggests, PGPR promotes plant growth and induces plant systemic resistance and tolerance via direct and indirect mechanisms. These mechanisms involve production of secondary metabolites, such as bioactive compounds (terpenoids, alkaloids, phenolics, and flavonoids), siderophores, hydrogen cyanide, organic acids, phosphatases, phytohormones (indole acetic acid, gibberellins, and abscisic acid), enzymes (protease, pectinase, cellulase, lipase, ACC deaminase, *etc.*), anticancer compounds, antibiotics, *etc.* (Pandey et al., 2021). Some examples of PGPR are *Gluconacetobacter azotocaptans*, *Pseudomonas putida*, and *Azospirillum lipoferum* (Mehnaz and Lazarovits, 2006); *Azotobacter chroococcum* (Kumar et al., 2014a); *Bacillus* spp. and *Pseudomonas* spp. (Baliah and Begum, 2015); *Pseudomonas chlororaphis* and *P. aurantiaca* (Shahid et al., 2017); *etc.*

1.3.1.2 Plant Endophytic Bacteria

Bacteria that live inside a plant host are known as endophytes and share beneficial relationships with their host. They are metabolically active to counter the plant defense system. Secondary metabolites produced by endophytes are antibiotics (lipopeptides, amino acid–rich peptides, cyclic cationic lipopeptides, and pigments), anticancer compounds (cyclic analogs, maytansinoids, and extracellular metabolites), antiviral compounds (flavonoids, saponins, and nanoparticles), and other compounds, such as terpenoids and alkaloids (Narayanan and Glick, 2022). Some examples of endophytes reported for producing secondary metabolites are *Pseudomonas viridiflava* EB273 (Miller et al., 1998), *Micromonospora lupini* (Igarashi et al., 2007), *Bacillus amyloliquefaciens* (Chen et al., 2013), *Actinomyces* sp. (Xu et al., 2014), *Aeromicrobium pontii* (Gos et al., 2017), *Burkholderia* sp. (Singh and Dubey, 2018), *Streptomyces cavourensis* YBQ59 (Vu et al., 2018), *Kytococcus schroeter* (Ghiasvand et al., 2020), *etc.*

1.3.1.3 Marine Bacteria

Marine habitats are a treasure house of microbial secondary metabolites. Different studies have reported bioactive secondary metabolites from marine bacteria belonging to diverse habitats, such as sediment, water, flora, and fauna. According to the literature survey between the year 1996 and 2020 by Srinivasan et al. (2021), the number of articles published were 834 (21%), 53 (1.3%), 148 (3.9%), 1,214 (31.8%), 3 (1%), and 1,565 (41%) related to antimicrobial activities of diverse bacterial phyla, *viz.*, Actinobacteria, Bacteroidetes, Cyanobacteria, Firmicutes, Planctomycetes, and Proteobacteria, respectively. It was estimated that about 23,000 bioactive secondary metabolites have been reported from marine microbes. Marine secondary metabolites possess various biological activities, like antimicrobial, antioxidant, antilarvicidal, anticancer, anti-inflammatory, enzyme inhibitory, *etc.* (Williams, 2009; Imhoff et al., 2011; Wang et al., 2020; Nawaz et al., 2021; Karthikeyan et al., 2022). Few examples of marine bacteria producing secondary metabolites are *Agrobacterium* (Mitova et al.,

2004); *Pseudoalteromonas* sp. (Shnit-Orland et al., 2012); *Streptomyces*, *Bacillus*, *Vibrio*, and *Pseudoalteromonas* (Raimundo et al., 2018); *Vibrio* sp. (Bibi et al., 2020); *Streptomyces* sp., *Salinispora tropica*, *Salinispora arenicola*, *Marinispora* (Petersen et al., 2020); *Maribacter*, *Aquimarina*, *Vagococcus*, and *Denitrobaculum* (Rodriguez et al., 2021); *etc.*

1.3.1.4 Extremophilic Bacteria

They are found in environments with harsh abiotic conditions, like high and low temperatures, anoxia, aridity, high salinity, acidity, alkalinity, high pressure, and intense irradiation. These microbes are repositories of important secondary metabolites. Sayed et al. (2020) reviewed secondary metabolites from microbes inhabiting extreme environments. The review reported that about 421 new secondary metabolites were identified from extremophilic microorganisms till the year 2018. These metabolites showed 25.6% anticancer, 9.6% antifungal, 33.6% antibacterial, 9.6% anti-inflammatory, 8% antiviral, and 13.6% other biological activities. Few examples of extremophilic bacteria producing secondary metabolites are *Streptomyces* sp. (Nachtigall et al., 2011), *Thermus filiformis* (Mandelli et al., 2012), *Bacillus* sp. (Son et al., 2016), *etc.*

Reports on bioactive secondary metabolites produced by various bacterial species isolated from diverse habitats are given in Table 1.1.

TABLE 1.1
Production of Secondary Metabolites by Different Bacterial Strains

Bacteria	Source	Secondary Metabolite	Bioactive Property	Reference
Bacillus subtilis I164	-	Bacillomycin F	Antifungal	Mhammedi et al., 1982
Bacillus subtilis A114	Freshwater lake	Iturin A_L	Antifungal	Winkelmann et al., 1983
Bacillus subtilis strain F-29–3	-	Fengycin	Antifungal	Vanittanakom et al., 1986
Bacillus coagulans I_4	Cattle feces	Coagulin	Bactericidal and bacteriolytic activity	Le Marrec et al., 2000
Vogesella indigofera	Water filter cartridge	Indigoidine (blue pigment)	Detection of heavy metal	Gu and Cheung, 2001
Bacillus subtilis 168	-	Bacilysocin	Antimicrobial	Tamehiro et al., 2002
Bacillus subtilis BBK-1	Salty conditions	Bacillomycin L, plipastatin, and surfactin	Biosurfactant	Roongsawang et al., 2002
Bacillus cereus 8A	Woodland soil	Cerein	Antibacterial	Bizani and Brandelli, 2002; Bizani et al., 2005
Pseudomonas putida BW11M1	Banana rhizosphere	Putidacin	Antibacterial	Parret et al., 2003
Bacillus subtilis GA1	Strawberry fruits	Lipopeptides	Antifungal	Touré et al., 2004
Pseudomonas tolaasii NCPPB2192	-	Tolaasin D	Antibacterial and antifungal	Bassarello et al., 2004
Bacillus subtilis strain M4	Soil	Fengycin-type lipopeptides	Antifungal	Ongena et al., 2005
Bacillus thuringiensis subsp. *kurstaki*strain BUPM4	Tunisian soil litter	Bacthurucin F4	Antibacterial	Kamoun et al., 2005
Bacillus thuringiensis	-	Thuricin S	Antibacterial	Chehimi et al., 2007
Brevibacillus laterosporus PNG276	Tissue of a tube worm collected off the coast of Loloata Island (Papua New Guinea)	Tauramamide (**6**), together with its methyl (**7**) and ethyl esters	Antibacterial	Desjardine et al., 2007
Bacillus subtilis B49	Tea rhizosphere	Bacillomycin D and fengycin	Antifungal	Ramarathnam et al., 2007
Hahella chejuensis KCTC 2396	Marine sediment	Prodiginines	Antibacterial	Kim et al., 2007
Pseudomonas stutzeri CMG1030	*Desmodema* spp. (Ribbonfish) intestinal tract	Zafrin	Antibacterial	Uzair et al., 2008
Pseudomonas fluorescens Pf-5	Soil	Rhizoxin analogs	Antifungal and antitumor	Loper et al., 2008
Bacillus subtilis 14B	Bitter almond rhizosphere	Bac 14B (bacteriocin)	Antibacterial	Hammami et al., 2009
Pseudomonas syringae pv. *syringae* B728a	-	Achromobactin	Iron scavenger	Berti and Thomas, 2009
Bacillus thuringiensis DPC 6431	Human feces	Thuricin CD	Antimicrobial	Rea et al., 2010
Bacillus subtilis SSE3	Shrimp shell waste	Subtulene A and iturin A	Antimicrobial	Thasana et al., 2010

TABLE 1.1 (Continued)
Production of Secondary Metabolites by Different Bacterial Strains

Bacteria	Source	Secondary Metabolite	Bioactive Property	Reference
Streptomyces sp.	Hyper-arid soil collected from the Atacama Desert	Atacamycins A–C	Antitumor	Nachtigall et al., 2011
Pedobacter	Samples collected from Doumer Island, Antarctica	Carotenoids	Antioxidant	Correa-Llantén et al., 2012
Thermus filiformis	-	Carotenoids	Antioxidant	Mandelli et al., 2012
Pseudomonas aurantiaca PBSt2	Sugarcane stems	Lahorenoic acid B (2),2,8-Dihydroxyphenazine (7), and phenazine-1-carboxylic acid (4)	2, 4, and 7 showing moderate antibacterial activity; 4 also exhibiting cytotoxic and antifungal properties	Mehnaz et al., 2013
Bacillus clausii GM17	*Ononis angustissima* rhizosphere	Bac GM17 (bacteriocin)	Antibacterial and antifungal	Mouloud et al., 2013
Pseudomonas aeruginosa strain PA14	Clinical isolate	Phenazine-1-carboxylic acid	Nematicidal	Cezairliyan et al., 2013
Bacillus amyloliquefaciens K103	Lemon	Bacillomycin L	Antifungal	Zhang et al., 2013
Collimonas fungivorans Ter 331 and *C. pratensis* Ter91	Soil	Volatile organic compounds	Antifungal	Garbeva et al., 2014
Pseudomonas stutzeri ST1302	Water resource (northeastern Thailand)	Pyrrolnitrin	Antifungal	Thongsri et al., 2014
Collimonas pratensis strain Ter91	Marram grass rhizosphere in sandy dune soils	Four monoterpenes (γ-terpinene, 1S-α-pinene, β-pinene, and β-myrcene)	Antibacterial	Song et al., 2015
Pseudomonas putida PUW5	*Oryza sativa* (rice) rhizosphere	5-methyl-phenazine-1-carboxyl acid	Antimicrobial and anticancer	Kennedy et al., 2015
Pseudomonas chlororaphis PA23	Soybean root tip	Pyrrolnitrin and HCN	Nematicidal	Nandi et al., 2015
Chromobacterium sp. NIIST (MTCC 5522)	Clay mine acidic sediment	Violacein	Antifungal	Sasidharan et al., 2015
Pseudomonas aeruginosa R_1 (KM924432) and *P. aeruginosa* U3 (KM603511)	Rice-cultivated soil and clinical specimen (urinary tract infection)	Pyocyanin	Antimicrobial	El-Fouly et al., 2015
Pseudomonas chlororaphis GP72	Green pepper rhizosphere	Phenazine-1-carboxylic acid and 2-hydroxyphenazine	Antifungal	Liu et al., 2007
Pseudomonas sp. CMR5c	Red cocoyam rhizosphere	Orfamide A	Antifungal	Ma et al., 2016
Hymenobacter sp. and *Chryseobacterium* sp.	Soil samples (Antarctica)	Carotenoid	Photo-sensitizers in dye-sensitized solar cells	Órdenes Aenishanslins et al., 2016
Bacillus subtilis MTCC10407 (JF834075)	Associated with brown seaweed *Sargassum myriocystum*	O-heterocyclic compounds	Antibacterial	Chakraborty et al., 2017a
Bacillus subtilis fmb60	Fermented compost of crop stalks	Aurantinins B, C, and D	Antibacterial	Yang et al., 2016
Bacillus sp. KCB14S006	Saltern (Incheon, Korea)	Iturin F1, iturin F2, iturin A9 (1–3), and iturin A8 (4),	Antifungal and cytotoxic activity	Son et al., 2016
Bacillus subtilis MTCC 10403	Associated with brown seaweed *Anthophycus longifolius*	Polyketide furanoterpenoids	Antibacterial	Chakraborty et al., 2017b
Bacillus amyloliquefaciens	Floor swabs	Bacocin	Antibacterial and antifungal	Brock et al., 2018
Co-culture of *Serratia* sp. and *Shewanella* sp.	-	Serratiochelin A	Antiproliferative and antibacterial	Schneider et al., 2020

1.3.1.5 Actinomycetes

Actinomycetes are the most explored and exploited bacterial group, which represents the largest group of bioactive microbial metabolites among all microorganisms.

Actinomycetes are filamentous, Gram-positive bacteria and a natural source of antibiotics, which are held responsible for the production of approximately 70% of the antibiotics used in the treatment of ailments (Bérdy, 2005). Morphologically, actinomycetes look like fungi because of their filament-like structure but are distinguished based on size, as actinomycete hyphae are much smaller than fungal hyphae. They can utilize a great variety of substrates found in soil, for example, chitin, cellulose, or hemicelluloses. The compounds produced by actinomycetes are commonly known as natural products, secondary or specialized metabolites.

Actinomycetes, also known as actinobacteria, can produce a wide array of microbial active compounds, and that makes them one of the most explored microorganisms among other prokaryotes. The secondary metabolic products produced by them are widely known for their role in several biological, physiological, and cellular processes. Actinomycetes are widely distributed among natural ecosystem habitats and are a source of secondary metabolites, for example, antibiotics, biopesticides, plant growth promoters, pharmacological compounds, pigments, enzymes, antitumor compounds, anti-inflammatory compounds, single-cell proteins, biosurfactants, *etc.* (Selim et al., 2021).

1.3.1.5.1 Distribution in Various Ecosystems

Capable of surviving in different types of habitats, actinomycetes are widely distributed in natural ecosystems. They are characterized by the formation of substrate and aerial mycelia on solid medium and have high GC DNA content (Goodfellow and Williams, 1983). Actinomycetes are soil-inhabiting organisms. Till 1964, only 10% of the actinomycetes were isolated from soil (Kuster and Williams, 1964). There is a need to screen undiscovered and capable actinobacteria that can prove effective in the production of antibiotics against resistant microorganisms. Several actinomycetes are found in rhizospheric soil. Rhizospheric streptomycetes protect plants by inhibiting fungal growth in roots, thus producing antifungal antibiotics (Gonzalez-Franco et al., 2009). Some plants fix nitrogen in their root nodules in combination with actinomycetes (*e.g.*, *Frankia* sp.). A novel antibiotic, "calcimycin," has been reported to be produced by *Frankia* sp. (Udwary et al., 2011). Certain actinomycetes isolated from hypersaline soils, like *Streptomyces alboflavus*, *Nocardia* sp., *Micromonospora* sp., *Streptomyces griseoflavus*, *etc.*, had been reported to show antifungal and antibacterial activities against *Fusarium solani*, *Aspergillus niger*, and *Cryptococcus* sp. (Karuppiah and Mustaffa, 2013; Mohan and Rajamanickam, 2018). Actinomycetes are widely present in the sediments of freshwater rivers (*e.g.*, *Streptomyces* and *Micromonospora*) and show antifungal activities against yeasts and molds (Rifaat, 2003). Many actinobacteria are found to be producing antibiotics with different chemical structures. Around 83 species of actinomycetes belonging to 28 genera had been reported to be isolated from marine habitats (Prave et al., 1987). There are possibilities of deep-sea studies on the diversity of marine actinomycetes. There are other reports of about 64 secondary metabolites and enzymes produced by marine streptomycetes which demonstrated high antimicrobial and anticancer activities (Mikami, 2007; Kumar et al., 2014b). The digestive tracts of insects harbor various symbiotic microbes that produce novel bioactive products. *Nocardiopsis* sp. has been reported to be isolated from the gut of a bee which expressed an antibiotic biosynthetic gene against *Bacillus* strains inside the bee gut and two other Gram-positive human pathogens (Kristufek et al., 1993). It was reported by Tan et al. (2009) that in the feces of goat, two dominant actinomycetes, *Oerskovia* and *Nocardiopsis*, produced a larger number of antifungal products. Other antibiotics, like monesin and flavomycin, are found to be produced by *Streptomyces* sp., which have growth-promoting activities. Many endophytic actinomycetes (*e.g.*, *Nocardia globerula*) were reported to produce bioactive compounds, which were used as antifungal agents against *Helminthosporium solani* silver scuff of potato. Another antibiotic, ansamycin, which is also produced by *Nocardia* sp., has been found to show antitumor activity (Snipes et al., 2007).

1.3.1.5.2 Secondary Metabolites

The order Actinomycetales is solely responsible for the production of more than 10,000 antimicrobial agents, which have diverse uses in the manufacturing of antibiotics, biopesticides, plant growth promoters, pharmacological compounds, pigments, enzymes, antitumor and anti-inflammatory compounds, single-cell proteins, biosurfactants, *etc.*

1.3.1.5.2.1 Antibiotics The most important secondary metabolite product to date has to be antibiotics. These are chemotherapeutic agents produced by microorganisms that can both inhibit the growth of harmful microbes and kill them. These are particularly defined as low-molecular-weight organic compounds produced by microorganisms. Maximum numbers of antibiotics are known to be produced by actinomycetes, with an extensive range and various types of activities. They have shown high intensity against several Gram-positive and Gram-negative bacteria. The highest number of antibiotics falls under *Streptomycetes* compared to bacteria and fungi. The class alone accounts for about 45% of the known antibiotics (Baltz, 2009). Some of the examples of antibiotics produced from *Streptomyces* spp. are cycloserine (*S. orchidaceous*), vancomycin (*S. orientalis*), neomycin and actinomycin (*S. fradiae*), amphotericin (*S. nodosus*), nystatin (*S. noursei*), rifampin (*S. mediterranei*), streptomycin (*S. griseus*), kanamycin (*S.

kanamyceticus), tobramycin (*S. tenebrarius*), spectinomycin (*S. spectabilis*), tetracycline (*S. vifidifaciens*), lincomycin and clindamycin (*S. lincolnensis*), oxytetracycline (*S. rimosus*), erythromycin (*S. erythraeus*), chloramphenicol (*S. vensuella*), chlortetracycline, dimethylchlor (*S. aureofaciens*), spiramycin (*S. ambofaciens*), avermycin (*S. avermitilis*), puromycin (*S. alboniger*), novabicin (*S. niveus*), platenmycin (*S. platensis*), daptomycin (*S. roseosporus*), ribostamycin (*S. ribosidificus*), cycloserine (*S. garyphalus*), viomycin (*S. vinaceus*), and cephalosporin (*S. clavuligerus*) (Hasani et al., 2014).

Other bioactive compounds, "polyketides," are an important natural product due to their varied uses in pharmaceuticals, for example, erythromycin, nystatin, and avermectin, having antibacterial, antifungal, and antiparasitic properties, respectively.

1.3.1.5.2.2 Biopesticides Some microorganisms or their by-products are generally biopesticidal agents that are used to inhibit or kill insects. Some actinomycetes have been reported to produce insecticidal compounds used to biologically control housefly, *Musca domestica*, and mosquito, *Culex quinquefasciatus* (Sharma et al., 2014).

1.3.1.5.2.3 Plant Growth–Promoting Hormones Many secondary metabolites produced by actinobacteria are being used as plant growth promoters, for example, auxins and gibberellins. Indole-3-acetic acid, an auxin, is mainly used for cell division, elongation, and differentiation in plants (Bloemberg and Lugtenberg, 2001).

1.3.1.5.2.4 Antitumor and Antiviral Compounds Certain actinomycetes secrete distinctive cytotoxic compounds that have antitumor properties against different malignant cell lines. Abyssomicin, actinofuranones A–B, metacycloprodigiosin, benzanthraquinone, butenolides, mechercharmycin, diphosphatidylglycerol, *etc.* are some examples of bioactive compounds derived from actinomycetes which have shown antibacterial, antitumor, and antiviral activities (Kumar et al., 2014b). Another product, "mitomycin," produced from *Streptomyces lavendulae*, has been reported to have aminobenzoquinone and aziridine rings in its functional groups and is generally utilized in the treatment of tumors (Sheldon et al., 1999; Mao et al., 1999). In another report by Angelova et al. (2006), *Streptomyces chromofuscus* produced PISC-2002 (protease inhibitor) from its culture supernatants, which played an important role as an antiviral agent against the influenza virus (A/Rostock/34 H7N7). Another species of *Streptomyces* secreted pimprinine, an extracellular alkaloid, which demonstrated significant antimicrobial, anticonvulsant, and antiviral activities against Enterovirus 71 (EV71) (Wei et al., 2014).

1.3.1.5.2.5 Pigments Pigments of varied colors are produced by actinomycetes that grow on natural and synthetic media, and the mycelia diffuse into it. These pigments are released by the bioactive natural compounds produced by actinomycetes (Wawrik et al., 2007). *Streptomyces hygroscopicus* is a pigment-producing actinomycete that releases antibiotic pigment 4-hydroxynitrobenzene and shows antibacterial activity against various drug-resistant pathogens *Bacillus subtilis* and *Shigella shiga* (Moore et al., 1999). Another pigment, "actinorhodin," produced by *S. coelicolor*, *S. violaceoruber*, and *S. lividans*, has applications in the food industry in making beverages, desserts, *etc.*

1.3.1.5.2.6 Enzymes *Streptomyces* produces important enzymes, for example, amylases, proteases, cellulases, *etc.*, which are exploited commercially in small and large-scale industries. Other enzymes, like L-glutaminase, L-asparaginase, and α-galactosidase, produced by marine Streptomycetes, also showed antitumor activities (Sahu et al., 2007).

1.3.1.5.2.7 Anti-Inflammatory Compounds 13527 Certain actinomycetes (*Micromonospora* sp.) are reported to produce anti-inflammatory compounds which have antimicrobial activities (Jayaprakashvel, 2012). In another report by Barbie and Kazmaier (2016), *Streptomyces arenicola* produced cyclomarin A and C, which showed antituberculosis and antimalarial activities.

1.3.1.5.2.8 Biosurfactants Actinomyces are known to have an effective role in the production of biosurfactants or bioemulsifiers. *Nocardia* sp. and *Streptomyces griseoflavus* produce trehalose dimycolates and many other biosurfactants which show biodegradability, less toxicity, and the ability to work under extreme conditions of temperature, pH, and salinity (Thampayak et al., 2008).

Some reports on secondary metabolites produced by actinomycetes are given in Table 1.2.

1.3.2 Fungi

In nature, fungi face multiple biotic and abiotic stress or factors, *viz.*, other more competent microorganisms, nutrient deprivation, pH, and temperature. They produce a vast number of SMs as a physiological response. Fungi can release their secondary metabolites from cells and excrete them into the environment, incorporate them into structural elements of the cell, or else, they can remain cell-bound. The term "extrolites" has been conferred as an alternative descriptor for SMs based primarily on the fact that their biological functions are assumed to be outwardly directed (Frisvad et al., 2007).

In the microbial world, fungi stand out as the most bounteous producers of bioactive compounds, with ~47% of the total. Fungal genomes analysis revealed that the class Ascomycetes has more genes of secondary metabolism as compared to Basidiomycetes, archeo-ascomycetes, and chytridiomycetes, whereas hemiascomycetes and zygomycetes

TABLE 1.2
Secondary Metabolites from Actinomycetes

Actinomycetes	Source	Secondary Metabolite	Bioactive Property	Reference
Streptomyces purpurogeneiscleroticus	Sea mud	Peri-hydroxyquinone derivative compound (antibiotic SS-228Y)	Antibacterial	Okazaki et al., 1975
S. albidoflavus	Corn seeds	Albaflavenone	Antibacterial	Gurtler et al., 1994
Streptomyces clavuligerus, *S. venezuelae*, and *S. kanamyceticus*	–	Cephamycin, chloramphenicol, and kanamycin	Antibacterial	Barrios-González et al., 2003
S. diastachromogenes and *S. nodosus*		Oligomycin and amphotericin	Antifungal	
S. antibioticus, *S. parvulus*, and *S. peucetius*		Actinomycin D and doxorubicin	Antitumor	
S. clavuligerus		Clavulanic acid	Enzyme inhibitor	
S. cinnamonensis and *S. fradiae*		Monensin and tylosin	Growth promoter	
S. *hygroscopicus*		Rapamycin	Immunosuppresive	
S. avermitilis and *S. hygroscopicus*		Avermectin and milbemycin	Insecticide and antiparasitic	
Marinispora sp. (NPS008920)	Lagoon sediment sample	2-alkylidene-5-alkyl-4-oxazolidinones, lipoxazolidinone A (9), B (10), and C (11)	Antimicrobial	Macherla et al., 2007
Micromonospora lupine 08	*Lupinus angustifolius* root nodules	Anthraquinones-Lupinacidins A (1) and B (2)	Antitumor	Igarashi et al., 2007
S. sannurensis CNQ-418	Marine sediment sample	Marinopyrroles A and B	Antibacterial activity against methicillin-resistant *Staphylococcus aureus* (MRSA)	Hughes et al., 2008
Marinispora sp. (NPS12745)	Mission Bay sediment samples	Lynamicins A-E (1–5)	Antimicrobial	McArthur et al., 2008
Nocardia sp. ALAA 2000	*Laurenica spectabilis* (marine red alga)	Ayamycin (4)	Antimicrobial	El-Gendy et al., 2008
Streptomyces strain AQBWWS1	Marine sponge *Callyspongia diffusa*	Carotenoid	Food-grade pigment	Dharmaraj et al., 2009
S. spinoverrucosus	Sand samples	Dibohemamines B–C	Anticancer	Fu et al., 2016
S. glaucescens NEAE-H	Soil samples (Egypt and Saudi Arabia)	Melanin	Anticancer and antioxidant agent	El-Naggar and El-Ewasy, 2017
Streptomyces sp. GG23	*Tenebrio molitor* (mealworm beetle) gut	Pentaminomycins C and D (1 and 2)	Autophagy-inducing activity	Hwang et al., 2020
Streptomyces sp. IB201691–2A	*Benedictia baicalensis* (endemic mollusk from Lake Baikal)	Baikalomycins	Moderate anticancer and antibacterial activities	Voitsekhovskaia et al., 2020

have a very low count (Collemare et al., 2008). *Microbes* are eco-friendly in use and have potential applications in medicine, agriculture, pharmaceutical, and other industries (Verma et al., 2017; Kour et al., 2020).

1.3.2.1 Secondary Metabolites in Industrial Applications

Microbes have been widely used in industrial biotechnology to manufacture various metabolites, such as butanol, ethanol, lactic acid, and riboflavin, in addition to biotransforming several chemicals to reduce pollution by means such as wastewater management, mycoremediation, bioremediation, and composting (Rana et al., 2019). Since fungi such as *Aspergillus*, *Penicillium*, and *Rhizopus* are tailored in absorptive mode to take up nutrition from host cells and produce extracellular enzymes, such as amylase, cellulases, chitinases, lipases, and proteases, there is a large class of secondary metabolites that originate from fungi, called terpenoids, which are being exploited by aromatic and flavoring industries. Many terpenoids are biologically active and have therapeutic potential against a variety of human diseases (Singh and Sharma, 2015).

Several industrial applications rely on the fungus enzyme laccase, including biosensors, dyes, phenols, and pesticides (Senthivelan et al., 2016). Fungal species *Periconia* sp. produces a thermotolerant secondary metabolite called β-glucosidase, which has high activity toward cellobiose and carboxymethylcellulose and also hydrolyzes rice straw into plain sugar. Moreover, this enzyme can convert lignocellulosic organic matter into biofuels (Wu et al., 2017).

1.3.2.1.1 Application in Pigment Industry

Some pre-eminent genera of fungi used for producing colors are *Aspergillus*, *Penicillium*, *Epicoccum*, *Lecanicillium*, and *Fusarium* species (Kumar et al., 2019). *P. aculeatum* produces pigments which are being used in soft drinks because of their stability at neutral pH (Mapari et al., 2009). Some other fungal species, *viz.*, *Aspergillus*, *Fusarium, Colletotrichum capsici*, *Hypoxylon fendleri*, and *Trichoderma*, have been used to produce anthraquinone pigment, which is used in dyestuff industries (Durán et al., 2002; Kusari et al., 2012; Ambika and Pandey, 2020).

1.3.2.1.2 Application in the Pharma and Medical Sector

Biomedicinal metabolites derived from fungi are well-known for their broad spectrum of pharmacological activities. Secondary metabolites produced by fungi have antioxidant, antimicrobial, and antitumor properties (Mishra et al., 2017; Uzma et al., 2018). Secondary metabolites can provide antifungal compounds like caspofungin (Macheleidt et al., 2016), antibacterial agents like penicillin, anticancer compounds like taxol (Yuan et al., 2006), lovastatin for lowering cholesterol, and immunosuppressant drugs like cyclosporine (Borel et al., 1994; Li and Vederas, 2009). Some phenolic metabolites of fungus species show a broad physiological spectrum, such as antiallergic, anti-inflammatory, antimicrobial, antiatherogenic, antithrombotic, cardioprotective, and vasodilatory activities (Balasundram et al., 2006; Pan et al., 2017). Another class of metabolites called alkaloids is secreted by *Aspergillus*, *Penicillium*, *Pestalotiopsis*, *etc*., having antimicrobial, anti-HIV, and antiparasitic activities (Devi et al., 2020). Some polyketides and flavonoids reported from *Exserohilum rostratum* and *Alternaria alternate* showed antagonistic properties toward phytopathogens (Pinheiro et al., 2017). Moreover, the extracts and fruiting bodies of fungi were found to be effective and economically feasible because of the fast-growing fruit bodies and mycelium. Some of the fungi that have pharmaceutical properties are *Pleurotus ostreatus*, *Agaricus brasiliensis*, *Ganoderma lucidum*, *Grifolafrondosa*, and *Termitomyces* and also reported as a source of vitamins, especially pyridoxine, flavin, and niacin; organic acids like ascorbate, fumarate, malate, and shikimate; carbohydrates like β-glucans; monoterpenoid and diterpenoid; proteins such as hydrophobins; and trace substances such as selenium (Chu et al., 2002; Ma et al., 2018). A few essential bioactive compounds found in fungi are being used in chemotherapy drugs or as components of drug formulations, which have brought out positive results. Some important reports of secondary metabolites produced from fungi/mushrooms are given in Table 1.3.

Over the past few years, many studies have been undertaken to characterize the biosynthetic pathways of fungal secondary metabolites. A search on NCBI (PubMed.gov) publications related to the keyword "fungal secondary metabolites" showed a significant increase in numbers from 1990 to 2022 (Figure 1.1). Research efforts in this field have led to the discovery of novel compounds, gene clusters, and key enzymes and have been greatly supported by the recent releases of fungal genome sequences (Collemare et al., 2008). Secondary metabolites have been rising in number over time. A wide range of secondary metabolites is being produced by slight variations in biosynthetic pathways that can have various applications in the field of pharmaceuticals, food, and agriculture.

1.3.2.2 Bioinformatics Approach for the Study of Fungal Secondary Metabolites

The sequencing and analysis of fungal genomics have become considerably simpler as a result of the advantages of bioinformatics techniques. For example, through the STITCH database, researchers can target a specific group of proteins that co-express and are crucial to fungi and their metabolism. Moreover, with the help of a bioinformatics tool, similar metabolites could be traced from different protein database sources, and interactions among them would be used for investigating the future potential.

The environment has a plethora of fungi, which are the explored and unexplored reservoirs of secondary metabolites. Secondary metabolites have plenty of applications in the field of pharmaceuticals, food, medicine, agriculture, and textiles as they exhibit antifungal, antibacterial, and anticancer activities and equally act as biocontrol agents against crop-deteriorating insects, pests, and pathogens. Some fungal species are also reported to be a potent source for natural pigments and colorants, having applications in the food and textile industries. Protocols for large-scale production could be optimized and validated for implementation in dye industries. Investigation of secondary metabolites from fungi has been rising, but that research is not being translated to the level of product development and patenting as well. It is also evident from studies and data available on online sources that product development is low from the identified secondary metabolites.

1.3.3 Algae

Algae are primitive unicellular or multicellular eukaryotic organisms that are primary producers, which means they take in solar energy and convert it into chemical energy for the production of organic compounds like sugars. The process of photosynthesis depends heavily on chlorophyll. These pigments facilitate the transfer of energy between the photosystems I and II reaction centers and the light source. Algal species can thrive and even flourish under several

TABLE 1.3
Secondary Metabolites from Fungi/Mushrooms

Fungus	Metabolites	Application	Reference
Pleurotuseryngii	Linoleic acid and oleic acid	Antibiotic, antitumor, and antiviral	Elisabeth et al., 2017
Lentinulaedodes	Ergosterol, lentionine, 2,4,6-tetrathiepane, dimethyldisulfide, demethylincisterol A3, lovastatin, and eritadenine	Antimicrobial, antitumor, antidiabetic, antiviral, and immunomodulator	Takahashi et al., 2020; Fukushima-Sakuno, 2020
Schizophyllum commune	Carotenoids, indole derivatives, flavonoids, and saponins	Antimicrobial, antitumor, antiviral, and anti-inflammatory	Tel-Çayan, 2019
Gloeophyllum trabeum	Sesquiterpenoids and (3Z,6Z,8E)-dodecatrien-1-ol	Antioxidant, anticholinergic, and esterase effect	Singh et al., 2021; Umezawa et al., 2020
Ganoderma lucidum	Polysaccharides (D-glucans, galactose, mannose, arabinose, rhamnose, and fructose) and triterpenoids	Antitumor, immunomodulatory, and antiangiogenic	Baby et al., 2015
Biatriospora sp.	Ascomycone B and 6-deoxyfusarubin	Cytotoxicity	Stodůlková et al., 2015
Fusarium fujikuroi	Apicidin F	Antimalarial	Von Bargen et al., 2013
Phaeolus schweinitzii	Hispidin	Antioxidant	Han et al., 2013
Pestalotiopsis sp.	Pestalotiopsone A	Antibacterial	Hemberger et al., 2013
Aspergillus sp.	Gliotoxin	Cytotoxic against human cancer cell lines HeLa	Nguyen et al., 2013
Botryosphaeria dothidea	Stemphyperylenol	Antifungal and cytotoxic against HCT116 cancer cell line	Xiao et al., 2014
Penicillium sp.	Verrucosidin	Antimycobacterial activity	Bu et al., 2015
Trichoderma spp.	T22 azaphilone, T39 butenolide, harzianolide, and harzianopyridone	Antagonistic activity against *Rhizoctonia solani*, *Pythium ultimum*, and *Gaeumannomyces graminis* in *in vitro* conditions	Vinale et al., 2006
Xylaria	Terpenoids, cytochalasins, mellein, alkaloids, and polyketides	Herbicide, biopesticide, and biocontrol	Macías-Rubalcava and Sánchez-Fernández, 2017
Monascus purpureus	Monascorubramine and rubropunctamine	Red color dye for textile	De Santis et al., 2005
Chlorociboria aeruginosa, *Scytalidium cuboideum*, and *S. ganodermophthorum*	Quinones	Green, red, and yellow colors for dying cotton, polyamide, polyester, and wool	Weber et al., 2014

challenging environmental conditions, including pH, temperature, osmotic pressure, salinity, exposure to ultraviolet light, and anaerobiosis (Shalaby et al., 2010). They are a substantial source of fatty acids, vitamins, minerals, and many important secondary metabolites.

1.3.3.1 Types of Algae

Algae are divided into two major categories, viz., microalgae and macroalgae.

1.3.3.1.1 Microalgae

Microalgae are a class of autotrophic microorganisms that thrive in freshwater, marine, and soil habitats and synthesize organic compounds by photosynthesis. Microalgae include diatoms (Bacillariophyta), dinoflagellates (Dinophyta), green and yellow-brown flagellates (Chlorophyta, Prasinophyta, Prymnesiophyta, Cryptophyta, and Rhaphidiophyta), as well as blue-green algae (Cyanophyta). These organisms play an important role in the productivity of oceans and are found at the base of the food chain. Microalgae are considered valuable bioresources due to their potential to accumulate proteins, carbohydrates, and other high-value metabolites inside the cell based on their ecological niche (Bonachela et al., 2011).

Microalgae can flourish in a range of habitats, such as terrestrial and aquatic habitats, or even in symbiosis with other organisms (Pandey et al., 2013). To thrive in these challenging conditions (light intensity, temperature, pH, and salinity), cyanobacteria produce a wide range of biologically active secondary metabolites, including carotenoids, essential fatty acids, polysaccharides, phycobiliproteins, amino acids, minerals, and several photoprotective compounds, like scytonemins and mycosporine-like amino acids (MAAs) (Kolanjinathan et al., 2014).

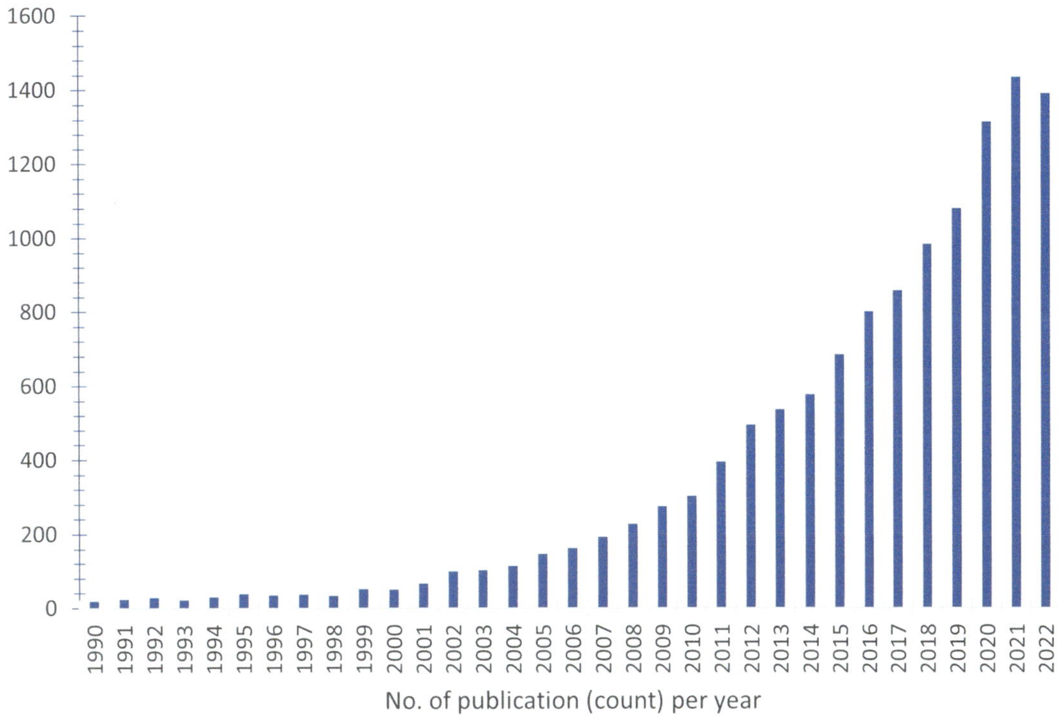

FIGURE 1.1 Number of reports on secondary metabolites from fungi, 1990–2022.

1.3.3.1.2 Macroalgae

The majority of macroalgae (seaweeds) are marine, multicellular, photosynthetic, eukaryotic, chlorophyll "a"–containing organisms that can be found from the intertidal zone to 300 m below the surface. Based on the photosynthetic pigment concentration, carbohydrate food reserve, cell wall constituents, and flagellar shape and orientation, more than 10,000 known macroalgal species are grouped into categories. Based on pigmentation and shape, the algae are grouped as Chlorophyta (green algae), Phaeophyta (brown algae), and Rhodophyta (red algae) (Rasmussen and Morrissey, 2007; Gupta and Abu-Ghannam, 2011).

1.3.3.2 Secondary Metabolites

Various species of algae produce bioactive and commercially valuable secondary metabolites, such as α- and β-carotene, phenolic or polyphenolic compounds, phycobiliproteins, and astaxanthin. Some of the highly valuable bioactive products isolated from algae are discussed in what follows.

1.3.3.2.1 Carotenoids

Carotenoids are essential for oxygenic photosynthesis; they also act as structural components that help stabilize protein folding in the photosynthetic system or as auxiliary pigments for capturing light. These fat-soluble accessory pigments trap light energy during photosynthesis. Due to their antioxidative properties, carotenoids shield the organism against high radiation and UV-induced photo-oxidation (Wertz et al., 2005; Joshi et al., 2018). Carotenoids have a wide range of applications, including those for cosmetic and pharmacological uses, feed supplements, nutraceuticals, and food coloring. The major carotenoids that are found in algae are beta-carotene, lutein, violaxanthin, neoxanthin, and zeaxanthin in green algae (Chlorophytes); alpha- and beta-carotene, lutein, and zeaxanthin in red seaweeds (Rhodophytes); and beta-carotene, violaxanthin, and fucoxanthin in brown algae (Phaeophytes) (Maeda et al., 2005).

1.3.3.2.2 Astaxanthin

Another important carotenoid pigment is astaxanthin. Astaxanthin is a distinct carotenoid produced by microalgae and has industrial uses. It is a keto-carotenoid that is mostly generated from the microalgae *Haematococcus pluvialis.* The microalga produces 4–5% astaxanthin per dry biomass (Sathasivam et al., 2017).

1.3.3.2.3 Phenolic Compounds

Phenolic compounds act as antioxidants by chelating metal ions, reducing free radicals. A few of the numerous compounds that make up the family of substances known as polyphenols include lignins, tocopherols, tannins, and phenolic acids. Phlorotannins are oligomers of phloroglucinol and are restricted to brown marine algae (Ragan and Karl-Werner, 1986).

1.3.3.2.4 Flavonoids

In algae, flavonoids are mostly found as glycosides and aglycones. Flavonoids are classified into two categories mainly based on the position of the benzenoid substituent (Zuiter, 2014):

- 2-phenylchromans (flavonoids), which contain flavanones, flavonols, flavones, flavan-3-ols, and anthocyanidins
- 3-phenylchromans (isoflavonoids), **which** contain isoflavones, isoflavans, and pterocarpans

1.3.3.2.5 Phlorotannins

Phlorotannins are important phenolic compounds that are mostly found in marine brown algae. These compounds include fucols, phlorethols, fucophlorethols, fuhalols, and halogenated and sulfated phlorotannins. Phlorotannin protects against oxidative degradation. The phenolic chemicals rutin, hesperidin, morin, caffeic acid, catechol, catechin, and epigallocatechin gallate have also been discovered in red macroalgae (Ragan and Glombitza, 1986).

1.3.3.2.6 Alkaloids

Alkaloids stand out as being of major importance in the development of novel drugs because they possess many of the pharmacological properties connected with the numerous complex chemical structures common in marine sources. There are a few known seaweed alkaloids, including caulerpin, a bisindole alkaloid with anti-inflammatory, anticancer, and growth-regulating effects (De Souza et al., 2009).

1.3.3.2.7 Sterols and Fatty Acids

The sterols found in plants are called phytosterols. Sterols are produced abundantly in a variety of microalgae species, and they make a considerable contribution to the production of phytosterols (Piironen et al., 2003). Microalgal sterols have beneficial impacts on health, including the prevention of Parkinson's disease, cancer, inflammation, and hypocholesterolemia. The three main sterol types found are stigmasterol, campesterol, and sitosterol (Abidi, 2001; Lagarda et al., 2006). Microalgae, including *Porphyridium cruentum*, *Arthrospira platensis*, *Odontella* sp., and *Isochrysis galbana*, produce useful fatty acids. Eicosapentaenoic acid (EPA) and docosahexaenoic acid (DHA) are two essential omega-3 polyunsaturated fatty acids used for treating inflammatory disorders, heart conditions, arthritis, asthma, headaches, and other conditions. A list of algae producing secondary metabolites is given in Table 1.4.

TABLE 1.4
Secondary Metabolites from Different Algal Species and Their Usage

Algae	Secondary Metabolite	Activity of Bioactive Compound/Uses	References
Microalgae			
Chlorella ellipsoidea	Violaxanthin	Anti-inflammatory and anticancer	Talero et al., 2015
Chlorella pyrenoidosa and *Chlorella protothecoides*	Lutein	AMD, atherosclerosis, and retinal neural damage	Talero et al., 2015
Chlorella vulgaris	Astaxanthin	Antitumoral	Talero et al., 2015
Chlorella zofingiensis	Astaxanthin	Antitumoral	Del Campo et al., 2004
	Lutein	AMD, atherosclerosis, and retinal neural damage	
Chlorococcum sp.	Astaxanthin	Antitumoral, antioxidant, and anti-inflammatory	Ma and Chen, 2001
Dunaliella bardawil	β-carotene	Antioxidant, provitamin A, AMD, liver fibrosis, and anti-inflammatory	Rabbani et al., 1998
Dunaliella salina	β-carotene	Antioxidant, provitamin A, AMD, liver fibrosis, and anti-inflammatory	Lamers et al. 2010
	Zeaxanthin	Antioxidant, maculopathy, cataracts, and anti-inflammatory	Fu et al., 2013; Bhosale and Bernstein, 2005
	Lutein	AMD, atherosclerosis, and retinal neural damage	Talero et al., 2015
Dunaliella sp.	Glutathione	Anticoagulant, antioxidant properties, anticancer activity, anti-Parkinson's disease, and detoxifying heavy metals	Li et al., 2004
Dunaliella tertiolecta	Violaxanthin	Anti-inflammatory and anticancer	Talero et al., 2015
	Lutein	AMD, atherosclerosis, and retinal neural damage	Barbosaa et al., 2023
	β-carotene	Antioxidant, provitamin A, AMD, liver fibrosis, and anti-inflammatory	Lamers et al., 2010;

TABLE 1.4 (Continued)
Secondary Metabolites from Different Algal Species and Their Usage

Algae	Secondary Metabolite	Activity of Bioactive Compound/Uses	References
Eustigmatos cf. *polyphem*	β-carotene	Antioxidant, provitamin A, AMD, liver fibrosis, and anti-inflammatory	Li et al., 2012
Haematococcus pluvialis	Astaxanthin	Antitumoral, antioxidant, and anti-inflammatory	Markou and Nerantzis, 2013
	Lutein	AMD, atherosclerosis, and retinal neural damage	Plaza et al., 2009
	Zeaxanthin	Antioxidant, maculopathy, cataracts, and anti-inflammatory	Singh and Dhar, 2011
	Canthaxanthin	Antioxidant property	
	β-carotene	Antioxidant, provitamin A, AMD, liver fibrosis, and anti-inflammatory	
	Oleic acid	Prevention of autoimmune diseases and anti-inflammatory	
Nostoc sp.	Cryptophycin	Antitumor	
Porphyridium cruentum	Zeaxanthin	Antioxidant, maculopathy, cataracts, and anti-inflammatory	Talero et al., 2015
Porphyridium sp.	Polysaccharides	Food, cosmetics, fabrics, stabilizers, and emulsifiers	Tannin-Spitz et al., 2005
Scenedesmus quadricauda	Polysaccharides	Food, cosmetics, fabrics, stabilizers, and emulsifiers	Mohamed, 2008
Spirulina platensis	Phenolic acids	Antimicrobial properties and antioxidant activity	Ibañez and Cifuentes, 2013
	Linolenic acid	Preventive against cardiovascular diseases	Plaza et al., 2009
	Palmitoleic acid	Antithrombotic activity	Singh and Dhar, 2011
Spirulina sp.	Polysaccharides	Food, cosmetics, fabrics, stabilizers, and emulsifiers	Amaro et al., 2013
Green Algae			
Ankistrodesmus angustus	Astaxanthin	Antitumoral	Borowitzka, 1988
Botryococcus braunii	Lutein	AMD, atherosclerosis, and retinal neural damage	Rao et al., 2006
	β-carotene	Antioxidant, provitamin A, AMD, liver fibrosis, and anti-inflammatory	Borowitzka, 1988
Chlamydobotrys stellata	Lutein	AMD, atherosclerosis, and retinal neural damage	Leya et al., 2009
	Canthaxanthin	Antioxidant property	
Chlamydocapsa spp.	Canthaxanthin	Antioxidant property	
	Astaxanthin	Antitumoral	
Chlamydomonas nivalis	Astaxanthin	Antitumoral	Remias et al., 2009
	Phenolic antioxidants	Antimicrobial properties and antioxidant activity	Duval et al., 1999
Chlamydomonas moewusii	Linoleic acid	Preventive against cardiovascular diseases	Arisz et al., 2000
	Hexadecatetraenoic acid	Anti-inflammatory	
Chlamydomonas acidophila	Lutein	AMD, atherosclerosis, and retinal neural damage	Garbayo et al., 2008
	β-carotene	Antioxidant, provitamin A, AMD, liver fibrosis, and anti-inflammatory	
Chlorella ellipsoidea	Violaxanthin, antherax anthin, and zeaxanthin	Protection of eye cells and antioxidant activity	Cha et al., 2008
Chlorella emersonii	Canthaxanthin	Antioxidant property	Bhosale and Bernstein, 2005
Chlorella fusca	Lutein	AMD, atherosclerosis, and retinal neural damage	Del Campo et al., 2000; Borowitzka, 1988
	Astaxanthin	Antitumoral	
	Canthaxanthin	Antioxidant property	

(Continued)

TABLE 1.4 (*Continued*)
Secondary Metabolites from Different Algal Species and Their Usage

Algae	Secondary Metabolite	Activity of Bioactive Compound/Uses	References
Chlorella protothecoides	Lutein	AMD, atherosclerosis, and retinal neural damage	Shi et al., 2006
Chlorella pyrenoidosa	Lutein	AMD, atherosclerosis, and retinal neural damage	Wu et al., 2007
Chlorella sorokiniana	Lutein	AMD, atherosclerosis, and retinal neural damage	Matsukawa et al., 2000
	α/β-carotene	Antioxidant, provitamin A, AMD, liver fibrosis, and anti-inflammatory	
Chlorella vulgaris	Lutein	AMD, atherosclerosis, and retinal neural damage	Cha et al., 2008; Gouveia et al., 1996; Mendes et al., 2003
	Canthaxanthin	Antioxidant property	
	Astaxanthin	Antitumoral	
Chlorella zofingiensis	Lutein	AMD, atherosclerosis, and retinal neural damage	Bhosale and Bernstein, 2005
	Astaxanthin	Antitumoral	Pelah et al., 2004
	Canthaxanthin	Antioxidant property	
Chlorococcum sp.	Astaxanthin	Antitumoral	Leya et al., 2009; Liu and Lee, 2000
	Canthaxanthin	Antioxidant property	Yuan et al., 2002
	Lutein	AMD, atherosclerosis, and retinal neural damage	
Chloromonas nivalis	Lutein	AMD, atherosclerosis, and retinal neural damage	Leya et al., 2009
	Zeaxanthin	Protection of eye cells and antioxidants	
Coccomyxa lacustris	Lutein	AMD, atherosclerosis, retinal neural damage	Del Campo et al., 2000
Coelastrella striolata	Canthaxanthin	Antioxidant	Abe et al., 2007
	Astaxanthin	Antitumoral	
Haematococcus pluvialis	Astaxanthin	Antitumoral	Domínguez-Bocanegra et al., 2004; Kang et al., 2005; Zhekisheva et al., 2005
Haematococcus lacustris	Canthaxanthin	Antioxidant property	Chattopadhyay et al., 2008
	Astaxanthin	Antitumoral	
Muriella aurantiaca	Lutein	AMD, atherosclerosis, and retinal neural damage	Del Campo et al., 2000
Muriellopsis sp.	Lutein	AMD, atherosclerosis, and retinal neural damage	Blanco et al., 2007; Del Campo et al., 2000
Neospongiococcum gelatinosum	Lutein	AMD, atherosclerosis, and retinal neural damage	Del Campo et al., 2000
Neospongiococcum spp.	Astaxanthin	Antitumoral	Borowitzka, 1988
	Canthaxanthin	Antioxidant property	
Platymonas subcordiformis	Lutein, violaxanthin	AMD, atherosclerosis, and retinal neural damage	Egeland et al., 1997
Platymonas striata	Lutein	AMD, atherosclerosis, and retinal neural damage	
Platymonas tetrahele	Lutein	AMD, atherosclerosis, and retinal neural damage	
	Neoxanthin	Antioxidant activity	
Pyramimonas obovata	β-carotene	Antioxidant, provitamin A, AMD, liver fibrosis, and anti-inflammatory	
Pyramimonas urceolata	Lutein	AMD, atherosclerosis, retinal neural damage	
Pyramimonas amylifera	β-carotene	Antioxidant, provitamin A, AMD, liver fibrosis, and anti-inflammatory	
	Siphonaxanthin	Antioxidant activity	

TABLE 1.4 (Continued)
Secondary Metabolites from Different Algal Species and Their Usage

Algae	Secondary Metabolite	Activity of Bioactive Compound/Uses	References
Scenedesmus obliquus	Astaxanthin	Antitumoral	Qin et al., 2008
Scenedesmus armatus	Lutein	AMD, atherosclerosis, and retinal neural damage	Ceron et al., 2008; Sánchez et al., 2008
Tetracystis intermedium and *T. tetrasporum*	Lutein	AMD, atherosclerosis, and retinal neural damage	Del Campo et al., 2000
Tetraselmis sp.	Eicosapentaenoic acid	Nutritional supplements and aquaculture feed	Patil et al., 2007; Reitan et al., 1994; Tzovenis et al., 2009
	α-Linolenic acid	Nutritional supplements	Patil et al., 2007; Reitan et al., 1994
	β-carotene	Antioxidant, provitamin A, AMD, liver fibrosis, and anti-inflammatory	Egeland et al., 1995
	Lutein	AMD, atherosclerosis, and retinal neural damage	
	Violaxanthinand neoxanthin	Antioxidant	
Tetraspora sp.	β-carotene	Antioxidant, provitamin A, AMD, liver fibrosis, and anti-inflammatory	
	Lutein	AMD, atherosclerosis, and retinal neural damage	
Red Algae			
Asparagopsis armata	Phenolic compounds, mycosporine, amino acids, and carotenoids	Antiaging	Leandro et al., 2019
Ceratodictyon spongiosum	Bromophenolic and peptide metabolites	Analgesic	Tan et al., 2000
Chondrococcus	Polyhalogenated monoterpenes	Antimicrobial and antitumor	Blunt et al., 2003
Corallina officinalis	Sulfated polysaccharides	Antioxidants	Benattouche et al., 2017
	Phenolic compounds, mycosporine, amino acids, and carotenoids	Antiaging	Leandro et al., 2019
Dichotomaria obtusata	Phenols	Analgesic	Vázquez et al., 2011
Gelidium lemaneiformis	Phenolic compounds	Antidiabetic	Chan et al., 2015
	Agarose	Skin whitening	Jin et al., 2017; Kim et al., 2017
Gelidium amansii	Phenolic compounds	Antidiabetic	Chan et al., 2015
Gelidium corneum	Phenolic compounds, mycosporine, amino acids, and carotenoids	Antiaging	Leandro et al., 2019
Gigartina skottsbergii	Carrageenans	Antiviral	Diogo et al., 2015
Gloiopeltis tenax	Phenols, sesquiterpenes, ketones, and sterols	Antimicrobial	Zheng et al., 2012
Gracilaria birdiae	Agar	Anticoagulant	Coura et al., 2015
Gracilaria changii,	Phenolic compounds	Antidiabetic	Chan et al., 2015
Gracilaria cornea	Sulfated polysaccharide	anti-inflammatory	Coura et al., 2015
	Agarose	Skin whitening	Jin et al., 2017; Kim et al., 2017
Gracilaria opuntia	Sulfated galactans	Antidiabetic	Makkar and Chakraborty, 2017
Gracilaria sp.	Agar	Anticancer and antioxidant	Cotas et al., 2020
Grateloupia indica	Sulfated galactans	Anticoagulant	Holtkamp et al., 2009

(Continued)

TABLE 1.4 (Continued)
Secondary Metabolites from Different Algal Species and Their Usage

Algae	Secondary Metabolite	Activity of Bioactive Compound/Uses	References
Jania rubens	N-hexadecanoic acid, docosanoic acid, 1,2,3-propanetriyl ester, hexanedioic acid, and dioctyl ester	Antimicrobial	El-Sheekh et al., 2016
Kappaphycus alvarezii	Carbohydrates, glycosides, and phenolic compounds	Antibacterial	Bhuyar et al., 2020
Laurencia obtusa	Polysaccharides	Antiviral	Gheda et al., 2016
	Polyphenols	Anticancer	Mhadhebi et al., 2014
Laurencia spp.	Halogenated biomolecules	Antimicrobial and cytotoxic potential	Brito et al., 2002
Mastocarpus stellatus	R-phycoerythrin	Antioxidant	Nguyen et al., 2017
Osmundea pinnatifida	Phenolic compounds	Antidiabetic	Chan et al., 2015
Palmaria palmata	R-phycoerythrin	Antioxidant	Fitzgerald et al., 2013
Plocamium	Polyhalogenated monoterpenes	Antimicrobial	Blunt et al., 2003
Polysiphonia urceolata	R-phycoerythrin	Antioxidant	Fitzgerald et al., 2013
	Bioactive peptides	Antioxidant	Fitzgerald et al., 2013
	Bromophenols, benzene acetamide, 2-piperidi-one, and *n*-hexadeca-noic acid	Antidiabetic	Lee and Jeon, 2013
Porphyra yezoensis	Agar	Anticoagulant	Coura et al., 2015
	Nori-peptides	Antihypertensive	Saito et al., 2002
Porphyra sp.	Porphyrans (anionic polysaccharides)	Anticancer	Jiang et al., 2012
Pterocladia capillacea	Polysaccharides	Antiviral	Gheda et al., 2016
Rhodomela confervoides	3-bromo-4,5-dihydroxy-benzaldehyde, bromophenols, 3-bromo-4,5-dihy-droxybenzoic acid, and methyl ester	Anticancer	Liu et al., 2011
	Bromophenols, benzene acetamide, 2-piperidi-one, and *n*-hexadeca-noic acid	Antidiabetic	Lee and Jeon, 2013
Solieria chordalis	Carrageenans	Antiviral	Boulho et al., 2017
Sphaerococcus coronopifolius	Brominated cyclic diterpenes	Antibacterial activity	Smyrniotopoulos et al., 2010
	Diterpene	Antifungal	Rodrigues et al., 2015
Symphyocladia latiuscula	Bromophenols	Antioxidant	Cotas et al., 2020
	Bromophenols, benzene acetamide, 2-piperidi-one, and *n*-hexadeca-noic acid	Antidiabetic	Lee and Jeon, 2013
Vidalia obtusaloba	Bromophenolic and peptide metabolites	Analgesic	Tan et al., 2000

TABLE 1.4 (Continued)
Secondary Metabolites from Different Algal Species and Their Usage

Algae	Secondary Metabolite	Activity of Bioactive Compound/Uses	References
Brown Algae			
Ascophyllum nodosum	Phlorotannin	Antibacterial, antioxidant, anti-inflammatory, antiproliferative, antitumor, antidiabetic, radioprotective, antiadipogenic, and anti-allergic effects	Agregán et al., 2017
	Flavonoids	Antitumor, antioxidant, anti-inflammatory, and antiviral	
Bifurcaria bifurcata	Phlorotannin	Antibacterial, antioxidant, anti-inflammatory, antiproliferative, antitumor, antidiabetics, radioprotective, antiadipogenic, and antiallergic effects	
	Alginate	Gel-forming and stabilizing agent	Aitouguinane et al., 2020
Caulocystis cephalornithos	6-tridecylsalicylic acid	Emollient (moisturizing ingredient)	Taki et al., 2020
Cystoseira tamariscifolia	Isololiolide	Cytotoxic and antitumor	Vizetto-Duarte et al., 2016
Dictyota ciliolata	Rutein	Antiapoptotic, antioxidant, and anti-inflammatory	Al-Saif et al., 2014
	Quercetin	Antioxidant, anti-inflammatory, and antifibrosis activities	De-Paula et al., 2018
	Palmitic acid	Anti-inflammatory	
	Diterpines	Antiplasmodial and antimalarial	
Dictyota pfaffii	Diterpenes	Anti-HIV-1	Pardo-Vargas et al., 2014
Ecklonia cava	8,8′-bieckol	Anti-inflammatory and anti-HIV-1 activity	Yang et al., 2014;
Ecklonia maxima	Eckol	Dietary supplements and food additives	Aremu et al., 2015
	Phloroglucinol	Anti-HIV	
Fucus vesiculosus	Phlorotannins	Antibacterial, antioxidant, anti-inflammatory, antiproliferative, antitumor, antidiabetics, radioprotective, antiadipogenic, and antiallergic effects	Agregán et al., 2017
Ishige okamurae	Diphlorethohydroxycarmalol (DPHC)	Antioxidant activity, antiadipogenic activity, anti-inflammatory activity	Kim et al., 2020
Sargassum ilicifolium	1-O-palmitoyl-2-O-stearoyl-3-O-b-Digalactopyranosyl glycerol	Acute coronary syndrome	Rahelivao et al., 2015
Sargassum polycystum	Fucoidan	Prebiotic and antitumor	Palanisamy et al., 2017
Sargassum thunbergii	Indole derivatives	Prevention of obesity or obesity-related diseases	Kang et al., 2017
Stochospermum marginatum	Spatane diterpenoids	Antitumor	Chinnababu et al., 2015
Turbinaria ornate	Apo-90-fucoxanthinone, 24-ketocholesterol, (22E)-3b-hydroxycholesta-5,22-dien-24-one, and saringosterol	Antitumoral and antimicrobial	Rahelivao et al., 2015

1.3.4 Archaea

Archaea are microorganisms that are abundant in nature but are mostly found in extreme environments with harsh conditions, such as high and low temperatures, high salinity, acidity, alkalinity, *etc.* Archaea produce secondary metabolites to thrive in extreme conditions and communicate/compete with other microorganisms. Secondary metabolites produced by archaea are archaeocins (halocins, sulfolobicins), diketopiperazines, acyl homoserine lactones, exopolysaccharides and polyhydroxyalkanoates, carotenoids, biosurfactants, phenazines, and organic solutes, which can be used in the pharma, food, and cosmetic industries (Charlesworth and Burns, 2015). Other secondary metabolites detected in archaeal genomes by the antiSMASH database are non-ribosomal peptides, bacteriocin, cyanobactin, lantipeptide, lassopeptide, terpene, ectoine, and siderophore (Wang and Lu, 2017). The survey of 203 archaea genome sequences by Wang et al. (2019) using the antiSMASH database confirmed that the most common BGCs in archaea genomes are bacteriocin (56/203) and terpene (49/203). Bacteriocins are mostly found in the families Thermoproteaceae, Sulfolobaceae, and Desulfurococcaceae, while terpene is found in the families Halobacteriaceae and Sulfolobaceae. Some examples of archaea species producing secondary metabolites are *Haloarcula mediterranea*, *Halobacterium noricense*, *Halobacterium volcanii*, *Halococcus saccharolyticus*, *Haloferax mediterranea*, *Halogeometricum borinquense*, *Halogranum amylolyticum*, *Halorubrum chaoviator*, *Haloterrigena hispanica*, *Halovivax* sp., *Methanosaeta harundinacea*, *Natrialba* sp., *Natronobacterium gregoryi*, *Sulfolobus solfataricus*, *etc.* (Yadav et al., 2019). Few reports on archaea producing secondary metabolites are given in Table 1.5.

1.4 CONCLUSION

For ages, microorganisms have been a gold mine of valuable natural products, especially secondary metabolites. The microbial ability to flourish in diverse environmental conditions has inspired researchers to exploit them for human benefits. To date, a large number of secondary metabolites from bacteria, fungi, and algae residing in

TABLE 1.5
Secondary Metabolites from Archaea

Archaea	Source	Secondary Metabolite	Biological Activity of Secondary Metabolite	Reference
Sulfolobus islandicus	Solfataric fields (Iceland)	Sulfolobicins	Antiarchaeal	Prangishvili et al., 2000
Haloferax gibbonsii	–	Halocin H6	Na^+/H^+ antiporter inhibitor in mammalian cells	Lequerica et al., 2006
Haloferax volcanii KPS1	Water sample from Kovalam solar saltern (India)	Halocin	Antibacterial	Kavitha et al., 2011
Haloterrigena hispanica	Fuente de Piedra saline lake (Spain)	Cyclo-(L-prolyl–L-valine) (diketopiperazine)	Activated N-acyl homoserine lactone (AHL) bioreporters	Tommonaro et al., 2012
Halomonas salifodinae MPM-TC	Water samples from Thamaraikulam solar salt works (India)	Fraction-III (F-III) contains 8 compounds: perfluorotributylamine, cyclopentane, 1-butyl-2-ethyl-, 1,10-Biphenyl]-3-amine, pyridine, 4-(phenylmethyl)-, hexadecane, 2-methyl, nonadecane, and phytol	Antiviral	Velmurugan et al., 2013
Haloarcula argentinensis (NR_112708), *Haloarcula japonica* (NR_112710), *Haloarcula vallismortis* (NR_112707), *Halorubrum tebenquichense* (HQ641750), and *Halobacterium* sp. (AB603514)	Water sample from Colorada Grande saltern (Argentina)	Halocins	Antiarchaeal	Nercessian et al., 2015
Halobacterium salinarum (DQ465019) and *Halobacterium piscisalsi* (JX067388)	Water sample from Salitral Negro saltern (Argentina)			
Halorussus salinus YJ-37-H	Marine solar saltern (China)	Archalan α (lanthipeptides)	Antiarchaeal	Liang et al., 2022

different habitats have been isolated and chemically characterized. They have been used in different industries, such as agriculture, pharmaceuticals, nutraceuticals, dyeing, *etc.*, but only a small fraction of microbial secondary metabolites have been explored. This is because these metabolites come from only 1% of cultured microbes and are expressed in low amounts under *in vitro* conditions. Modified conventional isolation techniques, along with metagenomic studies and bioinformatics tools, are now being used to solve these problems. Microorganisms from various environments, such as soil, marine, extreme, and microbes associated with flora and fauna, have been proven to be an excellent source of bioactive secondary metabolites. However, there are still a lot of untapped reservoirs of novel microbial secondary metabolites that need to be addressed, especially archaea. Additionally, microbes from polluted environments can be explored as they may provide new bioactive secondary metabolites, which is a pressing priority.

REFERENCES

Abe, K., H. Hattori, and M. Hirano. 2007. Accumulation and antioxidant activity of secondary carotenoids in the aerial microalga *Coelastrella striolata* var. *multistriata*. *Food Chemistry* 100: 656–61. https://doi.org/10.1016/j.foodchem.2005.10.026.

Abidi, S.L. 2001. Chromatographic analysis of plant sterols in foods and vegetable oils. *Journal of Chromatography A* 935: 173–201. https://doi.org/10.1016/S0021-9673(01)00946-3.

Agregán, R., P.E.S. Munekata, D. Franco, R. Dominguez, J. Carballo, and J.M. Lorenzo. 2017. Phenolic compounds from three brown seaweed species using LC-DAD–ESI-MS/MS. *Food Research International* 99: 979–85. https://doi.org/10.1016/j.foodres.2017.03.043.

Aitouguinane, M., S. Bouissil, A. Mouhoub, et al. 2020. Induction of natural defenses in tomato seedlings by using alginate and oligoalginates derivatives extracted from Moroccan brown algae. *Marine Drugs* 18 (October): 521. https://doi.org/10.3390/md18100521.

Al-Saif, S.S.A., N. Abdel-Raouf, H.A. El-Wazanani, and I.A. Aref. 2014. Antibacterial substances from marine algae isolated from Jeddah coast of Red Sea, Saudi Arabia. *Saudi Journal of Biological Sciences* 21: 57–64. https://doi.org/10.1016/j.sjbs.2013.06.001.

Amaro, H.M., R. Barros, A.C. Guedes, I. Sousa-Pinto, and F.X. Malcata. 2013. Microalgal compounds modulate carcinogenesis in the gastrointestinal tract. *Trends in Biotechnology* 31: 92–98. https://doi.org/10.1016/j.tibtech.2012.11.004.

Ambika, R.K., and A. Pandey. 2020. Exploring Ascomycetes fungus *Hypoxylon fendleri* as a source for natural textile dye. *Indian Forester* 146: 365–66. https://doi.org/10.36808/if/2020/v146i4/152438.

Angelova, L., M. Dalgalarrondo, I. Minkov, et al. 2006. Purification and characterization of a protease inhibitor from *Streptomyces chromofuscus* 34–1 with an antiviral activity. *Biochimica et Biophysica Acta* 1760: 1210–16. https://doi.org/10.1016/j.bbagen.2006.03.002.

Aremu, A.O., N.A. Masondo, K.R.R. Rengasamy, et al. 2015. Physiological role of phenolic biostimulants isolated from brown seaweed *Ecklonia maxima* on plant growth and development. *Planta* 241: 1313–24. https://doi.org/10.1007/s00425-015-2256-x.

Arisz, S.A., J.A. van Himbergen, A. Musgrave, H. van den Ende, and T. Munnik. 2000. Polar glycerolipids of *Chlamydomonas moewusii*. *Phytochemistry* 53: 265–70. https://doi.org/10.1016/S0031-9422(99)00505–1.

Baby, S., A.J. Johnson, and B. Govindan. 2015. Secondary metabolites from *Ganoderma*. *Phytochemistry* 114: 66–101. https://doi.org/10.1016/j.phytochem.2015.03.010.

Balasundram, N., K. Sundram, and S. Samman. 2006. Phenolic compounds in plants and agri-industrial by-products: Antioxidant activity, occurrence, and potential uses. *Food chemistry* 99: 191–203. https://doi.org/10.1016/j.foodchem.2005.07.042.

Baliah, N.T., and P.J. Begum. 2015. Isolation, identification and characterization of phosphate solubilizing bacteria (PSB) isolated from economically important crop plants. *International Journal of Current Microbiology and Applied Sciences* 4: 915–24. https://ijcmas.com/vol-4–3/N.%20Tensingh%20Baliah%20and%20P.%20Jemeema%20Begum.pdf.

Baltz, R.H. 2009. Daptomycin: Mechanisms of action and resistance, and biosynthetic engineering. *Current Opinion in Chemical Biology* 13: 144–51. https://doi.org/10.1016/j.cbpa.2009.02.031.

Barbie, P., and U. Kazmaier. 2016. Total synthesis of cyclomarin A, a marine cycloheptapeptide with anti-tuberculosis and anti-malaria activity. *Organic Letters* 18: 204–7. https://doi.org/10.1021/acs.orglett.5b03292.

Barbosaa, M., L.G. Inácioa, C. Afonsob, and P. Maranhão. 2023. The microalga *Dunaliella* and its applications: A review. *Applied Phycology* 4: 99–120. https://doi.org/10.1080/26388081.2023.2222318.

Barrios-González, J., F.J. Fernández, and A. Tomasini. 2003. Microbial secondary metabolites production and strain improvement. *Indian Journal of Biotechnology* 2: 322–33.

Bassarello, C., S. Lazzaroni, and G. Bifulco. 2004. Tolaasins A–E, five new lipodepsipeptides produced by *Pseudomonas tolaasii*. *Journal of Natural Products* 67: 811–16. https://doi.org/10.1021/np0303557.

Benattouche, Z., G.B. Raho, F. Sahnouni, A. Hariri, G. Bouhadi, and M. Benchohra. 2017. Antioxidant activities of sulfated polysaccharide obtained from red algae *Corallina officinalis*. *International Journal of Pharmacognosy* 4: 88–91. http://dx.doi.org/10.13040/IJPSR.0975–8232.

Bennett, J.W., and R. Bentley 1989. What's in a name?-microbial secondary metabolism. *Advances in Applied Microbiology* 34: 1–28. https://doi.org/10.1016/S0065-2164(08)70316–2.

Bérdy, J. 2005. Bioactive microbial metabolites. *The Journal of Antibiotics* 58: 1–26. https://doi.org/10.1038/ja.2005.1.

Bérdy, J. 2012. Thoughts and facts about antibiotics: Where we are now and where we are heading. *The Journal of Antibiotics* 65: 385–95. https://doi.org/10.1038/ja.2012.27.

Berti, A.D., and M.G. Thomas. 2009. Analysis of achromobactin biosynthesis by *Pseudomonas syringae* pv. *syringae* B728a. *Journal of Bacteriology* 191: 4594–604. https://doi.org/10.1128/JB.00457-09.

Bhosale, P., and P.S. Bernstein. 2005. Microbial xanthophylls. *Applied Microbiology and Biotechnology* 68: 445–55. https://doi.org/10.1007/s00253-005-0032-8.

Bhuyar, P., M.H. Rahim, S. Sundararaju, G.P. Maniam, and N. Govindan. 2020. Antioxidant and antibacterial activity of red seaweed *Kappaphycus alvarezii* against pathogenic bacteria. *Global Journal of Environmental Science and Management* 6: 47–58. https://doi.org/10.22034/GJESM.2020.01.04.

Bibi, F., M. Yasir, A. Al-Sofyani, M.I. Naseer, and E.I. Azhar. 2020. Antimicrobial activity of bacteria from marine sponge *Suberea mollis* and bioactive metabolites of *Vibrio* sp.

EA348. *Saudi Journal of Biological Sciences* 27: 1139–47. https://doi.org/10.1016/j.sjbs.2020.02.002.

Birkinshaw, J.H., J.H.V. Charles, A.C. Hetherington, and H. Raistrick. 1931. Part VI.-Quantitative examination by the carbon balance sheet method of the types of products formed from glucose by miscellaneous species of fungi. *Philosophical Transactions of the Royal Society of London* 220: 468–73. https://doi.org/10.1098/rstb.1931.0017.

Bizani, D., A.P.M. Dominguez, and A. Brandelli. 2005. Purification and partial chemical characterization of the antimicrobial peptide cerein 8A. *Letters in Applied Microbiology* 41: 269–73. https://doi.org/10.1111/j.1472-765X.2005.01748.x.

Bizani, D., and A. Brandelli. 2002. Characterization of a bacteriocin produced by a newly isolated *Bacillus* sp. strain 8A. *Journal of Applied Microbiology* 93: 512–19. https://doi.org/10.1046/j.1365-2672.2002.01720.x.

Blanco, A.M., J. Moreno, J.A.D. Campo, J. Rivas, and M.G. Guerrero. 2007. Outdoor cultivation of lutein-rich cells of *Muriellopsis* sp. in open ponds. *Applied Microbiology and Biotechnology* 73: 1259–66. https://doi.org/10.1007/s00253-006-0598-9.

Bloemberg, G.V., and B.J.J. Lugtenberg. 2001. Molecular basis of plant growth promotion and biocontrol by rhizobacteria. *Current Opinion in Plant Biology* 4: 343–50. https://doi.org/10.1016/S1369-5266(00)00183-7.

Blunt, J.W., B.R. Copp, M.M.H. Munro, P.T. Northcote, and M.R. Prinsep. 2003. Marine natural products. *Natural Product Reports* 20: 1–48. https://doi.org/10.1039/B207130B.

Bonachela, J.A., M. Raghib, and S.A. Levin. 2011. Dynamic model of flexible phytoplankton nutrient uptake. *Proceedings of the National Academy of Sciences* 108: 20633–38. https://doi.org/10.1073/pnas.1118012108.

Borel, J.F., C. Feurer, H.U. Gubler, and H. Stähelin. 1994. Biological effects of cyclosporine A: A new antilymphocytic agent. *Agents and Actions* 43: 179–86. https://doi.org/10.1007/bf01986686.

Borowitzka, M.A. 1988. Vitamins and fine chemicals from microalgae. In *Micro-algal biotechnology*, eds. M.A. Borowitzka and L.J. Borowitzka, 153–96. Cambridge: Cambridge University Press.

Boulho, R., C. Marty, Y. Freile-Pelegrín, D. Robledo, N. Bourgougnon, and G. Bedoux. 2017. Antiherpetic (HSV-1) activity of carrageenans from the red seaweed *Solieria chordalis* (Rhodophyta, Gigartinales) extracted by microwave-assisted extraction (MAE). *Journal of Applied Phycology* 29: 2219–28. https://doi.org/10.1007/s10811-017-1192-5.

Brito, I., M. Cueto, A.R. Díaz-Marrero, J. Darias, and A. San Martín. 2002. Oxachamigrenes, new halogenated sesquiterpenes from *Laurencia obtusa*. *Journal of Natural Products* 65: 946–48. https://doi.org/10.1021/np010580t.

Brock, S., J. Knadler, T. Ritter, and J.C. Baker. 2018. Characterization of a bacteriocin from *Bacillus amyloliquefaciens*. *International Journal of Current Microbiology and Applied Sciences* 7: 1492–503. https://doi.org/10.20546/ijcmas.2018.706.177.

Bu, Y.Y., H. Yamazaki, O. Takahashi, R. Kirikoshi, K. Ukai, and M. Namikoshi. 2016. Penicyrones A and B, an epimeric pair of α-pyrone-type polyketides produced by the marine-derived *Penicillium* sp. *The Journal of Antibiotics* 69: 57–61. https://doi.org/10.1038/ja.2015.82.

Bu'Lock, J.D. 1965. *The biosynthesis of natural products: An introduction to secondary metabolism*. London: McGraw-Hill.

Ceron, M.C., I. Campos, J.F. Sanchez, F.G. Acien, E. Molina, and J.M. Fernandez-Sevilla. 2008. Recovery of lutein from microalgae biomass: Development of a process for *Scenedesmus almeriensis* biomass. *Journal of Agricultural and Food Chemistry* 56: 11761–66. https://doi.org/10.1021/jf8025875.

Cezairliyan, B., N. Vinayavekhin, D. Grenfell-Lee, G.J. Yuen, A. Saghatelian, and F.M. Ausubel. 2013. Identification of *Pseudomonas aeruginosa* phenazines that kill *Caenorhabditis elegans*. *PLoS Pathogens* 9 (January): e1003101. https://doi.org/10.1371/journal.ppat.1003101.

Cha, K.H., S.Y. Koo, and D.U. Lee. 2008. Antiproliferative effects of carotenoids extracted from *Chlorella ellipsoidea* and *Chlorella vulgaris* on human colon cancer cells. *Journal of Agricultural and Food Chemistry* 56: 10521–26. https://doi.org/10.1021/jf802111x.

Chakraborty, K., B. Thilakan, R.D. Chakraborty, V.K. Raola, and M. Joy. 2017a. O-heterocyclic derivatives with antibacterial properties from marine bacterium *Bacillus subtilis* associated with seaweed, *Sargassum myriocystum*. *Applied Microbiology and Biotechnology* 101: 569–83. https://doi.org/10.1007/s00253-016-7810-3.

Chakraborty, K., B. Thilakan, and V.K. Raola. 2017b. Antimicrobial polyketide furanoterpenoids from seaweed-associated heterotrophic bacterium *Bacillus subtilis* MTCC 10403. *Phytochemistry* 142: 112–25. https://doi.org/10.1016/j.phytochem.2017.06.019.

Chan, P.T., P. Matanjun, S.M. Yasir, and T.S. Tan. 2015. Histopathological studies on liver, kidney and heart of normal and dietary induced hyperlipidaemic rats fed with tropical red seaweed *Gracilaria changii*. *Journal of Functional Foods* 17: 202–13. https://doi.org/10.1016/j.jff.2015.05.019.

Charlesworth, J.C., and B.P. Burns. 2015. Untapped resources: Biotechnological potential of peptides and secondary metabolites in archaea. *Archaea* 2015 (October): 82035. https://doi.org/10.1155/2015/282035.

Chattopadhyay, P., S. Chatterjee, and S.K. Sen. 2008. Biotechnological potential of natural food grade biocolorants. *African Journal of Biotechnology* 7: 2972–85. https://doi.org/10.5897/AJB08.433.

Chehimi, S., F. Delalande, S. Sablé, et al. 2007. Purification and partial amino acid sequence of thuricin S, a new anti-*Listeria* bacteriocin from *Bacillus thuringiensis*. *Canadian Journal of Microbiology* 53: 284–90. https://doi.org/10.1139/w06-116.

Chen, Y.T., Q. Yuan, L.T. Shan, M.A Lin, D.Q. Cheng, and C.Y. Li. 2013. Antitumor activity of bacterial exopolysaccharides from the endophyte *Bacillus amyloliquefaciens* sp. isolated from *Ophiopogon japonicus*. *Oncology Letters* 5: 1787–92. https://doi.org/10.3892/ol.2013.1284.

Chinnababu, B., S.P. Reddy, P.S. Rao, et al. 2015. Isolation, semi-synthesis and bio-evaluation of spatane derivatives from the brown algae *Stoechospermum marginatum*. *Bioorganic & Medicinal Chemistry Letters* 25: 2479–83. https://doi.org/10.1016/j.bmcl.2015.04.061.

Chu, K.K., S.S. Ho, and A.H. Chow. 2002. *Coriolus versicolor*: A medicinal mushroom with promising immunotherapeutic values. *The Journal of Clinical Pharmacology* 42: 976–84. https://doi.org/10.1177/009127000204200904.

Collemare, J., A. Billard, H.U. Böhnert, and M.H. Lebrun. 2008. Biosynthesis of secondary metabolites in the rice blast fungus *Magnaporthe grisea*: The role of hybrid PKS-NRPS in pathogenicity. *Mycological Research* 112: 207–15. https://doi.org/10.1016/j.mycres.2007.08.003.

Correa-Llantén, D.N., M.J. Amenábar, and J.M. Blamey. 2012. Antioxidant capacity of novel pigments from an Antarctic bacterium. *Journal of Microbiology* 50: 374–79. https://doi.org/10.1007/s12275-012-2029-1.

Cotas, J., A. Leandro, P. Monteiro, et al. 2020. Seaweed phenolics: From extraction to applications. *Marine Drugs* 18 (July): 384. https://doi.org/10.3390/md18080384.

Coura, C.O., R.B. Souza, J.A.G. Rodrigues, et al. 2015. Mechanisms involved in the anti-inflammatory action of a polysulfated fraction from *Gracilaria cornea* in rats. *PloS One*, 10 (March): e0119319. https://doi.org/10.1371/journal.pone.0119319.

De Santis, D., M. Moresi, A.M. Gallo, and M. Petruccioli. 2005. Assessment of the dyeing properties of pigments from *Monascus purpureus*. *Journal of Chemical Technology & Biotechnology* 80: 1072–79. https://doi.org/10.1002/jctb.1285.

De Souza, E.T., D.P. de Lira, A.C. de Queiroz, et al. 2009. The antinociceptive and anti-inflammatory activities of caulerpin, a bisindole alkaloid isolated from seaweeds of the genus *Caulerpa*. *Marine Drugs* 7: 689–704. https://doi.org/10.3390/md7040689.

Del Campo, J.A., J. Moreno, H. Rodríguez, M.A. Vargas, J. Rivas, and M.G. Guerrero. 2000. Carotenoid content of chlorophycean microalgae: Factors determining lutein accumulation in *Muriellopsis* sp. (Chlorophyta). *Journal of Biotechnology* 76: 51–59. https://doi.org/10.1016/S0168-1656(99)00178-9.

Del Campo, J.A., H. Rodríguez, J. Moreno, M.A. Vargas, J. Rivas, and M.G. Guerrero. 2004. Accumulation of astaxanthin and lutein in *Chlorella zofingiensis* (Chlorophyta). *Applied Microbiology and Biotechnology* 64(6): 848–54.

De-Paula, J.C., E.A.P. Lopes-Filho, F. Salgueiro, et al. 2018. Diterpenes content of the brown alga *Dictyota ciliolata* (Dictyotales, Phaeophyceae) and recognition of a Brazilian haplotype based on *psb*A sequences. *New Zealand Journal of Botany* 56: 415–29. https://doi.org/10.1080/0028825X.2018.1535441.

Desjardine, K., A. Pereira, H. Wright, T. Matainaho, M. Kelly, and R.J. Andersen. 2007. Tauramamide, a lipopeptide antibiotic produced in culture by *Brevibacillus laterosporus* isolated from a marine habitat: Structure elucidation and synthesis. *Journal of Natural Products* 70: 1850–53. https://doi.org/10.1021/np070209r.

Devi, R., T. Kaur, G. Guleria, et al. 2020. Fungal secondary metabolites and their biotechnological applications for human health. In *New and future developments in microbial biotechnology and bioengineering*, eds. A.A. Rastegari, A.N. Yadav, and N. Yadav, 147–61. Elsevier. https://doi.org/10.1016/B978-0-12-820528-0.00010-7.

Dharmaraj, S., B. Ashokkumar, and K. Dhevendaran. 2009. Food-grade pigments from *Streptomyces* sp. isolated from the marine sponge *Callyspongia diffusa*. *Food Research International* 42: 487–92. https://doi.org/10.1016/j.foodres.2009.02.006.

Diogo, J.V., S.G. Novo, M.J. González, M. Ciancia, and A.C. Bratanich. 2015. Antiviral activity of lambda-carrageenan prepared from red seaweed (*Gigartina skottsbergii*) against BoHV-1 and SuHV−1. *Research in Veterinary Science* 98: 142–44. https://doi.org/10.1016/j.rvsc.2014.11.010.

Domínguez-Bocanegra, A.R., I.G. Legarreta, F.M. Jeronimo, and A.T. Campocosio 2004. Influence of environmental and nutritional factors in the production of astaxanthin from *Haematococcus pluvialis*. *Bioresource Technology* 92: 209–14. https://doi.org/10.1016/j.biortech.2003.04.001.

Durán, N., M.F.S. Teixeira, R. De Conti, E. Esposito, 2002. Ecological-friendly pigments from fungi. *Critical Reviews in Food Science and Nutrition* 42: 53–66. https://doi.org/10.1080/10408690290825457.

Duval, B., K. Shetty, and W.H. Thomas. 1999. Phenolic compounds and antioxidant properties in the snow alga *Chlamydomonas nivalis* after exposure to UV light. *Journal of Applied Phycology* 11: 559–66. https://doi.org/10.1023/A:1008178208949.

Egeland, E.S., W. Eikrem, J. Throndsen, C. Wilhelm, M. Zapata, and S. Liaaen-Jensen. 1995. Carotenoids from further prasinophytes. *Biochemical Systematics and Ecology* 23: 747–55. https://doi.org/10.1016/0305-1978(95)00075-5.

Egeland, E.S., R.R. Guillard, and S. Liaaen-Jensen. 1997. Additional carotenoid prototype representatives and a general chemosystematic evaluation of carotenoids in Prasinophyceae (Chlorophyta). *Phytochemistry* 44: 1087–97. https://doi.org/10.1016/S0031-9422(96)00650-4.

El-Fouly, M.Z., A.M. Sharaf, A.A.M. Shahin, H.A. El-Bialy, and A.M.A. Omara. 2015. Biosynthesis of pyocyanin pigment by *Pseudomonas aeruginosa*. *Journal of Radiation Research and Applied Sciences* 8: 36–48. https://doi.org/10.1016/j.jrras.2014.10.007.

El-Gendy, M.M.A., U.W. Hawas, and M. Jaspars. 2008. Novel bioactive metabolites from a marine derived bacterium *Nocardia* sp. ALAA 2000. *The Journal of Antibiotics* 61: 379–86. https://doi.org/10.1038/ja.2008.53.

Elisabeth, W., J.M. Facchini, E.P. Alves, et al. 2017. A polysaccharide fraction extracted from *Pleurotus ostreatus* mycelial biomass inhibit Sarcoma 180 tumor. *Anais da Academia Brasileira de Ciências* 89: 2013–20. https://doi.org/10.1590/0001-3765201720150635.

El-Naggar, N.E.A., and S.M. El-Ewasy. 2017. Bioproduction, characterization, anticancer and antioxidant activities of extracellular melanin pigment produced by newly isolated microbial cell factories *Streptomyces glaucescens* NEAE-H. *Scientific Reports* 7 (February): 42129. https://doi.org/10.1038/srep42129.

El-Sheekh, M., S.M.M. El-Sabbagh, and B.M.A. El Samea. 2016. Control of some microbial skin diseases by some marine algal extract. *Journal of Agricultural Chemistry and Biotechnology* 7: 67–74. https://doi.org/10.21608/JACB.2016.40743.

Fitzgerald, C., E. Gallagher, P. O'Connor, et al. 2013. Development of a seaweed derived platelet activating factor acetylhydrolase (PAF-AH) inhibitory hydrolysate, synthesis of inhibitory peptides and assessment of their toxicity using the Zebrafish larvae assay. *Peptides* 50: 119–24. https://doi.org/10.1016/j.peptides.2013.10.006.

Fleming, A. 2001. On the antibacterial action of cultures of a *Penicillium*, with special reference to their use in the isolation of *B. influenza*. 1929. *Bulletin of World Health Organization* 79: 780–90. PMID: 11545337.

Foster, J.W. 1949. *Chemical activities of fungi*. New York: Academic Press.

Fouillaud, M., and L. Dufossé. 2022. Microbial secondary metabolism and biotechnology. *Microorganisms* 10 (January): 123. https://doi.org/10.3390%2Fmicroorganisms10010123.

Frisvad, J.C., T.O. Larsen, R. De Vries, et al. 2007. Secondary metabolite profiling, growth profiles and other tools for species recognition and important *Aspergillus mycotoxins*. *Studies in Mycology* 59: 31–37. https://doi.org/10.3114/sim.2007.59.04.

Fu, P., A. Legako, S. La, and J.B. MacMillan. 2016. Discovery, characterization, and analogue synthesis of bohemamine dimers generated by non-enzymatic biosynthesis. *Chemistry–A European Journal* 22: 3491–95. https://doi.org/10.1002/chem.201600024.

Fu, W., O. Guomundsson, G. Paglia, et al. 2013. Enhancement of carotenoid biosynthesis in the green microalga *Dunaliella salina* with light-emitting diodes and adaptive laboratory evolution. *Applied Microbiology and Biotechnology* 97: 2395–403. https://doi.org/10.1007/s00253-012-4502-5.

Fukushima-Sakuno, E. 2020. Bioactive small secondary metabolites from the mushrooms *Lentinula edodes* and *Flammulina velutipes*. *The Journal of Antibiotics* 73: 687–96. https://doi.org/10.1038/s41429-020-0354-x.

Gaden, E.L. Jr. 1959. Fermentation process kinetics. *Journal of Biochemical and Microbiological Technology and Engineering* 1: 413–29. https://doi.org/10.1002/jbmte.390010407.

Garbayo, I., M. Cuaresma, C. Vílchez, and J.M. Vega. 2008. Effect of abiotic stress on the production of lutein and β-carotene by *Chlamydomonas acidophila*. *Process Biochemistry* 43: 1158–61. https://doi.org/10.1016/j.procbio.2008.06.012.

Garbeva, P., C. Hordijk, S. Gerards, and W. De Boer. 2014. Volatiles produced by the mycophagous soil bacterium *Collimonas*. *FEMS Microbiology Ecology* 87: 639–49. https://doi.org/10.1111/1574-6941.12252.

Gaughran, E.R.L. 1969. Division of microbiology. From superstition to science: The history of a bacterium. *Translations of the New York Academy of Sciences* 31: 3–24. https://doi.org/10.1111/j.2164-0947.1969.tb02887.x.

Gheda, S.F., H.I. El-Adawi, and N.M. El-Deeb. 2016. Antiviral profile of brown and red seaweed polysaccharides against hepatitis C virus. *Iranian Journal of Pharmaceutical Research* 15: 483–91. https://doi.org/10.22037/ijpr.2016.1944.

Ghiasvand, M., A. Makhdoumi, M.M. Matin, and J. Vaezi. 2020. Exploring the bioactive compounds from endophytic bacteria of a medicinal plant: *Ephedra foliata* (Ephedrales: Ephedraceae). *Advances in Traditional Medicine* 20: 61–70. https://doi.org/10.1007/s13596-019-00410-z.

Gonzalez-Franco, A.C., L. Robles-Hernandez, A. Nuñez-Barrios, J.L. Strap, and D.L. Crawford. 2009. Molecular and cultural analysis of seasonal Actinomycetes in soils from *Artemisia tridentata* habitat. *International Journal of Experimental Botany* 78: 83–90. https://doi.org/10.32604/phyton.2009.78.083.

Goodfellow, M., and S.T. Williams. 1983. Ecology of actinomycetes. *Annual Review of Microbiology* 37: 189–216. https://doi.org/10.1146/annurev.mi.37.100183.001201.

Gos, F.M.W.R., D.C. Savi, K.A. Shaaban, et al. 2017. Antibacterial activity of endophytic actinomycetes isolated from the medicinal plant *Vochysia divergens* (Pantanal, Brazil). *Frontiers in Microbiology* 8 (September): 1642. https://doi.org/10.3389/fmicb.2017.01642.

Gouveia, L., V. Veloso, A. Reis, H. Fernandes, J. Novais, and J. Empis. 1996. Evolution of pigment composition in *Chlorella vulgaris*. *Bioresource Technology* 57: 157–63. https://doi.org/10.1016/0960-8524(96)00058-2.

Gu, J.D., and K.H. Cheung. 2001. Phenotypic expression of *Vogesella indigofera* upon exposure to hexavalent chromium, $Cr6^{+}$. *World Journal of Microbiology and Biotechnology* 17: 475–80. https://doi.org/10.1023/A:1011917409139.

Gupta, A., S. Jain, and N. Verma. 2022. *Industrial applications of soil microbes*, vol. 1. Bentham Science Publishers Pte Ltd. https://doi.org/10.2174/97898150399551220101.

Gupta, A., S. Jain, and N. Verma. 2024. *Industrial applications of soil microbes*, vol. 3. Bentham Science Publishers Pte Ltd. https://doi.org/10.2174/97898150797531240301.

Gupta, S., and N. Abu-Ghannam. 2011. Bioactive potential and possible health effects of edible brown seaweeds. *Trends in Food Science & Technology* 22: 315–26. https://doi.org/10.1016/j.tifs.2011.03.011.

Gürtler, H., R. Pedersen, U. Anthoni, et al. 1994. Albaflavenone, a sesquiterpene ketone with a zizaene skeleton produced by a streptomycete with a new rope morphology. *The Journal of Antibiotics* 47: 434–39. https://doi.org/10.7164/antibiotics.47.434.

Hammami, I., A. Rhouma, B. Jaouadi, A. Rebai, and X. Nesme. 2009. Optimization and biochemical characterization of a bacteriocin from a newly isolated *Bacillus subtilis* strain 14B for biocontrol of *Agrobacterium* spp. strains. *Letters in Applied Microbiology* 48: 253–60. https://doi.org/10.1111/j.1472-765X.2008.02524.x.

Han, J.J., L. Bao, L.W. He, et al. 2013. Phaeolschidins A–E, five hispidin derivatives with antioxidant activity from the fruiting body of *Phaeolus schweinitzii* collected in the Tibetan Plateau. *Journal of Natural Products* 76: 1448–53. https://doi.org/10.1021/np400234u.

Hasani, A., A. Kariminik, and K. Issazadeh. 2014. Streptomycetes: Characteristics and their antimicrobial activities. *International Journal of Advanced Biological and Biomedical Research* 2: 63–75. https://www.ijabbr.com/article_7033_7733c8235876d7ba635f6c831a916648.pdf.

Hemberger, Y., J. Xu, V. Wray, P. Proksch, J. Wu, and G. Bringmann. 2013. Pestalotiopens A and B: Stereochemically challenging flexible sesquiterpene-cyclopaldic acid hybrids from *Pestalotiopsis* sp. *Chemistry* 19: 15556–64. https://doi.org/10.1002/chem.201302204.

Holtkamp, A.D., S. Kelly, R. Ulber, and S. Lang. 2009. Fucoidans and fucoidanases – focus on techniques for molecular structure elucidation and modification of marine polysaccharides. *Applied Microbiology and Biotechnology* 82: 1–11. https://doi.org/10.1007/s00253-008-1790-x.

Hughes, C.C., A. Prieto-Davo, P.R. Jensen, and W. Fenical. 2008. The marinopyrroles, antibiotics of an unprecedented structure class from a marine *Streptomyces* sp. *Organic Letter* 10: 629–31. https://doi.org/10.1021/ol702952n.

Hwang, S., L.T.H.L. Le, S.I. Jo, J. Shin, M.J. Lee, and D.C. Oh. 2020. Pentaminomycins C–E: Cyclic pentapeptides as autophagy inducers from a mealworm beetle gut bacterium. *Microorganisms* 8 (September): 1390. https://doi.org/10.3390/microorganisms8091390.

Ibañez, E., and A. Cifuentes. 2013. Benefits of using algae as natural sources of functional ingredients. *Journal of the Science of Food and Agriculture* 93: 703–9. https://doi.org/10.1002/jsfa.6023.

Igarashi, Y., M.E. Trujillo, E. Martínez-Molina, et al. 2007. Antitumor anthraquinones from an endophytic actinomycete *Micromonospora lupini* sp. nov. *Bioorganic & Medicinal Chemistry Letters* 17: 3702–5. https://doi.org/10.1016/j.bmcl.2007.04.039.

Imhoff, J.F., A. Labes, and J. Wiese. 2011. Bio-mining the microbial treasures of the ocean: New natural products. *Biotechnology Advances* 29: 468–82. https://doi.org/10.1016/j.biotechadv.2011.03.001.

Jain, S., A. Gupta, and N. Verma. 2023. *Industrial applications of soil microbes*, vol. 2. Bentham Science Publishers Pte Ltd. https://doi.org/10.2174/97898150502641230201.

Jain, S., A. Gupta, and N. Verma. 2024. *Industrial applications of soil microbes*, vol. 4. Bentham Science Publishers Pte Ltd. https://doi.org/10.2174/97898151249961240401.

Jayaprakashvel, M. 2012. Therapeutically active biomolecules from marine actinomycetes. *Journal of Modern Biotechnology* 1: 1–7.

Jiang, Z., Y. Hama, K. Yamaguchi, and T. Oda. 2012. Inhibitory effect of sulphated polysaccharide porphyran on nitric oxide production in lipopolysaccharide-stimulated RAW264.7

macrophages. *The Journal of Biochemistry* 151: 65–74. https://doi.org/10.1093/jb/mvr115.

Jin, M., H. Liu, Y. Hou, et al. 2017. Preparation, characterization and alcoholic liver injury protective effects of algal oligosaccharides from *Gracilaria lemaneiformis*. *Food Research International* 100: 186–95. https://doi.org/10.1016/j.foodres.2017.08.032.

Joshi, S., R. Kumari, and V.N. Upasani. 2018. Applications of algae in cosmetics: An overview. *International Journal of Innovative Research in Science, Engineering and Technology* 7: 1269–78. https://doi.org/10.15680/IJIRSET.2018.0702038.

Kamoun, F., H. Mejdoub, H. Aouissaoui, J. Reinbolt, A. Hammami, and S. Jaoua. 2005. Purification, amino acid sequence and characterization of Bacthuricin F4, a new bacteriocin produced by *Bacillus thuringiensis*. *Journal of Applied Microbiology* 98: 881–88. https://doi.org/10.1111/j.1365-2672.2004.02513.x.

Kang, C.D., J.S. Lee, T.H. Park, and S.J. Sim. 2005. Comparison of heterotrophic and photoautotrophic induction on astaxanthin production by *Haematococcus pluvialis*. *Applied Microbiology and Biotechnology* 68: 237–41. https://doi.org/10.1007/s00253-005-1889-2.

Kang, M.C., Y. Ding, E.A. Kim, et al. 2017. Indole derivatives isolated from brown alga *Sargassum thunbergii* inhibit adipogenesis through AMPK activation in 3T3-L1 preadipocytes. *Marine Drugs* 15 (April): 119. https://doi.org/10.3390/md15040119.

Karthikeyan, A., A. Joseph, and B.G. Nair. 2022. Promising bioactive compounds from the marine environment and their potential effects on various diseases. *Journal of Genetic Engineering and Biotechnology* 20 (January): 14. https://doi.org/10.1186/s43141-021-00290-4.

Karuppiah, P., and M. Mustaffa. 2013. Antibacterial and antioxidant activities of *Musa* sp. leaf extracts against multidrug resistant clinical pathogens causing nosocomial infection. *Asian Pacific Journal of Tropical Biomedicine* 3: 737–42. https://doi.org/10.1016/S2221-1691(13)60148-3.

Kavitha, P., A.P. Lipton, A.R. Sarika, and M.S. Aishwarya. 2011. Growth characteristics and halocin production by a new isolate, *Haloferax volcanii* KPS1 from Kovalam solar saltern (India). *Research Journal of Biological Sciences* 6: 257–62. https://doi.org/10.3923/rjbsci.2011.257.262.

Kennedy, R.K., P.R. Naik, V. Veena, et al. 2015. 5-Methyl phenazine-1-carboxylic acid: A novel bioactive metabolite by a rhizosphere soil bacterium that exhibits potent antimicrobial and anticancer activities. *Chemico-Biological Interactions* 231: 71–82. https://doi.org/10.1016/j.cbi.2015.03.002.

Kim, D., J.S. Lee, Y.K. Park, et al. 2007. Biosynthesis of antibiotic prodiginines in the marine bacterium *Hahella chejuensis* KCTC 2396. *Journal of Applied Microbiology* 102: 937–44. https://doi.org/10.1111/j.1365-2672.2006.03172.x.

Kim, J.H., E.J. Yun, S. Yu, K.H. Kim, and N.J. Kang. 2017. Different levels of skin whitening activity among 3, 6-anhydro-l-galactose, agarooligosaccharides, and neoagarooligosaccharides. *Marine Drugs* 15 (October): 321. https://doi.org/10.3390/md15100321.

Kim, S.Y., G. Ahn, H.S. Kim, J.G. Je, K.N. Kim, and Y.J. Jeon. 2020. Diphlorethohydroxycarmalol (DPHC) isolated from the brown alga *Ishige okamurae* acts on inflammatory myopathy as an inhibitory agent of TNF-α. *Marine Drugs* 18 (October): 529. https://doi.org/10.3390/md18110529.

Kolanjinathan, K., P. Ganesh, and P. Saranraj. 2014. Pharmacological importance of seaweeds: A review. *World Journal of Fish and Marine Sciences* 6: 1–15. https://doi.org/10.5829/idosi.wjfms.2014.06.01.76195.

Kour, D., K.L. Rana, A.N. Yadav, et al. 2020. Microbial biofertilizers: Bioresources and eco-friendly technologies for agricultural and environmental sustainability. *Biocatalysis and Agricultural Biotechnology* 23 (January): 101487. https://doi.org/10.1016/j.bcab.2019.101487

Krištůfek, V., K. Ravasz, and V. Pižl. 1993. Actinomycetes communities in earthworm guts and surrounding soil. *Pedobiologia* 37: 379–84. https://dn720003.ca.archive.org/0/items/kristufek-et-al.-1993/Kri%C5%A1t%C5%AFfek%20et%20al.%201993.pdf.

Kumar, A., S. Prajapati, Nikhil, S. Nandan, and T.G. Neogi. 2019. Industrially important pigments from different groups of fungi. In *Recent advancement in white biotechnology through fungi. Fungal biology*, eds. A. Yadav, S. Singh, S. Mishra, and A. Gupta, 285–301. Switzerland Cham: Springer Nature. https://doi.org/10.1007/978-3-030-14846-1_10.

Kumar, A., R. Singh, D.D. Giri, P.K. Singh, and K.D. Pandey. 2014a. Effect of *Azotobacter chroococcum* CL13 inoculation on growth and curcumin content of turmeric (*Curcuma longa* L.). *International Journal of Current Microbiology and Applied Sciences* 3: 275–83. https://ijcmas.com/vol-3-9/Ajay%20Kumar,%20et%20al.pdf.

Kumar, R., K. Biswas, V. Soalnki, P. Kumar, and A. Tarafdar. 2014b. Actinomycetes: Potential bioresource for human welfare: A review. *Research Journal of Chemical and Environmental Sciences* 2: 5–16. https://aelsindia.com/june_2014.htm.

Kusari, S., C. Hertweck, and M. Spiteller. 2012. Chemical ecology of endophytic fungi: Origins of secondary metabolites. *Chemistry & Biology* 19: 792–98. https://doi.org/10.1016/j.chembiol.2012.06.004.

Kuster, E., and S.T. Williams. 1964. Production of hydrogen sulfide by streptomycetes and methods for its detection. *Applied Microbiology* 12: 46–52. https://pmc.ncbi.nlm.nih.gov/articles/PMC1058063/pdf/applmicro00351-0054.pdf.

Lagarda, M.J., G. García-Llatas, and R. Farré. 2006. Analysis of phytosterols in foods. *Journal of Pharmaceutical and Biomedical Analysis* 41: 1486–96. https://doi.org/10.1016/j.jpba.2006.02.052.

Lamers, P.P., C.C.W van de Laak, and P.S. Kaasenbrood. 2010. Carotenoid and fatty acid metabolism in light-stressed *Dunaliella salina*. *Biotechnology and Bioengineering* 106: 638–48. https://doi.org/10.1002/bit.22725.

Le Marrec, C., B. Hyronimus, P. Bressollier, B. Verneuil, and M.C. Urdaci. 2000. Biochemical and genetic characterization of coagulin, a new antilisterial bacteriocin in the pediocin family of bacteriocins, produced by *Bacillus coagulans* I(4). *Applied and Environmental Microbiology* 66: 5213–20. https://doi.org/10.1128/AEM.66.12.5213-5220.2000.

Leandro, A., L. Pereira, and A.M.M. Gonçalves. 2019. Diverse applications of marine macroalgae. *Marine Drugs* 18 (December): 17. https://doi.org/10.3390/md18010017.

Lee, S.H., and Y.J. Jeon. 2013. Anti-diabetic effects of brown algae derived phlorotannins, marine polyphenols through diverse mechanisms. *Fitoterapia* 86: 129–36. https://doi.org/10.1016/j.fitote.2013.02.013.

Lequerica, J.L., J.E. O'Connor, L. Such, et al. 2006. A halocin acting on Na^+/H^+ exchanger of Haloarchaea as a new type of inhibitor in NHE of mammals. *Journal of Physiology and Biochemistry* 62: 253–62. https://doi.org/10.1007/BF03165754.

Leya, T., A. Rahn, C. Lütz, and D. Remias. 2009. Response of arctic snow and permafrost algae to high light and nitrogen stress by changes in pigment composition and applied aspects for biotechnology. *FEMS Microbiology Ecology* 67: 432–43. https://doi.org/10.1111/j.1574-6941.2008.00641.x.

Li, J.W.H., and J.C. Vederas. 2009. Drug discovery and natural products: End of an era or an endless frontier? *Science* 325: 161–65. https://doi.org/10.1126/science.1168243.

Li, Y., G. Wei, and J. Chen. 2004. Glutathione: A review on biotechnological production. *Applied Microbiology and Biotechnology* 66: 233–42. https://doi.org/10.1007/s00253-004-1751-y.

Li, Z., X. Ma, A. Li, and C. Zhang. 2012. A novel potential source of β-carotene: *Eustigmatos* cf. *polyphem* (Eustigmatophyceae) and pilot β-carotene production in bubble column and flat panel photobioreactors. *Bioresource Technology* 117: 257–63. https://doi.org/10.1016/j.biortech.2012.04.069.

Liang, H., Z. Song, Z. Zhong, et al. 2022. Mining of the biosynthetic potential of archaea reveals diverse anti-archaeal lanthipeptides. *Research Square*: 1–21. https://doi.org/10.21203/rs.3.rs-1476892/v1.

Liu, B.H., and Y.K. Lee. 2000. Secondary carotenoids formation by the green alga *Chlorococcum* sp. *Journal of Applied Phycology* 12: 301–7. https://doi.org/10.1023/A:1008185212724.

Liu, H., Y. He, H. Jiang, H. Peng, X. Huang, X. Zhang, L.S. Thomashow, and Y. Xu. 2007. Characterization of a phenazine-producing strain *Pseudomonas chlororaphis* GP72 with broad-spectrum antifungal activity from green pepper rhizosphere. *Current Microbiology* 54: 302–06.

Liu, M., P.E. Hansen, and X. Lin. 2011. Bromophenols in marine algae and their bioactivities. *Marine Drugs* 9: 1273–92. https://doi.org/10.3390/md9071273.

Loper, J.E., M.D. Henkels, B.T. Shaffer, F.A. Valeriote, and H. Gross. 2008. Isolation and identification of rhizoxin analogs from *Pseudomonas fluorescens* Pf-5 by using a genomic mining strategy. *Applied and Environmental Microbiology* 74: 3085–93. https://doi.org/10.1128/AEM.02848-07.

Ma, G., W. Yang, L. Zhao, F. Pei, D. Fang, and Q. Hu. 2018. A critical review on the health promoting effects of mushrooms nutraceuticals. *Food Science and Human Wellness* 7: 125–33. https://doi.org/10.1016/j.fshw.2018.05.002.

Ma, R.Y.N., and F. Chen. 2001. Enhanced production of free trans-astaxanthin by oxidative stress in the cultures of the green microalga *Chlorococcum* sp. *Process Biochemistry* 36: 1175–79. https://doi.org/10.1016/S0032-9592(01)00157-1.

Ma, Z., N. Geudens, N.P. Kieu, et al. 2016. Biosynthesis, chemical structure, and structure-activity relationship of orfamide lipopeptides produced by *Pseudomonas protegens* and related species. *Frontiers in Microbiology* 7 (March): 382. https://doi.org/10.3389/fmicb.2016.00382.

Macheleidt, J., D.J. Mattern, J. Fischer, et al. 2016. Regulation and role of fungal secondary metabolites. *Annual Review of Genetics* 50: 371–92. https://doi.org/10.1146/annurev-genet-120215-035203.

Macherla, V.R., J. Liu, M. Sunga, et al. 2007. Lipoxazolidinones A, B, and C: Antibacterial 4-oxazolidinones from a marine actinomycete isolated from a Guam marine sediment. *The Journal of Natural Products* 70: 1454–57. https://doi.org/10.1021/np0702032.

Macías-Rubalcava, M.L., and R.E. Sánchez-Fernández. 2017. Secondary metabolites of endophytic *Xylaria* species with potential applications in medicine and agriculture. *World Journal of Microbiology and Biotechnology* 33 (November): 15. https://doi.org/10.1007/s11274-016-2174-5.

Maeda, H., M. Hosokawa, T. Sashima, K. Funayama, and K. Miyashita. 2005. Fucoxanthin from edible seaweed, *Undaria pinnatifida*, shows antiobesity effect through UCP1 expression in white adipose tissues. *Biochemical and Biophysical Research Communications* 332: 392–97. https://doi.org/10.1016/j.bbrc.2005.05.002.

Makkar, F., and K. Chakraborty. 2017. Antidiabetic and anti-inflammatory potential of sulphated polygalactans from red seaweeds *Kappaphycus alvarezii* and *Gracilaria opuntia*. *International Journal of Food Properties* 20: 1326–37. https://doi.org/10.1080/10942912.2016.1209216.

Mandelli, F., V.S. Miranda, E. Rodrigues, and A.Z. Mercadante. 2012. Identification of carotenoids with high antioxidant capacity produced by extremophile microorganisms. *World Journal of Microbiology and Biotechnology* 28: 1781–90. https://doi.org/10.1007/s11274-011-0993-y.

Mao, Y., M. Varoglu, and D.H. Sherman. 1999. Molecular characterization and analysis of the biosynthetic gene cluster for the antitumor antibiotic mitomycin C from *Streptomyces lavendulae* NRRL 2564. *Chemistry & Biology* 6: 251–63. https://doi.org/10.1016/S1074-5521(99)80040-4.

Mapari, S.A.S., A.S. Meyer, U. Thrane, and J.C. Frisvad. 2009. Identification of potentially safe promising fungal cell factories for the production of polyketide natural food colorants using chemotaxonomic rationale. *Microbial Cell Factories* 8: 1–15. https://doi.org/10.1186/1475-2859-8-24.

Markou, G., and E. Nerantzis. 2013. Microalgae for high-value compounds and biofuels production: A review with focus on cultivation under stress conditions. *Biotechnology Advances* 31: 1532–42. https://doi.org/10.1016/j.biotechadv.2013.07.011.

Matsukawa, R., M. Hotta, Y. Masuda, M. Chihara, and I. Karube. 2000. Antioxidants from carbon dioxide fixing *Chlorella sorokiniana*. *Journal of Applied Phycology* 12: 263–67. https://doi.org/10.1023/A:1008141414115.

McArthur, K.A., S.S. Mitchell, G. Tsueng, et al. 2008. Lynamicins A– E, chlorinated bisindole pyrrole antibiotics from a novel marine actinomycete. *Journal of Natural Products* 71: 1732–37. https://doi.org/10.1021/np800286d.

Mehnaz, S., and G. Lazarovits. 2006. Inoculation effects of *Pseudomonas putida*, *Gluconacetobacter azotocaptans*, and *Azospirillum lipoferum* on corn plant growth under greenhouse conditions. *Microbial Ecology* 51: 326–35. https://doi.org/10.1007/s00248-006-9039-7.

Mehnaz, S., R.S.Z. Saleem, B. Yameen, et al. 2013. Lahorenoic acids A–C, ortho-dialkyl-substituted aromatic acids from the biocontrol strain *Pseudomonas aurantiaca* PB-St2. *Journal of Natural Products* 76: 135–41. https://doi.org/10.1021/np3005166.

Mendes, R.L., B.P. Nobre, M.T. Cardoso, A.P. Pereira, and A.F. Palavra. 2003. Supercritical carbon dioxide extraction of compounds with pharmaceutical importance from microalgae. *Inorganica Chimica Acta* 356: 328–34. https://doi.org/10.1016/S0020-1693(03)00363-3.

Mhadhebi, L., A. Mhadhebi, J. Robert, and A. Bouraoui. 2014. Antioxidant, anti-inflammatory and antiproliferative effects of aqueous extracts of three mediterranean brown seaweeds of the genus *Cystoseira*. *Iranian Journal of Pharmaceutical Research* 13: 207–20. https://doi.org/10.22037/ijps.v10.41485.

Mhammedi, A., Peypoux, F., Besson, F., and Michel, G. 1982. Bacillomycin F, a new antibiotic of iturin group: Isolation and characterization. *The Journal of Antibiotics* 35: 306–11. https://doi.org/10.7164/antibiotics.35.306.

Mikami, Y. 2007. Biological work on medically important *Nocardia* species. *Actinomycetologica* 21: 46–51. https://doi.org/10.3209/saj.SAJ210107.

Miller, C.M., R.V. Miller, D. Garton-Kenny, et al. 1998. Ecomycins, unique antimycotics from *Pseudomonas Viridiflava*. *Journal of Applied Microbiology* 84: 937–44. https://doi.org/10.1046/j.1365-2672.1998.00415.x.

Mishra, V.K., A.K. Passari, P. Chandra, et al. 2017. Determination and production of antimicrobial compounds by *Aspergillus clavatonanicus* strain MJ31, an endophytic fungus from *Mirabilis jalapa* L. using UPLC-ESI-MS/MS and TD-GC-MS analysis. *PLoS One* 12 (October): e0186234. https://doi.org/10.1371/journal.pone.0186234.

Mitova, M., S. Popov, and S. De Rosa. 2004. Cyclic peptides from a *Ruegeria* strain of bacteria associated with the sponge *Suberites domuncula*. *Journal of Natural Products* 67: 1178–81. https://doi.org/10.1021/np049900+

Mohamed, Z.A. 2008. Polysaccharides as a protective response against microcystin-induced oxidative stress in *Chlorella vulgaris* and *Scenedesmus quadricauda* and their possible significance in the aquatic ecosystem. *Ecotoxicology* 17: 504–16. https://doi.org/10.1007/s10646-008-0204-2.

Mohan, K.D., and U. Rajamanickam. 2018. Biodiversity of actinomycetes and secondary metabolites. *International Journal of Sciences* 5: 22–27. https://innoriginal.com/index.php/iijs/issue/view/24.

Moore, B.S., J.A. Trischman, D. Seng, D. Kho, P.R. Jensen, and W. Fenical. 1999. Salinamides, antiinflammatory depsipeptides from a marine streptomycete. *Journal of Organometallic Chemistry* 64: 1145–50. https://doi.org/10.1021/jo9814391.

Mouloud, G., H. Daoud, J. Bassem, I. Laribi Atef, and B. Hani. 2013. New Bacteriocin from *Bacillus clausii* StrainGM17: Purification, Characterization, and Biological Activity. *Applied Biochemistry and Biotechnology* 171: 2186–200. https://doi.org/10.1007/s12010-013-0489-3.

Nachtigall, J., A. Kulik, S. Helaly, et al. 2011. Atacamycins A–C, 22-membered antitumor macrolactones produced by *Streptomyces* sp. C38. *The Journal of Antibiotics* 64: 775–80. https://doi.org/10.1038/ja.2011.96.

Nandi, M., C. Selin, A.K.C. Brassinga, et al. 2015. Pyrrolnitrin and hydrogen cyanide production by *Pseudomonas chlororaphis* strain PA23 exhibits nematicidal and repellent activity against *Caenorhabditis Elegans*. *PLoS One* 10 (April): e0123184. https://doi.org/10.1371/journal.pone.0123184.

Narayanan, Z., and B.R. Glick. 2022. Secondary metabolites produced by plant growth-promoting bacterial endophytes. *Microorganisms* 10 (October): 2008. https://doi.org/10.3390/microorganisms10102008.

Nawaz, A., R. Chaudhary, Z. Shah, et al. 2021. An overview on industrial and medical applications of bio-pigments synthesized by marine bacteria. *Microorganisms* 9 (December): 11. https://doi.org/10.3390/microorganisms9010011.

Nercessian, D., L. Di Meglio, R. De Castro, and R. Paggi. 2015. Exploring the multiple biotechnological potential of halophilic microorganisms isolated from two Argentinean salterns. *Extremophiles* 19: 1133–43. https://doi.org/10.1007/s00792-015-0785-7.

Nguyen, H.P.T., M. Morançais, J. Fleurence, and J. Dumay. 2017. *Mastocarpus stellatus* as a source of R-phycoerythrin: Optimization of enzyme assisted extraction using response surface methodology. *Journal of Applied Phycology* 29: 1563–70. https://doi.org/10.1007/s10811-016-1024-z.

Nguyen, V.T., J.S. Lee, Z.J. Qian, et al. 2013. Gliotoxin isolated from marine fungus *Aspergillus* sp. induces apoptosis of human cervical cancer and chondrosarcoma cells. *Marine Drugs* 12: 69–87. https://doi.org/10.3390/md12010069.

Okazaki, T., T. Kitahara, and Y. Okami. 1975. Studies on marine microorganisms. IV A new antibiotic SS-228 Y produced by *Chainia* isolated from shallow sea mud. *The Journal of Antibiotics* 28: 176–84. https://doi.org/10.7164/antibiotics.28.176.

Ongena, M., P. Jacques, Y. Touré, J. Destain, A. Jabrane, and P. Thonart. 2005. Involvement of fengycin-type lipopeptides in the multifaceted biocontrol potential of *Bacillus subtilis*. *Applied Microbiology and Biotechnology* 69: 29–38. https://doi.org/10.1007/s00253-005-1940-3.

Órdenes-Aenishanslins, N., G. Anziani-Ostuni, M. Vargas-Reyes, J. Alarcón, A. Tello, and J. Pérez-Donoso. 2016. Pigments from UV-resistant Antarctic bacteria as photosensitizers in dye sensitized solar cells. *Journal of Photochemistry and Photobiology B: Biology* 162: 707–14. https://doi.org/10.1016/j.jphotobiol.2016.08.004.

Palanisamy, S., M. Vinosha, T. Marudhupandi, P. Rajasekar, and N.M. Prabhu. 2017. Isolation of fucoidan from *Sargassum polycystum* brown algae: Structural characterization, *in vitro* antioxidant and anticancer activity. *International Journal of Biological Macromolecules* 102: 405–12. https://doi.org/10.1016/j.ijbiomac.2017.03.182.

Pan, F., T.J. Su, S.M. Cai, and W. Wu. 2017. Fungal endophyte-derived *Fritillaria unibracteata* var. *wabuensis*: Diversity, antioxidant capacities *in vitro* and relations to phenolic, flavonoid or saponin compounds. *Scientific Reports* 7 (February): 42008. https://doi.org/10.1038/srep42008.

Pandey, K.D., A.K. Patel, M. Singh, and A. Kumari. 2021. Secondary metabolites from bacteria and viruses. In *Natural bioactive compounds*, ed. R.P. Sinha and D.P. Häder, 19–40. London: Academic Press. https://doi.org/10.1016/C2019-0-00797-2.

Pandey, V.D., A. Pandey, and V. Sharma. 2013. Biotechnological applications of cyanobacterial phycobiliproteins. *International Journal of Current Microbiology and Applied Sciences* 2: 89–97. https://ijcmas.com/vol-2-9/V.D.%20Pandey,%20et%20al.pdf.

Pardo-Vargas, A., I.D.B. Oliveira, P.R.S. Stephens, et al. 2014. Dolabelladienols A–C, new diterpenes isolated from Brazilian brown alga *Dictyota pfaffii*. *Marine Drugs* 12: 4247–59. https://doi.org/10.3390/md12074247.

Parret, A.H., G. Schoofs, P. Proost, and R. De Mot. 2003. Plant lectin-like bacteriocin from a rhizosphere-colonizing *Pseudomonas* isolate. *Journal of Bacteriology* 185: 897–908. https://doi.org/10.1128/jb.185.3.897-908.2003.

Patil, V., T. Källqvist, E. Olsen, G. Vogt, and H.R. Gislerød. 2007. Fatty acid composition of 12 microalgae for possible use in aquaculture feed. *Aquaculture International* 15: 1–9. https://doi.org/10.1007/s10499-006-9060-3.

Pelah, D., A. Sintov, and E. Cohen. 2004. The effect of salt stress on the production of canthaxanthin and astaxanthin by *Chlorella zofingiensis* grown under limited light intensity. *World Journal of Microbiology and Biotechnology* 20: 483–86. https://doi.org/10.1023/B:WIBI.0000040398.93103.21.

Petersen, L.E., M.Y. Kellermann, and P.J. Schupp. 2020. Secondary metabolites of marine microbes: From natural products chemistry to chemical ecology. In *YOUMARES 9 – The oceans: Our research, our future*, eds. S. Jungblut, V. Liebich, and M. Bode-Dalby, 159–80. Cham: Springer. https://doi.org/10.1007/978-3-030-20389-4_8.

Piironen, V., J. Toivo, R. Puupponen-Pimiä, and A.M. Lampi. 2003. Plant sterols in vegetables, fruits and berries. *Journal of the Science of Food and Agriculture* 83: 330–37. https://doi.org/10.1002/jsfa.1316.

Pinheiro, E.A.A., J.R. Pina, A.O. Feitosa, et al. 2017. Bioprospecting of antimicrobial activity of extracts of endophytic fungi from *Bauhinia guianensis*. *Revista Argentina De Microbiologia* 49: 3–6. https://doi.org/10.1016/j.ram.2016.08.005.

Plaza, M., M. Herrero, A. Cifuentes, and E. Ibanez, 2009. Innovative natural functional ingredients from microalgae. *Journal of Agricultural and Food Chemistry* 57: 7159–70. https://doi.org/10.1021/jf901070g.

Prangishvili, D., I. Holz, E. Stieger, S. Nickell, J.K. Kristjansson, and W. Zillig, 2000. Sulfolobicins, specific proteinaceous toxins produced by strains of the extremely thermophilic archaeal genus *Sulfolobus*. *Journal of Bacteriology* 182: 2985–88. https://doi.org/10.1128/jb.182.10.2985-2988.2000.

Prave, P., U. Faust, W. Sitting, and D.A. Sukatsch. 1987. *Fundamentals of biotechnology.* Weinheim: VCH Verlagsgesellschaft.

Qin, S., G.X. Liu, and Z.Y. Hu. 2008. The accumulation and metabolism of astaxanthin in *Scenedesmus obliquus* (Chlorophyceae). *Process Biochemistry* 43: 795–802. https://doi.org/10.1016/j.procbio.2008.03.010.

Rabbani, S., P. Beyer, J.V. Lintig, P. Hugueney, and H. Kleinig. 1998. Induced β-carotene synthesis driven by triacylglycerol deposition in the unicellular alga *Dunaliella bardawil. Plant Physiology* 116: 1239–48. https://doi.org/10.1104/pp.116.4.1239.

Ragan, M.A. and K-W. Glombitza. 1986. Phlorotannins, brown algal polyphenols. In *Progress in Phycological Research*, eds. F.E. Round and D.J. Chapman, 129–241. Bristol: Biopress.

Rahelivao, M.P., M. Gruner, H. Andriamanantoanina, I. Bauer, and H.J. Knölker. 2015. Brown algae (Phaeophyceae) from the coast of Madagascar: Preliminary bioactivity studies and isolation of natural products. *Natural Products and Bioprospecting* 5: 223–35. https://doi.org/10.1007/s13659-015-0068-0.

Raimundo, I., S.G. Silva, R. Costa, and T. Keller-Costa. 2018. Bioactive secondary metabolites from octocoral-associated microbes – new chances for blue growth. *Marine Drugs* 16 (December): 485. https://doi.org/10.3390/md16120485.

Raistrick, H. 1949. *Proceedings of the royal society of London. Series A, mathematical and physical sciences.* Vol. 199. London: The Royal Society.

Ramarathnam, R., S. Bo, Y. Chen, W.G.D. Fernando, G. Xuewen, and T. De Kievit. 2007. Molecular and biochemical detection of fengycin-and bacillomycin D-producing *Bacillus* spp., antagonistic to fungal pathogens of canola and wheat. *Canadian Journal of Microbiology* 53: 901–11. https://doi.org/10.1139/W07-049.

Rana, K.L., D. Kour, I. Sheikh, et al. 2019. Biodiversity of endophytic fungi from diverse niches and their biotechnological applications. In *Advances in endophytic fungal research*, ed. B. Singh, 105–44. Cham: Springer. https://doi.org/10.1007/978-3-030-03589-1_6.

Rao, A.R., R. Sarada, V. Baskaran, and G.A. Ravishankar. 2006. Antioxidant activity of *Botryococcus braunii* extract elucidated *in vitro* models. *Journal of Agricultural and Food Chemistry* 54: 4593–99. https://doi.org/10.1021/jf060799j.

Rasmussen, R.S., and M.T. Morrissey. 2007. Marine biotechnology for production of food ingredients. *Advances in Food and Nutrition Research* 52: 237–92. https://doi.org/10.1016/S1043-4526(06)52005-4.

Rea, M.C., C.S. Sit, E. Clayton, et al. 2010. Thuricin CD, a post-translationally modified bacteriocin with a narrow spectrum of activity against *Clostridium difficile. Proceedings of the National Academy of Sciences* 107: 9352–57. https://doi.org/10.1073/pnas.0913554107.

Reitan, K.I., J.R. Rainuzzo, and Y. Olsen. 1994. Effect of nutrient limitation on fatty acid and lipid content of marine microalgae. *Journal of Phycology* 30: 972–79. https://doi.org/10.1111/j.0022-3646.1994.00972.x.

Remias, D., A. Holzinger, and C. Lütz. 2009. Physiology, ultrastructure and habitat of the ice alga *Mesotaenium berggrenii* (Zygnemaphyceae, Chlorophyta) from glaciers in the European Alps. *Phycologia* 48: 302–12. https://doi.org/10.2216/08-13.1.

Rifaat, H.M. 2003. The biodiversity of actinomycetes in the river Nile exhibiting antifungal activity. *Journal of Mediterranean Ecology* 4: 5–7. http://www.jmecology.com/wp-content/uploads/2014/03/01rifaat.pdf.

Rodrigues, D., C. Alves, A. Horta, S. Pinteus, J. Silva, G. Culioli, O.P. Thomas, and R. Pedrosa. 2015. Antitumor and antimicrobial potential of bromoditerpenes isolated from the red alga, *Sphaerococcus coronopifolius. Marine Drugs* 13(2): 713–26.

Rodriguez Jimenez, A., E. Dechamps, A. Giaux, et al. 2021. The sponges *Hymeniacidon perlevis* and *Halichondria panicea* are reservoirs of antibiotic-producing bacteria against multi-drug resistant *Staphylococcus aureus. Journal of Applied Microbiology* 131: 706–18. https://doi.org/10.1111/jam.14999.

Roongsawang, N., J. Thaniyavarn, S. Thaniyavarn, et al. 2002. Isolation and characterization of a halotolerant *Bacillus subtilis* BBK-1 which produces three kinds of lipopeptides: Bacillomycin L, plipastatin, and surfactin. *Extremophiles* 6: 499–506. https://doi.org/10.1007/s00792-002-0287-2.

Sahu, M.K., K. Sivakumar, E. Poorani, T. Thangaradjou, and L. Kannan. 2007. Studies on l-asparaginase enzyme of actinomycetes isolated from esturine fishes. *Journal of Environmental Biology* 28: 465–74. https://jeb.co.in/journal_issues/200704_apr07_supp/paper_19.pdf.

Saito, M., M. Kawai, H. Hagino, et al. 2002. P-488: Antihypertensive effect of Nori-peptides derived from red alga *Porphyra yezoensis* in hypertensive patients. *American Journal of Hypertension* 15: 210A. https://doi.org/10.1016/S0895-7061(02)02839-X.

Sánchez, J.F., J.M. Fernandez-Sevilla, F.G. Acien, M.C. Ceron, J. Perez-Parra, and E. Molina-Grima. 2008. Biomass and lutein productivity of *Scenedesmus almeriensis*: Influence of irradiance, dilution rate and temperature. *Applied Microbiology and Biotechnology* 79: 719–29. https://doi.org/10.1007/s00253-008-1494-2.

Sasidharan, A., N.K. Sasidharan, D.B. Amma, et al. 2015. Antifungal activity of violacein purified from a novel strain of *Chromobacterium* sp. NIIST (MTCC 5522). *Journal of Microbiology* 53: 694–701. https://doi.org/10.1007/s12275-015-5173-6.

Sathasivam, R., R. Radhakrishnan, A. Hashem, and E.F. Abd_Allah. 2017. Microalgae metabolites: A rich source for food and medicine. *Saudi Journal of Biological Sciences* 26: 709–22. https://doi.org/10.1016/j.sjbs.2017.11.003.

Sayed, A.M., M.H. Hassan, H.A. Alhadrami, H.M. Hassan, M. Goodfellow, and M.E. Rateb. 2020. Extreme environments: Microbiology leading to specialized metabolites. *Journal of Applied Microbiology* 128: 630–57. https://doi.org/10.1111/jam.14386.

Schneider, Y., M. Jenssen, J. Isaksson, K.O. Hansen, J.H. Andersen, and E.H. Hansen. 2020. Bioactivity of serratiochelin A, a siderophore isolated from a co-culture of *Serratia* sp. and *Shewanella* sp. *Microorganisms* 8 (July): 1042, 1–17. https://doi.org/10.3390/microorganisms8071042.

Selim, M.S.M., S.A. Abdelhamid, and S.S. Mohamed. 2021. Secondary metabolites and biodiversity of actinomycetes. *Journal of Genetic Engineering and Biotechnology* 19: 1–13. https://doi.org/10.1186/s43141-021-00156-9.

Senthivelan, T., J. Kanagaraj, and R.C. Panda. 2016. Recent trends in fungal laccase for various industrial applications: An eco-friendly approach-A review. *Biotechnology and Bioprocess Engineering* 21: 19–38. https://doi.org/10.1007/s12257-015-0278-7.

Shahid, I., M. Rizwan, D.N. Baig, R.S. Saleem, K.A. Malik, and S. Mehnaz. 2017. Secondary metabolites production and plant growth promotion by *Pseudomonas chlororaphis* and *P. aurantiaca* strains isolated from cactus, cotton, and para grass *Journal of Microbiology and Biotechnology* 27: 480–91. https://doi.org/10.4014/jmb.1601.01021.

Shalaby, E.A., S.M.M. Shanab, and V. Singh. 2010. Salt stress enhancement of antioxidant and antiviral efficiency of *Spirulina platensis*. *The Journal of Medicinal Plants Research* 4: 2622–32. https://doi.org/10.5897/JMPR09.300.

Sharma, M., P. Dangi, and M. Choudhary 2014. Actinomycetes: Source, identification, and their applications. *International Journal of Current Microbiology and Applied Sciences* 3: 801–32. https://www.ijcmas.com/vol-3-2/Mukesh%20Sharma.pdf.

Sheldon, P.J., Y. Mao, M. He, and D.H. Sherman. 1999. Mitomycin resistance in *Streptomyces lavendulae* includes a novel drug-binding-protein-dependent export system. *Journal of Bacteriology* 181: 2507–12. https://doi.org/10.1128/jb.181.8.2507-2512.1999.

Shnit-Orland, M., A. Sivan, A. Kushmaro. 2012. Antibacterial activity of *Pseudoalteromonas* in the coral holobiont. *Microbial Ecology* 64: 851–59. https://doi.org/10.1007/s00248-012-0086-y.

Shi, X., Z. Wu, and F. Chen. 2006. Kinetic modeling of lutein production by heterotrophic *Chlorella* at various pH and temperatures. *Molecular Nutrition & Food Research* 50(8): 763–68.

Singh, A., D.K. Singh, R.N. Kharwar, J.F. White, and S.K. Gond. 2021. Fungal endophytes as efficient sources of plant-derived bioactive compounds and their prospective applications in natural product drug discovery: Insights, avenues, and challenges. *Microorganisms 9* (January): 197, 1–47. https://doi.org/10.3390/microorganisms9010197.

Singh, B., and R.A. Sharma. 2015. Plant terpenes: Defense responses, phylogenetic analysis, regulation and clinical applications. *3 Biotech* 5: 129–51. https://doi.org/10.1007/s13205-014-0220-2.

Singh, N.K., and D.W. Dhar. 2011. Microalgae as second generation biofuel. A review. *Agronomy for Sustainable Development* 31: 605–29. https://doi.org/10.1007/s13593-011-0018-0.

Singh, R., and A.K. Dubey. 2018. Diversity and applications of endophytic actinobacteria of plants in special and other ecological niches. *Frontiers in Microbiology* 9 (August): 1767, 1–30. https://doi.org/10.3389/fmicb.2018.01767

Smyrniotopoulos, V., C. Vagias, M.M. Rahman, S. Gibbons, and V. Roussis. 2010. Ioniols I and II, tetracyclic diterpenes with antibacterial activity, from *Sphaerococcus coronopifoliusChemistry & Biodiversity* 7: 666–76. https://doi.org/10.1002/cbdv.200900026.

Snipes, C.E., D.O. Duebelbeis, M. Olson, et al. 2007. The ansacarbamitocins: Polar ansamitocin derivatives. *Journal of Natural Products* 70: 1578–81. https://doi.org/10.1021/np070275t.

Son, S., S.K. Ko, M. Jang, et al. 2016. New cyclic lipopeptides of the iturin class produced by saltern-derived *Bacillus* sp. KCB14S006. *Marine Drugs* 14 (April): 72, 1–12. https://doi.org/10.3390/md14040072.

Song, C., R. Schmidt, V. de Jager, et al. 2015. Exploring the genomic traits of fungus-feeding bacterial genus *Collimonas*. *Bmc Genomics* 16 (December): 1103, 1–17. https://doi.org/10.1186/s12864-015-2289-3.

Srinivasan, R., A. Kannappan, C. Shi, and X. Lin. 2021. Marine bacterial secondary metabolites: A treasure house for structurally unique and effective antimicrobial compounds. *Marine Drugs* 19 (September): 530, 1–36. https://doi.org/10.3390/md19100530.

Stodůlková, E., P. Man, M. Kuzma, et al. 2015. A highly diverse spectrum of naphthoquinone derivatives produced by the endophytic fungus *Biatriospora* sp. CCF 4378. *Folia Microbiologica* 60: 259–67. https://doi.org/10.1007/s12223-014-0366-7.

Takahashi, J.A., B.V.R. Barbosa, B.D.A. Martins, C.P. Guirlanda, and M.A.F. Moura. 2020. Use of the versatility of fungal metabolism to meet modern demands for healthy aging, functional foods, and sustainability. *Journal of Fungi* 6 (October): 223, 1–27. https://doi.org/10.3390/jof6040223.

Taki, A.C., R. Brkljača, T. Wang, et al. 2020. Natural compounds from the marine brown alga *Caulocystis cephalornithos* with potent *in vitro*-activity against the parasitic nematode *Haemonchus contortus*. *Pathogens* 9 (July): 550, 1–15. https://doi.org/10.3390/pathogens9070550.

Talero, E., S. García-Mauriño, J. Ávila-Román, A. Rodríguez-Luna, A. Alcaide, V. Motilva. 2015. Bioactive compounds isolated from microalgae in chronic inflammation and cancer. *Marine drugs* 13: 6152–209. https://doi.org/10.3390/md13106152.

Tamehiro, N., Y. Okamoto-Hosoya, S. Okamoto, et al. 2002. Bacilysocin, a novel phospholipid antibiotic produced by *Bacillus subtilis* 168. *Antimicrobial Agents and Chemotherapy* 46: 315–20. https://doi.org/10.1128/aac.46.2.315-320.2002.

Tan, H., Z. Deng, and L. Cao. 2009. Isolation and characterization of actinomycetes from healthy goat faeces. *Letters in Applied Microbiology* 49: 248–53. https://doi.org/10.1111/j.1472-765X.2009.02649.x.

Tan, L.T., R.T. Williamson, W.H. Gerwick, K.S. Watts, K. McGough, and R. Jacobs. 2000. Cis, cis-and trans, trans-ceratospongamide, new bioactive cyclic heptapeptides from the Indonesian red alga *Ceratodictyon spongiosum* and symbiotic sponge *Sigmadocia symbiotica*. *The Journal of Organic Chemistry* 65: 419–25. https://doi.org/10.1021/jo991165x.

Tannin-Spitz, T., M. Bergman, D. Van-Moppes, S. Grossman, and S. Arad. 2005. Antioxidant activity of the polysaccharide of the red microalga *Porphyridium* sp. *Journal of Applied Phycology* 17: 215–22. https://doi.org/10.1007/s10811-005-0679-7.

Tatum, E.L. 1944. The biochemistry of fungi. *Annual Review of Biochemistry* 13: 667–704. https://archive.org/details/sim_annual-review-of-biochemistry_1944_13/page/n6/mode/1up.

Tel-Çayan, G. 2019. Phenolic profiles, antioxidant, and anticholinesterase activities of three *Gloeophyllum* species with chemometric approach. *Journal of Food Biochemistry* 43 (April): e12790. https://doi.org/10.1111/jfbc.12790.

Thampayak, I., N. Cheeptham, W. Pathom-Aree, P. Leelapornpisid, and S. Lumyong. 2008. Isolation and identification of biosurfactant producing actinomycetes from soil. *Research Journal of Microbiology* 3: 499–507. https://scialert.net/abstract/?doi=jm.2008.499.507.

Thasana, N., B. Prapagdee, N. Rangkadilok, et al. 2010. *Bacillus subtilis* SSE4 produces subtulene A, a new lipopeptide antibiotic possessing an unusual C15 unsaturated β-amino acid. *FEBS Letters* 584: 3209–14. https://doi.org/10.1016/j.febslet.2010.06.005.

Thirumurugan, D., A. Cholarajan, S.S. Raja, and R. Vijayakumar. 2018. An introductory chapter: Secondary metabolites. In *Secondary metabolites-sources and applications*, eds. R. Vijayakumar and S.S.S. Raja, 3–21. *IntechOpen*. https://www.intechopen.com/chapters/62731.

Thongsri, Y., C. Aromdee, C. Yenjai, et al. 2014. Detection of diketopiperazine and pyrrolnitrin, compounds with anti-*Pythium insidiosum* activity, in a *Pseudomonas stutzeri* environmental strain. *Biomedical Papers of the Medical Faculty of the University Palacký Olomouc Czech Republic* 158: 378–83. https://doi.org/10.5507/bp.2012.090.

Thorn, C., and K.B. Raper.1945. *A manual of the aspergilli*. Baltimore: The Williams & Wilkins Company.

Tommonaro, G., G.R. Abbamondi, C. Iodice, K. Tait, and S. De Rosa. 2012. Diketopiperazines produced by the halophilic archaeon, *Haloterrigena hispanica*, activate AHL bioreporters. *Microbial Ecology* 63: 490–95. https://doi.org/10.1007/s00248-011-9980-y.

Touré, Y., M. Ongena, P. Jacques, A. Guiro, and P, Thonart. 2004. Role of lipopeptides produced by *Bacillus subtilis* GA1 in the reduction of grey mould disease caused by *Botrytis cinerea* on apple. *Journal of Applied Microbiology* 96: 1151–60. https://doi.org/10.1111/j.1365-2672.2004.02252.x

Tzovenis, I., E. Fountoulaki, N. Dolapsakis, et al. 2009. Screening for marine nanoplanktic microalgae from Greek coastal lagoons (Ionian Sea) for use in mariculture. *Journal of Applied Phycology* 21: 457–69. https://doi.org/10.1007/s10811-008-9393-6.

Udwary, D.W., E.A. Gontang, A.C. Jones, et al. 2011. Significant natural product biosynthetic potential of actinorhizal symbionts of the genus *Frankia*, as revealed by comparative genomic and proteomic analyses. *Applied Environmental Microbiology* 11: 3617–25. https://doi.org/10.1128/AEM.00038-11.

Ullah, A., A. Bano, and H.T. Janjua. 2020. Microbial secondary metabolites and defense of plant stress. In *Microbial services in restoration ecology*, eds. J.S. Singh and S.R. Vimal, 37–46. Elsevier. https://doi.org/10.1016/C2019-0-00997-1.

Umezawa, K., M. Niikura, Y. Kojima, B. Goodell, and M. Yoshida. 2020. Transcriptome analysis of the brown rot fungus *Gloeophyllum trabeum* during lignocellulose degradation. *PLoS One* 15 (December): e0243984. https://doi.org/10.1371/journal.pone.0243984.

Uzair, B., N. Ahmed, V.U. Ahmad, F.V. Mohammad, and D.H. Edwards. 2008. The isolation, purification and biological activity of a novel antibacterial compound produced by *Pseudomonas stutzeri*. *FEMS Microbiology Letters* 279: 243–50. https://doi.org/10.1111/j.1574-6968.2007.01036.x.

Uzma, F., M.C. Dhananjaya, A. Hashem, et al. 2018. Endophytic fungi – alternative sources of cytotoxic compounds: A review. *Frontiers in Pharmacology* 9 (April): 309, 1–37. https://doi.org/10.3389/fphar.2018.00309.

Vaishnav, P., and A.L. Demain. 2011. Unexpected applications of secondary metabolites. *Biotechnology Advances* 29: 223–29. https://doi.org/10.1016/j.biotechadv.2010.11.006.

Vanittanakom, N., W. Loeffler, U. Koch, U., and G. Jung. 1986. Fengycin-a novel antifungal lipopeptide antibiotic produced by *Bacillus subtilis* F-29–3. *The Journal of Antibiotics* 39: 888–901. https://doi.org/10.7164/antibiotics.39.888.

Vázquez, A.I.F., C.M.D. Sánchez, N.G. Delgado, A.M.S. Alfonso, Y.S. Ortega, and H.C. Sánchez. 2011. Anti-inflammatory and analgesic activities of red seaweed *Dichotomaria obtusata*. *Brazilian Journal of Pharmaceutical Sciences* 47: 111–18. http://dx.doi.org/10.1590/S1984-82502011000100014.

Velmurugan, S., K. Raman, V.T. Viji, et al. 2013. Screening and characterization of antimicrobial secondary metabolites from *Halomonas salifodinae* MPM-TC and its *in vivo* antiviral influence on Indian white shrimp *Fenneropenaeus indicus* against WSSV challenge. *Journal of King Saud University-Science* 25: 181–90. https://doi.org/10.1016/j.jksus.2013.03.002.

Verma, P., A.N. Yadav, V. Kumar, D.P. Singh, and A.K. Saxena. 2017. Beneficial plant-microbes interactions: Biodiversity of microbes from diverse extreme environments and its impact for crop improvement. In *Plant-microbe interactions in agro-ecological perspectives*, eds. D.P. Singh, H.B. Singh, and R. Prabha. 543-80. Singapore: Springer. https://doi.org/10.1007/978-981-10-6593-4_22.

Vinale, F., R. Marra, F. Scala, E.L. Ghisalberti, M. Lorito, and K. Sivasithamparam. 2006. Major secondary metabolites produced by two commercial *Trichoderma* strains active against different phytopathogens. *Letters in Applied Microbiology* 43: 143–48. https://doi.org/10.1111/j.1472-765X.2006.01939.x.

Vizetto-Duarte, C., L. Custodio, K.N. Gangadhar, et al. 2016. Isololiolide, a carotenoid metabolite isolated from the brown alga *Cystoseira tamariscifolia*, is cytotoxic and able to induce apoptosis in hepatocarcinoma cells through caspase-3 activation, decreased Bcl-2 levels, increased p53 expression and PARP cleavage. *Phytomedicine* 23: 550–57. https://doi.org/10.1016/j.phymed.2016.02.008.

Voitsekhovskaia, I., C. Paulus, C. Dahlem, et al. 2020. Baikalomycins A-C, new aquayamycin-type angucyclines isolated from Lake Baikal derived *Streptomyces* sp. IB201691–2A. *Microorganisms* 8 (May): 680, 1–21. https://doi.org/10.3390/microorganisms8050680.

Von Bargen, K.W., E.M. Niehaus, K. Bergander, R. Brun, B. Tudzynski, and H.U. Humpf. 2013. Structure elucidation and antimalarial activity of apicidin F: An apicidin-like compound produced by *Fusarium fujikuroi*. *Journal of Natural Products* 76: 2136–40. https://doi.org/10.1021/np4006053.

Vu, H.N.T., D.T. Nguyen, H.Q. Nguyen, et al. 2018. Antimicrobial and cytotoxic properties of bioactive metabolites produced by *Streptomyces cavourensis* YBQ59 isolated from *Cinnamomum cassia* Prels in Yen Bai Province of Vietnam. *Current Microbiology* 75: 1247–55. https://doi.org/10.1007/s00284-018-1517-x.

Wang, S., and Z. Lu. 2017. Secondary metabolites in archaea and extreme environments. In *Biocommunication of archaea*, ed. G. Witzany, 235–39. Cham: Springer. https://doi.org/10.1007/978-3-319-65536-9_14.

Wang, S., Z. Zheng, H. Zou, N. Li, and M. Wu. 2019. Characterization of the secondary metabolite biosynthetic gene clusters in archaea. *Computational Biology and Chemistry* 78: 165–69. https://doi.org/10.1016/j.compbiolchem.2018.11.019.

Wang, Y.N., L.H. Meng, and B.G. Wang. 2020. Progress in research on bioactive secondary metabolites from deep-sea derived microorganisms. *Marine Drugs* 18 (December): 614, 1–26. https://doi.org/10.3390/md18120614.

Wawrik, B., D. Kutliev, U.A. Abdivasievna, J.J. Kukor, G.J. Zylstra, and L. Kerkhof. 2007. Biogeography of actinomycete communities and type II polyketide synthase genes in soils collected in New Jersey and Central Asia. *Applied and Environmental Microbiology* 73: 2982–89. https://doi.org/10.1128/AEM.02611-06.

Weber, G., C. Hsiou-Lien, E. Hinsch, S. Freitas, and S. Robinson. 2014. Pigments extracted from the wood-staining fungi *Chlorociboria aeruginosa*, *Scytalidium cuboideum*, and *S. ganodermophthorum* show potential for use as textile dyes. *Coloration Technology* 130: 445–52. https://doi.org/10.1111/cote.12110.

Wei, Y., W. Fang, Z. Wan, et al. 2014. Antiviral effects against EV71 of pimprinine and its derivatives isolated from *Streptomyces* sp. *Virology Journal* 11 (November): 195, 1–14. https://doi.org/10.1186/s12985-014-0195-y.

Wertz, K., P.B. Hunziker, N. Seifert, et al. 2005. β-carotene interferes with ultraviolet light A-induced gene expression by multiple pathways. *Journal of Investigative Dermatology* 124: 428–34. https://doi.org/10.1111/j.0022-202X.2004.23593.x.

Williams, P.G. 2009. Panning for chemical gold: Marine bacteria as a source of new therapeutics. *Trends in Biotechnology* 27: 45–52. https://doi.org/10.1016/j.tibtech.2008.10.005.

Winkelmann, G., H. Allgaier, R. Lupp, and G. Jung.1983. Iturin AL-a new long chain iturin A possessing an unusual high content of C16-β-amino acids. *The Journal of Antibiotics* 36: 1451–57. https://doi.org/10.7164/antibiotics.36.1451.

Wu, W., R.W. Davis, M.B. Tran-Gyamfi, et al. 2017. Characterization of four endophytic fungi as potential consolidated bioprocessing hosts for conversion of lignocellulose into advanced biofuels. *Applied Microbiology and Biotechnology* 101: 2603–18. https://doi.org/10.1007/s00253-017-8091-1.

Wu, Z., S. Wu, and X. Shi. 2007. Supercritical fluid extraction and determination of lutein in heterotrophically cultivated *Chlorella pyrenoidosa*. *Journal of Food Process Engineering* 30: 174–85. https://doi.org/10.1111/j.1745-4530.2007.00102.x.

Xiao, J., Q. Zhang, Y.Q. Gao, J.J. Tang, A.L. Zhang, and J.M. Gao. 2014. Secondary metabolites from the endophytic *Botryosphaeria dothidea* of *Melia azedarach* and their antifungal, antibacterial, antioxidant, and cytotoxic activities. *Journal of Agricultural and Food Chemistry* 62: 3584–90. https://doi.org/10.1021/jf500054f.

Xu, D.B., W.W. Ye, Y. Han, Z.X. Deng, and K. Hong. 2014. Natural products from mangrove actinomycetes. *Marine Drugs* 12: 2590–13. https://doi.org/10.3390/md12052590.

Yadav, A.N., D. Kour, K.L. Rana, et al. 2019. Metabolic engineering to synthetic biology of secondary metabolites production. In *New and Future Developments in Microbial Biotechnology and Bioengineering*, eds. V.K. Gupta and A. Pandey, 279–320. Elsevier. https://doi.org/10.1016/B978-0-444-63504-4.00020-7.

Yang, J., X. Zhu, M. Cao, et al. 2016. Genomics-inspired discovery of three antibacterial active metabolites, aurantinins B, C, and D from compost-associated *Bacillus subtilis* fmb60. *Journal of Agricultural and Food Chemistry* 64: 8811–20. https://doi.org/10.1021/acs.jafc.6b04455.

Yang, Y.I., S.H. Jung, K.T. Lee, and J.H. Choi. 2014. 8, 8′-bieckol, isolated from edible brown algae, exerts its anti-inflammatory effects through inhibition of NF-κB signaling and ROS production in LPS-stimulated macrophages. *International Immunopharmacology* 23: 460–68. https://doi.org/10.1016/j.intimp.2014.09.019.

Yuan, J., A. Krämer, Y. Matthess, et al. 2006. Stable gene silencing of cyclin B1 in tumor cells increases susceptibility to taxol and leads to growth arrest *in vivo*. *Oncogene* 25: 1753–62. https://doi.org/10.1038/sj.onc.1209202.

Yuan, J.P., F. Chen, X. Liu, and X.Z. Li. 2002. Carotenoid composition in the green microalga *Chlorococcum*. *Food Chemistry* 76: 319–25. https://doi.org/10.1016/S0308-8146(01)00279-5.

Zhang, B., C. Dong, Q. Shang, Y. Cong, W. Kong, and P. Li. 2013. Purification and partial characterization of bacillomycin L produced by *Bacillus amyloliquefaciens* K103 from lemon. *Applied Biochemistry and Biotechnology* 171: 2262–72. https://doi.org/10.1007/s12010-013-0424-7.

Zhekisheva, M., A. Zarka, I. Khozin-Goldberg, Z. Cohen, and S. Boussiba. 2005. Inhibition of astaxanthin synthesis under high irradiance does not abolish triacylglycerol accumulation in the green alga *Haematococcus pluvialis* (Chlorophyceae). *Journal of Phycology* 41: 819–26. https://doi.org/10.1111/j.0022-3646.2005.05015.x.

Zheng, J., Y. Chen, F. Yao, W. Chen, and G. Shi. 2012. Chemical composition and antioxidant/antimicrobial activities in supercritical carbon dioxide fluid extract of *Gloiopeltis tenax*. *Marine Drugs* 10: 2634–47. https://doi.org/10.3390/md10122634.

Zuiter, A.S. 2014. Proanthocyanidin: Chemistry and biology: From phenolic compounds to proanthocyanidins. In *Reference module in chemistry, molecular sciences and chemical engineering*, ed. J. Reedijk. Elsevier Inc. https://doi.org/10.1016/B978-0-12-409547-2.11046-7.

2 Microbial Strains as a Factory for Valuable Secondary Metabolites

*Mukta Sharma
(School of Life Science and Technology, IIMT University, Meerut (UP), India; muktavats@yahoo.com)

Atul Singh
(School of Life Science and Technology, IIMT University, Meerut (UP), India; atul34867@gmail.com)

Anandveer Singh Sindu
(Department of Chemistry, Meerut College, Meerut (UP), India; dr.anandsindhu@yahoo.in)

*Corresponding Author

2.1 INTRODUCTION

Organic substances known as secondary metabolites perform critical roles in a number of biological processes but are not directly engaged in the reproduction and development of the organism. Microbial strains have long been known for their capacity to produce a wide variety of secondary metabolites. These metabolites often possess unique chemical structures and exhibit diverse biological activities, making them valuable for numerous applications, including pharmaceuticals, agriculture, and industrial processes.

Secondary metabolites are substances produced by some plant species and by strains of specific microbial species. They have complex chemical structures and come in a wide range of chemical forms. Even though antibiotics are the most commonly recognized secondary metabolite, there are numerous other secondary metabolites with a variety of biological roles, making them important for industry now or in the future.

Secondary metabolites are organic molecules that have a secondary role in an organism's regular growth, development, and reproduction (Anon, 2008). The defensive systems of many species usually substantially involve secondary metabolites (Stamp, 2003). Secondary metabolites are utilized by people as pharmaceuticals, flavors, and medications. Ecological competition, antibiotics, poisons, pigments, toxins, symbiosis agents, receptor antagonists and agonists, pheromones, insecticides, anticancer medications, immunomodulating compounds, and growth boosters for both plants and animals are a few examples of microbial secondary metabolites (Figure 2.1). Our society's finances, nutrition, and health are all significantly impacted by them (Gupta et al., 2022, 2024).

Metabolites are often generated during a later production stage (idiophase), which is a hallmark of secondary metabolism, rather than during the rapid development phase (trophophase). When one major nutritional source – nitrogen or phosphate or carbon – has run out, growth is constrained, and secondary metabolite production begins. When the culture medium's supply of glucose is depleted – for instance, *Penicillium chrysogenum* begins to consume lactose, a less quickly consumed sugar – this is when penicillin production begins.

Actinomycetes, particularly those belonging to the genus *Streptomyces*, and fungi are responsible for the majority of the economic significance.

2.2 BIOSYNTHETIC FAMILIES

Microbial strains are well-known for being significant factories that may create a variety of secondary metabolites (Jain et al., 2023, 2024). These metabolites, which are regularly produced by bacteria and fungi, exhibit a variety of biological functions and have various chemical compositions. Several biosynthetic groups of secondary metabolites frequently generated by microbial strains are listed in what follows.

2.2.1 POLYKETIDES

A broad class of secondary metabolites known as polyketides is distinguished by their polyketide backbone, which is created through the enzymatic synthesis of acetyl and malonyl-CoA units. Well-known drugs from this

DOI: 10.1201/9781003408833-2

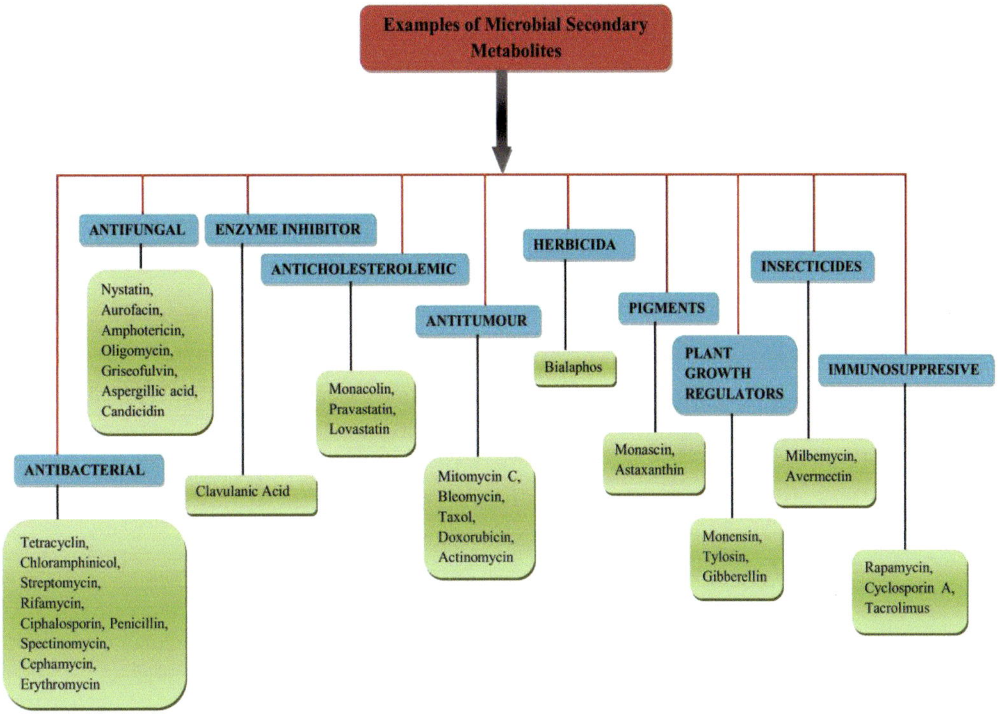

FIGURE 2.1 Examples of microbial secondary metabolites.

class include lovastatin, tetracycline, and erythromycin. Numerous bioactivities, including antibacterial, anticancer, and immunosuppressive characteristics, are possessed by polyketides.

2.2.2 Non-Ribosomal Peptides (NRPs)

Non-ribosomal peptide synthetases (NRPSs), rather than the ribosome, are responsible for NRP synthesis. In order to create complicated peptide structures, NRPSs, which are big multimodular enzymes, put together amino acid building blocks. NRPs include, for instance, the immunosuppressant cyclosporine A and antibiotics like vancomycin and penicillin.

2.2.3 Terpenoids

Isopentenyl diphosphate (IPP), the precursor molecule for terpenoids, is a broad and diversified class of secondary metabolites. They consist of substances like retinol, taxol, and artemisinin. Terpenoids are biological compounds that have a range of actions, including anti-inflammatory, antibacterial, and anticancer properties.

2.2.4 Alkaloids

Alkaloids are a group of secondary metabolites generated by bacteria, fungi, and plants that contain nitrogen. They frequently have various bioactivities and complicated structures. Vinblastine, caffeine, and morphine are a few examples of alkaloids made by microbial strains.

2.2.5 Phenazines

Some bacteria, including *Pseudomonas* and *Streptomyces* species, create phenazines, which are redox-active chemicals. They have broad-spectrum antibacterial activity and are crucial for microbial colonization and competition. It has been investigated whether phenazines have any promise for use in pharmacological and agricultural applications.

2.2.6 Terpenes

Terpenes are a large and diversified group of secondary metabolites. They are made from terpenoid precursors and frequently smell peculiar. Examples include paclitaxel, an anticancer medicine developed from the Pacific yew tree, and the antibacterial chemical carvacrol, which is found in oregano.

2.3 NOVEL BIOACTIVE COMPOUNDS

In the previous two decades, there has been a rapid development of key compounds used in a variety of industrial domains, most notably those of farming and agriculture, food, cosmetics, and pharmaceuticals, as discussed in Table 2.1.

TABLE 2.1
Biological Activity of Commercial Significance in Secondary Metabolites Produced by Microorganisms

S. N.	Activity	Producing Microorganism	Examples
1	Antibacterial	*Streptomyces aureofaciens*	Tetracycline
		Streptomyces venezuelae	Chloramphenicol
		Streptomyces griseus	Streptomycin
		Amycolatopsis mediterranei	Rifamycin
		Acremonium chrysogenum	Cephalosporin
		Penicillium chrysogenum	Penicillin
		Streptomyces spectablis	Spectinomycin
		Streptomyces clavuligerus	Cephamycin
		Streptomyces kanamyceticus	Kanamycin
		Saccharopolyspora erythraea	Erythromycin
2	Antifungal	*Streptomyces noursei*	Nystatin
		S. aureus	Aureofacin
		Streptomyces aureofaciens	Amphotericin
		Streptomyces nodosus	Oligomycin
		Streptomyces diastachromogenes	Griseofulvin
		Penicillumgriseofulvum	Aspergillic acid
		Aspergillus flavus	Candicidin
		Streptomyces griseus	
3	Enzyme inhibitors	*Streptomyces clavuligerus*	Clavulanic acid
4	Anticholesterolemics	*Monascus ruber*	Monacolin
		Penicillium citrinum	Pravastatin
		Streptomyces carbophilus	Lovastatin
		Aspergillus terreus	
5	Antitumorals	*Streptomyces lavendulae*	Mitomycin C
		Streptomyces peucetius	Bleomycin
		Taxomyces andreanae, plants	Taxol
		Streptomyces peucetius	Doxorubicin
		Streptomyces antibioticus	Actinomycin D
		S. parvulus	
6	Herbicidals	*Streptomyces hygroscopicus*	Bialaphos
7	Pigments	*Monacuspurpureus*	Monascin
		M. ruber	Astaxanthin
8	Growth promoters, plant growth regulators	*Streptomyces cinnamonensis*	Monensin
		Streptomyces fradiae	Tylosin
		Gibberella fujikuroi	Gibberellin
9	Insecticides and antiparasitic	*Streptomyces hygroscopicus*	Milbemycin
		Streptomyces avermitilis	Avermectin
10	Immunosuppressive	*Streptomyces hygroscopicus*	Rapamycin
		Tolypoclaudium infantum	Cyclosporin A
		Several *Streptomyces* spp.	Tacrolimus (FK-506)

Secondary metabolites produced by microorganisms, such as those that are cholesterolemic, anti-inflammatory, antiuterocontractants, hypotensive, and anticancer, are now being utilized more frequently to treat disorders that were previously solely managed by synthetic medications. Also, important insecticides, herbicides, pesticides, environmentally friendly herbicides, and plant growth regulators, as well as antiparasitic agents, are being employed in non-medical industries, like agriculture, with the help of new microbial metabolites.

Modern methods for locating microbial secondary metabolites are what have sparked this new era. Target-directed mode of action screens is displacing more traditional whole-cell assay techniques, like bioassays. New isolate culture broths are thus put to the test for their ability to bind to certain receptors either agonistically or antagonistically, or for important enzymatic activities. Understanding the biochemical and molecular details of many diseases is the foundation of this unique approach (Barrios-Gonzalez, 2003).

2.4 PRODUCTION

2.4.1 Solid-State Fermentation

Secondary metabolite generation has a significant potential using solid-state fermentation (Barrios-González, 1988; Tomasini et al., 1997; Robinson et al., 2001). Since ancient times, this method of fermentation has been employed in several Oriental nations to produce a variety of fermented foods from grains like rice or soybeans (Hesseltine, 1979). Several solid-state fermentation systems, some of which could be deemed unorthodox, have been developed over the past 15 years. Modern *solid-state fermentation*, according to the theory put forth by Lonsane et al. (1985), is the growth of a microbial culture on the exterior and interior of a solid matrix without the presence of free water. Depending on the solid phase utilized, there are now two different forms of solid-state fermentation (Barrios-González and Mejía, 1996):

1. One support substrate phase of a solid culture is made up of a substance that acts as both a support and a source of nutrients at the same time. Substrates for support are made of agricultural or even animal products or waste.
2. The two substrate support phases that make up the solid phase are inert support and liquid solution–injected support. Water and nutrients are held in reserve via inert support. An inert support can be made out of materials like polyurethane or sugarcane bagasse pith.

Solid-state fermentation allows fungi and actinomycetes to flourish because the conditions are comparable to those found in their native surroundings, such as soil and organic waste materials (Table 2.2).

The benefits of submerged fermentation over solid-state fermentation include the comparatively low energy requirements of the process and the direct delivery of oxygen to the bacteria. Frequently, secondary metabolites are created in much greater amounts, frequently in much shorter times, and frequently without the necessity for sterile conditions (Balakrishnan and Pandey, 1996; Ohno et al., 1993; Rosenblitt et al., 2000).

The parameters that must be managed for solid-state fermentation are the same as those for submerged fermentation. In this culture system, specific elements like beginning particle size, moisture content, and medium concentration must be changed. It has been demonstrated that the amounts of bagasse, water, and nutrients present have a considerable impact on penicillin synthesis in a solid-state fermentation–impregnated bagasse system. In this investigation, combinations that supported extremely high penicillin yields were found. While these conditions resulted in various growth phases, it's crucial to note that they were all beneficial for the fungus to collect nutrients slowly but sufficiently during the idiophase and support a slow but continuous respiratory activity throughout the production phase (Domínguez et al., 2000).

According to reports on enzyme synthesis, bacteria that create a lot of enzymes during submerged fermentation typically produce little during solid-state fermentation. High-yielding submerged fermentation strains are unlikely to function well in solid-state fermentation (Barrios-González, 1993). The necessity to create high-yielding strains specifically suited for solid-state fermentation is dictated by this circumstance. By employing hyperproducing strains created for submerged fermentation as parental strains, these specialized strains can be created more quickly.

According to several comparison studies contrasting submerged fermentation with solid-state fermentation, the

TABLE 2.2
Metabolites Produced by a Solid-State Fermentation System as Secondary Metabolites

S. N.	Metabolite	Substrate	Microorganism	Uses
1	Cephamycin C	Wheat straw	*Streptomyces clavuligerus*	Antibiotic
2	Cephalosporin	Rice grains	*Streptomyces* sp.	Antibiotic
3	Cyclosporin A	Wheat bran	*Tolypocladium inflatum*	Antibiotic
4	Ergot alkaloids	Sugarcane bagasse	*Claviceps fusiformis*	Medical
5	Gibberellic acid	Wheat bran, polyurethane, cassava	*Gibberella fujikuroi*	Vegetal hormone
6	Iturin	Soybean curd	*Bacillus subtilis*	Antifungal
7	Oxytetracycline	Sweet potato residue	*Streptomyces rimosus*	Antibiotic
8	Penicillin	Sugarcane bagasse	*Penicillum chrysogenum*	Antibiotic
9	Pyrazines	Wheat and soybean	*Aspergillus oryzae*	Aroma
10	Pigments	Rice grains	*Monascus purpureus*	Food and pharmaceuticals
11	Surfactin	Soybean curd	*Bacillus subtilis*	Surfactant, antibiotic
12	Tetracycline	Sweet potato residue	*Streptomyces viridifaciens*	Antibiotic

latter produces compounds in higher yields (Pandey et al., 2000), suggesting that some of these metabolites may be produced commercially by the latter. Solid-state fermentation is still not employed commercially in Western countries to create secondary metabolites due to scalability issues. Many of these issues have been researched, and some solutions have been put out. Some bioreactors have been created while considering these factors (Hardin et al., 2000).

Research focus was placed on the industrialization of this culture method in Southeast Asia, where solid-state fermentation is more prevalent. To create microbial enzymes and secondary metabolites for use in industry, Indian scientists developed the solid-state fermentation technique. The technology is housed in a solid-state bioreactor that also allows the fermentation product to be extracted from the solid matrix without having to open the reactor (Suryanarayan, 2003). Additionally, this reactor is run automatically. The first industrially produced secondary metabolite made by solid-state fermentation, lovastatin, has been made in this reactor since January 2001.

2.4.2 Liquid Fermentation

In the industry, batch or fed-batch culture submerged fermentation is widely used to produce secondary metabolites. Initially introduced to a growth medium in flasks, an improved strain of the producing microbe is then transferred to a relatively small fermenter, or "seed culture." A fermenter tank with a production medium that has a capacity between 30,000 and 200,000 L is used to inoculate this culture when it is undergoing rapid growth. Controlled variables include medium composition, temperature, pH, agitation, and aeration rate. Environmental manipulations work around the many regulatory processes that were described earlier. As a result, methionine is employed as an inducer in the fermentation of cephalosporins, phosphate is regulated in the fermentation of chlortetracyclines, and glucose is avoided in the fermentation of erythromycin or penicillin. Currently, antibiotics are fermented using slowly metabolized carbon sources, often lactose. When glucose is employed, it is often administered slowly and continuously to prevent catabolic regulation. It is possible to get around nitrogen or ammonium control by using nitrogen sources, like soybean meal. In certain circumstances, a precursor is introduced to boost the production of a certain desired metabolite. To speed up the production of cephamycin by *Streptomyces clavuligerus*, for instance, lysine is supplied as a cofactor and precursor (Khetan et al., 1999). Air must be provided at flow rates between 0.5 and 1.0 v/v per minute, and turbine impellers must be operated at a rate between 1 and 4 W/L. O_2 and CO_2 concentrations are often measured in exit gas. This can provide metabolic information to regulate how quickly nutrients and precursors are supplied.

Semisynthetic derivatives are created by further chemical modification of some natural antibiotics and other secondary metabolites.

2.5 REGULATION OF SECONDARY METABOLITES PRODUCTION

According to Adrio and Demain (2006), microorganisms can produce novel genetic traits through mutation and genetic recombination. It is now possible to alter the bacteria to create the desired color thanks to the development of cultivation technology (genetic engineering). Nutritional factors, feedback control, growth rate, enzyme inactivation, and enzyme stimulation all affect the synthesis of secondary metabolites. Microbial pigments are produced via basic biotechnological techniques, first by seeking out new sources, and then by increasing the yield of sources that have already been identified, either by strain enhancement or optimization (Venil et al., 2013).

The term "optimization" describes the optimum circumstances that promote microbial development and boost the production of secondary metabolites. These conditions include the influence of moisture content, incubation time, fermentation processes, carbon sources, pH, nitrogen sources, and ultraviolet light. Symbiosis, competence, conjugation, sporulation, biofilm formation, pathogenicity, and motility are a few physical processes that microorganisms engage in that might alter the generation of secondary metabolites.

Quorum-sensing technology permits cell communication by inducing the creation of chemical signals when the cell density reaches a predefined threshold. The set of regulatory procedures that result in chemical differentiation may be affected by signals generated by biosynthesis, either the addition of an inducer or the slowing of growth. A low-molecular-weight inducer serves as the signal's usual negative regulator, inhibiting the production of secondary metabolites (Demain, 1998).

Physical and chemical sources, such as UV light, pH, nitrogen, oxygen, antibiotics, and inorganic ions, can be used to improve strains (e.g., 4-nitro quinolone-1-oxide, hydroxylamine, nitrosoguanidine, methyl methane sulfonate, and ethyl methane sulfonate) (Simpson and Caten, 1979). Additionally, it may result in a change or modification to the chemical composition and pigment hue.

The development of microbial products has long benefited from genetic advancements. In metabolite-producing microbes, the following groups of genetic regulators are active. Key elements include permeability genes, which regulate the entry, exclusion, and excretion of products; structural genes, which code for the creation of products; regulatory genes, which regulate the expression of structural genes; resistance genes, which determine an organism's susceptibility to an antibiotic; and regulatory genes, which direct the pathways that produce precursors and cofactors (Strohl, 2001).

The main techniques used in molecular genetics tools to find effective transformation procedures, appropriate vectors, and biosynthetic pathways include the following:

1. A few strain advantages that can be improved include increased yields of desired metabolites, improved utilization of inexpensive carbon and nitrogen sources, removal of undesirable co-metabolites, modification of cellular morphology for increased mycelium production, and improved oxygen transfer in the fermenter.
2. It is simple to identify a fundamental understanding of product metabolism and pathway regulation that differs from that of ultimate interest for the choice of a particular attribute of the desired genotype.
3. Recombination between related species of microorganisms with genetic development results in the high synthesis of the newly found secondary metabolites from two strains.
4. Find a neutral location on the chromosome for targeted gene duplication so that genes can be put there without changing the strain's capacity for fermentation. The vector containing the antibiotic gene is cloned and integrated into the neutral site. Through homologous recombination, the gene is placed into the chromosomal neutral location after transformation.
5. Reduce the power and energy available to a route that competes for common intermediate crucial precursors, such as cofactors, using this method.
6. The bacterial strains' incorporation of new enzymes will result in the production of a new related product.
7. The generation of secondary metabolites in microorganisms is aided by the use of directorial biosynthesis, mutagenesis, combinatorial biosynthesis, and ribosome engineering.

2.6 VALUABLE STRAINS SCREENING AND CHARACTERIZATION

A recent study found that the coastal environment is a particularly rich biotope for the identification of novel microbes and subsequent significant chemical compounds.

It has been discovered that sponges and the bacteria that live inside of them produce a vast range of bioactive secondary metabolites. Numerous natural substances derived from marine species have already shown promise in the pharmaceutical industry as antiaging and/or life-extension agents. However, research on the biodiversity of the Indian Ocean's southwest is significantly less comprehensive, particularly in terms of antiaging approaches. Sad-Hassane et al. (2020) employed metagenomic analysis to examine the microbial diversity of the sea sponge *Scopalina hapalia*. The Proteobacteria phylum was the most numerous among the 26 bacterial phyla present in the sponge, which also contained two archaeal phyla. To produce secondary metabolites, 30 isolates from the family Thermoactinomycetaceae and the genera *Salinispora*, *Bacillus*, *Micromonospora*, *Rhodococcus*, *Chaetomium*, *Aspergillus*, and/or *Nigrospora* were grown. Sirtuin-1, tyrosinase, catalase, CDK-7, FYN kinase, and the proteasome were all discovered to be susceptible to crude extracts from particular microbial cultures. These results demonstrate the ability of marine sponge bacteria from the Indian Ocean to produce antiaging compounds.

In marine biotopes, where they are constantly under pressure from physicochemical factors (poor oxygenation, salinity, hydrostatic pressure), biotic competition, or predation (such as Nematodes and protozoans), marine epiphytic bacteria are also highly diverse and widespread. These marine species have evolved defense mechanisms to deal with challenges, such as the production of harmful bioactive chemicals. From these bacteria, several active metabolites resulting from the expression of particular genes have been discovered. The review focuses on the possibility that marine epiphytic bacteria could provide a fresh starting point for the creation of brand-new worm treatments (Salikin et al., 2020). We highlight novel ways to find novel strains, such as culture-independent, high-throughput screening.

A description of the processes used by marine bacteria to produce pigments and their uses is given in the review by Nawaz et al. (2021), which concentrates on pigments. Lutein or lycopene, prodigiosin, violacein, astaxanthin, and zeaxanthin are only a few of the compounds highlighted that are already appreciated at an industrial level.

Due to the variety of secondary metabolites, they are capable of producing, bacteria associated with insects are thought to have a role in the life cycle of their hosts. These strains are unquestionably among the researched sources of novel active substances. Numerous biosynthetic gene clusters may become active as a result of a strain's culture conditions changing, which could make it easier to find new bioactive substances. The OSMAC technique (one strain of many compounds) was created with the idea that certain microbial metabolites might not form under a specific set of physicochemical parameters but might do so if the conditions are changed. These two techniques significantly boosted the production of pentaminomycins C, D, and E, which are cyclic pentapeptides in the *Streptomyces* sp. culture. *Tenebrio molitor*, also known as the mealworm beetle, has actinobacteria GG-23 strain (Hwang et al., 2020). According to non-ribosomal peptide synthetase gene cluster analysis, the penicillin-binding protein type TE can catalyze macrocyclization in a non-canonical manner, and the low specificity of the adenylation domain in the NRPS module 1 is the cause of the compounds' uniqueness based on structural variations. The pentaminomycins C and D also showed potent autophagy-inducing action and offered in vitro cytoprotection against the oxidative stress brought on by menadione.

One cutting-edge field of research involves using metabolome technology to examine secondary metabolites produced by microbial co-cultures. A total of 23 isolates of the *Coscinoderma mathewsi* sponge from the Red Sea contained three actinomycete strains, each of which was identified as a unique species of the genera *Nocardia*, *Micromonospora*, and *Gordonia*. The results of this study (Shamikh et al., 2020) demonstrated that co-culturing two mycolic acid–containing actinomycetes, *Nocardia* sp. UA23 and *Gordonia* sp. UA19, enabled *Micromonospora* sp. UA17 to produce metabolites like neocopiamycin A, chicamycin B, and chlorocardicin that were lacking in the corresponding monocultures. This emphasizes how mycolic acid affects the stimulation of cryptic natural product biosynthetic pathways and demonstrates how quiet biosynthetic gene clusters can demonstrate their extraordinary capacities to produce secondary metabolites in unconventional culture environments.

Serratia marcescens, an opportunistic infection, produces siderophores, such as serratiochelins, which are specialized substances with a strong affinity for ferric iron. Pharmaceutical companies are interested in siderophores because they can be used to treat iron overload illnesses or make siderophore–antibiotic drug conjugates that make it easier for bacteria to absorb antibiotics. *Serratia* sp. and *Shewanella* sp. (a strain from the marine environment also active in the iron cycle) were co-cultured in the study by Schneider et al. (2020), and the unusual siderophore serratiochelin-A was recovered with high yields. The absence of this molecule in Serratia or Shewanella axenic cultures suggests that co-culturing, probably as a result of the two strains' competition for iron in the culture media, was responsible for the chemical's creation. Serratiochelin-A's antibacterial action was assessed and shown to be exclusive to *S. aureus*. The chemical also hurt the eukaryotic cell lines MRC5 and A2058.

After that, the creation of microbe consortia brings a limitless window of opportunity for research into the synthesis of fresh bioactive compounds.

2.7 STRAIN IMPROVEMENT

The art and science of altering and strengthening microbial strains to boost their metabolic capacity for biotechnological uses is known as strain augmentation. Since the microbial production strain is regarded as the brains of the fermentation business, improving it offers the best chance of cost reduction without a sizable capital investment (Parekh et al., 2000). A fermentation industry's ability to become and remain competitive also heavily depends on the ongoing development of the production strain. Increased yields of the target metabolite are frequently where improvement happens. However, it is also possible to enhance other strain features. Changing the cellular structure to one that is more favorable for the separation of the mycelium from the product and enhanced oxygen transmission in the fermenter is one example. Other ones include removing undesirable co-metabolites, using more affordable carbon and nitrogen sources, and increasing the usage of those sources.

One of two techniques can be used to improve strains:

1. Molecular genetics methods
2. Classical genetic methods

Each offers significant benefits, and in some circumstances, all these techniques can be used together to boost output.

2.8 FUTURE STRATEGIES AND PERSPECTIVES

Future alternatives to improve the use of microbes as regulated cell factories that produce high-value metabolites for our daily needs are abundant thanks to scientific and technical breakthroughs.

The discovery of new antibiotics and other secondary metabolites crucial to the medical industry, as well as other productive activities, is projected to be sped up by novel target-directed ways to identify microbial secondary metabolites. As part of the process known as combinatorial biosynthesis, new secondary metabolites will also be produced. The economically feasible synthesis of these compounds will rely heavily on the fermentation production process and the application of suitable strain augmentation techniques.

Natural microbiological products continue to be a crucial resource for therapeutic development with biological significance. Numerous research efforts have been started to determine the different sources that bacteria can use to produce secondary metabolites. Industrial waste utilization, optimizing medium elements, strain growth, and culture parameters are crucial methods for enhancing the bacterial system's functionality, which aids in raising output yield.

To changes in their environment, microbes can quickly adapt, and because it is simple to manipulate their genetic material, there are more options for obtaining secondary metabolites. The majority of natural colors are sufficiently complex that chemical synthesis will never be able to frequently produce them. Because it is simple to produce synthetic pigments in large quantities via fermentation, they are extremely rare yet have a high economic availability. These amines are toxic and carcinogenic and are not good for the environment. The effectiveness of microbial culture in the fiercely competitive field of commercial synthesis can be gauged by the fact that fermentation, as opposed to chemical synthesis, is utilized to produce even single molecules. Extraction is straightforward and more productive since microbial fermentation, the industrial production method for natural pigments, is cost-effective.

Soon, there will be many genetic and molecular genetics techniques available to boost fermentation product yields and other strain features. Some are cheap and straightforward and have only been utilized successfully in a few commercial applications, although they have huge theoretical promise, whereas others are more expensive and complex. Examples of the former include mutation and selection.

A particular strategy should be used depending on the cost-effectiveness of the biotechnological process and the genetic resources available for the target strain.

REFERENCES

Adrio, J.L., and A.L. Demain. 2006. Genetic improvement of processes yielding microbial products. *FEMS Microbiology Reviews* 30: 187–214. https://doi.org/10.1111/j.1574-6976.2005.00009.x.

Anonymous, 2008. Secondary metabolite. http://en.wikipedia.org/wiki/secondary_metabolite (Accessed July 21, 2023).

Balakrishnan, K., and A. Pandey. 1996. Production of biologically active secondary metabolites in solid state fermentation. *Journal of Scientific and Industrial Research* 55: 365–72.

Barrios-Gonzalez, J. 2003. Production of antibiotics. In *Concise encyclopedia of bioresource technology*, ed. A. Pandey. Binghamton, New York: The Haworth Press Inc.

Barrios-González, J., T.E. Castillo, and A. Mejía. 1993. Development of high penicillin producing strains for solid state fermentation. *Biotechnology Advances* 11: 525–37. https://doi.org/10.1016/0734-9750(93)90021-E.

Barrios-González, J., and A. Mejía. 1996. Production of secondary metabolites by solid state fermentation. *Biotechnology Annual Review* 2: 85–121. https://doi.org/10.1016/s1387-2656(08)70007-3.

Barrios-González, J., A. Tomasini, G. Viniegra-González, and L. López. 1988. Penicillin production by solid state fermentation. *Biotechnology Letters* 10: 793–98. https://doi.org/10.1007/BF01027575.

Demain, A.L. 1998. Induction of microbial secondary metabolism. *International Microbiology* 1: 259–64. PMID: 10943372

Domínguez, M., A. Mejía, and J. Barrios-González. 2000. Respiration studies of penicillin solid-state fermentation. *Journal of Bioscience and Bioengineering* 89: 409–13. https://doi.org/10.1016/S1389-1723(00)89088-X.

Gupta, A., S. Jain, and N. Verma. 2022. *Industrial applications of soil microbes*, vol. 1. Bentham Science Publishers Pte Ltd. https://doi.org/10.2174/97898150399551220101.

Gupta, A., S. Jain, and N. Verma. 2024. *Industrial applications of soil microbes*, vol. 3. Bentham Science Publishers Pte Ltd. https://doi.org/10.2174/97898150797531240301.

Hardin, M.T., D.A. Mitchell, and T. Howes. 2000. Approach to designing rotating drum bioreactors for solid-state fermentation on the basis of dimensionless design factors. *Biotechnology and Bioengineering* 67: 274–82. https://doi.org/10.1002/(SICI)1097-0290(20000205)67:3<274::AID-BIT3>3.0.CO;2-I.

Hassane, C.S., M. Fouillaud, G.L. Goff, et al. 2020. Microorganisms associated with the marine sponge *Scopalina hapalia*: A reservoir of bioactive molecules to slow down the aging process. *Microorganisms* 8 (August): 1262, 1–23. https://doi.org/10.3390/microorganisms8091262.

Hesseltine, C.W. 1979. Some important fermented foods of mid-Asia, the Middle East, and Africa. *Journal of the American Oil Chemists' Society* 56: 367–74. https://doi.org/10.1007/BF02671501.

Hwang, S., L.T.H.L. Le, S-II. Jo, J. Shin, M.J. Lee, and D.C. Oh. 2020. Pentaminomycins C–E: Cyclic pentapeptides as autophagy inducers from a mealworm beetle gut bacterium. *Microorganisms* 8 (September): 1390, 1–16. https://doi.org/10.3390/microorganisms8091390.

Jain, S., A. Gupta, and N. Verma. 2023. *Industrial applications of soil microbes*, vol. 2. Bentham Science Publishers Pte Ltd. https://doi.org/10.2174/97898150502641230201.

Jain, S., A. Gupta, and N. Verma. 2024. *Industrial applications of soil microbes*, vol. 4. Bentham Science Publishers Pte Ltd. https://doi.org/10.2174/97898151249961240401.

Khetan, A., L.H. Malmberg, Y.S. Kyung, D.H. Sherman, and W.S. Hu. 1999. Precursor and cofactor as a check valve for cephamycin biosynthesis in *Streptomyces clavuligerus*. *Biotechnology Progress* 15: 1020–27. https://doi.org/10.1021/bp990090f.

Lonsane, B.K., N.P. Ghildyal, S. Budiatman, and S.V. Ramakrishna. 1985. Engineering aspects of solid state fermentation. *Enzyme and Microbial Technology* 7: 258–65. https://doi.org/10.1016/0141-0229(85)90083-3.

Nawaz, A., R. Chaudhary, Z. Shah, et al. 2021. An overview on industrial and medical applications of bio-pigments synthesized by marine bacteria. *Microorganisms* 9 (December): 11, 1–25. https://doi.org/10.3390/microorganisms9010011.

Ohno, A., T. Ano, and M. Shoda. 1993. Production of the antifungal peptide antibiotic, iturin by *Bacillus subtilis* NB22 in solid state fermentation. *Journal of Fermentation and Bioengineering* 75: 23–27. https://doi.org/10.1016/0922-338X(93)90172-5.

Pandey, A.P., C.R. Soccol, and D. Mitchell. 2000. New developments in solid state fermentation: I-Bioprocesses and products. *Process Biochemistry* 35: 1153–69. https://doi.org/10.1016/S0032-9592(00)00152-7.

Parekh, S., V.A. Vinci, and R.J. Strobel. 2000. Improvement of microbial strains and fermentation processes. *Applied Microbiology and Biotechnology* 54: 287–301. https://doi.org/10.1007/s002530000403.

Robinson, T., D. Singh, and P. Nigam. 2001. Solid-state fermentation: A promising microbial technology for secondary metabolite production. *Applied Microbiology and Biotechnology* 55: 284–89. https://doi.org/10.1007/s002530000565.

Rosenblitt, A., E. Agosin, J. Delgado, R. Pérez-Correa. 2000. Solid substrate fermentation of *Monascus purpureus*: Growth, carbon balance, and consistency analysis. *Biotechnology Progress* 16: 152–62. https://doi.org/10.1002/btpr.5420160201.

Salikin, N.H., J. Nappi, M.E. Majzoub, and S. Egan. 2020. Combating parasitic nematode infections, newly discovered antinematode compounds from marine epiphytic bacteria. *Microorganisms* 8 (December): 1963, 1–19. https://doi.org/10.3390/microorganisms8121963.

Schneider, Y., M. Jenssen, J. Isaksson, K.Ø. Hansen, J.H. Andersen, and E.H. Hansen. 2020. Bioactivity of serratiochelin A, a siderophore isolated from a co-culture of *Serratia* sp. and *Shewanella* sp. *Microorganisms* 8 (July): 1042, 1–17. https://doi.org/10.3390/microorganisms8071042.

Shamikh, Y.I., A.A. El Shamy, Y. Gaber, et al. 2020. Actinomycetes from the red sea sponge *Coscinoderma mathewsi*: Isolation, diversity, and potential for bioactive compounds discovery. *Microorganisms* 8 (May): 783, 1–18. https://doi.org/10.3390/microorganisms8050783.

Simpson, I.N., and C.E. Caten. 1979. Induced quantitative variation for penicillin titre in clonal populations of *Aspergillus nidulans*. *Microbiology* 110: 1–12. https://doi.org/10.1099/00221287-110-1-1.

Stamp, N. 2003. Out of the quagmire of plant defense hypotheses. *The Quarterly Review of Biology* 78: 23–55. https://doi.org/10.1086/367580.

Strohl, W.R. 2001. Biochemical engineering of natural products biosynthesis pathways. *Metabolic Engineering* 3: 4–14. https://doi.org/10.1006/mben.2000.0172.

Suryanarayan, S. 2003. Current industrial practice in solid state fermentations for secondary metabolite production: The Biocon India experience. *Biochemical Engineering Journal* 13: 189–95. https://doi.org/10.1016/S1369-703X(02)00131-6.

Tomasini, A., C. Fajardo, and J. Barrios-González. 1997. Gibberellic acid production using different solid-state fermentation systems. *World Journal of Microbiology and Biotechnology* 13: 203–6. https://doi.org/10.1023/A:1018545932104.

Venil, C.K., Z.A. Zakaria, and W.A. Ahmad. 2013. Bacterial pigments and their applications. *Process Biochemistry* 48: 1065–79. https://doi.org/10.1016/j.procbio.2013.06.006.

3 Industrial Applications of Microbial Secondary Metabolites

Diversity, Classification, and Sources

*Moti Lal
(Department of Biotechnology, School of Sciences (SOS), ITM University, Gwalior, MP, India; motilal.bt@itmuniversity.ac.in)

Neelam
(Department of Microbiology, School of Sciences (SOS), ITM University, Gwalior, MP, India; neelam.sos@itmuniversity.ac.in)

*Corresponding Author

3.1 INTRODUCTION

Microbial secondary metabolites are a fascinating and diverse group of compounds produced by various microorganisms, including bacteria, fungi, and actinomycetes. Unlike primary metabolites, which are essential for the basic life functions of these microorganisms, secondary metabolites are not directly involved in growth or reproduction; instead, they serve a variety of ecological and physiological functions, often playing crucial roles in interactions with their surrounding environments (Jain et al., 2023, 2024). These compounds are characterized by their structural complexity and are known for their wide range of biological activities. Microbial secondary metabolites have been a valuable source of bioactive molecules, including antibiotics, antifungals, immunosuppressants, and anticancer agents, among many others. Their significance extends beyond medicine, as they also influence ecological relationships, aiding microorganisms in competition and cooperation within complex microbial communities. Approximately 60% of the world's population and about 80% of the population of developing countries depend on traditional medicine (secondary metabolites) for their primary health and care (Aharoni et al., 2004).

Understanding the biosynthesis, regulation, and functions of microbial secondary metabolites has not only revolutionized the field of pharmaceuticals but also offered insights into the intricate interplay of microorganisms in diverse ecosystems. Various new effective drugs are required day by day for the treatment of various cancers, drug-resistant bacteria, fungal infections, emerging viruses, and parasitic protozoan infections. Natural products have provided the basis for most of many new drugs, and their continuing roles in traditional health systems of many cultures reflect the bioactive properties of a wide variety of flora (Newman and Cragg, 2016). The importance of fungi in producing bioactive compounds was first reported by Fleming (1929) with the discovery of penicillin from *Penicillium notatum*. After that, there has been a particular focus on the discovery of natural products from fungi, especially during World War II, with large-scale production of secondary metabolites (Aly et al., 2010). Among common microorganisms, they represent themselves as promising sources of natural products that have the advantage of producing large quantities of bioactive compounds at a sustainable and reasonable cost by large-scale cultivation and fermentation of source organisms (Waites et al., 2001; Gupta et al., 2022, 2024). Microbacteria were isolated from soil samples in earlier times, but in the search for new sources of bioactive metabolites, especially those that are associated with higher plants, these bacteria have been proven to be a potential source of bioactive metabolites, producing a vast category of new biologically active secondary metabolites (Sandoval et al., 2004; Debbab et al., 2011). In recent times, *bioprospecting* is a term commonly used to clarify the discovery of fungal metabolites or different organisms of commercial importance from the world's biological community. The reality lies in the fact that tropical forests are rich in different and more species compared to temperate forests. Forests surrounded by tropical regions have many higher plants and different types of microdiversity. Therefore, sampling of a particular group of fungi from the defined area will be more revealing and fruitful in the search for the undescribed fungi for natural products. Approximately 300,000 vascular plant species have been estimated to serve as reserves of large numbers of microbes in tissue known as endophytes (Promputtha et al., 2007).

It has also often been observed that some fungi show similarity to photochemicals; rightly, the host's competence has changed the field of endophytic biology as well as the importance of producing secondary metabolites of endophytes. Based on different facts, the products obtained from

DOI: 10.1201/9781003408833-3

nature are of different classes, which have been discovered from different groups of endophytes. They have antibacterial, antifungal, antimalarial, antiparasitic, antituberculosis, and anticancer effects, among others. Even some unusual molecules have been isolated from these endophytes, such as antidiabetics, immunomodulators, insecticidal, herbicides, *etc.* (Kaul et al., 2014). Some data have been presented by Newman and Cragg (2016) following the data taken by the United States Food and Drug Administration (USFDA) from January 1981 to December 2014, in which different (total) numbers of different approved drug entries or medical entries are reported (Figure 3.1).

The different roles of natural products in the discovery of new therapeutic agents can be demonstrated by the analysis of the number and sources of bioactive agents. At least 200,000 natural metabolites with various bioactive properties have been confirmed to have been discovered (Selim et al., 2012). The discovery of natural products involves isolation, structural elucidation, and establishing of the biosynthetic pathway of the secondary metabolites. It remains a subject of great interest to scientists because of its different complexity and biosynthesis of isolated compounds (Horwitz et al., 1993). The world's first billion-dollar and most famous anticancer drug, Taxol, was separated from the yew tree in the early 1960s. This drug was approved by the US Food and Drug Administration in 1992 for the treatment of ovarian cancer. This medicine has also been found effective against lung, breast, head, and neck cancers and in the advanced forms of Kaposi's sarcoma (in people with AIDS). Taxol (mitotic inhibitor) has a unique mechanism of action for the breakdown of microtubules during cell division (Li et al., 1996). This has enforced the search for novel sources of Taxol from trees and the isolation of an endophytic fungus and bacteria (Zhou et al., 2007). However, the isolation of endophytic bacteria from plants is a relatively simple process, but the identification of Taxol-producing endophytic microbes is very difficult (Engelhard et al., 2000). Another surprising fact in terms of drug resistance is that the number of new antibacterial drugs receiving approval has decreased day by day over the years. Endophytes colonize host tissue internally, sometimes in very high numbers, identifying the symptoms of plant disease without harming the host.

3.2 SECONDARY BIOACTIVE METABOLITES

These antiproperty bioactive metabolites are organic compounds produced by bacteria, fungi, or plants, with no direct evidence of growth and production. Some specific types of secondary metabolites have been limited to an accumulated group of species within the phylogenetic group. It has often been observed that secondary metabolites play an important role in protecting the plant against external invasive elements, herbicides, and other interspersion interactions. For many centuries, humans have been using secondary metabolites as various drugs, flavor agents, dyes, and recreational drugs. Secondary metabolites have always been supportive of their host. They also

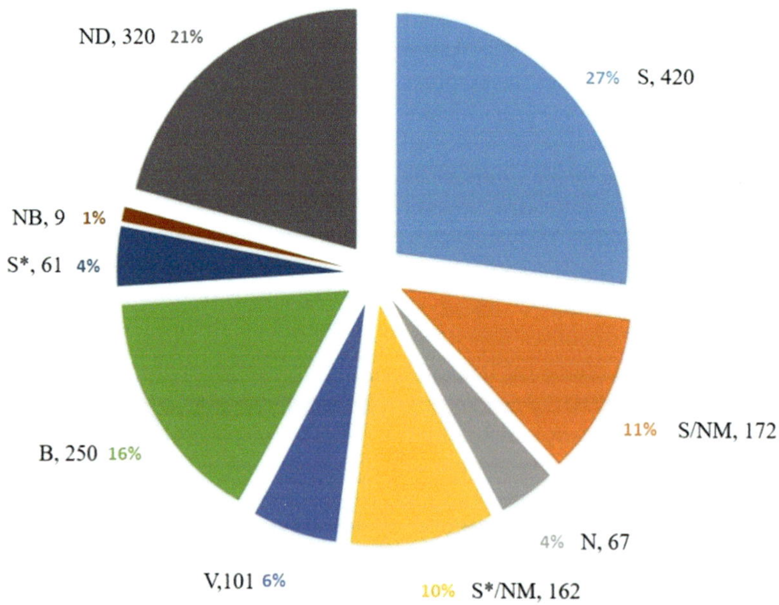

FIGURE 3.1 All new approved drugs (n = 1562), that is, therapeutic agents or drug entities approved by the USFDA from the past 34 years (January 1981–December 2014). Here, *B* is a biological macromolecule, *N* unaltered natural product, *NB* botanical drug (defined mixture), *ND* natural product derivative, *S* synthetic drug, *S** synthetic drug (NP pharmacophore), and *S*/NM* mimic of natural.

Source: Data obtained and modified from Newman and Cragg (2016).

play an important role in important functions, such as protection, competition, and species interaction. However, it has often been observed that hosts never use secondary bioactive metabolites to survive. A special type of specificity is found in each of the metabolites, which shows its characteristics. Typically, each of the metabolites is specific to an individual species of microbes (Pichersky and Gang, 2000). Research also shows that different species are affected in different ways by secondary metabolites. In a similar forest, four separate species of arboreal marsupial folivores reacted differently to a secondary metabolite in eucalyptus (Jensen et al., 2014). Several types of research have shown that bioactive metabolites also have a profound effect on various types of food consumed by humans. According to some research, some volatile bioactive metabolites may be responsible for human food diversity, which is based on various nutritional foods (Goff and Klee, 2006). It has often been observed that various bioactive metabolites help the plant obtain important nutrients, such as, for example, nitrogen. As an example, scientists observed that legumes use flavonoids to indicate a symbiotic relationship with nitrogen-fixing bacteria (*Rhizobium*) to increase their nitrogen (Croteau et al., 2012).

Therefore, many plants that use secondary metabolites are very high in nutrients and are highly beneficial for human consumption. According to Owen and Hundley (2004), endophyte bacteria act as chemical synthesizers inside the plant. Bacteria living in various medicinal plants contain a large amount of bioactive secondary metabolites, and these metabolites are used in the treatment of various diseases (Tejesvi et al., 2007; Kumar et al., 2014). From time to time, researchers try to get and identify various bioactive metabolites from different endophytes. Fermentation of endophytic bacteria with the potential for bioactive compound formation has some advantages, like dependable and reproducible output. These endophytes are grown in fermenters on a greater scale so that the bioactive compound can be properly made available to all people and thus commercially exploited to use different biosynthetic methods, under different culture conditions. Changes should be kept in mind so that we can produce various new compounds and their derivatives (Goutam et al., 2014). Newman and Cragg (2016) have presented 34 years of data obtained by the Food and Drug Administration of the United States (US-FDA), which reports the total number of drug entries (natural products) or therapeutic entries. This data is from January 1981 to December 2014 (Figure 3.2).

There is vitality, and a delicate equilibrium, between the activities of the primary and secondary metabolism, influenced by development, tissue differentiation, and the development of the cell or the body, and peripheral pressures. Therefore, *secondary metabolites* can also be defined as a heterogeneous group of natural metabolic products. This cannot be said to be necessary for the different vegetative growth of organisms. They are known as various compounds which show adverse conditions, such as symbiosis, metal and water transport, competition, self-defense, ecological interactions, signaling molecules, and acting as defensive compounds. The main reason for the secretions of secondary metabolism is to improve the health of mankind. It produces various compounds, such as antitumor agents, anticancer drugs, antiparasitic agents, antimalarials, enzyme inhibitors, growth enhancers of plants and animals, antibiotics, and immunomodulators.

3.3 CLASSIFICATION OF MICROBIAL SECONDARY METABOLITES

3.3.1 Polyketides

In terms of both structure and function, polyketides, a vast class of secondary metabolites, are known to exhibit extraordinary

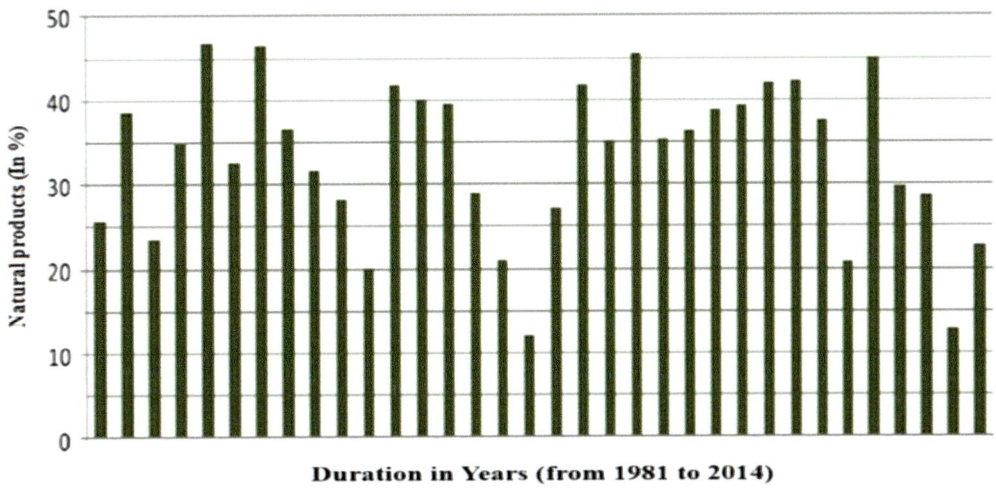

FIGURE 3.2 Percentage of natural products (*N*), that is, therapeutic agents or drug entities approved by the USFDA from the past 34 years (January 1981–December 2014).

Source: Blunt et al. (2017).

diversity (Moore and Hopke, 2001; Rokem et al., 2007). Numerous bioactivities, including those that are antibacterial (*e.g.*, tetracycline), antifungal (*e.g.*, amphotericin B), antiviral (*e.g.*, balticolid), immuneosuppressive (*e.g.*, rapamycin), anticholesterol (*e.g.*, lovastatin), and anti-inflammatory (*e.g.*, flavonoids), can be observed by polyketides (Katsuyama et al., 2007; Chopra and Roberts, 2001; Ghannoum and Rice, 1999). Some organisms can produce polyketides, such as bacteria (*e.g.*, tetracycline from *Streptomyces aureofaciens*), fungi (*e.g.*, lovastatin from *Phomopsis vexans*), and plants (*e.g.*, emodin from *Rheum palmatum*) (Hleba et al., 2018; Parthasarathy and Sathiyabama, 2015) (Figure 3.3).

Since the early 1940s, microorganisms have played a significant role in the history of antibiotics. Actinobacteria are one of the bacterial phyla that generate numerous significant antibiotics. *Streptomyces*, which has a filamentous morphology identical to fungi and has become a source of around two-thirds of all known natural antibiotics, is the most significant species. Polyketides make up a significant portion of the antibiotics that *Streptomyces* produces. Rapamycin (made by *Streptomyces hygroscopicus*), oleandomycin (manufactured by *Streptomyces antibioticus*), and actinorhodin (produced by *Streptomyces coelicolor* A3(2)) are a few examples of polyketides produced by *Streptomyces*.

3.3.2 Alkaloids

Alkaloids are a diverse group of naturally occurring organic compounds that are primarily found in plants, although some can also be found in fungi, animals, and microorganisms. They are characterized by their alkaline (basic) properties and often contain one or more nitrogen atoms in a heterocyclic ring structure. Alkaloids have a wide range of biological activities and are considered secondary metabolites because they are not essential for the basic growth and development of the producing organism (in contrast to primary metabolites like carbohydrates or amino acids). Instead, they often play important ecological roles as defense mechanisms against herbivores and pathogens or in interactions with other organisms. Some well-known alkaloids and their sources include the following.

3.3.2.1 Caffeine and Morphine

These are found in coffee beans, tea leaves, and cacao beans. Caffeine is a stimulant that affects the central nervous system. Morphine, derived from the opium poppy (*Papaver somniferum*), is a powerful pain reliever and is used in medicine as an analgesic.

3.3.2.2 Nicotine and Quinine

These are found in tobacco plants (*Nicotiana* spp.). Nicotine is a highly addictive compound and is the main reason for the addictive properties of tobacco products. Quinine is obtained from the bark of the cinchona tree (*Cinchona* spp.) and is used to treat and prevent malaria.

3.3.2.3 Atropine, Cocaine, and Codeine

These are present in plants like belladonna (*Atropa belladonna*) and jimsonweed (*Datura stramonium*). Atropine has various medicinal uses, including as a muscle relaxant and to

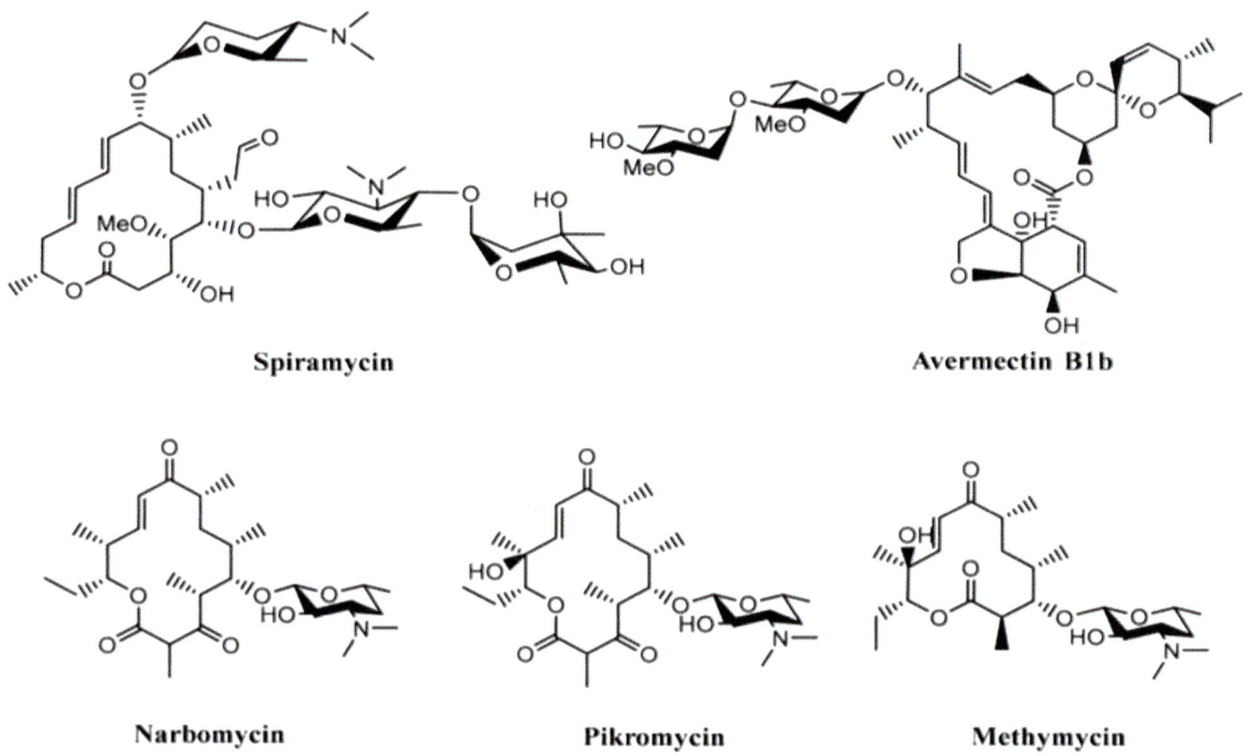

FIGURE 3.3 Some examples of polyketides.

Source: Risdian et al. (2019).

dilate pupils in eye exams. Another opium alkaloid, codeine, is used as a pain reliever and cough suppressant. Derived from coca plants (*Erythroxylum coca*), cocaine is a powerful stimulant with a history of recreational and medicinal use.

3.3.2.4 Vinblastine, Strychnine, Ephedrine, and Vincristine

These alkaloids are found in the Madagascar periwinkle (*Catharanthus roseus*) and are used in cancer chemotherapy. Strychnine, obtained from the seeds of the strychnine tree (*Strychnos nux-vomica*), is a potent neurotoxin. Ephedrine, obtained from various *Ephedra* spp., has shown stimulant and decongestant properties and has been used in traditional medicine.

3.3.3 Steroids

Steroids are another group of secondary bioactive metabolites, but unlike alkaloids, steroids are characterized by their distinct ring structure, which consists of four interconnected cycloalkane rings. Steroids are found in a wide range of organisms, including plants, animals, and fungi, and they serve various essential biological functions. Here are some key points about steroids as secondary metabolites.

3.3.3.1 Steroidal Alkaloids

Some steroids have alkaloid-like properties due to the presence of nitrogen-containing functional groups. For example, solanine and tomatine are steroidal alkaloids found in certain plants, like nightshades (*e.g.*, potatoes and tomatoes). These compounds can be toxic to humans if consumed in large amounts.

3.3.3.2 Plant Steroids

Plants also produce steroids, known as phytosterols. These compounds play important roles in the structure and function of plant cell membranes, and some steroids produced by plants have secondary metabolite functions. For example, steroidal glycoalkaloids found in some Solanaceae plants like potatoes and tomatoes serve as natural defense compounds against herbivores and pathogens.

3.3.4 Terpenoids

Terpenoids are a diverse group of secondary metabolites found in plants and various other organisms, including fungi and some animals. These compounds are derived from the basic hydrocarbon unit, isoprene (C_5H_8), and they play important roles in various biological processes, including defense against herbivores and pathogens, as well as serving as attractants for pollinators. Terpenoids are known for their wide range of bioactive properties, and they are often classified based on the number of isoprene units they contain.

3.3.4.1 Monoterpenoids, Sesquiterpenoids, and Diterpenoids

These contain two isoprene units ($C_{10}H_{16}$) and are commonly found in essential oils of plants. Examples include limonene and menthol. These contain three isoprene units ($C_{15}H_{24}$) and are also prevalent in essential oils. Examples include β-caryophyllene and curcumin. With four isoprene units ($C_{20}H_{32}$), diterpenoids have a wide range of biological activities and are found in various plant species. Some well-known diterpenoids include Taxol and gibberellins.

3.3.4.2 Triterpenoids and Tetraterpenoids

These have six isoprene units ($C_{30}H_{48}$) and are found in the sap of many trees and plants. They have various biological functions and include compounds like saponins and steroids. These contain eight isoprene units ($C_{40}H_{64}$) and are known for their role in photosynthesis. Carotenoids, such as beta-carotene and lycopene, fall into this category and serve as pigments in plants and certain microorganisms. Terpenoids have a wide range of biological activities and are essential in various ecological and physiological processes. They can serve as antioxidants, antimicrobials, antifungals, and even as signaling molecules in plant defense mechanisms. Additionally, many terpenoids have been used in traditional medicine for their therapeutic properties, and they are also of interest in the pharmaceutical and fragrance industries.

3.3.5 Quinones

Quinones are a class of secondary bioactive metabolites found in nature. They are characterized by their chemical structure, which includes a conjugated ring system with two carbonyl groups. Quinones play various roles in organisms, including plants, fungi, and some bacteria. These roles can range from defense mechanisms against pathogens and herbivores to serving as important intermediates in biochemical pathways. Quinones are compounds that contain a cyclic conjugated system of alternating double bonds and two carbonyl groups. The most common quinone structure is the 1,4-benzoquinone ring system.

3.3.5.1 Naphthoquinones and Anthraquinones

These quinones have a naphthalene ring structure and are found in plants and some fungi. Examples include juglone and plumbagin. These quinones have an anthracene ring structure and are often found in plants and fungi. They are known for their laxative properties and are used in traditional medicine. Examples include emodin and aloe-emodin.

3.3.5.2 Ubiquinone (Coenzyme Q) and Polyketide Quinones

Ubiquinone is an electron carrier in the electron transport chain of aerobic respiration. They are essential for energy production in cells. Some quinones are synthesized through polyketide pathways in microorganisms. These include compounds like doxorubicin, which is used as a chemotherapeutic agent.

3.3.6 Phenylpropanoids

Phenylpropanoids are a group of secondary bioactive metabolites that are derived from the phenylalanine pathway in plants. These compounds play a crucial role in plant growth, development, and defense mechanisms. Phenylpropanoids are also of significant interest in various industries and research fields due to their diverse biological activities and potential health benefits.

3.3.6.1 Phenolic Acids and Flavonoids

These compounds include ferulic acid, caffeic acid, and p-coumaric acid. They are often found in fruits, vegetables, and grains and are known for their antioxidant properties. They have potential health benefits due to their ability to neutralize free radicals. Flavonoids are a large group of phenylpropanoid derivatives, including quercetin, kaempferol, and naringenin. They are abundant in fruits, vegetables, tea, and red wine. Flavonoids have antioxidant, anti-inflammatory, and anticancer properties.

3.3.6.2 Lignans and Stilbenes

Lignans are found in seeds, whole grains, and certain vegetables. They are known for their potential role in reducing the risk of certain chronic diseases, including cancer and cardiovascular diseases. Resveratrol is a well-known stilbene found in red grapes and red wine. It is associated with various health benefits, including cardiovascular protection and potential antiaging effects.

3.3.6.3 Coumarins and Curcuminoids

Coumarins are found in plants such as cinnamon and sweet clover. Some coumarins have anticoagulant properties and are used in medicine. Curcumin is a phenylpropanoid compound found in turmeric. It is known for its anti-inflammatory and antioxidant properties and is used in traditional medicine. Phenylpropanoids not only are important for plants' defense against pathogens and environmental stressors but also have various applications in human health, including as dietary antioxidants, anti-inflammatory agents, and potential therapeutic compounds for treating various diseases. Research on phenylpropanoids and their bioactive properties continues to expand our understanding of their potential benefits to human health and their roles in plant biology.

3.3.7 Isocoumarins

Isocoumarins are indeed a group of secondary bioactive metabolites found in various organisms, including plants and microorganisms. They are characterized by their structural similarity to coumarins but have a distinct isomeric arrangement of atoms in the lactone ring. Isocoumarins often exhibit a wide range of biological activities and have been of interest in the fields of pharmacology and natural product chemistry. Some of their reported bioactive properties include the following.

3.3.7.1 Antimicrobial Activity and Antioxidant Activity

Isocoumarins have been investigated for their antimicrobial properties and have shown activity against various bacteria, fungi, and even viruses. Some isocoumarins have demonstrated antioxidant properties, which can help protect cells from oxidative damage.

3.3.7.2 Anticancer and Anti-Inflammatory Activity

Certain isocoumarins have exhibited potential anticancer properties, inhibiting the growth of cancer cells in laboratory studies. Isocoumarins have been studied for their anti-inflammatory effects, which can be valuable in treating inflammatory conditions.

3.3.7.3 Enzyme Inhibition and Neuroprotective Activity

Some isocoumarins can inhibit specific enzymes, making them potential candidates for drug development, especially in the context of enzyme-targeted therapies. Certain isocoumarins have shown promise in protecting nerve cells and may have applications in neurodegenerative disease research.

3.3.7.4 Anticoagulant Properties

Some isocoumarins have been explored for their ability to affect blood clotting, which could have implications in cardiovascular health.

3.4 SOURCES OF SECONDARY BIOACTIVE METABOLITES

Secondary bioactive metabolites are compounds produced by various organisms, including plants, microbes, and marine organisms, which serve specific ecological or defensive functions and often have therapeutic or pharmacological properties. These compounds can be derived from a variety of sources, and their production is influenced by environmental factors and genetic characteristics. Here are some common sources of secondary bioactive metabolites.

3.4.1 Plants

Many secondary metabolites are produced by plants as a defense mechanism against herbivores, pathogens, and environmental stressors. Examples include alkaloids (*e.g.*, caffeine, nicotine), flavonoids (*e.g.*, quercetin, resveratrol), and terpenoids (*e.g.*, taxol, artemisinin).

3.4.2 Microbes

Bacteria and fungi are prolific producers of secondary bioactive metabolites. They often synthesize these compounds as part of their competition for resources in their ecological niches or to defend against predators. Examples include antibiotics (*e.g.*, penicillin and streptomycin), mycotoxins (*e.g.*, aflatoxin), and immunosuppressive compounds (*e.g.*, cyclosporine).

3.4.3 Marine Organisms

Many marine organisms, such as sponges, corals, and certain algae, produce secondary metabolites with unique chemical structures and potent biological activities. These compounds can be used for defense against predation or to compete for space and resources in the marine ecosystem. Examples include marine alkaloids, polyketides, and peptides.

3.4.4 Insects and Animals

Some animals and insects have evolved to produce secondary metabolites for defense or communication purposes. For example, poison dart frogs produce alkaloid toxins on their skin, and certain beetles secrete defensive chemicals when threatened.

3.4.5 Endophytes

Endophytic fungi and bacteria that live inside plant tissues can also produce secondary metabolites. These compounds often contribute to the host plant's defense mechanisms. Endophytes have been a source of novel bioactive compounds with potential pharmaceutical applications.

3.5 DIVERSITY OF SECONDARY BIOACTIVE METABOLITES

The diversity of secondary bioactive metabolites is vast and encompasses a wide range of chemical structures and functions. Here are some key categories of secondary bioactive metabolites and examples of their diversity.

3.5.1 Glycosides

Some cardiac glycosides, like digitoxin, are found in foxglove plants, while saponins including ginsenosides are obtained from ginseng.

3.5.2 Pigments and Phycobiliproteins

Some pigments in organisms, such as melanin in humans and chlorophyll in plants, have bioactive properties beyond their coloration function, while pigments found in certain algae and cyanobacteria are used for photosynthesis and light capture.

3.5.3 Marine Natural Products and Microbial Metabolites

Many diverse secondary metabolites are isolated from marine organisms, including sponges, corals, and marine bacteria. These compounds often have unique structures and bioactivities. Bacteria and fungi produce a wide array of secondary metabolites with various bioactive properties, including antibiotics, immunosuppressants, and antifungals.

3.5.4 Insect Chemical Defenses and Bioactive Lipids

Insects produce a wide variety of bioactive compounds, such as alkaloids and terpenes, for defense against predators. These include prostaglandins, leukotrienes, and resolvins, which play crucial roles in inflammation and immune responses.

3.6 INDUSTRIAL APPLICATIONS OF MICROBIAL SECONDARY METABOLITES

3.6.1 Antibiotics Production

Many well-known antibiotics, such as penicillin, streptomycin, and tetracycline, are microbial secondary metabolites. These compounds have played a critical role in medicine by helping control bacterial infections (Table 3.1).

3.6.2 Antifungals and Antivirals Production

Microbial secondary metabolites also include compounds with antifungal and antiviral properties. These molecules can inhibit the growth and replication of fungi and viruses, providing potential sources for new drugs. Compounds like amphotericin B and nystatin, produced by *Streptomyces* species, are used as antifungal agents to treat fungal infections. Certain microbial secondary metabolites have antiviral properties and are used in the development of drugs to combat viral infections, such as HIV and influenza (Table 3.1).

3.6.3 Production of Anticancer Agents

Some microbial secondary metabolites exhibit anticancer properties and have been investigated for their potential in cancer treatment. For example, bleomycin is used in chemotherapy for certain types of cancer, and many anticancer drugs, such as doxorubicin and paclitaxel, are derived from microbial secondary metabolites which exhibit cytotoxic properties against cancer cells (Wall and Wani, 1996).

3.6.4 Production of Toxins

Not all microbial secondary metabolites are beneficial. Some can be toxic to other organisms, including humans. For instance, mycotoxins produced by molds can contaminate food and pose health risks.

TABLE 3.1
List of Some Volatile Antibiotics Produced by Microbes

Compounds	Biological Activity	Endophytic Bacteria	References
Bacillomycin	Antifungal and hemolytic	*B. subtilis* and *B. amyloliquefaciens*	Aranda et al., 2005
Nystatin	Antifungal	*Streptomyces noursei*	Fjaervik and Zotchev, 2005
KB425796-A	Antifungal	*Paenibacillus* sp. str. 530603	Kai et al., 2013
Bacilysocin	Antifungal	*B. subtilis* str.168	Tamehiro et al., 2002
Ecomycin	Antifungal	*Pseudomonas viridiflava*	Miller et al., 1998
Hormaomycin	Antibacterial	*Streptomyces* sp.	Bae et al., 2015
Munumbicin	Antibacterial	*Streptomyces* str. NRRL 30562	Castillo et al., 2002
Tetracycline	Antibacterial	*Streptomyces rimosus* and *S. aureofaciens*	Nelson, 2001
Bacteriocins	Antibacterial	*B. subtilis*	Sansinenea and Ortiz, 2011
Amicoumacin	Antibacterial and anti-inflammatory	*B. subtilis*	Pinchuk et al., 2002
Treponemycin	Antituberculous	*Streptomyces* str. MS-6–6	Yassien et al., 2015
Camptothecin	Anticancer	*Lysinibacillus* sp. and *B. cereus*	Shweta et al., 2013
Saadamycin	Antidermatophyte	*Streptomyces* sp. str. Hedaya48	Gendy and Bondkly, 2010
Strepturidin	Immunotherapy	*Streptomyces albus* str. DSM 40763	Pesic et al., 2014
Xiamycin	Anti-HIV activity	*Streptomyces* sp.	Ding et al., 2010
Thaxtomin A	Cellulose synthesis inhibitor	*Streptomyces scabies*	Francis et al., 2015

3.6.5 Symbiotic Relationships

Microbial secondary metabolites also play a role in symbiotic relationships between microorganisms and their hosts. For example, some bacteria in the root nodules of leguminous plants produce compounds called nodulation factors, which help establish a symbiotic relationship between the bacteria and the host plant (Ogbe et al., 2020).

3.7 BIOTECHNOLOGICAL APPLICATIONS OF SECONDARY METABOLITES

Microbial secondary metabolites have significant biotechnological applications, including drug discovery, bioremediation, and industrial processes. Researchers are continually exploring and harnessing their potential. Microbial secondary metabolites have a wide range of biotechnological applications due to their diverse chemical structures and biological activities. These compounds are produced by various microorganisms, such as bacteria, fungi, and actinomycetes, and often play crucial roles in their ecological functions and interactions. Here are some biotechnological applications of microbial secondary metabolites (Devi et al., 2020):

3.7.1 In Pharmaceuticals

Secondary metabolites from microorganisms have been a rich source of antibiotics, immunosuppressants, anticancer agents, and other pharmaceuticals. Examples include penicillin (from *Penicillium*), erythromycin (from *Saccharopolyspora erythraea*), and statins (from various fungi), among many others. Secondary metabolites have found numerous applications in the field of pharmaceuticals due to their diverse chemical structures and biological activities (Shankar and Sharma, 2022).

3.7.1.1 Drug Discovery and Development

Secondary metabolites are a valuable source of lead compounds for drug discovery. Natural products have been the basis for the development of many important drugs, such as antibiotics (*e.g.*, penicillin), anticancer agents (*e.g.*, paclitaxel), and immunosuppressants (*e.g.*, cyclosporine) (Verpoorte, 1998).

3.7.1.2 Pain Relievers

Alkaloids like morphine and codeine, found in the opium poppy, have been used as pain relievers for centuries.

3.7.1.3 Cardiovascular Drugs

Digitalis glycosides, extracted from the foxglove plant, have been used to treat heart conditions, such as congestive heart failure and atrial fibrillation (Vaishnav and Demain, 2011).

3.7.1.4 Anti-Inflammatory Agents

Compounds like salicylic acid (found in willow bark) and curcumin (from turmeric) have anti-inflammatory properties and are used to manage inflammatory conditions.

3.7.1.5 Herbal Medicines and Biodegradable Polymers

Traditional herbal medicines often rely on secondary metabolites for their therapeutic effects. These include compounds like ginsenosides from ginseng and artemisinin from sweet wormwood for malaria treatment. Microbial polyesters like polyhydroxyalkanoates (PHAs) are used in the production of biodegradable plastics, reducing environmental pollution.

3.7.1.6 Antiparasitic Drugs

Compounds like artemisinin and ivermectin have been used to treat parasitic infections, such as malaria and river blindness. Pharmaceutical researchers continue to explore natural product libraries to discover new drugs and therapeutic agents. Secondary metabolites remain a rich source of chemical diversity and biological activity, making them a valuable resource for drug development and health care.

3.7.2 Enzyme Inhibitors and Agriculture

Microbial secondary metabolites can serve as enzyme inhibitors, with applications in drug discovery and the development of enzyme-targeted therapies. Secondary metabolites can be used as biopesticides, herbicides, and plant growth–promoting compounds. For instance, the *Bacillus thuringiensis* toxin is used as a biological pesticide against certain insect pests.

3.7.3 Application in Bioremediation and Biocatalysts

Certain secondary metabolites can help in the bioremediation of polluted environments by breaking down or sequestering toxic compounds. Microbial secondary metabolites can be used as biocatalysts in various industrial processes, such as the production of biofuels and bioplastics.

3.7.9.1 Biofuels and Biological Control

Microbial secondary metabolites can enhance the production of biofuels by improving the growth and lipid content of biofuel-producing microorganisms. Some secondary metabolites can be used for the biological control of plant diseases, reducing the need for chemical pesticides. The discovery, isolation, and engineering of microbial secondary metabolites continue to be a vibrant area of research with significant implications for various industries, including health care, agriculture, and environmental management.

REFERENCES

Aharoni, A., A.P. Giri, F.W.A. Verstappen, et al. 2004. Gain and loss of fruit flavor compounds produced by wild and cultivated strawberry species. *The Plant Cell* 16: 3110–31. https://doi.org/10.1105/tpc.104.023895.

Aly, A.H., A. Debbab, J. Kjer, and P. Proksch. 2010. Fungal endophytes from higher plants: A prolific source of phytochemicals and other bioactive natural products. *Fungal Diversity* 41: 1–16. https://doi.org/10.1007/s13225-010-0034-4.

Aranda, F.J., J.A. Teruel, and A. Ortiz, 2005. Further aspects on the hemolytic activity of the antibiotic lipopeptide iturin A. *Biochimica et Biophysica Acta (BBA) – Biomembranes* 1713: 51–56. https://doi.org/10.1016/j.bbamem.2005.05.003.

Bae, M., B. Chung, K.B. Oh, J. Shin, and D.C. Oh. 2015. Hormaomycins B and C: New antibiotic cyclic depsipeptides from a marine mudflat-derived *Streptomyces* sp. *Marine Drugs* 13: 5187–200. https://doi.org/10.3390/md13085187.

Blunt, J.W., B.R. Copp, R.A. Keyzers, M.H.G. Munro, and M.R. Prinsep. 2017. Marine natural products. *Natural Product Reports* 34: 235–94. https://doi.org/10.1039/C6NP00124F.

Castillo, U.F., G.A. Strobel, E.J. Ford, et al. 2002. Munumbicins, wide-spectrum antibiotics produced by *Streptomyces* NRRL 30562, endophytic on *Kennedia nigriscans*. *Microbiology* 148: 2675–85. https://doi.org/10.1099/00221287-148-9-2675.

Chopra, I., and M. Roberts. 2001. Tetracycline antibiotics: Mode of action, applications, molecular biology, and epidemiology of bacterial resistance. *Microbiology and Molecular Biology Reviews* 65: 232–60. https://doi.org/10.1128/mmbr.65.2.232-260.2001.

Croteau, R., T.M. Kutchan, and N.G. Lewis. 2012. Natural products (secondary metabolites). In *Biochemistry & molecular biology of plants*, 2nd ed., eds. B. Buchanan, W. Gruissem, and R. Jones, 1250–319. Hoboken, NJ: Wiley. http://science.lecture.ub.ac.id/files/2012/04/plant-biosynthesis1.pdf.

Debbab, A., A.H. Aly, and P. Proksch. 2011. Bioactive secondary metabolites from endophytes and associated marine derived fungi. *Fungal Diversity* 49: 1–12. https://doi.org/10.1007/s13225-011-0114-0

Devi, K., T. Kaur, G. Guleria, et al. 2020. Fungal secondary metabolites and their biotechnological applications for human health. In *New and future developments in microbial biotechnology and bioengineering*, eds. A.A. Rastegari, A.N. Yadav, and N. Yadav, 147–61. Elsevier. https://doi.org/10.1016/B978-0-12-820528-0.00010-7.

Ding, L., J. Munch, H. Goerls, et al. 2010. Xiamycin, a pentacyclic indolo sesquiterpene with selective anti-HIV activity from a bacterial mangrove endophyte. *Bioorganic & Medicinal Chemistry Letters* 20: 6685–87. https://doi.org/10.1016/j.bmcl.2010.09.010.

Engelhard, M., T. Hurek, and B. Reinhold-Hurek. 2000. Preferential occurrence of diazotrophic endophytes, *Azoarcus* spp., in wild rice species and land races of *Oryza sativa* in comparison with modern races. *Environmental Microbiology* 2: 131–41. https://doi.org/10.1046/j.1462-2920.2000.00078.x.

Fjaervik, E., and S.B. Zotchev. 2005. Biosynthesis of the polyene macrolide antibiotic nystatin in *Streptomyces noursei*. *Applied Microbiology and Biotechnology* 67: 436–43. https://doi.org/10.1007/s00253-004-1802-4.

Fleming, A. 1929. On the antibacterial action of cultures of a *Penicillium* with special reference to their use in the isolation of *B. influenza*. *British Journal of Experimental Pathology* 10: 226–36. https://www.ncbi.nlm.nih.gov/pmc/articles/PMC2048009/.

Francis, I.M., S. Jourdan, S. Fanara, R. Loria, and S. Rigali. 2015. The cellobiose sensor CebR is the gatekeeper of *Streptomyces scabies* pathogenicity. *mBio* 6 (February): e02018, 1–6. https://doi.org/10.1128/mbio.02018-14.

Gendy, M.M., and A.M. Bondkly. 2010. Production and genetic improvement of a novel antimycotic agent, saadamycin, against dermatophytes and other clinical fungi from endophytic *Streptomyces* sp. Hedaya48. *Journal of Industrial Microbiology and Biotechnology* 37: 831–41. https://doi.org/10.1007/s10295-010-0729-2.

Ghannoum, M.A., and L.B Rice. 1999. Antifungal agents: Mode of action, mechanisms of resistance, and correlation of these mechanisms with bacterial resistance. *Clinical Microbiology Reviews* 12: 501–17. https://doi.org/10.1128/cmr.12.4.501.

Goff, S.A., and H.J. Klee. 2006. Plant volatile compounds: Sensory cues for health and nutritional value? *Science* 311: 815–19. https://www.science.org/doi/10.1126/science.1112614.

Goutam, M., C. Giriyapura, S.K. Mishra, and S. Gupta. 2014. Titanium allergy: A literature review. *Indian Journal of Dermatology* 59 (November): 630, 1–5. https://pubmed.ncbi.nlm.nih.gov/25484409/.

Gupta, A., S. Jain, and N. Verma. 2022. *Industrial applications of soil microbes*, vol. 1. Bentham Science Publishers Pte Ltd. https://doi.org/10.2174/97898150399551220101.

Gupta, A., S. Jain, and N. Verma. 2024. *Industrial applications of soil microbes*, vol. 3. Bentham Science Publishers Pte Ltd. https://doi.org/10.2174/97898150797531240301.

Hleba, L., I. Charousova, M. Cisarova, et al. 2018. Rapid identification of *Streptomyces* tetracycline producers by MALDI-TOF mass spectrometry. *Journal of Environmental Science and Health, Part A* 53: 1083–93. https://doi.org/10.1080/10934529.2018.1474581.

Horwitz, S.B., D. Cohen, S. Rao, I. Ringel, H.J. Shen, and C.P. Yang. 1993. Taxol: Mechanisms of action and resistance. *Journal of the National Cancer Institute* 15: 55–61. PMID: 7912530.

Jain, S., A. Gupta, and N. Verma. 2023. *Industrial applications of soil microbes*, vol. 2. Bentham Science Publishers Pte Ltd. https://doi.org/10.2174/97898150502641230201.

Jain, S., A. Gupta, and N. Verma. 2024. *Industrial applications of soil microbes*, vol. 4. Bentham Science Publishers Pte Ltd. https://doi.org/10.2174/97898151249961240401.

Jensen, L.M., I.R. Wallis, K.J. Marsh, B.D. Moore, N.L. Wiggins, and W.J. Foley. 2014. Four species of arboreal folivore show differential tolerance to a secondary metabolite. *Oecologia* 176: 251–58. https://doi.org/10.1007/s00442-014-2997-4.

Kai, H., M. Yamashita, S. Takase, et al. 2013. KB425796-A, a novel antifungal antibiotic produced by *Paenibacillus* sp. 530603. *The Journal of Antibiotics* 66: 465–71. https://doi.org/10.1038/ja.2013.63.

Katsuyama, Y., N. Funa, I. Miyahisa, and S. Horinouchi. 2007. Synthesis of unnatural flavonoids and stilbenes by exploiting the plant biosynthetic pathway in *Escherichia coli*. *Chemistry & Biology* 14: 613–21. https://doi.org/10.1016/j.chembiol.2007.05.004.

Kaul, S., M. Ahmed, T. Sharma, and M.K. Dhar. 2014. Unlocking the myriad benefits of endophytes: An overview. In *Microbial diversity and biotechnology in food security*, eds. R. Kharwar, R. Upadhyay, N. Dubey, and R. Raghuwanshi, 41–57. New Delhi: Springer. https://doi.org/10.1007/978-81-322-1801-2_4.

Kumar, A., P.K. Singh, R. Parihar, V. Dwivedi, S.C. Lakhotia, and S. Ganes. 2014. Decreased O-Linked GlcNAcylation protects from cytotoxicity mediated by Huntingtin Exon1 protein fragment. *Journal of Biological Chemistry* 289: 13543–53. https://doi.org/10.1074/jbc.M114.553321.

Li, J.Y., G. Strobel, R. Sidhu, W.M. Hess, and E.J. Ford. 1996. Endophytic taxol-producing fungi from bald cypress, *Taxodium distichum*. *Microbiology* 142: 2223–26. https://doi.org/10.1099/13500872-142-8-2223.

Miller, C.M., R.V. Miller, D. Garton-Kenny, et al. 1998. Ecomycins, unique antimycotics from *Pseudomonas viridiflava*. *Journal of Applied Microbiology* 84: 937–44. https://doi.org/10.1046/j.1365-2672.1998.00415.x.

Moore, B.S., and J.N. Hopke. 2001. Discovery of a new bacterial polyketide biosynthetic pathway. *ChemBioChem* 2: 35–38. https://doi.org/10.1002/1439-7633(20010105)2:1<35::AID-CBIC35>3.0.CO;2–1.

Nelson, M.L. 2001. The chemistry and cellular biology of the tetracyclines. In *Tetracyclin in biology, chemistry, and medicine*, eds. M. Nelson, W. Hillen, and R.A. Greenwald, 3–63. Basel: Birkhäuser. https://doi.org/10.1007/978-3-0348-8306-1_1.

Newman, D.J., and G.M. Cragg. 2016. Natural products as sources of new drugs from 1981 to 2014. *Journal of Natural Products* 79: 629–61. https://doi.org/10.1021/acs.jnatprod.5b01055.

Ogbe, A.A., J.F. Finnie, and J.V Staden. 2020. The role of endophytes in secondary metabolites accumulation in medicinal plants under abiotic stress. *South African Journal of Botany* 134: 126–34. https://doi.org/10.1016/j.sajb.2020.06.023.

Owen, N.L., and N. Hundley. 2004. Endophytes – The chemical synthesizers inside plants. *Science Progress* 87: 79–99. https://doi.org/10.3184/003685004783238553.

Parthasarathy, R., and M. Sathiyabama. 2015. Lovastatin-producing endophytic fungus isolated from a medicinal plant *Solanum xanthocarpum*. *Natural Product Research* 29: 2282–86. https://doi.org/10.1080/14786419.2015.1016938.

Pesic, A., B. Steinhaus, S. Kemper, et al. 2014. Isolation and structure elucidation of the nucleoside antibiotic strepturidin from *Streptomyces albus* DSM 40763. *The Journal of Antibiotics* 67: 471–77. https://doi.org/10.1038/ja.2014.16.

Pichersky, E., and D.R. Gang. 2000. Genetics and biochemistry of secondary metabolites in plants: An evolutionary perspective. *Trends in Plant Science* 5: 439–45. https://doi.org/10.1016/S1360-1385(00)01741-6.

Pinchuk, I.V., P. Bressollier, I.B. Sorokulova, B. Verneuil, and M.C. Urdaci. 2002. Amicoumacin antibiotic production and genetic diversity of *Bacillus subtilis* strains isolated from different habitats. *Research in Microbiology* 153: 269–76. https://doi.org/10.1016/S0923-2508(02)01320-7.

Promputtha, I., S. Lumyong, V. Dhanasekaran, E.H.C. McKenzie, K.D. Hyde, and R. Jeewon. 2007. A phylogenetic evaluation of whether endophytes become saprotrophs at host senescence. *Microbial Ecology* 53: 579–90. https://doi.org/10.1007/s00248-006-9117-x.

Risdian, C., T. Mozef, and J. Wink. 2019. Biosynthesis of polyketides in *Streptomyces*. *Microorganisms* 7 (May): 124, 1–18. https://doi.org/10.3390/microorganisms7050124.

Rokem, J.S., A.E. Lantz, and J. Nielsen. 2007. Systems biology of antibiotic production by microorganisms. *Natural Product Reports* 24: 1262–87. https://doi.org/10.1039/B617765B.

Sandoval, I.T., R.A. Davis, T.S. Bugni, G.P. Concepcion, M.K. Harper, and C.M. Ireland. 2004. Cytotoxic isoquinoline quinones from sponges of the genus *Petrosia*. *Natural Product Research* 18: 89–93. https://doi.org/10.1080/1057563031000122130.

Sansinenea, E., and A. Ortiz. 2011. Secondary metabolites of soil *Bacillus* spp. *Biotechnology Letters* 33: 1523–38. https://doi.org/10.1007/s10529-011-0617-5.

Selim, K.A., A.A. El-Beih, T.M. Abd El-Rahman, and A.I. El-Diwany. 2012. Biology of endophytic fungi. *Current Research in Environmental & Applied Mycology* 2: 31–82. https://www.creamjournal.org/pdf/Cream_2_1_3.pdf.

Shankar, A., and K.K. Sharma. 2022. Fungal secondary metabolites in food and pharmaceuticals in the era of multi-omics. *Applied Microbiology and Biotechnology* 106: 3465–88. https://doi.org/10.1007/s00253-022-11945-8.

Shweta, S., J.H. Bindu, J. Raghu, et al. 2013. Isolation of endophytic bacteria producing the anticancer alkaloid camptothecine from *Miquelia dentata* Bedd. (Icacinaceae). *Phytomedicine* 20: 913–17. https://doi.org/10.1016/j.phymed.2013.04.004.

Tamehiro, N., Y. Okamoto-Hosoya, and S. Okamoto. 2002. Bacilysocin, a novel phospholipid antibiotic produced by *Bacillus subtilis* 168. *Antimicrobial Agents and Chemotherapy* 46: 315–20. https://doi.org/10.1128/aac.46.2.315-320.2002.

Tejesvi, M.V., K.R. Kini, H.S. Prakash, V. Subbiah, and H.S. Shetty. 2007. Genetic diversity and antifungal activity of species of *Pestalotiopsis* isolated as endophytes from medicinal plants. *Fungal Diversity* 24: 37–54. https://www.fungaldiversity.org/fdp/sfdp/24-3.pdf.

Vaishnav, P., and A.L. Demain. 2011. Unexpected applications of secondary metabolites. *Biotechnology Advances* 29: 223–29. https://doi.org/10.1016/j.biotechadv.2010.11.006.

Verpoorte, R. 1998. Exploration of nature's chemodiversity: The role of secondary metabolites as leads in drug development. *Drug Discovery Today* 3: 232–38. https://doi.org/10.1016/S1359-6446(97)01167-7.

Waites, M.J., N.L. Morgan, J.S. Rockey, and G. Higton. 2001. *Industrial microbiology: An introduction*. Hoboken, NJ: Wiley-Blackwell. https://www.wiley.com/en-be/Industrial+Microbiology%3A+An+Introduction-p–9780632053070.

Wall, M.E., and M.C. Wani. 1996. Camptothecin and taxol: From discovery to clinic. *Journal of Ethnopharmacology* 51: 239–54. https://doi.org/10.1016/0378-8741(95)01367-9.

Yassien, M.A., H.M. Abdallah, M.A. El-Halawani, and A.A.M. Jiman-Fatani. 2015. Anti-tuberculous activity of treponemycin produced by a *Streptomyces* strain MS-6–6 isolated from Saudi Arabia. *Molecules* 20: 2576–90. https://doi.org/10.3390/molecules20022576.

Zhou, X., Z. Wang, K. Jiang, et al. 2007. Screening of taxol-producing endophytic fungi from *Taxus chinensis* var. *mairei*. *Applied Biochemistry and Microbiology* 43: 439–43. https://doi.org/10.1134/S000368380704014X.

4 Exploring Microbial Diversity for Secondary Metabolite Production

A Metagenomic Approach

**Ashwini A. Waoo*
(Department of Biotechnology, AKS University, Satna, MP, India; ashwiniwaoo@gmail.com)

C. Vyas
(Department of Biotechnology, AKS University, Satna, MP, India; mailcharu86@gmail.com)

B. A. Chopade
(Department of Biotechnology, AKS University, Satna, MP, India; bachopade@gmail.com)

*Corresponding Author

This chapter delves into the intricate realm of microbial diversity as a cornerstone for unlocking the vast potential of secondary metabolite (SM) production through a metagenomic approach. By employing advanced techniques in metagenomics, we navigate the complex genetic landscapes of microbial communities, uncovering hidden biosynthetic pathways and novel bioactive compounds. The exploration encompasses a thorough examination of metagenomic tools, computational analyses, and genetic reprogramming strategies, providing insights into how microbial diversity can be harnessed for enhanced SM discovery. As we embark on this metagenomic journey, the chapter highlights the transformative impact on our understanding of microbial ecosystems and the accelerated pace of uncovering valuable secondary metabolites with potential applications in medicine, agriculture, and biotechnology.

4.1 INTRODUCTION

Microorganisms affect nearly every aspect of human activity, both positively and negatively: They produce drugs that save lives (Demain and Fang, 2000), they aid in the digestion of food (Rowland et al., 2018), they cause infections (Morens et al., 2004), and they create bioterrorism weapons (Henderson, 1999). They also contribute to climate change (Bardgett et al., 2008). The ability of microbes to multiply into populations of billions and work together to alter vast quantities of molecules accounts for their immense influence on the macroscopic world. Many of the chemicals they emit are secondary metabolites; meaning, they are not necessary for growth, development, or division. However, it remains unclear why bacteria invest in secondary metabolites despite their significance (Cavalier-Smith, 1992).

Secondary metabolites (SM) allow for collective behaviors, such as cell–cell signaling, microbial warfare, pathogenicity (Dufour and Rao, 2011; Demain and Fang, 2000; Nicas and Iglewski, 1985), and nutrient scavenging. These collective features benefit a population's fitness (Maplestone et al., 1992), but they can also harm it by diverting precursors, energy, and cofactors away from basic primary metabolism (Yan et al., 2018; Jain et al., 2023, 2024).

Secondary metabolites are organic substances that are created by modifying the main metabolite synthesis. Secondary metabolites, unlike primary metabolites, do not participate in development, growth, or reproduction and are normally created at/near the end of the stationary phase of growth. Secondary metabolites have important roles in ecology, including in antibiotics, defense mechanisms, and pigment production (Gupta et al., 2022, 2024). Atropine and antibiotics, for example, erythromycin and bacitracin, are examples of secondary metabolites important in industrial microbiology. Atropine is a secondary metabolite generated from many plants that has essential clinical applications. Atropine is a competitive antagonist for acetylcholine receptors, specifically muscarinic receptors, and can be used to treat bradycardia.

Microbial secondary metabolites (SM) play crucial roles across industries: serving as pharmaceutical drugs, enhancing agricultural productivity, aiding in environmental cleanup, and contributing to various consumer products, as shown in Figure 4.1. Their diverse applications highlight their significance in medicine, agriculture, biotechnology,

DOI: 10.1201/9781003408833-4

and environmental science, making them invaluable resources for addressing societal challenges and driving innovation. From combating diseases to improving crop yields and promoting sustainability, microbial SM continues to shape advancements in science and industry.

Erythromycin, a widely used antibiotic produced from *Saccharopolyspora erythraea*, has a broad antibacterial range. It is mass-produced in industrial microbiology and is employed as an antibiotic and is typically taken orally. Finally, bacitracin is an antibiotic that is classed as a secondary metabolite. Bacitracin is an antibiotic produced from *Bacillus subtilis* organisms that is extensively used as a topical medication. Bacitracin is a non-ribosomal peptide synthetase that can manufacture peptides in nature; nonetheless, it is employed in medicine as an antibiotic.

In the intricate world of microbiology, the exploration of microbial diversity unveils a treasure trove of secondary metabolites, offering a myriad of potential applications across diverse industries. From pharmaceuticals to agriculture and biotechnology, these compounds play a pivotal role in shaping ecological interactions and driving innovation. This introductory paragraph sets the stage for delving into the fascinating realm of microbial diversity and its profound implications for secondary metabolite production, showcasing the multifaceted contributions of microorganisms to scientific advancement and societal well-being.

4.2 EXPLORING MICROBIAL SECONDARY METABOLITES

Microbial secondary metabolites, now known as specialized metabolites, frequently have odd structures, and many have shown significant implications on our society's health, nutrition, and economics (Monciardini et al., 2014). These chemicals are often low-molecular-mass by-products of secondary metabolism, which occurs independently of the primary microbial development phase. Different bacteria create metabolites in different ways. These are important for distinguishing between different microbe species. Microbial secondary metabolites are compounds that are not required for the growth of the generating cultures but are critical for human health.

Antibiotics, anticancer agents, cholesterol-lowering medications, and other medications are among them. They feature peculiar shapes. Its synthesis can be considerably altered by varying the kind and concentration of nutrients in the culture media. Among these, the effect of carbon sources as nutrients has been the topic of ongoing research for both businesses and research organizations. Various mechanisms have been postulated in fungi and bacteria to explain the detrimental effect of carbon catabolite on the production of secondary metabolites. Their expertise and modification have proven effective in either creating fermentation conditions or improving strains. Recent biochemical and molecular research has made significant progress in understanding these pathways.

Antibiotics, pigments, poisons, enzyme inhibitors, immunomodulators, ecological competition or symbiotic effectors, or substances with hormonal action or specific effects on lipid or carbohydrate metabolism are examples. Some have already proven to be effective antimicrobials, antivirals, antioxidants, antitumoral, contractants, vasorelaxants, and diuretics. Few are employed as colorants, insecticides, or growth stimulants (Fouillaud et al., 2016). For animals or plants, microorganisms are responsible for approximately 53% of FDA-approved natural product–based medications (Patridge et al., 2016).

Nutritive sources, growth circumstances, enzyme induction, feedback control, or inactivation all play a role in fine-tuning specialized metabolite production. Specific low-molecular-mass molecules, as well as transfer RNA factors and other factors, frequently influence their regulation.

During post-exponential development, gene products are generated. Clustered genes on chromosomal DNA rather than plasmid DNA are responsible for coding and specialized metabolite production. However, the linked pathways remain unresolved, opening up a new frontier for academic researchers in regulation, enzymology, and differentiation (Jeong et al., 2020). Omics sciences applied to industrial microbes, such as genomics, transcriptomics, and metabolomics, now provide new options for characterization, strain

Uses of Microbial Secondary Metabolites	Pharmaceuticals
	Agriculture
	Bioremediation
	Food and beverage industry
	Cosmetics
	Biofuels
	Research tools
	Animal and plant health

FIGURE 4.1 Applications of microbial secondary metabolites.

discovery, and development. As a result, there is significant potential for the development of new molecules for pharmacological, nutraceutical, dyeing, and other applications.

Microbial secondary metabolites, such as antibiotics, growth hormones, pigments, anticancer drugs, and others, are not required for microbial development and growth, but they have shown significant promise for animal and human health (Ruiz et al., 2010). Among the microorganisms that create the aforementioned compounds, as well as bacteria, fungi, and actinobacteria, produce a varied array of bioactive small molecules with substantial promise for use in medicine (O'Brien and Wright, 2011). These bioactive compounds are mostly created by the activation of cryptic gene clusters that are inactive under normal conditions; consequently, the expression of these clusters would be beneficial in the exploitation of microorganism chemical diversity (Pettit, 2011; Xu et al., 2019).

This will advance knowledge of secondary metabolites' capabilities and possibilities in business and medicine. One problem that needs to be solved is creating methods for understanding the workings of cryptic genes, their relationships, and how they relate to the production of bioactive compounds. To create novel bioactive chemicals, researchers also need to focus more on co-culturing various microbes that work well together.

Even though several studies on microbial metabolites have been reviewed and published recently (Passari et al., 2017; Zothanpuia et al., 2018; Overy et al., 2019), there is still much that needs to be known about how to boost the synthesis of bioactive secondary metabolites. It is imperative to concentrate on the vital components of microbial biodiversity and to extract and use marine, terrestrial, and endophytic microbial resources to produce secondary metabolites and their biological activities.

The production of new natural compounds from bacteria has significantly slowed down. Furthermore, most antibiotic screening programs uncover secondary metabolites that were previously unknown. It is believed that germs have lost their unique metabolic richness as a result. A global focus on the development of novel antimicrobials, novel treatment alternatives, and alternative antimicrobial therapies is necessary due to the rapid growth of multidrug-resistant pathogens. Over a million years of evolution, microbes developed resistance to drugs by lateral or vertical transmission. However, they also had the opportunity to adapt, neutralize, and fight these resistant infections by developing more potent, more effective secondary metabolites.

Since nature continues to be the most abundant and adaptable source of new antibiotics, real efforts are needed to utilize unknown and untapped microorganisms from natural environments to find secondary metabolites that have not yet been identified. In search of novel metabolites, this hope reignites the motivation to explore obscure gene clusters or new pathways in microorganisms.

The pharmaceutical sector is impacted by newly identified novel secondary metabolites of microbes. Amazingly, microorganisms produce natural products that are helpful for human health. Presently, half of all drugs available on the market are natural products. Since the discovery of penicillin by Alexander Fleming 85 years ago, antibiotics have been crucial in the treatment of infectious diseases, saving countless lives.

Novel chemicals possessing hypocholesterolemic, anticancer, and immunosuppressive properties have been introduced to treat various disorders. Novel antibiotics and secondary metabolites that are no longer produced by giant pharmaceutical companies are being created by smaller biotechnology companies and academic labs. Certain contemporary technologies, like genome mining and metagenomics, enable the creation and identification of novel compounds with a range of uses, including therapeutic substitutes needed to address serious issues like antibiotic resistance, the SARS-CoV-2 pandemic, and other newly emerging illnesses.

Advances in microbial methods have been relevant in allowing uncultured microorganisms to flourish as potential sources of new chemical compounds, as the secondary metabolites found so far originate from 1% of cultured microbes (Newman, 2016). Furthermore, metagenomics opens up a vast, previously unexplored reservoir of genetic variation and metabolic diversity.

The identification of novel compounds from microbial secondary metabolites can be effectively achieved using metagenomics and metabolomics approaches. *Metagenomics* allows for the exploration of microbial genetic material from diverse environments, uncovering novel genes responsible for secondary metabolite production (Nam et al., 2023). Meanwhile, *metabolomics* provides a detailed analysis of metabolites through techniques like mass spectrometry and nuclear magnetic resonance, revealing new compounds and their functions (Wang et al., 2010). Integrating these methods enhances the discovery of unique metabolites with potential applications across various fields. For further reading, check out Shah and Newgard (2015).

Another possibility is to search for novel and more efficient compounds by screening microbes from unusual sources. Among the sources were oceans, hot springs, caverns, high-salinity areas or seas, and arid soils (Demain et al., 2019). The fascinating idea of finding new microbial products from a largely untapped pool could be made possible by this metagenomics technique (Kowalchuk et al., 2007; Trindade et al., 2015). Genome mining (Martinez-Klimova et al., 2017) presents an additional intriguing possibility: It enables the identification of the expression of particular secondary metabolites generated that are distinct to particular environments. As a result, there are significant issues in obtaining cryptic clusters in the lab, which calls for the employment of alternative methods.

4.3 SCREENING AND CHARACTERIZATION OF VALUABLE STRAINS

Researchers have found that the ocean is a very productive biotope for identifying new bacteria and, consequently, characterizing profitable new chemicals. It has been found that sponges and their microbiota produce a diverse array of advantageous secondary metabolites. In the field of medicine, several naturally occurring substances derived from marine animals have already demonstrated their capacity to delay aging and/or lengthen lifespan.

However, the biodiversity of the Indian Ocean's southwest has received far less attention, particularly in terms of antiaging measures. In the article published by Hassane et al. (2020), the microbial diversity of the sea sponge *Scopalina hapalia* was investigated utilizing metagenomic analysis. The sponge had 26 bacterial and 2 archaeal phyla, with the *Proteobacteria* being the most abundant. Thirty isolates from the genera *Bacillus*, *Rhodococcus*, *Micromonospora*, *Aspergillus*, *Salinispora*, *Chaetomium*, *Nigrospora*, and/or linked to the family *Thermoactinomycetaceae* were cultured to produce secondary metabolites. It was shown that crude extracts from several microbial cultures exhibited activity against several enzymes, including elastase, tyrosinase, catalase, sirtuin 1, CDK7 (cyclin-dependent kinase 7), Fyn kinase, and the proteasome. These results show that microorganisms associated with a marine sponge in the Indian Ocean are capable of producing antiaging compounds. Salikin et al. (2020) emphasized the potential of marine epiphytic bacteria to offer a new platform for the development of innovative antinematode drugs. Novel strain discovery approaches that are still under development are highlighted, such as high-throughput screening that is not dependent on culture.

Nawaz et al. (2021) gave a summary of pigments in their review, with marine bacteria that synthesize biopigments, as well as their applications. It highlights some compounds, including lutein, lycopene, astaxanthin, zeaxanthin, and violacein, that are already well-known to the business. These strains are undoubtedly one of the underutilized sources of novel active chemicals. The OSMAC (one strain, many compounds) strategy is based on the notion that certain microbial metabolites may not be generated under specific physicochemical circumstances but may occur when conditions change.

Analyzing secondary metabolites obtained from microbial co-cultures using metabolome technology is a new area of research. The rich microbiota collected from the Red Sea sponge *Coscinoderma mathewsi* (23 isolates) included three actinomycete strains that were identified as novel species of *Micromonospora*, *Nocardia*, and *Gordonia*. The potential for marine fungi to manufacture anticancer chemicals has been assessed (Deshmukh et al., 2018), and the capacity of the sea-derived *Streptomyces helimycini* to make actinomycins has been described (Zhu et al., 2017). In an intriguing study, it was demonstrated that co-cultivating fungi and bacteria resulted in the creation of novel secondary metabolites. There is an increasing interest in discovering novel sources for the study of novel microbial communities with the potential to produce bioactive natural compounds. As a result, the bacterial and fungal populations taken from the *Aquilaria malaccensis* tree and soil increased agarospirol production within three months of artificial infection (Chhipa and Kaushik, 2017).

The study conducted by Alenezi et al. (2017) highlighted a direct correlation between the development of a novel gramicidin and the biological activities of *Aneurinibacillus* in migrant isolates. Rao et al. (2017) focused on the importance of pigments made from bacteria and fungi and their various uses in both industry and health. Li et al. (2017) generated somalimycin, a new antimycin-type depsipeptide, using a mutant strain of *Streptomyces somaliensis* that was isolated from deep-sea waters.

Similarly, Thgersen et al. (2016) discovered that the marine bacteria *Pseudoalteromonas luteoviolacea* produces the potentially antibacterial compounds indolmycin and violacein when its maeA mutant is present. Important prospects are also endophytic microorganisms, which have been found to create secondary metabolites (Mefteh et al., 2017). Thus, in contrast to healthy plants, they showed that plants subjected to biotic stress generated unique endophytes with different bioactivities.

According to Sharma et al. (2017), dietary components of turmeric extracts enhanced the formation of cryptic and bioactive metabolites with antibacterial and antioxidant properties by the endophytic fungus *Colletotrichum gloeosporioides*. Furthermore, Abdel-Azeem et al. (2016) found that the endophytic fungus *Chaetomium globosum*, which was isolated from Egyptian medicinal herbs, had antirheumatoid activity. Given the societal ramifications, the hunt for bioactive characteristics in biobased compounds requires significant effort from research teams.

Aurantiochytrium sp., a marine microalga, is thought to be a viable source of docosahexaenoic acid production of DHA. DHA is an n-3 long-chain polyunsaturated fatty acid that is essential for cellular functions involved in health maintenance. UV mutagenesis was used by Liu et al. (2020) to develop a competitive *Aurantiochytrium* sp. strain with increased DHA synthesis (1.90-fold greater than the wild strain). The transcriptomes of the mutant and parent strains were compared to identify the main genes associated with increased DHA buildup. Gene expression was manipulated to control DHA yields in *Aurantiochytrium* sp.

Thus, genetic enhancement of microbial strains is merely scratching the surface of its enormous potential. Biofilms, which are made up of exopolysaccharides released by bacteria, defend them against attack. The use of natural antimicrobial substances is a traditional method for removing biofilms. This strategy could be applied to combat other biofilm-producing bacteria (Argüelles et al., 2020).

According to the hypothesis put forth by Hamed et al. (2020), the use of antimicrobial textiles could significantly reduce the risk of nosocomial infections caused by endophytic fungi that are isolated from marine animals and gathered from saline settings. The antibacterial and antioxidant properties of 32 fungal isolates were evaluated against a panel of harmful bacteria and fungi. The 21 strains' crude ethyl acetate extracts first showed antioxidant or antibacterial properties. MCT CD (monochlorotriazinyl cyclodextrin) grafting was used to create core-shaped hydrophobic voids in cellulosic textiles. Furthermore, the three most active fungal extracts (*Aspergillus terreus* 7S4, *Aspergillus calidoustus* M113, and *Alternaria alternata* 13A) were successfully incorporated into the hydrophobic cavities. According to the results, *Aspergillus calidoustus* strain M113 showed the most promising improvement in antimicrobial functionality and the second best in UV protection of this innovative generation of fabrics.

Streptomyces is a well-studied genus well recognized for producing bioactive chemicals. However, strains from different backgrounds can provide new natural compounds. *Streptomyces* sp. strain IB201691–2A from the Lake Baikal–endemic mollusk is one example.

Benedictia baicalensis synthesizes three novel angucyclines (baikalomycins A–C), as well as rabelomycin and 5-hydroxy-rabelomycin, in high quantities (Voitsekhovskaia et al., 2020). Baikalomycins A–C displayed varied degrees of antitumor activity. Rabelomycin and 5-hydroxy-rabelomycin were similarly found to have antiproliferative properties. The gene cluster responsible for baikalomycin biosynthesis was discovered, cloned, and expressed in *S. albus* J1074. The glycosyltransferase functions involved in the manufacture of these original chemicals were identified using heterologous expression and deletion assays.

Furthermore, a specific antibacterial activity against *Staphylococcus* sp. was discovered, as well as an intriguing antiproliferative capability against the human breast cancer MCF7 cell line. This extract yielded a total of 22 metabolites.

Two main pure chemicals, including a novel polyhydroxyacid, called their acid, were identified. Kheiric acid demonstrated excellent suppression of methicillin-resistant *Staphylococcus aureus* (MRSA). As a result, endophytes deserve additional attention as a source of novel bioactive chemicals.

4.4 STRATEGIES FOR ACTIVATING CRYPTIC GENE CLUSTERS FOR SM BIOSYNTHESIS

The current catastrophic situation demands immediate international cooperation in the search for new antimicrobials. The current situation will take us back to a time before antibiotics if left untreated. In the post-genomic era, several new strategies, like turning on mysterious gene clusters in microbes, are renewing a recent belief in natural selection.

Many complex secondary metabolites had not been identified before. The genes that drive these biosynthetic clusters either are phenotypically cryptic (silent) under ordinary laboratory conditions or express themselves at very low levels. In recent decades, the prevalence of multidrug-resistant bacteria has increased, presenting a serious global health risk. Product research may reveal novel secondary metabolites with therapeutic and other uses that have not yet been identified. The deciphering of this flexible endogenous genetic pool might soon offer a more plausible explanation for the evolution of antibiotics.

Strategies for activating cryptic gene clusters for secondary metabolite biosynthesis involve a range of innovative approaches aimed at unveiling the hidden potential of microbial genomes. One such strategy involves the use of bioinformatics tools to identify cryptic gene clusters within microbial genomes based on sequence homology and biosynthetic gene cluster architecture. Once identified, various genetic engineering techniques, such as gene knockout, overexpression, or promoter engineering, can be employed to activate these silent gene clusters and trigger the production of secondary metabolites. Additionally, co-culturing techniques and environmental manipulation have shown promise in inducing the expression of cryptic gene clusters, leading to the discovery of novel bioactive compounds with pharmaceutical or industrial applications.

Davies (2011) states that microbes' parvome, or small bioactive compounds, are thought to constitute an endless supply of secondary metabolites. Many industrial and medical metabolite chemicals have been found to date (Bentley, 1997; Bérdy, 2012). In pursuit of new secondary metabolites, a vast body of research conducted over the past 80 years has tested a large number of microorganisms, primarily actinomycetes (53%) and fungal (30%) species. A considerable number of secondary metabolites (17%) are produced by the cyanobacteria, *Pseudomonas*, and the myxobacteria *Bacillus* (Bérdy, 2005).

Interestingly, because secondary metabolites have a distinct molecular structure and lack scaffolds found in other chemical libraries, using microbial potential to extract them offers an additional benefit (Feher and Schmidt, 2003). These bacteria's unique secondary metabolism has been applied to the manufacture and screening of commercial metabolites.

Furthermore, advancements in omics technologies, including genomics, transcriptomics, proteomics, and metabolomics, have provided valuable insights into the regulatory mechanisms governing cryptic gene expression. By elucidating the intricate interplay between microbial physiology and environmental cues, researchers can devise targeted strategies to unlock the biosynthetic potential encoded within cryptic gene clusters. Overall, these multifaceted approaches hold great promise for expanding the chemical diversity of microbial secondary metabolites and accelerating the discovery of bioactive compounds for various biomedical, agricultural, and biotechnological applications.

4.5 METAGENOMICS FOR BACTERIAL SM PRODUCTION

Researchers can access most of the genetic material of bacteria that are resistant to growth by using metagenomics, which allows searching in the broad biochemical array found in the genomes of microorganisms found in environmental samples (Fernı́andez-Arrojo et al., 2010). In this way, unique functional gene clusters involved in the synthesis of bioactive compounds may be found by directly cloning or sequencing the DNA extracted from bacteria (Figure 4.2). The activity-based metagenomic technique facilitates the identification of the targeted genes encoding the necessary activities through the construction and subsequent screening of metagenomic libraries (Alves et al., 2017; Hertweck, 2009).

Metagenomics has thus been successfully applied to find novel pathways coding for bioactive chemicals in a range of contexts, as evidenced by numerous examples in the literature (Banik and Brady, 2010; Feng et al., 2011; Felczykowska et al., 2014; Iqbal et al., 2014; Jackson et al., 2015; Guo et al., 2017). Additionally, additional gene clusters implicated in the generation of secondary metabolites have been discovered by metagenome analysis (Chang and Brady, 2013; Cacho et al., 2014; Weber et al., 2015; Blin et al., 2017). Metabolism has thus developed into a suitable methodological instrument for enhancing and developing natural products (Santana-Pereira and Liles, 2017; Westmann et al., 2017).

The genes responsible for secondary metabolites are arranged into operons or clusters, which lead to the well-organized production of chemicals through a series of sequential stages carried out by an assembly of functionally related enzymes (Cimermancic et al., 2014).

The entire metabolic transformation repertoire, combined with finely tuned catalytic synergy control, would enable the synthesis and design of new complex bioactive molecules. Ren et al. (2017) suggested utilizing expression cassettes to facilitate a plug-and-play route-remodeling methodology for high-throughput pathway construction in *E. coli* and *S. cerevisiae*.

A successful construction of 96 pathways for combinatorial carotenoid biosynthesis was achieved as a proof of concept. By way of gene cluster remodeling from *Streptomyces* sp. strain NRRL F-525 using the production host *S. lividans* for expression, Freestone et al. (2017) discovered a new phosphonoacetic acid. Concurrently, efforts are underway to optimize chemical biosynthesis by employing developing algorithms that consider a broad spectrum of genetic and nongenetic characteristics (Zhou et al., 2015; Bhatia et al., 2017).

However, the recovery of natural products from metagenomes is now severely hampered by the use of *E. coli* as a host, which is phylogenetically distant from bacteria that are native producers of natural products (Guazzaroni et al., 2015). Therefore, the most abundant producers of a wide range of secondary metabolites, which are widely used in a variety of medicinal applications, are the Gram-positive bacteria of the phylum Actinobacteria (Bentley et al., 2002; Ohnishi et al., 2008; Barka et al., 2016).

Among the main obstacles preventing *E. coli* from being used as a host for natural products (NPs) discovery in metagenomic libraries are transporters, regulatory signals, appropriate protein folding, cofactor availability, and low precursor availability for secondary metabolite synthesis (Lewin et al., 2017). Actinobacteria members are being used in several systems biology and metabolic engineering studies to create sophisticated heterologous gene expression platforms for NP production. For instance, endogenous gene clusters have been removed in several *S. coelicolor* strains that have undergone genetic modification to enhance the release of secondary metabolites (Gomez-Escribano and Bibb, 2012), deleting transcriptional factors or altering natural promoters (Smanski et al., 2012; Du et al., 2012).

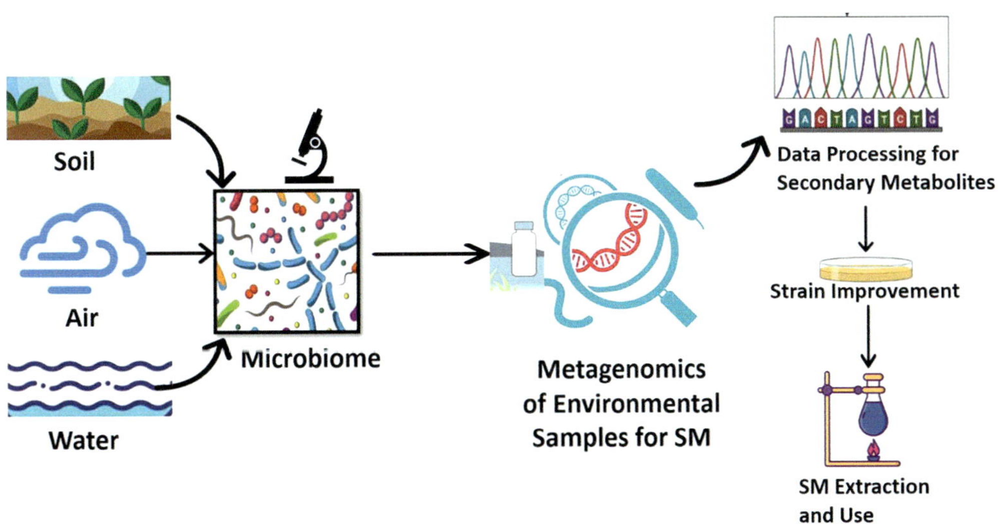

FIGURE 4.2 Metagenomics strategies for novel SM biosynthesis.

Additionally, by carefully reducing nonessential genes from the genome of the industrial bacteria *Streptomyces avermitilis*, a versatile model host for heterologous NP expression was created (Komatsu et al., 2010). Additionally, *Actinobacteria* have already been shown to have a variety of genetic tools for genome editing and genetic modification, including the widely recognized CRISPR/Cas system (Gust et al., 2004; Du et al., 2012; Jones et al., 2013; Cobb et al., 2015; Huang et al., 2015; Du et al., 2015; Tong et al., 2015).

In conclusion, the utilization of metagenomics for bacterial secondary metabolite (SM) production offers a transformative approach in biotechnological and pharmaceutical industries. By tapping into the vast genetic reservoir of unculturable bacteria, metagenomics provides unprecedented access to diverse biosynthetic gene clusters, facilitating the discovery of novel bioactive compounds. This method improves our knowledge of microbial ecology and evolution while also hastening the identification of promising therapeutic candidates. Moving forward, continued advancements in metagenomic technologies and bioinformatics tools hold promise for unlocking even greater microbial diversity and harnessing its potential for sustainable SM production.

4.6 FUTURE PERSPECTIVES AND CHALLENGES

Today, genetic modifications provide scientists with amazing tools to promote the synthesis of interesting compounds in host animals other than their original producers. Reliance on the climatic and spatial constraints inherent in plant cultures is obviated by using bacteria as productive hosts. Sunlight can be employed as an energy source for long-term cultivation and production operations since the molecules can be produced at reasonably high growth rates in inexpensive media.

Hilgers et al. (2021) improved the plant sesquiterpenoid pathway leading to caryophyllene by utilizing the heterologous host *Rhodobacter capsulatus*. The highest yield recorded was 139 +/− 31 mg/L. The antiphytopathogenic characteristics of sesquiterpenes have led to the testing of the bioactivity of caryophyllene and its oxygenated derivative against a range of phytopathogenic fungi. The chemicals had a significant growth suppression effect on the two phytopathogenic bacteria, *Sclerotinia sclerotiorum* and *Fusarium oxysporum*, while leaving other bacteria, including some that promote plant growth, unharmed. Therefore, the ability of *Rhodobacter capsulatus* to produce caryophyllene and caryophyllene oxide offers a possible method for producing natural compounds to control some plant pathogenic fungi in the production of crops.

Microbiological product extraction typically requires a lot of solvents, which goes against the notion of environmentally sustainable practices. Lebeau et al. (2020) looked at compounds of ergosterol and azaphilone red pigments. A marine-derived species of *Talaromyces* is reported, and its products are discussed. A novel approach to extracting fungal pigments is developed, utilizing solvents that are safe for the environment along with high pressure. The prospective uses of these fungal pigments in the creation of novel pharmaceuticals may make them interesting.

Nevertheless, to maximize the chemicals' extraction, enormous efforts are needed. Encouraging the production and use of microbial natural chemicals while introducing additional pollutants into ecosystems might be detrimental.

4.7 CONCLUSION

Numerous taxa of bacteria harbor the gene, which is clustered for the production of secondary metabolites with potential biological uses. The development of creative techniques to build natural bioactive chemical producers to increase production and monitor/discover novel metabolites has been greatly aided by the advent of postgenomic tools. There is the use of biosensors. It is important to remember that many computational tools and approaches for the development of novel regulatory elements and routes for the synthesis of SM have emerged with the advent of synthetic biology.

Utilizing powerful computational tools and metagenomic methods holds great promise for accelerating the discovery of secondary metabolites (SM) and novel bioactive compounds. These advancements enable more accurate pathway design and reprogramming of bacteria for SM production, potentially providing unique genetic pieces. Assembling synthetic gene clusters and further investigating these approaches, in conjunction with high-throughput screening and engineering techniques, is expected to significantly increase the rate of SM discovery, facilitating the identification of valuable bioactive compounds.

REFERENCES

Abdel-Azeem, A.M., S.M. Zaki, W.F. Khalil, N.A. Makhlouf, and L.M. Farghaly. 2016. Anti-rheumatoid activity of secondary metabolites produced by endophytic *Chaetomium globosum*. *Frontiers in Microbiology* 7 (September): 1477, 1–11. https://doi.org/10.3389/fmicb.2016.01477.

Alenezi, F.N., I. Rekik, A.C. Bouket, et al. 2017. Increased biological activity of *Aneurinibacillus migulanus* strains correlates with the production of new gramicidin secondary metabolites. *Frontiers in Microbiology* 8 (April): 517, 1–11. https://doi.org/10.3389/fmicb.2017.00517.

Alves, L.F., R. Silva-Rocha, and M.E. Guazzaroni. 2017. Enhancing metagenomic approaches through synthetic biology. In *Functional metagenomics: Tools and applications*, eds. T. Charles, M. Liles, and A. Sessitsch, 75–94. Cham: Springer. https://doi.org/10.1007/978-3-319-61510-3_5.

Argüelles, A., R. Sánchez-Fresneda, J.P. Guirao-Abad, et al. 2020. Novel bi-factorial strategy against *Candida albicans* viability using carnosic acid and propolis: Synergistic antifungal action. *Microorganisms* 8 (May): 749, 1–11. https://doi.org/10.3390/microorganisms8050749.

Banik, J.J., and S.F. Brady. 2010. Recent application of metagenomic approaches toward the discovery of antimicrobials and other bioactive small molecules. *Current Opinion in Microbiology* 13: 603–9. https://doi.org/10.1016/j.mib.2010.08.012.

Bardgett, R.D., C. Freeman, and N.J. Ostle. 2008. Microbial contributions to climate change through carbon cycle feedbacks. *The ISME Journal* 2: 805–14. https://doi.org/10.1038/ismej.2008.58.

Barka, E.A., P. Vatsa, L. Sanchez, et al. 2016. Taxonomy, physiology, and natural products of actinobacteria. *Microbiology and Molecular Biology Reviews* 80 (March): 1, 1–43. https://doi.org/10.1128/mmbr.00019-15.

Bentley, R. 1997. Microbial secondary metabolites play important roles in medicine; and prospects for the discovery of new drugs. *Perspectives in Biology and Medicine* 40: 364–94. https://muse.jhu.edu/article/401288.

Bentley, S.D., K.F. Chater, A.M. Cerdeno-Tarraga, et al. 2002. Complete genome sequence of the model actinomycete *Streptomyces coelicolor* A3(2). *Nature* 417: 141–47. https://doi.org/10.1038/417141a.

Bérdy, J. 2005. Bioactive microbial metabolites. *The Journal of Antibiotics* 58: 1–26. https://doi.org/10.1038/ja.2005.1.

Bérdy, J. 2012. Thoughts and facts about antibiotics: Where we are now and where we are heading. *The Journal of Antibiotics* 65: 385–95. https://doi.org/10.1038/ja.2012.27.

Bhatia, S.P., M.J. Smanski, C.A. Voigt, and D.M. Densmore. 2017. Genetic design via combinatorial constraint specification. *ACS Synthetic Biology* 6: 2130–35. https://doi.org/10.1021/acssynbio.7b00154.

Blin, K., M.H. Medema, R. Kottmann, S.Y. Lee, and T. Weber. 2017. The antiSMASH database, a comprehensive database of microbial secondary metabolite biosynthetic gene clusters. *Nucleic Acids Research* 45: D555–59. https://doi.org/10.1093/nar/gkw960.

Cacho, R.A., Y. Tang, and Y.H. Chooi. 2014. Next-generation sequencing approach for connecting secondary metabolites to biosynthetic gene clusters in fungi. *Frontiers in Microbiology* 5 (January): 774, 1–16. https://doi.org/10.3389/fmicb.2014.00774.

Cavalier-Smith, T. 1992. Origins of secondary metabolism. In *Secondary metabolites: Their function and evolution: Ciba foundation symposium 171*, eds. D.J. Chadwick, and J. Whelan, 64–87. Hoboken, NJ: Wiley. https://doi.org/10.1002/9780470514344.ch5.

Chang, F.Y., and S.F. Brady. 2013. Discovery of indolotryptoline antiproliferative agents by homology-guided metagenomic screening. *Proceedings of the National Academy of Sciences of the United States of America* 110: 2478–83. https://doi.org/10.1073/pnas.1218073110.

Chhipa, H., and N. Kaushik. 2017. Fungal and bacterial diversity isolated from *Aquilaria malaccensis* tree and soil, induces agarospirol formation within 3 months after artificial infection. *Frontiers in Microbiology* 8 (July): 273540, 1–12. https://doi.org/10.3389/fmicb.2017.01286.

Cimermancic, P., M.H. Medema, J. Claesen, et al. 2014. Insights into secondary metabolism from a global analysis of prokaryotic biosynthetic gene clusters. *Cell* 158: 412–21. https://doi.org/10.1016/j.cell.2014.06.034.

Cobb, R.E., Y. Wang, and H. Zhao. 2015. High-efficiency multiplex genome editing of *streptomyces* species using an engineered CRISPR/Cas System. *ACS Synthetic Biology* 4: 723–28. https://doi.org/10.1021/sb500351f.

Davies, J. 2011. How to discover new antibiotics: Harvesting the parvome. *Current Opinion in Chemical Biology* 15: 5–10. https://doi.org/10.1016/j.cbpa.2010.11.001.

Demain, A.L., and A. Fang. 2000. The natural functions of secondary metabolites. In *History of modern biotechnology I. Advances in biochemical engineering/biotechnology*, ed. A. Fiechter, 69: 1–39. Berlin, Heidelberg: Springer. https://doi.org/10.1007/3-540-44964-7_1.

Demain, A.L., B. Gomez-Ortiz, B. Ruiz-Villafan, R. Rodriguez-Sanoja, and S. Sanchez. 2019. Recent findings of molecules with anti-infective activity: Screening of non-conventional sources. *Current Opinion in Pharmacology* 48: 40–47. https://doi.org/10.1016/j.coph.2019.04.003.

Deshmukh, S.K., V. Prakash, and N. Ranjan. 2018. Marine fungi: A source of potential anticancer compounds. *Frontiers in Microbiology* 8 (January): 274495, 1–24. https://doi.org/10.3389/fmicb.2017.02536.

Du, D., L. Wang, Y. Tian, H. Liu, H. Tan, and G. Niu. 2015. Genome engineering and direct cloning of antibiotic gene clusters via phage ϕbT1 integrase-mediated site-specific recombination in *Streptomyces*. *Scientific Reports* 5, 8740 (March): 1–8. https://doi.org/10.1038/srep08740.

Du, L., R.H. Liu, L. Ying, and G.R. Zhao. 2012. An efficient intergeneric conjugation of DNA from *Escherichia coli* to mycelia of the lincomycin-producer *Streptomyces lincolnensis*. *International Journal of Molecular Sciences* 13: 4797–806. https://doi.org/10.3390/ijms13044797.

Dufour, N., and R.P. Rao. 2011. Secondary metabolites and other small molecules as intercellular pathogenic signals. *FEMS Microbiology Letters* 314: 10–17. https://doi.org/10.1111/j.1574-6968.2010.02154.x.

Feher, M., and J.M. Schmidt. 2003. Property distributions: Differences between drugs, natural products, and molecules from combinatorial chemistry. *Journal of Chemical Information and Modeling* 43: 218–27. https://doi.org/10.1021/ci0200467.

Felczykowska, A., A. Dydecka, and M. Bohdanowicz, et al. 2014. The use of fosmid metagenomic libraries in preliminary screening for various biological activities. *Microbial Cell Factories* 13 (July): 105, 1–7. https://doi.org/10.1186/s12934-014-0105-4.

Feng, Z., D. Kallifidas, and S.F. Brady. 2011. Functional analysis of environmental DNA-derived type II polyketide synthases reveals structurally diverse secondary metabolites. *Proceedings of the National Academy of Sciences* 108: 12629–34. https://doi.org/10.1073/pnas.1103921108.

Ferníandez-Arrojo, L., M.E. Guazzaroni, N. López-Cortés, A. Beloqui, and M. Ferrer. 2010. Metagenomic era for biocatalyst identification. *Current Opinion in Biotechnology* 21: 725–33. https://doi.org/10.1016/j.copbio.2010.09.006.

Fouillaud, M., M. Venkatachalam, E. Girard-Valenciennes, Y. Caro, and L. Dufossé. 2016. Anthraquinones and derivatives from marine-derived fungi: Structural diversity and selected biological activities. *Marine Drugs* 14 (March): 64, 1–64. https://doi.org/10.3390/md14040064.

Freestone, T.S., K.S. Ju, B. Wang, and H. Zhao. 2017. Discovery of a phosphonoacetic acid derived natural product by pathway refactoring. *ACS Synthetic Biology* 6: 217–23. https://doi.org/10.1021/acssynbio.6b00299.

Gomez-Escribano, J.P., and M.J. Bibb. 2012. *Streptomyces coelicolor* as an expression host for heterologous gene clusters. *Methods in Enzymology* 517: 279–300. https://doi.org/10.1016/B978-0-12-404634-4.00014-0.

Guazzaroni, M.E., R. Silva-Rocha, and R.J. Ward. 2015. Synthetic biology approaches to improve biocatalyst identification in metagenomic library screening. *Microbial Biotechnology* 8: 52–64. https://doi.org/10.1111/1751-7915.12146.

Guo, C.J., F.Y. Chang, T.P. Wyche, et al. 2017. Discovery of reactive microbiota-derived metabolites that inhibit host proteases. *Cell* 168: 517–26. https://doi.org/10.1016/j.cell.2016.12.021.

Gupta, A., S. Jain, and N. Verma. 2022. *Industrial applications of soil microbes*, vol. 1. Bentham Science Publishers Pte Ltd. https://doi.org/10.2174/97898150399551220101.

Gupta, A., S. Jain, and N. Verma. 2024. *Industrial applications of soil microbes*, vol. 3. Bentham Science Publishers Pte Ltd. https://doi.org/10.2174/97898150797531240301.

Gust, B., G. Chandra, D. Jakimowicz, T. Yuqing, C.J. Bruton, and K.F. Chater. 2004. Red-mediated genetic manipulation of antibiotic-producing *Streptomyces*. *Advances in Applied Microbiology* 54: 107–28. https://doi.org/10.1016/S0065-2164(04)54004-2.

Hamed, A.A., S. Soldatou, M.M. Qader, et al. 2020. Screening fungal endophytes derived from under-explored Egyptian marine habitats for antimicrobial and antioxidant properties in factionalised textiles. *Microorganisms* 8 (October): 1617, 1–19. https://doi.org/10.3390/microorganisms8101617.

Hassane, C.S., M. Fouillaud, G. Le Goff, et al. 2020. Microorganisms associated with the marine sponge *Scopalina hapalia*: A reservoir of bioactive molecules to slow down the aging process. *Microorganisms* 8 (August): 1262, 1–23. https://doi.org/10.3390/microorganisms8091262.

Henderson, D.A. 1999. The looming threat of bioterrorism. *Science* 283: 1279–82. https://doi.org/10. 1126/science.283.5406.1279.

Hertweck, C. 2009. The biosynthetic logic of polyketide diversity. *Angewandte Chemie International Edition* 48: 4688–716. https://doi.org/10.1002/anie.200806121.

Hilgers, F., S.S. Habash, A. Loeschcke, et al. 2021. Heterologous production of β-caryophyllene and evaluation of its activity against plant pathogenic fungi. *Microorganisms* 9 (January): 168, 1–19. https://doi.org/10.3390/microorganisms9010168.

Huang, H., G. Zheng, W. Jiang, H. Hu, and Y. Lu. 2015. One step high-efficiency CRISPR/Cas9-mediated genome editing in *Streptomyces*. *Acta Biochimica et Biophysica Sinica* 47: 231–43. https://doi.org/10.1093/abbs/gmv007.

Iqbal, H.A., J.W. Craig, and S.F. Brady. 2014. Antibacterial enzymes from the functional screening of metagenomic libraries hosted in *Ralstonia metallidurans*. *FEMS Microbiology Letters* 354: 19–26. https://doi.org/10.1111/1574-6968.12431.

Jackson, S.A., E. Borchert, F. O'Gara, and A.D.W. Dobson. 2015. Metagenomics for the discovery of novel biosurfactants of environmental interest from marine ecosystems. *Current Opinion in Biotechnology* 33: 176–82. https://doi.org/10.1016/j.copbio.2015.03.004.

Jain, S., A. Gupta, and N. Verma. 2023. *Industrial applications of soil microbes*, vol. 2. Bentham Science Publishers Pte Ltd. https://doi.org/10.2174/97898150502641230201.

Jain, S., A. Gupta, and N. Verma. 2024. *Industrial applications of soil microbes*, vol. 4. Bentham Science Publishers Pte Ltd. https://doi.org/10.2174/97898151249961240401.

Jeong, Y., S.H. Cho, H. Lee, et al. 2020. Current status and future strategies to increase secondary metabolite production from cyanobacteria. *Microorganisms* 8 (November): 1849, 1–24. https://doi.org/10.3390/microorganisms8121849.

Jones, A.C., B. Gust, A. Kulik, L. Heide, M.J. Buttner, and M.J. Bibb. 2013. Phage P1-derived artificial chromosomes facilitate heterologous expression of the FK506 gene cluster. *PLoS ONE* 8 (July): e69319, 1–9. https://doi.org/10.1371/journal.pone.0069319.

Komatsu, M., T. Uchiyama, S. Omura, D.E. Cane, and H. Ikeda. 2010. Genome-minimized *Streptomyces* host for the heterologous expression of secondary metabolism. *Proceedings of the National Academy of Sciences* 107: 2646–51. https://doi.org/10.1073/pnas.0914833107.

Kowalchuk, G.A., A.G.C.L. Speksnijder, K. Zhang, R.M. Goodman, and J.A. van Veen. 2007. Finding the needles in the metagenome haystack. *Microbial Ecology* 53: 475–85. https://doi.org/10.1007/s00248-006-9201-2.

Lebeau, J., T. Petit, M. Fouillaud, L. Dufossé, and Y. Caro. 2020. Alternative extraction and characterization of nitrogen-containing azaphilone red pigments and ergosterol derivatives from the marine-derived fungal *Talaromyces* sp. 30570 strains with industrial relevance. *Microorganisms* 8 (December): 1920, 1–20. https://doi.org/10.3390/microorganisms8121920.

Lewin, A., R. Lale, and A. Wentzel. 2017. Expression platforms for functional metagenomics: Emerging technology options beyond *Escherichia coli*. In *Functional metagenomics: Tools and applications,* eds. T.C. Charles, M.R. Liles, and A. Sessitsch, 13–44. Cham: Springer. https://doi.org/10.1007/978-3-319-61510-3_2.

Li, H., H. Huang, L. Hou, J. Ju, and W. Li. 2017. Discovery of antimycin-type depsipeptides from a *wbl* gene mutant strain of deepsea-derived *Streptomyces somaliensis* SCSIO ZH66 and their effects on pro-inflammatory cytokine production. *Frontiers in Microbiology* 8 (April): 678. https://doi.org/10.3389/fmicb.2017.00678.

Liu, L., Z. Hu, S. Li, et al. 2020. Comparative transcriptomic analysis uncovers genes responsible for the DHA enhancement in the mutant *Aurantiochytrium* sp. *Microorganisms* 8 (April): 529, 1–19. https://doi.org/10.3390/microorganisms8040529.

Maplestone, R.A., M.J. Stone, and D.H. Williams. 1992. The evolutionary role of secondary metabolites – A review. *Gene* 115: 151–57. https://doi.org/10.1016/0378-1119(92)90553-2.

Martinez-Klimova, E., S. Centeno-Leija, and S. Sanchez. 2017. Modern approaches to genome mining for the development of new anti-infectives: *In silico* gene prediction and experimental metabolomics. In *Frontiers in clinical drug research-antiinfectives*, ed. A.U. Rahman, 3–48. Sharjah: Bentham Science Publishers. https://www.eurekaselect.com/chapter/10207.

Mefteh, F.B., A. Daoud, A.C. Bouket, et al. 2017. Fungal root microbiome from healthy and brittle leaf diseased date palm trees (*Phoenix dactylifera* L.) reveals a hidden untapped arsenal of antibacterial and broad spectrum antifungal secondary metabolites. *Frontiers in Microbiology* 8 (February): 253126, 1–18. https://doi.org/10.3389/fmicb.2017.00307.

Monciardini, P., M. Iorio, S. Maffioli, M. Sosio, and S. Donadio. 2014. Discovering new bioactive molecules from microbial sources. *Microbial Biotechnology* 7: 209–20. https://doi.org/10.1111/1751-7915.12123.

Morens, D.M., G.K. Folkers, and A.S. Fauci. 2004. The challenge of emerging and re-emerging infectious diseases. *Nature* 430: 242–49. https://doi.org/10.1038/nature02759.

Nam, N.N., H.D.K. Do, K.T.L. Trinh, and N.Y. Lee. 2023. Metagenomics: An effective approach for exploring microbial diversity and functions. *Foods* 12 (May): 2140, 1–23. https://doi.org/10.3390/foods12112140.

Nawaz, A., R. Chaudhary, Z. Shah, et al. 2021. An overview on industrial and medical applications of bio-pigments synthesized by marine bacteria. *Microorganisms* 9 (December): 11, 1–25. https://doi.org/10.3390/microorganisms9010011.

Newman, D.J. 2016. Predominately uncultured microbes as sources of bioactive agents. *Frontiers in Microbiology* 7 (November): 1832, 1–15. https://doi.org/10.3389/fmicb.2016.01832.

Nicas, T.I., and B.H. Iglewski. 1985. The contribution of exoproducts to virulence of *Pseudomonas aeruginosa*. *Canadian Journal of Microbiology* 31: 387–92. https://doi.org/10.1139/m85-074.

O'Brien, J., and G.D. Wright. 2011. An ecological perspective of microbial secondary metabolism. *Current Opinion in Biotechnology* 22: 552–58. https://doi.org/10.1016/j.copbio.2011.03.010.

Ohnishi, Y., J. Ishikawa, H. Hara, et al. 2008. Genome sequence of the streptomycin-producing microorganism *Streptomyces griseus* IFO 13350. *Journal of Bacteriology* 190: 4050–60. https://doi.org/10.1128/jb.00204-08.

Overy, D.P., T. Rämä, R. Oosterhuis, A.K. Walker, and K.L. Pang. 2019. The neglected marine fungi, *Sensu stricto*, and their isolation for natural products discovery. *Marine Drugs* 17 (January): 42, 1–20. https://doi.org/10.3390/md17010042.

Passari, A.K., V.K. Mishra, G. Singh, et al. 2017. Insights into the functionality of endophytic actinobacteria with a focus on their biosynthetic potential and secondary metabolites production. *Scientific Reports* 7 (September): 11809, 1–17. https://doi.org/10.1038/s41598-017-12235-4.

Patridge, E., P. Gareiss, M.S. Kinch, and D. Hoyer. 2016. An analysis of FDA-approved drugs: Natural products and their derivatives. *Drug Discovery Today* 21, 204–7. https://doi.org/10.1016/j.drudis.2015.01.009.

Pettit, R.K. 2011. Small-molecule elicitation of microbial secondary metabolites. *Microbial Biotechnology* 4: 471–78. https://doi.org/10.1111/j.1751-7915.2010.00196.x.

Rao, M.P.N., M. Xiao, and W. Li. 2017. Fungal and bacterial pigments: Secondary metabolites with wide applications. *Frontiers in Microbiology* 8 (June): 250699, 1–13. https://doi.org/10.3389/fmicb.2017.01113.

Ren, H., P. Hu, and H. Zhao. 2017. A plug-and-play pathway refactoring workflow for natural product research in *Escherichia coli* and *Saccharomyces cerevisiae*. *Biotechnology and Bioengineering* 114: 1847–54. https://doi.org/10.1002/bit.26309.

Rowland, I., G. Gibson, A. Heinken, et al. 2018. Gut microbiota functions metabolism of nutrients and other food components. *European Journal of Nutrition* 57: 1–24. https://doi.org/10.1007/s00394-017-1445-8.

Ruiz, B., A. Chávez, A. Forero, et al. 2010. Production of microbial secondary metabolites: Regulation by the carbon source. *Critical Reviews in Microbiology* 36: 146–67. https://doi.org/10.3109/10408410903489576.

Salikin, N.H., J. Nappi, M.E. Majzoub, and S. Egan. 2020. Combating parasitic nematode infections, newly discovered antinematode compounds from marine epiphytic bacteria. *Microorganisms* 8 (December): 1963, 1–19. https://doi.org/10.3390/microorganisms8121963.

Santana-Pereira, A.L.R., and M.R. Liles. 2017. Challenges and opportunities in discovery of secondary metabolites using a functional metagenomic approach. In *Functional metagenomics: Tools and applications*, eds. T. Charles, M. Liles, and A. Sessitsch, 119–38. Cham: Springer. https://doi.org/10.1007/978-3-319-61510-3_7.

Shah, S.H., and C.B. Newgard. 2015. Integrated metabolomics and genomics: Systems approaches to biomarkers and mechanisms of cardiovascular disease. *Circulation Cardiovascular Genetics* 8: 410–19. https://doi.org/10.1161/CIRCGENETICS.114.000223.

Sharma, V.K., J. Kumar, D.K. Singh, et al. 2017. Induction of cryptic and bioactive metabolites through natural dietary components in an endophytic fungus *Colletotrichum gloeosporioides* (Penz.) Sacc. *Frontiers in Microbiology* 8 (June): 252837, 1–9. https://doi.org/10.3389/fmicb.2017.01126.

Smanski, M.J., J. Casper, R.M. Peterson, Z. Yu, S.R. Rajski, and B. Shen. 2012. Expression of the platencin biosynthetic gene cluster in heterologous hosts yielding new platencin congeners. *Journal of Natural Products* 75: 2158–67. https://doi.org/10.1021/np3005985.

Thgersen, M.S., M.W. Delpin, J. Melchiorsen, et al. 2016. Production of the bioactive compounds violacein and indolmycin is conditional in a maeA mutant of *Pseudoalteromonas luteoviolacea* S4054 lacking the malic enzyme. *Frontiers in Microbiology* 7 (September): 1461, 1–11. https://doi.org/10.3389/fmicb.2016.01461.

Tong, Y., P. Charusanti, L. Zhang, T. Weber, and S.Y. Lee. 2015. CRISPR-Cas9 based engineering of actinomycetal genomes. *ACS Synthetic Biology* 4: 1020–29. https://doi.org/10.1021/acssynbio.5b00038.

Trindade, M., L.J. Van Zyl, J. Navarro-Fernandez, and A. Abd Elrazak. 2015. Targeted metagenomics as a tool to tap into marine natural product diversity for the discovery and production of drug candidates. *Frontiers in Microbiology* 6 (August): 890, 1–14. https://doi.org/10.3389/fmicb.2015.00890.

Voitsekhovskaia, I., C. Paulus, C. Dahlem, et al. 2020. Baikalomycins A-C, new aquayamycin-type angucyclines isolated from lake baikal derived *Streptomyces* sp. IB201691–2A. *Microorganisms* 8 (May): 680, 1–21. https://doi.org/10.3390/microorganisms8050680.

Wang, J.H., J. Byun, and S. Pennathur. 2010. Analytical approaches to metabolomics and applications to systems biology. *Seminars in Nephrology* 30: 500–11. https://doi.org/10.1016/j.semnephrol.2010.07.007.

Weber, T., K. Blin, S. Duddela, et al. 2015. AntiSMASH 3.0-A comprehensive resource for the genome mining of biosynthetic gene clusters. *Nucleic Acids Research* 43: W237–43. https://doi.org/10.1093/nar/gkv437.

Westmann, C.A., L.D.F. Alves, R. Silva-Rocha, and M.E. Guazzaroni. 2017. Assessing and characterizing the repertoire of constitutive promoter elements in soil metagenomic libraries in *Escherichia coli*. *bioRxiv* (January): 211367, 2–26. https://doi.org/10.1101/211367.

Xu, F., Y. Wu, C. Zhang, et al. 2019. A genetics-free method for high-throughput discovery of cryptic microbial metabolites. *Nature Chemical Biology* 15: 161–68. https://doi.org/10.1038/s41589-018-0193-2.

Yan, Q., L.D. Lopes, B.T. Shaffer, et al. 2018. Secondary metabolism and interspecific competition affect the accumulation of spontaneous mutants in the grass-gas regulatory system in *Pseudomonas protegens*. *MBio* 9 (January): e01845–17, 1–18. https://doi.org/10.1128/mBio.01845-17.

Zhou, H., B. Vonk, J.A. Roubos, R.A.L. Bovenberg, and C.A. Voigt. 2015. Algorithmic co-optimization of genetic constructs and growth conditions: Application to 6-ACA, a potential nylon-6 precursor. *Nucleic Acids Research* 43: 10560–70. https://doi.org/10.1093/nar/gkv1071.

Zhu, J., X. Xu, M. Yuan, H. Wu, Z. Ma, and W. Wu. 2017. Optimum O_2:CH_4 ratio promotes the synergy between aerobic methanotrophs and denitrifiers to enhance nitrogen removal. *Frontiers in Microbiology* 8 (June): 250939, 1–10. https://doi.org/10.3389/fmicb.2017.01112.

Zothanpuia, A.K. Passari, V.V. Leo, et al. 2018. Bioprospection of actinobacteria derived from freshwater sediments for their potential to produce antimicrobial compounds. *Microbial Cell Factories* 17 (May): 68, 1–14. https://doi.org/10.1186/s12934-018-0912-0.

5 Biomolecular Microbial Metabolism and Microbial Secondary Metabolites Production Pathways

Yogitha N. Sagar
(Department of Biotechnology, School of Applied Sciences, REVA University, Bengaluru, KA, India; yogithansagar262@gmail.com)

Nitya Mankal
(Department of Biotechnology, School of Applied Sciences, REVA University, Bengaluru, KA, India; nityamankal09@gmail.com)

Senthilkumar R.
(Department of Biotechnology, School of Applied Sciences, REVA University, Bengaluru, KA, India; senthilanal@yahoo.com)

**Ankita Chatterjee*
(Department of Biotechnology, School of Applied Sciences, REVA University, Bengaluru, KA, India; ankita.chatterjee@reva.edu.in)

*Corresponding Author

5.1 CONCEPT AND INTRODUCTION OF SECONDARY METABOLITES

Metabolites are intermediate or by-products of the metabolic process. These are the biomolecules that are smaller in size and can be produced by industries for medicinal purposes. Metabolites are of two types, which are primary metabolites and secondary metabolites. Primary metabolites are used to build up cofactors, precursors, and biological polymers that help in bacterial proliferation and growth. These include nucleotides, amino acids, and some end products of fermentation, such as alcohol and organic acids. Secondary metabolites are predominantly organic compounds not associated with the growth, development, and reproduction of the microorganisms, which are extracellular predominant (Jain et al., 2023, 2024). These are produced throughout bacterial growth but mostly occur in late exponential and stationary phases and have high potential in health-care activities, such as enzyme inhibitors, antitumors, immunosuppressants, *etc.* (Seyedsayamdost, 2019). Bacterial reproduction takes place by binary fission. Five different types of growth patterns are observed to be involved in bacterial growth: (1) growth cycle, (2) biphasic growth, (3) maintenance of cells in exponential phase, (4) synchronous growth, and (5) bacterial growth *in vivo*. Bacterial growth includes four different phases, namely, lag, log, exponential, and death phase. During the lag phase, the bacteria adapt themselves to the growing conditions and are not able to divide but mature. The duration of the phase depends on the type of bacteria, environmental factors, *etc.* The *log phase* (8 h) is expanded as a logarithmic phase. The number of new bacteria growing per unit of time is proportional to the present population. The *stationary phase* occurs due to the depletion of a nutrient and the formation of inhibitory products. In the *decline phase*, bacteria run out of nutrients, there is accumulation of toxic products, and autolytic enzymes are increased (Maier and Pepper, 2015).

Secondary metabolites have major potential. Microorganisms are recognized as one of the novel and valuable assets in the food industry, as well as biologically active compounds. They are used in the synthesis of various products with diverse applications across the food, agricultural, and pharmaceutical industries (Verardo et al., 2020; Gupta et al., 2022, 2024). For example, prebiotics exhibit their effects via multiple mechanisms, along with the production of vitamins, by interacting with the host's immune system, preventing pathogen adhesion to the host cells, and influencing the morphology of the intestine. By the modulation of the intestinal microbiota, the mechanisms are mediated. The criteria are fulfilled by a wide range of dietary compounds. At present, the most promising dietary fibers that exert prebiotic properties are non-digestible oligosaccharides composed of 3–9 sugar monomers (Sorour et al., 2017). Nisin, a bacitracin,

DOI: 10.1201/9781003408833-5

acts as a biopreservative. Rapamycin, an antifungal agent, is isolated from *Actinomycetes*, residing in soil, and is used to prevent organ rejection in transplant recipients.

In 2002, amrubicin hydrochloride, which is an anticancer compound, was isolated from *Streptomyces peucetius*. Additionally, a new class of antibiotics called pumalicidins A, B, C, D, E, F, and G was procured from the culture broth of *Bacillus pumilus*. Related compounds to Bacillomycin D, which were isolated from *Bacillus subtilis* (B38), were found to possess antioxidative and DNA-protective properties. Moreover, surfactin, which is produced by *Bacillus subtilis* CYS191, has been observed to induce apoptosis in human breast cancer cells (MCF-7) by the induction of oxidative stress (Mushtaq et al., 2018; Taghinasab and Jabaji, 2020). Norsolorinic acid, procured from *Aspergillus* spp., has been reported to induce apoptosis in breast cancer cells (MCF-7) as well as human bladder cancer cells (T-24) (Nataraj et al. 2020; Taghinasab and Jabaji, 2020). The ethanolic extracts from *Fomitopsis pinicola* have been found to exhibit cytotoxic effects on several cancer cell lines, including human hepatoma, colorectal, lung, and breast cancer cells. Additionally, these extracts have shown synergistic effects when they are used in combination with cisplatin *in vivo*.

Streptomycetes are abundant producers of bioactive compounds. Among these, macrolide polyenes, like nystatin, are derived from *Streptomyces*, which are found as antifungals. Approximately two-thirds of bioactive compounds originate from this group and show diverse clinical potential against various organisms, such as bacteria, fungi, and parasites. Actinomycetes are well-known for their ability to produce a wide range of antibiotics, including peptides/glycopeptides, angucyclinones, tetracyclines, phenazines, macrolides, anthraquinones, polyenes, anthracyclines, β-lactams, piercidins, octaketides, benzoxazolophenanthridines, heptadecaglycosides, and lactones. This diversity in antibiotic production shows the versatility of actinomycetes as a valuable source for various types of antimicrobial compounds (Pham et al., 2019; Abdel-Razek et al., 2020).

These secondary metabolites are characterized by their structures and chemical properties and serve as exceptional therapeutics for addressing a wide range of diseases. In addition to natural products, the progress in recombinant DNA technology has led to the development of an extensive range of biopharmaceutical products, which include recombinant proteins. These advancements made breakthroughs in the treatment of various medical illnesses and conditions by offering new possibilities for improved health care (Vrancken and Anné, 2009).

5.2 SECONDARY METABOLITES AS BIOACTIVE COMPOUNDS

Microbial secondary metabolites play a wide-ranging role as active ingredients in medicine, agriculture, and food additives (Figure 5.1). Metabolism encloses the essential chemical transformations that sustain life within cells, with small molecules known as metabolites serving as intermediates. Among these metabolites, primary microbial metabolites (such as ethanol and amino acids) are essential for the development of organisms and for reproduction. On the other hand, secondary metabolites (including pigments and antibiotics), which are produced by the microorganisms, benefit living organisms by enhancing their metabolic processes as well, and are advantageous for human usage. These microbial secondary metabolites find utility in various sectors, such as agriculture, medicine, food industry,

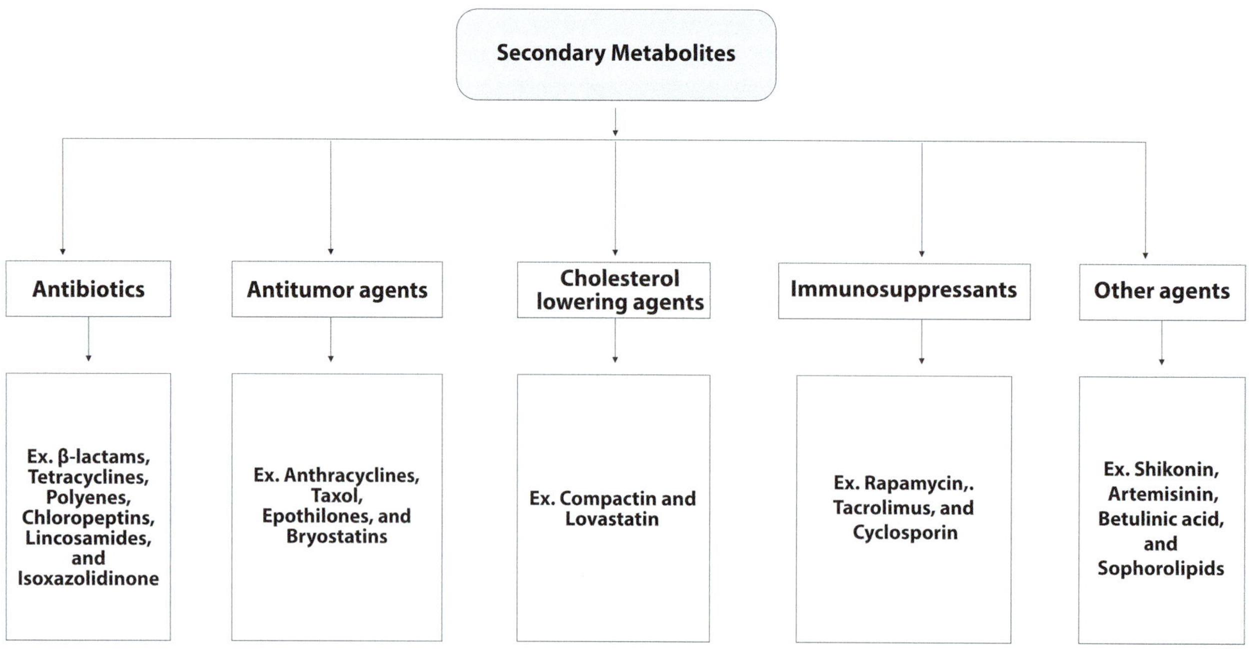

FIGURE 5.1 Therapeutic applications of secondary metabolites.

Source: Vaishnav and Demain (2011).

chemical industry, and other related fields, ultimately contributing to human welfare and diverse biological processes (Berdy, 2005; Demain, 1980; Desire et al., 2014).

The collection of endophytic bacteria from medicinal plants offers a promising avenue for discovering novel bioactive compounds that can effectively target the proteins of SARS-CoV-2. This approach holds significant potential in expediting drug innovation and development efforts (Akter et al., 2022). From the diverse range of studies on bioactive compounds from microbes and secondary metabolites, it can be deduced that these microorganisms belong to various phyla, including both bacteria and fungi, such as Rhodophyta, Proteobacteria, Haptophyta, Cyanobacteria, Crenarchaeota, Chlorophyta, Bacteroidetes, Ascomycota, and Actinobacteria (Table 5.1). Among these, the most prevalent genera are *Streptomyces*, *Aspergillus*, *Bacillus*, *Acremonium*, *Penicillium*, *Pseudomonas*, *Fusarium*, *Halobacterium*, and *Amycolatopsis* (Ruiz et al., 2010; Barrios-Gonzalez et al., 2003).

The global scientific community has reported the discovery of novel and potentially valuable secondary metabolites synthesized by microorganisms across a wide range of habitats. These habitats include microbiomes associated with plants (such as rhizospheric, epiphytic, and endophytic environments), as well as microbiomes found in extreme habitats (including halophilic, thermophilic, and acidophilic/alkaliphilic conditions). These secondary metabolites have been identified in diverse species belonging to various genera, such as *Streptomyces*, *Rhodococcus*, *Oceanicola*, *Fodinicurvata*, *Bacillus*, *Aeromicrobium*, and *Actinomadura* (Yadav et al., 2019). The remarkable diversity of these Actinomycetes and their ability to produce substances that are novel hence position this class prominently. They are responsible for generating approximately 50% of the investigated bioactive secondary metabolites, including notable contributions in the fields of antibiotics, anticancer agents, anti-inflammatory agents, and enzymes (Selim et al., 2021). Certain soil bacteria, such as *Serratia* spp., exhibit antagonistic effects against worms, and the zeamine molecule, produced by the bacteria, demonstrates effectiveness against yeasts and other kinds of biological systems. Prodigiosin, a renowned pigment synthesized by the marine bacterium *Vibrio ruber*, possesses a broad spectrum of antimicrobial activity and induces autolytic activity in specifically targeted cells like *Bacillus subtilis*.

Lantibiotics, which are class I bacteriocins produced by Gram-positive bacteria, can be bioengineered to enhance their effectiveness against a wider range of bacterial strains and improve their stability during gastric transit by rendering them protease-resistant. These findings offer new possibilities for antibiotic therapy at a time when antibiotic resistance is considered a significant threat to public health (World Health Organization, 2014).

Probiotic strains can produce a range of therapeutic molecules, including amino acids, vitamins, bacteriocins, enzymes, immunomodulatory compounds, and short-chain fatty acids. These bioactive substances contribute to the potential health benefits associated with probiotics (Indira et al., 2019). *Aurantiochytrium* sp., a marine microalga, holds great promise as a valuable source to produce docosahexaenoic acid (DHA). DHA, an n-3 long-chain polyunsaturated fatty acid, plays a vital role in various cellular

TABLE 5.1
Secondary Metabolites Secreted by Microorganisms

Sl. No.	Secondary Metabolites	Sources	Reference
1	Penicillin G	*Penicillium rubens*	Bills and Gloer., 2016
2	Cephalosporin C	*Acremonium chrysogenum*	Bills and Gloer., 2016
3	Pleuromutilin	*Clitopilus passeckerianus*	Bills and Gloer., 2016
4	Fusidic acid	*Acremonium fusidioides*	Bills and Gloer., 2016
5	Enfumafungin	*Hormonema carpetanum*	Bills and Gloer., 2016
6	Nostodione	*Nostoc* sp.	Burja et al., 2001
7	Lyngbyatoxin	*Moorea* sp.	Edwards and Gerwick, 2004
8	Nodularin	*Nodularia* sp.	Moffitt and Neilan, 2004
9	Cryptophycins	*Nostoc* sp.	Magarvey et al., 2006
10	Tubercidin	*Tolypothrix* sp.	Banker and Carmeli, 1998
11	Cycloserine	*S. orchidaceus*	Selim et al., 2021
12	Neomycin	*S. fradiae*	Selim et al., 2021
13	Tetracycline	*S. viridifaciens*	Selim et al., 2021
14	Erythromycin	*S. erythraeus*	Selim et al., 2021
15	Novobiocin	*S. niveus*	Selim et al., 2021
16	β-carotene	*Dunaliella salina*	Sreenikethanam et al., 2022
17	Phycocyanin	*Spirulina platensis*	Sreenikethanam et al., 2022
18	Fucoxanthin	*Phaeodacty lumtricornutum*	Sreenikethanam et al., 2022
19	Zeaxanthin	*Dunalilella salina*	Sreenikethanam et al., 2022
20	Astaxanthin	*Chlorella vulgaris*	Sreenikethanam et al., 2022

processes essential for maintaining overall health. The study conducted by Liu et al. (2020) further explores the potential and benefits associated with *Aurantiochytrium* sp. in DHA production. Endophytic *Penicillium* sp. and *Fusarium moniliforme* have exhibited antimicrobial activity against multidrug-resistant (MDR) human pathogens, including *Staphylococcus aureus*, *Pseudomonas aeruginosa*, *Klebsiella pneumoniae*, and *Salmonella typhi*. These findings highlight the potential of these endophytic fungi in combating MDR pathogens and offer promising avenues for further research and development of novel antimicrobial agents (Das et al., 2023).

5.3 METABOLIC PATHWAYS

5.3.1 The Embden–Meyerhof–Parnas (EMP) Pathway

The EMP pathway, which is a glycolytic pathway, is used by eukaryotes which refill the TCA cycle. Ethanol and carbon dioxide are synthesized by fungi and bacteria. Various gases, like fatty acids and alcohol, are produced. The key enzyme of the EMP pathway is 6-phosphofructokinase (PFK). Two ATP molecules are required to activate the initial reactions of the EMP pathway, whereas the later stages of the EMP pathway involve the phosphoregulation of the triose phosphate molecule. Later, these are oxidized further and catalyzed by glyceraldehyde dehydrogenase. The EMP pathway yields 4 ATP and 2 pyruvate molecules (Wolfe, 2015). Many bacteria, for example, *Escherichia coli* and *Salmonella enterica* serovar Typhimurium (S. Tm), use different types of sugars as the source of energy. Glucose breakdown initiates with the phosphorylation step, via hexokinase (Hxk), (ATP: d-hexose phosphotransferase). This step consumes one ATP molecule and generates glucose 6-phosphate (G6P). Accordingly, G6P undergoes isomerization into fructose 6-phosphate (F6P) through the action of glucose phosphate isomerase Pgi-(d-glucose-6-phosphate isomerase). This isomerization reaction is a reversible step. The subsequent step, which is phosphorylation, requires additional ATP to convert F6P into fructose 1,6-bisphosphate (F16BP or FDP). This is catalyzed by the crucial enzyme, which is phosphofructokinase (Pfk) (ATP: d-fructose-6-phosphate 1 phosphotransferase), which regulates the flow via the EMP pathway. Pfk can act as a limiting factor for the EMP pathway and has an allosteric property, which is inhibited by PEP. The latter step involves the cleavage of 1,6-diphosphate (F16BP or FDP) into two triose phosphate molecules. This reaction is catalyzed by fructose diphosphate aldolase (Fba) (fructose-1,6-bisphosphate: d-glyceraldehyde-3-phosphate-lyase). The cleavage of FDP produces glyceraldehyde 3-phosphate (GAP) and dihydroxyacetone phosphate (DHAP), which can again interconvert through the reversible action of triose phosphate isomerase (Tpi) (d-glyceraldehyde-3-phosphate ketol-isomerase).

Step 6 involves an oxidation and phosphorylation process which is catalyzed by glyceraldehyde-phosphate dehydrogenase (GAPDH) (d-glyceraldehyde-3-phosphate: NAD oxidoreductase). This step generates NADH and the high-energy molecule 1,3-bisphospho-d-glycerate (1,3BPG). In step 7, 1,3BPG releases one phosphate group, resulting in the production of ATP. This reaction is facilitated by phosphoglycerate kinase (Pgk) (ATP: 3-phospho-d-glycerate 1-phospho-transferase), which yields 3-phosphoglycerate (3 PG). The conversion of 3 PG to 2-phosphoglycerate (2 PG) occurs in step 8 and is catalyzed by phosphoglycerate mutase (Pgm) (2, 3-diphospho-d-glycerate: 2-phospho-d-glycerate phosphotransferase).

Step 9 is mediated by enolase (Eno) (2-phospho-d-glycerate hydro-lyase) and results in the formation of phosphoenol pyruvate (PEP). This step includes an intracellular electron shift, known as the intramolecular oxidation-reduction reaction. PEP is energetically more approving compared to 2 PG. In the final step, which is step 10 of the EMP pathway, pyruvate kinase (Pyk) (ATP: pyruvate phosphotransferase) catalyzes the reaction that produces pyruvate (PYR). During this reaction, the phosphate group of PEP is transferred to ADP, resulting in the generation of ATP. Pyk serves as the second allosteric enzyme in the EMP pathway (Shimizu, 2013).

Escherichia coli, a bacterium, can thrive on a simple medium composed of glucose and mineral salts. It can synthesize all the required components for cell growth using the provided medium. Glucose is metabolized via the EMP pathway and the hexose monophosphate (HMP) pathway. The end product of glucose metabolism, pyruvate, undergoes oxidative decarboxylation to form acetyl-CoA, which again enters the tricarboxylic acid (TCA) cycle for oxidation. Twelve intermediates are compounds produced during these pathways that serve as carbon skeletons for the biosynthesis of essential cellular components. In eukaryotes, glucose is metabolized via the EMP pathway, which generates ATP, pyruvate, and NADH. The HMP pathway produces metabolic intermediates that are not readily available from the EMP pathway, such as pentose-5-phosphate, erythrose-4-phosphate, and NADPH. Most prokaryotes follow similar metabolic mechanisms, although some prokaryotes have unique glucose metabolism pathways, such as the Entner–Doudoroff (ED) pathway and the phosphoketolase (PK) pathway.

Certain prokaryotes possess genes for both the ED and EMP pathways and are expressed in various prokaryotes, including thermophilic bacteria (*e.g.*, *Thermotoga maritima*), thermophilic archaea (*e.g.*, *Thermoproteus tenax*), and halophilic archaea (*e.g.*, *Halococcus saccharolyticus*). *Escherichia coli* metabolizes glucose primarily via the EMP pathway, where gluconate is oxidized through the ED pathway. Modified versions of both EMP and ED pathways are commonly observed in archaea (Kim and Gadd, 2008). Archaea exhibit unique characteristics in their EMP pathway. Instead of using NADH, they utilize ferredoxin as the reducing agent. Furthermore, they have various enzymes, such as pyruvate ferredoxin oxidoreductase, non-orthologous phosphoglycerate mutases (PGMs), and non-phosphorylating glyceraldehyde-3-phosphate (GAP) dehydrogenase.

For instance, pyruvate ferredoxin oxidoreductase helps in the conversion of pyruvate to acetyl-CoA and carbon dioxide. These enzymes are not present in bacteria and eukaryotes. In contrast, the ED pathway demonstrates Archaea's ability to use glucose and galactose as energy sources. Archaea exhibit two variations of the ED pathway: a non-phosphorylative version and a semi-phosphorylative version for carbohydrate catabolism. Hence, EMP pathway is well-conserved in bacteria and eukaryotes, the ED pathway is exclusive to prokaryotes, and archaea have developed their own variations of these pathways (Alnomasy and Al-Harbi, 2017).

5.3.2 The Entner–Doudoroff Pathway

The ED pathway is found in various bacteria and serves as a significant pathway for glucose catabolism, mainly in aerobic conditions. The pathway was identified in *P. saccharophila* by Entner and Doudoroff in 1952. They observed that glucose was oxidized by this organism, resulting in the production of two pyruvate molecules, with one of the carboxyl carbons in pyruvate originating from carbon 1 of glucose. Further investigations conducted by MacGee and Doudoroff in 1954 showed the isolation and characterization of the sodium salt of 2-keto-3-deoxy-6-phosphogluconate, confirming the pathway. Similar studies carried out by Kovachevich and Wood in 1955 demonstrated the presence of the same pathway in *Pseudomonas fluorescens*. The pathway is called the oxidative branch of the pentose phosphate pathway (PPP) (Spector, 2009). The ED pathway works directly for glucose conversion to pyruvic acid when compared to the glycolysis pathway. It begins with a similar step to glycolysis, initially with phosphorylation, followed by an oxidative process that converts the compound into phosphogluconic acid. Further dehydration takes place, leading to the formation of keto-deoxy-phosphoglucose acid. In the final step, pyruvic acid and glyceraldehyde phosphate are produced, with the latter able to be converted to pyruvic acid as well. The ED pathway generates two molecules of NADPH and one molecule of ATP from each glucose molecule, which are utilized by cellular biosynthetic reactions. This pathway is present in certain Gram-negative bacteria, including *Pseudomonas*, *Rhizobium*, and *Agrobacterium*. However, it is absent in Gram-positive bacteria (Elbeshbishy, 2014). While the ED pathway is naturally active in *Zymomonas mobilis*, it can also be induced in other bacteria, like *E. coli*, depending on various genetic and culture conditions (Shimizu, 2013).

Fungi have the potential for the development of nutraceuticals that can help in NTCD (non-toxigenic *Clostridioides difficile*). These are used as natural food additives, like stabilizers, colorants, *etc.* They have a lower risk of health concerns when compared to synthetic food additives. Metabolic engineering is a modern approach utilized for the derivation of additives from fungi. Agroindustrial by-products are used for the sustainable fermentation of the fungus (Takahashi et al., 2020). Fungi are used in the production of pigments, lipids, glycolipids, vitamins, *etc.* These are used in biomineralization for their antimicrobial activities. They also play an important role as disease suppressors, as plant growth promoters, and mainly in making mycoproteins (Solomon et al., 2019). The strictly fermentative bacterium *Zymomonas mobilis* utilizes the ED pathway exclusively under anaerobic conditions, and this serves as the sole catabolic pathway for glucose and other sugars to generate energy. *Z. mobilis* employs the nonoxidative branch of the PPP and an incomplete TCA cycle to synthesize essential metabolic precursors and intermediates. Notably, the pyruvate generated via the ED pathway is ultimately converted into ethanol in *Z. mobilis*. Due to the lower energy production efficiency of the ED pathway and the absence of electron transport system enzymes, *Z. mobilis* relies on the rapid uptake and metabolism of sugars present in its environment to achieve and sustain significant growth rates.

The Entner–Doudoroff (ED) pathway is found in various bacteria and can serve as a major pathway for glucose catabolism under aerobic conditions. It branches off from the oxidative branch of the PPP. The pathway begins with the phosphorylation of D-glucose by ATP-dependent hexokinase, resulting in the formation of glucose-6-phosphate. Subsequently, Glc-6P dehydrogenase (Glc-6PDH) and 6-phosphogluconolactonase act on glucose-6-phosphate to convert it into 6-phosphogluconate, generating one $NADPH/H^+$ in the process. The next step involves the conversion of 6-phosphogluconate to 2-keto-3-deoxy-6-phosphogluconate (KDPG) through the dehydration reaction catalyzed by 6-phosphogluconate dehydratase. KDPG is later cleaved by KDPG aldolase into pyruvate and GAP (glyceraldehyde-3-phosphate). The GAP was later converted to pyruvate through the triose phosphate portion of the EMP pathway, resulting in the production of two ATP and one $NADH/H^+$. In summary, a single glucose molecule is catabolized via the ED pathway. It is degraded into two pyruvates, yielding a net of one ATP plus one $NADPH/H^+$ and one $NADH/H^+$, depending on the specific microorganism. In significance, the ED pathway provides half the net amount of energy in the form of ATP compared to the EMP pathway, when a single glucose molecule is catabolized into two pyruvates (Spector, 2009).

5.3.3 The Pentose Phosphate Pathway

The pentose phosphate pathway is an important pathway for glucose catabolism. PPP includes two main branches, which are oxidative and non-oxidative. The oxidative branch produces NADPH for anabolic processes. The non-oxidative branches of glucose help in the production of ribose 5 phosphate (R-5P), a five-carbon sugar. R-5P can also serve as a precursor for the generation of other important intermediates, like glyceraldehyde 3-phosphate (GA-3P) and fructose 6-phosphate (F-6P). Non-oxidative PPP helps in the central metabolic role and acts as a precursor to RNA ribose 5-phosphate and erythrose 4–4-phosphate for amino acids. This can occur non-enzymatically or by the interconversion of glucose 6-phosphate to 6-phosphogluconate. Although the oxidative branch of PPP is newer and more blooming than the non-oxidative branch, the oxidative branch plays a major role in eukaryotes. They are active and help in converting the glucose 6-phosphate into ribulose 5-phosphate with the reaction of G6PDH (glucose-6-phosphate dehydrogenase), 6-6-phosphogluconolactonase (6PGL) with the increase in specificity of the enzyme. With these reactions, 2 NADPH are yielded. Ribulose 5-phosphate is then transformed into ribose 5-phosphate by ribose 5-phosphate isomerase (RPI) or to xylulose 5-phosphate by ribulose 5-phosphate epimerase (RPE) by entering into the non-oxidative branch. The backbone of RNA and DNA is formed by ribose 5-phosphate. For the synthesis of histidine, erythrose 4–4-phosphate helps act as a precursor. The completion of the pathway involves the RPI (ribose 5-phosphate isomerase) and RPE (ribulose 5-phosphate epimerase) reactions. These reactions play a vital role in converting ribose 5-phosphate, xylulose 5-phosphate, glyceraldehyde 3-phosphate, and fructose 6-phosphate, which are intermediates in glycolysis/gluconeogenesis, through the rearrangement of monophosphate sugars. The enzymatic transformations are facilitated by transketolase (TKL) and transaldolase (TAL), which are responsible for interconversion reactions involving multiple substrates at the core of the non-oxidative pentose phosphate pathway (PPP). Transketolase (TKL) plays a role in the non-oxidative pentose phosphate pathway (PPP) by using xylulose 5-phosphate as a ketose donor and ribose 5-phosphate or erythrose 4-phosphate as aldose acceptors. This enzymatic process forms the glyceraldehyde 3-phosphate and sedoheptulose 7-phosphate or fructose 6-phosphate as aldose and ketose products, respectively. TKL facilitates the transfer of two-carbon fragments, known as "activated glycolaldehyde," for the interconversion of monosaccharides. The activity of TKL relies on the cofactor thiamine diphosphate, which is bound at the interface of the TKL homodimer. On the other hand, transaldolase (TAL) is responsible for transferring the three-carbon fragment dihydroxyacetone between sugar phosphates of varying lengths, forming a Schiff base at a lysine residue in the active site. TAL utilizes ketose sugar phosphates, fructose 6-phosphate, and sedoheptulose 7-phosphate as donor substrates, while aldose sugar uses phosphates such as glyceraldehyde 3-phosphate and erythrose 4-phosphate as acceptor substrates. TAL and TKL act as a connection between glycolysis and the PPP by sharing intermediate metabolites, such as fructose 6-phosphate and glyceraldehyde 3-phosphate. They also interact with sedoheptulose 7-phosphate, which can be synthesized by other enzymes, like sedoheptulokinase (SHPK) in mammals, sedoheptulose 1,7-bisphosphatase (SH17BPase) in yeast, and sedoheptulose 7-phosphate isomerase (SHI) in bacteria. Sedoheptulose 7-phosphate serves as an entry and exit point into/from the non-oxidative PPP, connecting it with open-chain and polyol sugar metabolism, as well as bacterial lipopolysaccharide biosynthesis. These connections with glycolysis, amino acid biosynthesis, and open-chain sugar metabolism establish the PPP as a central component within the metabolic network.

In the majority of the organisms, such as fungi and metazoans, the PPP primarily takes place in the cytosol, contributing to both the cytoplasmic metabolite pool and the redox cofactor pool. However, exceptions are in where the pathway is divided between the cytosol and other organelles, such as the plastid in plants or peroxisomes and glycosomes in parasitic protozoa. Additionally, a portion of the PPP may also occur in the endoplasmic reticulum (ER). Microsomes, which are vesicles derived from the ER during cell homogenization, contain several PPP enzymes, including hexose 6-phosphate dehydrogenase (H6PDH), an enzyme that is similar to G6PDH. H6PDH plays a vital role in providing NADPH to luminal reductases and exhibits a broader range of substrates compared to G6PDH. It is nonselective regarding the nucleotide cofactor, accommodating both NAD^+ and $NADP^+$. The concentration of reduced NADP(H) in the endoplasmic lumen indicates a preference for glucose 6-phosphate and $NADP^+$ under physiological conditions. Although the PPP predominantly occurs in the cytosol, these alternative organelle localizations, including the endoplasmic reticulum, are of significant importance and contribute to the overall functionality of the pathway. The pentose phosphate pathway (PPP) plays a central role in the metabolism of both single- and multicellular organisms.

Despite being evolutionarily ancient, the PPP exhibits a remarkable level of flexibility, making it an attractive target for biotechnology and medicine. Its primary biochemical function is the biosynthesis of nucleic acid and amino acid sugar phosphate precursors. This crucial function is closely linked to the generation of NADPH, which serves as a vital provider of biochemical reducing equivalents and helps maintain redox homeostasis. The PPP has high flexibility and adaptability, allowing it to respond to varying nutrient availability and stress conditions. This adaptive nature enables the pathway to coordinate its functions and meet the metabolic demands of cells in a constantly changing environment. The PPP holds significant importance in biotechnology, as its flexibility can be harnessed to regulate NADPH production.

In medical research, the activity of the PPP is influenced by bacterial and eukaryotic parasites during infection, stem cell differentiation, redox homeostasis in cancer cells, and energy metabolism in neurons. Despite 80 years of detailed research in both basic and medical fields, the intricate regulation of the PPP remains incompletely understood. Unraveling this complex regulation is essential for comprehending metabolic adaptations and their impact on cellular and organismal physiology (Stincone et al., 2015; Bertels et al., 2021).

5.4 METABOLIC PATHWAY IN BACTERIA

In bacteria, glucose can be catabolically processed by many pathways, in which glycolysis is the most common route associated with anaerobic or fermentative metabolism in bacteria and yeasts. Additionally, minor heterofermentative pathways, such as the phosphoketolase pathway, are present in bacteria. Furthermore, two other glucose-catabolizing pathways are present in bacteria: the oxidative pentose phosphate pathway (hexose monophosphate shunt) and the ED pathway. The Entner–Doudoroff pathway is predominantly found in aerobic bacteria. Certain organisms are highly oxidative, like *Azotobacter*, and most *Pseudomonas* species use the ED pathway for glucose catabolism due to the absence of phosphofructokinase. These organisms are unable to synthesize fructose 1,6-diphosphate, a crucial intermediate compound in the glycolytic pathway. Phosphofructokinase is sensitive to O_2 and, therefore, does not function in obligate aerobes. Additionally, bacteria lacking aldolase, an enzyme that splits fructose-1,6-diphosphate into two triose phosphate compounds, cannot sustain a functional glycolytic pathway. Although the ED pathway is typically associated with obligate aerobic bacteria, it is also present in the facultative anaerobe *Zymomonas mobilis*. This organism carries out glucose dissimilation to ethanol, making it a significant player in bacterial alcoholic fermentation reactions. Bacteria that inhabit the rhizosphere and internal tissues of plants can synthesize diverse types of secondary metabolites. These metabolites play a crucial role in enhancing plant growth through various mechanisms, including biocontrol. Some of these metabolites are produced as a result of interactions between plants and microbes, while others are exclusively synthesized by endophytic bacteria.

The secondary metabolites produced by rhizospheric and endophytic bacteria are valued and untapped resources of therapeutic agents that can be used in the treatment of multidrug resistance and cancer. Exploiting these metabolites holds significant potential for human health, agroindustry, and environmental applications. Additionally, the study of prophage-induced production of secondary metabolites by lysogenic bacteria and cyanobacteria represents an old yet underexplored research field in microbiology (Pandey et al., 2021). In a simplified model of the rhizosphere involving *Bacillus cereus*, *Flavobacterium johnsoniae*, and *Pseudomonas koreensis*, it has been studied that the dynamics of secondary metabolism are influenced by the composition and the interactions between species. Comparative metatranscriptomics and meta-metabolomics analyses have revealed differences in the abundance of transcripts from biosynthetic gene clusters (BGCs) and molecular features in monocultures, dual cultures, and a tripartite community. In both two- and three-member co-cultures, *P. koreensis* has been observed to change the expression of BGCs responsible for producing zwittermicin, petrobactin, and other secondary metabolites in *B. cereus* and *F. johnsoniae*. Interestingly, the BGC transcriptional response in *P. koreensis* itself is minimal when exposed to the community. The pairwise and tripartite co-cultures with *P. koreensis* exhibited distinct molecular features, which appear to be derivatives of lokisin, indicating the occurrence of metabolic interactions and exchanges between species. By deleting the BGC responsible for koreenceine, another metabolite produced by *P. koreensis*, notable changes in transcript and metabolite profiles were observed throughout the community. This included significant up-regulation of the petrobactin and bacillibactin BGCs in *B. cereus*, suggesting that koreenceine represses the production of siderophores. These findings from the model community highlight how the expression of bacterial BGCs and the resulting chemical outputs are dependent on the specific identity and biosynthetic capabilities of the co-cultured partners. It suggests that the composition of the community and interactions within the microbiome potentially shape the regulation of secondary metabolism in natural environments (Chevrette et al., 2022). For example, the utilization of engineered lactic acid bacteria strains has proven beneficial in enhancing the production of various organic acids, like γ-aminobutyric acid and lactic acid, as well as exopolysaccharides, vitamins, and other valuable products. These engineered strains can also be employed to express enzymes involved in the degradation of proteins and polysaccharides. Notably, *Levilactobacillus brevis*, *Lactococcus lactis*, *Lactiplantibacillus plantarum*, and other lactic acid strains have shown remarkable advancements in metabolic engineering.

The advancement of modern biotechnology, including multiomics technology and gene editing techniques, has facilitated extensive research on lactic acid bacteria metabolism. These technologies have significantly contributed to unraveling the metabolic pathways and characteristics of lactic acid bacteria. Such research insights hold promise in guiding the application of lactic acid bacteria in the food industry and the human health sector (Wang et al., 2021).

5.5 METABOLIC PATHWAY IN FUNGI

Fungi have a high potential to convert monomers from plant polysaccharides (de Vries et al., 2001). Conversion of complex networks makes these fungi attractive cell factories. D-glucose is a major constituent of xyloglucan, xylan, and pectin. D-xylose and L-arabinose are the most abundant monosaccharides found in nature, after D-glucose (Seiboth and Metz, 2011). PCP, which is a pentose catabolic pathway, is composed of two interconnected oxidoreductive pathway branches. Both of the pentoses will undergo reactions like oxidation, reduction, and phosphorylation to form D-D-xylulose-5-phosphate; later, they are metabolized via the pentose phosphate path (Witteveen et al., 1989). Secondary metabolite products are controlled by transcription factors, which are encoded by genes. These genes also help in physiological events and environmental conditions, like temperature, nutrition, pH, *etc.* (Yu and Keller, 2005; Hoffmeister and Keller, 2007). Gene cluster contains a transcription factor which regulates secondary metabolism, and thus, this cluster acts on gene within the cluster (Fox and Howlett, 2008). Fungi have the potential for the development of nutraceuticals that can help in the NTCD. These are used as natural food additives, like stabilizers, colorants, *etc.* These have a lower risk of health concerns when compared to synthetic food additives.

Metabolic engineering is a modern approach utilized for the derivation of additives from fungi. Agroindustrial by-products are used for the sustainable fermentation of the fungus (Takahashi et al. 2020). Fungi are used in the production of pigments, lipids, glycolipids, vitamins, *etc.* These are mainly used in biomineralization as they possess antimicrobial activities. These also play an important role as disease suppressors, plant growth promoters, and also mainly in making mycoproteins (Solomon et al.,2019). The investigation of a large number of secondary metabolite clusters, many of which include knowledge about the encoded enzymes, combined with the availability of fungal genome sequences in databases, has sparked a significant surge in genetic mining studies across various fungi. This progress has been facilitated by computer prediction programs like SMURF, antiSMASH, and MIDDAS, enabling the identification of new clusters with their biosynthetic functions tentatively assumed until experimental validation can be conducted. A current research challenge in this field involves assigning metabolites to newly discovered clusters, which requires a combination of genetic, chemical, and biochemical methods. Notably, there are substantial differences in the presence of secondary metabolite clusters, not only between different genera, but also among different species within the same genus. Remarkable examples of this versatility can be found in *Aspergillus*, *Penicillium*, and *Fusarium*.

The identification of novel metabolites is an exciting area of study, given its immense biotechnological potential, as the majority of these compounds are still waiting to be discovered. In this context, the manipulation of secondary metabolite pathways, leveraging the combinatorial possibilities offered by the modular organization of polyketide synthase (PKS) and non-ribosomal peptide synthetase (NRPS) enzymes, holds particular promise, and significant progress has already been made in this direction. In conclusion, fungal secondary metabolism represents a continuously growing and evolving field, and future discoveries hold the potential to unlock unprecedented opportunities in medicine and biotechnology (Avalos and Limón, 2022).

5.6 METABOLIC PATHWAY IN ALGAE

The biochemical details of glycolysis in algae reveal a striking similarity to the process observed in higher plants rather than metazoans. In algae, both ATP-dependent enzymes, such as phosphofructokinase (PFK), and PPi-dependent enzymes, such as PFP, are present and capable of converting fructose-6-phosphate to fructose-1,6-bisphosphate. Within the chlorophyte (trebouxophycean) green microalga *Selenastrum*, the pathway beyond phosphoenolpyruvate exhibits phenotypic modification in response to low phosphate availability. This modification involves bypassing the pyruvate kinase reaction. Instead, pyruvate is produced either through the action of phosphoenolpyruvate phosphatase or through the sequential involvement of phosphoenolpyruvate carboxylase, malic dehydrogenase, and malic enzyme (Raven, 1984; Falkowski et al., 2013; Turpin et al., 1988).

Oxidative pentose phosphate pathway. Like higher plants, green algae possess enzymes in the oxidative pentose phosphate pathway in both their cytosol and their plastids (Buchanan et al., 2015). A study by Raven (1984) suggested that the absence of specific enzymes related to glycolysis and the oxidative pentose phosphate pathway in Euglena's plastids could be linked to the lack of polysaccharides in those plastids. Additionally, Euglena's inability to utilize nitrate as a nitrogen source might correlate with the absence of specific oxidative pentose phosphate pathway enzymes in its plastids, as there is no need for reductant (NADPH) in the dark for nitrite reduction, which typically occurs in the plastids of green algae and higher plants. While there is limited information available regarding the localization of oxidative pentose phosphate cycle enzymes in other algal divisions, recent research has demonstrated that isolated diatom (Heterokontophyta) plastids can exhibit light-dependent oxygen evolution in the presence of nitrate, indicating the presence of nitrite reductase activity in these plastids (Wittpoth et al., 1998).

The tricarboxylic acid cycle is a metabolic pathway that facilitates the complete oxidation of pyruvate, generated by glycolysis, to CO_2 under aerobic conditions. This process involves the reoxidation of NADH and reduced flavoprotein generated within the cycle through the oxidative phosphorylation sequence (Buchanan et al., 2015). Apart from its role in energy production, the tricarboxylic acid cycle is essential for providing carbon skeletons required for biosynthesis. Intermediates of this cycle, such as 2-oxoglutarate, serve as crucial precursors for the synthesis of glutamate and its family of amino acids and amides, as well as algal tetrapyrroles (except for mitochondrial tetrapyrroles in *Euglena*, which are synthesized using succinate and glycine, derived from another tricarboxylic acid intermediate). Notably, 2-oxoglutarate is uniquely synthesized by the tricarboxylic acid cycle. The photolithotrophic metabolism of certain algae reaches its utmost extent in obligate photolithotrophs, such as numerous cyanobacteria and eukaryotes. These organisms cannot be cultivated in the absence of light using any of the organic carbon substrates that have been tested thus far (Beardall and Raven, 2012).

Certain algae deviate from strict photolithotrophy due to impairments in downstream metabolic processes, leading to their dependency on external sources of specific growth substances or vitamins. Among these substances, biotin, thiamin, and vitamin B12 play prominent roles. Vitamin B12, for instance, primarily acts as a cofactor for vitamin B12-dependent methionine synthase in algal metabolism. While some cyanobacteria are capable of synthesizing vitamin B12, eukaryotic algae seem to lack the necessary genes for a functional pathway to biosynthesize vitamin B12. Some eukaryotic algae possess a vitamin B12–independent methionine synthase and therefore do not require dietary vitamin B12. On the other hand, certain algae rely on obtaining vitamin B12 from associated biota, such as bacteria, as they possess only the vitamin B12–dependent methionine synthase (Helliwell et al., 2011).

Algae possess a wide range of important macro- and micromolecules that are specific to their biology. These molecules encompass various functions, including light-harvesting pigment–protein complexes, redox catalysts involved in photosynthesis, enzymes and transporters for inorganic carbon acquisition, enzymes associated with glycolate metabolism, and polyglucan organic carbon stores that serve as metabolic buffers between organic carbon production or acquisition and downstream metabolism. Furthermore, algae produce alga-specific end products, such as extracellular structural materials (*e.g.*, agarose, agaropectin, alginic acid, and carrageenans), as well as antibiophage compounds and compatible solutes that are widespread among algae but synthesized through alga-specific pathways. An example of such a molecule is dimethyl sulfoniopropionate (DMSP). These diverse macro- and micromolecules play essential roles in the biology and physiology of algae. Algal cells contain a range of low-molecular-weight solutes, known as "secondary metabolites" or "antibiophage compounds," with diverse biosynthetic origins. These compounds encompass various substances, including phlorotannins, found in brown algae; halocarbons, present in different types of algae; and potentially the toxins released by certain cyanobacteria and dinoflagellates. Additionally, compounds like DMSP (dimethyl sulfoniopropionate) and MAAs (mycosporine-like amino acids) serve as both storage molecules and secondary metabolites, playing significant roles in algal responses to stress (Beardall and Raven, 2012). Several algal species, including *Chlorella*, *Haematococcus*, *Dunaliella*, and *Spirulina*, present a diverse array of secondary metabolites that hold promise for the creation of healthier food products (Orejuela-Escobar et al., 2021).

Certain groups of macroalgae possess notable properties and potential cosmetic applications, particularly in terms of polysaccharides, pigments, and phenolic compounds. While all three types of macroalgae (red, green, and brown) have been investigated for their compounds, brown and red macroalgae, in general, exhibit greater diversity. The compounds derived from macroalgae can serve as additives, excipients, or even active ingredients in cosmetic formulations. Numerous studies have highlighted the intriguing properties of these compounds. While some macroalgae compounds are already utilized as thickening or gelling agents, their untapped potential remains significant, particularly in the realm of sunscreens. Their photoprotective and antiphotoaging effects make them promising candidates for the sunscreen industry (Lourenço-Lopes et al., 2020).

Specific microalgae species contain notable quantities of wax, which consists of esters formed by long-chain fatty acids and alcohol. One such example is *Euglena gracilis*, a eukaryotic microalga that, when grown in an anaerobic environment, undergoes a conversion of stored polysaccharides into wax. This wax accumulates inside the cells. For instance, by initially cultivating *E. gracilis* in a heterotrophic manner and subsequently transitioning to anaerobic conditions for several days, waxes can be obtained, accounting for up to 50% of the dry weight (Park et al., 2022).

5.7 METABOLIC PATHWAY OF ACTINOMYCETES

Actinomycetes are widely found in both natural and man-made environments and play a crucial role in the decomposition of organic matter. Additionally, they are recognized as a valuable reservoir of bioactive secondary metabolites (Berdy, 2012). Actinomycetes exhibit adaptability to diverse habitats and have a broad distribution in natural ecosystems. They are classified as Gram-positive bacteria and are characterized by the formation of aerial and substrate mycelium on solid media. Furthermore, actinomycetes possess spores with distinct surface features and have a high guanine + cytosine (G + C) content in their DNA (Goodfellow and Williams, 1983). The bioactive compounds synthesized by actinomycetes demonstrate specificity against *Bacillus* strains found in bees, as well as two drug-resistant, Gram-positive human pathogens. Among these actinomycetes, *Nocardiopsis alba* is an uncommon species isolated from the gut of honeybees. *N. alba* produces phenazine, which contributes to its ability to survive temporarily under anaerobic conditions commonly encountered in the honeybee gut (Park et al., 2022; Teske et al., 2003). Marine actinobacteria offer a promising source of novel therapeutics, providing the necessary medications to aid us in our ongoing battle against drug-resistant infectious diseases for the foreseeable future. These actinobacteria present a wealth of untapped potential and hold the key to developing new classes of therapeutics that can address the growing challenge of drug resistance (Vimal et al., 2009).

Marine sponges play a fundamental role in deep-sea ecosystems, forming essential components of these communities. Through intricate associations with diverse microorganisms, they have established unique relationships that hold ecological and biological significance. Recent studies have revealed the presence of novel actinomycete taxa in both the Great Barrier Reef and Mediterranean sponges. These actinomycetes exhibit the ability to produce distinct bioactive compounds, further highlighting the remarkable diversity and potential of these marine organisms (Selvan et al., 2012). Secondary metabolism in actinomycetes typically occurs during the late growth phase of microorganisms. This temporal pattern of secondary metabolism is primarily governed by genetic factors, but its expression can be significantly influenced by environmental conditions. Factors such as nutrient depletion, introduction of specific inducers, reduced growth rates, or signals from neighboring organisms in the soil, such as γ-butyrolactones produced by various Streptomyces and Actinomycetes, can trigger secondary metabolite production. These environmental manipulations can stimulate the activation of specific biosynthetic pathways, leading to the synthesis of diverse bioactive compounds (Bibb, 2005).

Analysis of genomic sequence data has uncovered the existence of a significant portion of potentially dormant biosynthetic gene clusters within actinomycete genomes. These gene clusters are responsible for encoding secondary metabolites; however, they remain inactive and undetected under typical fermentation conditions (Abdelmohsen et al., 2015). In typical laboratory culture conditions, a considerable number of secondary metabolite biosynthetic gene clusters (smBGCs) remain inactive or concealed. To awaken these dormant smBGCs, various techniques have been devised, such as culture-based and genetic engineering–based strategies. Among these approaches, co-culture stands out as a promising method to stimulate the production of novel secondary metabolites from actinomycetes. By replicating an ecological environment, co-culture creates conditions that can activate cryptic smBGCs within actinomycetes, offering potential for the discovery of previously unknown secondary metabolites (Kim et al., 2021). Nevertheless, the co-culture technique frequently lacks reproducibility and is not suitable for large-scale cultivation aimed at abundant production of specific target secondary metabolites. However, the stimuli identified for inducing secondary metabolite production in co-culture studies can still be harnessed for industrial applications through single-culture approaches. Therefore, achieving a thorough and precise comprehension of the underlying mechanisms in co-culture is of utmost importance (Lee et al., 2020).

Till date, numerous co-culture studies have effectively uncovered new secondary metabolites derived from actinomycetes. However, the specific mechanisms underlying these interactions remain largely unexplored. It is crucial to conduct more extensive and comprehensive investigations into the induction mechanisms involved in co-culture. This deeper understanding is necessary to grasp the intricate regulation of secondary metabolism and to guide the development of genetic engineering–based strategies aimed at inducing or enhancing the production of specific target secondary metabolites (Kim et al., 2021).

5.8 MODIFICATION OF EXISTING PATHWAY TO ENHANCE PRODUCTION

Streptomyces spp. can generate over 30 secondary metabolites, but the majority of the biosynthetic gene clusters responsible for their production remain cryptic or silent. The expression of these genes is highly conditional and relies on specific culture conditions. Various methods have been developed to activate these gene clusters and unlock their potential. Goadsporin is a chemical compound that is extracted from *Streptomyces* sp. TP-A0584. It can stimulate secondary metabolism and sporulation in numerous *Streptomyces* species. Another approach, known as combined culture, involves co-culturing *Streptomyces* with mycolic acid–containing bacteria. This method induces notable alterations in secondary metabolism, with approximately 90% of *Streptomyces* species exhibiting changes compared to pure culture conditions. Both goadsporin and combined-culture techniques hold potential applications in the discovery of natural product–based drugs (Onaka, 2017).

Over the past few years, there has been a noticeable rise in the quantity of microorganisms that can undergo metabolic engineering using the principles of systems biology and the advanced molecular tools of synthetic biology (Lee and Kim, 2015; Trinh et al., 2015; Jullesson et al., 2015; Chubukov et al., 2016; Smanski et al., 2016). The elimination of competing internal pathways, the precise regulation of gene expression and substrate transport, and the redirection of small molecules at specific points in the network have become key strategies. However, the fundamental biochemical processes that drive microbial cell factories are often overlooked and rarely modified as a whole. Only a few endeavors have sought to reshape the central enzymatic processes completely (Antonovsky et al., 2016). For instance, *Pseudomonas putida*, a microorganism found in soil, serves as an ideal host for robust redox reactions. However, this bacterium lacks the activity of 6-phosphofructo-1-kinase (Pfk), rendering the EMP pathway nonfunctional. Simply introducing Pfk into the biochemical network of *P. putida* proved insufficient to activate hexose catabolism via the EMP pathway. Furthermore, it led to adverse effects on growth and reduced resistance to oxidative stress. Therefore, redirecting glucose metabolism toward pyruvate in *P. putida*, while ensuring sufficient NAD(P)H production and maximum ATP yield, necessitates a multifaceted approach. This approach involves incorporating additional glycolytic genes or specific combinations, resulting in the reconfiguration of the entire EMP pathway from *Escherichia coli* into a modular framework known as GlucoBricks. These GlucoBricks allow for the complete or partial implementation of the glycolytic pathway in various Gram-negative bacteria, including *P. putida* and others (Chan et al., 2005).

Approaches that seek to modify the entire metabolome of a target strain and induce pleiotropic effects to activate any pathway present within the strain have been developed. These methodologies are typically straightforward technically, making them well-suited for scaling up to high-throughput systems. However, one drawback of these techniques is that they often sacrifice the benefits offered by next-generation sequencing and genome mining. As a result, they are unable to specifically target the most intriguing gene clusters for activation (Baral et al., 2018). Ribosome engineering is a crucial mechanism in bacteria that enables them to adapt to environmental stresses, such as amino acid depletion. This adaptation is known as the "stringent response" and is regulated by the temporary buildup of guanosine tetraphosphate (ppGpp) within the cell (Ochi, 1987). Histones dynamically modulate the packaging and accessibility of DNA in response to physiological and developmental cues (Jenuwein and Allis, 2001; Cedar and Bergman, 2009). Secondary metabolism gene clusters are typically located in clustered arrangements within subtelomeric regions of the genome. These gene clusters are co-regulated, and their transcription is significantly influenced by histone acetylation and methylation (Yu and Keller, 2005; Gacek and Strauss, 2012). In *A. nidulans*, the inactivation of the histone deacetylase (HDAC) hdaA led to a decrease in the expression of secondary metabolic pathways. Conversely, the loss of function of cclA, an ortholog of bre2 responsible for histone H3 lysine 4 methylation, resulted in the activation of cryptic secondary metabolic clusters. This activation led to the production of monodictyphenone, emodin, and their derivatives (Shwab et al., 2007; Bok et al., 2009).

Regulatory genes play a crucial role in categorizing secondary metabolism pathways and manipulating specific segments or fragments of these pathways in various organisms, including bacteria, fungi, and algae. One notable example is the Fus3 MAPK pathway in fungi, which serves as a controlling regulator for both sexual growth and secondary metabolism, demonstrating its coordinated role in resilient fungal species. Exploring the distinct functionalities of *Aspergillus* Fus3 cells in signal transduction to regulate fungal growth and secondary metabolism holds promising prospects for future research. The microbial synthesis pathways of melatonin share similarities with those found in animals, but they differ from those observed in plants. Currently, the utilization of microorganism fermentation to produce melatonin is gaining traction, and the focus is shifting toward exploring and modifying the biosynthetic pathway of melatonin in microorganisms. This approach offers numerous advantages over traditional chemical synthesis methods. One of the essential precursors for melatonin synthesis is L-tryptophan, which can be produced through a four-step reaction. However, enzymes involved in melatonin synthesis exhibit low specificity and catalytic efficiency. Enhancing enzyme activity and increasing metabolic flow to promote microbial melatonin production can be achieved through site-directed mutation, directed evolution, or promoting cofactor synthesis. Overall, advancements can be made to improve the current status and overcome bottlenecks in melatonin biosynthesis, providing valuable insights for future microbial modifications.

Metagenomics is an important and widely employed technique that allows researchers to directly access the genetic material of microbes, bypassing the need for laboratory cultivation through the isolation of environmental DNA. This approach has led to the discovery of numerous bioactive compounds, including violacein, terragins, and turbomycins, along with their unique configurations. However, there are certain drawbacks associated with this approach. These include difficulties in effectively capturing complete gene sequences and incompatibility of specific gene expression components within their respective hosts. The technique of pathway engineering can be employed to enhance the production of desired compounds by utilizing gene deletion, gene expression repetition, and the introduction of new genes into the native host. For example, in *S. coelicolor*, the production of actinorhodin can be increased by 20 times through the overexpression of multiple copies of the actinorhodin gene clusters. Similarly, in *S. hygroscopicus*, the production of validomycin A can be boosted by 34% through the overexpression of multiple copies of the validomycin A gene clusters. Gene deletion is also beneficial for removing competing pathways that divert useful precursor metabolites. These competing pathways may consume cellular resources that are crucial for improving the yields of the desired products. Similar approaches have been used to improve the secretion productivity and capacity of natural products.

Research has shown that deleting specific genes associated with protein degradation can increase the secretion of protease-sensitive human growth hormones (hGH). By manipulating enzymes or metabolic pathways involved in the supply of precursor molecules, it is possible to achieve an enhanced availability of these molecules. For example, in the synthesis of polyketides, methylmalonyl-CoA and malonyl-CoA are commonly used biosynthetic precursors that are readily available metabolically. Studies have shown that the precursor molecule required for the synthesis of FK506 is methyl malonyl-CoA. By increasing the concentration of methyl malonyl-CoA, such as through the supply of methyl oleate, the production of FK506 can be significantly enhanced. In fact, in the microbial species *S. clavuligerus*, this approach has led to a 2.5-fold increase in the formation of FK506 (Iwai et al., 1992; Shukla, 2015; Sekurova et al., 2019).

5.9 CONCLUSION AND FUTURE PERSPECTIVES

It can be concluded that secondary metabolites are not deemed necessary for the growth and development of microorganisms; they play a crucial role in enabling a wide range of survival functions in nature. These molecules facilitate antagonistic and symbiotic interactions among diverse species, spanning from single-cell to multicellular organisms (Mandwal, 2022). Given the ongoing and critical need for novel pharmaceutical agents to combat conditions such as cancers, cardiac disorders, pests, cytotoxic effects, mosquitoes, infectious diseases, and autoimmune disorders affecting both animals and plants, the changing climate further exacerbates the occurrence of these events. In light of this, secondary metabolites possessing significant biological activity are increasingly viewed as viable alternatives to synthetic drugs and other commercially valuable compounds. In the ongoing development of synthetic microbial consortia, numerous challenges lie ahead. Improvements are needed in several areas, including gaining a deeper understanding of natural complex communities, estab lishing orthogonal cell-to-cell signaling, minimizing the occurrence of mutants in genetically modified consortia, and effectively controlling the population composition.

The manipulation of ecological interactions offers a means to regulate the composition of synthetic microbial consortia. As these interactions become more finely controlled, the potential for increased complexity within these consortia expands. Valuable insights can be gleaned from studying natural complex communities like gut or soil microbiomes, and this knowledge can inform the design of synthetic microbial consortia. Conversely, the control achieved in engineered microbial consortia will contribute to a better understanding of the intricate interactions occurring within natural communities (Duncker et al., 2021). Advancements in experimental techniques have opened up new possibilities in understanding biological networks and their intricate interactions through the utilization of high-throughput data from genomics, transcriptomics, and proteomics. Additionally, the emergence of innovative technologies like gene shuffling holds great promise for the future of metabolic engineering. The availability of kinetic and biochemical information regarding the pathway leading to a desired product enables a quantitative analysis that can significantly enhance the chances of success in metabolic engineering by identifying the factors limiting the rate of production.

Looking ahead, the field of metabolic engineering is likely to encompass not only the relaxation of rate-controlling steps in structural genes but also the deregulation and increased flow of precursor biosynthesis in secondary metabolism. However, metabolic flux analysis remains the most commonly employed approach. The future prosperity of the pharmaceutical industry relies on the identification or development of novel compounds with unique activities, specifically targeting particular end points. Recent accomplishments in the metabolic engineering of certain bacteria signify the initial steps toward sustainable biotechnological production of specialty fine and bulk chemicals. With ongoing genome, transcriptome, proteome, and other systems biology analyses, it is expected that an increasing number of targets for strain development will be discovered. The current achievements and observed trends point toward a promising future where microbial biotechnology will have an ever-growing impact on chemical production (Kumar and Prasad, 2011). The engineering of genetic circuits has allowed for the precise control of transcription in operons responsible for producing and secreting recombinant spider silk monomers in *Salmonella*. As synthetic biology progresses, the development of well-characterized components and increasingly complex devices will simplify the design, modelling, synthesis, and analysis of synthetic metabolic pathways.

In addition to enhancing metabolic control through gene expression regulation, such as engineered promoters and tunable intergenic regions, as well as enzyme colocalization through modular protein scaffolds or synthetic organelles, the enzymes themselves can be modified through directed evolution or computational design. Recent research demonstrated the creation of chimeric enzymes with improved properties by exchanging catalytic domains from three fungal cellulases. Furthermore, engineered enzymes with novel catalytic properties could enable the implementation of non-natural metabolic pathways. However, the rational engineering of enzymes remains challenging and requires a case-by-case approach.

Therefore, it would be advantageous for the metabolic engineering community to operate within the framework of synthetic biology, focusing on generating and populating families of genetic parts that are scalable, measurable, and amenable to modelling. The concurrent construction and characterization of these parts will facilitate model-based pathway design in the future, akin to building complex electrical circuits in silico using well-modelled components. High-throughput screening technologies have emerged as powerful platforms for quantifying the performance of components derived from combinatorially generated libraries (McArthur and Fong, 2009). Co-cultivation engineering in microbial biosynthesis represents a transformative approach for balancing metabolic pathways. It offers a broader range of possibilities for fine-tuning complex metabolic pathways and can be tailored to efficiently produce various bioproducts.

Compared to mono-cultivation systems, co-cultivation engineering provides several advantages, including robustness, modularity, enhanced tolerance (where toxic intermediates or waste produced by one partner are consumed or degraded by the other partner), and increased productivity. By harnessing the metabolic capabilities and resources of each co-cultivation partner, specific cofactors and precursors can be supplied, resulting in improved conversion yields of modularized biosynthetic pathways. This strategy enables the production of more complex compounds with higher productivity by distributing the metabolic workload among consortium members. Although co-cultivation engineering is an emerging research field within metabolic engineering, it is still in its early stages. Most recent studies on co-cultivation engineering involving microbial consortia have focused on utilizing two constituent strains/species to achieve their engineering objectives. While synthetic microbial consortia hold immense potential, there are inherent challenges that require addressing through the application of synthetic biology approaches. It is anticipated that shortly, co-cultivations comprising multiple specialized members, or polycultures, will be developed and utilized to meet the demands of more intricate biosynthetic pathways (Jawed et al., 2019).

REFERENCES

Abdelmohsen, U.R., T. Grkovic, S. Balasubramanian, M.S. Kamel, R.J. Quinn, and U. Hentschel. 2015. Elicitation of secondary metabolism in actinomycetes. *Biotechnology Advances* 33: 798–811. https://doi.org/10.1016/j.biotechadv.2015.06.003.

Abdel-Razek, A.S., M.E. El-Naggar, A. Allam, O.M. Morsy, and S.I. Othman. 2020. Microbial natural products in drug discovery. *Processes* 8 (April): 470, 1–19. https://doi.org/10.3390/pr8040470.

Akter, Y., R. Barua, M. Nasir Uddin, A.F.M. Sanaullah, and L.W. Marzan. 2022. Bioactive potentiality of secondary metabolites from endophytic bacteria against SARS-COV-2: An in-silico approach. *PLoS One* 17 (August): e0269962, 1–37. https://doi.org/10.1371/journal.pone.0269962.

Alnomasy, S.F., and Y.S. Al-Harbi. 2017. Insights into glucose metabolism inarchaea and bacteria: Comparison study of Embden-Meyerhof-Parnas (EMP) and Entner Doudoroff

(ED) pathways. *Advances in Biotechnology & Microbiology* 6 (August): 555679, 1–3. https://doi.org/10.19080/AIBM.2017.06.555679.

Antonovsky, N., S. Gleizer, E. Noor, et al. 2016. Sugar synthesis from CO2 in *Escherichia coli*. *Cell* 166: 115–25. https://doi.org/10.1016/j.cell.2016.05.064.

Avalos, J., and M.C. Limón. 2022. Fungal secondary metabolism. *Encyclopedia* 2: 1–13. https://doi.org/10.3390/encyclopedia2010001.

Banker, R., and S. Carmeli. 1998. Tenuecyclamides A–D, cyclic hexapeptides from the cyanobacterium *Nostoc spongiaeforme* var. *tenue*. *Journal of Natural Products* 61: 1248–51. https://doi.org/10.1021/np980138j.

Baral, B., A. Akhgari, and M. Metsä-Ketelä. 2018. Activation of microbial secondary metabolic pathways: Avenues and challenges. *Synthetic and Systems Biotechnology* 3: 163–78. https://doi.org/10.1016/j.synbio.2018.09.001

Barrios-Gonzalez, J., F.J. Fernandez, and A. Tomasini. 2003. Microbial secondary metabolites production and strain improvement. *Indian journal of Biotechnology* 2: 322–33.

Beardall, J., and J.A. Raven. 2012. *Algal metabolism*. In eLS, John Wiley & Sons, Ltd. https://doi.org/10.1002/9780470015902.a0000321.pub2.

Berdy, J. 2005. Bioactive microbial metabolites. *The Journal of Antibiotics* 58: 1–26. https://doi.org/10.1038/ja.2005.1.

Berdy, J. 2012. Thoughts and facts about antibiotics: where we are now and where we are heading. *The Journal of Antibiotics* 65: 385–95. https://doi.org/10.1038/ja.2012.27.

Bertels, L.K., L.F. Murillo, and J.J. Heinisch. 2021. The pentose phosphate pathway in yeasts–more than a poor cousin of glycolysis. *Biomolecules* 11 (May): 725, 1–20. https://doi.org/10.3390/biom11050725.

Bibb, M.J. 2005. Regulation of secondary metabolism in *Streptomycetes*. *Current Opinion in Microbiology* 8: 208–15. https://doi.org/10.1016/j.mib.2005.02.016.

Bills, G.F., and J.B. Gloer. 2016. Biologically active secondary metabolites from the fungi. *Microbiology Spectrum* 4 (November): FUNK-0009, 1–32. https://doi.org/10.1128/microbiolspec.funk-0009-2016.

Bok, J.W., Y.M. Chiang, E. Szewczyk, et al. 2009. Chromatin-level regulation of biosynthetic gene clusters. *Nature Chemical Biology* 5: 462–64. https://doi.org/10.1038/nchembio.177.

Buchanan, B.B., W. Gruissem, and R.L. Jones. 2015. Senescence and cell death. In *Biochemistry and molecular biology of plants*, 2nd ed., ed. B.B. Buchanan, W. Gruissem, and R.L Jones, 925–82. John Wiley & Sons.

Burja, A.M., B. Banaigs, E. Abou-Mansour, J.G. Burgess, and P.C. Wright. 2001. Marine cyanobacteria – A prolific source of natural products. *Tetrahedron* 57: 9347–77. https://doi.org/10.1016/S0040-4020(01)00931–0.

Cedar, H., and Y. Bergman. 2009. Linking DNA methylation and histone modification: patterns and paradigms. *Nature Reviews Genetics* 10: 295–304. https://doi.org/10.1038/nrg2540.

Chan, L.Y., S. Kosuri, and D. Endy. 2005. Refactoring bacteriophage T7. *Molecular Systems Biology* 1 (September): 1–10. https://doi.org/10.1038/msb4100025.

Chevrette, M.G., C.S. Thomas, A. Hurley, et al. 2022. Microbiome composition modulates secondary metabolism in a multispecies bacterial community. *Proceedings of the National Academy of Sciences* 119 (October): e2212930119, 1–10. https://doi.org/10.1073/pnas.2212930119.

Chubukov, V., A. Mukhopadhyay, C.J. Petzold, J.D. Keasling, and H.G. Martín. 2016. Synthetic and systems biology for microbial production of commodity chemicals. *NPJ Systems Biology and Applications* 2 (April): 16009, 1–11. https://doi.org/10.1038/npjsba.2016.9.

Das, P., B. Sarkar, and S. Mandal. 2023. Endophytic microbes: A potential source of bioactive metabolites with therapeutic values. In *Endophytic association: What, why and how*, eds. M. Shah and D. Deka, 435–57. Elsevier Inc., Academic Press. https://doi.org/10.1016/B978-0-323-91245-7.00004-3.

de Vries, R.P., and J. Visser. 2001. *Aspergillus* enzymes involved in degradation of plant cell wall polysaccharides. *Microbiology and Molecular Biology Reviews* 65: 497–522. https://doi.org/10.1128/mmbr.65.4.497-522.2001.

Demain, A.L. 1980. Microbial production of primary metabolites. *Naturwissenschaften* 67: 582–87. https://doi.org/10.1007/BF00396537.

Desire, M.H., F. Bernard, M.R. Forsah, C.T. Assang, and O.N. Denis. 2014. Enzymes and qualitative phytochemical screening of endophytic fungi isolated from *Lantana camara* Linn. leaves. *Journal of Applied Biology and Biotechnology* 2: 01–06. https://doi.org/10.7324/JABB.2014.2601.

Duncker, K.E., Z.A. Holmes, and L. You. 2021. Engineered microbial consortia: strategies and applications. *Microbial Cell Factories* 20 (November): 211, 1–13. https://doi.org/10.1186/s12934-021-01699-9.

Edwards, D.J., and W.H. Gerwick. 2004. Lyngbyatoxin biosynthesis: Sequence of biosynthetic gene cluster and identification of a novel aromatic prenyltransferase. *Journal of the American Chemical Society* 126: 11432–33. https://doi.org/10.1021/ja047876g.

Elbeshbishy, E. 2014. Metabolic pathways release of energy (Anaerobic). *Encyclopedia of food microbiology*, 2nd ed., eds. C.E. Batt and M.-L. Tortorello, 588–601. London: Elsevier. https://doi.org/10.1016/B978-0-12-384730-0.00198-1.

Falkowski, P.G., and J.A. Raven. 2013. *Aquatic photosynthesis*, 2nd ed. Princeton University Press.

Fox, E.M., and B.J. Howlett. 2008. Secondary metabolism: Regulation and role in fungal biology. *Current Opinion in Microbiology* 11: 481–87. https://doi.org/10.1016/j.mib.2008.10.007.

Gacek, A., and J. Strauss. 2012. The chromatin code of fungal secondary metabolite gene clusters. *Applied Microbiology and Biotechnology* 95: 1389–404. https://doi.org/10.1007/s00253-012-4208-8.

Goodfellow, M., and S.T. Williams. 1983. Ecology of actinomycetes. *Annual Review of Microbiology* 37: 189–216. https://doi.org/10.1146/annurev.mi.37.100183.001201.

Gupta, A., S. Jain, and N. Verma. 2022. *Industrial applications of soil microbes*, vol. 1. Bentham Science Publishers Pte Ltd. https://doi.org/10.2174/97898150399551220101.

Gupta, A., S. Jain, and N. Verma. 2024. *Industrial applications of soil microbes*, vol. 3. Bentham Science Publishers Pte Ltd. https://doi.org/10.2174/97898150797531240301.

Helliwell, K.E., G.L. Wheeler, K.C. Leptos, R.E. Goldstein, and A.G. Smith. 2011. Insights into the evolution of vitamin B12 auxotrophy from sequenced algal genomes. *Molecular Biology and Evolution* 28: 2921–33. https://doi.org/10.1093/molbev/msr124.

Hoffmeister, D., and N.P. Keller. 2007. Natural products of filamentous fungi: Enzymes, genes, and their regulation. *Natural Product Reports* 24: 393–416. https://doi.org/10.1039/B603084J.

Indira, M., T.C. Venkateswarulu, K.A. Peele, M.N. Bobby, and S. Krupanidhi. 2019. Bioactive molecules of probiotic bacteria and their mechanism of action: A review. *3 Biotech* 9 (July): 306. https://doi.org/10.1007/s13205-019-1841-2.

Iwai, Y., and Y. Takahashi. 1992. Selection of microbial sources of bioactive compounds. In *The search for bioactive compounds*

from microorganisms, ed. S. Ōmura, 281–302. New York, NY: Springer. https://doi.org/10.1007/978-1-4612-4412-7_15.
Jain, S., A. Gupta, and N. Verma. 2023. *Industrial applications of soil microbes*, vol. 2. Bentham Science Publishers Pte Ltd. https://doi.org/10.2174/97898150502641230201.
Jain, S., A. Gupta, and N. Verma. 2024. *Industrial applications of soil microbes*, vol. 4. Bentham Science Publishers Pte Ltd. https://doi.org/10.2174/97898151249961240401.
Jawed, K., S.S. Yazdani, and M.A. Koffas, 2019. Advances in the development and application of microbial consortia for metabolic engineering. *Metabolic Engineering Communications* 9 (December): e00095, 1–12. https://doi.org/10.1016/j.mec.2019.e00095.
Jenuwein, T., and C.D. Allis. 2001. Translating the histone code. *Science* 293: 1074–80. https://doi.org/10.1126/science.1063127.
Jullesson, D., F. David, B. Pfleger, and J. Nielsen. 2015. Impact of synthetic biology and metabolic engineering on industrial production of fine chemicals. *Biotechnology Advances* 33: 1395–402. https://doi.org/10.1016/j.biotechadv.2015.02.011.
Kim, B.H., and G.M. Gadd. 2008. Glycolysis. In *Bacterial physiology and metabolism*, ed. B.H. Kim, and G.M. Gadd, 60–84. Cambridge: Cambridge University Press. https://doi.org/10.1017/CBO9780511790461.005.
Kim, J.H., N. Lee, S. Hwang, et al. 2021. Discovery of novel secondary metabolites encoded in actinomycete genomes through coculture. *Journal of Industrial Microbiology and Biotechnology* 48 (January): kuaa001, 1–16. https://doi.org/10.1093/jimb/kuaa001.
Kumar, R.R., and S. Prasad. 2011. Metabolic engineering of bacteria. *Indian Journal of Microbiology* 51: 403–9. https://doi.org/10.1007/s12088-011-0172-8.
Lee, N., W. Kim, J. Chung, et al. 2020. Iron competition triggers antibiotic biosynthesis in *Streptomyces coelicolor* during coculture with *Myxococcus xanthus*. *The ISME Journal*, 14: 1111–24. https://doi.org/10.1038/s41396-020-0594-6.
Lee, S.Y., and H.U. Kim. 2015. Systems strategies for developing industrial microbial strains. *Nature Biotechnology* 33: 1061–72. https://doi.org/10.1038/nbt.3365.
Liu, L., Z. Hu, S. Li, et al. 2020. Comparative transcriptomic analysis uncovers genes responsible for the DHA enhancement in the mutant *Aurantiochytrium* sp. *Microorganisms* 8 (April): 529, 1–19. https://doi.org/10.3390/microorganisms8040529.
Lourenço-Lopes, C., M. Fraga-Corral, C. Jimenez-Lopez, et al. 2020. Metabolites from macroalgae and its applications in the cosmetic industry: A circular economy approach. *Resources* 9 (August): 101, 1–30. https://doi.org/10.3390/resources9090101.
Magarvey, N.A., Z.Q. Beck, T. Golakoti, et al. 2006. Biosynthetic characterization and chemoenzymatic assembly of the cryptophycins. potent anticancer agents from *Nostoc* cyanobionts. *ACS Chemical Biology* 1: 766–79. https://doi.org/10.1021/cb6004307.
Maier, R.M., and I.L. Pepper. 2015. Bacterial growth. In *Environmental microbiology*, 3rd ed., eds. I.L. Pepper, C.P. Gerba, and T.J. Gentry, 37–56. Elsevier Inc., Academic Press. https://doi.org/10.1016/B978-0-12-394626-3.00003-X.
Mandwal, A. 2022. Diverse survival functions of secondary metabolites in nature. In *Secondary metabolites: Trends and reviews*, eds. R. Vijayakumar and S.S.S. Raja, 13. https://doi.org/10.5772/intechopen.101977.
McArthur, G.H., and S.S. Fong. 2009. Toward engineering synthetic microbial metabolism. *Journal of Biomedicine and Biotechnology* 2010 (December): 459760, 1–10. https://doi.org/10.1155/2010/459760.
Moffitt, M.C., and B.A. Neilan. 2004. Characterization of the nodularin synthetase gene cluster and proposed theory of the evolution of cyanobacterial hepatotoxins. *Applied and Environmental Microbiology* 70: 6353–62. https://doi.org/10.1128/AEM.70.11.6353-6362.2004.
Mushtaq, S., B.H. Abbasi, B. Uzair, and R. Abbasi. 2018. Natural products as reservoirs of novel therapeutic agents. *EXCLI Journal* 17: 420–51. https://doi.org/10.17179/excli2018–1174.
Nataraj, B.H., S.A. Ali, P.V. Behare, and H. Yadav. 2020. Postbiotics-parabiotics: The new horizons in microbial biotherapy and functional foods. *Microbial Cell Factories* 19 (August): 168, 1–22. https://doi.org/10.1186/s12934-020-01426-w.
Ochi, K. 1987. Metabolic initiation of differentiation and secondary metabolism by *Streptomyces griseus*: significance of the stringent response (ppGpp) and GTP content in relation to A factor. *Journal of Bacteriology* 169: 3608–16. https://doi.org/10.1128/jb.169.8.3608-3616.1987.
Onaka, H. 2017. Novel antibiotic screening methods to awaken silent or cryptic secondary metabolic pathways in actinomycetes. *The Journal of Antibiotics* 70: 865–70. https://doi.org/10.1038/ja.2017.51.
Orejuela-Escobar, L., A. Gualle, V. Ochoa-Herrera, and G.P. Philippidis. 2021. Prospects of microalgae for biomaterial production and environmental applications at biorefineries. *Sustainability* 13 (March): 3063, 1–19. https://doi.org/10.3390/su13063063.
Pandey, K.D., A.K. Patel, M. Singh, Vandana, A. Kumari, and Jalaluddin. 2021. Secondary metabolites from bacteria and viruses. *Natural Bioactive Compounds Technological Advancements*, ed. R.P. Sinha and D.P Häder, 19–40. Elsevier Inc., Academic Press. https://doi.org/10.1016/C2019-0-00797-2.
Park, Y.H., S.I. Han, B. Oh, et al. 2022. Microalgal secondary metabolite productions as a component of biorefinery: A review. *Bioresource Technology* 344 (January): 126206. https://doi.org/10.1016/j.biortech.2021.126206.
Pham, J.V., M.A. Yilma, A. Feliz, et al. 2019. A review of the microbial production of bioactive natural products and biologics. *Frontiers in Microbiology* 10 (June): 1404, 1–27. https://doi.org/10.3389/fmicb.2019.01404.
Raven, J.A. 1984. *Energetics and transport in aquatic plants*, 587. University of California: A.R. Liss. https://books.google.co.in/books?id=vr8dAQAAIAAJ.
Ruiz, B., A. Chávez, A. Forero, et al. 2010. Production of microbial secondary metabolites: Regulation by the carbon source. *Critical Reviews in Microbiology* 36: 146–67. https://doi.org/10.3109/10408410903489576.
Seiboth, B., and B. Metz. 2011. Fungal arabinan and L-arabinose metabolism. *Applied Microbiology and Biotechnology* 89: 1665–73. https://doi.org/10.1007/s00253-010-3071-8.
Sekurova, O.N., O. Schneider, and S.B. Zotchev. 2019. Novel bioactive natural products from bacteria via bioprospecting, genome mining and metabolic engineering. *Microbial Biotechnology* 12: 828–44. https://doi.org/10.1111/1751-7915.13398.
Selim, M.S.M., S.A. Abdelhamid, and S.S. Mohamed. 2021. Secondary metabolites and biodiversity of actinomycetes. *Journal of Genetic Engineering and Biotechnology* 19 (December): 72, 1–13. https://doi.org/10.1186/s43141-021-00156-9.

Selvan, G.P., S. Ravikumar, A. Ramu, and P. Neelakandan. 2012. Antagonistic activity of marine sponge associated *Streptomyces* sp. against isolated fish pathogens. *Asian Pacific Journal of Tropical Disease* 2: S724–28. https://doi.org/10.1016/S2222-1808(12)60252-7.

Seyedsayamdost, M.R. 2019. Toward a global picture of bacterial secondary metabolism. *Journal of Industrial Microbiology and Biotechnology* 46: 301–11. https://doi.org/10.1007/s10295-019-02136-y.

Shimizu, K. 2013. *Bacterial cellular metabolic systems: Metabolic regulation of a cell system with 13C-metabolic flux analysis*. Elsevier, Woodhead Publishing.

Shukla, R. 2015. Studies on bioactive compounds from different microorganisms. *International Journal of Scientific & Engineering Research* 6: 1225–33.

Shwab, E.K., J.W. Bok, M. Tribus, J. Galehr, S. Graessle, and N.P. Keller. 2007. Histone deacetylase activity regulates chemical diversity in *Aspergillus*. *Eukaryotic Cell* 6: 1656–64. https://doi.org/10.1128/ec.00186-07.

Smanski, M.J., H. Zhou, J. Claesen, B. Shen, M.A. Fischbach, and C.A. Voigt. 2016. Synthetic biology to access and expand nature's chemical diversity. *Nature Reviews Microbiology* 14: 135–49. https://doi.org/10.1038/nrmicro.2015.24.

Solomon, L., V.P. Tomii, and A.A. Dick. 2019. Importance of fungi in the petroleum, agro-allied, agriculture and pharmaceutical industries. *New York Science Journal* 12: 8–15. https://doi.org/10.7537/marsnys120519.02.

Sorour, N.M., A.A. Tayel, R.N. Abbas, and O.M. Abonama. 2017. Microbial biosynthesis of health-promoting food ingredients. In *Food biosynthesis, handbook of food bioengineering*, ed. A.M. Grumezescu and A.M. Holban, 55–93. Elsevier Inc., Academic Press. https://doi.org/10.1016/B978-0-12-811372-1.00002-6.

Spector, M.P. 2009. Metabolism, central (intermediary). In *Encyclopedia of microbiology*, 3rd ed., ed. M. Schaechter, 242–64. Elsevier Inc., Academic Press. https://doi.org/10.1016/B978-012373944-5.00078-X.

Sreenikethanam, A., S. Raj, R.J. Banu, P. Gugulothu, and A.K. Bajhaiya. 2022. Genetic engineering of microalgae for secondary metabolite production: Recent developments, challenges, and future prospects. *Frontiers in Bioengineering and Biotechnology* 10 (March): 836056, 1–14. https://doi.org/10.3389/fbioe.2022.836056.

Stincone, A., A. Prigione, T. Cramer, et al. 2015. The return of metabolism: biochemistry and physiology of the pentose phosphate pathway. *Biological Reviews* 90: 927–63. https://doi.org/10.1111/brv.12140.

Taghinasab, M., and S. Jabaji. 2020. Cannabis microbiome and the role of endophytes in modulating the production of secondary metabolites: An overview. *Microorganisms* 8 (March): 355, 1–16. https://doi.org/10.3390/microorganisms8030355.

Takahashi, J.A., B.V.R. Barbosa, B.A. Martins, C.P. Guirlanda, and M.A.F. Moura. 2020. Use of the versatility of fungal metabolism to meet modern demands for healthy aging, functional foods, and sustainability. *Journal of Fungi* 6 (October): 223, 1–27. https://doi.org/10.3390/jof6040223.

Teske, A., A. Dhillon, and M.L. Sogin. 2003. Genomic markers of ancient anaerobic microbial pathways: sulfate reduction, methanogenesis, and methane oxidation. *The Biological Bulletin* 204: 186–91. https://doi.org/10.2307/1543556.

Trinh, C.T., Y. Liu, and D.J. Conner. 2015. Rational design of efficient modular cells. *Metabolic Engineering* 32: 220–31. https://doi.org/10.1016/j.ymben.2015.10.005.

Turpin, D.H., I.R. Elrifi, D.G. Birch, H.G. Weger, and J.J. Holmes. 1988. Interactions between photosynthesis, respiration, and nitrogen assimilation in microalgae. *Canadian Journal of Botany* 66: 2083–97. https://doi.org/10.1139/b88-286.

Vaishnav, P., and A.L. Demain. 2011. Unexpected applications of secondary metabolites. *Biotechnology Advances* 29: 223–29. https://doi.org/10.1016/j.biotechadv.2010.11.006.

Verardo, V., A.M. Gómez-Caravaca, and G. Tabanelli. 2020. Bioactive components in fermented foods and food by-products. *Foods* 9 (February): 153, 1–3. https://doi.org/10.3390/foods9020153.

Vimal, V., B.M. Rajan, and K. Kannabiran. 2009. Antimicrobial activity of marine actinomycete, *Nocardiopsis* sp. VITSVK 5 (FJ973467). *Asian Journal of Medical Sciences* 1: 57–63.

Vrancken, K., and J. Anné. 2009. Secretory production of recombinant proteins by *Streptomyces*. *Future Microbiology* 4: 181–88. https://doi.org/10.2217/17460913.4.2.181.

Wang, Y., J. Wu, M. Lv, et al. 2021. Metabolism characteristics of lactic acid bacteria and the expanding applications in food industry. *Frontiers in Bioengineering and Biotechnology* 9 (May): 612285, 1–19. https://doi.org/10.3389/fbioe.2021.612285.

Witteveen, C.F.B., R. Busink, P.V.D. Vondervoort, C. Dijkema, K. Swart, and J. Visser. 1989. L-arabinose and D-xylose catabolism in *Aspergillus niger*. *Microbiology* 135: 2163–71. https://doi.org/10.1099/00221287-135-8-2163.

Wittpoth, C., P.G. Kroth, K. Weyrauch, K.V. Kowallik, and H. Strotmann. 1998. Functional characterization of isolated plastids from two marine diatoms. *Planta* 206: 79–85. https://doi.org/10.1007/s004250050376.

Wolfe, A.J. 2015. Glycolysis for microbiome generation. *Microbiology Sspectrum* 3 (June): MBP-0014, 1–12. https://doi.org/10.1128/microbiolspec.mbp-0014-2014.

World Health Organization. 2014. *WHO's first global report on antibiotic resistance reveals serious, worldwide threat to public health*. https://www.who.int/southeastasia/news/detail/30-04-2014-who-s-first-global-report-on-antibiotic-resistance-reveals-serious-worldwide-threat-to-public-health.

Yadav, A.N., D. Kour, K.L. Rana, et al. 2019. Metabolic engineering to synthetic biology of secondary metabolites production. In *New and future developments in microbial biotechnology and bioengineering*, eds. V.K. Gupta and A. Pandey, 279–320. Elsevier. https://doi.org/10.1016/C2014-0-00306-1.

Yu, J.H., and N. Keller. 2005. Regulation of secondary metabolism in filamentous fungi. *Annual Review of Phytopathology* 43: 437–58. https://doi.org/10.1146/annurev.phyto.43.040204.140214.

6 Enzymes for Microbial Secondary Metabolite Production

*H. Pandey
(Department of Food Processing Technology, College of Food Processing Technology and Bio-Energy, Anand Agricultural University, Anand 388110, Guj., India; hridyeshpandey@gmail.com)

R. Prakasha
(Department of Food Processing Technology, College of Food Processing Technology and Bio-Energy, Anand Agricultural University, Anand 388110, Guj., India; prakashgowdar017@gmail.com)

6.1 INTRODUCTION

Microbial secondary metabolites are low-molecular-weight compounds produced mainly by bacteria, fungi, and actinomycetes during the late phases of the growth cycle (idiophase) (Kumar et al., 2021). Although these compounds are not directly involved in the normal growth, development, or reproduction of an organism, they play an important role in biological activities, especially in the fields of pharmaceuticals, cosmetics, food, and agricultural farming (Gupta et al., 2022, 2024). Microbial secondary metabolites include antibiotics, pigments, toxins, enzyme inhibitors, immune-modulating agents, antitumor agents, insecticides, plant growth regulators, and environmentally friendly herbicides and pesticides (Dixit and Kumari, 2023). They have a major impact on the health, nutrition, and economics of our society. These compounds are typically produced by liquid-submerged fermentation, but some of them can be advantageously produced by solid-state fermentation (Jain et al., 2023, 2024).

Enzymes play a crucial role in the biosynthesis of microbial secondary metabolites, as they catalyze the chemical reactions that transform precursor molecules into final products. In many cases, enzymes are also responsible for the modification and processing of the final products to produce compounds with specific bioactivities. There are several types of enzymes that can be used in this context, including oxidoreductases, transferases, and hydrolases (Figure 6.1). The selection of enzymes for microbial secondary metabolite production will depend on various factors, such as the specific compound being produced, the microbial strain being used, and the desired production conditions, such as temperature, pH, etc. However, by carefully selecting and optimizing the use of enzymes, it is possible to enhance the yield and purity of microbial secondary metabolites.

6.2 ENZYMES AND SECONDARY METABOLITE PRODUCTION

Secondary metabolites are produced from primary metabolite pools and central metabolic pathways using specific precursors for particular metabolites. The biosynthesis of secondary metabolites involves various pathways, including those for the production of aromatic compounds, isoprenes, oligosaccharides, peptides, polyketides, and ß-lactam rings (Ruiz et al., 2010). The steps involved in secondary metabolites include the synthesis of metabolite by the different reactions of multifunctional enzymes that are divided into two categories, namely, modular and iterative enzymes (Table 6.1), which are then modified with the help of modifying enzymes if desired. All these enzymes act as multifunctional enzymes, which may be modular or iterative, including synthases and modifying enzymes (Figure 6.1).

Synthases are responsible for assembling the basic framework of secondary metabolites using specific substrates. The most commonly encountered synthetic enzymes are PKSs, NRPSs, and TPSs (Wang et al., 2023). In contrast, modifying enzymes, including CYP450s, are essential for the abundance and activity of secondary metabolites. CYP450s participate in the post-modification process and play a crucial role in the regulation of these metabolites. With their multifunctional properties, CYP450s can efficiently utilize energy and substrates in both *in vivo* and *in vitro* settings, leading to a rapid and precise response when needed and facilitating the organism's vital activities.

DOI: 10.1201/9781003408833-6

TABLE 6.1
Comparison of Modular and Iterative Enzymes

	Modular Enzymes	Iterative Enzymes
Definition	Enzymes consist of multiple independent functional domains, each of which is responsible for a distinct step in the biosynthetic pathway.	Enzymes catalyze the same reaction repeatedly in a single active site, having only one functional module.
Functions	1. Responsible for the synthesis of complex natural products that require multiple enzymatic steps. 2. Each functional domain of a modular enzyme is responsible for a specific chemical reaction in the biosynthetic pathway, allowing for the precise control of product formation.	Responsible for the repeated addition of a specific chemical group to a growing molecule, leading to the formation of a homogenous product.
Domains	Consists of multiple domains, such as acyltransferases, ketosynthases, and thioesterases, each of which is responsible for a distinct step in the biosynthetic pathway.	Consists of a single domain that catalyzes the same reaction repeatedly.
Advantages	1. Easily modified and evolved by rearranging the domains to create new enzymatic activities. 2. Ability to engineer the biosynthetic pathway to produce novel compounds.	1. Higher degree of specificity for their substrate. 2. Simplicity and efficiency in catalyzing the same reaction repeatedly lead to high yields of a homogenous product.
Disadvantages	Complexity and the difficulty of engineering the biosynthetic pathway to produce novel compounds.	Limited versatility and difficulty of controlling the product formation beyond the repetitive addition of a specific chemical group.
Examples	Type-I modular PKSs are associated with the production of highly reduced metabolites, such as macrolide antibiotics.	Bacterial type-II iterative PKSs are involved in the biosynthesis of aromatic polyketides.

Source: Ōmura et al. (2001); Wang et al. (2023).

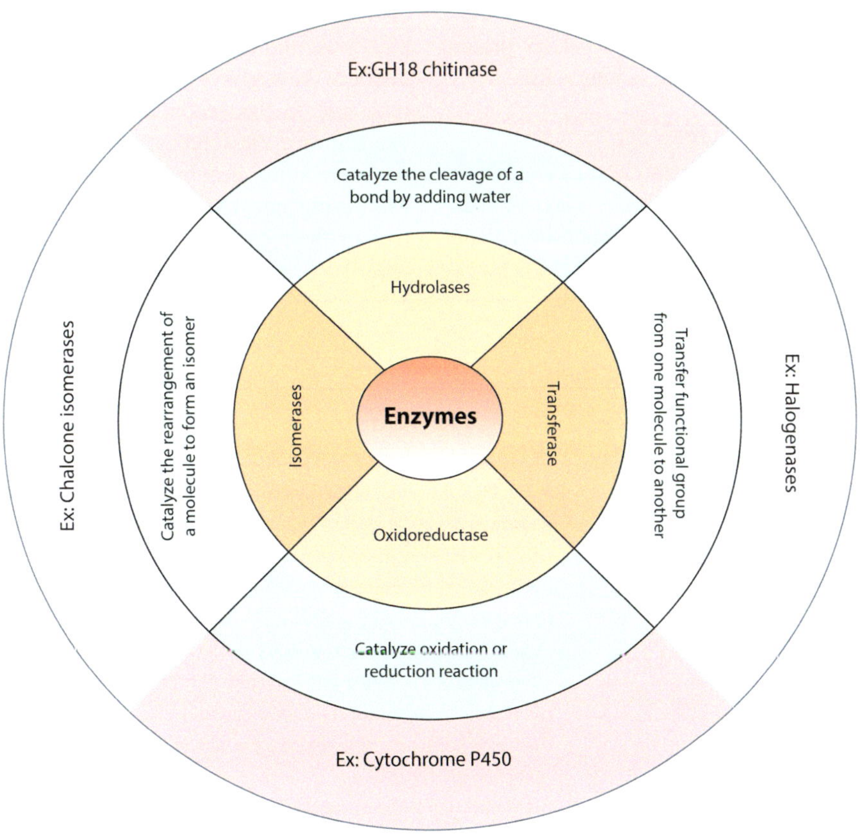

FIGURE 6.1 Classification of enzymes and reactions catalyzed by them.

6.3 MAJOR CLASS OF ENZYMES FOR METABOLITE PRODUCTION

6.3.1 Synthases and Post-Modifying Enzymes

6.3.1.1 Polyketide Synthases (PKS)

PKS enzymes are large, multifunctional enzymes that contain several distinct domains, each with a specific role in the biosynthetic pathway. These domains include ketosynthase, acyltransferase, acyl carrier protein, and ketoreductase, among others. PKS enzymes are responsible for the assembly of polyketide chains, which are the backbone of many secondary metabolites. Polyketides are synthesized by the condensation of small carboxylic acid molecules, such as malonyl-CoA, into long carbon chains, which are then modified by a series of enzymatic reactions to produce the final product as shown in Table 6.2.

6.3.1.2 Non-Ribosomal Peptide Synthetase (NRPS)

NRPS enzymes are another group of important enzymes involved in the biosynthesis of microbial secondary metabolites. NRPS enzymes are responsible for the assembly of non-ribosomal peptides. Non-ribosomal peptides are synthesized through a modular enzymatic process that involves the stepwise addition of amino acids to a growing peptide chain. Unlike ribosomal peptides, which are synthesized by ribosomes in a linear fashion, non-ribosomal peptides are synthesized by large, multifunctional NRPS enzymes that contain several distinct domains, each with a specific role in the biosynthetic pathway. The domains present in NRPS enzymes include adenylation (A), peptidyl carrier protein (PCP), and condensation (C) domains, among others.

6.3.1.3 Terpene Synthases (TPSs)

TPSs are a family of enzymes that catalyze the biosynthesis of terpenes, which are a large and diverse class of natural compounds found in many plants and microorganisms. TPSs typically use isoprenyl diphosphate substrates to generate a wide range of structurally diverse terpenoid products that are responsible for catalyzing the cyclization of linear terpene precursors to form various terpenoids. One example of a microbial secondary metabolite that is produced using terpene synthases is the antifungal compound pleuromutilin. It is a cyclic diterpenoid that is produced by the fungus *Clitopilus passeckerianus* and exhibits potent antibacterial and antifungal activity (Xu et al., 2018).

6.3.2 Modifying and Post-Modifying Enzymes

Post-modifying enzymes are a group of enzymes that modify the intermediate metabolites produced by primary biosynthetic enzymes in the biosynthesis of microbial secondary metabolites. These enzymes are responsible for introducing further modifications into the initial metabolite, leading to the formation of a structurally diverse final natural product. Post-modifying enzymes catalyze a wide range of chemical reactions, including methylation, acylation, glycosylation, phosphorylation, and oxidation, among others. Understanding the role and mechanism of post-modifying enzymes is important for the rational design and engineering of microbial secondary metabolites with improved biological activities.

For example, in the biosynthesis of the antibiotic penicillin in the fungus *Penicillium chrysogenum* (Martín et al., 2023), modifying enzymes are responsible for the acylation of the penicillin core structure, leading to the formation of different penicillin derivatives with different biological activities, and in the case of antibiotic erythromycin production, these glycosyltransferase enzymes transfer a sugar molecule to the secondary metabolite; this modification can improve the solubility, stability, and bioavailability of the metabolite, improving its pharmacokinetic properties. Modifying enzymes are often highly specific and selective in their catalytic activity, and their activity is often

TABLE 6.2
Synthases with Their Domains and Functions

Enzyme	Domains	Function
Polyketide synthases	Ketosynthase	Condensation of the building blocks into the polyketide chain
	Acyl transferase	Selection and loading of the appropriate building block onto the acyl carrier protein domain
	Ketoreductase	Reduction of ketone groups in the polyketide chain
	Acyl carrier protein	Carriage of the polyketide chain between the different domains of the PKS enzyme
Non-ribosomal peptide synthetase	Adenylation	Recognition and activation of specific amino acids
	Peptidyl carrier protein	Carriage of the growing peptide chain between the different domains of the NRPS enzyme
	Condensation	Formation of peptide bonds between the activated amino acids, which results in the formation of a non-ribosomal peptide
Terpene synthases	α domain	Binding and recognition of the precursor molecule
	β domain	Control of the stereochemistry of the product molecule and also responsible for catalysis and product release
	αβ domain	Biosynthesis of C15 terpenoids
	βγ domain	Biosynthesis of C10 and C15 terpenoids

tightly regulated in order to ensure the correct biosynthesis of the desired secondary metabolite. In addition, modifying enzymes often work in concert with other biosynthetic enzymes, such as polyketide synthases and non-ribosomal peptide synthetases, to produce structurally complex and diverse secondary metabolites.

6.3.3 Cytochrome P450s (CYP450s)

CYP450s enzymes are involved in the biosynthesis of many secondary metabolites produced by microorganisms, including antibiotics, antifungal agents, and anticancer drugs. These enzymes catalyze a wide range of reactions, including hydroxylation, epoxidation, dealkylation, oxidation, and other chemical modifications of intermediate molecules, leading to the formation of complex and diverse secondary metabolites. For example, CYP107D1 from *Streptomyces griseus* can catalyze the oxidative C-C bond cleavage of the natural product pleiotropic regulator AprX in the biosynthesis of the antibiotic apramycin (Rudolf et al., 2017). Similarly, the biosynthesis of the antibiotic erythromycin in the soil bacterium *Saccharopolyspora erythraea* involves a series of CYP450-catalyzed reactions (Cho et al., 2019). These reactions result in the formation of the macrolide ring structure and the addition of various functional groups to the molecule, leading to the production of the final compound.

6.4 ENZYME MODIFICATION TO INCREASE YIELD

6.4.1 Immobilization

Enzyme immobilization involves fixing the enzyme onto a solid support, which improves the enzyme's stability and activity. Immobilized enzymes can be used repeatedly, reducing the cost of enzyme use. Immobilization of enzymes can also improve the selectivity of the enzyme, which can lead to higher yields of the desired product.

6.4.2 Genetic Modification

The genetic modification of enzymes involves changing the enzyme's DNA sequence to alter its activity or specificity. This modification can increase the yield of the desired product by improving the enzyme's activity or by changing the enzyme's specificity to produce the desired metabolite.

6.4.3 Directed Evolution

Directed evolution involves generating a large number of enzyme variants with random mutations and selecting the variants with improved activity. This method can be used to identify enzymes with improved activity, specificity, or stability, which can lead to higher yields of the desired metabolite.

6.4.4 Rational Design

Rational design involves using computer-aided modelling and simulations to predict the effect of specific mutations on enzyme activity or specificity. This method can be used to design enzymes with improved activity or specificity, which can lead to higher yields of the desired metabolite.

6.4.5 Cofactor Engineering

Cofactors are molecules that are required for enzyme activity. Cofactor engineering involves modifying the enzyme or the production host to increase the availability of cofactors, which can improve enzyme activity and increase the yield of the desired metabolite.

6.4.6 Metabolic Engineering

Metabolic engineering involves modifying the metabolic pathways of the production host to increase the availability of precursors required for the production of the desired metabolite. This modification can increase the yield of the desired metabolite by providing the necessary building blocks for the production of the metabolite.

6.5 METHODOLOGY FOR THE PRODUCTION OF SECONDARY METABOLITES THROUGH FERMENTATION

6.5.1 Strain Selection

The first step in producing microbial secondary metabolites is selecting the appropriate microbial strain. The strain should have a high yield of the desired secondary metabolite and be stable during the fermentation process. Strain selection is typically done through screening of various strains from natural sources, such as soil or plant samples, or the genetic engineering of existing strains.

6.5.2 Medium Preparation

Once the strain is selected, the next step is to prepare a suitable growth medium. The medium should provide all the nutrients necessary for the growth of the microorganism and the production of the secondary metabolite. The medium should have the proper pH, temperature, and aeration conditions to optimize the growth of the microorganism. The medium can be a complex medium, which contains a variety of nutrients, or a defined medium, which has a specific set of nutrients. The medium can also be optimized through various techniques, such as statistical experimental designs or computational modelling.

6.5.3 Inoculation

The selected strain is then inoculated into the prepared medium. The inoculum is typically prepared by growing the strain in a smaller-scale culture and then transferring it to the larger production scale. The inoculation can be done using batch, fed-batch, or continuous culture methods, depending on the requirements of the strain and the fermentation process.

6.5.4 Enzyme Addition

Enzymes are added to the fermentation process to catalyze specific reactions that are necessary for the production of the secondary metabolite. Enzymes can be added at different stages of the fermentation process, depending on their function. For example, enzymes can be added to catalyze the biosynthesis of precursor molecules, modify the metabolic pathways of the production host, or improve the efficiency of downstream processing. Enzyme addition can be done through various methods, such as adding the enzyme directly to the fermentation broth or using genetically engineered strains that produce the enzyme *in situ*.

6.5.5 Harvesting and Purification

After the fermentation process is complete, the culture broth is harvested, and the secondary metabolite is purified. The purification process involves various techniques, including filtration, centrifugation, chromatography, and crystallization. The goal of purification is to isolate the secondary metabolite from the fermentation broth and obtain it in a pure form for downstream processing.

6.6 APPLICATION OF ENZYMES IN MICROBIAL SECONDARY METABOLITE PRODUCTION

6.6.1 Antibiotic Compounds

Antibiotic compounds such as penicillin and cephalosporins are produced through fermentation processes using enzymes. Penicillin production through fermentation involves the utilization of penicillin acylase (PAC) enzymes, which play a crucial role in various aspects of the process. The most widespread use of PACs is in the production of 6-aminopenicillanic acid (6-APA) from both penicillin G and penicillin V. Immobilized PAC enzymes are derived from *Escherichia coli*, *Bacillus megaterium*, and *Enterococcus faecalis*. Reactions are conducted at a scale exceeding 5,000 L under controlled conditions, with the pH being controlled around 8.0 or slowly ramped up to as high as 8.5, depending on the specific catalyst employed. The process aims to minimize exposure to high temperatures (>30°C) and alkaline pH (>8) to prevent enzyme inactivation and maintain the high product yield of the relatively unstable 6-APA compound. PACs are also extensively employed in the large-scale production of semisynthetic penicillins and cephalosporins, involving the condensation of a suitable D-amino acid derivative with the β-lactam nucleus in a PAC-catalyzed reaction. These reactions occur at low pH (<7.0) and employ nucleophiles such as 6-APA combined with free acids. Kinetically controlled synthesis using PAC involves acyl group transfer reactions, where activated acids, esters, or amides act as the acylating agents. The optimization of pH, the addition of suitable solvents, and the use of high concentrations of acyl donor and nucleus are employed to enhance reaction efficiency. Similarly, cephalosporin synthesis involves enzymes like deacetoxycephalosporin C synthase (DAOCS) and acylase. Several different antibiotics, including amoxicillin, ampicillin, cephalexin, and cefazolin, are produced using these controlled strategies. PACs also find application in peptide synthesis. This wide range of applications highlights the versatility and significance of PAC enzymes in the production of penicillin and related compounds through fermentation.

6.6.2 Antitumor Compounds

Enzymes play a crucial role in the production of antitumor compounds through fermentation. Harnessing the power of these biological catalysts can enhance the synthesis of valuable antitumor compounds that are challenging to extract from natural sources or expensive to chemically produce. Through genetic engineering and metabolic engineering approaches, enzymes can be optimized and introduced into microbial or yeast hosts, allowing for the scalable production of antitumor compounds through fermentation. Paclitaxel, a valuable antitumor compound, is produced using metabolic engineering in microbes. The primary objective is to generate substantial quantities of paclitaxel, which is challenging to extract from natural sources and economically prohibitive to synthesize chemically (Shankar, 2019). To accomplish this, foreign genes encode biosynthetic enzymes introduced from the early stages of the taxoid biosynthetic pathway and the isoprenoid pathway into *Saccharomyces cerevisiae*. One key enzyme in this process is taxadiene synthase, derived from *Taxus chinensis*, responsible for converting geranylgeranyl diphosphate into taxadiene, a vital precursor in paclitaxel synthesis. However, expressing taxadiene synthase alone did not yield sufficient taxadiene due to the limited availability of geranylgeranyl diphosphate. To overcome this hurdle, researchers co-expressed geranylgeranyl diphosphate synthase and a truncated version of 3-hydroxyl-3-methylglutaryl-CoA reductase (HMG-CoA reductase) to enhance precursor levels and circumvent feedback inhibition. Furthermore, they introduced a mutant regulatory protein, UPC2–1, to enhance steroid uptake. These modifications resulted in a remarkable 50% increase in taxadiene production. Further enhancements were achieved by replacing the geranylgeranyl diphosphate synthase with a non-competing counterpart from *Sulfolobus acidocaldarius* and optimizing the codons of taxadiene synthase for efficient expression. As a result, taxadiene production escalated by an impressive 40-fold to reach 8.77 ± 0.85 mg/L, accompanied by significant quantities of geranylgeraniol (Engels et al., 2008).

Similarly, other antitumor compounds, such as eribulin, bleomycin, vinblastine, and camptothecin, have also undergone genetic modification to enhance their production yields. For instance, eribulin, derived from the marine sponge *Halichondria okadai*, has been genetically engineered to enhance its production yield (Shetty and Gupta, 2014). This modification involves the utilization of enzymes such as halogenases, methyltransferases, and acyltransferases, which play crucial roles in the biosynthesis of eribulin (Li and Lou, 2018). Similarly, bleomycin, a potent antitumor antibiotic produced by *Streptomyces* bacteria, has been subjected to genetic manipulation to improve its yield. The biosynthesis of bleomycin relies on enzymes such as non-ribosomal peptide synthetases and oxygenases (Edwards, 2001).

Furthermore, vinblastine and vincristine, essential antitumor alkaloids extracted from the Madagascar periwinkle plant (*Catharanthus roseus*) (Iskandar and Iriawati, 2016), have had their biosynthetic pathway genes cloned and inserted into microorganisms like yeast and bacteria. These engineered microorganisms, through fermentation, enable the production of vinblastine and vincristine on a larger scale. The modification of vinblastine and vincristine biosynthesis involves the involvement of enzymes like methyltransferases, prenyltransferases, and acetyltransferases (El-Sayed and Verpoorte, 2007; Zhu et al., 2014). Another notable antitumor compound, camptothecin, naturally occurring in the Chinese tree *Camptotheca acuminata*, has been successfully biosynthesized through genetic engineering in various organisms, including *Escherichia coli* and *Streptomyces*. The biosynthesis of camptothecin relies on the activity of enzymes such as acetyltransferases, cyclases, and glycosyltransferases (Chen et al., 2016).

Tailored triacylglycerol (lipase enzymes) is responsible for the hydrolysis of triacylglycerol into free fatty acids and glycerols. Depending on the regiospecificity of enzymes, monoacylglycerols and diacylglycerols are produced. Using hydrolysis, esterification, and interesterification of acylglycerols, a change in the melting and physicochemical properties of fat is obtained. This type of fat found application in the industry as a milk fat substitute, as low-calorie triacylglycerols, as cocoa and bakery fat substitute, and to enrich specific fatty acids in milk.

6.6.3 Antioxidant and Antimicrobial Compounds

Most of the natural antioxidants are polyhydroxy compounds and are hydrophilic in nature, which limits solubility in the lipophilic phase. Enzyme modification of antioxidants like flavonoids and phenolic compounds improves lipophilicity. Lipophilic antioxidants are synthesized by reaction with lipophilic alcohols in the case of phenolic acids and fatty acids in flavonoids. Further lyophilization enhances bioavailability and some biological properties. Lipolysis is done through esterification by lipase enzymes. Zieniuk et al. (2021) studied the biosynthesis of different esters of phenolic compounds and their antioxidant and antimicrobial properties. They found the antibacterial effect of the ester of 3-(4-hydroxyphenyl) propanoic acid against *L. monocytogenes*. One more interesting notice was that increasing the alkyl chain length of the ester decreased the minimum inhibitory concentration (MIC) and minimum bacterial concentration (MBC). Another compound, p-methoxy cinnamic acid, is a well-known antidiabetic, anticancer, and hepato- and neuroprotective compound. The compound is particularly present in coffee, turmeric, sugarcane, peanut, and rice bran and is also a bioactive ingredient of *Scrophularia buergeriana* and *Kaempferia galanga* herbs. The bioavailability of the compound is very low due to rapid metabolism. hence, Rychlicka and Gliszczyńska (2020) modified the compound using an enzymatic interesterification process. The egg yolk phosphatidylcholine was modified with p-methoxycinnamic acid by an enzyme lipase B from *Candida antarctica* and Novozym 435 as a biocatalyst. This method of phenolipid production yields 32% p-methoxycinnamoylated lysophosphatidylcholine and 3% p-methoxycinnamoylated phosphatidylcholine.

Kojic acid, also known as 5-hydroxy-2-(hydroxymethyl) pyran-4-one, is derived as a secondary metabolite from certain fungi belonging to the *Aspergillus* genus. It is widely utilized in the cosmetics industry as a skin-lightening agent. This compound exhibits remarkable antioxidant properties, aiding in the protection of the skin against damage caused by harmful free radicals. Kojic acid has been proven to effectively diminish hyperpigmentation and dark spots, resulting in a more uniform and radiant complexion. Its versatility is evident in its incorporation into various skincare products, including creams, lotions, serums, and masks. Lajis et al. (2017) synthesized a 7-O-kojic acid monopalmitate (7-O-KAP) from kojic acid and palmitic acid using immobilized lipase Novozyme (N435). 7-O-KAP has better depigmentation, antioxidant, and reduced peroxide activity compared to kojic acid. The author reported that 7-O-KAP has high chemical stability; furthermore, the *Danio rerio* embryo assay reveals that the compound exhibited greater hydrophobicity compared to kojic acid. Additionally, it displayed enhanced cell membrane permeability, leading to decreased tyrosinase activity and inhibition of melanin formation.

6.7 ADVANTAGES

6.7.1 High Specificity

Enzymes are highly specific in their action, which means they can catalyze specific reactions without interfering with other metabolic pathways. This specificity ensures that the desired product is produced with high yield and purity.

6.7.2 Enhanced Yield

Enzymes can increase the yield of the desired product by accelerating the rate-limiting step in the biosynthesis pathway. This increased yield can reduce the cost of production and increase the profitability of the process.

6.7.3 Reduced Environmental Impact

Enzymatic reactions require milder conditions than chemical reactions, which reduces the environmental impact of the process. Additionally, enzymes can be reused, reducing the waste generated during the process.

6.7.4 Reduced Production Time

Enzymatic reactions are often faster than chemical reactions, which reduces the production time and increases the efficiency of the process.

6.8 DISADVANTAGES

6.8.1 Cost

Enzymes can be expensive to produce, which increases the cost of production. Additionally, enzymes may require specific conditions, such as specific pH or temperature ranges, which can add to the cost of the process.

6.8.2 Sensitivity

Enzymes can be sensitive to environmental factors, such as pH, temperature, and substrate concentration. Any deviation from the optimal conditions can reduce the activity of the enzyme, resulting in reduced yield and purity of the desired product.

6.8.3 Limited Substrate Range

Enzymes are often specific to a particular substrate, which limits their use in metabolite production. This specificity may require the use of multiple enzymes or pathways to produce the desired metabolite, which can increase the complexity and cost of the process.

6.8.4 Potential Contamination

Enzymes can be contaminated with other microorganisms or chemicals, which can affect the purity and yield of the desired product. Additionally, enzymes can be deactivated by contaminants, reducing the activity of the enzyme and reducing the yield of the desired product.

6.9 CURRENT CHALLENGES AND LIMITATIONS IN UTILIZING ENZYMES FOR SECONDARY METABOLITE PRODUCTION

6.9.1 Limited Knowledge of Enzyme Functions

Understanding the precise functions and mechanisms of enzymes involved in secondary metabolite production remains a challenge. Many enzymes in these pathways have not been fully characterized, making it difficult to manipulate their activities or optimize production.

6.9.2 Low Enzyme Yields

Some enzymes may have low expression levels or unstable structures, leading to low yields in enzyme production. This limitation hinders their practical application in industrial settings where high enzyme quantities are required.

6.9.3 Substrate Specificity and Promiscuity

Enzymes involved in secondary metabolite biosynthesis often exhibit narrow substrate specificity, limiting their ability to catalyze reactions with different substrates. Additionally, some enzymes may have promiscuous activities, catalyzing side reactions that reduce the desired secondary metabolite yield.

6.9.4 Metabolic Regulation and Feedback Inhibition

Enzyme activities in secondary metabolite biosynthesis pathways are often tightly regulated, leading to challenges in optimizing their expression and activity. Feedback inhibition by end products or intermediate metabolites can further complicate enzyme manipulation.

6.10 EMERGING TECHNOLOGIES AND APPROACHES TO OVERCOME THESE CHALLENGES

6.10.1 Metagenomics and Functional Genomics

Advancements in metagenomics and functional genomics enable the exploration of diverse microbial communities for novel enzymes involved in secondary metabolite production. These approaches can help identify new enzymes with unique functions and expand the pool of enzymes available for biotechnological applications.

6.10.2 Synthetic Biology and Enzyme Engineering

Synthetic biology techniques allow the engineering of enzymes to improve their catalytic efficiency, substrate specificity, and stability. Directed evolution, rational design, and protein engineering strategies can be employed to optimize enzymes for enhanced secondary metabolite production.

6.10.3 Systems Biology and Computational Modelling

Systems biology approaches combined with computational modelling can provide insights into the complex interactions and regulation of enzyme networks involved in secondary metabolite biosynthesis. This knowledge can aid in the identification of key enzyme targets for manipulation and the optimization of metabolic pathways.

6.10.4 Bioprocess Optimization and Fermentation Strategies

Advances in bioprocess engineering, such as optimization of fermentation conditions, bioreactor design, and culture media formulation, can improve enzyme production yields. Continuous fermentation and integrated bioprocessing approaches can also enhance the overall efficiency of secondary metabolite production.

6.11 FUTURE DIRECTIONS AND POTENTIAL ADVANCEMENTS IN THE FIELD

6.11.1 Expanding Enzyme Toolbox

Continued exploration of microbial diversity, especially in extreme environments, can uncover novel enzymes with unique functions and broader substrate specificity. This expansion of the enzyme toolbox will enable the production of a wider range of secondary metabolites.

6.11.2 CRISPR-Based Genome Editing

CRISPR-Cas9 and related technologies have revolutionized genome editing and hold promise for precise manipulation of enzyme-encoding genes in various organisms. This approach can facilitate the development of high-yielding strains and the engineering of metabolic pathways for enhanced secondary metabolite production.

6.11.3 Multiomics Integration

Integrating data from genomics, transcriptomics, proteomics, and metabolomics can provide a holistic understanding of enzyme networks and their regulation in secondary metabolite biosynthesis. This integration will enable more accurate predictions and targeted manipulation of enzymes for improved production.

6.11.4 Biocatalysis and Enzymatic Synthesis

Enzymes have significant potential for the production of secondary metabolites through biocatalysis and enzymatic synthesis. Advancements in enzyme immobilization, enzyme recycling, and reaction engineering can enable the efficient and sustainable synthesis of complex secondary metabolites.

6.11.5 Bioreactor Design and Process Intensification

Innovations in bioreactor design, process monitoring, and control systems can enhance secondary metabolite production efficiency. The integration of online sensing technologies, advanced monitoring techniques, and intelligent process control can optimize enzyme activities and substrate utilization in real time.

By addressing these challenges and exploring emerging technologies, the field of enzyme utilization for secondary metabolite production holds great promise. Continued research and development efforts can lead to the discovery of novel enzymes, improved enzyme engineering strategies, and more efficient bioprocesses, paving the way for the sustainable production of valuable secondary metabolites with diverse applications in pharmaceuticals, agriculture, and other industries.

6.12 CONCLUSION

Enzymes are important in biosynthesis, modification, and large-scale production of microbial secondary metabolites, which are vital to pharmaceutical, agricultural, food, and cosmetic industries. The unique specificity of enzymes enables the formation of structurally diverse and biologically potent compounds, including antibiotics, antitumor agents, antioxidants, and antimicrobial molecules. The enzymes offer high selectivity, efficiency, and eco-friendly processing. While challenges persist in terms of production costs, stability under industrial conditions, and limited substrate versatility.

Recent advancements in enzyme immobilization, genetic engineering, rational design, and metabolic engineering have expanded opportunities to enhance yield and tailor metabolite structures for improved bioactivity. Integration of emerging tools such as CRISPR-based genome editing, synthetic biology, metagenomics, and multi-omics promises to accelerate the discovery of novel enzymes and optimize their performance for industrial applications. Moreover, innovations in bioprocess engineering and bioreactor design are likely to address existing challenges of scale-up and process efficiency.

The future of microbial secondary metabolite production lies in combining biological innovation with engineering solutions to create sustainable, high-yielding, and economically viable processes. By harnessing the catalytic versatility of enzymes and integrating modern biotechnological strategies, researchers can unlock new possibilities in drug discovery, crop protection, and health-related products.

REFERENCES

Chen, F., W. Li, L. Jiang, et al. 2016. Functional characterization of a geraniol synthase-encoding gene from *Camptotheca acuminata* and its application in production of geraniol in *Escherichia coli*. *Journal of Industrial Microbiology and Biotechnology* 43: 1281–92. https://doi.org/10.1007/s10295-016-1802-2.

Cho, M.A., S. Han, Y.R. Lim, V. Kim, H. Kim, and D. Kim. 2019. *Streptomyces* cytochrome P450 enzymes and their roles in the biosynthesis of macrolide therapeutic agents. *Biomolecules & Therapeutics* 27: 127–33. https://doi.org/10.4062/biomolther.2018.183.

Dixit, R., and M. Kumari. 2023. Microbial metabolites in plant disease management. In *Microbial biomolecules: Emerging approach in agriculture, pharmaceuticals and environment management*, eds. A. Kumar, M. Bilal, L.F.R. Ferreira, M. Kumari, 159–79. Elsevier Inc and Academic Press. https://doi.org/10.1016/C2021-0-02184-2.

Edwards, D.J. 2001. *Bleomycin biosynthesis, insights into a hybrid nonribosomal peptide synthetase/polyketide synthase system*. Davis: University of California. https://books.google.co.in/books?id=N_DhjK1KcYQC.

El-Sayed, M., and R. Verpoorte. 2007. *Catharanthus* terpenoid indole alkaloids: Biosynthesis and regulation. *Phytochemistry Reviews* 6: 277–305. https://doi.org/10.1007/s11101-006-9047-8.

Engels, B., P. Dahm, and S. Jennewein. 2008. Metabolic engineering of taxadiene biosynthesis in yeast as a first step towards taxol (paclitaxel) production. *Metabolic Engineering* 10: 201–06. https://doi.org/10.1016/j.ymben.2008.03.001.

Gupta, A., S. Jain, and N. Verma. 2022. *Industrial applications of soil microbes*, vol. 1. Bentham Science Publishers Pte Ltd. https://doi.org/10.2174/97898150399551220101.

Gupta, A., S. Jain, and N. Verma. 2024. *Industrial applications of soil microbes*, vol. 3. Bentham Science Publishers Pte Ltd. https://doi.org/10.2174/97898150797531240301.

Iskandar, N.N., and I. Iriawati. 2016. Vinblastine and vincristine production on Madagascar Periwinkle (*Catharanthus roseus* (L.) G. Don) callus culture treated with polethylene glycol. *Makara Journal of Science* 20: 7–16. https://scholarhub.ui.ac.id/science/vol20/iss1/2/.

Jain, S., A. Gupta, and N. Verma. 2023. *Industrial applications of soil microbes*, vol. 2. Bentham Science Publishers Pte Ltd. https://doi.org/10.2174/97898150502641230201.

Jain, S., A. Gupta, and N. Verma. 2024. *Industrial applications of soil microbes*, vol. 4. Bentham Science Publishers Pte Ltd. https://doi.org/10.2174/97898151249961240401.

Kumar, V., V. Ahluwalia, S. Saran, J. Kumar, A.K. Patel, and R.R. Singhania. 2021. Recent developments on solid-state fermentation for production of microbial secondary metabolites: Challenges and solutions. *Bioresource Technology* 323 (March): 124566. https://doi.org/10.1016/j.biortech.2020.124566.

Lajis, A.F.B., M. Hamid, S. Ahmad, and A.B. Arif. 2017. Lipase-catalyzed synthesis of kojic acid derivative in bioreactors and the analysis of its depigmenting and antioxidant activities. *Cosmetics* 4 (July): 22, 1–18. https://doi.org/10.3390/cosmetics4030022.

Li, G., and H.X. Lou. 2018. Strategies to diversify natural products for drug discovery. *Medicinal Research Reviews* 38: 1255–94. https://doi.org/10.1002/med.21474.

Martín, J.F., C. García-Estrada, and P. Liras. 2023. Insights into the molecular mechanisms of β-lactam antibiotic synthesizing and modifying enzymes in fungi. In *Biotechnology of microbial enzymes*, ed. G. Brahmachari, 199–228. Elsevier Inc and Academic Press. https://doi.org/10.1016/B978-0-443-19059-9.00015-3.

Ōmura, S., H. Ikeda, J. Ishikawa, et al. 2001. Genome sequence of an industrial microorganism *Streptomyces avermitilis*: Deducing the ability of producing secondary metabolites. *Proceedings of the National Academy of Sciences* 98: 12215–20. https://doi.org/10.1073/pnas.211433198.

Rudolf, J.D., C.Y. Chang, M. Ma, and B. Shen. 2017. Cytochromes P450 for natural product biosynthesis in *Streptomyces*: Sequence, structure, and function. *Natural Product Reports* 34: 1141–72. https://doi.org/10.1039/C7NP00034K.

Ruiz, B., A. Chávez, A. Forero, et al. 2010. Production of microbial secondary metabolites: Regulation by the carbon source. *Critical Reviews in Microbiology* 36: 146–67. https://doi.org/10.3109/10408410903489576.

Rychlicka, M., and A. Gliszczyńska. 2020. Interesterification of egg-yolk phosphatidylcholine with p-methoxycinnamic acid catalyzed by immobilized lipase B from *Candida antarctica*. *Catalysts* 10 (October): 1181, 1–17. https://doi.org/10.3390/catal10101181.

Shankar, N.B. 2019. Developments in taxol production through endophytic fungal biotechnology: A review. *Oriental Pharmacy and Experimental Medicine* 19: 1–13. https://doi.org/10.1007/s13596-018-0352-8.

Shetty, N., and S. Gupta. 2014. Eribulin drug review. *South Asian Journal of Cancer* 3: 57–59. https://www.thieme-connect.de/products/ejournals/abstract/10.4103/2278-330X.126527.

Wang, J.T., T.T. Shi, L. Ding, J. Xie, and P.J. Zhao. 2023. Multifunctional enzymes in microbial secondary metabolic processes. *Catalysts* 13 (March): 581, 1–21. https://doi.org/10.3390/catal13030581.

Xu, M., M. Jia, Y.J. Hong, et al. 2018. Premutilin synthase: Ring rearrangement by a class II diterpene cyclase. *Organic Letters* 20: 1200–02. https://doi.org/10.1021/acs.orglett.8b00121.

Zhu, X., X. Zeng, C. Sun, and S. Chen. 2014. Biosynthetic pathway of terpenoid indole alkaloids in *Catharanthus roseus*. *Frontiers of Medicine* 8: 285–93. https://doi.org/10.1007/s11684-014-0350-2.

Zieniuk, C.B., E. Białecka-Florjańczyk, and A. Fabiszewska. 2021. Anti-*Listerial* effect of 4-hydroxyphenylpropanoic acid esters synthesized by lipase-catalyzed esterification. *Proceedings* 70 (November): 17, 1–5. https://doi.org/10.3390/foods_2020-07700.

7 Sustainable Agriculture and Microbial Secondary Metabolites

Shalini Fagna
(Department of Botany, University of Rajasthan, Jaipur 302004, Rajasthan, India; shalini.fagna@gmail.com)

Apexa Pareek
(Department of Botany, University of Rajasthan, Jaipur 302004, Rajasthan, India; pareekapeksha34@gmail.com)

Payal Lodha
(Department of Botany, University of Rajasthan, Jaipur 302004, Rajasthan, India; payallodha10@yahoo.in)

7.1 INTRODUCTION

There is concurrently a growing concern about the effects of synthetic plant protection agents on human and animal health, as well as pollution caused by inorganic fertilizer residues (Buzek and Malmström, 2009; Rees et al., 2013; Björk and Kauppinen-Räisänen, 2016). Chemical fertilizers are crucial in ensuring that the world's population's rising food needs are met. To produce crops with the highest yields, nitrogen (N), phosphate (P), and potassium (K) are the three main commercial fertilizer types that are being employed (McGuire, 2015). However, the increasing use of chemical fertilizers in agriculture harms ecosystems. Chemical fertilizers are not absorbed sufficiently by plants; thus, they eventually seep into water bodies, where they produce eutrophication.

Pesticides may change the microbial populations in soil that are connected with plants (Meena et al., 2020). Finding ecologically acceptable alternatives is, therefore, becoming crucial. Plants' development and health can be improved by using microbiological tools like bacteria and fungi. Through eco-friendly methods, microbiological technologies have considerable potential to increase crop yields (Mahanty et al., 2016).

Microbial inoculants are attractive tools for sustainable agriculture since they can, at their best, both boost plant health and growth and also enhance nutrient availability and uptake (Gupta et al., 2022, 2024). In this chapter, we will give a brief overview of microbial secondary metabolites that are used in microbiological tools and discuss their formulation, current state for use in crop production, and ecologically friendly agriculture.

7.2 SECONDARY METABOLITES

Specialized substances that plant species make as a result of their ecological adaptation are known as secondary metabolites. These bioactive substances are mostly produced by soil microbiomes and plant microbiomes for defensive purposes (Berdy, 2005). Microbial communities in the phyllosphere are extremely complex and comprise both cultured and naturally occurring bacteria (Müller and Ruppel, 2014; Jain et al., 2023, 2024).

The scientific community has divided these substances into three major categories based on their presumed functions: primary metabolites, which are directly necessary for plant growth; secondary (or specialized) metabolites, which mediate interactions between plants and their environment; and hormones, which control organismal functions and metabolism (Erb and Kliebenstein, 2020).

These secondary metabolites are divided into terpenoids, alkaloids, polyketides, non-ribosomal peptides, and cytochalasins due to their metabolic origin (Yadav et al., 2019a).

Secondary metabolites, which are organic substances that are not necessary for the fundamental development and reproduction of the organism but frequently play crucial roles in survival and interactions with the environment, are known to be produced in large quantities by a wide range of microorganisms (Berdy, 2005). The diverse secondary metabolites created by microbes include antibiotics, antifungal, antiviral, anticancer, immunomodulatory, pigments, and substances that promote plant development.

The use of secondary metabolites spans a wide range of industries, including pharmaceuticals, such as antibiotics, anticancer drugs, and immunosuppressants; agrochemicals, such as pesticides, insecticides, and antifeedants; biofuels, such as squalene and oleoresin; and dietary supplements, such as carotenoids and flavonoids (Giddings and Newman, 2013).

7.3 MICROBIAL SECONDARY METABOLITES AND THEIR BIOACTIVITY

Many secondary metabolites produced by microbes have properties such as antibiotic, antifungal, antiviral,

DOI: 10.1201/9781003408833-7

anticancer, immune-modulating, and pigment use and also include plant growth–promoting substances.

The discovery of novel and potential secondary metabolites produced by microbes has been reported from a variety of habitats, including plant-related microbiomes (rhizospheric, epiphytic, and endophytic) and microbiomes of extreme habitats (halophilic, psychrophilic/thermophilic, acidophilic/alkaliphilic, and xerophilic).

These microbiomes belong to diverse species of different genera, such as *Streptosporangium*, *Streptomyces*, *Rhodococcus*, *Paenibacillus*, *Oceanicola*, *Marinobacter*, *Fodinicurvata*, *Bacillus*, *Agrococcus*, *Aeromicrobium*, and *Actinomadura* (Yadav et al., 2019b; Rastegari et al., 2019).

Extremophilic microbiomes, which include xerophiles, psychrophiles/thermophiles, halophiles, and acidophiles/alkaliphiles, produce a variety of secondary metabolites with potential biotechnological applications in the following genera: *Streptomyces*, *Rhodococcus*, *Phaeobacter*, *Oceanicola*, *Oceanibulbus*, *Muscodor*, *Micromonospora*, and *Kitasatosporia* (Chaabouni et al., 2012; Yadav et al., 2019a; Vemireddy et al., 2020).

7.3.1 Bacterial Secondary Metabolites

Bacteria are known to produce a wide array of secondary metabolites, which are organic compounds that are not directly involved in the growth, development, or reproduction of the organism. These metabolites often have diverse biological activities and are of great interest for various applications in medicine, agriculture, and industry, as shown in Table 7.1.

7.3.2 Fungal Secondary Metabolites

In natural conditions, fungi must contend with a variety of biotic and abiotic stresses, including other microbes, nutrient deficiency, pH, and temperature. They generate a large number of secondary metabolites (SMs) as one physiological response (Macheleidt et al., 2016), as shown in Table 7.2.

7.3.3 Algal Secondary Metabolites

It is known that the photoautotrophic microorganisms known as "microalgae" – which include microeukaryotes and cyanobacteria – produce a variety of secondary metabolites with different biological functions (Leflaive and Ten-Hage, 2007), as shown in Table 7.3.

TABLE 7.1
Some Commercial Bacterial Secondary Metabolites

Secondary Metabolites	Compound Produced	Bacterial Biostain	References
Antibacterial products (alkaloids and angucyclines)	Streptomycin	*Streptomyces* sp.	Sivalingam et al., 2019
Antifungal compounds	1. Nystatin 2. Amphotericin	1. Streptomyces noursei 2. Streptomyces nodosus	Gray et al., 2012
Antitumor compounds	Eribulin	*Saccharomonospora*	Jordan and Wilson, 2004
Immunomodulators	Muramyl peptides	Various bacteria	Martinon and Tschopp, 2007

TABLE 7.2
Some Fungal Secondary Metabolites and Their Bioactivity

Secondary Metabolites	Producer Fungal Stain	Bioactivity	References
Cyclosporine A	*Tolypocladium*	Immunosuppressant	Survase et al., 2011
Aflatoxin	*Aspergillus*	Mycotoxin	Klich, 2007
Ergot alkaloids	*Claviceps purpurea*	Infection of cereal crops and cause of ergotism	Schardl et al., 2006
Bikaverin	*Fusarium*	Antibiotic	Limón et al., 2010
Lovastatin	*Aspergillus*	HMG-CoA reductase inhibitor	Mulder et al., 2015

TABLE 7.3
Algal Secondary Metabolites and Their Bioactivity

Secondary Metabolites	Algal Biostain	Bioactivity	References
Phlorotannins	Phaeophyceae	Antioxidant, anti-inflammatory, and anticancer properties	Kim et al., 2012
Fucoxanthin	Phaeophyceae	Antiobesity, antidiabetic, and anticancer properties	Maeda et al., 2009
Mycosporine-like amino acids	Rhodophyta and Chlorophyta		Singh et al., 2008

7.4 ISOLATION OF MICROBIAL SECONDARY METABOLITES

The isolation of microbial secondary metabolites typically involves a combination of extraction, fractionation, and purification techniques. Some commonly used methods are as follows.

7.4.1 Solid-Phase Extraction (SPE)

SPE is a method that uses solid sorbents to concentrate and selectively extract target molecules from complicated mixtures. It entails putting the sample through a solid-phase disk or cartridge, where the desired analytes are kept and the undesired ones are eliminated. For the isolation of microbial secondary metabolites, SPE has been widely used.

A fermentation process on a solid support that has low moisture content and takes place in a non-septic and natural condition is known as solid-state or substrate fermentation (Robinson et al., 2001). These substances, which are generated on an industrial scale for use in agriculture and the treatment of disease, are connected to the stationary phase of microbial development.

The microorganisms that are mostly used to produce secondary metabolites have a mycelial shape that is well-suited to development on a stable support. Because efficient metabolite production necessitates very viscous liquid media, which can obstruct oxygen transmission, this can also have a negative impact on product formation in liquid medium (Ates et al., 1997).

In solid-state fermentation settings, wild-type strains of bacteria and fungi typically outperform genetically engineered microorganisms, requiring less energy and less money (Barrios-González et al., 1993).

In comparison to solid-state fermentation, the scale-up from the laboratory to the industrial fermenter is significantly simpler, with easier monitoring and control of the parameters with this method.

7.4.2 Liquid–Liquid Extraction (LLE)

Compounds can be extracted using the traditional LLE method if they are soluble in two immiscible liquids. The desired components are extracted from the sample by combining it with an appropriate organic solvent, which is followed by the separation of the two liquid phases. In order to extract microbial secondary metabolites from fermentation broths, LLE is frequently used as a first step (Salvachúa et al., 2015).

7.4.3 High-Performance Liquid Chromatography (HPLC)

HPLC is a widely used technique for separating, identifying, and purifying compounds. It employs a liquid mobile phase that passes through a column packed with a stationary phase. Components in the sample are separated based on their interactions with the stationary phase and detected using various detectors.

7.4.4 Preparative Liquid Chromatography

An approach for the large-scale separation and purification of chemicals is preparative liquid chromatography. Utilizing an analytical HPLC system with a bigger column and more sample loading space is required. When more microbial secondary metabolites are needed, preparative liquid chromatography is used.

7.4.5 Supercritical Fluid Extraction (SFE)

Since SFE uses supercritical solvents, which have different physicochemical properties than conventional extraction methods, such as density, diffusivity, viscosity, and dielectric constant, it has various operational advantages over those methods. Supercritical fluids offer improved transport qualities compared to liquids because of their low viscosity and relatively high diffusivity. They can also quickly diffuse through solid materials, which can lead to faster extraction rates. The ability to alter a supercritical fluid's density by adjusting its pressure and temperature is one of its key qualities. Since density and solubility are connected, the solvent strength of the fluid can be changed by varying the extraction pressure. The use of solvents generally recognized as safe (GRAS), the higher efficiency of the extraction process in terms of increasing yields and shorter extraction times, and the potential for direct coupling with analytical chromatographic techniques like gas chromatography (GC) or supercritical fluid chromatography (SFC) are additional benefits when compared to other extraction techniques (Herrero et al., 2006).

7.5 ROLE OF MICROBIAL SECONDARY METABOLITES IN SUSTAINABLE AGRICULTURE

High crop yields and soil health could be achieved through the use of microbial bioactive compounds and secondary metabolites for sustainable agriculture. For instance, the ability of bacteria from the genus *Azospirillum* and *Rhizobia* to either fix atmospheric nitrogen or to produce phytohormones like indole-3-acetic acid (IAA) and cytokinins has been examined (Morel et al., 2012).

Aggressive biosurfactants are widely used in agricultural soils nowadays. Biosurfactants are thought to be less harmful to the environment (Sachdev and Cameotra, 2013).

The development of biosurfactants from ecological isolates has promise for the agrochemical and agricultural sectors. Biosurfactants are produced by several phyllospheric, endophytic, and rhizospheric bacteria (Yadav, 2020).

Through soil restoration, biosurfactants can be effectively used to raise the quality of agricultural land. The dangerous chemical surfactants now employed in the multi-billion-dollar pesticide industry could be replaced by these biomolecules (Kumar et al., 2021).

The applications of microbial secondary metabolites in sustainable agriculture are numerous. They can be utilized as biopesticides, plant growth boosters, biofertilizers, biocontrol agents, and to induce systemic resistance (ISR). The following are some examples of microbial secondary metabolites' specialized uses in sustainable agriculture.

7.5.1 Biofertilizers

Microbial-originated secondary metabolites can increase plant nutrient availability and uptake, enhancing soil fertility and crop output. They are able to mobilize nutrients *via* fixing atmospheric nitrogen, phosphate solubilization, and the production of organic acids (Vessey, 2003).

Biofertilizer is a substance that contains living microorganisms or their by-products. Most biofertilizers colonize either around rhizosphere or in the tissue of the plant and stimulate the development of plants.

Biofertilizers are often bacteria or fungi that can fix nitrogen, solubilize phosphate, oxidize sulfur, produce plant hormones, or break down organic molecules (Verma et al., 2018). For instance, *Pseudomonas fluorescens* K-34 generates organic acids, which may be in charge of the plant's release of phosphate. IAA, a plant hormone, can also be produced by *P. fluorescens* K-34, *P. fluorescens* 1773/K, *P. trivalis* BIHB 745, and *Bacillus circulans* (Parani and Saha, 2012). Overall, biofertilizers cycle nutrients and ensure that crops grow and develop to their full potential (Bhardwaj et al., 2014).

7.5.2 Biocontrol Agents

By definition, biological control uses other living organisms as a non-chemical way to manage plant diseases. Nowadays, microbes such as bacteria, fungi, and their products are being used as biocontrol agents.

A microbe's ability to exert biocontrol can be attributed to its ability to produce antibiotic compounds, enzymes that can break down fungal cell walls, iron-depleting rhizosphere bacteria, induced systemic resistance, and competition for rhizosphere niches with pathogens (Parani and Saha, 2012). They prevent the spread of disease, stop pathogen growth and activity, and support long-term disease management (Compant et al., 2005).

The mechanism most frequently linked with the capacity for biocontrol is the production of one or more antibiotics. The ability to produce antifungal enzymes, such as chitinases, 1,3-glucanases, proteases, or lipases, with the ability to lyse fungal cells, is exhibited by several biocontrol strains. The growth of fungal infections can be stopped by the synthesis of low-molecular-mass siderophores that chelate iron in the soil close to roots (Köhl et al., 2019). For instance, *Pseudomonas trivalis* strain BIHB 745 can create siderophores (Parani and Saha, 2012). *P. fluorescens* has been shown to produce the siderophores pyochelin and pyoverdine (Sharma et al., 2017). By outcompeting phytopathogens for nutrients, many biocontrol strains can defend the host plant. They assist the plant by occupying rhizosphere niches and preventing diseases from infecting it (Kloepper et al., 2004).

7.5.3 Plant Growth–Promoting Rhizobacteria (PGPR)

Plant growth–promoting rhizobacteria (PGPR) are a kind of bacteria that are being used as biofertilizers and biocontrol agents (Pii et al., 2015; Paterson et al., 2017). Many plant species have PGPR that colonize the rhizosphere and have positive effects on the host, such as improved plant development and decreased susceptibility to illnesses brought on by nematodes, fungus, bacteria, and viruses (Kloepper et al., 2004).

The genera *Alcaligenes*, *Arthrobacter*, *Azospirillum*, *Azotobacter*, *Bacillus*, *Burkholderia*, *Serratia*, *Enterobacter*, *Klebsiella*, *Pseudomonas*, and *Rhizobium* are home to the bulk of the most well-known PGPRs (Kloepper et al., 2004). Increased shoot and root weights, root growth, yield, leaf area, chlorophyll content, nutrient uptake, protein content, hydraulic activity, resistance to abiotic stress, and delayed senescence are only a few advantages of PGPR (Adesemoye and Kloepper, 2009). Thus, PGPRs are frequently used as biofertilizers. *Azotobacter chroococcum* and *A. vinelandii*, for instance, are utilized as nitrogen fertilizers all throughout the world. For phosphorus fertilization, *Bacillus megaterium*, *B. amyloliquefaciens* IT45, and *Pseudomonas fluorescens* strains are utilized. *Frateuria aurantia*–based biofertilizers are available to provide potassium nutrition for crops. As fertilizers with zinc, sulfur, and silicate, strains such as *Thiobacillus thiooxidans*, *Delfia acidovorans*, and *Bacillus* spp. are commonly used in Asia (Mącik et al., 2020).

7.5.4 Biopesticides

Chemical pesticides can be replaced with microbial secondary metabolites, which are more environmentally benign. They have antibacterial, antifungal, and insecticidal qualities that efficiently combat plant diseases and pests while causing the least harm to the environment (Palaniyandi et al., 2013).

Streptomyces species are effective biocontrol agents against soil- and foliar-borne fungal diseases brought on by significant agricultural pathogens like *Phytophthora cinnamomi* (Shimizu et al., 2000). *Colletotrichum musae*, *Fusarium oxysporum* (Taechowisan et al., 2003), *Botryts cinerea* (Lian et al., 2017), *Colletotrichum gloeosporioides* (Palaniyandi et al., 2013), *Fusarium moniliforme* (Kaur et al., 2019), *Rhizoctonia solani*, *Fusarium solani*, *Fusariumv erticillioides*, *Alternaria alternata* (Al-Askar et al., 2011), *Sclerotinia sclerotiorum* (Baniasadi et al., 2009), *Macrophomina phaseolina* (Shrivastava et al., 2017), *Corynespora cassiicola* (Devi and Rao, 2017), *Pythium aphanidermatum* (Loliam et al., 2013), and

Penicillium digitatum are used as biocontrol agents (Maldonado et al., 2010).

7.5.5 Induced Systemic Resistance (ISR)

Plants can develop systemic resistance to diseases and pests as a result of secondary metabolites produced by microbes. They trigger the plant's natural defensive mechanisms, offering long-lasting defense (Van Wees et al., 2008).

Biological (insects, diseases, or microorganisms) or chemical inducers can cause biotic stress, and plants can respond by developing induced resistance, which is controlled by a network of interconnected signaling pathways, with plant hormones playing a key regulatory function. The term "induced systemic resistance" (ISR) refers to the fact that induced resistance is exhibited not only locally at the infection site but also systemically in plant components that are physically far from the infection. Used as biocontrol agents that cause systemic resistance in plants are several bacterial species from *Pseudomonads*, bacilli, actinomycetes, and fungi from different species of non-pathogenic *Fusarium*, *Trichoderma*, and *Streptomyces* (Choudhary and Johri, 2009).

7.6 CONCLUSION AND FUTURE PROSPECTS

Chemical contamination is currently causing harm to the world on all fronts. Effective solutions must be identified and quickly put into action in order to guarantee that the future generation lives in a healthy environment. Future sustainable food production will require the development of microbial agents as well as their greater use.

Both microbial secondary metabolites are likely to see an increase in demand, which offers a solid foundation for ongoing research and development. The uniformity of the products in field conditions is the main issue with the present tools. The existing techniques are all based on a small number of well-known mechanisms of action, but interactions between microbes and plants are more complex than that and require further study. A solid foundation for creating dependable tools to improve plant health and growth in various field situations will likely come from a thorough study of each microbial strain, their methods of action, and the elements of a healthy plant microbiome. The development of technologies for microbial culture and life preservation should therefore receive more attention.

REFERENCES

Adesemoye, A.O., and J.W. Kloepper. 2009. Plant–microbes interactions in enhanced fertilizer-use efficiency. *Applied Microbiology and Biotechnology* 85: 1–12. https://doi.org/10.1007/s00253-009-2196-0.

Al-Askar, A.A., W.M. Abdul Khair, and Y.M. Rashad. 2011. In vitro antifungal activity of *Streptomyces spororaveus* RDS28 against some phytopathogenic fungi. *African Journal of Agricultural Research* 6: 2835–42. https://doi.org/10.5897/AJAR11.320.

Ates, S., M. Elibol, and F. Mavituna. 1997. Production of actinorhodin by *Streptomyces coelicolor* in batch and fed-batch cultures. *Process Biochemistry* 32: 273–78. https://doi.org/10.1016/S0032-9592(96)00071-4.

Baniasadi, F., G.H.S. Bonjar, A. Baghizadeh, et al. 2009. Biological control of *Sclerotinia sclerotiorum*, causal agent of sunflower head and stem rot disease, by use of soil borne actinomycetes isolates. *American Journal of Agricultural and Biological Sciences* 4: 146–51. https://doi.org/10.3844/ajabssp.2009.146.151.

Barrios-González, J., T.E. Castillo, and A. Mejía. 1993. Development of high penicillin producing strains for solid state fermentation. *Biotechnology Advances* 11: 525–37. https://doi.org/10.1016/0734-9750(93)90021-E.

Berdy, J. 2005. Bioactive microbial metabolites. *The Journal of Antibiotics* 58: 1–26. https://doi.org/10.1038/ja.2005.1.

Bhardwaj, D., M.W. Ansari, R.K. Sahoo, and N. Tuteja. 2014. Biofertilizers function as key player in sustainable agriculture by improving soil fertility, plant tolerance and crop productivity. *Microbial Cell Factories* 13 (May): 66, 1–10. https://doi.org/10.1186/1475-2859-13-66.

Björk, P., and H. Kauppinen-Räisänen. 2016. Local food: A source for destination attraction. *International Journal of Contemporary Hospitality Management* 28: 177–94. https://doi.org/10.1108/IJCHM-05-2014-0214.

Buzek, J., and C. Malmström. 2009. Regulation (EC) No 1107/2009 of the European parliament and of the council of 21 October 2009 concerning the placing of plant protection products on the market and repealing council directives. *Official Journal of the European Union* 309: 1–50. http://data.europa.eu/eli/reg/2009/1107/oj.

Chaabouni, I., A. Guesmi, and A. Cherif. 2012. Secondary metabolites of *Bacillus*: Potentials in biotechnology. In *Bacillus thuringiensis biotechnology*, ed. E. Sansinenea, 347–66 Dordrecht, Netherlands: Springer, 347–66. https://doi.org/10.1007/978-94-007-3021-2_18.

Choudhary, D.K., and B.N. Johri. 2009. Interactions of *Bacillus spp*. and plants–with special reference to induced systemic resistance (ISR). *Microbiological Research* 164: 493–513. https://doi.org/10.1016/j.micres.2008.08.007.

Compant, S., B. Duffy, J. Nowak, C. Clément, and E.A. Barka. 2005. Use of plant growth-promoting bacteria for biocontrol of plant diseases: Principles, mechanisms of action, and future prospects. *Applied and Environmental Microbiology* 71: 4951–59. https://doi.org/10.1128/AEM.71.9.4951-4959.2005.

Devi, S.S., and K.V.B. Rao. 2017. Exploration of antimicrobial compounds from *Streptomyces* S9 against a phytopathogen, *Corynespora cassiicola* (Berk & Curtis). *Journal of Biopesticides* 10: 1–9. http://www.jbiopest.com/users/LW8/efiles/vol_10_1_01-09.pdf.

Erb, M., and D.J. Kliebenstein. 2020. Plant secondary metabolites as defenses, regulators, and primary metabolites: The blurred functional trichotomy. *Plant Physiology* 184: 39–52. https://doi.org/10.1104/pp.20.00433.

Giddings, L.A., and D.J. Newman. 2013. Microbial natural products: Molecular blueprints for antitumor drugs. *Journal of Industrial Microbiology and Biotechnology* 40: 1181–210. https://doi.org/10.1007/s10295-013-1331-1.

Gray, K.C., D.S. Palacios, I. Dailey, et al. 2012. Amphotericin primarily kills yeast by simply binding ergosterol. *Proceedings of the National Academy of Sciences* 109: 2234–39. https://doi.org/10.1073/pnas.1117280109.

Gupta, A., S. Jain, and N. Verma. 2022. *Industrial applications of soil microbes*, vol. 1. Bentham Science Publishers Pte Ltd. https://doi.org/10.2174/97898150399551220101.

Gupta, A., S. Jain, and N. Verma. 2024. *Industrial applications of soil microbes*, vol. 3. Bentham Science Publishers Pte Ltd. https://doi.org/10.2174/97898150797531240301.

Herrero, M., A. Cifuentes, and E. Ibañez. 2006. Sub- and supercritical fluid extraction of functional ingredients from different natural sources: Plants, food-by-products, algae and microalgae: A review. *Food Chemistry* 98: 136–48. https://doi.org/10.1016/j.foodchem.2005.05.058.

Jain, S., A. Gupta, and N. Verma. 2023. *Industrial applications of soil microbes*, vol. 2. Bentham Science Publishers Pte Ltd. https://doi.org/10.2174/97898150502641230201.

Jain, S., A. Gupta, and N. Verma. 2024. *Industrial applications of soil microbes*, vol. 4. Bentham Science Publishers Pte Ltd. https://doi.org/10.2174/97898151249961240401.

Jordan, M.A., and L. Wilson. 2004. Microtubules as a target for anticancer drugs. *Nature Reviews Cancer* 4: 253–65. https://doi.org/10.1038/nrc1317.

Kaur, T., R. Rani, and R.K. Manhas. 2019. Biocontrol and plant growth promoting potential of phylogenetically new *Streptomyces sp*. MR14 of rhizospheric origin. *AMB Express* 9 (August): 125, 1–14. https://doi.org/10.1186/s13568-019-0849-7.

Kim, Y.J., E.H. Kim, and K.B. Hahm.2012. Oxidative stress in inflammation-based gastrointestinal tract diseases: Challenges and opportunities. *Journal of Gastroenterology and Hepatology Foundation* 27: 1004–10. https://doi.org/10.1111/j.1440-1746.2012.07108.x.

Klich, M.A. 2007. *Aspergillus flavus*: The major producer of aflatoxin. *Molecular Plant Pathology* 8: 713–22. https://doi.org/10.1111/j.1364-3703.2007.00436.x.

Kloepper, J.W., C.M. Ryu, and S. Zhang. 2004. Induced systemic resistance and promotion of plant growth by *Bacillus* spp. *Phytopathology* 94: 1259–66. https://doi.org/10.1094/PHYTO.2004.94.11.1259.

Köhl, J., R. Kolnaar, and W.J. Ravensberg. 2019. Mode of action of microbial biological control agents against plant diseases: Relevance beyond efficacy. *Frontiers in Plant Science* 10 (July): 845, 1–19. https://doi.org/10.3389/fpls.2019.00845.

Kumar, M., A.N. Yadav, R. Saxena, D. Paul, and R.S. Tomar. 2021. Biodiversity of pesticides degrading microbial communities and their environmental impact. *Biocatalysis and Agricultural Biotechnology* 31 (January): 101883. https://doi.org/10.1016/j.bcab.2020.101883.

Leflaive, J.P., and L. Ten-Hage. 2007. Algal and cyanobacterial secondary metabolites in freshwaters: A comparison of allelopathic compounds and toxins. *Freshwater Biology* 52: 199–14. https://doi.org/10.1111/j.1365-2427.2006.01689.x.

Lian, Q., J. Zhang, L. Gan, Q. Ma, Z. Zong, and Y. Wang. 2017. The biocontrol efficacy of *Streptomyces pratensis* LMM15 on *Botrytis cinerea* in tomato. *BioMed Research International* 2017 (November): 9486794, 1–11. https://doi.org/10.1155/2017/9486794.

Limón, M.C., R. Rodríguez-Ortiz, and J. Avalos. 2010. Bikaverin production and applications. *Applied Microbiology and Biotechnology* 87: 21–29. https://doi.org/10.1007/s00253-010-2551-1.

Loliam, B., T. Morinaga, and S. Chaiyanan. 2013. Biocontrol of *Pythium aphanidermatum* by the cellulolytic actinomycetes *Streptomyces rubrolavendulae* S4. *Science Asia* 39: 584–90. https://www.scienceasia.org/2013.39.n6/scias39_584.pdf.

Macheleidt, J., D.J. Mattern, J. Fischer, et al. 2016. Regulation and role of fungal secondary metabolites. *Annual Review of Genetics* 50: 371–92. https://doi.org/10.1146/annurev-genet-120215-035203.

Mącik, M., A. Gryta, and M. Frąc. 2020. Biofertilizers in agriculture: An overview on concepts, strategies and effects on soil microorganisms. *Advances in Agronomy* 162: 31–87. https://doi.org/10.1016/bs.agron.2020.02.001.

Maeda, H., M. Hosokawa, T. Sashima, K. Murakami-Funayama, and K. Miyashita. 2009. Anti-obesity and anti-diabetic effects of fucoxanthin on diet-induced obesity conditions in a murine model. *Molecular Medicine Reports* 2: 897–902. https://doi.org/10.3892/mmr_00000189.

Mahanty, T., S. Bhattacharjee, M. Goswami, et al. 2016. Biofertilizers: A potential approach for sustainable agriculture development. *Environmental Science and Pollution Research* 24: 3315–35. https://doi.org/10.1007/s11356-016-8104-0.

Maldonado, M.C., C.E.F. Orosco, M. Gordillo, and A.R. Navarro. 2010. *In vivo* and *in vitro* antagonism of *Streptomyces* sp. RO3 against *Penicillium digitatum* and *Geotrichum candidum*. *African Journal of Microbiology Research* 4: 2451–56. http://hdl.handle.net/11336/77469.

Martinon, F., and J. Tschopp. 2007. Inflammatory caspases and inflammasomes. Master switches of inflammation. *Cell Death & Differentiation* 14: 10–22. https://doi.org/10.1038/sj.cdd.4402038.

McGuire, S. 2015. The state of food insecurity in the world 2015: Meeting the 2015 international hunger targets: Taking stock of uneven progress. *The Food and Agriculture Organization Rome* 6: 623–24. http://www.fao.org/3/a-i4646e.pdf.

Meena, R.S., S. Kumar, R. Datta, et al. 2020. Impact of agrochemicals on soil microbiota and management: A review. *Land* 9 (January): 34, 1–21. https://www.mdpi.com/2073-445X/9/2/34/pdf.

Morel M.A., V. Braña, and S. Castro-Sowinski. 2012. Legume crops, importance and use of bacterial inoculation to increase the production. In *Crop plant*, ed. A.K. Goyal, 217–40. InTechOpen. https://www.intechopen.com/chapters/35625.

Mulder, K.C.L., F. Mulinari, O.L. Franco, M.S.F. Soares, B.S. Magalhães, and N.S. Parachin. 2015. Lovastatin production: From molecular basis to industrial process optimization. *Biotechnology Advances* 33: 648–65. https://doi.org/10.1016/j.biotechadv.2015.04.001.

Müller, T., and S. Ruppel. 2014. Progress in cultivation-independent phyllosphere microbiology. *FEMS Microbiology Ecology* 87: 2–17. https://doi.org/10.1111/1574-6941.12198.

Palaniyandi, S.A., S.H. Yang, L. Zhang, and J.W. Suh. 2013. Effects of actinobacteria on plant disease suppression and growth promotion. *Applied Microbiology and Biotechnology* 97: 9621–36. https://doi.org/10.1007/s00253-013-5206-1.

Parani, K., and B.K. Saha. 2012. Prospects of using phosphate solubilizing *Pseudomonas* as bio fertilizer. *European Journal of Biological Sciences* 4: 40–44. http://idosi.org/ejbs/4(2)12/2.pdf.

Paterson, J., G. Jahanshah, Y. Li, Q. Wang, S. Mehnaz, and H. Gross. 2017. The contribution of genome mining strategies to the under-standing of active principles of PGPR strains. *FEMS Microbiology Ecology* 93 (December): fiw249, 1–12. https://doi.org/10.1093/femsec/fiw249.

Pii, Y., T. Mimmo, N. Tomasi, R. Terzano, S. Cesco, and C. Crecchio. 2015. Microbial interactions in the rhizosphere: Beneficial influences of plant growth-promoting rhizobacteria on nutrient acquisition process. A review. *Biology and Fertility of Soils* 51: 403–15. https://doi.org/10.1007/s00374-015-0996-1.

Rastegari, A.A., A.N. Yadav, N. Yadav, and N.T. Sarshari. 2019. Bioengineering of secondary metabolites. In

New and future developments in microbial biotechnology and bioengineering, eds. V.K. Gupta and A. Pandey, 55–68. Amsterdam: Elsevier. https://doi.org/10.1016/B978-0-444-63504-4.00004-9.

Rees, R.M., J.A. Baddeley, A. Bhogal, et al. 2013. Nitrous oxide mitigation in UK agriculture. *Soil Science and Plant Nutrition* 59: 3–15. https://doi.org/10.1080/00380768.2012.733869.

Robinson, T., D. Singh, and P. Nigam. 2001. Solid-state fermentation: A promising microbial technology for secondary metabolite production. *Applied Microbiology and Biotechnology* 55: 284–89. https://doi.org/10.1007/s002530000565.

Sachdev, D.P., and S.S. Cameotra. 2013. Biosurfactants in agriculture. *Applied Microbiology and Biotechnology* 97: 1005–16. https://doi.org/10.1007/s00253-012-4641-8.

Salvachúa, D., E.M. Karp, C.T. Nimlos, D.R. Vardon, and G.T. Beckham. 2015. Towards lignin consolidated bioprocessing: Simultaneous lignin depolymerization and product generation by bacteria. *Green Chemistry* 17: 4951–67. https://doi.org/10.1039/C5GC01165E.

Schardl, C.L., D.G. Panaccione, and P. Tudzynski. 2006. Ergot alkaloids – biology and molecular biology. In *The alkaloids: Chemistry and biology*, ed. G.A. Cordell, 63: 45–86. Elsevier: Academic Press. https://doi.org/10.1016/S1099-4831(06)63002–2.

Sharma, P., L. Bora, K. Puzari, et al. 2017. Review on bacterial blight of rice caused by *Xanthomonas oryzae* pv. *oryzae*: Different management approaches and role of *Pseudomonas fluorescens* as a potential biocontrol agent. *International Journal of Current Microbiology and Applied Sciences* 6: 982–1005. https://doi.org/10.20546/ijcmas.2017.603.117.

Shimizu, M., Y. Nakagawa, Y. Sato, et al. 2000. Studies on endophytic actinomycetes (I) *Streptomyces sp.* isolated from rhododendron and its antifungal activity. *Journal of General Plant Pathology* 66: 360–66. https://doi.org/10.1007/PL00012978.

Shrivastava, P., R. Kumar, and M.S. Yandigeri. 2017. *In vitro* biocontrol activity of halotolerant *Streptomyces aureofaciens* K20: A potent antagonist against *Macrophomina phaseolina* (Tassi) Goid. *Saudi Journal of Biological Sciences* 24: 192–99. https://doi.org/10.1016/j.sjbs.2015.12.004.

Singh, S.P., M. Klisch, R.P. Sinha, and D.P. Häder. 2008. Effects of abiotic stressors on synthesis of the mycosporine-like amino acid shinorine in the cyanobacterium *Anabaena variabilis* PCC 7937. *Photochemistry and Photobiology* 84: 1500–5. https://doi.org/10.1111/j.1751-1097.2008.00376.x.

Sivalingam P., K. Hong, J. Pote, and K. Prabakar. 2019. Extreme environment Streptomyces: Potential sources for new antibacterial and anticancer drug leads? *International Journal of Microbiology* 2019 (July): 5283948, 1–20. https://doi.org/10.1155/2019/5283948.

Survase, S.A., L.D. Kagliwal, U.S. Annapure, and R.S. Singhal. 2011. Cyclosporin A-A review on fermentative production, downstream processing and pharmacological applications. *Biotechnology Advances* 29: 418–35. https://doi.org/10.1016/j.biotechadv.2011.03.004.

Taechowisan, T., J.F. Peberdy, and S. Lumyong. 2003. Isolation of endophytic actinomycetes from selected plants and their antifungal activity. *World Journal of Microbiology and Biotechnology* 19: 381–85. https://doi.org/10.1023/A:1023901107182.

Van Wees, S.C.M., S. Van der Ent, and C.M.J. Pieterse. 2008. Plant immune responses triggered by beneficial microbes. *Current Opinion in Plant Biology* 11: 443–48. https://doi.org/10.1016/j.pbi.2008.05.005.

Vemireddy, B., A. Madasi, A. Ajmeera, and K.R. Vanteru. 2020. Distribution and diversity of endophytic fungi associated with three medicinal tree species from Eturnagaram Wildlife Sanctuary, TS, India. *Journal of Applied Biology & Biotechnology* 8: 7–12. https://doi.org/10.7324/JABB.2020.80602.

Verma, M., J. Mishra, and N.K. Arora. 2018. Plant growth-promoting rhizobacteria: Diversity and applications. In *Environmental biotechnology: For sustainable future*, eds. R. Sobti, N. Arora and R. Kothari, 129–73. Singapore: Springer. https://doi.org/10.1007/978-981-10-7284-0_6.

Vessey, J.K. 2003. Plant growth promoting rhizobacteria as biofertilizers. *Plant and Soil* 255: 571–86. https://doi.org/10.1023/A:1026037216893.

Yadav, A.N. 2020. Plant microbiomes for sustainable agriculture: Current research and future challenges. In *Plant microbiomes for sustainable agriculture*, eds. A.N. Yadav, J. Singh, A.A. Rastegari, and N. Yadav, 475–82. Cham: Springer. https://doi.org/10.1007/978-3-030-38453-1_16.

Yadav, A.N., D. Kour, K.L. Rana, et al. 2019a. Metabolic engineering to synthetic biology of secondary metabolites production. In *New and future developments in microbial biotechnology and bioengineering*, eds. V.K. Gupta and A. Pandey, 279–320. Amsterdam: Elsevier. https://doi.org/10.1016/B978-0-444-63504-4.00020-7.

Yadav, A.N., N. Yadav, S.G. Sachan, and A.K. Saxena. 2019b. Biodiversity of psychrotrophic microbes and their biotechnological applications. *Journal of Applied Biology & Biotechnology* 7: 99–108. https://doi.org/10.7324/JABB.2019.70415.

8 Beneficial Microbe-Induced Secondary Metabolites

Interactions with Plants and Soil Microbiome

*Anandkumar Naorem
(ICAR – Central Arid Zone Research Institute, Jodhpur, Rajasthan 342003, India; naoremanand@gmail.com)

M. Chandra Surya Rao
(Department of Agricultural and Horticultural Sciences, School of Agriculture and Food Technology, Vignan's Foundation for Science, Technology, and Research, Vadlamudi, Guntur, Andhra Pradesh 522213, India; suryahorti@gmail.com)

Devi Shidayaichenbi
(Dr. Rajendra Prasad Central Agricultural University, Pusa 848125, Bihar, India; shidayaish@gmail.com)

Saranya R.
(ICAR – Central Arid Zone Research Institute, RRS-Jaisalmer, Rajasthan 345001, India; vrsaran92@gmail.com)

8.1 INTRODUCTION

The vital function that soils, an intricate and constantly morphing combination of organic and inorganic components, play in sustaining life on our planet cannot be overstated. Comprising a slender stratum of non-living elements, it occupies the interface where the Earth's solid terrain interfaces with the enveloping atmosphere. Sand, silt, and clay are just a few of the several types of particles that cohabit in this layer. These particles create an incredibly diverse and rich ecosystem of pores and aggregates where gases, nutrients, and water can travel. Soil is the foundation for plant growth, serving as a medium for root anchorage, a source of nutrients, and a location for water retention.

Soil is an extremely complex and heterogeneous environment in which a wide range of microorganisms can coexist on a small scale. Hence, soil can be considered one of the most intricate biological environments, demonstrating an extraordinary capacity for the capture, alteration, and discharge of substances. Soil characteristics, both in terms of their physical and chemical attributes, are significantly molded by the interplay of living organisms and non-living factors that define the soil milieu. These factors encompass the presence of flora and fauna, climatic conditions, land contours, and geological aspects. One of the key characteristics of soil is its ability to support a diverse range of microorganisms, including bacteria, fungi, viruses, and protozoa. Microbes fulfill a vital function in nutrient recycling and the well-being of soil, as they decompose organic material and liberate vital nutrients necessary for plant development (Jain et al., 2023, 2024). The abundance of microorganisms in soil is truly astonishing, with approximations extending up to 8,300,000 bacterial varieties per gram of soil (Roesch et al., 2007).

Microbes dwelling in the soil serve as valuable assets, facilitating an extensive array of transformative procedures and playing a substantial role in enhancing soil richness (Mader et al., 2002). For example, they perform a vital function in ensuring the accessibility of phosphorus, a fundamental nutrient crucial for the development of plants. Soil bacteria and fungi help in the breakdown of organic matter, releasing phosphorus, which is then taken up by plants (Richardson, 2001). Additionally, some soil bacteria can directly solubilize phosphorus, making it more available to plants. By enhancing the availability of phosphorus in the soil, soil microorganisms can increase plant growth and yield (Gupta et al., 2022, 2024).

In addition, soil microbes contribute significantly to the process of biodegradation, wherein organic contaminants undergo breakdown and conversion into benign compounds (Ahemad et al., 2011). Numerous soil microorganisms and fungi are capable of breaking down complex organic substances, including insecticides, herbicides, and petroleum products. By breaking down these pollutants, soil microorganisms help remove them from the environment, reducing

DOI: 10.1201/9781003408833-8

their impact on human and environmental health. Beyond their capacity for biotransformation, soil microorganisms serve as a significant source of intricate organic compounds. These compounds are produced through the metabolic activities of soil microorganisms and are important for soil health and fertility. For example, some soil bacteria produce exopolysaccharides, which help improve soil structure and water-holding capacity. Other microorganisms produce enzymes that help break down organic matter, releasing nutrients that are then available to plants.

8.2 PHYSIOLOGY OF RHIZOBACTERIA THAT STIMULATE PLANT GROWTH

The co-evolution between plants and microorganisms has resulted in certain bacteria adapting to become facultative intracellular endophytes (Bulgarelli et al., 2013). One of these free-living bacteria, plant growth–promoting rhizobacteria (PGPR), has been employed to increase biotic and abiotic stress tolerance as well as water and nutrient intake. However, the molecular underpinnings of the mechanisms by which plants and bacteria interact, leading to physiological changes, are frequently not fully understood. Due in part to newly developed "omics" techniques, the mode(s) of action by which bacteria display beneficial actions are starting to be discovered.

8.2.1 Communication between Plant Roots and PGPR: A Mutual Dialogue

The exchange of signals between plant roots and bacteria is one way that PGPR promotes plant growth. Phytohormones are important regulators of plant growth and development, serving as molecular signals in reaction to environmental elements that, in the absence of intervention, might restrict plant development or prove fatal (Fahad et al., 2015). Numerous rhizosphere bacteria have a reputation for releasing hormones that facilitate root absorption or regulate the plant's hormonal equilibrium to enhance growth and stress resilience. PGPR strains that generate auxins like indole-3-acetic acid (IAA) have demonstrated their ability to trigger alterations in gene expression related to hormones, defense mechanisms, and cell wall components. This, in turn, results in longer roots, greater root biomass, and reduced stomatal size and density (Hong et al., 1991; Ruzzi and Aroca, 2015; Omer et al., 2004). Additionally, PGPR can create gibberellins and cytokinins, which might increase the plant's production of root exudate and perhaps increase the amount of PGPR present in the plant (Gupta et al., 2015). Enhancement of shoot development can result from PGPR production of cytokinins, whilst changes to root architecture can result from these hormones' interactions with auxins (Jha and Saraf, 2015).

For certain PGPR, the gaseous hormone ethylene plays a significant part in enhancing plant stress tolerance. In order to prevent ethylene levels from rising to a point where they would inhibit plant development, PGPR secretes 1-aminocyclopropane-1-carboxylase (ACC) deaminase (Glick, 2014). Inoculating plants with PGPR, which produces ACC deaminase, has been found to increase their ability to withstand stress in several investigations, as was the case with *Camelina sativa* (Ahemad and Kibret, 2014; Heydarian et al., 2016).

Numerous factors come into play when it comes to influencing plant productivity and the availability of nutrients. One of these elements, the rhizosphere's population of soil microbes, stands out as being particularly important. The rhizosphere represents the soil surrounding plant roots, where intricate interactions unfold among the roots themselves, the soil, and a multitude of microorganisms. Within this community of soil microorganisms, which encompasses bacteria, fungi, archaea, and viruses, lies a critical role in nutrient cycling, soil structure maintenance, and overall plant well-being. These microorganisms inhabiting the rhizosphere engage in intricate interactions with both plant roots and the soil itself, giving rise to a complex web of relationships that ultimately bolster plant growth and productivity.

One of the most important ways that microbes enhance plant growth is by making nutrients more available to the plant. Microbes in the rhizosphere can break down organic matter in the soil and release essential nutrients, like nitrogen, phosphorus, and potassium, which plants need to grow. In turn, plants secrete root exudates containing sugars, amino acids, and organic acids, which act as substrates and signaling molecules for the microbes. This creates a positive feedback loop, where the microbes and plants support each other's growth. The microbial community in the rhizosphere also plays a crucial role in maintaining soil structure. Microbes produce polysaccharides and other extracellular substances that bind soil particles together, leading to the formation of stable aggregates. These aggregates improve soil aeration, water-holding capacity, and nutrient availability, all of which are essential for plant growth.

Along with maintaining the structure of the soil and biogeochemical cycles, the microbial population in the rhizosphere shields plants against harmful pathogens. Antibiotics and other secondary metabolites that hinder the development of plant pathogens can be produced by microbes. By generating hormones that promote plant growth and development, such as auxins, cytokinins, and gibberellins, they can also encourage plant growth. Scientists have been fascinated by the complicated relationship between plants and their environment for ages. Numerous studies have been carried out over time to comprehend how plants interact with their surroundings. Although certain research has delved into the intricacies of singular plant–microbe connections, it's essential to recognize that these interactions are significantly more intricate, involving a wide spectrum of microbes that frequently collaborate to produce synergistic outcomes (Mendes et al., 2011).

One of the key players in this complex relationship is the soil microbiome. The soil microbiome comprises a diverse collection of microorganisms, encompassing bacteria, fungi, and protozoa, all of which engage with plant roots through various interactions. These microorganisms can have either positive or negative effects on plant health and productivity. Consistently, plant roots release a complex, wide range of compounds into the surrounding soil, establishing intricate interactions with neighboring organisms. This multifaceted process appears to confer distinct advantages throughout various stages of their growth and development. These emissions, scientifically recognized as root exudates, encompass a diverse array of organic and inorganic molecules, functioning as signaling agents, sources of energy, and mechanisms for actively deterring competing microorganisms.

According to recent studies, a plant's root exudates' phytochemical content varies throughout its lifetime. This means that plants can tailor the composition of their root exudates to meet their changing needs as they grow and develop. For example, young plants may produce root exudates that are high in sugars to attract beneficial microbes that can help them establish themselves in the soil. To encourage growth and development, the plant may start creating root exudates that are rich in amino acids and other nutrients as it ages. The soil microbiome plays an important role in plant health and productivity by interacting with these root exudates. Some microorganisms can break down complex organic compounds in the soil and make them available to the plant as nutrients. Other microorganisms can form beneficial associations with plant roots, providing them with protection against pathogens or helping them take up nutrients more efficiently.

Not all soil microbes, meanwhile, are helpful to plants. By competing with plants for nutrients or by releasing toxins that might hurt the plant, some microbes can hinder plant growth and development. In order to increase plant health and productivity, it is essential to comprehend how each soil microbiome component interacts with and influences the others.

The rhizosphere represents the soil area enveloping and shaped by plant roots. It forms a thriving environment teeming with nutrients and a diverse microbial community, where a multitude of microorganisms engage in interactions amongst themselves and with the plants. Research has revealed robust connections between the microbial communities in the rhizosphere and the intricate blend of compounds known as root exudates, emanating from plant roots and released into the soil.

Studies have shown that changes in root exudate composition result in significant modifications of the soil microbial community. For instance, a study by Broeckling et al. (2008) revealed that rhizosphere microbial communities respond differently to different root exudate compositions. In another study, Badri and Vivanco (2009) demonstrated that alterations in root exudate composition resulted in changes in microbial community composition and function.

Research findings indicate that alterations in the composition of root exudates can lead to notable transformations in the soil microbial community. For example, in a study conducted by Broeckling et al. (2008), it was observed that rhizosphere microbial communities displayed varying responses to distinct root exudate compositions. Similarly, Badri and Vivanco (2009) demonstrated in another study that changes in root exudate composition brought about shifts in both the composition and the functional attributes of the microbial community.

The makeup and levels of root exudates hinge on a range of factors, encompassing both soil-related and biological influences. Soil-related factors, such as soil type, pH levels, and nutrient availability, play a crucial role in shaping root exudate characteristics. The existence and activity of soil microorganisms also play a significant role in this complex interplay. Flores et al. (1999) found that changes in soil microbial communities resulted in significant alterations in the composition of root exudates. The dynamic and reciprocal relationships between plants and soil bacteria are highlighted by these studies. According to Schmülling et al. (1988), plant development can be influenced by soil microbes in both positive and negative ways. *Agrobacterium rhizogenes*, for example, is a soil microbe that can influence and manipulate plant development to form hairy roots (Ortíz-Castro et al., 2009). Similar to this, soil microorganisms like plant growth–promoting rhizobacteria (PGPRs) can affect plant development by secreting volatile chemical compounds or hormones like auxin and cytokinin. A group of advantageous microbes known as PGPRs colonize the rhizosphere and improve plant growth and productivity. By causing systemic resistance against pathogens, these microbes can also improve plant health.

Plants have developed defense mechanisms to safeguard themselves against pathogens and pests. An example of such a mechanism is age-related resistance (ARR), observed in various plant species including bean, maize, soybean, cabbage, cowpea, cotton, and *Arabidopsis*. ARR signifies that a plant's vulnerability to pathogens and pests fluctuates as it matures. To illustrate, younger plants might exhibit greater susceptibility to a specific pathogen, whereas mature plants may display increased resistance to the same pathogen (Develey-Riviere and Galiana, 2007). One instance of ARR can be found in maize, where mature plants exhibit the opposite of what younger plants do in terms of susceptibility to common rust (*Puccinia sorghi*) (Abedon and Tracy, 1996). This shows that the degree to which plants grow influences their susceptibility to infections and pests.

It is crucial to grasp the impact of plant growth stages on the interactions with microbes in the rhizosphere, which is the soil zone encompassing plant roots. In this region, plants release a diverse array of compounds that exert a profound influence on the composition of the microbial community. The rhizosphere microbiome plays a vital role in facilitating nutrient uptake and fostering plant growth. Notably, as outlined in studies by Malagoli et al. (2004), plants have been shown to have heightened demand for nitrogen (N) during later stages of development.

Rhizobia and legumes' symbiotic relationship is a prime illustration of plant–microbe interactions. Only when the plant is subjected to N-limiting circumstances is this association observed (Davidson and Robson, 1986). The release of flavones and flavanols, which start the rhizobia–legume symbiosis, is also boosted in N-limiting circumstances (Coronado et al., 1995). This shows that plants have some degree of control over the composition and operation of the rhizosphere microbiome.

Studies have demonstrated that the rhizosphere microbiome is influenced by plant growth. For instance, a study of the potato rhizosphere discovered that immature potato plants had microbial communities that were reliant on cultivar, but as the plants grew older, these distinctions disappeared (Inceoğlu et al., 2011). The rhizosphere microbial communities of soybean plants were also shown to be influenced by plant development, with early reproductive growth stages creating more complex microbial communities than late-stage soybean plants.

The rhizosphere, the region encompassing plant roots, is greatly affected by the release of organic compounds known as root exudates. The type and quantity of these exudates are subject to variation based on plant species and environmental factors. Recent investigations have underscored the pivotal role of root exudates in molding the microbial community within the rhizosphere. The connection between root exudates and rhizosphere microbial composition is gaining substantial support. As reported in studies by Badri et al. (2008) and Chaparro et al. (2012), numerous chemicals present in root exudates serve as substrates, chemotactic agents, or signaling molecules, orchestrating shifts in microbial composition. Furthermore, research by Badri and Vivanco (2009) and Neal et al. (2012) has highlighted that root exudates can shape the structure of rhizosphere microbial communities. Root exudates function as substrates, signals, and even antimicrobials in the rhizosphere, influencing the relative abundance of microbial taxa. For example, specific root exudates can stimulate the growth of beneficial bacteria, fostering plant growth while suppressing pathogenic microorganisms. In contrast, some root exudates can directly inhibit the growth of pathogenic microorganisms. However, much remains unknown about the functional capabilities of the majority of these rhizosphere microorganisms, and our comprehension of the relationship between root exudation and microbiome functioning remains limited. Nevertheless, the manipulation of the rhizosphere microbiome through the use of root exudates holds significant promise as a strategy for enhancing plant growth, bolstering disease resistance, and improving soil fertility and productivity.

As a result, the rhizosphere's microbes engage in severe competition for nutrients received from plants. Due to this rivalry, microbes may engage in a bloody struggle for supremacy and survival. To reach their host or to multiply sufficiently on their host before infecting host tissue and eluding the rhizosphere combat zone, soilborne pathogens, including fungi, bacteria, and viruses, typically depend on saprophytic development in the rhizosphere. The microbial population of the soil that a disease infects affects how successful it is.

Every natural soil has some degree of pathogen suppression capacity. This can be seen by contrasting the degree of disease that develops after a pathogen is injected into pasteurized soils, which are devoid of microorganisms, and non-pasteurized soils, which are rich in microorganisms. This phenomenon is known as general disease suppression and is attributed to the total microbial activity in the soil.

The root microbiome is an essential part of the soil microbial community that aids in the prevention of disease. The rhizosphere's microbes can compete with soilborne pathogens for nutrients, produce antibiotics and other secondary metabolites that can stop pathogen growth, and boost plant defenses. Plant health and disease susceptibility can be significantly influenced by the makeup and variety of the root microbiome.

In a favorable soil, organic additives like compost, manure, and cover crops can promote the activity of microbial communities. This increases the diversity and number of advantageous bacteria in the rhizosphere, which enhances the environment's ability to suppress illness in general. Additionally, boosting soil structure, water-holding capacity, and nutrient availability, these additions can also increase plant health and production (Hoitink and Boehm, 1999).

8.3 INTERACTIONS OF PGPR-INDUCED SECONDARY METABOLITES WITH SOIL MICROBES

8.3.1 Modulation of Soil Microbial Community Structure

Soil microorganisms exist as the most complex structure and are associated with variable levels of biological organization, having genetic variability and evenness in communities. They work for an enhanced ecosystem function, and the quality of the ecosystem depends on its diversity rather than the total number of microbial species. The soil microbial community is composed of billions of organisms, which predominantly consist of bacteria, fungi, actinomycetes, and algae (Islam and Wright, 2004). Among them, bacteria are the most diverse group (20,000 species gram^{-1} soil) of soil microorganisms (Brady et al., 2002), followed by fungi. The existence of a soil microbe relies on the accessibility of the sources of energy and carbon supplied by plants. The heterotrophic soil bacteria immediately break down the simple compounds of plants, such as starch/amylum, sucrose, and protein, and then fungi and actinomycetes become active to act on the remaining compounds or by-products. These soil microorganisms are associated with antagonistic (predation, parasitism), competitive, and mutualistic interactions between them and with the roots of higher plants and exist together as a community in the soil. Some beneficial associations of soil microorganisms with higher plants are free-living, symbiotic, and asymbiotic.

The responsibilities and activities of this microbial community are more active in the rhizospheric zone, in contrast to the soil bulk.

The soil microorganisms interact among themselves within a food web cycle, becoming prey or predators of each other. The initiation of this cycle is decomposition, mineralization, and immobilization of the plant's residues and exudates, mostly by bacteria. The plants, while growing, release various metabolites (primary and secondary) as exudates. Both plants and soil microorganisms release common metabolites. The primary (1°) metabolites include carbohydrates, proteins, hormones, lipids (fats, oils, fatty acids), *etc.*, whereas the secondary metabolites include terpenoids, phenolics, alkaloid groups, *etc.* The secondary (2°) metabolites are derived from primary metabolites. These metabolites are the source of maintaining a biological hotspot in the rhizospheric zone of plants. The primary metabolites are the food sources for various soil microorganisms and induce the food web cycle, whereas, the secondary metabolites, especially of soil microbes, play a key role in ecological functions, that is, plant growth–promoting rhizobacteria (PGPR), induced systemic resistance (ISR), growth inhibition of infectious agents, *etc.* (Tyc et al., 2017).

The PGPR interacts with plants through chemical signaling agents, such as quorum sensing (QS), plant growth hormones (PGH), and volatile organic compounds (VOCs). These signaling molecules are the tools for communicating between soil microorganisms and the soil microorganisms with higher plants. PGPR–higher plant interactions symbiotically exist. The PGPR protects the plants from various stresses (biotic and abiotic), supplies growth hormones, and defends against pests and diseases, and in return, the host plant provides carbon and energy sources for survival. The soil microbial communities are more sustained on the plant and the surrounding environment than the interactions among them.

8.3.2 Stimulation of Beneficial Soil Microbes

The soil inhabitants that are symbiotically associated with higher roots of the plant, maintaining a soil nutrient cycle through the decomposition and mineralization processes of organic matter, inducing plant growth hormones, and are active as biocontrol agents are termed "beneficial soil microorganisms (BSMs)." The soil microbial functions that are bound to the plant's health improvement and are eco-friendly with the surrounding environment are BSMs. It includes plant growth–promoting microorganisms (PGPMs), especially comprising soil bacteria and fungi, which facilitate the agroecosystem maintenance that helps plant growth and enhances soil productivity (Santos et al., 2019; Rouphael and Colla, 2020). The PGPR is more concerned with the agroecological system regarding soil nutrient improvement; enhancement of crop nutrient availability; resistance to plant stresses, such as biotic and abiotic; and betterment of plant water absorption (Lobo et al., 2019; Goswami and Deka, 2020).

The encouragement of the BSMs population in the rhizospheric zone depends on the optimum supply of food sources, water, and aeration. It positively impacts soil health in several ways: the addition of organic manure (farmyard manure, vermicompost, green manure, compost), maintenance of optimum soil water and soil air regimes, crop residue management (soil incorporation, mulching), avoidance of soil physical hindrances, and no utilization of pesticides. Many of these BSMs naturally occur in the soil. However, under certain circumstances, it is advantageous to proliferate the microbial populations either by soil inoculation as biofertilizers or by the application of agricultural operation techniques that enhance their abundance and functions.

In the present era, PGPR is popular as a bioinoculant or biofertilizer for atmospheric nitrogen (N_2) fixation; solubilization-mineralization of phosphorus (P), potassium (K), and iron (Fe); enhancement of adaptation against plant stresses; elimination of pest and diseases problems; and enhancement of crop yield and nutritional quality (Koskey et al., 2021). They are axenically cultured in a microbial cell growth medium containing yeast extract, meat extract, malt extract, polypeptone compound, casamino acids, rumen juice, corn steep liquor, cane molasses, *etc.* (Ameyama et al., 1984). The culture medium contains easily available nutrients, energy derivatives, minerals, buffering agents, and a solidifying agent that maintains microbial development, differentiation, and population proliferation. It is the strategy for producing biofertilizers axenically. The biofertilizer may be composed of an inoculant or a combination of inoculants. Before the production of biofertilizers, an axenic compatibility test among the inoculants and influences on plant growth are necessary to understand, and a positive result indicates that the microbial secondary metabolites are in a synergistic relationship with each other.

8.4 Mechanisms of Action of PGPR-Induced Secondary Metabolites

8.4.1 Direct Interactions with Plant and Soil Microbes

There are two types of soil microbes, epiphytes and endophytes. The soil microbes that colonize outside the plant parts are epiphytes, whereas the microbes that inhabit inside the plant parts are endophytes (Nadarajah et al., 2021). They are mostly colonized near the roots, the rhizospheric zone, and interact with the plant roots through their exudates (Dong et al., 2019; Sivakumar et al., 2020). Plant exudates may be amino acids, organic acids, flavonoids, sugars, nutrients, *etc.* that act as chemical signals to regulate the density of soil microbes and induce plant–microbe interaction (Egamberdieva et al., 2017; Compant et al., 2020). The interaction of plants and soil microbes can be categorized as symbiosis and parasitism. The beneficial interactions, that is, the symbiotic interaction, include rhizobacteria–root, mycorrhizal–roots, actinobacteria–root, and other microbes–root interactions.

Among these, the interaction of rhizobacteria with the roots of higher plants (legumes) is the widely explored relationship.

There are no records that soil microorganisms and plants possess sense organs. It became a trigger point to understand how they recognize each other for interactions since a host plant allows a specific species of soil microbe for a specific function, say, *Rhizobium sp.* for nodule formation. The plant exudates as secondary metabolites become a chemical signal agent and are released to attract soil microbes for interaction. In the soil system, each species of microbes has specific signaling agents with the plant's roots and *vis-à-vis* (Figure 8.1). For example, the legume root exudates flavonoid secondary metabolites act as a plant-signaling agent to attract Rhizobia soil microbes and act as the precursor of *Rhizobial Nod*-LCOs factor, which is again detected by the plant, recognizing each other and initiating the interaction pathway leading to nodule formation (Vishwakarma et al., 2020). Likewise, the PGPR detects the plant's signaling secondary metabolite, that is, the LuxR-binding plant compounds QS (quorum sensing) mimics, by its secondary metabolites, such as QS molecules, antimicrobial VOCs (volatile organic compounds), and phytohormones, and recognizes each other for interaction. These interactions respond to soil microbes for rhizomicrobiome shaping, symbiosis, mutualism, *etc.* and respond to plants for hormonal balance, recognition priming, plant gene expression, metabolism, stress defense, induction of systemic defense, *etc.* (Venturi and Keel, 2016). However, the soil microbial population density is a factor for plant–microbe interactions. The quorum-sensing (QS) molecule is the signaling network between microbial cells to a cell for monitoring the population density. It disintegrates when it is not detected or has less microbial population density to respond to the function (Moghaddam et al., 2014). The microbial QS molecule (AHL-Acyl homoserine lactone) mimics plant AHL for interactions and responses to plants for stress, hormone, defense, *etc.*, and virulence, biofilm, motility, protein folding, *etc.* to microbes (Bauer and Mathesius, 2004). The microbial secondary metabolites as phytohormones include auxin, cytokinin, gibberellin, abscisic acids, jasmonates, and ethylene. The metabolites auxin and cytokinin of *Rhizobium* sp. help induce nodule organogenesis and enhance the atmospheric N_2-fixation in legumes (Boivin et al., 2016). Moreover, bacterial secondary metabolite as auxin (IAA – indole acetic acid) indirectly responds to reducing plant stresses by lowering ethylene toxicity and simultaneously increasing the plant's growth improvement (Glick, 2014). The VOCs (butanediol and acetoin) of PGPR are chemically recognized and communicate with the VOCs of plant roots and respond to the plant for biocontrol, bioprotection, phytostimulation, biotolerance, enhancement of plant productivity, and flavor (Ortíz-Castro et al., 2009). Beyond the aforementioned microbial secondary metabolites, many remain to be discovered. The exploration of microbial secondary metabolites of PGPR is still limited, and various mechanisms and pathways of interaction with plants are unknown. This becomes a restriction to enhancing plant growth and development while maintaining a sustainable environment.

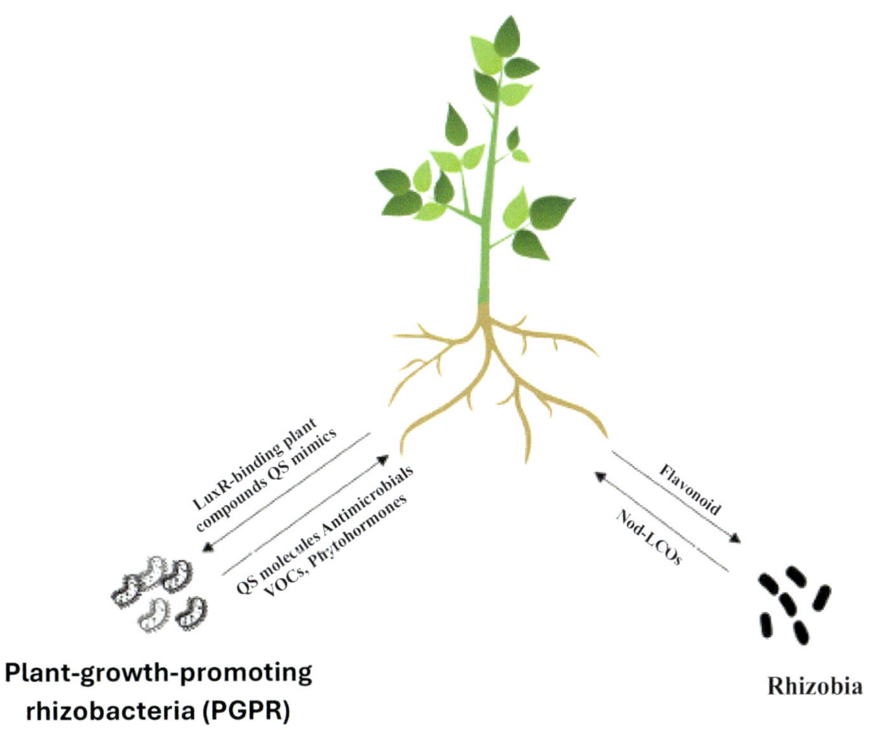

FIGURE 8.1 Chemical signaling agents to recognize and communicate between soil microorganisms and plants.

8.4.2 Interactions of PGPR-Induced Secondary Metabolites with Plants

8.4.2.1 Antagonism against Pathogens

Induced resistance is the defense mechanism of plants that is induced by biological infection or chemical stimuli. It is classified into two broad basic forms: (1) systemic acquired resistance (SAR) and (2) induced systemic resistance (ISR). Generally, SAR is the kind of resistance that is activated through pathogen infection, which depends on the salicylic acid (SA) signaling pathway, which is characterized by over-expression of PR (pathogenesis-related) proteins. On the other side, ISR is the resistance, which is triggered by PGPR or non-pathogenic biocontrol organisms (van Loon et al., 1998), and requires signaling compounds ethylene and jasmonic acid. These signal molecules trigger the accumulation of defense-associated enzymes (polyphenol oxidase (PPO), peroxidase (PO), phenylalanine ammonia lyase (PAL)) and other substances (chalcone synthase, phytoalexin, phenolic compounds, *etc.*) for the creation of chemical and structural barriers against pathogen invasion. Polyphenol oxidase (PPO) and peroxidase (PO) are enzymes that accelerate the synthesis of lignin and some oxidative phenols, which help in the development of defense-related barriers in the plant cell structures in resistance to pathogens (Avdiushko et al., 1993). Other enzymes involved in the production of phytoalexins or phenolic compounds include chalcone synthase, phenylalanine ammonia lyase (PAL), and tyrosine ammonia lyase (TAL) (Gundlach et al., 1992; Creelman et al., 1992). According to research by Ramamoorthy et al. (2002), PAL is an important enzyme of the phenylpropanoid pathway that results in the synthesis of lignin, phenolic compounds, and phytoalexins, which help form the structural and chemical barriers that protect host plants from pathogen attack. PPO catalyzes the production of defense compounds such as tannin, which is lethal to pathogenic microbes (Chen et al., 2000), as well as the synthesis of oxidative phenols, which help inhibit pathogen invasion in the plant cell (Avdiushko et al., 1993). Sarkar et al. (2014) observed that *Pseudomonas* Cgr and S1 strains enhanced PPO activity in the consortial mode, as well as PPO function in chickpea when challenged with *Sclerotinia sclerotiorum* and subjected to salt stress. Similarly, in cucumber plants, the endophytic bacteria of wheat, *Pseudomonas aeruginosa* PW09, alleviate biotic and abiotic stresses caused by *Sclerotium rolfsii* and 150 mM NaCl, respectively (Pandey et al., 2012). PGPR and their metabolites have played a critical role in the reduction of chickpea wilt (*Fusarium oxysporum* f. sp. *ciceris*) in greenhouse and field studies.

8.5 APPLICATIONS OF PGPR-INDUCED SECONDARY METABOLITES IN AGRICULTURE

8.5.1 Disease Management

PGPRs are known to produce biologically active secondary metabolites that affect host–pathogen interactions (Kumar et al., 2017). Siderophores, bacteriocins, and antibiotics are the most powerful and well-known mechanisms an antagonist might use to limit or prevent pathogenic growth and proliferation.

8.5.2 Antibiotics

Antibiotics are a broad and heterogeneous class of low-molecular-weight chemical substances synthesized by a diverse group of microbes. These are toxic to the development or metabolic activities of other bacteria at very low concentrations. The antibiotics, *viz.*, iturin, surfactin, fengycin, pyoluteorin, pyrrolnitrin, phenazins, 2,4-diacetylphloroglucinol, mycosubtilin, subtilosin, sublancin, bacilysin, viscosinamide, tensin, amphisin, polymyxin, circulin and colistin, xanthobaccin, zwittermicin A, oligomycin A, oomycin A, and kanosamine have been isolated from various PGPR (Beneduzi et al., 2012). The two most significant PGPR genera, *Bacillus* and *Pseudomonas*, are the subject of extensive research on the effectiveness of their metabolites in managing different diseases.

Pseudomonas has been successfully used for many years as a biocontrol agent for managing plant diseases. It is an agile organism that can produce a staggering number of secondary metabolites with antimicrobial properties. Many antibiotic metabolites produced by *Pseudomonas* spp., which include phenazine 1-carboxylic acid, pyrrolnitrin, 2,4-diacetylphloroglucinol, hydrogen cyanide, phenazine-1-carboxamide, oomycin A, pyocyanin, pyoluteorin, anthranilate, viscosinamide, gluconic acid, and ammonia, have been reported to manage many soilborne plant pathogenic organisms. After taking into account the method of root colonization by pseudomonads, both abiotic and biotic variables affecting colonization, bacterial characteristics, and the genes that influence rhizosphere competence, and disease suppression mechanisms, Weller (2007) concluded that *Pseudomonas* spp. have great potential as biocontrol agents in the management of soilborne diseases.

Bacillus is the second-most extensively used PGPR for the biological control of plant diseases, particularly soilborne diseases, after *Pseudomonas*. It produces a variety of physiologically active metabolites with antibiotic characteristics, such as lipopeptides and polyketides. The lipopeptide families fengycin and iturin are accountable for the antagonistic effect of *B. subtilis* on *Podosphaera fusca*. Furthermore, fengycin and surfactin lipopeptides operate like bacterial impulses to provoke immunological responses in plant cells through the progress of ISR (Ongena et al., 2007). The development of the bacilysin and difficidin polyketides by *B. amyloliquefaciens* is responsible for the successful management of fire blight disease. Therefore, *Bacillus* spp. take part in the biological management of fungal and bacterial diseases of crop plants (Prashanth, 2007; Shafi et al., 2017).

8.5.3 Bacteriocins

A key attribute that bacteriocins vary from conventional antibiotics is that they frequently have a very limited killing range and are only harmful to bacteria that are closely related to the developing strain. It is identified in important plant pathogenic bacterial species, such as *Pseudomonas*, *Xanthomonas*, *Agrobacterium*, and *Pectobacterium* (Rooney et al., 2020). Based on their size, these bacteriocins are classified as small peptide bacteriocins, colicin-like bacteriocins, tailocins (phage-tail-like bacteriocins), and lectin-like bacteriocins (Grinter et al., 2012a) (Table 8.1). The use of bacteriocins against bacterial diseases of plants is limited; however, it is owing to a lack of research studies of bacteriosin diversity and the mode of action of their associated proteins on plant pathogenic bacteria, rather than a lack of potential. Agrocin 84 (1.4 kDa), a bacteriocin-like chemical substance produced in certain non-pathogenic *Agrobacterium tumefaciens* strains, was characterized and utilized to treat crown gall diseases that are produced by pathogenic strains of *A. tumefaciens* (Ellis et al., 1979). Because of the efficacy of this bacteriocin as a method for controlling crown gall disease, Bio-Care Technology developed an agrocin 84–synthesizing strain of *A. tumefaciens*, which was properly branded and sold in the name of Nogall (Jones et al., 1988). *P. fluorescens* SF4c tailocins show antibacterial action against several plant pathogenic bacteria. Príncipe et al. (2018) evaluated the efficiency of foliar sprays of these bacteriocins in controlling *Xanthomonas vesicatoria* (Xcv Bv5–4a), causing bacterial-spot disease in tomatoes. The severity and incidence indexes of disease were lowered by 44% and 36%, respectively. In another study, Lavermicocca et al. (2002) demonstrated the application of an unidentified bacteriocin produced by *P. syringae* pv. *ciccaronei* to prevent olive knot disease. This study found that applying crude bacteriocin to stem wounds before infection with the pathogen *P. syringae* pv. *savastanoi* reduced knot formation by 60–80%. Bacteriocins produced by *Pectobacterium* strains containing Ctv and CGE are commercialized in Japan under the brand name "Biokeeper" to treat bacterial soft rot disease in potatoes (Chuang et al., 1999).

8.5.4 Siderophore

PGPR reduces pathogen growth through the production of siderophores, which bind the iron (III) present near the plant roots, restricting pathogen development near the root. Plants, on the other hand, are unaffected by such territorial soil depletion of iron since they can tolerate much lower iron concentrations (1,000-fold less) than bacteria. Different types of bacterial siderophores have been found, primarily hydroxamates, carboxylates, and catecholates. Among the studied bacterial siderophores are those produced by pseudomonads, which have a high affinity for the ferric ion. For example, the potent siderophore pyoverdin can prevent the growth of bacteria and fungi in environments that lack iron (Kloepper et al., 1980). In iron-deficient soil, a pseudobactin siderophore produced by *P. putida* B10 strain inhibited the formation of iron chelators by microbes; however, this suppression was eliminated when the soil was restored with iron, a condition that inhibits the production of iron chelators by microorganisms (Kloepper et al., 1980). According to Scher and Baker (1982), the deprivation of ferric iron by *P. putida*–producing siderophores is the mechanism by which the *Fusarium* wilt pathogens of cucumber, radish, and flax are suppressed. The pathogen *Botrytis cinerea* is similarly suppressed by fluorescent *Pseudomonas* that secretes the siderophore pyoverdine by depleting iron (Trapet et al., 2016). In another study, the root rot disease is suppressed by siderophores produced by *Bacillus* spp. (MR-11(2) and MRF) and fluorescent *Pseudomonas* EM85 isolated from the maize rhizosphere (Pal et al., 2001). A rhizobacterium, *Streptomyces* sp. IISRBPAct1, has been shown to limit the growth of *S. rolfsii* and protect black pepper plants by synthesizing siderophores, as well as to minimize foot rot occurrence by up to 80% (Thampi and Bhai, 2017).

8.5.5 Other Secondary Metabolites

8.5.5.1 Hydrolytic Enzymes

Several PGPR with hyperparasitic activity combat pathogens by secreting enzymes that break down cell walls, like chitinases, glucanases, cellulases, and proteases (Santoyo et al., 2021). In this case, *Botrytis cinerea*, a pathogenic fungus, fails to germinate or extend its germ tube when *Serratia plymuthica* secretes chitinase (Frankowski et al., 2001). *Streptomyces* sp. 385 and *Paenibacillus* sp. 300

TABLE 8.1
Classes of Bacteriocin and Their Examples

Type	Name of Bacteriocin	Producing Organism	Reference
Colicin-like bacteriocins	Carocin S1K, Carocin D, Carocin S1 and S2, and Pectocins M1 and M2	*Pectobacterium carotovorum*	Atanaskovic et al., 2020,Grinter et al., 2012b
	Syringacin	*P. syringae*	Hockett et al., 2017
Lectin-like bacteriocins	Putidacin L1 (PL1)	*P. putida*	Parret et al., 2003
	Pyocin L1	*P. aeruginosa*	McCaughey et al., 2014
Tailocins	Tailocin	*P. fluorescens* SF4c	Príncipe et al., 2018
	Carotovoricin (Ctv) Er	*P. carotovorum IAM 1068*	Yamada et al., 2006
	Carotovoricin CGE	*P. carotovorum* CGE234-M403	

both produce β-1,3-glucanase and chitinase to lyse the cell walls of wilt-causing fungus *F. oxysporum* f. sp. *cucumerinum*, as demonstrated by Singh et al. (1999). Fungal pathogens such as *F. oxysporum*, *F. solani*, and *R. solani* are all inhibited by the enzymes chitinase and 1,4-β-glucanase produced by *Bacillus* sp. BPR7 (Kumar et al., 2012). The mycelial growth of *B. cinerea* was hindered by the enzyme chitinase produced by *Bacillus thuringiensis* UM96 (Martínez-Absalón et al., 2014), whereas the three enzymes chitinase, glucanase, and protease produced by *B. atrophaeus* JZB120050 also inhibited the growth of *B. cinerea* (Ni et al., 2018).

Another group of hydrolytic enzymes, that is, lactonases and acylases, is involved in the degradation of signaling molecules in bacteria; this phenomenon is known as signal interference. Most plant pathogenic bacteria can elicit pathogenicity/virulence factors at the highest bacterial population density through the recognition of signaling molecules or quorum-sensing (QS) molecules (also known as autoinducers), such as N-acylhomoserine lactones (AHLs) (Helman and Chernin, 2015). The interference with QS mediated by the enzymatic breakdown of AHLs is the basis for signal interference as a biocontrol approach. For example, *Ralstonia* sp. acylases break the AHL amide (Lin et al., 2003), while *Bacillus thuringiensis* lactonases hydrolyze the AHL lactone ring (Dong et al., 2002). By inhibiting AHL-quorum sensing, *Rhodococcus erythropolis* and *B. simplex* decreased the disease-producing ability of *A. tumefaciens* (Uroz et al., 2003) and *E. amylovora* (Hanano et al., 2014).

8.5.5.2 Lipopeptides (LP)

Lipopeptides (LPs) are oligopeptides with a lipid tail linked to them. Lipopeptides are synthesized by various fungi and bacteria, including *Streptomyces*, *Bacillus*, and *Pseudomonas*. LPs have received a lot of interest because of their antibacterial, anticancer, cytotoxic, immunosuppressor, and surface-sterilizing activities (Gross and Loper, 2009). The main mechanism of action postulated for LPs is the development of a pore in cell membranes, resulting in a discrepancy in transmembrane ion fluxes and cell death (Baltz, 2009). LPs have been noted for their inhibitory properties by preventing the growth of fungi with increased branching, inflated hyphae, reduced hydrophobicity of hyphae, decreased intracellular pH, changed mitochondrial organization, and impaired activities of mitochondria and esterases, all of which contribute to the inhibitory effect on fungi (Thrane et al., 1999). Lipopeptides developed by *Pseudomonas* spp. possess the ability to lyse *Pythium* and *Phytophthora* spp. zoospores. This effect has been well characterized in viscosin-containing LPs (Raaijmakers et al., 2010). The antiviral efficacy of a viscosin group of LPs toward enveloped viruses was also reported by Nybroe and Sørensen (2004). NAHL-based quorum sensing is involved with the synthesis of cyclic LPs in some phytopathogenic *Pseudomonas* strains, such as *P. fluorescens* 5064. Non-pathogenic *Pseudomonas* strains enhance plant defenses by producing LPs resistant to oomycetes and other pathogenic fungi. The application of obtained massetolide-A, a product of *P. fluorescens*, to tomato root systems led to plant resistance to the *P. infestans* infestation. (Tran et al., 2007).

8.6 LIMITATIONS

8.6.1 Standardization Challenges

Plant growth–promoting rhizobacteria (PGPR) are a class of advantageous microbes that colonize the roots of plants and improve their health and growth in a variety of ways. In agriculture, the use of PGPR has drawn a lot of interest as an environmentally benign strategy to increase crop yield, cut back on chemical inputs, and support sustainable farming. However, a significant obstacle to the development and marketing of PGPR products has been the lack of standardization and quality control. Here are some restrictions and difficulties brought on by the absence of PGPR standards and quality control.

The first stage in creating a beneficial plant–microbe interaction is for the microbe to colonize the plant roots. This complex process necessitates the capacity of bacteria to compete in the rhizosphere soil for a suitable niche. The climatic conditions required for a particular crop determine the effectiveness of PGPR field application (Mahanty et al., 2017). To optimize the efficiency of the used PGPR strain, it is strongly encouraged to identify region-specific microbial strains. Inadequate amounts of PGPRs are employed regularly as a rhizosphere inoculum for host plants, preventing effective rhizosphere colonization in a natural setting. These PGPRs are not combined with an appropriate carrier. To ensure that the environment is ideal for the PGPR strain's survival and growth in the rhizosphere, it is crucial to carefully choose the right one.

8.6.2 Lack of Defined Criteria for PGPR

There are no universally accepted definition or criteria for PGPR, which makes it difficult to compare the efficacy of different PGPR products. Different strains of bacteria can have different modes of action, and without a clear definition, it is challenging to ensure that the product is consistent in its efficacy.

8.6.3 Heterogeneous Nature of PGPR

PGPRs are a broad set of microorganisms that have different needs for development, modes of operation, and types of chemicals they create. As a result, creating a standardized procedure for the manufacturing and quality assurance of PGPR products is challenging.

8.6.4 Lack of Standardized Production Protocols

Small-scale manufacturers who employ various culturing and formulation techniques frequently create PGPR products. The goods' effectiveness and quality may differ as a consequence of all this.

8.6.5 Limited Knowledge on the Safety of PGPR

The safety of PGPR products is not well understood, especially when it comes to their application in food crops. It is challenging to control and standardize the manufacture and use of these items due to the lack of safety information.

8.6.6 Lack of Standardization in Testing

Standardized protocols for testing the efficacy of PGPR products are lacking. This makes equating the results of different studies difficult, which can stymie the development and commercialization of these products.

8.6.7 Complexity of Field Trials

Conducting field trials to test the efficacy of PGPR products is a complex process that involves several variables, such as soil type, environmental conditions, and crop variety. These variables can have an impact on trial results, determining the efficiency of the commodities.

8.6.8 Quality Control Challenges

Because of the heterogeneity of the microorganisms and the lack of standardized production protocols, quality control of PGPR products is difficult. This might lead to discrepancies in product quality and efficacy, affecting their commercial viability.

Overall, the lack of standardization and quality control of PGPR products poses significant challenges to the development and commercialization of these products. Addressing these challenges requires the establishment of clear criteria for PGPR, the development of standardized protocols for production and testing, and the generation of safety data.

8.7 CONCLUSION

The study of plant growth–promoting rhizobacteria (PGPR)–induced secondary metabolites has emerged as a transformative approach to sustainable agriculture. These bioactive compounds play a crucial role in enhancing soil fertility, improving plant growth, and suppressing plant pathogens, thereby reducing reliance on chemical fertilizers and pesticides. The intricate interactions between PGPR, secondary metabolites, and the soil microbiome create a dynamic and beneficial ecosystem that supports plant health and productivity.

Key findings from this research emphasize that PGPR-mediated production of secondary metabolites such as antibiotics, lipopeptides, siderophores, and volatile organic compounds significantly influences plant–microbe interactions. These metabolites enhance plant resistance through induced systemic resistance (ISR) and systemic acquired resistance (SAR), modulating the plant immune response against biotic and abiotic stresses. Furthermore, these metabolites contribute to nutrient solubilization, biofilm formation, and quorum sensing regulation, which collectively improve microbial community dynamics in the rhizosphere.

Additionally, PGPR-induced secondary metabolites have shown immense potential in disease management strategies by suppressing phytopathogens through antagonism, competition, and direct inhibition. The successful application of PGPR as biocontrol agents and biofertilizers underlines their relevance in modern agricultural practices, particularly in organic and precision farming systems. The modulation of the rhizosphere microbiome through PGPR interactions also enhances soil health, leading to long-term agricultural sustainability.

Despite these advancements, challenges persist in the standardization and commercialization of PGPR-based products. Variability in field performance due to environmental factors, host plant specificity, and microbial competition limits their widespread application. Future research should focus on multistrain consortia development, elucidation of molecular mechanisms governing plant–microbe interactions, and optimization of formulation strategies to ensure consistency and efficacy. The integration of omics technologies, synthetic biology, and bioinformatics tools will further refine our understanding of PGPR-induced secondary metabolites, paving the way for innovative and scalable biotechnological applications in sustainable agriculture.

REFERENCES

Abedon, B.G., and W.F. Tracy. 1996. Corngrass 1 of maize (*Zea mays* L.) delays development of adult plant resistance to common rust (*Puccinia sorghi* Schw.) and European corn borer (*Ostrinia nubilalis* Hubner). *Journal of Heredity* 87: 219–23. https://doi.org/10.1093/oxfordjournals.jhered.a022988Ahemad, M., and M.S. Khan. 2011. Plant growth promoting rhizobacteria: Recent advancements. *Insight Microbiology* 1: 39–54. https://doi.org/10.5567/IMICRO-IK.2011.39.54.

Ahemad, M., and M. Kibret. 2014. Mechanisms and applications of plant growth promoting rhizobacteria: Current perspective. *Journal of King saud University-science* 26: 1–20. https://doi.org/10.1016/j.jksus.2013.05.001.

Ameyama, M., E. Shinagawa, K. Matsushita, and O. Adachi. 1984. Growth stimulating substance for microorganisms produced by *Escherichia coli* causing the reduction of the lag phase in microbial growth and identity of the substance with pyrroloquinoline quinone. *Agricultural and Biological Chemistry* 48: 3099–07. https://www.tandfonline.com/doi/abs/10.1080/00021369.1984.10866645.

Atanaskovic, I., K. Mosbahi, C. Sharp, et al. 2020. Targeted killing of *Pseudomonas aeruginosa* by pyocin G occurs *via* the hemin transporter Hur. *Journal of Molecular Biology* 432: 3869–80. https://doi.org/10.1016/j.jmb. 2020.04.020.

Avdiushko, S.A., X.S. Ye, and J. Kuc. 1993. Detection of several enzymatic activities in leaf prints cucumber plant.

Physiological and Molecular Plant Pathology 42: 441–54. https://doi.org/10.1006/pmpp.1993.1033sl.
Badri, D.V., V.M. Loyola-Vargas, C.D. Broeckling, et al. 2008. Altered profile of secondary metabolites in the root exudates of *Arabidopsis* ATP-binding cassette transporter mutants. *Plant Physiology* 146: 762–71. https://doi.org/10.1104/pp.107.109587.
Badri, D.V., and J.M. Vivanco. 2009. Regulation and function of root exudates. *Plant Cell and Environment* 32: 666–81. https://doi.org/10.1111/j.1365-3040.2008.01926.x.
Baltz, R.H. 2009. Daptomycin: Mechanisms of action and resistance, and biosynthetic engineering. *Current Opinion in Chemical Biology* 13: 144–51. https://doi.org/10.1016/j.cbpa.2009.02.031.
Bauer, W.D., and U. Mathesius. 2004. Plant responses to bacterial quorum sensing signals. *Current Opinion in Plant Biology* 7: 429–33. https://doi.org/10.1016/j.pbi.2004.05.008.
Beneduzi, A., A. Ambrosini, and L.M.P. Passaglia. 2012. Plant growth-promoting rhizobacteria (PGPR): Their potential as antagonists and biocontrol agents. *Genetics and Molecular Biology* 35: 1044–51. https://doi.org/10.1590/S1415-47572012000600020
Boivin, S., C. Fonouni-Farde, and F. Frugier. 2016. How auxin and cytokinin phytohormones modulate root microbe interactions. *Frontiers in Plant Science* 7 (August): 1240, 1–12. https://www.frontiers.org/articles/10.3 389/fpls.2016.01240/full.
Brady, N.C., and R.R. Weil. 2002. Organisms and ecology of the soil. In *Nature and properties of soils*, 13th ed., 449–97. NJ: Prentice Hall.
Broeckling, C.D., A.K. Broz, J. Bergelson, D.K. Manter, and J.M. Vivanco. 2008. Root exudates regulate soil fungal community composition and diversity. *Applied Environmental Microbiology* 74: 738–44. https://doi.org/10.1128/AEM.02188-07.
Bulgarelli, D., K. Schlaeppi, S. Spaepen, E.V.L.V. Themaat, and P. Schulze-Lefert. 2013. Structure and functions of the bacterial microbiota of plants. *Annual Review of Plant Biology* 64: 807–38. https://doi.org/10.1146/annurev-arplant-050312-120106.
Chaparro, J.M., A.M. Sheflin, D.K. Manter, and J.M. Vivanco. 2012. Manipulating the soil microbiome to increase soil health and plant fertility. *Biology and Fertility of Soils* 48: 489–99. https://doi.org/10.1007/s00374-012-0691-4.
Chen, C., R.R. Belanger, N. Benhamou, and T.C. Paulitz. 2000. Defence enzymes induced in cucumber roots by treatment with plant growth-promoting rhizobacteria (PGPR) and *Pythium aphanidermatum*. *Physiological and Molecular Plant Pathology* 56: 13–23. https://doi.org/10.1006/pmpp.1999.0243.
Chuang, D.Y., A.G. Kyeremeh, Y. Gunji, Y. Takahara, Y. Ehara, and T. Kikumoto. 1999. Identification and cloning of an *Erwinia carotovora* subsp. *carotovora* bacteriocin regulator gene by insertional mutagenesis. *Journal of Bacteriology* 181: 1953–57. https://doi.org/10.1128/jb.181.6.1953-1957.1999.
Compant, S., M.C. Cambon, C. Vacher, B. Mitter, A. Samad, and A. Sessitsch. 2020. The plant endosphere world-bacterial life within plants. *Environmental Microbiology* 23: 1812–29. https://doi.org/10.1111/1462-2920.15240.
Coronado, C., J.A.S. Zuanazzi, C. Sallaud, et al. 1995. Alfalfa root flavonoid production is nitrogen regulated. *Plant Physiology* 108: 533–42. https://doi.org/10.1104/pp.108.2.533.
Creelman, R.A., M.L. Tierney, and J.E. Mullet, 1992. Jasmonic acid/methyl jasmonate accumulation in wounded soybean hypocotyls and modulate wound gene expression. *Proceedings of National Academy of Sciences USA* 89: 4938–41. https://doi.org/10.1073/pnas.89.11.4938.
Davidson, I.A., and M.J. Robson. 1986. Effect of contrasting patterns of nitrate application on the nitrate uptake, N2-fixation, nodulation and growth of white clover. *Annals of Botany* 57: 331–38. https://doi.org/10.1093/oxfordjournals.aob.a087114.
Develey-Riviere, M.P., and E. Galiana. 2007. Resistance to pathogens and host developmental stage: A multifaceted relationship within the plant kingdom. *New Phytologist* 175: 405–16. https://doi.org/10.1111/j.1469-8137.2007.02130.x.
Dong, C.J., L.L. Wang, Q. Li., and Q.M. Shang. 2019. Bacterial communities in the rhizosphere, phyllosphere and endosphere of tomato plants. *PLoS One* 14 (November): e0223847, 1–17. https://doi.org/10.1371/journal.pone.0223847.
Dong, Y.H., A.R. Gusti, Q. Zhang, J.L. Xu, and L.H. Zhang. 2002. Identification of quorum-quenching N-acyl homoserine lactonases from *Bacillus* species. *Applied and Environmental Microbiology* 68: 1754–59. https://doi.org/10.1128/AEM.68.4.1754-1759.2002
Egamberdieva, D., S.J. Wirth., A.A. Alqarawi, E.F. Abd_Allah, and A. Hashem. 2017. Phytohormones and beneficial microbes: Essential components for plants to balance stress and fitness. *Frontiers in Microbiology* 8 (October): 2104, 1–14. https://doi.org/10.3389/fmicb.2017.02104.
Ellis, J.G., A. Kerr, M. Van Montagu, and J. Schell. 1979. *Agrobacterium*: Genetic studies on agrocin 84 production and the biological control of crown gall. *Physiological Plant Pathology* 15: 311–16. https://doi.org/10.1016/0048-4059(79)90082–1.
Fahad, S., S. Hussain, A. Bano, et al. 2015. Potential role of phytohormones and plant growth-promoting rhizobacteria in abiotic stresses: Consequences for changing environment. *Environmental Science and Pollution Research* 22: 4907–21. https://doi.org/10.1007/s11356-014-3754-2.
Flores, H.E., J.M. Vivanco, and V.M. Loyola-Vargas. 1999. 'Radicle' biochemistry: The biology of root-specific metabolism. *Trends in Plant Science* 4: 220–26. https://doi.org/10.1016/S1360-1385 (99)01411–9.
Frankowski, J., M. Lorito, F. Scala, R. Schmid, G. Berg, and H. Bahl. 2001. Purification and properties of two chitinolytic enzymes of *Serratia plymuthica* HRO-C48. *Archives of Microbiology* 176: 421–26. https://doi.org/10.1007/s002030100347.
Glick, B.R. 2014. Bacteria with ACC deaminase can promote plant growth and help to feed the world. *Microbiological Research* 169: 30–39. https://doi.org/10.1016/j.micres.2013.09.009.
Goswami, M., and S. Deka. 2020. Plant growth-promoting rhizobacteria – alleviators of abiotic stresses in soil: A review. *Pedosphere* 30: 40–61. https://doi.org/10.1016/S1002-0160(19)60839–8.
Grinter, R., J. Milner, and D. Walker. 2012a. Bacteriocins active against plant pathogenic bacteria. *Biochemical Society Transactions* 40: 1498–02. https://doi.org/10.1042/bst20120206.
Grinter, R., J. Milner, and D. Walker. 2012b. Ferredoxin containing bacteriocins suggest a novel mechanism of iron uptake in *Pectobacterium* spp. *PLoS One* 7 (March): e33033, 1–9. https://doi.org/10.1371/journal.pone.0033033.
Gross, H., and J.E. Loper. 2009. Genomics of secondary metabolite production by *Pseudomonas* spp. *Natural Product Reports* 26: 1408–46. https://doi.org/10.1039/B817075B.
Gundlach, H., M.J. Muller, T.M. Kutchan, and M.H. Zenk. 1992. Jasmonic acid is a signal transducer in elicitor-induced plant cell cultures. *Proceedings of National Academy of Sciences of USA* 89: 2389–93. https://doi.org/10.1073/pnas.89.6.2389.

Gupta, A., S. Jain, and N. Verma. 2022. *Industrial applications of soil microbes*, vol. 1. Bentham Science Publishers Pte Ltd. https://doi.org/10.2174/97898150399551220101.

Gupta, A., S. Jain, and N. Verma. 2024. *Industrial applications of soil microbes*, vol. 3. Bentham Science Publishers Pte Ltd. https://doi.org/10.2174/97898150797531240301.

Gupta, G., S.S. Parihar, N.K. Ahirwar, S.K. Snehi, and V. Singh. 2015. Plant growth promoting rhizobacteria (PGPR): Current and future prospects for development of sustainable agriculture. *Journal of Microbial and Biochemical Technology* 7: 96–102. http://dx.doi.org/10.4172/1948-5948.1000188.

Hanano, A., M. Harba, M. Al-Ali, and H. Ammouneh. 2014. Silencing of *Erwinia amylovora* sy69 AHL-quorum sensing by a *Bacillus simplex* AHL-inducible aiiA gene encoding a zinc-dependent *N*-acyl-homoserine lactonase. *Plant Pathology* 63: 773–83. https://doi.org/10.1111/ppa.12142.

Helman, Y., and L. Chernin. 2015. Silencing the mob: Disrupting quorum sensing as a means to fight plant disease. *Molecular Plant Pathology* 16: 316–29. https://doi.org/10.1111/mpp.12180.

Heydarian, Z., M. Yu, M. Gruber, B.R. Glick, R. Zhou, and D.D. Hegedus. 2016. Inoculation of soil with plant growth promoting bacteria producing 1-aminocyclopropane-1-carboxylate deaminase or expression of the corresponding acdS gene in transgenic plants increases salinity tolerance in *Camelina sativa*. *Frontiers in Microbiology* 7 (December): 1966, 1–17. https://doi.org/10.3389/fmicb.2016.01966.

Hockett, K.L., M. Clark, S. Scott, and D.A. Baltrus. 2017. Conditionally redundant bacteriocin targeting by *Pseudomonas syringae*. *bioRxiv*: 1–25. https://doi.org/10.1101/167593.

Hoitink, H., and M. Boehm. 1999. Biocontrol within the context of soil microbial communities: A substrate-dependent phenomenon. *Annual Review of Phytopatho*logy 37: 427–46. https://doi.org/10.1146/annurev.phyto.37.1.427.

Hong, Y.W., B.R. Glick, and J.J. Pasternak. 1991. Plant microbial interaction under gnotobiotic conditions: A scanning electron microscope study. *Current Microbiology* 23: 111–14. https://doi.org/10.1007/BF02092259.

İnceoğlu, Ö., W.A. Al-Soud, J.F. Salles, A.V. Semenov, and J.D. van Elsas. 2011. Comparative analysis of bacterial communities in a potato field as determined by pyrosequencing. *PloS One* 6 (August): e23321, 1–11. https://doi.org/10.1371/journal.pone.0023321.

Islam, K.R., and S.R. Wright. 2004. Microbial Biomass Measurement Methods. In *Encyclopedia of soil science*, ed. R. Lal. New York: Marcel Dekker, Inc.

Jain, S., A. Gupta, and N. Verma. 2023. *Industrial applications of soil microbes*, vol. 2. Bentham Science Publishers Pte Ltd. https://doi.org/10.2174/97898150502641230201.

Jain, S., A. Gupta, and N. Verma. 2024. *Industrial applications of soil microbes*, vol. 4. Bentham Science Publishers Pte Ltd. https://doi.org/10.2174/97898151249961240401.

Jha, C.K., and M. Saraf. 2015. Plant growth promoting rhizobacteria (PGPR): A review. *E3 Journal of Agricultural Research and Development* 5: 108–19. https://e3journals.org/cms/articles/1434021343_Chaitanya%20and%20Meenu.pdf.

Jones, D.A., M.H. Ryder., B.G. Clare, S.K. Farrand, and A. Kerr. 1988. Construction of a Tra-deletion mutant of pAgK84 to safeguard the biological control of crown gall. *Molecular Genetics and Genomics* 212: 207–14. https://doi.org/10.1007/BF00334686.

Kloepper, J.W., J. Leong, M. Teintze, and M.N. Schroth. 1980. *Pseudomonas* siderophores: A mechanism explaining disease suppressive soils. *Current Microbiology* 4: 317–20. https://doi.org/10.1007/BF02602840.

Koskey, G., S.W. Mburu., R. Awino., E.M. Njeru., and J.M. Maingi. 2021. Potential use of beneficial microorganisms for soil amelioration, phytopathogen biocontrol, and sustainable crop production in smallholder agroecosystems. *Frontiers in Sustainable Food Systems* 5 (April): 606308, 1–21. https://doi.org/10.3389/fsufs.2021.606308.

Kumar, A., H. Verma, V.K. Singh, et al. 2017. Role of *Pseudomonas* sp. in sustainable agriculture and disease management. In *Agriculturally important microbes for sustainable agriculture*, eds. V.S. Meena, P.K. Mishra, J.K. Bisht, and A. Pattanayak, 195–215. Singapore: Springer.

Kumar, P., R.C. Dubey, and D.K. Maheshwari. 2012. *Bacillus* strains isolated from rhizosphere showed plant growth promoting and antagonistic activity against phytopathogens. *Microbiological Research* 167: 493–99. https://doi.org/10.1016/j.micres.2012.05.002.

Lavermicocca, P., S.L. Lonigro, F. Valerio, A. Evidente, and A. Visconti. 2002. Reduction of olive knot disease by a bacteriocin from *Pseudomonas syringae* pv. *ciccaronei*. *Applied and Environmental Microbiology* 68: 1403–07. https://doi.org/10.1128/AEM.68.3.1403-1407.2002.

Lin, Y.-H., J.-L. Xu, J. Hu, et. al. 2003. Acyl-homoserine lactone acylase from *Ralstonia* strain XJ12B represents a novel and potent class of quorum-quenching enzymes. *Molecular Microbiology* 47: 849–60. https://doi.org/10.1046/j.1365-2958.2003.03351.x.

Lobo, C.B., M.S.J. Tomás., E. Viruel., M.A. Ferrero., and M.E. Lucca. 2019. Development of low-cost formulations of plant growth-promoting bacteria to be used as inoculants in beneficial agricultural technologies. *Microbiological Research* 219: 12–25. https://doi.org/10.1016/j.micres.2018.10.012.

Mader, P., A. Fliessbach, D. Dubois, L. Gunst, P. Fried, and U. Niggli. 2002. Soil fertility and biodiversity in organic farming. *Science* 296: 1694–97. https://www.science.org/doi/10.1126/science.1071148.

Mahanty, T., S. Bhattacharjee, M. Goswami, et al. 2017. Biofertilizers: A potential approach for sustainable agriculture development. *Environmental Science and Pollution Research* 24: 3315–35. https://doi.org/10.1007/s11356-016-8104-0.

Malagoli, P., P. Laine, E.L. Deunff, L. Rossato, B. Ney, and A. Ourry. 2004. Modeling nitrogen uptake in oilseed rape cv. Capitol during a growth cycle using influx kinetics of root nitrate transport systems and field experimental data. *Plant Physiology* 134: 388–400. https://doi.org/10.1104/pp.103.029538.

Martínez-Absalón, S., D. Rojas-Solís, R. Hernández-León, et al. 2014. Potential use and mode of action of the new strain *Bacillus thuringiensis* UM96 for the biological control of the grey mould phytopathogen *Botrytis cinerea*. *Biocontrol Science and Technology* 24: 1349–62. https://doi.org/10.1080/09583157.2014.940846.

McCaughey, L.C., R. Grinter, I. Josts, et al. 2014. Lectin-like bacteriocins from *Pseudomonas* spp. utilise D-rhamnose containing lipopolysaccharide as a cellular receptor. *PLoS Pathogens* 10 (February): e1003898, 1–15. https://doi.org/10.1371/journal.ppat.1003898.

Mendes, R., M. Kruijt, I. de Bruijn, et al. 2011. Deciphering the rhizosphere microbiome for disease-suppressive bacteria. *Science* 332: 1097–100. https://www.science.org/doi/10.1126/science.1203980.

Moghaddam, M.M., S. Khodi, and A. Mirhosseini. 2014. Quorum sensing in bacteria and a glance on *Pseudomonas aeruginosa*. *Clinical Microbiology* 3 (June): 1000156, 1–11. http://dx.doi.org/10.4172/2327-5073.1000156.

Nadarajah, K., and N.S.N. Abdul Rahman. 2021. Plant–microbe interaction: Aboveground to belowground, from the good to the bad. *International Journal of Molecular Sciences* 22 (September): 10388, 1–35. https://www.mdpi.com/1422-0067/22/19/10388.

Neal, A.L., S. Ahmad, R. Gordon-Weeks, and J. Ton. 2012. Benzoxazinoids in root exudates of maize attract *Pseudomonas putida* to the rhizosphere. *PLoS One* 7 (April): e35498, 1–10. https://doi.org/10.1371/journal.pone.0035498.

Ni, M., Q. Wu, J. Wang, et al. 2018. Identification and comprehensive evaluation of a novel biocontrol agent *Bacillus atrophaeus* JZB120050. *Journal of Environmental Science and Health – Part B Pesticides, Food Contaminants, and Agricultural Wastes* 53: 777–85. https://doi.org/10.1080/03601234.2018.1505072.

Nybroe, O., and J. Sørensen. 2004. Production of cyclic lipopeptides by fluorescent pseudomonads. In *Pseudomonas: Biosynthesis of macromolecules and molecular metabolism*, ed. J.L. Ramos, 147–72. Boston, MA: Springer. https://doi.org/10.1007/978-1-4419-9088-4_5.

Omer, Z.S., R. Tombolini, A. Broberg, and B. Gerhardson. 2004. Indole-3-acetic acid production by pink-pigmented facultative methylotrophic bacteria. *Plant Growth Regul*ation 43: 93–96. https://doi.org/10.1023/B:GROW.0000038360.09079.ad.

Ongena, M., E. Jourdan, A. Adam, et al. 2007. Surfactin and fengycin lipopeptides of *Bacillus subtilis* as elicitors of induced systemic resistance in plants. *Environmental Microbiology* 9: 1084–90. https://doi.org/10.1111/j.1462-2920.2006.01202.x

Ortíz-Castro, R., H.A. Contreras-Cornejo, L. Macías-Rodríguez, and J. López-Bucio. 2009. The role of microbial signals in plant growth and development. *Plant Signaling & Behavior* 4: 701–12. https://doi.org/10.4161/psb.4.8.9047.

Pal, K.K., K.V.B.R. Tilak, A.K. Saxena, R. Dey, and C.S. Singh. 2001. Suppression of maize root diseases caused by *Macrophomina phaseolina*, *Fusarium moniliforme* and *Fusarium graminearum* by plant growth promoting rhizobacteria. *Microbiological Research* 156: 209–23. https://doi.org/10.1078/0944-5013-00103.

Pandey, P.K., S.K. Yadav, A. Singh, B.K. Sarma, A. Mishra, and H.B. Singh. 2012. Cross-species alleviation of biotic and abiotic stresses by the endophyte *Pseudomonas aeruginosa* PW09. *Journal of Phytopathology* 160: 532–39. https://doi.org/10.1111/j.1439-0434.2012.01941.x.

Parret, A.H.A., G. Schoofs, P. Proost, and R. De Mot. 2003. Plant lectin-like bacteriocin from a rhizosphere-colonizing *Pseudomonas* isolate. *Journal of Bacteriology* 185: 897–08. https://doi.org/10.1128/jb.185.3.897-908.2003.

Prashanth, S. 2007. *Biological control of Macrophomina root rot and plant growth promotion in groundnut by Bacillus licheniformis MML2501, and azole compound producing Rhizobacterium*. Ph.D. thesis, University of Madras, Madras, India.

Príncipe, A., M. Fernandez, M. Torasso, A. Godino, and S. Fischer. 2018. Effectiveness of tailocins produced by *Pseudomonas fluorescens* SF4c in controlling the bacterial-spot disease in tomatoes caused by *Xanthomonas vesicatoria*. *Microbiological Research* 212–13: 94–102. https://doi.org/10.1016/j.micres.2018.05.010.

Raaijmakers, J.M., I. de Bruijn, O. Nybroe, and M. Ongena. 2010. Natural functions of lipopeptides from *Bacillus* and *Pseudomonas*: More than surfactants and antibiotics. *FEMS Microbiology Reviews* 34: 1037–62. https://doi.org/10.1111/j.1574-6976.2010.00221.x.

Ramamoorthy, V., T. Raguchander, and R. Samiyappan. 2002. Induction of defense-related proteins in tomato roots treated with *Pseudomonas fluorescens* Pf1 and *Fusarium oxysporum* f. sp. *lycopersici*. *Plant and Soil* 239: 55–68. https://doi.org/10.1023/A:1014904815352.

Richardson A.E. 2001. Prospects for using soil microorganisms to improve the acquisition of phosphorus by plants. *Australian Journal of Plant Physiology* 28: 897–906. https://www.publish.csiro.au/fp/PP01093.

Roesch, L., R. Fulthorpe, A. Riva, et al. 2007. Pyrosequencing enumerates and contrasts soil microbial diversity. *International Society for Microbial Ecology Journal* 1: 283–90. https://doi.org/10.1038/ismej.2007.53.

Rooney, W.M., R.W. Grinter, A. Correia, J. Parkhill, D.C. Walker, and J.J. Milner. 2020. Engineering bacteriocin-mediated resistance against the plant pathogen *Pseudomonas syringae*. *Plant Biotechnology Journal* 18: 1296–06. https://doi.org/10.1111/pbi.13294.

Rouphael, Y., and G. Colla. 2020. Editorial: Biostimulants in agriculture. *Frontiers in Plant Science* 11 (February): 40, 1–7. https://doi.org/10.3389/fpls.2020.00040.

Ruzzi, M., and R. Aroca. 2015. Plant growth-promoting rhizobacteria act as biostimulants in horticulture. *Scientia Horticulturae* 196: 124–34. https://doi.org/10.1016/j.scienta.2015.08.042.

Santos, M.S., M.A. Nogueira, and M. Hungria. 2019. Microbial inoculants: Reviewing the past, discussing the present and previewing an outstanding future for the use of beneficial bacteria in agriculture. *AMB Express* 9 (December): 205, 1–22. https://doi.org/10.1186/s13568-019-0932-0.

Santoyo, G., C.A. Urtis-Flores, P.D. Loeza-Lara, M.D.C. Orozco-Mosqueda, and B.R. Glick. 2021. Rhizosphere colonization determinants by plant growth-promoting rhizobacteria (PGPR). *Biology* 10 (May): 475, 1–18. https://doi.org/10.3390/biology10060475.

Sarkar, A., J.S. Patel, S. Yadav, B.K. Sarma, J.S. Srivastava, and H.B. Singh. 2014. Studies on Rhizosphere-bacteria mediated biotic and abiotic stress tolerance in chickpea (*Cicer arietinum* L.). *Vegetos* 27: 158–69. https://www.vegetosindia.org/article/studies-on-rhizosphere-bacteria-mediated-biotic-and-abiotic-stress-tolerance-in-chickpea-cicer-arietinum-l.

Scher, F.M., and R. Baker.1982. Effect of *Pseudomonas putida* and a synthetic iron chelator on induction of soil suppressiveness to *Fusarium* wilt pathogens. *Phytopathology* 72: 1567–73. https://doi.org/10.1094/phyto-72-1567.

Schmülling, T., J. Schell, and A. Spena. 1988. Single genes from *Agrobacterium rhizogenes* influence plant development. *The EMBO Journal* 7: 2621–29. https://doi.org/10.1002/j.1460-2075.1988.tb03114.x.

Shafi, J., H. Tian, and M. Ji. 2017. *Bacillus* species as versatile weapons for plant pathogens: A review. *Biotechnology & Biotechnological Equipment* 31: 446–59. https://doi.org/10.1080/13102818.2017.1286950.

Singh, P.P., Y.C. Shin, C.S. Park, and Y.R. Chung. 1999. Biological control of *Fusarium* wilt of cucumber by chitinolytic bacteria. *Phytopathology* 89: 92–99. https://doi.org/10.1094/PHYTO.1999.89.1.92.

Sivakumar, N., R. Sathishkumar, G. Selvakumar, R. Shyamkumar, and K. Arjunekumar. 2020. Phyllospheric microbiomes: Diversity, ecological significance, and biotechnological applications. In *Plant microbiomes for sustainable agriculture: Sustainable development and biodiversity*, eds. A. Yadav, J. Singh, A. Rastegari, and N. Yadav, Vol. 25, 113–72. Cham: Springer. https://doi.org/10.1007/978-3-030-38453-1_5.

Thampi, A., and R.S. Bhai. 2017. Rhizosphere actinobacteria for combating *Phytophthora capsici* and *Sclerotium rolfsii*, the major soil borne pathogens of black pepper (*Piper nigrum* L.). *Biological Control* 109: 1–13. https://doi.org/10.1016/j.biocontrol.2017.03.006.

Thrane, C., S. Olsson, T.H. Nielson, and J. Sorensen. 1999. Vital fluorescent strains for detection of stress in *Pythium ultimum* and *Rhizoctonia solani* challenged with viscosinamide from *Pseudomonas fluorescens* DR54. *FEMS Microbiology Ecology* 30: 11–23. https://doi.org/10.1016/S0168-6496(99)00034–3.

Tran, H., A. Ficke, T. Asiimwe, M. Hofte, and J.M. Raaijmakers. 2007. Role of the cyclic lipopeptide massetolide a in biological control of *Phytophthora infestans* and in colonization of tomato plants by *Pseudomonas fluorescens*. *New Phytologist* 175: 731–42. https://doi.org/10.1111/j.1469-8137.2007.02138.x.

Trapet, P., L. Avoscan, A. Klinguer, et al. 2016. The *Pseudomonas fluorescens* siderophore pyoverdine weakens *Arabidopsis thaliana* defense in favor of growth in iron-deficient conditions. *Plant Physiology* 171: 675–93. https://doi.org/10.1104/pp.15.01537.

Tyc, O., C. Song, J.S. Dickschat, M. Vos, and P. Garbeva. 2017. The ecological role of volatile and soluble secondary metabolites produced by soil bacteria. *Trends in Microbiology* 25: 280–92. http://doi.org/10.1016/j.tim.2016.12.002.

Uroz, S., C.D'Angelo-Picard, A. Carlier, et al. 2003. Novel bacteria degrading *N*-acylhomoserine lactones and their use as quenchers of quorum-sensing-regulated functions of plant-pathogenic bacteria. *Microbiology* 149: 1981–89. https://doi.org/10.1099/mic.0.26375-0.

van Loon, L.C., P.A.H.M. Bakker, and C.M.J. Pieterse. 1998. Systemic resistance induced by rhizosphere bacteria. *Annual Review of Phytopathology* 36: 453–83. https://doi.org/10.1146/annurev.phyto.36.1.453.

Venturi, V., and C. Keel. 2016. Signaling in the rhizosphere. *Trends in Plant Science* 21: 187–98. https://doi.org/10.1016/j.tplants.2016.01.005.

Vishwakarma, K., N. Kumar, C. Shandilya, S. Mohapatra, S. Bhayana, and A. Varma. 2020. Revisiting plant – microbe interactions and microbial consortia application for enhancing sustainable agriculture: A review. *Frontiers in Microbiology* 11 (December): 3195, 1–21. https://doi.org/10.3389/fmicb.2020.560406.

Weller, D.M. 2007. *Pseudomonas* biocontrol agents of soilborne pathogens: Looking back over 30 years. *Phytopathology* 97: 250–56. https://doi.org/10.1094/phyto-97-2-0250.

Yamada, K., M. Hirota, Y. Niimi, et al. 2006. Nucleotide sequences and organization of the genes for carotovoricin (Ctv) from *Erwinia carotovora* indicate that Ctv evolved from the same ancestor as *Salmonella typhi* prophage. *Bioscience, Biotechnology, and Biochemistry* 70: 2236–47. https://doi.org/10.1271/bbb.60

9 Viral Secondary Metabolites and Their Applications

*T. Attri
(Dr. Y. S. Parmar University of Horticulture and Forestry, Nauni, Solan, India; attritarushi947@gmail.com)

A. Sharma
(College of Horticulture and Forestry, Dr. Y. S. Parmar University of Horticulture and Forestry (Solan), Neri, Hamirpur, India; amit.sharma3178@gmail.com)

M. Sharma
(College of Horticulture and Forestry, Dr. Y. S. Parmar University of Horticulture and Forestry (Solan), Neri, Hamirpur, India; dmonicasharma@gmail.com)

9.1 INTRODUCTION

Metabolism is a very crucial process for any living entity for the functionality of different physiological processes, and such processes lead to the production of two types of intermediary organic compounds, which are primary and secondary metabolites. Primary metabolites, *viz*., carbohydrates, proteins, and fats, have a direct role in growth and reproduction, whereas secondary metabolites (belonging to three categories: terpenes, phenols and nitrogen, and carbon-containing compounds) are the organic compounds produced by the different microorganisms as well as plants when they are under environmental stress (Gupta et al., 2022, 2024). Secondary metabolites at a certain concentration render their hosts with defense strategies. But at higher concentrations, they also have an adverse effect on the plants by inhibiting their growth and reproduction.

The virus is one of the pathogens that can be considered a smart infective particle that hijacks the cellular machinery of the host (Bazzini et al., 2011). Viruses are the entities that cause severe infection in different hosts, such as humans, animals, and plants. They acquire the building blocks for replication by entering the host. This ultimately leads to different kinds of alterations in the host. The host transcription and translation processes are controlled by the host for their resource utilization. Viruses require a substantial amount of host resources for their synthesis, which is achieved by the action of the shutdown of host genes (Thivierge et al., 2005). Viral infection leads to the production of different secondary metabolites in the hosts, which act as antiviral compounds, bioactive molecules, *etc*. By understanding the altered pathways of the host in viral infections, the knowledge to develop therapeutic targets to treat viral infections can be understood (Brunton et al., 2013; Noto et al., 2014).

Innate immunity, which is the first line of defense, is the key process during infection. To understand the role of the innate immunity system, we need to uncover all the central metabolic cycles going on in the host. While viruses are often associated with causing diseases, they possess a hidden world of biochemical wonders. Among the plethora of genetic information they carry, viruses also harbor an impressive array of secondary metabolites – small organic compounds with diverse chemical structures and biological activities. These viral secondary metabolites have garnered increasing attention from scientists due to their potential applications in medicine, biotechnology, and environmental management. In this chapter, we will explore the fascinating world of viral secondary metabolites, their origins and functions, and the promising avenues they offer for various fields of research.

9.2 CLASSIFICATION OF SECONDARY METABOLITES

Secondary metabolites are mainly classified into three different groups.

9.2.1 Terpenes

These are the compounds that are derived from IPP (isopentenyl diphosphate), which is a 5-carbon precursor. Terpenes and terpenoids are also the primary constituents of the essential oils of many types of plants and flowers. In plants, terpenes and terpenoids are important mediators of ecological interactions.

DOI: 10.1201/9781003408833-9

9.2.2 Nitrogen-Containing and Sulfur-Containing Compounds

These compounds mainly include alkaloids and flavonoids.

9.2.2.1 Alkaloids

These are biosynthesized from amino acids.

9.2.2.2 Flavonoids

Compounds produced from the acid polyketide pathway.

9.2.3 Phenolics

These are the compounds that are formed from the shikimic acid pathway.

9.3 BIOSYNTHESIS OF SECONDARY METABOLITES

Biosynthetic pathways for secondary metabolites are complex biochemical processes that organisms, including plants, fungi, and bacteria, use to produce a wide range of compounds that are not essential for basic growth and reproduction but often play crucial roles in adaptation, defense, and communication (Jain et al., 2023 and 2024). The biosynthesis of different metabolites has been described through the flowchart given in Figure 9.1.

9.4 VIRAL SECONDARY METABOLITES

Viruses are obligate intracellular parasites that rely on host cells for their replication and survival. During their infection

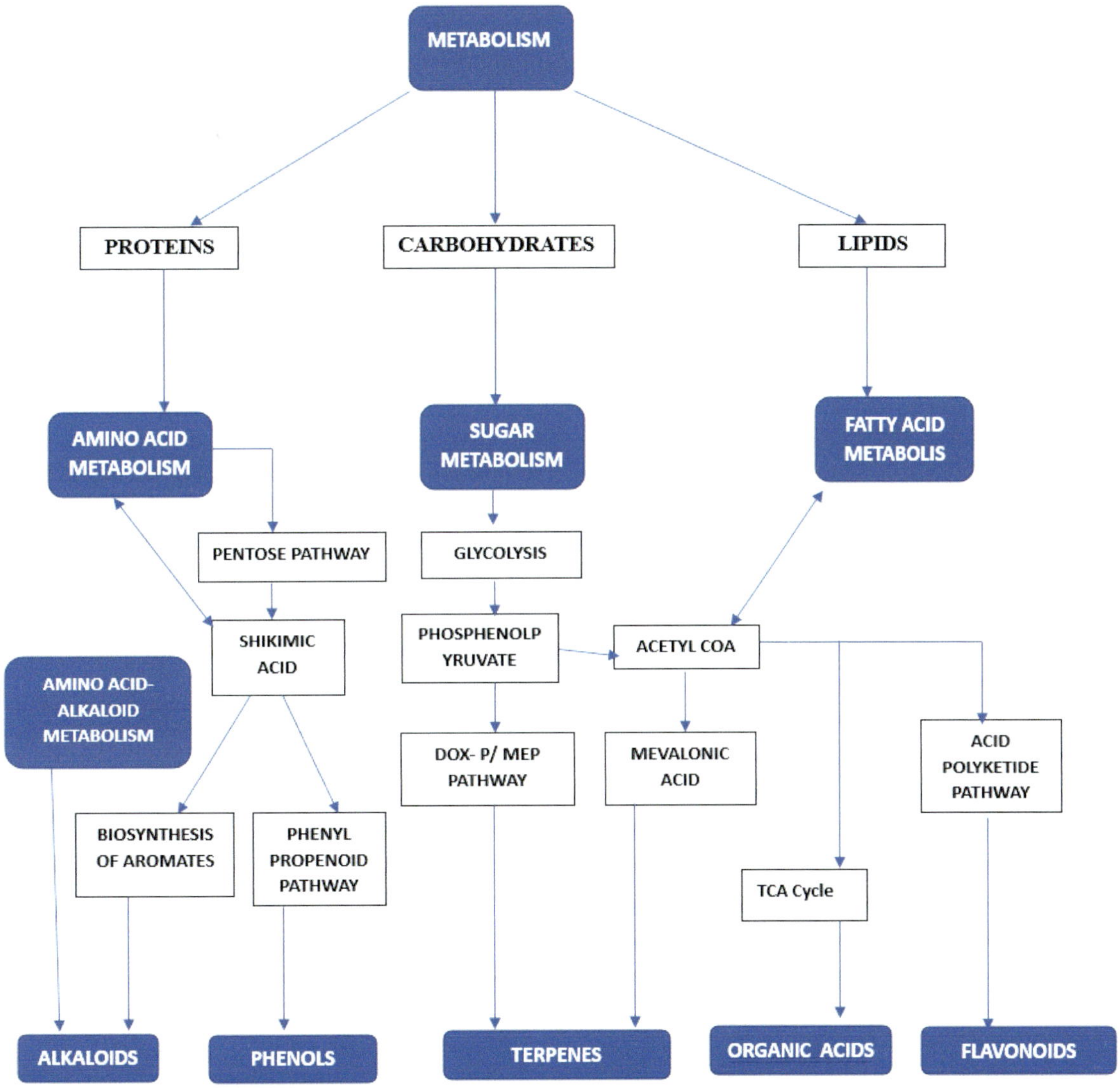

FIGURE 9.1 Biosynthetic pathways for the production of secondary metabolites.

Source: Modified from Mishra et al. (2020); Pandey et al. (2016).

cycle, viruses can manipulate the metabolic machinery of the host cell, leading to alterations in the synthesis and accumulation of various compounds. Among these alterations are the production of secondary metabolites, which are not essential for the virus's replication but contribute to the viral life cycle in other ways. Viral secondary metabolites often mirror or augment the host cell's natural secondary metabolite pathways, suggesting an intriguing co-evolution between viruses and their hosts.

Viral secondary metabolites are compounds produced by viruses or their hosts (fungi, bacteria, and plants) that have biological effects on viral replication or host defense. Some plant secondary metabolites also have antiviral properties against different types of viruses, such as terpenoids, phenols, polyphenols, alkaloids, and glycyrrhizin. These compounds can interfere with various stages of the viral life cycle, such as attachment, entry, replication, assembly, and release.

9.5 TYPES OF VIRAL SECONDARY METABOLITES

9.5.1 Antiviral Defense Compounds

Some viral secondary metabolites are thought to play a role in the defense mechanisms of the infected host cell. These compounds can impede viral replication or limit the spread of the infection to neighboring cells. By studying these compounds, scientists hope to uncover new strategies for combating viral infections.

9.5.2 Immunomodulatory Agents

Viruses can produce secondary metabolites that affect the immune response of the host. These compounds may suppress or stimulate specific immune pathways, influencing the severity and outcome of viral infections. Understanding these interactions could lead to the development of novel immunotherapies.

9.5.3 Enzyme Inhibitors

Certain viral secondary metabolites act as enzyme inhibitors, disrupting essential cellular processes within the host cell. By identifying and characterizing these compounds, researchers can gain insights into potential targets for drug development.

9.5.4 Bioactive Molecules

Viruses can synthesize a wide range of small bioactive molecules with diverse pharmacological properties. These molecules may have applications in drug discovery, agriculture, and environmental management.

9.6 EFFECT OF VIRAL INFECTION ON THE HOST METABOLISM

9.6.1 In Animals

Viral infections in humans and animals result in the disruption of various metabolic pathways and master protein regulators of metabolism. The effects of viruses on secondary metabolites in animals are illustrated in Table 9.1.

9.6.1.1 Glucose Metabolism

By modifying the signaling pathways, viruses alter glucose metabolism to boost available energy and encourage their replication. The Warburg effect, which is how most viruses achieve this, is what causes aerobic glycolysis (Sanchez and Lagunoff, 2015). Even when oxygen is present, the lactate dehydrogenase (LDH) enzyme ferments pyruvate into lactate as part of the Warburg effect. Hexokinase (HK), PFK, pyruvate kinase, and LDH are the primary regulators of aerobic glycolysis, along with glucose transporters (GLUTs) and hexokinase (HK). According to Liberti and Locasale (2016), aerobic glycolysis results in a buildup of more lactic acid, fewer glycolytic intermediates that contribute to the TCA cycle, and increased glucose intake. Then, to promote

TABLE 9.1
Metabolic Alteration in Animals

S. No.	Virus	Metabolic Effect	Reference
1	VACV and DENV	Alteration of glycolysis, FAS, and glutaminolysis	Kelly and O'Neill, 2015
2	HCMV	De novo pyrimidine biosynthetic flux in host cells	Zhu et al., 2015
3	HCMV and influenza A	Inhibition of FAS, suppression of replication of HCMV, and influenza A	Zhu et al., 2015
4	HCMV	Infection requirement of long-chain FA metabolism: acyl-CoA synthetases and fatty acid elongases	Koyuncu et al., 2013
5	KSHV	Inhibition of FAS and a decrease in the number of lytic enveloped viruses–induced apoptosis in infected cells	Delgado et al., 2012
6	SARS COV-A	Increase in glucose carbon entry into the TCA cycle via increased pyruvate carboxylase expression, and reduction of oxidative glutamine metabolism	Kumar et al., 2020
7	EBV, VZV, poliovirus, and HCV	Inhibition of FAS and a lower number of infectious viruses	Noto et al., 2014

its replication, the infected cells of the host cause greater expression of the glycolytic enzymes HK1, LDH, and glucose-6-phosphate dehydrogenase. It was discovered that stopping aerobic glycolysis and pyruvate dehydrogenase kinase inhibition prevented viral multiplication. Examples include the ISKNV (infectious spleen and kidney necrosis virus).

9.6.1.2 Glutaminolysis

In the absence of pyruvate, glutamine is used in the process of glutaminolysis to produce TCA cycle intermediates. The TCA cycle is restored when glutamine is transformed into glutamate by glutaminase (GLS), which is then transformed into ketoglutarate by glutamate dehydrogenase (GDH). This results in the generation of additional intermediates, such as citrate, which aid in the synthesis of fatty acids. Due to the TCA cycle's production of NADPH, which provides electrons to the *ETC*, glutamine can potentially function as a substitute energy source. There is a dependency on glutamine as a result of the separation between glycolysis and the TCA cycle seen in viral life cycles that promote aerobic glycolysis. Take HSV-1 and influenza as examples. For their best replication, an infection also increases mitochondrial GLS-1 activity (Thai et al., 2015). A glutamine-free environment reduces viral replication by 75% in cells infected with KSHV (Sanchez et al., 2017).

9.6.1.3 Lipid Metabolism

To speed up their replication cycle, viruses modify the formation of LDs (lipid droplets) by utilizing their components, such as triglycerides, cholesterol esters, and the highly dynamic membrane trafficking system (Herker et al., 2012). For instance, SREBP-1 and peroxisome proliferator-activated receptor are up-regulated in SARS-CoV-2-infected primary human monocytes, human lung epithelial cell lines (A549), and human lung microvascular endothelial cell lines, resulting in an increase in LDs in the cytoplasm. These two transcription factors are essential for increasing fatty acid transporters (CD36) and lipogenic enzymes, respectively. DGAT-1, a crucial regulator of TAG synthesis and LD formation, is also modulated by SARS-CoV-2 to increase protein trafficking to LDs, which in turn increases lipid biogenesis. A922500's inhibition of DGAT-1 lessens the cell death and inflammation caused by SARS-COV-2 in primary monocytes.

9.6.1.4 Fatty Acid Synthesis

Virus-infected HF cells show a substantial increase in all rate limiting enzymes for lipid biosynthesis, namely, ACC1, ATP-citrate lyase, and FASN (Yu et al., 2012). In HF cells, mature SREBP-1 was detected in the nucleus of the cell after 24 h of HCMV infection, resulting in fatty acid synthesis and the accumulation of lipid droplets (LDs) in the cytoplasm, which assist in the replication and production of the lipid bilayer required by the dsDNA virus, for example, HCMV (human cytomegalovirus).

9.6.2 In Plants

Plant viruses exhibit a wide host range. For example, tobacco mosaic virus alone infects different hosts, such as tobacco, tomato, and many solanaceous crops. More than 1,000 plant species are infected by cucumber mosaic virus. Virus infection affects different metabolic processes in host plants, for example, carbohydrate metabolism by affecting genes involved in glycolysis/glucogenesis and starch metabolism. There are different examples of secondary metabolites that are affected due to viral infection in plants, as given in Table 9.2. The affected metabolic pathways are responsible for decreasing soluble solids in virus-infected plants (Blanco-Ulate et al., 2017). Normal functioning of the grape plant is regulated due to viral infection, which results in:

1. *Up-regulation of photosynthesis-related pathways.* Both thylakoid electron transfer (PSII proteins) and the Calvin cycle (ribulose-1,5-bisphosphate carboxylase/oxygenase and fructose-bisphosphate aldolase) are up-regulated.
2. *Enrichment of invertase/pectin methyl esterase (PME) inhibitors among the up-regulated genes in virus-infected plants.* These inhibitors control PME activity at early stages of berry development in the grape crop and appear to function as regulators of berry enlargement and softening. (Lionetti et al., 2015)

Infection with *Poppy mosaic virus* resulted in the accumulation of five alkaloids, *viz.*, papaverine, thebaine, narcotine, codeine, and morphine, in some poppy genotypes. The infection of saffron by *Saffron latent virus* affects the spice quality of saffron by altering the concentration and proportions of crocetin, esters, picrocrocin, safranal, and kaempferols (Parizad et al., 2019). The infection of *Passiflora edulis* with *Cucumber mosaic virus* showed an increase in the total polyphenol (26–28%) and flavonoid (48–58%) content (Lan et al., 2020).

The virus infection changes the phytohormone levels depending on the combination of virus–host interaction. Viruses can even manipulate the biosynthesis of plant hormones to help in the suppression of plant defense and to improve the replication and spread of the viruses. The transcriptional suppression of secondary metabolic pathways, in particular the biosynthesis of phenylpropanoids, stilbenoids, and lignin, occurs in host plants after the viral infection. This affects the anthocyanin content. For example, virus-infected grapevines undergo reduced metabolic flux in the core phenylpropanoid pathway and alterations in flavonoid and anthocyanin biosynthesis, observed due to lower anthocyanin content. Similarly, in virus-infected grape plants, genes involved in nucleic acid metabolism showed higher expression in RNA processing. The alterations in RNA metabolic pathways due to viral infection are responsible for resistance and susceptibility in plants (Moutinho-Pereira et al., 2012).

TABLE 9.2
Metabolic Alterations in Plants

S. No.	Virus	Host	Secondary Metabolites	References
1	Cucumber mosaic virus	*Passiflora edulis*	Polyphenols and flavonoids	Lan et al., 2020
2	Grapevine leafroll–associated virus 3	*Vitis vinifera*	Flavanols (myricetin, kaempferol, and quercetin derivatives) and hydroxycinnamic acids (which include derivatives of caffeic acid)	Montero et al., 2016
3	Grapevine red blotch–associated virus	*Vitis vinifera*	Flavonoid and anthocyanin biosynthesis (malvidin-3-O-glucoside, petunidin-3-O-glucoside, delphinidin-3-Oglucoside, pelargodin-3-O-glucoside, and cyanidin-3-O-glucoside), phenylpropanoid metabolism, and isoflavonoid metabolism	Blanco-Ulate et al., 2017
4	Hop mosaic virus and hop latent virus	*Humulus lupulus*	α- and β-bitter acids, polyphenols, and essential oils	Pethybridge et al., 2002; Jelínek et al., 2012
5	Poppy mosaic virus	*Papaver somniferum* var. *Sampada*	Alkaloids: papaverine, narcotine, codeine, morphine, and thebaine	Zaim et al., 2014a, 2014b
6	Prunus necrotic ringspot virus (PNRSV-A and PNRSV-I)	*Humulus lupulus*	α- and β-bitter acids, polyphenols, and essential oils	Pethybridge et al., 2002; Jelínek et al., 2012
7	Saffron latent virus	*Crocus sativa*	Crocetin, esters, picrocrocin, safranal, and kaempferols	Parizad et al., 2019
8	Telosma mosaic virus	*Passifora edulis*	Phenols	Chen et al., 2018
9	Tobacco mosaic virus	*Nicotiana tobaccums*	Scopoletin and its glucoside scopolin	Chong et al., 2002; Costet et al., 2002
10	Turnip crinkle virus and cauliflower mosaic virus	*Arabidopsis thaliana*	Camalexin	Callaway et al., 1996; Dempsey et al., 1997
11	Potato virus Y	*Solanum tuberosum*	Valine, malate, and chlorogenate	Kogovsek et al., 2016

9.7 INDUSTRIAL APPLICATION OF VIRAL SECONDARY METABOLITES

Various therapeutics can be made by studying the metabolic effects of different viruses on their hosts. Viral infections can be controlled by a prophylactic strategy and/or drug therapy. There is a need to discover novel strategies/drugs that are less toxic, selective, and stable under *in vivo* conditions. Table 9.3 describes the different metabolites from different hosts which are produced upon viral infection and can be used as bioactive molecules for drug development against different viruses, which are described in Figure 9.2.

9.7.1 Medical Applications

The exploration of viral secondary metabolites has significant implications for medicine. As antiviral resistance becomes a growing concern, viral secondary metabolites may offer alternative therapeutic options. By analyzing the chemical diversity of viral secondary metabolites, scientists can identify potential new antiviral compounds or enhance existing ones.

Furthermore, viral secondary metabolites can play a role in developing novel immunomodulatory therapies. For instance, certain compounds might be harnessed to boost the immune response against infections or dampen hyperactive immune reactions in autoimmune diseases.

9.7.2 Biotechnological Uses

Viral secondary metabolites hold great promise in biotechnology as well. Enzyme inhibitors discovered from viral sources can be utilized in various industrial processes, including food production, biofuel synthesis, and waste management. Additionally, bioactive molecules from viruses might find applications in agricultural practices, serving as natural pesticides or growth stimulants for crops.

9.7.3 Environmental Significance

Studying viral secondary metabolites also sheds light on their ecological functions. Some compounds might contribute to the virus–host interactions in natural ecosystems, influencing microbial diversity and population dynamics. Understanding these roles is essential for comprehending the broader ecological impacts of viral infections.

TABLE 9.3

Secondary Metabolites from Different Hosts Act as Antiviral Defense/Bioactive Compounds

S. No.	Host	Antiviral Compound	Group	Targeted Virus	Mechanism of Inhibition	Reference
Fungi						
1	*Penicillium sclerotiorum*	Sclerotiorin	Polyketone	HIV1	HIV-1 integrase and protease	Arunpanichlert et al., 2010
2	*Aspergillus terreus*	Asperterrestide A	Peptide	H1N1 and H3N2	NR	Nong et al., 2014
3	*Xylaria* sp.	Integric acid	Terpenoid	HIV	HIV integrase	Singh et al., 1999
4	*Stachybotrys* sp.	Stachybogrisephenone	Pyrone	EV71	Replication of EV-71	Qin et al., 2015
5	*Stachybotrys* sp.	Stachyflin and acetylstachyflin	Alkaloid	H1N1	Fusion of viral envelope and endosome	Minagawa et al., 2002
6	*Chaetomium* sp.	Isocochliodinol and didemethylasterriquinone D	Quinone	HIV	HIV protease	Sekita, 1983
7	*Saccharomyces cerevisiae*	Silybin	Polyphenol	HCV	Inhibition of penetration GMM	Yang et al., 2020
Bacteria and Cyanobacteria						
1	*Spirulina platensis*	Calcium spirulan	Polysaccharide	HSV-1, hCMV, H1N1, HIV-1, measles virus, and mumps virus	Inhibition of replication	Hayashi et al., 1996
2	*Nostoc ellipsosporum*	Cyanovirin-N	Peptide	HIV	Inhibition through binding to the envelope protein gp120	Botos et al., 2002
3	*Bacillus pumilus*	Pumilacidins	Peptide	HSV1	Not discovered	Naruse et al., 1989
4	*E. coli*	Resveratrol	Polyphenols	RSV	Immune response GMM	Lim et al., 2011
5	*Streptomyces* sp.	Xiamycin	Terpenoid	HIV	HIV infection	Ding et al., 2010
Microalgae						
1	*Porphyridium purpureum*	Sulfated exopolysaccharide	Polysaccharide	VACV	Viral entry	Radonić et al., 2011
2	*Rhodella reticulata*	Exopolysaccharide	Polysaccharide	Murine sarcoma and leukemia viruses	Inhibit early steps in the virus replication cycle	Talyshinsky et al., 2002
3	*Coccomixa* sp.	Monogalactosyl diacylglyceride	Galactolipid	HSV 2	Inhibition of viral replication	Hayashi et al., 2019

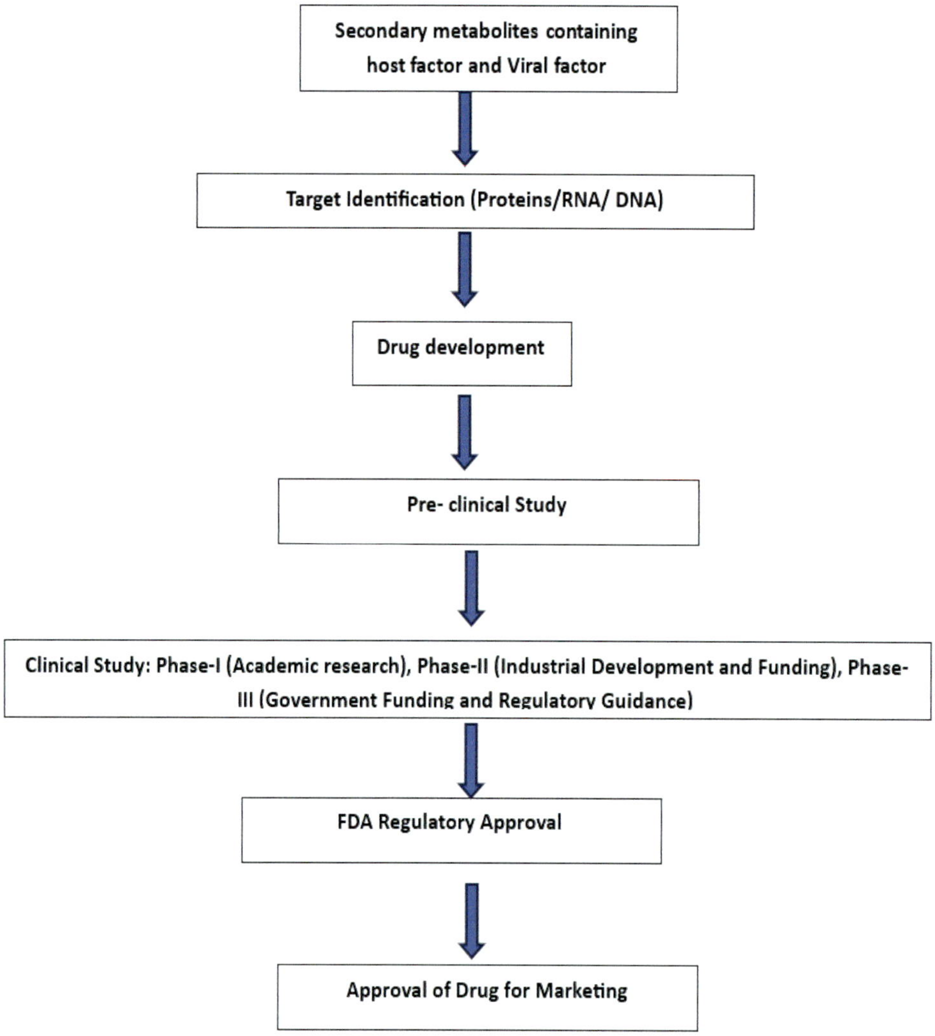

FIGURE 9.2 Drug development from metabolites.

Source: Raihan et al. (2021).

9.8 CHALLENGES AND FUTURE PROSPECTS

While the field of viral secondary metabolites holds tremendous potential, there are challenges to overcome. Identifying and characterizing these compounds can be technically demanding due to their low concentrations within host cells and the vast diversity of viruses that remain uncultured and uncharacterized.

Despite these challenges, advancements in sequencing technologies, metagenomics, and synthetic biology are revolutionizing the study of viral secondary metabolites. Collaborative efforts between virologists, microbiologists, chemists, and biotechnologists are essential to unlock the full potential of these fascinating compounds.

9.9 CONCLUSION

Viral secondary metabolites represent a captivating intersection of virology and natural product chemistry. They offer unique insights into the intricate relationship between viruses and their hosts, as well as potential solutions to challenges in medicine, biotechnology, and environmental science. As research in this field advances, we can expect to harness the power of viral secondary metabolites for the benefit of human health, industry, and the environment.

REFERENCES

Arunpanichlert, J., V. Rukachaisirikul, Y. Sukpondma, et al. 2010. Azaphilone and isocoumarin derivatives from the endophytic fungus *Penicillium sclerotiorum* PSU-A13. *Chemical and Pharmaceutical Bulletin* 58: 1033–36. https://doi.org/10.1248/cpb.58.1033.

Bazzini, A.A., C.A. Manacorda, T. Tohge, et al. 2011. Metabolic and miRNA profiling of *TMV* infected plants reveals biphasic temporal changes. *PLoS One* 6 (December): e28466, 1–14. https://doi.org/10.1371/journal.pone.0028466.

Blanco-Ulate, B., H. Hopfer, R. Figueroa-Balderas, et al. 2017. Red blotch disease alters grape berry development and metabolism by interfering with the transcriptional and hormonal regulation of ripening. *Journal of Experimental Botany* 68: 1225–38. https://doi.org/10.1093/jxb/erw506.

Botos, I., B.R. O'Keefe, S.R. Shenoy, et al. 2002. Structures of the complexes of a potent anti-HIV protein cyanovirin-N and high mannose oligosaccharides. *Journal of Biological Chemistry* 277: 34336–42. https://doi.org/10.1074/jbc.M205909200.

Brunton, J., S. Steele, B. Ziehr, N. Moorman, and T. Kawula. 2013. Feeding uninvited guests: mTOR and AMPK set the table for intracellular pathogens. *PLoS Pathogens* 9 (October): e1003552, 1–4. https://doi.org/10.1371/journal.ppat.1003552.

Callaway, A., W. Liu, V. Andrianov, et al. 1996. Characterization of cauliflower mosaic virus (CaMV) resistance in virus-resistant ecotypes of *Arabidopsis*. *Molecular Plant-Microbe Interactions* 9: 810–18. https://doi.org/10.1094/mpmi-9-0810.

Chen, S., N. Yu, S. Yang, B. Zhong, and H. Lan. 2018. Identification of *Telosma mosaic virus* infection in *Passifora edulis* and its impact on phytochemical contents. *Virology Journal* 15 (November): 168, 1–8. https://doi. org/10.1186/s12985-018-1084-6.

Chong, J., R. Baltz, C. Schmitt, R. Befa, B. Fritig, and P. Saindrenan. 2002. Downregulation of a pathogen-responsive tobacco UDP Glc:phenylpropanoid glucosyltransferase reduces scopoletin glucoside accumulation, enhances oxidative stress, and weakens virus resistance. *Plant Cell* 14: 1093–07. https://doi.org/10.1105/tpc.010436.

Costet, L., B. Fritig, and S. Kaufmann. 2002. Scopoletin expression in elicitor-treated and Tobacco mosaic virus-infected tobacco plants. *Physiologia Plantarum* 115: 228–35. https://doi.org/10.103 4/j.1399–3054.2002. 1150208.x.

Delgado, T., E.L. Sanchez, R. Camarda, and M. Lagunoff. 2012. Global metabolic profiling of infection by an oncogenic virus: KSHV induces and requires lipogenesis for survival of latent infection. *PLoS Pathogens* 8 (August): e1002866, 1–11. https://doi.org/10.1371/journal.ppat.1002866.

Dempsey, D.A., M.S. Pathirana, K.K. Wobbe, and D.F. Klessig. 1997. Identification of an *Arabidopsis* locus required for resistance to Turnip crinkle virus. *The Plant Journal* 11: 301–11. https://doi.org/10.1046/j.1365-313x.1997.11020301.x.

Ding, L., J. Münch, H. Goerls, et al. 2010. Xiamycin, a pentacyclic indolosesquiterpene with selective anti-HIV activity from a bacterial mangrove endophyte. *Bioorganic & Medicinal Chemistry Letters* 20: 6685–87. https://doi.org/10.1016/j.bmcl.2010.09.010.

Gupta, A., S. Jain, and N. Verma. 2022. *Industrial applications of soil microbes*, vol. 1. Bentham Science Publishers Pte Ltd. https://doi.org/10.2174/97898150399551220101.

Gupta, A., S. Jain, and N. Verma. 2024. *Industrial applications of soil microbes*, vol. 3. Bentham Science Publishers Pte Ltd. https://doi.org/10.2174/97898150797531240301.

Hayashi, K., T. Hayashi, and I. Kojima. 1996. A natural sulfated polysaccharide, calcium spirulan, isolated from *Spirulina platensis*: *In vitro* and *Ex vivo* evaluation of anti-herpes simplex virus and anti-human immunodeficiency virus activities. *AIDS Research and Human Retroviruses* 12: 1463–71. https://doi.org/10.1089/aid.1996.12.1463.

Hayashi, K., J.B. Lee, K. Atsumi, et al. 2019. *In vitro* and *in vivo* anti-herpes simplex virus activity of monogalactosyl diacylglyceride from *Coccomyxa* sp. KJ (IPOD FERM BP22254), a green microalga. *PLoS One* 14 (July): e0219305, 1–14. https://doi.org/10.1371/journal.pone.0219305.

Herker, E., and M. Ott. 2012. Emerging role of lipid droplets in host/pathogen interactions. *Journal of Biological Chemistry* 287: 2280–87. https://doi.org/10.1074/jbc.R111.300202

Jain, S., A. Gupta, and N. Verma. 2023. *Industrial applications of soil microbes*, vol. 2. Bentham Science Publishers Pte Ltd. https://doi.org/10.2174/97898150502641230201.

Jain, S., A. Gupta, and N. Verma. 2024. *Industrial applications of soil microbes*, vol. 4. Bentham Science Publishers Pte Ltd. https://doi.org/10.2174/97898151249961240401.

Jelínek, L., M. Dolečková, M. Karabín, T. Hudcová, B. Kotlíková, and P. Dostálek. 2012. Influence of growing area, plant age, and virus infection on the contents of hop secondary metabolites. *Czech Journal of Food Sciences* 30: 541–47. https://doi.org/10.17221/50/2012-CJFS.

Kelly, B., and L.A. O'Neill. 2015. Metabolic reprogramming in macrophages and dendritic cells in innate immunity. *Cell Research* 25: 771–84. https://doi.org/10.1038/cr.2015.68.

Kogovsek, P., M. Pompe-Novak, M. Petek, L. Fragner, W. Weckwerth, and K. Gruden. 2016. Primary metabolism, phenylpropanoids and antioxidant pathways are regulated in potato as a response to Potato virus Y Infection. *PLoS One* 11 (January): e0146135, 1–20. https://doi.org/10.1371/journal.pone.0146135.

Koyuncu, E., J.G. Purdy, J.D. Rabinowitz, and T. Shenk. 2013. Saturated very long chain fatty acids are required for the production of infectious human cytomegalovirus progeny. *PLoS Pathogens* 9 (May): e1003333, 1–15. https://doi.org/10.1371/journal.ppat.1003333.

Kumar, N., S, Sharma, and R. Kumar. 2020. Host-directed antiviral therapy. *Clinical Microbiology Reviews* 33 (July): e00168–19, 1–36. https://doi.org/10.1128/CMR.00168-19

Lan, H., B. Lai, P. Zhao, et al. 2020. *Cucumber mosaic virus* infection modulated the phytochemical contents of *Passifora edulis*. *Microbial Pathogenesis* 138 (January): 103828. https://doi.org/10.1016/j.micpath.2019.103828.

Liberti, M.V., and J.W. Locasale. 2016. The Warburg effect: How does it benefit cancer cells? *Trends in Biochemical Sciences* 41: 211–18. https://doi.org/10.1016/j.tibs.2015.12.001.

Lim, C.G., Z.L. Fowler, T. Hueller, S. Schaffer, and M.A.G. Koffas. 2011. High-yield resveratrol production in engineered *Escherichia coli*. *Applied and Environmental Microbiology* 77: 3451–60. https://doi.org/10.1128/AEM.02186-10.

Lionetti, V., A. Raiola, B. Mattei, and D. Bellincampi. 2015. The grapevine VvPMEI1 gene encodes a novel functional pectin methylesterase inhibitor associated to grape berry development. *PLoS One* 10 (July): e0133810, 1–19. https://doi.org/10.1371/journal.pone.0133810.

Minagawa, K., S. Kouzuki, J. Yoshimoto, et al. 2002. Stachyflin and acetylstachyflin, novel anti-influenza a virus substances, produced by *Stachybotrys* sp. RF-7260 I. Isolation, structure elucidation and biological activities. *Journal of Antibiotics* 55: 155–64. https://doi.org/10.7164/antibiotics.55.155.

Mishra, J., R. Srivastava, P.K. Trivedi, and P.C. Verma. 2020. Effect of virus infection on the secondary metabolite production and phytohormone biosynthesis in plants. *3 Biotech* 10 (November): 547, 1–16. https://doi.org/10.1007/s13205-020-02541-6.

Montero, R., M.L. Pérez-Bueno, M. Barón, et al. 2016. Alterations in primary and secondary metabolism in *Vitis vinifera* "Malvasía de Banyalbufar" upon infection with *Grapevine leaf roll associated virus*. *Physiologia Plantarum* 157: 442–52. https://doi. org/10.1111/ppl.12440.

Moutinho-Pereira, J., C.M. Correia, and B. Gonçalves. 2012. Impacts of leafroll-associated viruses (GLRaV-1 and -3) on the physiology of the Portuguese grapevine cultivar 'Touringa Nacional' growing under field conditions. *Annals of Applied Biology* 3: 237–49. https://doi.org/10.1111/J.1744-7348.2012.00536.X.

Naruse, N., O. Tenmyo, K.Tomita, et al. 1989. Lanthiopeptin, a new peptide antibiotic. Production, isolation and properties of lanthiopeptin. *Journal of Antibiotics* 42: 837–45. https://doi.org/10.7164/antibiotics.42.837.

Nong, X-H., Y-F. Wang, X-Y. Zhang, M-P. Zhou, X-Y. Xu, and S-H. Qi. 2014. Territrem and butyrolactone derivatives from a marine-derived fungus *Aspergillus terreus*. *Marine* Drugs 12: 6113–24. https://doi.org/10.3390/md12126113.

Noto, A., A. Dessi, M. Puddu, M. Mussap, and V. Fanos. 2014. Metabolomics technology and their application to the study of the viral infection. *Journal of Maternal-Fetal & Neonatal Medicine* 27: 53–57. https://doi.org/10.3109/14767058.2014.955963.

Pandey, V., R. Srivastava, N. Akhtar, J. Mishra, P. Mishra, P.C. Verma. 2016. Expression of *Withania somnifera* steroidal glucosyltransferase gene enhances withanolide content in hairy roots. *Plant Molecular Biology Reporter* 34: 681–89. https://doi.org/10.1007/s1110 5–015-0955-x.

Parizad, S., A. Dizadji, M.K. Habibi, et al. 2019. The effects of geographical origin and virus infection on the saffron (*Crocus sativus* L.) quality. *Food Chemistry* 295: 387–94. https://https://doi.org/10.1016/j.foodchem.2019.05.116.

Pethybridge, S.J., C.R. Wilson, F.S. Hay, G.W. Leggett, and L.J. Sherrif. 2002. Effect of viruses on agronomic and brewing characteristics of four hop (*Humulus lupulus*) cultivars in Australia. *Annals of Applied Biology* 140: 97–105. https://doi.org/10.1111/j.1744-7348.2002. tb00161.

Qin, C., X. Lin, X. Lu, et al. 2015. Sesquiterpenoids and xanthones derivatives produced by sponge-derived fungus *Stachybotry* sp. HH1 ZSDS1F1–2. *Journal of Antibiotics* 68: 121–25. https://doi.org/10.1038/ja.2014.97.

Radonić, A., S. Thulke, J. Achenbach, et al. 2011. Anionic polysaccharides from phototrophic microorganisms exhibit antiviral activities to Vaccinia virus. *Journal of Antivirals & Antiretrovirals* 2: 51–55. https://doi.org/10.4172/jaa.1000023.

Raihan, T., M.F. Rabbee, P. Roy, S. Choudhury, K.-H. Baek, and A.K. Azad. 2021. Microbial metabolites: The emerging hotspot of antiviral compounds as potential candidates to avert viral pandemic alike COVID–19. *Frontiers in Molecular Biosciences* 8 (September): 732256, 1–31. https://doi.org/10.3389/fmolb.2021.732256.

Sanchez, E.L., and M. Lagunoff. 2015. Viral activation of cellular metabolism. *Virology*, 479–80: 609–18. https://doi.org/10.1016/j.virol.2015.02.038.

Sanchez, E.L., T.H. Pulliam, T.A. Dimaio, A.B. Thalhofer, T. Delgado, and M. Lagunoff. 2017. Glycolysis, glutaminolysis, and fatty acid synthesis are required for distinct stages of Kaposi's sarcoma-associated herpesvirus lytic replication. *Journal of Virology* 91 (May): e02237–16, 1–15. https://doi.org/10.1128/jvi.02237-16.

Sekita, S. 1983. Isocochliodinol and neocochliodinol, bis (3-indolyl)-benzoquinones from *Chaetomium* Spp. *Chemical & Pharmaceutical Bulletin* 31: 2998–3001. https://doi.org/10.1248/cpb.31.2998.

Singh, S.B., D. Zink, J. Polishook, et al. 1999. Structure and absolute stereochemistry of HIV-1 integrase inhibitor integric acid. A novel eremophilane sesquiterpenoid produced by a *Xylaria* sp. *Tetrahedron Letters* 40: 8775–79. https://doi.org/10.1016/S0040-4039(99)01878-X.

Talyshinsky, M.M., Y.Y. Souprun, and M.M. Huleihel. 2002. Antiviral activity of red microalgal polysaccharides against retroviruses. *Cancer Cell International* 2 (July): 8, 1–7. https://doi.org/10.1186/1475-2867-2-8.

Thai, M., S.K. Thaker, J. Feng, et al. 2015. MYC-induced reprogramming of glutamine catabolism supports optimal virus replication. *Nature Communications* 6 (November): 8873, 1–9. https://doi.org/10.1038/ncomms9873.

Thivierge, K., V. Nicaise, P.J. Dufresne, et al. 2005. Plant virus RNAs. Coordinated recruitment of conserved host functions by (+) ssRNA viruses during early infection events. *Plant Physiology* 138: 1822–27. https://doi. org/10.1104/pp.105.064105.

Yang, J., J. Liang, L. Shao, et al. 2020. Green production of silybin and isosilybin by merging metabolic engineering approaches and enzymatic catalysis. *Metabolic Engineering* 59: 44–52. https://doi. org/10.1016/j.ymben.2020.01.007.

Yu, Y., T.G. Maguire, and J.C. Alwine. 2012. Human cytomegalovirus infection induces adipocyte-like lipogenesis through activation of sterol regulatory element binding protein. *Journal of Virology* 86: 2942–49. https://doi.org/10.1128/jvi.06467-11.

Zaim, M., R.K. Lal, R.K. Verma, and R. Pandey. 2014a. Studies on effect of Poppy mosaic virus infection on poppy produce and some secondary metabolites. *Acta Horticulturae* 1036: 151–55. https://doi.org/10.17660/ActaHortic.2014.1036.16.

Zaim, M., R.K. Verma, R. Pandey, and R.K. Lal. 2014b. Genotype-dependent response of an RNA virus infection on selected pharmaceutically important alkaloids in *Papaver somniferum*. *Journal of Herbs, Spices and Medicinal Plants* 20: 124–31. https://doi.org/10.1080/10496475.2013.840817.

Zhu, L., Q. Zhao, T. Yang, W. Ding, and Y. Zhao. 2015. Cellular metabolism and macrophage functional polarization. *International Reviews of Immunology* 34: 82–100. https://doi.org/10.3109/08830185.2014.969421.

10 Various Bacterial Secondary Metabolites and Their Roles in Basic Metabolic Processes

Jyoti Lodhi
(Govt Motilal Vigyan Mahavidhayalaya, Affiliated to Barkatullah University, Bhopal, Madhya Pradesh 462001, India; lodhijyoti71@gmail.com)

**Monu Ashutosh Shukla*
(Drug Chemistry Research Centre, Indore, Madhya Pradesh 452003, India
(Email: drugchemistryresearchcentre@gmail.com)

10.1 INTRODUCTION

Secondary metabolites (SMs) are naturally occurring substances that are primarily produced by bacteria, fungi, and plants. They are low-molecular-weight compounds with a range of chemical and biological properties. Metabolites which consist of lipids, amino acids, carbohydrates, and nucleic acids are known as primary metabolites. Secondary metabolites, on the other hand, are substances that are not necessary for the growth and reproduction of organisms (Jain et al., 2023, 2024. On the other hand, far from being secondary, the phrase "specialized metabolites" have begun to be used to refer to them. Nowadays, it is acknowledged that SMs serve crucial roles in the survival of the species because SMs control interactions in their environment (Mosunova et al., 2021). Usually, only tiny molecules are included when using the word "metabolite." Metabolites serve a variety of purposes, including those of fuel, structure, signaling, stimulating, and inhibiting effects on enzymes, independent catalytic activity (often as an enzyme cofactor), defense, and connections with other species, as shown in Figure 10.1 (Zhao et al., 2005). In contrast to being non-essential for the growth and development of microorganisms, microbial secondary metabolites, such as antibiotics, pigments, growth hormones, anticancer medicines, *etc*., have demonstrated a significant potential for enhancing human and animal health (Gupta et al., 2022, 2024).

Among the microorganisms that create the aforementioned substances, bacteria and fungi produce a wide range of bioactive small molecules with great potential for application in medicine. Chemical structure (such as having rings or containing a sugar), composition (such as whether they contain nitrogen or not), solubility in different solvents, or the mechanism by which they are synthesized (such as the production of tannins through the phenylpropanoid pathway) may all be used to categorize secondary metabolites. Typically, they are categorized by their biosynthesis pathways. Phenolics, terpenes and steroids, alkaloids and flavonoids are the three main categories of big molecules that are often taken into account. Some of them have extremely severe consequences (Overy et al., 2019).

All spheres of life have produced secondary metabolites, at least to this point. Those made by bacteria were some of the first to be found and were the targets of the earliest engineering attempts. Fungi and plants are recognized as two of the eukaryotic kingdom's most abundant producers of secondary metabolites. The fact that animals do not make as many of these substances is noteworthy. This is probably because these species have alternative means of interacting with their surroundings, such as locomotion. Although the number of instances described is rather modest, secondary metabolites have also been extracted from Archaea. This might be partly explained by the difficulties in cultivating a large number of species from this domain (Harborne, 1999).

Primary and secondary bacterial metabolites are frequently separated into two groups. The building blocks, cofactors, and biological polymers that bacteria need to develop and proliferate are called primary metabolites. These are the sets of chemicals that provide life to bacteria. The interaction between a bacterium and its environment is made easier by secondary metabolites, which are primarily extracellular, low-molecular-weight substances (Bourgaud et al., 2001). Secondary metabolites are of interest for both their ecological and biogeochemical effects, as well as their potential for use in medicine and biotechnology (O'Connor, 2015). Since the commercial significance of secondary metabolites has increased in recent years, there has been a lot of interest in the prospect of modifying the synthesis of bioactive plant metabolites utilizing tissue culture techniques. Explants such as leaves, stems, roots, and meristems can be used to develop plant cell and tissue culture technologies regularly under sterile circumstances for both

DOI: 10.1201/9781003408833-10

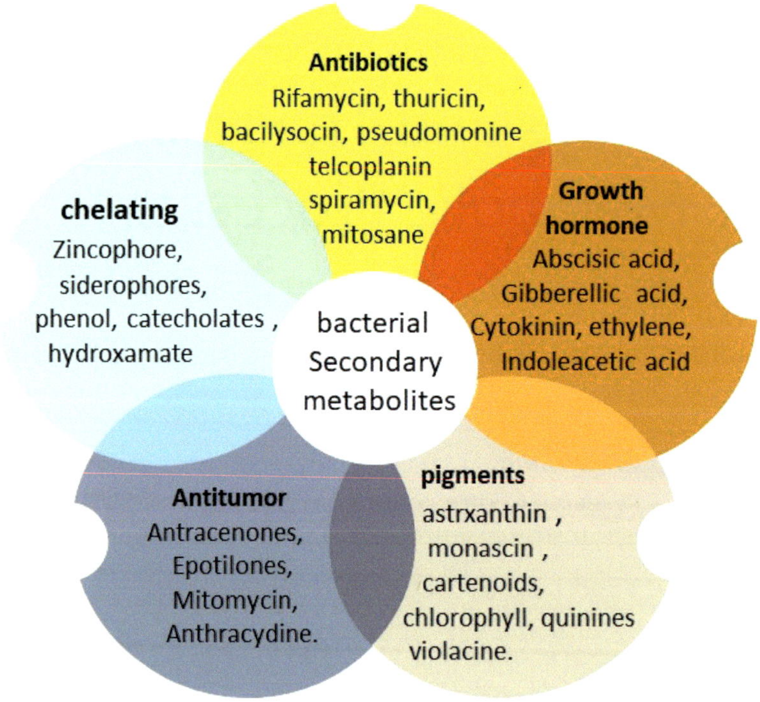

FIGURE 10.1 Schematic view of bacterial secondary metabolites.

the multiplication of cells and the extraction of secondary metabolites. It has been observed that both industrial and medicinal plants may produce secondary metabolites *in vitro* in plant cell suspension cultures (Baltz, 2017).

We still need to clarify the origins and selection processes involved in the spread of the many mechanisms of antibiotic resistance that we are now detecting in clinical and environmental contexts, in the context of a larger picture of microbial ecology. Findings on the ecology of antibiotic resistance genes *in situ* might have a significant influence on clinical microbiology and provide a great environment for researching how characteristics adapt in bacterial populations (Cragg and Newman, 2013).

10.2 ANTIBIOTICS

10.2.1 Antibiotics as Secondary Metabolites Produced by Bacteria

The term "antibiotic" has its roots in the word "antibiose," which Vuillemin (1889) initially used in his article to characterize the antagonistic interaction between various microorganisms (for instance, fungus vs. bacteria; bacteria vs. protozoa) (Sharanabasappa et al., 2007). Molecular compounds referred to as antibiotics are used to treat and prevent bacterial infections. *Efficacious therapies* for the treatment of several medical conditions, such as pneumonia, typhoid fever, or gonorrhea, were lacking before their debut in clinical practice (Pegadraju et al., 2021). Penicillin was the first organic antibiotic to be successfully isolated by Fleming and colleagues (1929). The scientists discovered a Petri plate containing colonies of *Staphylococcus* infected with a mold, eventually identified as *Penicillium notatum*, when they returned from vacation. Fleming was astounded that no germs could develop around the mold, but he was also astute enough to think that the occurrence was noteworthy. In 1945, Dorothy Hodgkin determined the beta-lactam composition of penicillin (Simeis and Serra, 2021), settling the long-running argument between Robert Robinson, who advocated a thiazolidine-oxazolone structure, and several eminent chemists, including Chain, Abrahams, and Woodward, who thought it was a beta-lactam (Aminov, 2010). This significant invention made it possible to create semisynthetic derivatives that could overcome penicillin resistance.

The discovery of streptomycin, the first drug that effectively treated TB, came about as a result of the screening of streptomycetes. Over the following 20 years, the pharmaceutical industry largely embraced this "Waksman platform," leading to the production of the major families of antibiotics. As they discovered novel treatments for infectious illnesses, corporations like Merck, Pfizer, and Eli Lilly, among others, saw their riches rise. However, after 20 years of success, overmining of soil streptomycetes (and other actinomycetes) led to declining returns, leading to the rediscovery of known chemicals and the platform's collapse. Although it was becoming harder to use currently available drugs, altering antibiotics led to the production of active analogues. For a while, this strategy was mostly productive and produced efficient treatments (Curtis and Jones, 2007).

The majority of the time, antibiotics in poultry are given to the entire flock and are used for growth promotion, illness prevention, and therapy. Antibiotic growth promoters are now legal in China and Brazil but were outlawed in the EU in 2006 and the United States in 2017 (Fischbach and Walsh, 2009). All major nations that produce chickens are allowed to use antibiotics to prevent illness. Treatment with antibiotics is used for intestinal infections, such as colibacillosis, necrotic enteritis, and other illnesses typically brought on by *Salmonella*, *E. coli*, or *Clostridium* spp. These illnesses are a major risk for poultry and result in significant financial losses (Poole and Sheffield, 2013). Given their significance for both human benefit and natural systems, several extensive initiatives to find novel bacterial metabolites have been initiated over the past century with great success. Louis Pasteur said that using microbial conflict would give "perhaps the greatest hopes for therapeutics" as early as 1876. Waksman and others employed systematic screening techniques in the middle of the twentieth century to find soil microorganisms that generated antibiotics. These early efforts helped usher in the "golden age of antibiotic discovery," which produced numerous new classes of antibiotics that were crucial in the fight against infectious illnesses (Kingston, 2008).

This chapter, looking at various antibiotics produced by microorganisms (Table 10.1), explores the application of this area. The search for immunosuppressive, anticancer, and other therapeutically important compounds continued with similar initiatives, which have continuously driven both the industrial and academic drug discovery pipelines: More than 68% of the almost 3,000 FDA-approved medications from 1981 to 2019 are either bacterial natural products or are influenced by their chemistry (Katz and Baltz, 2016). Benjamin M. Duggar discovered the first member of one of the most significant groups of antibiotics, the tetracyclines, from *Streptomyces aureofaciens* in 1945. It was initially called *aureomycin* because of its golden hue, but it is now more often referred to as chlortetracycline (Newman and Cragg, 2012). It was authorized for therapeutic use as a broad-spectrum antibiotic against both Gram-positive and Gram-negative bacteria in 1948 (Mak et al., 2014).

10.2.2 Inhibitory Roles of Antibiotics

Antibiotics, antimicrobial compounds with the ability to stop the development of germs or kill them, are frequently used to treat bacterial infections in both humans and animals as well as in non-medical settings. Antibiotics are manufactured worldwide yearly in quantities between 100,000 and 200,000 t, with more than 1 billion t produced since 1940 (Tang, 2010). The term "minimal inhibitory concentration" (MIC) refers to the lowest antibiotic concentration that, when used in therapeutic antimicrobial medication doses, stops observable microorganism growth following incubation with bacteriological medium. The minimum antibiotic concentration that promotes selection of a resistant mutant over wild-type cells is known as the minimal selective concentration (MSC). In sub-inhibitory doses (concentrations below the MIC), the susceptible bacterial cells continue to develop at a decreased pace. According to estimates, the MSC for a variety of different microorganisms falls between 1/4 and 1/230 of the MIC values (Czekalski et al., 2014).

10.2.3 Antibiotic-Producing Bacteria and Their Mechanisms of Action

Antibiotic-producing bacteria have a wide range of complex defense mechanisms to protect themselves from their antibiotics. They frequently have many systems operating at once to provide total defense against the physiologically active substances they create. It's interesting to note that the genes responsible for antibiotic biosynthesis are usually always grouped with the genes responsible for self-resistance, and

TABLE 10.1
Some Antibiotics Produced by Microorganisms and Their Applications

S. No.	Antibiotics	Producer Microorganism	Application	Reference
1	Penicillin	*Penicillium chrysogenum*	Medical and pharmaceutical	Etebu and Arikekpar, 2016
2	Tetracycline	*S. aureofaciens*, *S. avellaneus*, *S. lusitanus*, and *S. viridifaciens*	Pharmaceutical, antibiotics therapies, and chemotherapeutic	Robinson et al., 2001
3	Chlortetracycline	*S. lusitanus*, *S. aureofaciens*, and *S. lividans*	Pharmaceuticals	Robinson et al., 2001
4	Oxytetracycline	*S. alboflavus*, *S. albofaciens*, *S. erumpens*, *S. griseus*, *S. platensis*, *S. rimosus* subsp. *Rimosus*, and *S. varsoviensis*	Pharmaceuticals	Robinson et al., 2001
5	Streptomycin	*Streptomyces griseus*	Pharmaceuticals	Peterson and Kaur, 2018
6	Cephalosporin	*Cephalosporium acremonium*	Pharma industries	Etebu and Arikekpar, 2016

their expression is co-regulated. The section that follows provides detailed examples for each of the key biochemical groups of self-defense systems that are present in producing organisms (Andersson and Hughes, 2012). In particular, chloramphenicol, β-lactams, and aminoglycoside antibiotics (such as kanamycin, gentamycin, and streptomycin) are frequently targeted for antibiotic modification, which is a method for making an antibiotic ineffective. Numerous aminoglycoside modification enzymes (AMEs), such as N-acetyl transferases (AAC), O-phosphotransferases (APH), and O-adenyltransferases (ANT), which acetylate, phosphorylate, or adenylylate the antibiotic, are known to occur in producer bacteria.

10.2.4 Impact of Antibiotics on Bacterial Communities and Antibiotic Resistance

This research examines how antibiotic resistance affects the world's health and focuses on initiatives to address this challenging issue (Velez and Sloand, 2016). Antibiotic resistance in bacterial pathogens is a global problem that is linked to high morbidity and death rates (Akova, 2016). Gram-positive and Gram-negative bacteria with multidrug resistance have caused illnesses that are difficult to cure or are even untreatable with traditional antimicrobials. Broad-spectrum antibiotics are often and usually unnecessarily utilized in health-care settings due to the absence of early identification of pathogenic bacteria and their antimicrobial sensitivity patterns in patients with bacteremia and other severe diseases (Wesgate et al., 2016). Modern times have changed this concept to also include antimicrobials that are partially or entirely manufactured synthetically. While some medicines can eradicate other germs, others can just stop their development. Bacteriostatic substances are those that prevent the development of bacteria, whereas bactericidal substances kill germs (Walsh, 2003). Although the term "antibiotic" commonly refers to a substance that fights bacteria, antibiotic substances are classified as either antibacterials, antifungals, or antivirals depending on the type of microorganisms they are effective against (Brooks et al., 2004).

10.3 SECONDARY METABOLITES AS ANTIFUNGAL DRUGS

Many other secondary metabolites, including polyketides, terpenes, and non-ribosomal peptides, are being screened from fungi (Keller et al., 2005). These secondary metabolites do not directly contribute to the growth, development, reproduction, or energy production of the organism, but they have shown biological activity that helps to survive in the ecological niche it has filled. Most secondary metabolites have biological actions that have the potential to be useful in medication development (Dias et al., 2012). Some of the pharmaceutically and commercially significant secondary metabolites include lovastatin, a medication that lowers cholesterol, and penicillin, a β-lactam antibiotic (Boruta, 2018). The most popular secondary metabolite of fungal origin is penicillin, the original broad-spectrum antibiotic. Penicillin revolutionized the profession of medicine, changed the direction of pharmaceutical research, altered the outcome of World War II, and saved countless lives. Alexander Fleming discovered *Penicillium notatum* as the origin of this drug. The treatment of bacterial infections significantly improved as a result of the discovery of penicillin in both medical and scientific fields (de la Fuente-Nunez, 2019). The inventory of secondary metabolites from different fungi strains, which reveal certain characteristics, was one of several factors that greatly influenced the unanticipated finding (Houbraken et al., 2011).

10.3.1 Role of Antifungal Metabolites in Inhibiting Fungal Growth and Preventing Infections

An antifungal activity is the prevention or the death of fungi. It serves as a medicinal preparation. A medicine called an *antifungal* is used to treat fungal diseases, including mycosis (including fungal infections in humans), athlete's foot, dermatophytosis, candidiasis, and many other severe systemic infections.

10.3.2 Bacteria Producing Antifungal Compounds and Their Modes of Action

The development of secure, broad-spectrum antifungal medicines is made more challenging by both. There are several antifungal medicines in the market; out of several, amphotericin B and a few other substances are fungistatic. Screening massive libraries of artificially made small compounds or naturally occurring products has historically been the most popular method for discovering antifungal chemicals (Pianalto and Alspaugh, 2016). Screening of natural compounds led to the discovery of two groups of antifungal drugs that are now approved: echinocandins and polyenes. A few instances of this method have been recorded recently. Lesional psoriasis scale preparations yielded psoriasin, a tiny fungicidal protein that was potent against a variety of filamentous fungi but not against *Candida albicans*, including *Trichophyton rubrum* and *A. fumigatus* (Hein et al., 2015). To cause fungal apoptosis, psoriasin chelates free intracellular zinc. As a result, a novel strategy to discover new antifungal drugs may be to use selective fungal cell–penetrating zinc chelators. Amphotericin B continues to be the only polyene medication of choice for the treatment of invasive fungal infections despite the discovery of more than 200 polyenes with antifungal action since the 1950s (Zotchev, 2003). The antimetabolite 5-flucytosine was initially developed and utilized in the treatment of cancer. Due to its lackluster antineoplastic effect, 5-FC is increasingly used in combinational antifungal treatment (Vermes et al., 2000). The first-generation triazoles are fluconazole (FLC) and itraconazole (ITC), but the second-generation triazoles have better pharmacological characteristics and include voriconazole (VOR), posaconazole (POS), and isavuconazole (ISV) (Ping et al., 2013). The echinocandins are the most recent class of peptide antifungals, and their members are capofungin (CSF), micafungin (MCF), and anidulafungin (ANF).

10.3.3 Applications of Antifungal Metabolites in Agriculture, Medicine, and Industry

Cyclosporin A (derived from *Trichoderma polysporum*) was first prescribed as an antifungal drug, which is now well-known for its potent immunosuppressive properties (Keller et al., 2005). When someone has had surgery for an organ transplant, it is utilized to stop rejection. Secondary metabolites may also serve as hormones, such as the gibberellins from *Gibberella fujikuroi*, which are used to produce seedless grapes (Gong et al., 2004); to increase vegetable yield, speed up the malting of barley, and improve the quality of the resulting beer; to regulate flowering, seed germination, and stem elongation; and to speed up the harvest of lettuce and sugar beet seed crops (Golan et al., 2011). *Gibberella* sp. (*Gibberella zeae*) also produces the estrogen zearelanone, which is utilized to boost growth and feed effectiveness in sheep and cattle (Wurzbacher et al., 2010). As is evident with the aflatoxins that are responsible for several illnesses, secondary metabolites can have adverse consequences. These include both acute and chronic illnesses, such as cancer, immune system suppression, and in extreme circumstances, death (Fox and Howlett, 2008).

10.4 ANTIVIRAL COMPOUNDS

10.4.1 Overview of Antiviral Secondary Metabolites Produced by Bacteria

Since early in the history of human civilization, infectious diseases have been well-known. Different microorganisms (bacteria, viruses, and fungi) cause infectious diseases (Balloux and van Dorp, 2017). In contrast to the complex structures of fungi, helminths, and protozoa, viral structure is simple and consists of a protein coat, nucleic acid, and occasionally, a lipid envelope. Additionally, viruses are obligatory intracellular pathogens because they replicate via the host's cellular infrastructure. These traits make it challenging to build medications that have selective toxicity against viruses (Kumar et al., 2021). The development of antiviral medications is a laborious process that involves several steps, including target discovery and screening, lead generation and optimization, clinical investigations, and drug registration, among others (Saxena et al., 2010). A high-throughput screening of libraries, including already-approved medications or random molecules, is the most efficient method for finding antiviral medicines quickly. Potential drug candidates can be found by such research, which can also offer useful information for lead optimization techniques like QSAR (quantitative structure–activity relationship). However, the cost-effectiveness and scope of these initiatives are frequently constrained by the libraries of available compounds or fragments.

Natural products have long been a significant source of inspiration in the search for novel antivirals due to their various and distinctive structures and methods of action. One of them is the nucleoside spongouridine that the sea sponges *Cryptotethya crypta* generate (Riccio et al., 2020). The two antiviral medication classes with the most prevalent modes of action are those that target the virus and the host (Lou et al., 2014). Antivirals that target specific viruses concentrate on blocking key transcription and replication enzymes, such as polymerases and proteases (Kiser and Flexner, 2013), or on directly inactivating viral structural proteins (Yu et al., 2013). Contrarily, host-targeting antivirals concentrate on the suppression of cyclophilins, which are known to be significant cellular factors that are hijacked by some viruses during the replication cycle (Lou et al., 2014), the use of immunomodulators like interferons (Raziky et al., 2013) and gamma globulins (Buttinelli et al., 2003).

10.4.2 Discussion on Mechanisms of Action against Viral Infections

In vitro anti-RSV, but not anti-HSV-1 or anti–adenovirus serotype 7 (DNA virus), was found in *Myrcianthes cisplatensis* (Kott et al., 1998). Other medicinal plants, such as *Nepeta coerulea*, *Nepeta nepetella*, *Nepeta tuberosa*, *Sanguisorba minor* ssp. *magnolii*, and *Dittrichia viscose*, on the other hand, demonstrated definite antiviral activity against DNA and RNA viruses, including HSV-1 and VSV, as well as poliovirus type 1 in the case of *Dittrichia viscose* (Abad et al., 2000). Several viruses, including smallpox (DNA), chicken pox (DNA), poxvirus (DNA), poliomyelitis (RNA), and herpes viruses (DNA), have been discovered to be susceptible to the *Azadirachta indica* leaf extract. Inactivating extracellular viruses such as HSV, horse herpes, pseudorabies, and influenza virus was possible with an extract from the cactus plant *Opuntia streptacantha*. In preclinical and clinical testing against the influenza virus (RNA) and HSV (DNA), the *Bergenia ligulata*, *Nerium indicum*, and *Holoptelia integrifolia* plants demonstrated significant antiviral activity (Rajbhandari et al., 2001).

10.4.3 Antiviral Metabolites Produced by Bacteria and Their Potential Applications

Galactan sulfate, a polysaccharide that was isolated by Witvrouw et al. (1994) from *Agardhiella tenera*, demonstrated efficacy against HIV-1 and HIV-2 with IC50 values of 0.5 and 0.05 g/L, respectively. The methanolic extracts of various Indian red seaweeds were studied by Devi et al. (2008) for their antioxidant and antibacterial properties. *Scinaia hatei*, an Indian red seaweed, contained the antiviral component sulfated xylomannan, which suppressed HSV-1 and HSV-2 (IC50 = 0.5–1.4 g/mL) (Mandal et al., 2008). Water-soluble polysaccharidic extracts of *Boergeseniella thuyoides* and *Sphaerococcus coronopifolius* from the Moroccan coast demonstrated antiviral activity against viruses, including HIV and HSV-1 (Bouhlal et al., 2011). From *Ceramium rubrum*, a plant isolated from the Black Sea, was identified a bioactive substance. *In vitro* and *in vivo* inhibition of the types A and B influenza viruses by this metabolite was followed by a decrease in cytopathogenic effects (Serkedjieva, 2004). The recent COVID-19 pandemic event demonstrated the need for further funding to be put toward the quest for new marine natural products with antiviral activity, as well as the exploration of the vast maritime ecosystem, because many organisms are still unexplored (Liu et al., 2020).

Microorganisms, particularly those that are easily cultivable, have been deemed new and potential sources of novel bioactives in light of adopting eco-friendly and eco-sustainable approaches to drug development (in keeping with the European concept of a Blue Economy based on marine resources). In actuality, extensive collection techniques are necessary for isolation from macroorganisms. Numerous solutions, including chemical synthesis and heterologous expression, and manufacturing, have been considered to address this issue (Lauritano et al., 2019). Additionally, current research has concentrated on developing novel culturing techniques to cultivate "uncultivable organisms" to isolate novel chemicals from them (Lauritano and Ianora, 2018). The ocean is a vast, unexplored supply of natural chemicals with antiviral and other potential bioactivities that are beneficial to the environment and human health.

10.4.4 Challenges and Opportunities in the Development of Bacterial Antiviral Therapies

The potential of marine creatures as medicines must be fully realized, which presents numerous difficulties. Lack of access to deep-sea regions is a significant obstacle to gathering sufficient samples for further study and the development of numerous incredibly effective novel marine actives with simpler synthesis methods and tailored characteristics. It is generally known that the marine environment is a source of rich natural products and that it offers a wide range of different chemical structures with promising bioactivities that can be used as therapeutic agents (Martins et al., 2014).

10.5 IRON CHELATORS AND SIDEROPHORES

10.5.1 Role of Iron in Bacterial Metabolism and Its Limited Availability in the Environment

The majority of metabolic enzymes involved in biological processes like respiration, DNA biosynthesis, gene control, and the tricarboxylic acid (TCA) cycle depend on iron, the most prevalent transition metal in the host. Iron is essentially inaccessible within animal hosts due to its limited solubility under aerobic circumstances at neutral pH. Iron is chelated by proteins, including ferritin, hemoglobin, and transferrin, which might limit the availability of this vital metal to invasive infections. Host cells have the ability to deplete metals from inside phagosomes in addition to extracellular metal limitation mechanisms. In certain conditions, where the availability of iron is limited, bacteria produce and release high-affinity iron-chelating molecules called siderophores. These molecules help the bacteria survive by removing iron from the host (Hood and Skaar, 2012). Since iron is both necessary for life and potentially poisonous to most living things, maintaining iron homeostasis precisely is crucial to existence. As a result, many organisms have evolved robust homeostatic mechanisms that keep intracellular iron concentrations within ranges that are not harmful to cells (Baichoo and Helmann, 2002).

10.5.2 Iron Acquisition via Siderophores

The acquisition of sufficient amounts of iron is made possible by strictly controlled bacterial iron homeostasis, which also limits the toxicity caused by reactive oxygen species produced as a result of iron. As a result, only environments with low levels of available iron can express iron acquisition systems. The main transcription factor of iron homeostasis in a variety of bacterial taxa, ferric uptake regulator (Fur), senses changes in intracellular iron 6 concentrations (Sheldon and Heinrichs, 2015).

10.5.3 Siderophores as Regulators of Microbial Communities

The composition and operation of many microbial communities are influenced by the availability of iron, a key environmental element. Given that most bacteria need iron as a growth factor and that different bacterial taxa have diverse abilities to scavenge for this micronutrient, this is not surprising. Dietary interventions that alter iron concentrations in the colon cause compositional changes to the intestinal microbiota as well as significant alterations in the fecal metabolome (Lee et al., 2017).

10.5.4 Introduction to Siderophores and Other Iron-Chelating Secondary Metabolites

In the Earth's crust, iron is the fourth most common metal and is necessary for practically all life forms. Iron plays a role in a variety of metabolic processes. For instance, it plays a crucial role as a cofactor in numerous enzymes and is involved in the transfer of oxygen, respiration, nitrogen fixation, photosynthesis, and the production of DNA, fatty acids, and amino acids, among other processes (Serrano, 2017). One of the most prevalent techniques established by bacteria, fungi, and plants is the synthesis of siderophores, or "iron carriers" in Greek. Natural iron chelators known as siderophores have modest molecular weights

(400–2,000 Da) and a strong affinity for ferric iron (Ahmed and Holmström, 2014). They are created when there is an iron deficiency to scavenge iron by creating soluble iron (III) complexes. They give living things access to insoluble forms of iron, and in pathogenic microbes, they play a role in iron theft from host proteins (Abergel et al., 2006).

10.5.5 Importance of Siderophores in Iron Acquisition and Bacterial Survival

Targets for siderophores' iron-scavenging activity include extracellular iron complexed with host proteins, including transferrin and lactoferrin. Small chemicals called siderophores, which are released by bacteria, have an extraordinarily high affinity for iron. Staphyloferrin A and staphyloferrin B, which are produced by *S. aureus* separately but share many characteristics, are two different siderophores (Friedman et al., 2006). We hypothesized that siderophore-mediated iron acquisition gives *S. epidermidis* a competitive advantage within the host, given that it is a significant contributor to bloodstream infections resulting from indwelling medical device contamination (Kleinschmidt et al., 2015). Using a murine *in vivo* model for *S. epidermidis* bacteremia, we found that groups of mice infected with the siderophore biosynthetic 1sfa mutant had significantly lower bacterial loads compared to those infected with the wild-type strain. This difference was evident both in the bloodstream and in specific organs, such as the liver and kidneys, six hours after infection. To sum up, our *in vivo* results show that siderophore-mediated iron uptake is crucial for *S. epidermidis* survival inside the host, and they also provide strong evidence for this claim.

10.5.6 Examples of Siderophore-Producing Bacteria and Their Implications in Host–Pathogen Interactions

Several significant infections' pathogeneses have been linked to siderophores (Madigan et al., 2015). When *Staphylococcus epidermidis* biofilm cells come into contact with human blood, a study by França et al. (2014) found that the transcription of certain genes, such as htsC and fhuC, which are thought to be components of the iron ABC transporter system, and a siderophore homolog locus called sfaABCD increases. These genes are thought to be involved in iron acquisition. Iron may be important during *S. epidermidis* infections, according to the theory that resulted from this (Oliveira et al., 2021).

10.6 QUORUM-SENSING MOLECULES

10.6.1 Explanation of Quorum Sensing and Its Significance in Bacterial Communication

The universe is a collection of numerous intricate mechanisms that connect microbes. To adapt, cohabit, and conjugate (reproduce), bacteria must work together under a variety of situations. This defense mechanism – metabolite synthesis – is the result of this cooperation. Typical defense mechanisms, also known as pheromones or auto-inducers, include iron chelators, antibiotics, and signal molecules that aid in microbial communication (Williams et al., 2007).

10.6.2 Introduction to Quorum-Sensing Molecules as Secondary Metabolites

Quorum sensing (QS) is a microbial communication and gene expression regulatory phenomenon. Since the 1960s, it has been understood that unicellular bacteria do not exist in isolation from their surroundings, but rather that there is cell-to-cell communication that enables the population to regulate its behavior (Nealson et al., 1970). Small diffusible chemicals produced by the microbial population serve as "autoinducers" by accumulating during microbial growth and mediating this complex cell-to-cell signaling (Williams et al., 2012). This phenomenon was originally investigated in bacteria (Miller and Bassler, 2001), and it is now recognized to play a role in the control of a variety of biological activities, including pathogenesis, symbiosis, competence, conjugation, nutrient uptake, morphological differentiation, generation of secondary metabolites, development of biofilms, *etc.* (Bandara et al., 2012).

10.6.3 Roles of Quorum-Sensing Molecules in Regulating Various Metabolic Processes

Bacteria interact using signaling molecules referred to as autoinducers through the quorum-sensing system, which is reliant on population density (Bassler and Losick, 2006). Bacteria typically make autoinducers during quorum sensing, and these molecules build up in the environment as cell density rises. Their receptors, which are found on the cell surface or in the cytoplasm, determine the function of these autoinducers. A specific quorum-sensing gene's transcription is activated by an autoinducer's activation of its cognate receptor (Rutherford and Bassler, 2012). This process gives bacteria a way to coordinate their social behavior, communicate with one another, and control gene expression in response to population density. Another sort of quorum sensing is present in Gram-positive bacteria, where autoinducers are returned to the bacterial cytoplasm, where they interact with a particular transcription factor to control gene expression (Rutherford and Bassler, 2012).

10.6.4 Examples of Quorum-Sensing Molecules and Their Impact on Bacterial Behavior and Virulence

Quorum sensing (QS) is a molecular communication system that many bacteria, including *P. aeruginosa*, employ to coordinate the expression of specific genes and exhibit group behavior. Among QS-regulated features, the synthesis of virulence factors such as pyocyanin and proteases (Hentzer et al., 2003), motility (Daniels et al., 2004), and biofilm formation are involved in the progression of infections (Whiteley et al., 2001). Many examples have been documented and discuss the reduction of virulence factor secretion (e.g., siderophores, proteases, rhamnolipids, *etc.*) (Diggle et al., 2007; Welsh and Blackwell, 2016) and or biofilm development, notably employing QSI molecules (Hentzer et al., 2003). However, the majority of these papers focus on studying the model strains PAO1 and PA14 of *P. aeruginosa*, and only a small number of reports describe how clinical isolates react to QS disruption, even though wild isolates typically include QS gene alterations (Ciofu et al., 2010).

10.7 TOXINS AND VIRULENCE FACTORS

10.7.1 Overview of Bacterial Toxins and Virulence Factors as Secondary Metabolites

Pathogenic bacteria are endowed with a variety of chemicals and mechanisms that function as virulence factors and circumvent the host defense mechanisms to start disease processes. In addition to metabolic systems and cell surface structures, bacteria can produce a variety of chemicals, including enzymes, toxins, and exopolysaccharides. Many of these secreted biomolecules can penetrate the host cells and regulate or harm the intracellular signaling or structures (Leitão, 2020). Each of these bacterial macromolecules eventually contributes to pathogenicity, which may make therapy more challenging. The connection between several bacterial virulence factors and carcinogenesis is a fascinating feature. It is believed that *E. coli*, which induces chronic inflammation in inflammatory bowel disease, may increase the risk of developing colon cancer (Khan et al., 2017). Notably, *E. coli*, *Klebsiella* spp., and another Gram-negative bacillus, *Pseudomonas aeruginosa*, have all been isolated from cancer patients (Bhat et al., 2021; Islas-Muñoz et al., 2018; Perez et al., 2014).

10.7.2 Discussion on the Roles in Pathogenesis and Host Interactions

One of the most fascinating elements of life is the continual interaction between hosts and pathogens. These interactions have evolved over millions of years; hosts build defense systems against pathogenic incursions, while pathogens get around these new lines of defense (Morens and Fauci, 2012). Proteomics has increased the number of tools available for studying pathogen infections by integrating with other biochemical, molecular biology, and omics methodologies. To develop a systems-level knowledge of the infection process, genomic, transcriptomic, and metabolomic research offers orthogonal data that supports proteome analyses. Here, we examine the state of the art in proteomics techniques to clarify host–pathogen protein interaction networks, changes in the structure and content of the host cell proteome, and infection-induced post-translational regulation.

10.7.3 Examples of Toxin-Producing Bacteria and the Diseases They Cause

Escherichia coli is a part of the natural flora in the human stomach; however, some strains are harmful. Internal pathogenic *E. coli* can be divided into six groups based on their pathotypes: enterotoxigenic (ETEC), enteropathogenic (EPEC), enteroaggregative (EAEC), enteroinvasive (EIEC), and diffusely adherent (DAEC) (Kaper et al., 2004). *Shigella dysenteriae* serotype 1 and STEC both express the family of bacterial exotoxins known as Stxs (Fraser et al., 1994). These toxins are the main virulence factors for bloody diarrheal disease, which can progress to potentially fatal systemic sequelae, like acute renal failure syndrome (also known as hemolytic uremic syndrome, HUS), as well as abnormalities of the central nervous system (Tarr et al., 2005). Hemorrhagic colitis, infantile diarrhea, and hemolytic-uremic syndrome are only a few of the disorders that are brought on by bacteria that carry Stx-encoding phages when they are identified in people. Even though all isoforms of Stx2 and all classes of the protein have the same functional properties, only a few of them are frequently linked to more severe human disease. However, it has not yet been conclusively determined how important each Stx class and type is to the pathogenesis of humans. *E. coli* O157:H7 is the most frequently discovered Shiga toxin–producing *E. coli* (STEC) strain. STEC O157:H7 developed from a progenitor that was similar to O55:H7 (Feng et al., 1998).

10.7.4 Importance of Understanding Bacterial Toxins for Developing Therapeutics and Vaccines

Antimicrobials have made it possible to treat infections that may be fatal, helping countless patients around the world by saving their lives and enhancing their health. The ability to prevent and treat many kinds of infectious diseases that were once curable has been compromised by the rising quantity and global dispersion of drug-resistant microorganisms, which is one of the biggest health challenges. Since antibiotic resistance has greatly increased over the past few decades, investigation into the resistance mechanism in *P. aeruginosa* has become vital. Despite the high level of interest, the variability and complexity of drug resistance, as well as the paucity of a thorough understanding of pathogenic mechanisms in *P. aeruginosa*, are holding back the development of new antibiotics and other therapeutic approaches for *P. aeruginosa* infections. The need for coordinated efforts has increased as did the need for developing efficient therapeutic techniques (such as phage therapy, immunotherapy, gene editing therapy, antimicrobial peptides, and vaccine therapies) to combat *P. aeruginosa* invasion (Chin et al., 2018).

Lack of knowledge regarding immune defense and *S. aureus*–countering mechanisms is the main challenge facing vaccine development. Around 20–80% of human anterior nares are colonized by *S. aureus*, one of the major causes of bacteremia and a significant burden of morbidity from skin infections. Developing a vaccine is made more difficult by ignorance of the immune system's main actors and the *S. aureus* master immune responses (Browne et al., 2014).

10.8 BIOACTIVE COMPOUNDS

10.8.1 Introduction to Bioactive Secondary Metabolites with Pharmaceutical Potential

Almost all biosphere-wide processes can be impacted by microorganisms, which can also change their immediate surroundings. Drug-resistant pathogenic microorganisms and the advent of novel diseases are just two examples of the difficulties the rapidly changing microbial population has placed on humans (Chang et al., 2015; Bengtsson-Palme, 2018). For instance, antibiotics and anticancer drugs are being discovered at a decreasingly rapid rate (Lam, 2007; Hesterkamp, 2015). In addition, several medicinal plants are rare, grow slowly, are situated in odd habitats or at high elevations, and are constantly threatened by humans (Kala et al., 2006; Chen et al., 2016). The pharmaceutical industry has discovered over the past several years that endophytic fungi are a rich source of naturally occurring bioactive compounds with potential advantages (Higginbotham et al., 2013; Mousa and Raizada, 2013). The production of several of these secondary metabolites by their various hosts is intriguingly similar, and they may be used to replace important plant-based chemicals (Kumara et al., 2012). Endophyte research is still in its infancy despite several reports of bioactive substances from fungal endophytes; commercial synthesis of such bioactive chemicals has met with minimal success.

10.8.2 Discussion on Their Diverse Biological Activities, Such as Anticancer or Immunomodulatory Effects

The biological activity of the d-glucan group of polysaccharides is influenced by various physicochemical factors, including solubility, molecular structure, molecular weight, and branching (Synytsya and Novák, 2013). Due to their diversity, these polysaccharides have a wide range of uses in the food, biomedical, pharmaceutical, and cosmetic industries (Cunha et al., 2012; Xu et al., 2013; Pentikäinen et al., 2014). In biology, polysaccharides can be utilized for biosensors, tissue engineering, skin restoration for burn sufferers, enzyme immobilization, and controlled medication release. The substances significantly inhibited mitogenic responses in mice spleen cells due to the lectin, concanavalin A, and lipopolysaccharides, which have been shown to play an immunomodulatory role by inhibiting lymphocyte activity or shifting the kinetic parameters of immune responses (Basch et al., 2003). Momordicin induces surface membrane immunoglobulin activity, which activates and boosts B cell proliferation. It also increases the expression of the humoral immune component of B cell subsets known as CD86 (cell activation target point). Additionally, after 96 hours of co-culture, it might cause spleen cells to release significant levels of nonspecific immunoglobulin IgM, which can influence immunological control (Huang et al., 2008).

10.8.3 Examples of Bacteria Producing Bioactive Compounds and Their Applications in Drug Discovery

One of the most promising anticarcinogens created or synthesized to date is Taxol (paclitaxel), a complex diterpene alkaloid generated by the endophyte *Metarhizium anisopliae* found in the bark of the *Taxus* tree. Nothapodytes foetida's camptothecin is well-known for its cytotoxic and antifungal effects (Zhang et al., 2009; Nair and Padmavathy, 2014). An inhibitor of cholinesterase is hupA, a compound derived from *Hupezia serrata*. It has been claimed that lignans, such as cathartics, emetics, and cholagogues, isolated from endophytic *Podophyllum hexandrum*, have anticancer properties. *P. emodi* resins have potent anticancer properties, including etoposide and teniposide. Bactericidal activity is known to exist in several substances, including ampicillin, catechin, oxacillin, gallic acid, and cephalexin (Konuklugil, 1995; Akiyama et al., 2001). According to reports, endophytes create a variety of bioactive metabolites in a single plant or microbe, serving as a great source of medications for treating a variety of ailments and having potential applications in the food, cosmetics, agriculture, and medical industries. The functional groups of these secondary metabolites include alkaloids, benzopyranones, chinones, flavonoids, phenolic acids, quinones, steroids, saponins, tannins, terpenoids, tetralones, xanthones, and many more (Strobel and Daisy, 2003; Jalgaonwala et al., 2011; Godstime et al., 2014; Shukla et al., 2014).

10.9 EMERGING RESEARCH ON BACTERIAL SECONDARY METABOLITES AND THEIR ROLES IN METABOLIC PROCESSES

We are approaching a "new golden age" or renaissance age for the discovery and development of natural products, according to the combination and interaction of these three factors, as well as improvements in bioinformatics and mass spectrum investigations of secondary metabolites (Boddy, 2014; Medema et al., 2014; Weber et al., 2015). For full use of these developments, it is essential to have reliable genetic techniques on hand that can quickly express otherwise cryptic SMGCs and enhance product titers for isolation, characterization, product development, and production. The following genetic, molecular genetics, and synthetic biology techniques are currently being used to activate or improve SM production: (1) chemical, physical, and transposon mutagenesis; (2) recombination by protoplast fusion; (3) targeted gene/gene cluster duplications and higher-order amplifications; (4) overexpression of positive regulators; (5) disruption of negative regulators; (6) modulation of transcription by mutations in rpoB or hrdB; and (7) modulation (Weber et al., 2015). Although this combination of methodologies was frequently utilized in various optimization and modelling processes, it is uncommon to see them used in microbial fermentation to produce primary and secondary products. Accordingly, Shafi et al. (2018) combined response surface methodology with an artificial neural network to optimize temperature, pH, gelatine percentage, incubation time, and agitation speed to increase the biomass of *Bacillus* sp. and its antifungal efficacy against *Botrytis cinerea*.

10.10 CONCLUSION

In conclusion, there is a very good chance that techniques for genetically modifying actinomycetes to increase the synthesis of secondary metabolites will continue to progress. This is encouraging for the future use of "biosynthetic dark matter" in crop protection, animal health, and medicine development for humans. The free flow of concepts and genetic techniques between academic and corporate researchers has improved the existing state of the art. Success in the future depends on maintaining the global collaboration that has come to represent the actinomycete research community. In particular, for biosynthetically "gifted" germs, we describe the major difficulties in establishing a global, complete view of bacterial secondary metabolomes. We will cover current methodological developments that can address these issues. Applications of these technologies in conjunction with current developments will provide a full image of the world's secondary metabolomes, which will, in turn, provide insight into the biology, chemistry, and enzymology of natural products and facilitate the drug discovery process.

REFERENCES

Abad, M.J., J.A. Guerra, P. Bermejo, A. Irurzun, and L. Carrasco. 2000. Search for antiviral activity in higher plant extracts. *Phytotherapy Research* 14: 604–07. https://doi.org/10.1002/1099-1573(200012)14:8<604::AID-PTR678>3.0.CO;2-L.

Abergel, R.J., M.K. Wilson, J.E. Arceneaux, et al. 2006. Anthrax pathogen evades the mammalian immune system through stealth siderophore production. *Proceedings of the National Academy of Sciences* 103: 18499–03. https://doi.org/10.1073/pnas.0607055103.

Ahmed, E., and S.J.M. Holmström. 2014. Siderophores in environmental research: Roles and applications. *Microbial Biotechnology* 7: 196–208. https://doi.org/10.1111/1751-7915.12117.

Akiyama, H., K. Fujii, O. Yamasaki, T. Oono, and K. Iwatsuki. 2001. Antibacterial action of several tannins against *Staphylococcus aureus*. *Journal of Antimicrobial Chemotherapy* 48: 487–91. https://doi.org/10.1093/jac/48.4.487.

Akova, M. 2016. Epidemiology of antimicrobial resistance in bloodstream infections. *Virulence* 7: 252–66. https://doi.org/10.1080/21505594.2016.1159366.

Aminov, R.I. 2010. A brief history of the antibiotic era: Lessons learned and challenges for the future. *Frontiers in Microbiology* 1 (December): 134, 1–7. https://doi.org/10.3389/fmicb.2010.00134.

Andersson, D.I., and D. Hughes. 2012. Evolution of antibiotic resistance at non-lethal drug concentrations. *Drug Resistance Updates* 15: 162–72. https://doi.org/10.1016/j.drup.2012.03.005.

Baichoo, N., and J.D. Helmann. 2002. Recognition of DNA by Fur: A reinterpretation of the Fur box consensus sequence. *Journal of Bacteriology* 184: 5826–32. https://doi.org/10.1128/jb.184.21.5826-5832.2002.

Balloux, F., and L. van Dorp. 2017. Q&A: What are pathogens, and what have they done to and for us? *BMC Biology* 15 (October): 91, 1–6. https://doi.org/10.1186/s12915-017-0433-z.

Baltz, R.H. 2017. Gifted microbes for genome mining and natural product discovery. *Journal of Industrial Microbiology and Biotechnology* 44: 573–88. https://doi.org/10.1007/s10295-016-1815-x.

Bandara, H.M.H.N., O.L.T. Lam, L.J. Jin, and L. Samaranayake. 2012. Microbial chemical signaling: A current perspective. *Critical Reviews in Microbiology* 38: 217–49. https://doi.org/10.3109/1040841X.2011.652065.

Basch, E., S. Gabardi, and C. Ulbricht. 2003. Bitter melon (*Momordica charantia*): A review of efficacy and safety. *American Journal of Health-System Pharmacy* 60: 356–59. https://doi.org/10.1093/ajhp/60.4.356.

Bassler, B.L., and R. Losick. 2006. Bacterially speaking. *Cell* 125: 237–46. https://doi.org/10.1016/j.cell.2006.04.001.

Bengtsson-Palme, J. 2018. The diversity of uncharacterized antibiotic resistance genes can be predicted from known gene variants – but not always. *Microbiome* 6 (July): 125, 1–12. https://doi.org/10.1186/s40168-018-0508-2.

Bhat, S., S. Muthunatarajan, S.S. Mulki, A.K. Bhat, and K.H. Kotian. 2021. Bacterial infection among cancer patients: Analysis of isolates and antibiotic sensitivity pattern. *International Journal of Microbiology* 1 (January): 8883700, 1–7. https://doi.org/10.1155/2021/8883700.

Boddy, C.N. 2014. Bioinformatics tools for genome mining of polyketide and non-ribosomal peptides. *Journal of Industrial Microbiology and Biotechnology* 41: 443–50. https://doi.org/10.1007/s10295-013-1368-1.

Boruta, T. 2018. Uncovering the repertoire of fungal secondary metabolites: From Fleming's laboratory to the international space station. *Bioengineered* 9: 12–16. https://doi.org/10.1080/21655979.2017.1341022.

Bouhlal, R., C. Haslin, J.C. Chermann, et al. 2011. Antiviral activities of sulfated polysaccharides isolated from *Sphaerococcus coronopifolius* (Rhodophytha, Gigartinales) and *Boergeseniella thuyoides* (Rhodophyta, Ceramiales). *Marine Drugs* 9: 1187–09. https://doi.org/10.3390/md9071187.

Bourgaud, F., A. Gravot, S. Milesi, and E. Gontier. 2001. Production of plant secondary metabolites: A historical perspective. *Plant science* 161: 839–51. https://doi.org/10.1016/S0168-9452(01)00490–3.

Brooks, G.F., J.S. Butel, and S.A. Morse. 2004. *Jawetz, Melnick, & Adelberg's medical microbiology*. New York, N.Y.: Lange Medical Books/McGraw-Hill, Medical Pub. Division.

Browne, N., F. Hackenberg, W. Streciwilk, M. Tacke, and K. Kavanagh. 2014. Assessment of *in vivo* antimicrobial activity of the carbene silver (I) acetate derivative SBC3 using *Galleria mellonella* larvae. *BioMetals* 27: 745–52. https://doi.org/10.1007/s10534-014-9766-z.

Buttinelli, G., V. Donati, S. Fiore, et al. 2003. Nucleotide variation in Sabin type 2 poliovirus from an immunodeficient patient with poliomyelitis. *Journal of General Virology* 84: 1215–21. https://doi.org/10.1099/vir.0.18974-0.

Chang, H.H., T. Cohen, Y.H. Grad, W.P. Hanage, T.F. O'Brien, and M. Lipsitch. 2015. Origin and proliferation of multiple-drug resistance in bacterial pathogens. *Microbiology and Molecular Biology Reviews* 79: 101–16. https://doi.org/10.1128/mmbr.00039-14.

Chen, S.L., H. Yu, H.M. Luo, Q. Wu, C.F. Li, and A. Steinmetz. 2016. Conservation and sustainable use of medicinal plants: Problems, progress, and prospects. *Chinese Medicine* 11 (July): 37, 1–10. https://doi.org/10.1186/s13020-016-0108-7.

Chin, W., G. Zhong, Q. Pu, et al. 2018. A macromolecular approach to eradicate multidrug resistant bacterial infections while mitigating drug resistance onset. *Nature Communications* 9 (March): 917, 1–14. https://doi.org/10.1038/s41467-018-03325-6.

Ciofu, O., L.F. Mandsberg, T. Bjarnsholt, T. Wassermann, and N. Høiby. 2010. Genetic adaptation of *Pseudomonas aeruginosa* during chronic lung infection of patients with cystic fibrosis: Strong and weak mutators with heterogeneous genetic backgrounds emerge in *mucA* and/or *lasR* mutants. *Microbiology* 156: 1108–19. https://doi.org/10.1099/mic.0.033993-0.

Cragg, G.M., and D.J. Newman. 2013. Natural products: A continuing source of novel drug leads. *Biochimica et Biophysica Acta (BBA)-General Subjects* 1830: 3670–95. https://doi.org/10.1016/j.bbagen.2013.02.008.

Cunha, M.A., J.A. Turmina, R.C. Ivanov, et al. 2012. Lasiodiplodan, an exocellular (1→6)-β-d-glucan from *Lasiodiplodia theobromae* MMPI: Production on glucose, fermentation kinetics, rheology and anti-proliferative activity. *Journal of Industrial Microbiology and Biotechnology* 39: 1179–88. https://doi.org/10.1007/s10295-012-1112-2.

Curtis, R., and J. Jones. 2007. Robert Robinson and penicillin: An unnoticed document in the saga of its structure. *Journal of Peptide Science* 13: 769–75. https://doi.org/10.1002/psc.888.

Czekalski, N., E.G. Díez, and H. Bürgmann. 2014. Wastewater as a point source of antibiotic-resistance genes in the sediment of a freshwater lake. *The ISME Journal* 8: 1381–90. https://doi.org/10.1038/ismej.2014.8.

Daniels, R., J. Vanderleyden, and J. Michiels. 2004. Quorum sensing and swarming migration in bacteria. *FEMS Microbiology Reviews* 28: 261–89. https://doi.org/10.1016/j.femsre.2003.09.004.

de la Fuente-Nunez, C. 2019. Toward autonomous antibiotic discovery. *Msystems* 4 (June): e00151–19, 1–5. https://doi.org/10.1128/msystems.00151-19.

Devi, K.P., N. Suganthy, P. Kesika, and S.K. Pandian. 2008. Bioprotective properties of seaweeds: *In vitro* evaluation of antioxidant activity and antimicrobial activity against food borne bacteria in relation to polyphenolic content. *BMC Complementary and Alternative Medicine* 8 (July): 38, 1–11. https://doi.org/10.1186/1472-6882-8-38.

Dias, D.A., S. Urban, and U. Roessner. 2012. A historical overview of natural products in drug discovery. *Metabolites* 2: 303–36. https://doi.org/10.3390/metabo2020303.

Diggle, S.P., S. Matthijs, V.J. Wright, et al. 2007. The *Pseudomonas aeruginosa* 4-quinolone signal molecules HHQ and PQS play multifunctional roles in quorum sensing and iron entrapment. *Chemistry & Biology* 14: 87–96. https://doi.org/10.1016/j.chembiol.2006.11.014.

Etebu, E., and I. Arikekpar. 2016. Antibiotics: Classification and mechanisms of action with emphasis on molecular perspectives. *International Journal of Applied Microbiology and Biotechnology Research* 4: 90–101.

Feng, P., K.A. Lampel, H. Karch, and T.S. Whittam. 1998. Genotypic and phenotypic changes in the emergence of *Escherichia coli* O157:H7. *The Journal of Infectious Diseases* 177: 1750–53. https://doi.org/10.1086/517438.

Fischbach, M.A., and C.T. Walsh. 2009. Antibiotics for emerging pathogens. *Science* 325: 1089–93. https://www.science.org/doi/10.1126/science.1176667.

Fleming, A. 1929. On the antibacterial action of cultures of a penicillium, with special reference to their use in the isolation of *B. influenzae*. *British Journal of Experimental Pathology* 10: 226–36. https://www.ncbi.nlm.nih.gov/pmc/articles/PMC2048009/pdf/brjexppathol00255–0037.pdf.

Fox, E.M., and B.J. Howlett. 2008. Secondary metabolism: Regulation and role in fungal biology. *Current Opinion in Microbiology* 11: 481–87. https://doi.org/10.1016/j.mib.2008.10.007.

França, A., V. Carvalhais, T. Maira-Litrán, M. Vilanova, N. Cerca, and G. Pier. 2014. Alterations in the *Staphylococcus epidermidis* biofilm transcriptome following interaction with whole human blood. *Pathogens and Disease* 70: 444–48. https://doi.org/10.1111/2049-632X.12130.

Fraser, M.E., M.M. Chernaia, Y.V. Kozlov, and M.N.G. James. 1994. Crystal structure of the holotoxino from *Shigella dysenteriae* at 2.5 Å resolution. *Nature Structural Biology* 1: 59–64. https://doi.org/10.1038/nsb0194-59.

Friedman, D.B., D.L. Stauff, G. Pishchany, C.W. Whitwell, V.J. Torres, and E.P. Skaar. 2006. *Staphylococcus aureus* redirects central metabolism to increase iron availability. *PLoS Pathogens* 2: 777–89. https://doi.org/10.1371/journal.ppat.0020087.

Godstime, C., O. Enwa Felix, O. Jewo Augustina, and O. Eze Christopher. 2014. Mechanisms of antimicrobial actions of phytochemicals against enteric pathogens–A review. *Research Journal of Pharmaceutical, Biological and Chemical Sciences* 2: 77–85.

Golan, D.E., A.H. Tashjian, and E.J. Armstrong. 2011. *Principles of pharmacology: The pathophysiologic basis of drug therapy*. Philadelphia: Lippincott Williams & Wilkins.

Gong, Y., A. Hounsa, S. Egal, et al. 2004. Postweaning exposure to aflatoxin results in impaired child growth: A longitudinal study in Benin, West Africa. *Environmental Health Perspectives* 112: 1334–38. https://doi.org/10.1289/ehp.6954.

Gupta, A., S. Jain, and N. Verma. 2022. *Industrial applications of soil microbes*, vol. 1. Bentham Science Publishers Pte Ltd. https://doi.org/10.2174/97898150399551220101.

Gupta, A., S. Jain, and N. Verma. 2024. *Industrial applications of soil microbes*, vol. 3. Bentham Science Publishers Pte Ltd. https://doi.org/10.2174/97898150797531240301.

Harborne, J.B. 1999. Classes and functions of secondary products from plants. In *Chemicals from plants, perspectives on plant secondary products*, eds. N.J. Walton and D.E. Brown, 1–25. Singapore: World Scientific Publishing Co Pte Ltd. https://doi.org/10.1142/9789812817273_0001.

Hein, K.Z., H. Takahashi, T. Tsumori, et al. 2015. Disulphide-reduced psoriasin is a human apoptosis-inducing broad-spectrum fungicide. *Proceedings of the National Academy of Sciences* 112: 13039–44. https://doi.org/10.1073/pnas.1511197112.

Hentzer, M., H. Wu, J.B. Andersen, et al. 2003. Attenuation of *Pseudomonas aeruginosa* virulence by quorum sensing inhibitors. *The EMBO Journal* 22: 3803–15. https://doi.org/10.1093/emboj/cdg366.

Hesterkamp, T. 2015. Antibiotics clinical development and pipeline. In *How to overcome the antibiotic crisis: Facts, challenges, technologies and future perspectives*, eds. M. Stadler, and P. Dersch, 447–74. Cham: Springer. https://doi.org/10.1007/82_2015_451.

Higginbotham, S.J., A.E. Arnold, A. Ibañez, C. Spadafora, P.D. Coley, and T.A. Kursar. 2013. Bioactivity of fungal endophytes as a function of endophyte taxonomy and the taxonomy and distribution of their host plants. *PLoS One* 8 (September): e73192, 1–11. https://doi.org/10.1371/journal.pone.0073192.

Hood, M.I., and E.P. Skaar. 2012. Nutritional immunity: Transition metals at the pathogen–host interface. *Nature Reviews Microbiology* 10: 525–37. https://doi.org/10.1038/nrmicro2836.

Houbraken, J., J.C. Frisvad, and R.A. Samson. 2011. Fleming's penicillin producing strain is not *Penicillium chrysogenum* but *P. rubens*. *IMA Fungus* 2: 87–95. https://doi.org/10.5598/imafungus.2011.02.01.12.

Huang, L., T. Adachi, Y. Shimizu, et al. 2008. Characterization of lectin isolated from *Momordica charantia* seed as a B cell activator. *Immunology Letters* 121: 148–56. https://doi.org/10.1016/j.imlet.2008.10.004.

Islas-Muñoz, B., P. Volkow-Fernández, C. Ibanes-Gutiérrez, A. Villamar-Ramírez, D. Vilar-Compte, and P. Cornejo-Juárez. 2018. Bloodstream infections in cancer patients. Risk factors associated with mortality. *International Journal of Infectious Diseases* 71: 59–64. https://doi.org/10.1016/j.ijid.2018.03.022.

Jain, S., A. Gupta, and N. Verma. 2023. *Industrial applications of soil microbes*, vol. 2. Bentham Science Publishers Pte Ltd. https://doi.org/10.2174/97898150502641230201.

Jain, S., A. Gupta, and N. Verma. 2024. *Industrial applications of soil microbes*, vol. 4. Bentham Science Publishers Pte Ltd. https://doi.org/10.2174/97898151249961240401.

Jalgaonwala, R.E., B.V. Mohite, and R.T. Mahajan. 2011. A review: Natural products from plant associated endophytic fungi. *Journal of Microbiology and Biotechnology Research* 1: 21–32.

Kala, C.P., P.P. Dhyani, and B.S. Sajwan. 2006. Developing the medicinal plants sector in northern India: Challenges and opportunities. *Journal of Ethnobiology and Ethnomedicine* 2 (August): 32, 1–15. https://doi.org/10.1186/1746-4269-2-32.

Kaper, J.B., J.P. Nataro, and H.L.T. Mobley. 2004. Pathogenic *Escherichia coli*. *Nature Reviews Microbiology* 2: 123–40. https://doi.org/10.1038/nrmicro818.

Katz, L., and R.H. Baltz. 2016. Natural product discovery: Past, present, and future. *Journal of Industrial Microbiology and Biotechnology* 43: 155–76. https://doi.org/10.1007/s10295-015-1723-5.

Keller, N.P., G. Turner, and J.W. Bennett. 2005. Fungal secondary metabolism – from biochemistry to genomics. *Nature Reviews Microbiology* 3: 937–47. https://doi.org/10.1038/nrmicro1286.

Khan, A.A., Z. Khan, A. Malik, et al. 2017. Colorectal cancer-inflammatory bowel disease nexus and felony of *Escherichia coli*. *Life Sciences* 180: 60–67. https://doi.org/10.1016/j.lfs.2017.05.016.

Kingston, W. 2008. Irish contributions to the origins of antibiotics. *Irish Journal of Medical Science* 177: 87–92. https://doi.org/10.1007/s11845-008-0139-x.

Kiser, J.J., and C. Flexner. 2013. Direct-acting antiviral agents for Hepatitis C virus infection. *Annual Review of Pharmacology and Toxicology* 53: 427–49. https://doi.org/10.1146/annurev-pharmtox-011112-140254.

Kleinschmidt, S., F. Huygens, J. Faoagali, I.U. Rathnayake, and L.M. Hafner. 2015. *Staphylococcus epidermidis* as a cause of bacteremia. *Future Microbiology* 10: 1859–79. https://doi.org/10.2217/fmb.15.98.

Konuklugil, B. 1995. The importance of aryltetralin (*Podophyllum*) lignans and their distribution in the plant kingdom. *Journal of Faculty of Pharmacy of Ankara University* 24: 109–25. https://dergipark.org.tr/tr/download/article-file/425010.

Kott, V., L. Barbini, M. Cruanes, et al. 1998. Antiviral activity in Argentine medicinal plants. *Journal of Ethnopharmacology* 64: 79–84. https://doi.org/10.1016/S0378-8741(98)00098–1.

Kumar, A., S.K. Parida, I.C. Giri, and A. Kumar. 2021. Development and characterization of novel analytical approach for antiviral drugs. *World Journal of Pharmaceutical Research* 10: 490–500. https://doi.org/10.20959/wjpr202110–21238.

Kumara, P.M., S. Zuehlke, V. Priti, et al. 2012. *Fusarium proliferatum*, an endophytic fungus from *Dysoxylum binectariferum* Hook.f, produces rohitukine, a chromane alkaloid possessing anti-cancer activity. *Antonie Van Leeuwenhoek* 101: 323–29. https://doi.org/10.1007/s10482-011-9638-2.

Lam, K.S. 2007. New aspects of natural products in drug discovery. *Trends in Microbiology* 15: 279–89. https://doi.org/10.1016/j.tim.2007.04.001.

Lauritano, C., M.I. Ferrante, and A. Rogato. 2019. Marine natural products from microalgae: An-omics overview. *Marine Drugs* 17 (May): 269, 1–18. https://doi.org/10.3390/md17050269.

Lauritano, C., and A. Ianora. 2018. Grand challenges in marine biotechnology: Overview of recent EU-funded projects. In *Grand challenges in marine biotechnology*, eds. P.H. Rampelotto and A. Trincone, 425–49. Cham: Springer. https://doi.org/10.1007/978-3-319-69075-9_11.

Lee, T., T. Clavel, K. Smirnov, et al. 2017. Oral versus intravenous iron replacement therapy distinctly alters the gut microbiota and metabolome in patients with IBD. *Gut Microbiota* 66: 863–71. https://doi.org/10.1136/gutjnl-2015-309940.

Leitão, J.H. 2020. Microbial virulence factors. *International Journal of Molecular Sciences* 21 (July): 5320, 1–6. https://doi.org/10.3390/ijms21155320.

Liu, C., Q. Zhou, Y. Li, et al. 2020. Research and development on therapeutic agents and vaccines for COVID-19 and related human corona virus diseases. *ACS Central Science* 6: 315–31. https://doi.org/10.1021/acscentsci.0c00272.

Lou, Z., Y. Sun, and Z. Rao. 2014. Current progress in antiviral strategies. *Trends in Pharmacological Sciences* 35: 86–102. https://doi.org/10.1016/j.tips.2013.11.006.

Madigan, C.A., A.J. Martinot, J.R. Wei, et al. 2015. Lipidomic analysis links mycobactin synthase K to iron uptake and virulence in *M. tuberculosis*. *PLoS Pathogens* 11 (March): e1004792, 1–21. https://doi.org/10.1371/journal.ppat.1004792.

Mak, S., Y. Xu, and J.R. Nodwell. 2014. The expression of antibiotic resistance genes in antibiotic-producing bacteria. *Molecular Microbiology* 93: 391–02. https://doi.org/10.1111/mmi.12689.

Mandal, P., C.A. Pujol, M.J. Carlucci, K. Chattopadhyay, E.B. Damonte, and B. Ray. 2008. Anti-herpetic activity of a sulfated xylomannan from *Scinaia hatei*. *Phytochemistry* 69: 2193–99. https://doi.org/10.1016/j.phytochem.2008.05.004.

Martins, A., H. Vieira, H. Gaspar, and S. Santos. 2014. Marketed marine natural products in the pharmaceutical and cosmeceutical industries: Tips for success. *Marine Drugs* 12: 1066–01. https://doi.org/10.3390/md12021066.

Medema, M.H., P. Cimermancic, A. Sali, E. Takano, and M.A. Fischbach. 2014. A systematic computational analysis of biosynthetic gene cluster evolution: Lessons for engineering biosynthesis. *PLoS Computational Biology* 10 (December): e1004016, 1–12. https://doi.org/10.1371/journal.pcbi.1004016.

Miller, M.B., and B.L. Bassler. 2001. Quorum sensing in bacteria. *Annual Reviews in Microbiology* 55: 165–99. https://doi.org/10.1146/annurev.micro.55.1.165.

Morens, D.M., and A.S. Fauci. 2012. Emerging infectious diseases in 2012: 20 years after the institute of medicine report. *MBio* 3 (December): e00494–12, 1–4. https://doi.org/10.1128/mbio.00494-12.

Mosunova, O., J.C. Navarro-Muñoz, and J. Collemare. 2021. The biosynthesis of fungal secondary metabolites: From fundamentals to biotechnological applications. *Encyclopedia of Mycology* 2: 458–76. https://doi.org/10.1016/b978-0-12-809633-8.21072-8.

Mousa, W.K., and M.N. Raizada. 2013. The diversity of antimicrobial secondary metabolites produced by fungal endophytes: An interdisciplinary perspective. *Frontiers in Microbiology* 4 (March): 65, 1–18. https://doi.org/10.3389/fmicb.2013.00065.

Nair, D.N., and S. Padmavathy. 2014. Impact of endophytic microorganisms on plants, environment and humans. *The Scientific World Journal* 1 (January): 250693, 1–11. https://doi.org/10.1155/2014/250693.

Nealson, K.H., T. Platt, and J.W. Hastings. 1970. Cellular control of the synthesis and activity of the bacterial luminescent system. *Journal of Bacteriology* 104: 313–22. https://doi.org/10.1128/jb.104.1.313-322.1970.

Newman, D.J., and G.M. Cragg. 2012. Natural products as sources of new drugs over the 30 years from 1981 to 2010. *Journal of Natural Products* 75: 311–35. https://doi.org/10.1021/np200906s.

O'Connor, S.E. 2015. Engineering of secondary metabolism. *Annual Review of Genetics* 49: 71–94. https://doi.org/10.1146/annurev-genet-120213-092053.

Oliveira, F., H. Rohde, M. Vilanova, and N. Cerca. 2021. The emerging role of iron acquisition in biofilm-associated infections. *Trends in Microbiology* 29: 772–75. https://doi.org/10.1016/j.tim.2021.02.009.

Overy, D.P., T. Rämä, R. Oosterhuis, A.K. Walker, and K.L. Pang. 2019. The neglected marine fungi, *Sensu stricto*, and their isolation for natural products' discovery. *Marine Drugs* 17 (January): 42, 1–20. https://doi.org/10.3390/md17010042.

Pegadraju, H., J.A. Thomas, J.A. Mutha, and R. Reshmi. 2021. Recapitulation of antibiotic resistance and MRSA. *International Research Journal of Modernization in Engineering Technology and Science* 3: 857–68. https://www.irjmets.com/uploadedfiles/paper/volume3/issue_4_april_2021/8284/1628083340.pdf.

Pentikäinen, S., L. Karhunen, L. Flander, et al. 2014. Enrichment of biscuits and juice with oat β-glucan enhances postprandial satiety. *Appetite* 75: 150–56. https://doi.org/10.1016/j.appet.2014.01.002.

Perez, F., J. Adachi, and R.A. Bonomo. 2014. Antibiotic-resistant Gram-negative bacterial infections in patients with cancer. *Clinical Infectious Diseases* 59: S335–39. https://doi.org/10.1093/cid/ciu612.

Peterson, E., and P. Kaur. 2018. Antibiotic resistance mechanisms in bacteria: Relationships between resistance determinants of antibiotic producers, environmental bacteria, and clinical pathogens. *Frontiers in Microbiology* 9 (November): 2928, 1–21. https://doi.org/10.3389/fmicb.2018.02928.

Pianalto, K.M., and J.A. Alspaugh. 2016. New horizons in antifungal therapy. *Journal of Fungi* 2 (October): 26, 1–24. https://doi.org/10.3390/jof2040026.

Ping, B., Y. Zhu, Y. Gao, C. Yue, and B. Wu. 2013. Second-versus first-generation azoles for antifungal prophylaxis in hematology patients: A systematic review and meta-analysis. *Annals of Hematology* 92: 831–39. https://doi.org/10.1007/s00277-013-1693-5.

Poole, T.L., and C.L. Sheffield. 2013. Use and misuse of antimicrobial drugs in poultry and livestock: Mechanisms of antimicrobial resistance. *Pakistan Veterinary Journal* 33: 266–71. https://api.semanticscholar.org/CorpusID:85770498.

Rajbhandari, M., U. Wegner, M. Jülich, T. Schoepke, and R. Mentel. 2001. Screening of Nepalese medicinal plants for antiviral activity. *Journal of Ethnopharmacology* 74: 251–55. https://doi.org/10.1016/S0378-8741(00)00374–3.

Raziky, M.E., W.F. Fathalah, W.A. El-Akel, et al. 2013. The effect of peginterferon alpha-2a vs. peginterferon alpha-2b in treatment of naive chronic HCV genotype-4 patients: A single centre Egyptian study. *Hepatitis Monthly* 13 (May): e10069, 1–8. https://doi.org/10.5812/hepatmon.10069.

Riccio, G., N. Ruocco, M. Mutalipassi, et al. 2020. Ten-year research update review: Antiviral activities from marine organisms. *Biomolecules* 10 (July): 1007, 1–36. https://doi.org/10.3390/biom10071007.

Robinson, T., D. Singh, and P. Nigam. 2001. Solid-state fermentation: A promising microbial technology for secondary metabolite production. *Applied Microbiology and Biotechnology* 55: 284–89. https://doi.org/10.1007/s002530000565.

Rutherford, S.T., and B.L. Bassler. 2012. Bacterial quorum sensing: Its role in virulence and possibilities for its control. *Cold Spring Harbor Perspectives in Medicine* 2 (September): a012427, 1–26. https://perspectivesinmedicine.cshlp.org/content/2/11/a012427.

Saxena, S.K., S. Saxena, R. Saxena, M.L.A. Swamy, A. Gupta, and M.P.N. Nair. 2010. Emerging trends, challenges and prospects in antiviral therapeutics and drug development for infectious diseases. *Electronic Journal of Biology* 6: 26–31. https://ejbio.imedpub.com/articles/emerging-trends-challenges-and-prospects-in-antiviraltherapeutics-and-drug-development-for-infectious-diseases.php?aid=5859.

Serkedjieva, J. 2004. Antiviral activity of the red marine alga *Ceramium rubrum*. *Phytotherapy Research* 18: 480–83. https://doi.org/10.1002/ptr.1458.

Serrano, L.O.D. 2017. Biotechnology of siderophores in high-impact scientific fields. *Biomolecular Concepts* 8: 169–78. https://doi.org/10.1515/bmc-2017-0016.

Shafi, J., Z. Sun, M. Ji, Z. Gu, and W. Ahmad. 2018. ANN and RSM based modelling for optimization of cell dry mass of *Bacillus* sp. strain B67 and its antifungal activity against *Botrytis cinerea*. *Biotechnology & Biotechnological Equipment* 32: 58–68. https://doi.org/10.1080/13102818.2017.1379359.

Sharanabasappa, G.K., M.K. Santosh, D. Shaila, Y.N. Seetharam, and I. Sanjeevarao. 2007. Phytochemical studies on *Bauhinia racemosa* lam. *Bauhinia purpurea* Linn. and *Hardwickia binata* roxb. *E-Journal of Chemistry* 4: 21–31. https://doi.org/10.1155/2007/874721.

Sheldon, J.R., and D.E. Heinrichs. 2015. Recent developments in understanding the iron acquisition strategies of gram positive pathogens. *FEMS Microbiology Reviews* 39: 592–30. https://doi.org/10.1093/femsre/fuv009.

Shukla, S.T., P.V. Habbu, V.H. Kulkarni, K.S. Jagadish, A.R. Pandey, and V.N. Sutariya. 2014. Endophytic microbes: A novel source for biologically/pharmacologically active secondary metabolites. *Asian Journal of Pharmacology and Toxicology* 2: 1–16.

Simeis, D.D., and S. Serra. 2021. Actinomycetes: A never-ending source of bioactive compounds – An overview on antibiotics production. *Antibiotics* 10 (April): 483, 1–32. https://doi.org/10.3390/antibiotics10050483.

Strobel, G., and B. Daisy. 2003. Bioprospecting for microbial endophytes and their natural products. *Microbiology and Molecular Biology Reviews* 67: 491–02. https://doi.org/10.1128/mmbr.67.4.491-502.2003.

Synytsya, A., and M. Novák. 2013. Structural diversity of fungal glucans. *Carbohydrate Polymers* 92: 792–09. https://doi.org/10.1016/j.carbpol.2012.09.077.

Tarr, P.I., C.A. Gordon, and W.L. Chandler. 2005. Shiga-toxin-producing *Escherichia coli* and haemolytic uraemic syndrome. *The Lancet* 365: 1073–86. https://doi.org/10.1016/S0140-6736(05)71144–2.

Velez, R., and E. Sloand. 2016. Combating antibiotic resistance, mitigating future threats and ongoing initiatives. *Journal of Clinical Nursing* 25: 1886–89. https://doi.org/10.1111/jocn.13246.

Vermes, A., H.J. Guchelaar, and J. Dankert. 2000. Flucytosine: A review of its pharmacology, clinical indications, pharmacokinetics, toxicity and drug interactions. *Journal of Antimicrobial Chemotherapy* 46: 171–79. https://doi.org/10.1093/jac/46.2.171.

Vuillemin, J.P. 1889. *Antibiose et symbiose*. Association française pour l'avancement des sciences Paris, Secretariat de l'Association. https://books.google.co.in/books?id=XB1ktAEACAAJ.

Walsh, C. 2003. *Antibiotics: Actions, origins, resistance*. Washington: American Society for Microbiology (ASM). https://www.cabidigitallibrary.org/doi/full/10.5555/20043133125.

Tang, J.C. 2010. Research of antibiotics pollution in soil environments and its ecological toxicity. *Journal of Agro-Environment Science* 29: 261–66. https://api.semanticscholar.org/CorpusID:130520430.

Weber, T., P. Charusanti, E.M. Musiol-Kroll, et al. 2015. Metabolic engineering of antibiotic factories: New tools for antibiotic production in actinomycetes. *Trends in Biotechnology* 33: 15–26. https://doi.org/10.1016/j.tibtech.2014.10.009.

Welsh, M.A., and H.E. Blackwell. 2016. Chemical genetics reveals environment-specific roles for quorum sensing circuits in *Pseudomonas aeruginosa*. *Cell Chemical Biology* 23: 361–69. https://doi.org/10.1016/j.chembiol.2016.01.006.

Wesgate, R., P. Grasha, and J.Y. Maillard. 2016. Use of a predictive protocol to measure the antimicrobial resistance risks associated with biocidal product usage. *American Journal of Infection Control* 44: 458–64. https://doi.org/10.1016/j.ajic.2015.11.009.

Whiteley, M., M.G. Bangera, R.E. Bumgarner, et al. 2001. Gene expression in *Pseudomonas aeruginosa* biofilms. *Nature* 413: 860–64. https://doi.org/10.1038/35101627.

Williams, H.E., J.C.P. Steele, M.O. Clements, and T. Keshavarz. 2012. γ-heptalactone is an endogenously produced quorum-sensing molecule regulating growth and secondary metabolite production by *Aspergillus nidulans*. *Applied Microbiology and Biotechnology* 96: 773–81. https://doi.org/10.1007/s00253-012-4065-5.

Williams, P., K. Winzer, W.C. Chan, and M. Camara. 2007. Look who's talking: Communication and quorum sensing in the bacterial world. *Philosophical Transactions of the Royal Society B: Biological Sciences* 362: 1119–34. https://doi.org/10.1098/rstb.2007.2039.

Witvrouw, M., J.A. Este, M.Q. Mateu, et al. 1994. Activity of a sulfated polysaccharide extracted from the red seaweed *Aghardhiella tenera* against human immunodeficiency virus and other enveloped viruses. *Antiviral Chemistry & Chemotherapy* 5: 297–03. https://doi.org/10.1177/095632029400500503.

Wurzbacher, C.M., F. Bärlocher, and H.P. Grossart. 2010. Fungi in lake ecosystems. *Aquatic Microbial Ecology* 59: 125–49. https://www.int-res.com/articles/ame_oa/a059p125.pdf.

Xu, J., G.E. Inglett, D. Chen, and S.X. Liu. 2013. Viscoelastic properties of oat β-glucan-rich aqueous dispersions. *Food Chemistry* 138: 186–91. https://doi.org/10.1016/j.foodchem.2012.10.049.

Yu, F., L. Lu, L. Du, X. Zhu, A.K. Debnath, and S. Jiang. 2013. Approaches for identification of HIV-1 entry inhibitors targeting gp41 pocket. *Viruses* 5: 127–49. https://doi.org/10.3390/v5010127.

Zhang, P., P.P. Zhou, and L.J. Yu. 2009. An endophytic taxol-producing fungus from *Taxus media*, *Cladosporium cladosporioides* MD2. *Current Microbiology* 59: 227–32. https://doi.org/10.1007/s00284-008-9270-1.

Zhao, J., L.C. Davis, and R. Verpoorte. 2005. Elicitor signal transduction leading to production of plant secondary metabolites. *Biotechnology Advances* 23: 283–33. https://doi.org/10.1016/j.biotechadv.2005.01.003.

Zotchev, S.B. 2003. Polyene macrolide antibiotics and their applications in human therapy. *Current Medicinal Chemistry* 10: 211–23. http://dx.doi.org/10.2174/0929867033368448.

11 Potential Strategies for Fungal Secondary Metabolites-Based Mycoherbicide Development to Control Invasive Weeds

*A. K. Singh
(Mycology Research Laboratory, Department of BioScience, R. D. University, Jabalpur, Madhya Pradesh; email: drajay2009@gmail.com)

A. K. Pandey
(Mycology Research Laboratory, Department of BioScience, R. D. University, Jabalpur, Madhya Pradesh; akpmycol@yahoo.co.in)

11.1 INTRODUCTION

Weed control is one of the most important steps in agricultural production. The sales of herbicides rank first among the global market share of pesticides. However, the use of a large number of chemical herbicides has caused environmental problems such as water quality degradation and soil pollution, especially the constant occurrence of resistant weeds, which has brought great challenges to agricultural production. According to statistics, as of 2018, a total of 495 cases (biotypes) of resistant weeds have been reported worldwide, involving 255 types of weeds. In China, from the first report of weed resistance in 1992 to 2018, a total of 44 resistance cases have been reported, and the trend of resistant weeds is not optimistic. This technology is showing its age with the ever-growing evolution of resistance to available commercial herbicides and even resistance by non-target site mechanisms of resistance to herbicides yet to be introduced (Gaines et al., 2020).

The number of herbicides being introduced with new molecular targets and/or new chemical classes for which no resistance exists is few (Qu et al., 2021). Natural products have been the inspiration for insecticides and fungicides with new molecular targets, but to a much lesser extent with herbicides (Lorsbach et al., 2019; Sparks and Duke, 2021). This may be partially due to the relatively reduced effort of herbicide discovery after glyphosate dominated the herbicide market, reducing the overall value of that market significantly (Duke, 2012). The fungicides and insecticides markets have not experienced a similar phenomenon. Nevertheless, information on the molecular targets of many natural phytotoxins indicates that natural products could be the basis for new herbicides to help cope with the growing herbicide resistance problem (Dayan and Duke, 2014).

With the increase of global environmental awareness and the need for sustainable agricultural development, the development of bio-herbicides with good environmental compatibility, high selectivity, and strong biological activity is urgently needed. Microbial bioherbicides, as defined by the United States Environmental Protection Agency, are preparations from killed or live microbes to kill weeds. Their utilization carries different problems and risks than those of chemical herbicides. Although there have been significant improvements in the adoption of microbial bio-fungicides and bioinsecticides in the past 50 years (Marrone, 2019; Jain et al., 2023, 2024), there have been relatively few advancements in this area with microbial bioherbicides. We briefly discuss these problems and how they might be overcome.

The growing popularity of and political pressure for organic agriculture and food grown without synthetic pesticides is also providing an impetus to discover and develop weed management technologies that can meet these criteria. The demands for such products are not necessarily based on unequivocal information that such organic or more natural farming is more environmentally or toxicologically safe than conventional farming (Gupta et al., 2022, 2024). For example, some studies have shown organic grains to have more mycotoxins than those from conventional farms (Kuzdraliński et al., 2013). Some mycotoxins (*e.g.*, fumonisins) produced by microbial pathogens that infect crops are orders of magnitude more toxic to humans than most pesticide residues. Nevertheless, there is good reason for intensifying research to provide safe weed management technologies to meet the demands of farmers, consumers, and regulatory bodies through the use of natural phytotoxins and microbial bioherbicides.

The early stages link discovery to proof-of-concept and platform technology development. Basic assessments of

DOI: 10.1201/9781003408833-11

biology, environment, biochemistry, and small-scale fermentation and formulation are conducted under laboratory, greenhouse, and field conditions, with emphasis placed on characterization, safety, and practicality of the bioherbicide organism (Boyetchko and Svircev, 2009; Bailey et al., 2010). The importance of regulatory and market considerations cannot be underestimated since they will dictate the success or failure of commercialization (Bailey et al., 2010). Although these two areas are not necessarily a research consideration to scientists, they are equally important because defining the field of use and the market will pre-determine the experimentation and scientific data required to register the product. Another key feature of the innovation chain is that it encourages development of novel technology platforms that can be expanded to other potential applications and crop pests, whether this is for multiple use patterns, application methods, broadening host targets, new production systems, or be integrated into other crop production systems (Boyetchko and Rosskopf, 2006; Hynes and Boyetchko, 2006). Although a great deal of science and understanding of microbial physiology and biochemistry are involved, there is as much art required in this aspect of product development.

The later stages in the innovation chain (*i.e.*, application development and technology transfer) test the robustness of the earlier decisions by actively working with various stakeholders involved in biopesticide development. These include the industry partner, other collaborators, and regulators to develop the data required for the registration package. Moving into commercial scale-up can still affect the final stages of product development, often requiring significant investment back into technology development to move from the bench to pilot scale to commercial manufacture. Commercial scale-up, registration, and technology adoption is usually directed by an industry partner, but the champion of the project is usually the lead scientist or inventor who should work collaboratively on the project with industry until technology adoption has occurred. The corporate memory or "know-how" and intuition of the scientist, along with the scientific expertise, should not be overlooked by the industry partner. Experience shows that it can ease the transition phases and add potential new product value.

The successful development of a bioherbicide is a combination of science, art, entrepreneurship, and serendipity, which can take over10–15 years to complete. This review is meant to discuss the development of mycoherbicides for the control of some noxious weeds of Madhya Pradesh.

11.2 STRATEGIC FRAMEWORK FOR BIOHERBICIDE DEVELOPMENT

The cost and consistency of the microbial organisms with varying environmental conditions, efficient mass production (*i.e.*, fermentation), formulation and delivery systems, and prolonged shelf-life have always been a challenge to the development and commercialization of bioherbicide products (Boyetchko and Peng, 2003; Boyetchko, 2005; Hynes and Boyetchko, 2006). Past experiences have shown extensive complexities within pest/pathogen systems, which have affected product development. Research must tackle these complexities with more innovative approaches by addressing biological, environmental, and technological issues to achieve success (Boyetchko et al., 2002; Boyetchko and Rosskopf, 2006).

Contrary to popular belief, many government and university researchers believe that once a microbial organism has been discovered in the laboratory, the final product is imminent, and a bioherbicide product is one where all the platform technologies involved in product development are considered (Boyetchko and Peng, 2003). The taxonomy, biological characterization, mode of action, and efficacy are among the factors that are core to the selection of the bioherbicide organism, but the platform technologies, including fermentation, formulation, and application/delivery systems, are integral to the actual product itself. Discovery of a promising organism as the active ingredient is often the easiest part of the battle, but it is the selection and development of the appropriate fermentation system (*e.g.*, liquid/submerged vs. solid-state) for economic scale-up in combination with the most suitable formulation (*e.g.*, liquid/spray application vs. granule/seed treatment/soil application) and the type of application method (*e.g.*, foliar- vs. soil-applied) that determines the bioherbicide product and its performance (Boyetchko and Rosskopf, 2006; Hynes and Boyetchko, 2006). Moreover, the fermentation process can affect shelf life and stability in the formulation phase, while formulation ingredients and downstream processing will influence delivery and application. Conversely, the type of application method required will often dictate the most appropriate formulation required to achieve effective weed control under field conditions. These mycoherbicide technologies are often interrelated, and any minor changes to each of these processes will have an impact on the other, thus potentially leading to significant improvements in bioherbicide performance in the field.

11.3 BIOLOGY OF MYCOHERBICIDES

For several years of bioprospecting for weed pathogens from their host plants was employed in various regions. Various microbial preparations were tested as bioherbicides, but they were either low phytotoxic or non-phytotoxic. Finally, several novel strains of weeds have shown phytotoxicity toward their host plants, and it was found that the cell-free liquid, in which the mold had grown, could inhibit weed growth.

11.3.1 Fungal Diversity

The industrial microbiology field has utilized only a very minor portion of nature's fungal arsenal for the discovery of useful molecules. The reason is the inability of microbiologists to culture the vast majority of microbes in nature. It is estimated that only 5% of fungi have been cultivated.

11.3.2 Bioherbicidal Agents

Although the early emphasis on fungal metabolite discovery was mainly devoted to herbicides that some of these had been quietly exploited in the past, and that such broadening of scope should be expanded. Thus, a broad screening of herbicidal active molecules for herbicidal activity against weeds was proposed to yield new and useful products for weed control and to isolate new bioactive compounds. A large number of *in vitro* laboratory tests were developed to help detect, isolate, and purify useful compounds.

11.3.3 Laboratory Fermentation Improvement

Fungi isolated from nature usually produce extremely low levels of metabolites. Overproduction must be achieved in the laboratory before pilot plant scale-up is attempted. Knowledge of fungal physiology is crucial for success since the environment presented to the fungi is of the utmost importance. Virtually every industrial fermentation process runs at its optimum rate and efficiency when a nutritional limitation is imposed. Metabolic processes are controlled by sources of carbon, nitrogen, phosphorus, metals, induction, feedback regulation, growth rate, and enzyme decay. The areas of physiology that must be considered to develop successful industrial processes are nutrition, growth and death, transport, energy, building blocks, polymer synthesis, regulation of enzyme synthesis, action and degradation, as well as cellular differentiation.

When fungi from nature are brought into the laboratory, they are isolated away from contaminating microorganisms by streaking on an agar surface in a Petri dish. When purified, the original colony is often found to be made up of a mixed population of cells, all of the members belonging to the same strain but possibly showing somewhat different colonial morphologies. The individual isolates are then tested for production ability, and the best strain becomes the subject of a multi-year fermentation development and mutagenesis program. Throughout the life of the commercial product, and fermentation parts of the improvement program go hand-in-hand,

Fermentation development in the laboratory occurs as follows: first, the carbon, nitrogen, inorganic, and, if necessary, complex nutrients supporting growth must be determined, and then modifications in the medium must be made to support product biosynthesis.

11.3.4 Process Development in Pilot Plant

One of the most important developments is large-scale fermentations. It was the finding that culture morphology has a major effect on fermenter performance by filamentous organisms, especially molds. Generally speaking, the long filamentous type of growth is undesirable because it leads to high culture viscosity, thereby reducing oxygen transfer into the bulk culture and leading to oxygen starvation.

11.3.4.1 Formulation

The word formulation is described as a mixture of an active ingredient (a.i.) and the appropriate or compatible inert ingredients. Adjuvants are the most popular inert ingredients commonly found in bioherbicide formulations. Adjuvants are well known to improve the efficacy of bioherbicides by altering their physicochemical properties. Surfactants, emulsifiers, and hydrophilic polymers are examples of adjuvants important for improving the effectiveness of bioherbicides. Bioherbicide compounds are generally applied to weeds in the form of an emulsion, which can increase weed control stability and effectiveness (Boyette et al., 2016). When surfactants were mixed properly with an aqueous suspension of a bacterial pathogen, the bacteria effectively invaded plant leaves while also expanding their host range (Deveau et al., 2018). Silwet L-77 is a non-ionic surfactant that enables bacterial cells and spores to quickly penetrate weed tissue. While hydrophilic polymers are readily miscible with water, emulsifiers are employed to blend water (hydrophilic) and oil (hydrophobic) components in a formulation into a stable emulsion. Unfortunately, the cost of bioherbicides sometimes increases because of the high cost of the adjuvant used in the formulation. Additionally, a careful selection of surfactants must be taken into consideration, since certain ingredients used in herbicide formulations can be toxic to humans (Castro et al., 2013).

For example, while pathogens can efficiently control a variety of weeds, they can also create some undesired toxins that could harm mammals and avian (Kremer, 2005). A *Cynara cardunculus* plant extract demonstrated a phytotoxic effect on *P. minor*, *T. incarnatum*, and *Silybum marianum* during both pre-emergence (Rial et al., 2016; Scavo et al., 2019) and post-emergence (Kaab et al., 2020).

In one study, an herbicide formulation was produced with a combination of *C. cardunculus* plant extract, non-ionic surfactants, and vegetable oil. The vegetable oil and surfactants increased the adsorption ability of polar molecules, dissolved cuticular fatty acids, and thereby improved the penetration of hydrophilic active substances (Hazrati et al., 2017).

11.3.4.2 Spray Preparation

Water is one of the main inputs of spray preparation. The amount of water applied per hectare is closely connected to spray coverage and bioherbicide performance. The water quality used for bioherbicide mixing can influence the effectiveness of bioherbicide usage (Bailey, 2014). Water of poor quality can reduce agricultural chemicals, increase chemical breakdown in the spray water (hydrolysis), block spray lines or nozzles, as well as reduce the uniformity of chemical application (Schilder, 2008).

11.3.5 Application Methods

Application methods have a significant impact on bioherbicide efficacy. Spray droplet size, distribution, retention, volume, and types of equipment are all factors involved in the application method of bioherbicides, which influence their efficacy (Peng and Wolf, 2008). Bioherbicides' effectiveness is also influenced by the delivery method (Harding and Raizada, 2015). Weeds' morphological characteristics, leaf surface characteristics, and the type of adjuvants are also important factors that can influence the efficacy of a bioherbicide. Byer et al. (2006) reported that a smaller droplet size of *Colletotrichum truncatum* was able to control *Matricaria perforata* Mérat with greater efficacy. The use of dual nozzle sprayers has a major effect on bioherbicide efficacy (Hintz, 2007). The application of bioherbicide with different nozzles was used in the biocontrol of diseases in *Amaranthus tuberculatus* (Moq.) Sauer (Doll et al., 2005).

Other factors that influence bioherbicide efficacy include bioherbicide host spectrum, which may be broad or targeted to specific species, and the nature of the formulation. Broad-spectrum bioherbicides tend to show different levels of efficacy in different locations/areas. It has been suggested that combining a bioherbicide with several pathogens would enhance its potential. When *Alternaria crassa* was combined with plant filtrates and fruit pectin, it showed broad-spectrum characteristics (Yandoc-Ables et al., 2007). Chandramohan and *Crotalaria spectabilis* Roth. and *Senna obtusifolia* (L.) H.S. Irwin and Barney were successfully controlled using a mixture of three pathogens, namely *Phomopsis amaranthicola*, *A. cassiae*, and *Colletotrichum dematium* (Chandramohan and Charudattan, 2003). In Florida, three pathogens, *viz*., *Exserohilum longirostratum*, *Drechslera gigantea*, and *Exserohilum rostratum*, were properly mixed and applied in citrus groves to control weeds (Chandramohan and Charudattan, 2003).

11.4 REGULATIONS

Regulatory frameworks and product registrations are used worldwide to guide the commercial development of microbial products. When developing new microbial products, the requisite regulatory framework varies by country, the product's characteristics, and its intended usage. These national and international regulations must be taken into account during every part of the product development cycle, including its earliest stages, as certain regulations also outline where natural microbes can and cannot be harvested. Interestingly, the regulations about cell-free broth, crude broth, and biocontrol strains, while similar, may differ in certain parts of the world. Nevertheless, regulatory cycles for the development of new biochemical and biocontrol products are generally streamlined and well-articulated. As a result, microbial products are an appealing and cost-effective choice when taking an integrated, systems-level approach toward crop productivity and agricultural weed management.

11.5 MODE OF ACTION OF BIOHERBICIDE ON WEED GROWTH

11.5.1 Action on Seed Germination

Seed germination is considered an important factor in plant development and productivity. During germination, the physiological, biochemical, and morphological changes are strongly linked with the survival rate of seedlings. Plant-based bioherbicides can inhibit seed germination by blocking the hydrolysis of nutrient reserves and cell division (Irshad and Cheema, 2004). The inhibition of the germination process by plant extracts involves osmotic effects on imbibition rates, which eventually inhibit germination and, in particular, cell elongation (El-Mergawi and Al-Humaid, 2019). The phytotoxicity of microbial extracts, residue, or mulch may affect weed germination and growth. Microbial extract releases biochemical substances into the soil that can hinder seed germination and the seedling growth of weeds. Aslani et al. (2014) reported that leaf extract of *Tinospora tuberculata* Baume inhibited the germination and growth of *E. crus-galli*, although it also showed phytotoxicity in the crops (*O. sativa*, *Daucus carota* L., *L. sativa*, *Cucumis sativus* L., and *S. lycopersicum*). Likewise, the extracts of *S. nigrum*, *C. album*, and *Matricaria chamomilla* L. exhibited suppressive action on the germination and seedling growth of *H. vulgare*, *Phaseolus vulgaris* L., *Cicer arietinum* L., *Zea mays* L., *Allium cepa* L., *Capsicum annuum* L., *S. lycopersicum*, and *Triticum aestivum* L. (Kadioglu et al., 2005). The seed germination of *C. arietinum*, *T. aestivum*, *B. nigra*, and *Lens culinaris* L. was significantly reduced by the application of *Butea monosperma* L. (Prasad et al., 2016).

11.5.2 Action on Shoot and Root

Shoot length is one of the significant growth parameters considered for the growth and development of plants. Generally, shoot growth is less sensitive to phytotoxic plant extracts compared to radical growth (Motmainna et al., 2021a). The greater sensitivity of radical growth to plant extracts is due to the radical being the first organ to be exposed to the phytotoxic substances and having a more permeable tissue than other organs and/or a low mitotic division in the root apical meristem (Motmainna et al., 2021a). Moreover, phytotoxic substances can affect genes responsible for the cellular characterization of radial tissues and the endoderm, inhibiting their development. The leaf extract of *D. stramonium* at 25–100% concentration significantly reduced the shoot and root growth of *Cenchrus ciliaris* L. and *Neonotonia wightii* Wight and Arn. (Elisante et al., 2013). Motmainna et al. (2021b) reported that the methanol extracts of *P. hysterophorus*, *Cleome rutidosperma* DC, and *Borreria alata* (Aubl.) DC showed a promising inhibitory effect in the shoot and root growth of *Oryza sativa* f. *spontanea* Roshev (weedy rice).

11.5.3 Action on Leaf Area

The leaf area of a plant is an important parameter for assessing growth. It is a variable that relates to the atmospheric condition of the plant through the process, such as transpiration, respiration, and photosynthesis. It is a fundamentally essential tool used in scientific disciplines such as agronomy, plant physiology, entomology, ecology, plant pathology, and many others. The cuticle is present on the upper and lower surfaces of the leaf that line the sub-stomatal cavities. The epidermal surfaces of the plant are covered by a cuticle to protect against water loss and desiccation. Herbicide movement or absorption into leaves depends on the spray retention of a herbicide on the leaf surface and the diffusion through the cuticle. In wheat, the leaf area index was significantly affected by sorghum extract at different application times (Rab et al., 2016). The leaf area index decreased with each increase in concentration, and a lower leaf area index (3.00) was recorded with 1:5 concentrations. Sorghum water extract applied at tillering gave a higher leaf area index, whereas sorghum water extract applied at 50% emergence + 50% tillering resulted in a reduced leaf area index (Rab et al., 2016).

11.5.4 Action on Stomatal Conductance

Stomata are small pores that absorb CO_2 and remove water vapor from the top and/or bottom of a leaf. When plant stomata open, water escapes through transpiration, and carbon dioxide is taken up via photosynthesis. The connection between the plant hydraulic system and photosynthesis is governed by a gaseous exchange and certain biochemical processes (Ewers, 2013). The measurement of the gaseous exchange is important, as it is linked to photosynthesis and transpiration, which may be manipulated by environmental elements such as carbon dioxide, light, humidity, temperature, and wind speed (Tardieu, 2013). The rhizome yield of *Zingiber officinale* Roscoe was significantly low when treated with *Tamarindus indica* L. (512 g/plant) leaves and mulched *Mangifera indica* L. (521 g/plant), attributed to drastic decreases in stomatal conductance, leaf production, rhizome thickness, rhizome spread, root length and spread, tiller production, and plant growth (Abraham et al., 2016)

11.5.5 Action on Chlorophyll Pigment

Chlorophyll molecules play an important role in photosynthesis and serve as a green pigment embedded in photosynthetic membranes, which is why a chlorophyll reduction normally results in a decrease in photosynthesis. Many plant extracts have been reported to affect the chlorophyll content directly or indirectly. Phenolic compounds such as *p*-coumarin, ferulic, and *o*-hydroxyphenyl acetic acids have been shown to inhibit chlorophyll biosynthesis and stimulate its degradation within 1h after treatment. A crude powder of *M. polymorpha* significantly decreased the chlorophyll content and total photosynthetic pigments of the recipient species (Najafpour, 2011). Plant extracts inhibit certain enzymes associated with chlorophyll and ultimately affect the integrity of chloroplasts and thylakoid membranes.

11.5.6 Action on Photosynthesis

Plant extracts affect protein metabolism in plants, with the result that protein binding in chlorophyll a/b is reduced by twofold. They also affect photosynthesis by suppressing chlorophyll synthesis. Oxygen-evolving enhancer protein 1 (OEE1) plays an important role in increasing the thioredoxin movement, defends the tetra-manganese cluster, and helps to release oxygen by splitting water (Algandaby and El-Darier, 2018). Plant-based bioherbicides can affect the gas and nutrient exchange of weeds by reducing OEE1 biosynthesis (Lee et al., 2015). Magnesium (Mg) is an essential part of chlorophyll, involved in carbon fixation and a variety of biochemical processes. Bioherbicides from plant extracts can reduce the Mg concentration in weed species, which has a major effect on chlorophyll synthesis (Shaul, 2002). Photosynthesis-inhibiting herbicides like bentazon can limit electron transfer and carbohydrate synthesis (de Sousa et al., 2014).

11.5.7 Action on Plant Hormones

Plant hormones are essential to plant growth through different metabolic activities. Gibberellins (GA) are a plant hormone that assists in the development of plant shoot growth (Radhakrishnan et al., 2013). The allelochemicals isolated from the seed of *Sicyos angulatus* L., a broadleaf weed, contain a high amount of 2-linoleoyl glycerol, which can inhibit the gibberellin pathway (Lee et al., 2015). On the other hand, 2-linoleoyl glycerol can stimulate the accumulation of abscisic acid (ABA), jasmonic acid (JA), and salicylic acid (SA). Increased ABA production causes stomatal closing, catalyzes plant senescence, and lowers the rate of photosynthesis, further inhibiting plant growth and development (Grossmann, 2003). Moreover, higher accumulations of JA could greatly reduce photosynthesis. Janda et al. (2012) also reported that higher SA accumulations inhibit photosynthesis

11.6 MARKET EXPECTATIONS AND PLACEMENT OF MYCOHERBICIDE PRODUCTS

Worldwide markets for mycoherbicides are increasing since the chemical herbicide market is based on a broader spectrum than the selective one. Broad-spectrum chemical herbicides are damaging grazing land in some cases crops due to drift, and herbicide resistance to weeds is opening a selective herbicide market worldwide. In the world pesticide market, herbicides constitute about 25–30% market.

The market dependence on herbicides continues to be there due to the acute shortage of farm labor. It will not be surprising that the share of herbicides in total pesticides used in India is going to increase several fold, as manual labor is not available and weed management through mechanization is not economical, and it is limited to farmers of large holdings only.

11.7 FUTURE DIRECTION OF MYCOHERBICIDES

Despite the recent developments in the research of bioherbicides, there is still much room for scientists to explore new solutions and improve existing ones. In addition, studies on the crop selectivity, appropriate formulation, and mechanism of action of this promising bioherbicide are still lacking. Hence, this opens up more opportunities for researchers and industries to venture into the advancement of this product as an alternative to the current synthetic herbicides in both the ornamental and agricultural fields. Moving forward, more emphasis should be given to advancing the formulation of bioherbicide active substances, which is a key factor influencing the efficacy of a bioherbicide product. In particular, techniques employing multiple surfactants or nano-formulation could be considered to enhance the penetration and absorption of active herbicidal compounds in plants. Moreover, we should look into isolating and identifying new herbicidal compounds from microbes and, upon validating their bioactivity under laboratory and field conditions, recommending them for herbicide development for sustainable agriculture. Due to their easy degradability and high cost, the direct use of cell-free culture extracts as bioherbicides in the field is not feasible in many cases. This could be resolved if the chemical industries elaborated appropriate formulations of the bioherbicides to improve their activity and synthesized the relevant herbicidal compounds whenever this is more convenient than obtaining them from nature. Researchers working on bioherbicides shall continue to identify new and potential species containing a wide range of phytotoxic compounds acting at the molecular or cellular level of target plants and investigate the enzymatic activity related to the mechanism of inhibition and tissue injury of these compounds.

11.8 CONCLUSION

Although only a few studies have sought to elucidate the physiological changes in weeds following bioherbicide application, the investigation of how bioherbicides impact essential biochemical elements like photosynthesis, antioxidants, nutrients, and hormones is well documented: weed populations are suppressed by disrupting regular cell activity and the secretion of toxic metabolites from the bioherbicidal agent. Cell division, nutrient uptake, pigment synthesis, and plant growth–promoting regulators are all inhibited, while the germination and growth of weeds are controlled by stress-mediated hormones, the irregular activation of antioxidants, and other metabolites. On top of the available information, this review suggests that more biomolecular studies are needed to elucidate the mechanism of biological interaction between weeds and bioherbicides.

The developed natural biological herbicide portfolio was derived from a series of fungal extracts, featuring new sites of and modes of action. The first new market entrant to address both organic and industrial agriculture. The beneficial parts of this product are

1. Thriving fields of crops with no resistant weeds
2. Healthier soils sustaining increased production
3. A bioherbicide that gives back to the environment
4. Novel tools to support farmers' needs

This biological herbicide enters the plant cells through the cell wall and disrupt membrane function by releasing large amounts of free radicals, peroxides, and ions into the cytoplasm, resulting in discoloration of plant tissues, rapid wilting due to loss of turgor pressure, and ultimately results in cell death.

REFERENCES

Abraham, E., J. John, and P.S. Pillai. 2016. Allelopathic effect of leaf loppings of homestead trees on ginger (*Zingiber officinale* Roscoe). *Journal of Tropical Agriculture* 54: 60–65. https://jtropag.kau.in/index.php/ojs2/article/view/371.

Algandaby, M.M., and S.M. El-Darier. 2018. Management of the noxious weed, *Medicago polymorpha*, L. *via* allelopathy of some medicinal plants from Taif region, Saudi Arabia. *Saudi Journal of Biological Sciences* 25: 1339–47. https://doi.org/10.1016/j.sjbs.2016.02.013.

Aslani, F., A.S. Juraimi, M.S. Ahmad-Hamdani, et al. 2014. Allelopathic effect of methanol extracts from *Tinospora tuberculata* on selected crops and rice weeds. *Acta Agriculturae Scandinavica, Section B – Soil & Plant Science* 64: 165–77. https://doi.org/10.1080/09064710.2014.898784.

Bailey, K.L. 2014. The bioherbicide approach to weed control using plant pathogens. In *Integrated pest management: Current concepts and ecological perspectives*, ed. D.P. Abrol, 245–66. San Diego: Academic Press, Elsevier. https://doi.org/10.1016/B978-0-12-398529-3.00014-2.

Bailey, K.L., S.M. Boyetchko, and T. Längle. 2010. Social and economic drivers shaping the future of biological control: A Canadian perspective on the factors affecting the development and use of microbial biopesticides. *Biological Control* 52: 221–29. https://doi.org/10.1016/j.biocontrol.2009.05.003.

Boyetchko, S.M. 2005. Biological herbicides in the future. In *Weed management in transition: Topics in Canadian weed science*, Vol. 2, ed. J.A. Ivany, 29–47. Sainte-Anne-de-Bellevue, Québec: Canadian Weed Science Society – Société canadienne de malherbologie. https://www.weedscience.ca/wp-content/uploads/2021/04/Vol-2-Halifax_Full_book–8.pdf.

Boyetchko, S.M., and G. Peng. 2003. Challenges and strategies for development of mycoherbicides. In *Fungal biotechnology in agricultural, food, and environmental applications*, ed. D.K. Arora, 111–21. Boca Raton: CRC Press. https://doi.org/10.1201/9780203913369.

Boyetchko, S.M., and E.N. Rosskopf. 2006. Strategies for developing bioherbicides for sustainable weed management. In *Handbook of sustainable weed management*, eds. H.P. Singh, D.R. Batish, and R.K. Kohli, 393–430. Boca Raton: CRC Press. https://doi.org/10.1201/9781482293593.

Boyetchko, S.M., E.N. Rosskopf, A.J. Caesar, and R. Charudattan. 2002. Biological weed control with pathogens: Search for candidates to applications. In *Applied mycology and biotechnology, agriculture and food production*, eds. G.G. Khachatourians and D.K. Arora, 239–74. Netherlands: Elsevier. https://doi.org/10.1016/S1874-5334(02)80013–2.

Boyetchko, S.M., and A.M. Svircev 2009. Biopesticides: Strategies for discovery, development, and adoption. *Agriculture and Agri-Food Canada* (10733): 1–4. https://publications.gc.ca/collection_2010/agr/A52-120-2009-eng.pdf

Boyette, C.D., R.E. Hoagland, and K.C. Stetina. 2016. Efficacy improvement of a bioherbicidal fungus using a formulation-based approach. *American Journal of Plant Sciences* 7: 2349–58. https://doi.org/10.4236/ajps.2016.716206.

Byer, K.N., G. Peng, T.M. Wolf, and B.C. Caldwell. 2006. Spray retention and its effect on weed control by mycoherbicides. *Biological Control* 37: 307–13. https://doi.org/10.1016/j.biocontrol.2005.12.002.

Castro, M.J.L., C. Ojeda, and A.F. Cirelli. 2013. Surfactants in agriculture. In *Green materials for energy, products and depollution*, eds. E. Lichtfouse, J. Schwarzbauer, and D. Robert, 287–334. Dordrecht, Germany: Springer. https://doi.org/10.1007/978-94-007-6836-9_7.

Chandramohan, S., and R. Charudattan. 2003. A multiple-pathogen system for bioherbicidal control of several weeds. *Biocontrol Science and Technology* 13: 199–205. https://doi.org/10.1080/0958315021000073466.

Dayan, F.E., and S.O. Duke. 2014. Natural compounds as next generation herbicides. *Plant Physiology* 166: 1090–105. https://doi.org/10.1104/pp.114.239061.

de Sousa, C.P., M.E. de Farias, A.A. Schock, and M.A. Bacarin. 2014. Photosynthesis of soybean under the action of a photosystem II-inhibiting herbicide. *Acta Physiologiae Plantarum* 36: 3051–62. https://doi.org/10.1007/s11738-014-1675-9.

Deveau, A., G. Bonito, J. Uehling, et al. 2018. Bacterial–fungal interactions: Ecology, mechanisms and challenges. *FEMS Microbiology Review* 42: 335–52. https://doi.org/10.1093/femsre/fuy008.

Doll, D.A., P.E. Sojka, and S.G. Hallett. 2005. Effect of nozzle type and pressure on the efficacy of spray applications of the bioherbicidal fungus *Microsphaeropsis amaranthi*. *Weed Technology* 19: 918–23. https://doi.org/10.1614/WT-04-352R2.1

Duke, S.O. 2012. Why have no new herbicide modes of action appeared in recent years? *Pest Management Science* 68: 505–12. https://doi.org/10.1002/ps.2333.

Elisante, F., M.T. Tarimo, and P.A. Ndakidemi. 2013. Allelopathic effect of seed and leaf aqueous extracts of *Datura stramonium* on leaf chlorophyll content, shoot and root elongation of *Cenchrus ciliaris* and *Neonotonia wightii*. *American Journal of Plant Sciences* 4: 2332–39. https://doi.org/10.4236/ajps.2013.412289.

El-Mergawi, R.A., and A.I. Al-Humaid. 2019. Searching for natural herbicides in methanol extracts of eight plant species. *Bulletin of the National Research Center* 43 (February): 22, 1–6. https://doi.org/10.1186/s42269-019-0063-4.

Ewers, B.E. 2013. Understanding stomatal conductance responses to long-term environmental changes: A Bayesian framework that combines patterns and processes. *Tree Physiology* 33: 119–22. https://doi.org/10.1093/treephys/tpt008.

Gaines, T.A., S.O. Duke, S. Morran, et al. 2020. Mechanisms of evolved herbicide resistance. *Journal of Biological Chemistry* 295: 10307–30. https://doi.org/10.1074/jbc.REV120.013572.

Grossmann, K. 2003. Mediation of herbicide effects by hormone interactions. *Journal of Plant Growth Regulation* 22: 109–22. https://doi.org/10.1007/s00344-003-0020-0.

Gupta, A., S. Jain, and N. Verma. 2022. *Industrial applications of soil microbes*, vol. 1. Bentham Science Publishers Pte Ltd. https://doi.org/10.2174/97898150399551220101.

Gupta, A., S. Jain, and N. Verma. 2024. *Industrial applications of soil microbes*, vol. 3. Bentham Science Publishers Pte Ltd. https://doi.org/10.2174/97898150797531240301.

Harding, D.P., and M.N. Raizada. 2015. Controlling weeds with fungi, bacteria and viruses: A review. *Frontiers in Plant Science* 6 (August): 659, 1–14. https://doi.org/10.3389/fpls.2015.00659.

Hazrati, H., M.J. Saharkhiz, M. Niakousari, and M. Moein. 2017. Natural herbicide activity of *Satureja hortensis* L. essential oil nanoemulsion on the seed germination and morphophysiological features of two important weed species. *Ecotoxicology and Environmental Safety* 142: 423–30. https://doi.org/10.1016/j.ecoenv.2017.04.041.

Hintz, W. 2007. Development of *Chondrostereum purpureum* as a mycoherbicide for deciduous brush control. In *Biological control: A global perspective*, eds. C. Vincent, M.S. Goettel, and G. Lazarovits, 284–90. Wallingford, UK: CABI Digital Library. https://doi.org/10.1079/9781845932657.0284.

Hynes, R.K., and S.M. Boyetchko. 2006. Research initiatives in the art and science of biopesticide formulations. *Soil Biology and Biochemistry* 38: 845–49. https://doi.org/10.1016/j.soilbio.2005.07.003.

Irshad, A., and Z.A. Cheema. 2004. Influence of some plant water extracts on the germination and seedling growth of barnyard grass (*Echinochola crus-galli* (L.) Beauv). *Pakistan Journal of Scientific and Industrial Research* 43: 222–26. https://v2.pjsir.org/index.php/biological-sciences/article/view/1523.

Jain, S., A. Gupta, and N. Verma. 2023. *Industrial applications of soil microbes*, vol. 2. Bentham Science Publishers Pte Ltd. https://doi.org/10.2174/97898150502641230201.

Jain, S., A. Gupta, and N. Verma. 2024. *Industrial applications of soil microbes*, vol. 4. Bentham Science Publishers Pte Ltd. https://doi.org/10.2174/97898151249961240401.

Janda, K., E. Hideg, G. Szalai, L. Kovács, and T. Janda. 2012. Salicylic acid may indirectly influences the photosynthetic electron transport. *Journal of Plant Physiology* 169: 971–78. https://doi.org/10.1016/j.jplph.2012.02.020.

Kaab, S.B., I.B. Rebey, M. Hanafi, et al. 2020. Screening of Tunisian plant extracts for herbicidal activity and formulation of a bioherbicide based on *Cynara cardunculus*. *South African Journal of Botany* 128: 67–76. https://doi.org/10.1016/j.sajb.2019.10.018.

Kadioglu, I., Y. Yanar, and U. Asav. 2005. Allelopathic effects of weeds extracts against seed germination of some plants. *Journal of Environmental Biology* 26: 169–73. PMID: 16161968.

Kremer, R.J. 2005. The role of bioherbicides in weed management. *Biopesticides International* 1: 127–41. https://www.ars.usda.gov/ARSUserFiles/50701000/cswq-0294–193032.pdf.

Kuzdraliński, A., E. Solarska, and J. Mazurkiewicz. 2013. Mycotoxin content of organic and conventional oats from southeastern Poland. *Food Control* 33: 68–72. http://dx.doi.org/10.1016/j.foodcont.2013.02.014.

Lee, S.M., R. Radhakrishnan, S.M. Kang, et al. 2015. Phytotoxic mechanisms of bur cucumber seed extracts on lettuce with special reference to analysis of chloroplast proteins, phytohormones, and nutritional elements. *Ecotoxicology and Environmental Safety* 122: 230–37. https://doi.org/10.1016/j.ecoenv.2015.07.015.

Lorsbach, B.A., T.C. Sparks, R.M. Cicchillo, N.V. Garizi, D.R. Hahn, and K.G. Meyer. 2019. Natural products: As strategic lead generation approach in crop protection discovery. *Pest Management Science* 75: 2301–09. https://doi.org/10.1002/ps.5350.

Marrone, P.G. 2019. Pesticidal natural products: Status and future potential. *Pest Management Science* 75: 2325–40. https://doi.org/10.1002/ps.5433.

Motmainna, M., A.S. Juraimi, M.K. Uddin, N.B. Asib, A.K.M.M Islam, and M. Hasan. 2021a. Bioherbicidal properties of *Parthenium hysterophorus*, *Cleome rutidosperma* and *Borreria alata* extracts on selected crop and weed species. *Agronomy* 11 (March): 643, 1–24. https://doi.org/10.3390/agronomy11040643.

Motmainna, M., A.S.B. Juraimi, M.K. Uddin, N.B. Asib, A.K.M.M. Islam, and M. Hasan. 2021b. Allelopathic potential of Malaysian invasive weed species on weedy rice (*Oryza sativa* f. *spontanea* Roshev). *Allelopathy Journal* 53: 53–68. http://dx.doi.org/10.26651/allelo.j/2021-53-1-1327.

Najafpour, M.M. 2011. Oxygen evolving complex in photosystem II: Better than excellent. *Dalton Transactions* 40: 9076–84. https://doi.org/10.1039/C1DT10746A.

Peng, G., and T.M. Wolf. 2008. Spray retention and its potential impact on bioherbicide efficacy. *Pest Technology* 2: 70–80. http://www.globalsciencebooks.info/Online/GSBOnline/images/0812/PT_2(1&2)/PT_2(2)70-80o.pdf.

Prasad, R., V.D. Tripathi, P. Singh, et al. 2016. Allelopathic potential of *Butea monosperma* L.: Effect of aqueous leaf extract on seed germination and seedling growth of winter season (rabi) crops. *Indian Journal of Agroforestry* 18: 63–69.

Qu, R.Y., B. He, J.F. Yang, et al. 2021. Where are the new herbicides? *Pest Management Science* 77: 2620–25. https://doi.org/10.1002/ps.6285.

Rab, A., S.K. Khalil, M. Asim, et al. 2016. Response of sorghum (*Sorghum bicolor* L.) extract type, concentration and application time to weeds weight, grain and biomass yield of wheat. *Pure Applied Biology* 5: 847–55. http://dx.doi.org/10.19045/bspab.2016.50106.

Radhakrishnan, R., A.L. Khan, and I.J. Lee. 2013. Endophytic fungal pre-treatments to seeds alleviate salinity stress effects in soybean plants. *Journal of Microbiology* 51: 850–57. https://doi.org/10.1007/s12275-013-3168-8.

Rial, C., B.F. García, R.M. Varela, A. Torres, J.M.G. Molinillo, and F.A. Macías. 2016. The joint action of sesquiterpene lactones from leaves as an explanation for the activity of *Cynara cardunculus*. *Journal of Agricultural and Food Chemistry* 64: 6416–24. https://doi.org/10.1021/acs.jafc.6b02678.

Scavo, A., C. Rial, J.M.G. Molinillo, R.M. Varela, G. Mauromicale, and F.A. Macias. 2019. The extraction procedure improves the allelopathic activity of cardoon (*Cynara cardunculus* var. altilis) leaf allelochemicals. *Industrial Crops and Products* 128: 479–87. https://doi.org/10.1016/j.indcrop.2018.11.053.

Schilder, A. 2008. *Effect of water pH on the stability of pesticides*. East Lansing, MI: Michigan State University Extension. https://www.canr.msu.edu/news/effect_of_water_ph_on_the_stability_of_pesticides.

Shaul, O. 2002. Magnesium transport and function in plants: The tip of the iceberg. *Biometals* 15: 307–21. https://doi.org/10.1023/A:1016091118585.

Sparks, T.C., and S.O. Duke. 2021. Structure simplification of natural products as a lead generation approach in agrochemical discovery. *Journal of Agricultural and Food Chemistry* 69: 8324–46. https://doi.org/10.1021/acs.jafc.1c02616.

Tardieu, F. 2013. Plant response to environmental conditions: Assessing potential production, water demand, and negative effects of water deficit. *Frontiers in Physiology* 4 (February): 17, 1–11. https://doi.org/10.3389/fphys.2013.00017.

Yandoc-Ables, C.B., E.N. Rosskopf, and R. Charudattan. 2007. Plant pathogens at work: Improving weed control efficacy. *Plant Health Progress* 8 (August): 1, 1–13. https://doi.org/10.1094/PHP-2007-0822-02-RV.

12 Host-Specific Fungal Toxins
Status and Perspectives

Jaya Singh
(Biodiversity, Conservation and Rural Biotechnology Centre, Jabalpur (MP), India; dr.jayasingh@yahoo.com)

**Saurabh Gupta*
(Department of Biotechnology, Sri Guru Tegh Bahadur Khalsa College, Jabalpur (MP), India; gupta.saurabh89@yahoo.com)

12.1 INTRODUCTION

Plants encounter a multitude of fungal spores, but only a small portion of them can cause infection. Generally, plants possess natural defense mechanisms that work against most fungi, except for a few specialized parasites. A precise recognition of these parasites by plants has long fascinated plant pathologists. Host recognition is seen as a particular occurrence that prompts a swift and decisive response from the host, either aiding or hindering the pathogen's further development (Kohmoto and Otani, 1991). A mechanism that determines such a practical specificity comprises three basic processes: (1) spores of a fungal parasite release, before penetration, a host recognition factor (host-specific toxin) as a signal; (2) the released signal factor selectively binds to receptor sites in host cells; and (3) the signal is transduced in such a way as to immobilize the host's reactions and condition the cells to produce an "accessible state" for fungal parasite penetration, before cell death. Recent comprehensive research has provided evidence supporting this theoretical framework, particularly through the examination of host-specific toxins (HSTs) and a thorough assessment of their significance in the initial stages of plant penetration and colonization by pathogens (Kohmoto and Otani, 1991; Nishimura et al., 1987; Otani et al., 1995).

The understanding of toxins' role in plant diseases caused by pathogenic microorganisms can be traced back to the latter part of the nineteenth century. Toxins are defined as microbial metabolites that can harm plants even at very low concentrations. According to Scheffer and Briggs (1981), toxins are microbial products, excluding enzymes, that visibly harm plant tissues and are reasonably believed to contribute to disease progression. Most of the plant pathogenic microbes produce toxins in culture and their hosts. In many cases, especially if they are produced during the early stages of plant disease development, these compounds play a key role in pathogenesis and reproduce some or even all of the symptoms of the disease caused by the producer organisms. Most of the toxins are small molecules and have a variety of structures; they can diffuse from the site of infection into the surrounding tissues or are translocated within the plant *via* the apoplast. The potency of plant pathogens, and occasionally their ability to cause disease, can hinge on their capacity to produce one or more toxins and deploy them as chemical weaponry (Durbin, 1983).

The recognition of host-specific toxins as pivotal factors in pathogenicity was initially observed in two diseases: black spot disease in Japanese pears and Victoria blight in oats, dating back to the 1930s and 1940s. However, it wasn't until 1964 that the term "host-specific toxin" (HST) was coined by Pringle and Scheffer (1964). HSTs are known to affect only plants of a genotype susceptible to the pathogen and are the only known agents of specificity in any plant/microbe interaction. At low (physiological) concentrations, they reproduce the symptoms of the natural infection. The most important attribute of HSTs is that they are agents of virulence or pathogenicity: a fungus that makes an HST cause more disease on its host than one that is otherwise identical but does not make the HST. Mutants of the pathogen lacking the genes for toxicity (TOX-), that is, unable to produce toxin, are not virulent (Daly and Deverall, 1983; Gilchrist et al., 1985; Nakatsuka et al., 1986; Kono and Suzuki, 1988; Nishimura and Kohmoto, 1983; Nishimura et al., 1987; Scheffer, 1983; Navarre et al., 1995; Wolpert et al., 1985; Yoder and Turgeon, 1985). Therefore, host-specific toxins (HSTs) appear to play a crucial role in both the pathogenicity of the disease and the range of hosts it can infect. Conversely, when a plant shows insensitivity to an HST, it gains increased resistance against the organism producing it as shown in Table 12.1.

Sensitivity to specific HSTs in host plants is typically governed by a single gene, although this sensitivity can manifest as dominant, semi-dominant, recessive, or cytoplasmic inheritance. While hundreds of genes related to disease specificity (resistance genes in plants and pathogenicity genes in pathogens) have been identified, the biochemical and physiological mechanisms controlled by these specificity genes remain largely unknown. The basis of specificity has been elucidated for only two HSTs.

DOI: 10.1201/9781003408833-12

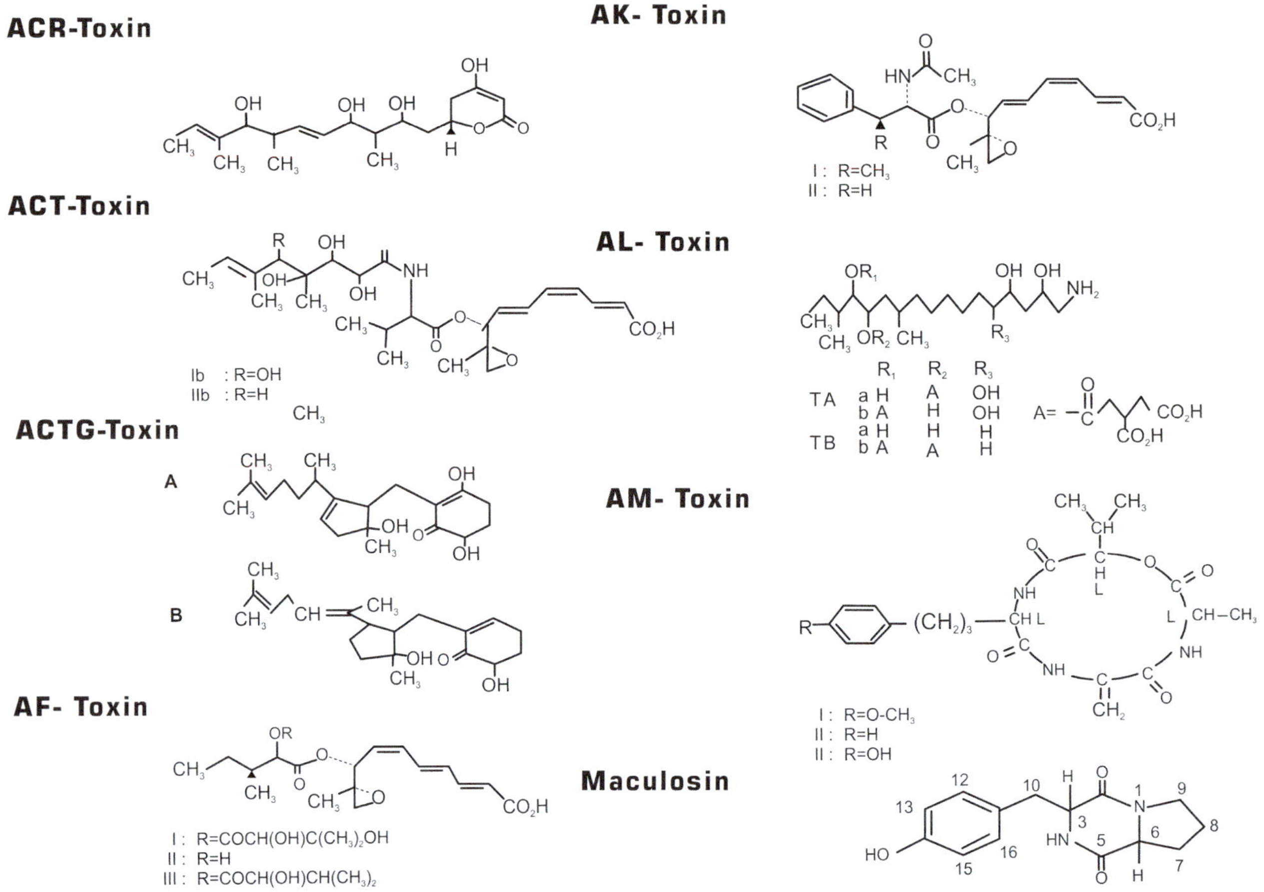

FIGURE 12.1 Structures of host-specific toxins (HSTs) produced by pathotypes of *Alternaria alternata*.

The unique reaction to T-toxin produced by *Cochliobolus heterostrophus* is facilitated by a novel protein known as URF 13, encoded by the mitochondrial genome of maize and situated within the inner mitochondrial membrane of plants that are sensitive to the toxin (Dewey et al., 1988). The specificity of HG-toxin, produced by the maize pathogen *Cochliobolus carbonum,* is not related to its site of action but to differential detoxification; resistance is dominant (Johal and Briggs, 1992; Meeley et al., 1992). HST produced by *Cochliobolus victoriae* is related to differential binding to a membrane-localized protein (Wolpert and Macko, 1989). Recent progress on the chemistry and biology of HSTs indicates that they are ordinary natural products. The chemical structures, biosynthetic pathways, and molecular genetics of HSTs are similar to those of other microbial secondary metabolites (Jain et al., 2023, 2024). The majority of secondary metabolites lack identified biological functions; however, it is their role as agents of virulence and specificity in microbe-plant interactions that renders host-specific toxins (HSTs) significant in the realms of natural products biology and, notably, plant pathology (Gupta et al., 2022, 2024).

In this chapter, the focus is primarily on the structure, production (including genetic and molecular mechanisms), function, application, and ecological consequences of HST(s).

12.2 STRUCTURE AND ACTIVITY

The known HSTs, except Tomas et al. (1990), are of low-molecular-weight secondary metabolites, like any other ordinary natural products, including cyclic peptides, terpenoids, polyketides, and compounds of unknown biogenesis with great diversity in their chemistry and biological effects (Figures 12.1 and 12.2). Most HSTs from any single pathogen occur as families of closely related compounds. More than 19 HSTs have been reported: Most of them are produced by *A. alternate* and *Cochliobolus* (Nishimura et al., 1987; Scheffer and Livingston, 1984).

12.3 HST PRODUCED BY PATHOTYPES OF *ALTERNARIA*

Currently, it is known that ten or more pathogenic species of *Alternaria* produce HSTs, which have varied effects on the structural components of the host, including the plasma membrane, chloroplast, and mitochondria. These toxins can be classified into three groups based on their primary site of action:

1. ACT-, AF-, and AK-toxins share a common epoxy-decatrienoic acid structure and primarily affect the plasma membrane of susceptible cells.

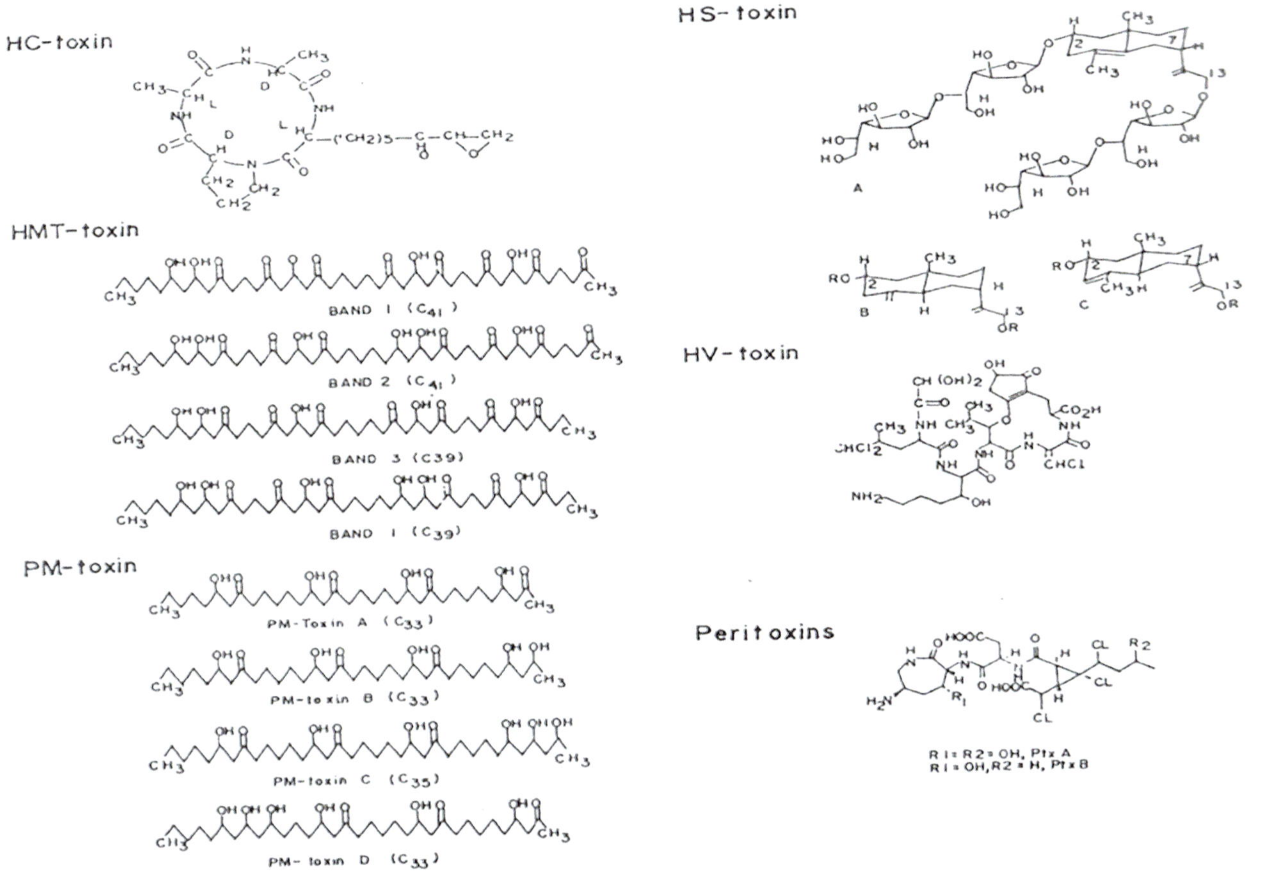

FIGURE 12.2 Structures of host-specific toxins (HSTs) produced by species of *Cochliobolus*, *Phyllosticta*, and *Periconia*.

2. ACR (L)-toxin induces changes in mitochondria, including swelling, reduced electron density of the matrix, vesiculation of cristae, increased NADH oxidation rate, and cessation of malate oxidation.
3. AM-toxin appears to have an early impact on both chloroplasts and plasma membranes.

Despite their diverse structural features and physiological responses, these toxins are deemed essential for colonization of host tissue by the pathogens.

12.3.1 AK-, AF-, and ACT-Toxins

These toxins are structurally related and have in common the 8-substituted-9, 10-epoxy-9-methyl decatrienoic acid structure (Figure 12.1).

Nakashima et al. (1982) isolated two toxins, that is, AK toxin I and II in crystalline form from *A. alternata*, specifically the Japanese pear pathotype (Figure 12.1). AK-toxin I has a significant impact on only 9 cultivars, such as "Nijisseikei" and related ones, even at a remarkably low concentration of 10^{-9} M (Nakatsuka et al., 1989). Interestingly, 36 other cultivars tested remained insensitive to AK-toxin I, even at a concentration as high as 10^{-4} M (Nakatsuka et al., 1989). This "all or nothing" type of selectivity in toxin action perfectly aligns with the host-or-nonhost response observed in pear cultivars. AK-toxin II exhibits similar selectivity, albeit with a much lower level of activity, approximately 1/20th that of AK-toxin I.

Otani et al. (1995) prepared a scheme for the mechanism of AK-toxin action in susceptible pear as shown in Figure 12.3. AK-toxin has pleiotropic effects on susceptible cells after specific recognition at sites on plasma membranes, which is an important event for the induction of accessibility. AK-toxin is detrimental to the defense mechanism of pears by suppressing resistance-inducing factor (a polysaccharide of about 40 kDa) and papilla formation beneath the appressorium of germinating spores of the pathogen, which finally pave the way for penetration/pathogenesis.

AF-toxins I, II, and III were isolated by Nakatsuka et al. (1986) from *A. alternata* pathotype strawberry, causing black spot disease only on Monoka-16 cultivars. However, in laboratory tests, it also affects certain cultivars of Japanese pears, showing the same host range as the Japanese pear type (AK-toxin producer) of *A. alternata* (Maekawa et al., 1984). AF-toxin I causes leaf necrosis of both strawberry and Japanese pear, while AF-toxin II causes necrosis of pear only (Maekawa et al., 1984). AF-toxin III is toxic to strawberry and less toxic to pear (Namiki et al., 1986). The decatrienoic acid moiety in the major AF-toxins has a2E,

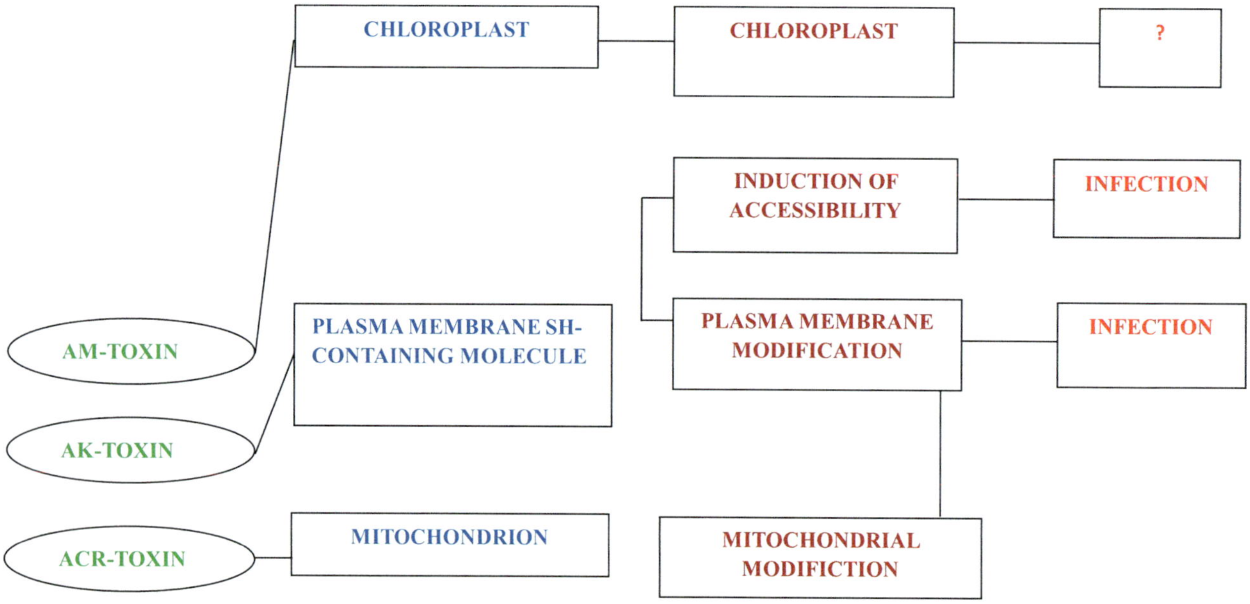

FIGURE 12.3 Brief mechanism of action of AK-, ACR-, and AM-toxins in susceptible plants.

Source: Redrawn and modified after Otani et al. (1995).

4E, 6Z geometry, while that of the AK-toxins has 2E, 4Z, 6E. In addition, the strawberry pathogen produces small amounts of stereoisomers which are different with respect to the trienoic acid structure. Among the three isomers of AF-toxin I examined for toxicity, that with the 2E, 4E, and 6Z geometry shows the highest toxicity to the strawberry leaves. The three isomers of AF-toxin II, characterized by a free hydroxy group at 1he 2'-position of the 8-acyl moiety, show no toxicity to strawberry but are toxic to pear. Moreover, some 2'-0-acyl derivatives (acetyl, propionyl, and isovaleryl) prepared from AF-toxin II by acylation are as toxic as AF-toxin I or III on both strawberry and pear leaves (Kohmoto and Otani, 1991). Nishimura and Nakatsuka (1989) suggested that the 2'-hydroxy group represents a feature of the molecule essential only for toxicity to pear. Its role may be the same as that played by the amide NH at the 2'-position of the AK-toxin. Interestingly, the 2E, 42-6E isomer of AF-toxin isolated from culture filtrates exhibited high toxicity to pear, comparable to that of AK-toxin I, which has the same stereochemistry. The free w-epoxy-decatrienoic acid itself has no toxicity for any of the plants tested.

ACTG-toxins were isolated from *A. alternata* tangerine pathotype causing disease on citrus varieties "Dancy" tangerine, "Emperor" mandal'm and "Murcott" tangor (Kono et al., 1986; Itch et al., 1989). Both ACTG-toxins A and B exhibit toxicity to susceptible plants at a concentration as low as 0.1 µg/ml. additionally, another group of host-specific toxins, ACT-toxins, which differ significantly in chemical composition from ACTG-toxins but are closely related to AK- and AF-toxins (Figure 12.1), were also isolated from tangerine pathotype (Itch et al., 1989; Nakatsuka et al., 1989). ACT-toxins are produced during spore germination, which is an important event for successful pathogenesis. The ACT-toxin lb is toxic at a concentration of 10 ng/ml only to species, cultivars, and hybrid lines susceptible to the *A. alternata* Tangerine pathotype. There is a close correlation between disease susceptibility and toxin sensitivity. The w-epoxydecatrienoic moiety of ACT-toxin lb has the same stereochemistry (2E, 4Z, 6E) as that of AK-toxin. ACT-toxin le, which is produced in minor amounts, contains the 2E, 4E, 6E isomer and is less active. Another HST, ACT-toxin IIb, which is the 5'-deoxy derivative of ACT toxin lb, causes a 2,000-fold reduction in toxicity to tangerines, while it increases the toxicity to Japanese pears up to 10-fold, thus reaching a potency comparable to that of AK-toxin (Kohmoto and Otani, 1991). Moreover, the sensitivity spectrum in Japanese pear cultivars to toxin lb and IIb is the same as AJ-toxin. AK-toxin, however, is not toxic to Tangerines and Mandarins. A collaborative investigation into the structure-activity relationships among three closely related groups of host-specific toxins (HSTs) – AK, AF, and ACT – indicates that the susceptible Japanese pear demonstrates a lower level of specificity in recognizing signal molecules at potential receptor sites compared to susceptible strawberry, tangerine, and mandarin varieties.

12.3.2 ACR (L)-Toxins

A. alternata, causing brown spots on rough lemon, produces multiple HSTs (Figure 12.1), the major one designated as ACRL toxin I (Gardener et al., 1985). The toxin induces ultra-structural changes, including swelling of mitochondria, reduction in numbers and vesiculation of cristae, and a decrease in the electron density of the matrix. After affecting mitochondria, the toxin induces a rapid increase in electrolyte loss from leaf tissues (plasma membrane disorder) and eventually causes veinal necrosis (Figure 12.3), which is an essential event for accessibility.

12.3.3 AM-Toxin

The apple pathotype of *A. alternata,* causing the blotch of apple, produces multiple HSTs named AM-toxin I, II, and III. AM-toxin I is a cyclic depsipeptide, and AM-toxin II and III are derivatives of AM-toxin I (Okuno et al., 1974; Otani et al., 1995). Two sites of action of the toxin have been identified: chloroplasts, where the toxin induces dissociation and vesiculation of grana lamellae and inhibits photosynthetic CO_2 fixation; plasma membrane, where the toxin causes modification and an increase of electrolyte loss (Figure 12.3). AM-toxin predominantly affects both chloroplasts and plasma membranes in susceptible cells. While chloroplast sites may contribute to host-specific action of AM-toxin, dysfunction of the plasma membrane is closely associated with the light-inhibiting process that results in necrosis and susceptibility to fungal infection (Otani et al., 1995).

12.3.4 AAL-Toxin

A. alternata f. sp. *Lycopersici* (stem canker strain) produces an HST, ML-toxin both in host plant tissue as well as defined culture medium (Gilchrist et al., 1996). The toxin consists of two related esters of 1,2,3-propanetricarboxylic acid esterified to 1-amino-11, 15-dimethylheptadeca-2, 4, 5, 13, 14-pentol. Interveinal localized cell death in leaflets of plants with natural stem infection, a diagnostic symptom of the disease, is reproduced by phytotoxin recovered from cell-free culture filtrates of the pathogen. Moreover, only cultivars of tomato susceptible to the pathogen are highly sensitive to the toxin (20 nM). This toxin is a potent inhibitor of ceramide synthase at 20 nM with an IC50 in the range of 35–40 nM. A net decrease in sucrose influx to treated leaves occurs. Accumulation of ethylene precursor 1-aminocyclopropane-1-carboxylic acid (ACC) followed by a burst in ethylene within 12 hours, occurs after exposure to AAL-toxin and before the death of the plant tissue. Genetic resistance to the pathogen and the AAL-toxins is regulated in tomato by a single Mendelian gene (Ase), which is inherited as a dominant to resistance to the pathogen but expresses incomplete dominance to Ile toxin (Clouse and Gilchrist, 1987). The stem canker disease is currently controlled in commercial cultivation by the dominant Ase gene.

12.3.5 Maculosin

Maculosin, produced by the *A. alternata* spotted knapweed pathotype, is a remarkable host-specific toxin responsible for leaf necrosis in *Centaurea maculosa.* It belongs to the chemical class of diketopiperazines and is a cyclic dipeptide. Maculosin induces necrosis on knapweed leaves at a concentration of 10 μM, while other grasses, dicots, and *Centaurea* species remain unaffected. This makes maculosin the first phytotoxin produced by a weed pathogen to exhibit such a high degree of specificity. Maculosin contains a phenolic moiety, which is not uncommon in phytotoxins. Studies on diketopiperazines containing proline suggest that L, L compounds assume an extended conformation, making both the aromatic moiety and its hydroxyl group accessible for interaction with a biological site (Stierele et al., 1989).

12.4 HST Produced by Various Species of *Cochliobolus*

12.4.1 HV-Toxin (Victorin)

Victoria blight of oats is caused by the fungus. *C. victoriae* was first described by Meehan and Murphy (1946). Only oat lines containing Victoria-type resistance to crown rust are susceptible to *C. victoriae.* The novel discovery made is that the cell-free *C. victoriae* filtrates contain a toxic substance that not only reproduces the symptoms of the fungus on susceptible plants but also exhibits the same specificity as the pathogen. The most abundant toxic component purified from fungal culture filtrates is victorin C (molecular weight 814 corresponding to empirical formula $C_{31}H_{45}O_{13}N_6C_{13}$) is composed of a cyclic combination of the six components: glyoxylic acid, 5. 5-dichloroleucine, erythro-B-hydroxyleucine, vitalanine (2-alanyl-3, 5-dihydroxy-⇑2 – cyclopentenone-1), threo-B-hydroxylysine, and alpha – amino-B-chloro-acrylic acid (Macko et al., 1985; Wolpert et al., 1985, 1988). An ether linkage between a hydroxyl group of the victalanine and the hydroxy leucine residue forms the 12-membered macrocyclic ring. Kono et al. (1986a) identified a form of victorin and designated as HV-toxin M (molecular weight 797 corresponding to empirical formula C_{31} H_{46} O_{11} N_7 C_{13}), which is identical to victorin C, except that the glyoxylic acid residue of victorin C is replaced by a glycine residue. Additional toxic compounds identified from the culture filtrates were victorin B, D, and E (Macko et al., 1985; Wolpert et al., 1985; Navarre et al., 1995). Thus, victorin is a collection of closely related cyclized pentapeptides (Figure 12.2).

Experimental evidence suggests that the nature and substitution pattern on the 12-membered macrocyclic ring are important for association with the active site. Removal of the glyoxylic acid residue results in the complete loss of toxic activity, indicating that it is essential for recognition of the active site. However, aldehyde is important for the toxic activity of victorin, more because it can bind spontaneously with proteins (possible active site) covalently (Navarre et al., 1995).

12.4.2 HMT-Toxin

The fungus *Helminthosporium maydis* race T, which sparked a severe corn blight epidemic in the USA during

1970–1971, produces a host-specific toxin known as HMT-toxin. This toxin impacts corn plants carrying Texas male sterile cytoplasm, but does not affect corn with normal fertile cytoplasm. HMT-toxin is a complex comprising chemically related compounds (Figure 12.2) and it influences various physiological processes in Tms-com, including ion leakage, dark CO_2 fixation, coleoptile elongation, and mitochondrial oxidation (Kono, 1989).

12.5 HST PRODUCED BY *PHYLLOSTICTA MAYDIS*

12.5.1 PM-Toxin

Phyllosticta maydis, the causal agent of yellow leaf blight in corn, produces a host-specific toxin known as PM-toxin, which affects corn carrying Texas male sterile cytoplasm but not corn with normal fertile cytoplasm (Yoder, 1973). PM-toxin comprises four major components (A, B, C, D), all of which are equally active and host-specific, similar to HMT-toxin produced by *Helminthosporium maydis* race T. The absolute configurations of PM-toxins A, B, C, and D have been determined (Figure 12.2) (Danko et al., 1984; Kono et al., 1985). PM-toxin B does not affect resistant corn even at a concentration of 10^{-5} M, but each purified component of PM-toxin exhibits high toxicity (ranging from 10^{-8} to 10^{-9}M) on tissues and mitochondria obtained from susceptible corn varieties, akin to HMT-toxin isolated from the unrelated plant pathogen *H. maydis* race T (Danko et al., 1984). The oxygen groups in HMT (between C-8 and C-36) and PM toxins (C-6 and C-33) are located in similar regions on straight carbon chains (Figure 12.2). This structural resemblance is considered a factor explaining the similarity in their biological activities.

12.6 HST PRODUCED BY *PERICONIA CIRCINATA*

Destructive outbreaks of root and crown rot disease in Sorghum caused by *Periconia circinata* occurred in the southern plains of the USA in the mid-1920s (Leukel and Johnson, 1948). The affected plants were stunted and produced little or no grain; the leaves became chlorotic and eventually necrotic; the roots and crown developed a dark-red discoloration that frequently progressed up the stalk through the vascular tissues. However, the fungus is widespread and exists as a saprophyte on the roots of many other species, including sorghum in South Africa. Pathogenic fungal species did not coevolve with their host near the center of origin of sorghum (northeastern Africa) but emerged only after the introduction of sorghum into regions where a particular and unique toxin-producing ecotype occurred (USA).

P. circinata produces peritoxins A and B (Dunkle and Macko, 1995), which exhibit equal specific toxicities. These toxins are low-molecular-weight peptides with molecular weights of 575 and 558, respectively. They contain a residue of aspartate, an E-lactam form of 5-amino-3-hydroxy-lysine, and a chlorinated C10 residue (Figure 12.2). Additionally, two other structurally related but inactive analogs, circinatin and 7-chlorocircinatin, have been discovered alongside peritoxin production and are potentially metabolic precursors of the toxin. Interestingly, tox-strain does not produce these two precursors, indicating that the block in toxin biosynthesis occurs before circinatin in tox-strains. These strains seemingly differ from TOX+ strains in several metabolic steps (Dunkle and Macko, 1995).

Dunkle and Macko (1995) developed a model (Figure 12.4) to show ht effects of peritoxin on susceptible sorghum genotype. According to their hypothesis, the dominant allele at the Pc locus encodes a toxin-sensitive site, a proteinaceous receptor on or near the cell surface, which reacts with peritoxin to initiate a series of responses leading to development and expression of toxin-induced disease symptoms. The initial interaction is sensitive to proteases, phospholipase D, and perhaps heat shock and is prevented by biotinylation of the putative toxin-receptor protain enhanced synthesis of a specific group of proteins, the P16 proteins, is induced, perhaps as a consequence of other perturbations, but is not directly linked to or responsible for symptom development. The synthesis of the P16 protein is stimulated by various factors, including treatment with a fungal elicitor, which does not cause disease symptoms, as well as exposure to Hg^{2+} and Cd^{2+} ions, which do replicate the disease symptoms. The toxin triggers a range of symptoms, such as mitosis inhibition, electrolyte leakage, hindered primary root elongation and seedling growth, as well as necrosis and wilting.

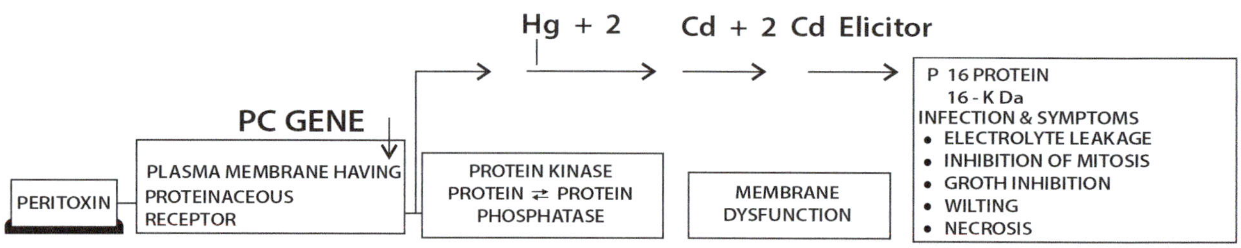

FIGURE 12.4 Possible mechanism of action of peritoxin on susceptible sorghum genotypes.

Source: Modified after Dunkle and Macko (1995).

12.7 SUMMARY OF ACTIVITY

The primary action sites of host-specific toxins (HSTs) are typically associated with three target organelles: the plasma membrane, mitochondria, and chloroplasts. In susceptible genotypes, HSTs induce immediate dysfunction in plasma membranes upon application. This often leads to a rapid and dose-dependent increase in electrolyte loss from cells, accompanied by invagination of their membranes – a common effect observed with PC-, HV-, HS-, PM-, AK-, AF-, and ACT-toxins. However, HG-toxin stands as an exception, as it enhances ion uptake (Kohmoto and Otani, 1991). AF-, AK-, HV-, and HS-toxins depolarize the membrane of susceptible genotypes but have little effect on the membrane potential of resistant genotypes; conversely, HG-toxin hyperpolarizes the membrane of susceptible maize.

Susceptible host cell mitochondria are affected by ACR-, AL-, AT-, HMT-, and PM-toxins. For instance, HMT-toxin impacts the mitochondria of Texas male-sterile (Tms) but not male-fertile (N) cytoplasm maize, exhibiting symptoms such as swelling, reduction in cristae numbers, vesiculation of cristae, a decrease in matrix electron density, an increase in NADH oxidation rate, and inhibition of malate oxidation. Additionally, HMT-toxin increases the permeability to NAD+ of the inner mitochondrial membrane in Tms, but not N, cytoplasm. In contrast, AT- and AL-toxins induce ultrastructural changes in mitochondria of susceptible cultivars within 24 hours. However, detailed investigations regarding the effects of HSTs on chloroplasts are still going on. Apart from the primary sites of action, several secondary sites also exhibit effects of HSTs.

12.8 GENETICS OF HST PRODUCTION AND SPECIFICITY

In most cases, pathogenicity and resistance are controlled by single Mendelian loci in host/pathogen interactions. Single gene resistance to particular species, races, pathovars and strains of viruses, nematodes, bacteria, and fungi has been identified in most major crop plants. It has also been shown in several fungi and some bacteria that specific pathogenicity is inherited monogenically. Though monogenic inheritance excludes complex molecules as mediators of specificity, primary gene products, that is, protein, must do so. This is because complex molecules require many genes contributing to the specificity of a particular disease interaction. Plant secondary metabolites require many individual enzymatic steps and genes. However, in microorganisms, the tight clustering of secondary metabolite genes is very common. Gene clusters control biosynthesis of secondary metabolites in both bacteria and fungi (Smith et al., 1990).

Very little is known about the genetic control of toxin production by plant pathogenic fungi. The production of HSTs, known to play a role in disease development, has been genetically analyzed in three species of *Cochliobolus*, wherein single but different genes control their production. These genes are called TOX 1, TOX 2, and TOX 3, and control production of T-toxin, HG-toxin, and victorin in *C. heterostrophus*, *C. carbonum*, and *C. victoriae*, respectively (Bronson, 1991). In *C. heterostrophus*, a single genetic locus with two alleles has been identified controlling the production of HMT-toxin. This locus appears to be at or near the breakpoint of chromosome rearrangement. Single genetic loci have also been identified controlling the production of HG-toxin by *C. carbonum* and HV-toxin by *C. victoriae*. How do the single genes single-handedly control the production of these complex secondary metabolites? More likely, it is the TOX genes that encode multifunctional enzymes or are gene clusters. For example, the Mendelian locus TOX 2 in *C. carbonum* that controls HC-toxin biosynthesis is a cluster of at least 54 kb encoding in part two copies of a large multifunctional enzyme. Since non-toxin-producing isolates of *C. carbonum* lack DNA homologous to TOX 2, the entire region segregates as a single Mendelian locus (Panaccione et al., 1992). Based on various studies, particularly on TOX 2, it is concluded that the mediation of specificity by complex secondary metabolites is completely compatible with monogenic inheritance of specificity (Walton and Panaccione, 1993).

"Receptors" play a pivotal role in plant–pathogen interactions and are believed to be encoded by plant genes associated with resistance or susceptibility. In the case of host-specific toxins (HSTs), it is proposed that the presence of the appropriate receptor is necessary for the toxin to exert its effects. Since HSTs contribute to pathogenicity, the absence of the receptors renders the plant resistant (Gardener and Scheffer, 1973). This hypothesis aligns with the dominant sensitivity observed, for instance, to victorin. In disease models, where pathogen avirulence dominates over virulence, it is suggested that active host defenses are activated upon binding of a pathogen signal (a specific elicitor) to a specific receptor. In the absence of the receptor, recognition fails, leading to disease. A gene that encodes a protein capable of deactivating an HST or transducing an elicitor of plant defenses would confer dominant resistance. Conversely, a gene encoding a protein that activates a toxin (such as activation of tabtoxin by plant peptidases) or a receptor that transduces a suppressor of plant defenses would result in recessive resistance.

In case of HV-toxin (victorin), sensitivity to victorin, and thus susceptibility to the fungus is controlled by a dominant allele at the Vb locus. Furthermore, the Vb gene is either closely linked or identical to the Pc-2 gene, which confers resistance to the crown rust pathogen *Puccinia coronata*. Navarre et al. (1995) have identified two proteins, referred to as victorin binding protein (VBP), which bind biologically active radio-labeled victorin *in vivo*. A protein, 100-kDa VBP, binds victorin *in vivo* only in toxin-sensitive, susceptible genotypes. The other VBP, a 15 kDa protein, binds victorin *in vivo* in both susceptible and resistant genotypes. Wolpert et al. (1994) characterized the 100 kDa VBP and have recently identified a 15 kDa VBP *in vitro*. Both these proteins are components of the multienzyme complex, glycine decarboxylase, which is the site of action of victorin, or a component of the complex, very likely regulates the specificity of the oat plant's response to victorin and therefore to *C. victoriae*.

12.9 MOLECULAR MECHANISM OF HST PRODUCTION

It is important to understand the nature of the pathogen genes that control the synthesis of HSTs, which, in turn, might allow plant pathologists and breeders to develop more effective disease control strategies. The molecular genetics of the biosynthesis of HSTs, coupled with the cloning of the TOX gene, has been studied only in the case of HG-toxin by *Cochliobolus carbonum*. Races 1 and 2 of *C. carbonum* differ by a single Mendelian gene for pathogenicity on maize lacking the Hm gene for resistance (Scheffer et al., 1967). The ability to produce HG-toxin segregates with the pathogenicity gene called TOX 2 (Yoder et al., 1989).

The elucidation of the structure of HG-toxin as a cyclic tetrapeptide, cyclo (D-prolyl-L-alanyl-D-alanyl-L-Aeo), where Aeo stands for 2-amino-9, 1 O-epoxy-8-oxodecanoic acid, suggests that the compound might be synthesized like other small microbial peptides. At least two enzymes, HGC-toxin synthetase 1 (HTS-1) and HG-toxin synthetase 2 (HTS-2), participate in the synthesis of HG-toxin, and therefore, TOX 2 is a gene cluster (Walton and Panaccione, 1993). Gene including HTS-1 has three characteristics: First, it is part of a contiguous region of DNA, 22 kb in length, that has absolutely no homology to DNA from any non-HG-toxin-producing race of *C. carbonum* or species of *Cochliobolus* tested. Second, this 22 kb DNA piece is flanked on either side by moderately reiterated DNA (Panaccione et al., 1992). Elements from 5' and 3' flanks have homology to one another as well as to repeated sequences from other races of *C. carbonum* and *C. victoriae*. Third, the entire 22 kb race 1-unique region, as well as at least 6 kb of repeated DNA on the 5' flank and 2 kb on 3' flank, are duplicated, and the duplicate copies are genetically linked. These findings have interesting implications for the evolution of race-specific pathogenicity in *C. carbonum*. Lack of DNA homology between races 1 and 2 indicates that biosynthetic capability might at one time have been common to the species and was subsequently deleted in certain isolates, to create race 2, or it might have been recently acquired by the species, to create race 1. Disruption and sequencing of the gene for HTS-1 indicate that HTS-1 and HTS-2 are potentially part of a single polypeptide. A striking feature of HTS 1 is its apparent lack of introns.

TOX 2, despite being estimated to be at least 56 kb in size and potentially large enough to match the size of the chromosome it resides on, remains poorly understood. Little information is available about TOX 2, including its known repetitive DNA flanking the 22 kb tox+ – unique DNA, which is duplicated and exhibits tox+ – unique RFLPs. There is ongoing research in molecular plant pathology aimed at unraveling the mysteries surrounding both HTS 1 and TOX 2.

12.10 EVOLUTION AND ECOLOGICAL PERSPECTIVES

Fungal pathogens of plants demonstrate a remarkable ability to adapt to changes in agriculture. Particularly illuminating are fungi within a group whose species produce host-specific toxins (HSTs) as crucial factors in causing disease. There is compelling evidence supporting the hypothesis that these fungi, originally opportunistic saprophytes, evolved from low-grade pathogens through the acquisition of a single gene responsible for HST production (Scheffer, 1991). Consequently, these fungi became virulent and host-specialized. The best-known model cases belong to the genera *Cochliobolus* and *Alternaria*. The first striking example of the significance of single gene changes was that of *Victoria* blight of oats in North America (Scheffer, 1991). The causal agent, *C. victoriae,* apparently evolved from *C. sativus*. The single gene in *C. victoriae* that controls pathogenicity and toxin production made possible a plant disease epidemic of continental proportions. Of course, a prerequisite for a destructive epidemic was the extensive planting of a crop with genetically uniform most of the oat crops in the mid-USA contained a single gene (Vb), making the plant susceptible to *C. victoriae*, sensitive to its toxin. Crops with genetic homogeneity, such as self-fertilized crops or vegetatively propagated ones, are especially vulnerable. The ecological consequences of HST production by a fungus are exemplified by the leaf blight disease of maize caused by *C. heterostrophus* race T. The older form of the fungus (race O) is probably the parental type (Scheffer, 1989). Race T is believed to have resulted from a mutation or some other genetic rearrangement. The ecological consequences became apparent when the maize crop in North America became genetically uniform for Tms cytoplasm (maternally inherited) and when weather conditions favored disease spread in 1970. Tms maize contains a receptor protein in the mitochondrion that binds HMT tom (Braun et al., 1990), leading to disruption of the respiratory process; maize lacking Tms cytoplasm lacks this protein and is insensitive to the toxin. Genetic uniformity of the maize crop resulted in selection and population explosion of a fungal strain (race T) that could exploit the situation. Thus, a minor, geographical restricted fungus was transformed into a virulent pathogen that invaded the total range of maize in North America.

The ability of *Periconia circinata* to produce toxins provides a clear advantage to the species in regions where susceptible sorghum genotypes are cultivated. Odvody and Dunkle (1984) analyzed strains of *P. circinata* isolated from soil and from roots of susceptible and resistant genotypes to determine the proportion of TOX+ and TOX– phenotypes and to investigate their pathogenic potential on several sorghum lines and hybrids.

Their findings revealed that the proportion of TOX+ strains isolated from the roots of susceptible plants was two to three times higher than those found in the soil (34% of isolates from roots compared to 13% from soil). This suggests that continuous cultivation of susceptible genotypes would lead to an enrichment of the population with toxin-producing strains. However, regardless of toxin production, none of the isolates affected other types of sorghum.

In contrast to pathogens like *Cochliobolus* and *Alternaria*, which develop new races or pathotypes through the acquisition of new genetic information leading to toxin production, *P. circinata* does not exhibit such diversity. TOX+ strains of *P. circinata* are likely to function as weak pathogens until inoculum levels reach damaging thresholds during monoculture of milo sorghum. It is suggested that *P. circinata* primarily behaves as a saprophyte that becomes an opportunistic pathogen, relying heavily on its host-selective toxins to stress potential hosts and overcome or bypass resistance mechanisms. This is unlike the majority of toxin-producing fungi, which are typically pathogens that can also act as opportunistic saprophytes.

P. circinata, causing sorghum grain blight, is an example of evolution with reference to seasonal incidence of disease (Scheffer, 1991). Sorghum is a high-temperature plant, growing well at 35°C. When plants are held at this temperature for several hours, they become insensitive to the toxin, perhaps because a protein receptor is deranged. When the temperature is lowered to 20°C, the plants regain sensitivity in two to three days, perhaps because new receptors are synthesized (Bronson and Scheffer, 1977).

This heat effect on toxin sensitivity fits with disease observations in the field and greenhouse. The ecological point of significance is that the seasonal incidence of several plant diseases is determined by heat-induced tolerance of plant tissues to toxins. The parallel between toxin sensitivity and disease incidence is further proof of the key roles of selective toxins in these diseases.

12.11 HST IN BREEDING AND SELECTION OF RESISTANT PLANT

Conventional plant breeding for resistance to pathogens has been successful in many cases. However, especially in vegetatively propagated crop plants, it remains slow and cumbersome, often struggling to keep pace with pathogen adaptation. This is compounded by a lack of genetic variability in cultivated varieties. Toxins, because of their role in disease development, have been proposed as convenient markers for early screening of resistant genotypes and as selective agents for *in vitro* selection (Upadhyay et al., 1990; Stierle et al., 1991; Novak et al., 1993). Until now, toxins and crude extracts have been mainly used as tools for early screening of segregating populations within classical as well as mutation breeding programs, or for direct *in vitro* selection of tolerant cells and subsequent regeneration of putative resistant plants. Toxin sensitivity tests may indeed be of real value to the breeder, as suggested by Yoder (1981). However, their efficacy depends on using the appropriate test conditions, which may include assessing root growth, leaf necrosis and chlorosis, protoplast and cell survival, as well as evaluating cotyledons and calli. Ensuring the use of the most effective differentiating conditions is crucial for obtaining reliable results.

A major drawback in selecting plants for resistance is the lack of genetic variability for this character in cultivated crops. Plant cells and tissue cultures are known to be highly variable and show a somewhat different mutation spectrum from that observed *in vivo* (Buiatti, 1989) and therefore have been suggested as a potentially useful source of mutants resistant to pathogens. Selection, in plant species where a reliable regeneration technique has been developed may be carried out at the level of the whole plant using all the known screening techniques. However, a more efficient approach would be to screen for novel mutants *in vitro*, before plant regeneration. Indeed, the use of toxin sensitivity tests can enable a scale of selection that would be challenging to achieve through other means. This approach can lead to significant savings in terms of space, time, and money. For instance, it has been estimated that a breeder working with seedlings can only screen around 100,000 individuals for resistance each year, whereas protoplasts can be screened *in vitro* in a single experiment (MacDonald and Ingram, 1986). Therefore, there are compelling reasons to develop efficient *in vitro* screening procedures (Buiatti and Ingram, 1991; Novak et al., 1993).

Gengenbach et al. (1977) and Brettell et al. (1980) used selection of cell cultures to select maize cell variants tolerant to HMT-toxin. Such variants were obtained from susceptible Tms corn lines and have been shown to maintain a high level of tolerance to HMT-toxin even after several months of culture on toxin-tree medium. Furthermore, the tolerance expressed by cell cultures was correlated with a tolerance of isolated mitochondria of HMT-toxin. Most significantly, plants regenerated from tolerant Tms cultures were fully resistant. The character was transmitted to the progeny through the female parent and appeared to be located on the mitochondrial genome (Brettell et al., 1980; Dixon, 1986; Gengenbach et al., 1977). Rines and Luke (1985) regenerated oat plants on media containing victorin, the toxin of *C. victoriae*, from calluses of the genotype VB/vb susceptible to the pathogen. Nine out of the eleven regenerated plants were shown to be resistant to the pathogen, and the resistance was found to be heritable. Likewise, banana cultivars have been successfully screened for resistance using tetralone and fijiensin produced by *Mycosphaerella fijiensis* var. *difformis*, causing black Sigatoka disease of banana and plantains (Novak et al., 1993; Stierle et al., 1991; Upadhyay et al., 1990). The utilization of toxins, such as fijiensin and tetralone, during mutation breeding, especially in the phases of callus growth and regeneration, could enhance the development of resistant or tolerant cultivars. *In vitro* selection methods demonstrate synchrony regarding resistance or susceptibility traits observed in the field. This approach offers a promising strategy for efficiently

TABLE 12.1
Host-Specific Toxins, Their Sources, Range of Activities, Genetics, and Target Sites

Disease	Pathogen Species/ Pathotype (Synonym)	Toxin (Synonym)	Host Range (Susceptible cultivar)	Genetics (Dominance)	Target Site
Alternaria blotch of apple	*Alternaria alternata* apple pathotype (*A. mali*)	AM-toxin I, II, and III	Apple (Red Gold, Starking)	Multiple genes (susceptible)	Chloroplast and plasma membrane
Alternaria leaf spot of pigeon pea	*A. tenuissima*	ATC-toxin	Pigeon pea	-	-
Alternaria stem canker of tomato	*A. alternata*/tomato pathotype (*A. alternata* f. sp. *lycopersicai*)	AL (or AAL)-toxin (I and II or Ta and Tb	Tomato (Earlypak 7, First)	Single (homo) gene (susceptible)	Mitochondrion ACTase
Black leaf spot of strawberry	*A. alternata*/strawberry pathotype	AF-toxin I, II, and III	Strawberry (Morioka-16)	Single (hetero) gene (susceptible)	Plasma membrane
Black spot of Japanese pear	*A. alternata*/Japanese pear pathotype (*A. kikuchiana*)	AK-toxin I and II	Japanese pear (Nijisseiki)	Single (hetero) gene (susceptible)	Plasma membrane
Brown spot of rough lemon	*A. alternata*/rough lemon pathotype (*A. citri*)	ACR(L)-toxin I	Rough lemon	-	Mitochondria
Brown spot of tangerine	*A. alternata*/tangerine pathotype (*A. citri*)	ACT-toxin I and II; ACTG-toxin A and B	Tangerine (Dancy)	-	Plasma membrane
Brown spot of tobacco	*A. alternata*/tobacco pathotype (*A. longipes*)	AT-toxin	Tobacco	-	Mitochondria
Leaf spot/necrosis of knapweed	*A. alternata* knapweed pathotype	Maculosin	Knapweed	-	-
Blackspot of rapeseed	*A. brassicae*	Destruxin B	Rapeseed	-	Cell wall/plasma membrane
Eyespot of sugarcane	*Bipolaris sacchari* (*Helminthosporium*)	HS-toxin A, B, and C	Sugarcane (51-NG97)	-	Plasma membrane
Milo disease of sorghum	*Periconia circinata*	PC-toxin (Peritoxin A and B)	Grain sorghum (giant milo)	Single gene (susceptible)	Plasma membrane
Northern leaf spot of maize	*Cochliobolus carbonum* (*B. carbonum*) race 1	HC-toxin I, II, and III	Maize (K-44, K-61)	Multiple genes (resistant)	Plasma membrane
Southern leaf spot of maize	*C. heterostrophus* (*B. maydis*) race T	HMT-toxin band 1, 2, 3 and 1'	Maize (Tms cytoplasm)	Cytoplasmic	Mitochondria
Target leaf spot of tomato	*Corynespora cassiicola*	CC-toxin	Tomato (Ife-1)	-	-
Victoria blight of oats	*Cochliobolus victoriae* (*B. victoriae*)	HV-toxin (victorin C)	Oats (Victoria)	Single (hetero) gene (susceptible)	Plasma membrane
Yellow leaf blight of maize	*Phyllosticta maydis*	PM-toxin A, B, C, and D	Maize (Tms cytoplasm)	Cytoplasmic	Mitochondrion

developing crop varieties with enhanced resistance to pathogens or tolerance to environmental stresses.

12.12 CONCLUSION

Certain fungal groups utilize host-specific toxins (HSTs) as crucial factors in host recognition and pathogenicity during spore germination on the plant surface. These toxins, once released, specifically bind to putative receptors on the host cells, initiating signaling mechanisms that result in pleiotropic effects on the cells. Among these effects, some of the most significant ones include the suppression or negation of the host defense system. This enables the fungal pathogens to effectively establish infection and exploit host resources for their own growth and reproduction. The pathogenicity/specificity in the pathogen and susceptibility in the plant have been reported to be controlled by a single gene each or gene clusters as single Mendelian loci. Despite involvement of HSTs in many diseases, it cannot be generalized that all or even the majority of diseases may prove to involve HSTs or HST-like molecules, including elicitors. The molecular mechanism of genes controlling physiological and biochemical reactions leading to pathogenicity/specificity and resistance/susceptibility is yet to be investigated. Recombinant DNA technology may help in developing transgenic cultivars lacking toxin-binding material, thereby evolving resistant germplasm. The sole pathway to achieve a breakthrough in comprehending host–pathogen interactions, particularly those involving host-specific toxins (HSTs), alongside related biotechnological advancements, is through the molecular characterization of genes crucial to this interaction.

REFERENCES

Braun, C.J., J.N. Siedow, and C.S. Levings III. 1990. Fungal toxins bind to the URF_{13} protein in maize mitochondria and *Escherichia coli*. *The Plant Cell* 2: 153–61. https://doi.org/10.1105/tpc.2.2.153.

Brettell, R.I.S., E. Thomas, and D.S. Ingram. 1980. Reversion of Texas male sterile cytoplasm maize in culture to give fertile, T-toxin resistant plants. *Theoretical and Applied Genetics* 57: 55–58. https://doi.org/10.1007/BF00277764.

Bronson, C.R. 1991. The genetics of phytotoxin production by plant pathogenic fungi. *Experientia* 47: 771–76. https://doi.org/10.1007/BF01922456.

Bronson, C.R., and R.P. Scheffer. 1977. Heat and aging induced tolerance of sorghum and oat tissue to host selective toxins. *Phytopathology* 67: 1232–38. https://www.apsnet.org/publications/phytopathology/backissues/Documents/1977Articles/Phyto67n10_1232.pdf.

Buiatti, M. 1989. Use of cell and tissue cultures for mutation breeding. *Vorträge für Pflanzenzüchtung* 16: 179–200. https://www.cabidigitallibrary.org/doi/full/10.5555/19901612464.

Buiatti, M., and D.S. Ingram. 1991. Phytotoxins as tools in breeding and selection of disease resistant plants. *Experientia* 47: 811–19. https://doi.org/10.1007/BF01922461.

Clouse, S.D., and D.G. Gilchrist. 1987. Interaction of the asc locus in F8 paired lines of tomato with *Altemaria altemata* f. sp. *lycopersici* and AAL-toxin. *Phytopathology* 77: 80–82. https://www.apsnet.org/publications/phytopathology/backissues/Documents/1987Articles/Phyto77n01_80.PDF.

Daly, J.M., and B.J. Deverall. 1983. *Toxins and plant pathogenesis.* Australia: Academic Press, p. 181.

Danko, S.J., Y. Kono, J.M. Daly, Y. Suzuki, S. Takeuchi, and D.A. McCrery. 1984. Structure and biological activity of a host-specific toxin produced by the fungal corn pathogen *Phyllosticta maydis. Biochemistry* 23: 759–66. https://doi.org/10.1021/bi00299a027.

Dewey, R.E., J.N. Siedow, D.H. Timothy, and C.S. Levings III. 1988. A 13-kilo dalton maize mitochondrial protein in *E. coli* confers sensitivity to *Bipolaris maydis* toxin. *Science* 239: 293–95. https://www.science.org/doi/10.1126/science.3276005.

Dixon, R.A. 1986. The phytoalexin response: Elicitation, signalling and control of host gene expression. *Biological Reviews* 61: 239–91. https://doi.org/10.1111/j.1469-185X.1986.tb00719.x.

Dunkle, L.D., and V. Macko. 1995. Peritoxins and their effects on Sorghum. *Canadian Journal of Botany* 73: S444–52. https://doi.org/10.1139/b95-281.

Durbin, R.D. 1983. The biochemistry of fungal and bacterial toxins and their modes of action. In *Biochemical plant pathology*, ed: J.A. Callow, 137–62. New York, Chichester: John Wiley & Sons.

Gardener, J.M., Y. Kono, J.H. Tatum, Y. Suzuki, and S. Takeuchi. 1985. Plant pathotoxins from *Alternaria citri*: The major toxin specific for rough lemon plants. *Phytochemistry* 24: 2861–67. https://doi.org/10.1016/0031-9422(85)80015-7.

Gardener, J.M., and R.P. Scheffer. 1973. Effects of cycloheximide and sulfhydryl-binding compounds on sensitivity of oat tissues to *Helminthosporium victoriae* toxin. *Physiological Plant Pathology* 3: 147–57. https://doi.org/10.1016/0048-4059(73)90031-3.

Gengenbach, B.G., C.E. Green, and C.M. Donovan. 1977. Inheritance of selected pathotoxin resistance in maize plants regenerated from cell cultures. *Proceedings of the National Academy of Sciences (PNAS)* 74: 5113–17. https://doi.org/10.1073/pnas.74.11.5113.

Gilchrist, D.G., R.M. Bostock, and H. Wang. 1996. Sphingosine-related mycotoxins in plant and animal diseases. *Canadian Journal of Botany* 73: 459–67. https://doi.org/10.1139/b95-283.

Gilchrist, D.G., S.D. Clouse, B.L. McFarland, and A.N. Martensen. 1985. Phytotoxins as molecular determinants of pathogenicity and virulence. In *Molecular genetics of filamentous fungi,* ed. W.E. Timberlake, 405–20. New York: Alan A. Liss Inc.

Gupta, A., S. Jain, and N. Verma. 2022. *Industrial applications of soil microbes*, vol. 1. Bentham Science Publishers Pte Ltd. https://doi.org/10.2174/97898150399551220101.

Gupta, A., S. Jain, and N. Verma. 2024. *Industrial applications of soil microbes*, vol. 3. Bentham Science Publishers Pte Ltd. https://doi.org/10.2174/97898150797531240301.

Itch, Y., K. Kohmoto, H. Otani, et al. 1989. Isolation and biological activities of host specific toxins, ACT-toxin produced by the tangerine pathotype of *Alternaria alternata. Annals of the Phytopathological Society of Japan* 55: 482.

Jain, S., A. Gupta, and N. Verma. 2023. *Industrial applications of soil microbes*, vol. 2. Bentham Science Publishers Pte Ltd. https://doi.org/10.2174/97898150502641230201.

Jain, S., A. Gupta, and N. Verma. 2024. *Industrial applications of soil microbes*, vol. 4. Bentham Science Publishers Pte Ltd. https://doi.org/10.2174/97898151249961240401.

Johal, G.S., and S.P. Briggs. 1992. Reductase activity encoded by the *HM1* disease resistance gene in maize. *Science* 258: 985–87. https://www.science.org/doi/10.1126/science.1359642.

Kohmoto, K., and H. Otani. 1991. Host recognition by toxigenic plant pathogens. *Experientia* 47: 755–64. https://doi.org/10.1007/BF01922454.

Kono, Y. 1989. Structural studies on host specific pathotoxins in corn blight and citrus brown spot diseases. In *Phytotoxins and plant pathogenesis*, eds. A. Graniti, A.O. Durbin, and A. Ballio, 7–21. Berlin, Heidelberg: Springer. https://doi.org/10.1007/978-3-642-73178-5_2.

Kono, Y., J.M. Gardner, Y. Suzuki, and S. Takeuchi. 1986. Studies on host selective toxins produced by pathotype of *Alternaria citri* causing brown spot disease of mandarins. *Agricultural and Biological Chemistry* 50: 1597–606. https://doi.org/10.1080/00021369.1986.10867585.

Kono, Y., T. Kinoshita, S. Takeuchi, and J. Daly. 1986a. Structure of HV-toxin M, a HST-related compound produced by *Helminthosporium victoriae. Agricultural and Biological Chemistry* 50: 2689–91. https://api.semanticscholar.org/CorpusID:83995546.

Kono, Y., and Y. Suzuki. 1988. Host-specific toxins. In: *Bioactive natural products,* ed. S. Shibata, 73. Japan: Ishiyaku-Shuppan.

Kono, Y., Y. Suzuki, S. Takeuchi, H.W. Knoche, and J.M. Daly. 1985. Studies on the host-specific pathotoxins produced by *H. maydis*, race T and *P. maydis*: Absolute configuration of PM-toxins and HMT-toxins. *Agricultural and Biological Chemistry* 49: 559–62. https://doi.org/10.1080/00021369.1985.10866765.

Leukel, R.W., and A.G. Johnson. 1948. *Periconia circinata*, the cause of milo disease. *Science* 107: 93–94. https://www.science.org/doi/10.1126/science.107.2769.93.

MacDonald, M.V., and D.S. Ingram. 1986. Towards the selection *in vitro* for resistance to *Alternaria brassicicola* (SCHW.) wilts., in *Brassica napus* ssp. *oleifera* (METZG.) Sinsk., winter oilseed rape. *New Phytologist* 104: 621–29. https://doi.org/10.1111/j.1469-8137.1986.tb00662.x.

Macko, V., T.J. Wolpert, W. Acklin, et al. 1985. Characterization of victorin C, the major host-selective toxin from *Cochliobolus victoriae*: Structure of degradation products. *Experientia* 41: 1366–70. https://doi.org/10.1007/BF01949993.

Maekawa, N., M. Yamamoto, S. Nishimura, K. Kohmoto, M. Kuwada, and Y. Watanabe. 1984. Studies on host-specific AF-toxins produced by *Alternaria alternata* strawberry pathotype causing *Alternaria* black spot of strawberry. *Japanese Journal of Phytopathology* 50: 600–09. https://doi.org/10.3186/jjphytopath.50.600.

Meehan, F., and H.C. Murphy. 1946. A new *Helminthosporium* blight of oat. *Science* 104: 413–14. https://www.science.org/doi/10.1126/science.104.2705.413.

Meeley, R.B., G.S. Johal, S.P. Briggs, and J.D. Walton. 1992. A biochemical phenotype for a disease resistance gene of maize. *The Plant Cell* 4: 71– https://doi.org/10.1105/tpc.4.1.71.

Nakashima, T., T. Ueno, and H. Fukami. 1982. Structure elucidation of AK-toxins, host-specific phytotoxic metabolites produced by *Alternaria kikuchiana* Tanaka. *Tetrahedron Letters* 23: 4469–72. https://doi.org/10.1016/S0040-4039(00)85630–0.

Nakatsuka, S., T. Goto, H. Itoh, et al. 1989. Chemical studies on structure and host specificities of ACT-toxins produced by *Altemaria altemata* tangerine pathotype causing citrus brown spot disease. Papers of 32nd Symposium on the chemistry of Natural Products, Nagoya, 671–76.

Nakatsuka, S., T. Goto, K. Kohmoto, and S. Nishimura. 1986. Host specific phytotoxins. In *Natural products and biological activities*, eds. H. lmura, T. Gote, T. Murachi, and T. Nakajima, 11–18. Tokyo: University of Tokyo Press.

Namiki, F., M. Yamamoto, S. Nishimura, et al. 1986. Studies on host specific AF-toxins produced by *Alternaria alternata* strawberry pathotype causing *Alternaria* black spot of strawberry. (4) Protective effect of AF-toxin II on AF-toxin I-induced toxic action and fungal infection. *Phytopathological Society of Japan* 52: 428–36. https://web.archive.org/web/20181102152200id_/https://www.jstage.jst.go.jp/article/jjphytopath1918/52/3/52_3_428/_pdf.

Navarre, D.A., J.M. Lorang, D.L. Moore, and T.J. Wolpert. 1995. Molecular interactions of victorin and oats. *Canadian Journal of Botany* 73: S475–82. https://doi.org/10.1139/b95-285.

Nishimura, S., and K. Kohmoto. 1983. Host-specific toxins and chemical structures from *Alternaria* species. *Annual Review of Phytopathology* 21: 87–116. https://doi.org/10.1146/annurev.py.21.090183.000511.

Nishimura, S., and S. Nakatsuka. 1989. Trends in host selective toxin research in Japan. In *Host-specific toxins: Recognition and specificity factors in plant disease*, eds. K. Kohmoto and R.D. Durbin, 19–31. Tottori: Tottori University Press.

Nishimura, S., C.P. Vance, and N. Doke. 1987. *Molecular determinants of plant diseases*. Tokyo: Japan Scientific Societies Press.

Novak, F.J., H. Brunner, R. Afza, et al. 1993. Improvement of *Musa* through biotechnology and mutation breeding. In *Workshop on biotechnology applications for banana and plantain improvement, San José, Costa Rica*, 143–58. Montpellier, France: INIBAP. https://www.musalit.org/seeMore.php?id=2151.

Odvody, G.N., and L.D. Dunkle. 1984. Sorghum root rots, a critical review. In *Proceedings of the consultative group discussion on research needs and strategies for control of sorghum root and stalk rot diseases*, ed. L.K. Mughogho, 43–48. Bellagio, Italy: ICRISAT. https://oar.icrisat.org/804/1/RA_00066.pdf.

Okuno, T., Y. Ishita, K. Sawai, and T. Matsumoto. 1974. Characterization of alternariolide, a host specific toxin produced by *Alternaria mali* Roberts. *Chemistry Letters* 3: 635–38. https://doi.org/10.1246/cl.1974.635.

Otani, H., K. Kohmoto, and M. Kodama. 1995. *Alternaria* toxins and their effects on host plants. *Canadian Journal of Botany* 73: S453–58. https://doi.org/10.1139/b95-282

Panaccione, D.G., J.S. Scott-Craig, J.A. Pocard, and J.D. Walton. 1992. A cyclic peptide synthetase gene required for pathogenicity of the fungus *Cochliobolus carbonum* on maize. *Proceedings of the National Academy of Sciences (PNAS)* 89: 6590–94. https://doi.org/10.1073/pnas.89.14.6590.

Pringle, R.B., and R.P. Scheffer. 1964. Host-specific plant toxins. *Annual Review of Phytopathology* 2: 133–56. https://doi.org/10.1146/annurev.py.02.090164.001025.

Rines, H.W., and H.H. Luke. 1985. Selection and regeneration of toxin-insensitive plants from tissue cultures of oats (*Avena sativa*) susceptible to *Helminthosporium victoriae*. *Theoretical and Applied Genetics* 71: 16–21. https://doi.org/10.1007/BF00278247.

Scheffer, R.P. 1983. Toxins as chemical determinants of plant disease. In *Toxin and plant pathogenesis*, eds. J.M. Daly and B.J. Deverall, 1–40. Sydney: Academic Press.

Scheffer, R.P. 1989. Ecological consequences of toxin production by *Cochliobolus* and related fungi. In *Phytotoxins and plant pathogenesis*, eds. A. Graniti, R.D. Durbin, and A. Ballio, 285–300. Heidelberg, Berlin: Springer. https://doi.org/10.1007/978-3-642-73178-5_22.

Scheffer, R.P. 1991. Role of toxins in evolution and ecology of plant pathogenic fungi. *Experientia* 47: 804–11. https://doi.org/10.1007/BF01922460.

Scheffer, R.P., and S.P. Briggs. 1981. Introduction: A perspective of toxin studies in plant pathology. In *Toxin in plant diseases*, ed. R.D. Durbin, 1–20. New York: Academic Press.

Scheffer, R.P., and R.S. Livingston. 1984. Host-Selective toxins and their role in plant diseases. *Science* 223: 17–21. https://www.science.org/doi/10.1126/science.223.4631.17.

Scheffer, R.P., R.R. Nelson, and A.J. Ullstrup. 1967. Inheritance of toxin production and pathogenicity in *Cochliobolus carbonum* and *Cochliobolus victoriae*. *Phytopathology* 57: 1288–91.

Smith, D.J., M.K. Burnham, J.H. Bull, et al. 1990. Beta-lactam antibiotic biosynthetic genes have been conserved in clusters in prokaryotes and eukaryotes. *The EMBO Journal* 9: 741–47. https://doi.org/10.1002/j.1460-2075.1990.tb08168.x.

Stierele, A.C., J.H. Cardellina II, and G.A. Strobel. 1989. Maculosin, a host-specific phytotoxin for spotted knapweed from *Alternaria alternata*. *Proceedings of the National Academy of Sciences (PNAS)* 85: 8008–11. https://doi.org/10.1073/pnas.85.21.8008.

Stierle, A.A., R. Upadhyay, J. Hershenhorn, G.A. Strobel, and G. Molina. 1991. The phytotoxins of *Mycosphaerella fijiensis*, the causative agent of black sigatoka disease of bananas and plantains. *Experientia* 47: 853–59. https://doi.org/10.1007/BF01922472.

Tomas, A., G. Feng, G. Reeck, W. Bockus, and J. Leach. 1990. Purification of a cultivar-specific toxin from *Pyrenophora tritici-repentis*, causal agent of tan spot of wheat. *Molecular Plant-microbe Interactions* 3: 221–24. https://www.apsnet.org/publications/mpmi/backissues/Documents/1990Articles/Microbe03_221.pdf.

Upadhyay, R.K., G.A. Strobel, S.J. Coval, and J. Clardy. 1990. Fijiensin, the first phytotoxin from *Mycosphaerella fijiensis*, the causative agent of black sigatoka disease. *Experientia* 46: 982–84. https://doi.org/10.1007/BF01939396.

Walton, J.D., and D.G. Panaccione. 1993. Host-selective toxins and disease specificity: Perspectives and progress.

Annual Review of Phytopathology 31: 275–03. https://doi.org/10.1146/annurev.py.31.090193.001423.

Wolpert, T.J., and V. Macko. 1989. Specific binding of victorin to a 100-kDa protein from oats. *Proceedings of the National Academy of Sciences (PNAS)* 86: 4092–96. https://doi.org/10.1073/pnas.86.11.4092.

Wolpert, T.J., V. Macko, W. Acklin, and D. Arigoni. 1988. Molecular features affecting the biological activity of the host-selective toxins from *Cochliobolus victoriae*. *Plant Physiology* 88: 37–41. https://doi.org/10.1104/pp.88.1.37.

Wolpert, T.J., V. Macko, W. Acklin, et al. 1985. Structure of victorin C, the major host-selective toxin from *Cochliobolus victoriae*. *Experientia* 41: 1524–29. https://doi.org/10.1007/BF01964789.

Wolpert, T.J., D.A. Navarre, D.L. Moore, and V. Macko. 1994. Identification of the 100-kD victorin binding protein from oats. *The Plant Cell* 6: 1145–55. https://www.ncbi.nlm.nih.gov/pmc/articles/PMC160508/pdf/061145.pdf.

Yoder, O.C. 1973. A selective toxin produced by *Phyllosticta maydis*. *Phytopathology* 63: 1361–66. https://www.apsnet.org/publications/phytopathology/backissues/Documents/1973Articles/Phyto63n11_1361.PDF.

Yoder, O.C. 1981. Assay. In *Toxins in plant disease*, ed. R.D. Durbin, 45–78. Academic Press. https://doi.org/10.1016/B978-0-12-225050-7.50008-6.

Yoder, O.C., and B.G. Turgeon. 1985. Molecular bases of fungal pathogenicity to plants. In *Gene manipulation in fungi*, eds. J.W. Bennett and L.L. Lasure, 417–48. Academic Press.

Yoder, O.C., B.G. Turgeon, L.M. Ciuffetti, and W. Schafer. 1989. Genetic analysis of toxin production by fungi. In *Phytotoxins and plant pathogenesis*, eds. A. Graniti, R.D. Durbin, and A. Ballio, 43–60. Heidelberg, Berlin: Springer. https://doi.org/10.1007/978-3-642-73178-5_5

13 Mycorrhizal Secondary Metabolites
An Overview

Avinash Kumar Chaurasia
(School of Biotechnology, Institute of Science, BHU, Varanasi 221005, India; avinash11vns@gmail.com)

Ritika Kumari Singh
(School of Biotechnology, Institute of Science, BHU, Varanasi 221005, India; ritika_singh123@yahoo.co.in)

Riden Saxena
(Department of Obstetrics and Gynecology, Institute of Medical Sciences, BHU, Varanasi 221005, India; Department of Microbiology, Vivek University, Bijnor, 246701, India; ridensaxena@gmail.com)

Arvind Kumar
(School of Biotechnology, Institute of Science, BHU, Varanasi 221005, India; arvindk@bhu.ac.in)

**Kriti Shrinet*
(School of Biotechnology, Institute of Science, BHU, Varanasi 221005, India; School of Biotechnology, IFTM University, Moradabad 244102, India; kritishrinet@gmail.com)

13.1 INTRODUCTION

Throughout history, medicinal plants have played a significant role in providing therapeutic substances for the treatment of a wide range of ailments and diseases, including but not limited to diarrhea, fever, colds, and malaria. Mycorrhiza refers to a mutualistic relationship established between specific fungi and the root systems of plants (Dambisya and Tindimwebwa, 2003). In contemporary times, these organisms additionally serve as a valuable resource for the advancement of novel pharmaceuticals aimed at treating significant ailments, including cancer (Beik et al., 2020). The therapeutic efficacy of these substances is frequently attributed to the abundance and diversity of bioactive compounds found in the secondary metabolic pathways, including flavonoids, alkaloids, phenolics, and terpenoids (Hussein and El-Anssary, 2019). Currently, a significant proportion, approximately 80%, of individuals residing in developing nations rely exclusively on herbal medications as their primary source of health care (Ekor, 2014). Furthermore, more than a quarter, specifically 25%, of pharmaceutical drugs prescribed in developed countries have been sourced from medicinal plants obtained through rough collection (Hamilton, 2004). Various techniques, including the extraction of compounds from medicinal plants and other synthetic chemistry, combinatorial chemistry, natural resources, and molecular modeling, have been employed in the process of drug discovery (Lombardino and Lowe III, 2004). Nevertheless, it is worth noting that natural products, specifically medicinal plants, continue to hold significant value as a primary reservoir for novel pharmaceuticals, potential drug candidates, and novel chemical entities (Butler, 2004). This is mostly because they are more culturally acceptable, work better with the human body, and can adapt to it better than synthetic chemicals (Garg et al., 2021). Based on the findings of the International Union for Conservation of Nature and the World Wildlife Fund (Chen et al., 2016), it has been determined that approximately 80,000 species of flowering plants are utilized for their medicinal properties.

Plants in natural habitats establish relationships with a wide range of beneficial microorganisms, including endophytic and symbiotic bacteria and fungi (Jain et al., 2023, 2024). These microorganisms play a crucial part in promoting herbal plant well-being, growth, and production, as well as encouraging the synthesis of metabolites (de Vries et al., 2020). The arbuscular mycorrhizal fungus (AMF) is a group of soil microorganisms that establish mutualistic interactions with more than 70% of vascular floras (Brundrett and Tedersoo, 2018). Arbuscular mycorrhizas

DOI: 10.1201/9781003408833-13

display a characteristic trait whereby arbuscules, intricately branching structures, develop in the root's cortical cells of the host plants (Coleman et al., 2004). The arbuscules function as sites for bidirectional transport, enabling the transfer of nutrients from the fungal cell to the host plant cell, and carbon-based substances are carried in the reverse manner. The formation of AMF symbiosis requires a reciprocal recognition process between the two interacting entities. Lipo-chitooligosaccharides, referred to as Myc factors, are recognized by the plant in response to signaling molecules, notably strigolactones, that are excreted by the roots (Akiyama and Hayashi, 2006). After the establishment of mutual recognition, AMF hyphae commence the development of a hyphopodium on the outermost layer of the root, subsequently establishing colonization in the cortical portion of the root. Concurrently, fungal hyphae proliferate into the surrounding soil, forming an extensive extraradical mycelium, which accounts for a varying proportion of 9 to 55% of the total microbial biomass present in the soil (Olsson et al., 1999). The presence of a concentrated extraradical mycelium significantly enhances the capability of plant roots to access water and essential inorganic mineral nutrients such as calcium, nitrogen, sulfur, zinc, phosphorus, potassium, and copper. This often leads to an increase in plant biomass and improved crop quality (Noceto et al., 2021). Furthermore, the extra radical mycelium has been observed to alter the composition of the soil matrix (Chen et al., 2018). Consequently, this process leads to an improvement in the overall soil quality and fertility (Thirkell et al., 2017). The capacity of arbuscular mycorrhizal fungi (AMF) to augment plant resistance or tolerance to diverse stressors, such as drought, salt, nutrient deficiency, heavy metals, severe temperatures, pests, and diseases, has garnered significant recognition (Abdel-Salam et al., 2018). Furthermore, it has been observed that these benefits have the potential to exert both quantitative and qualitative influences and promote the synthesis of bioactive secondary metabolites in the host (Kaur and Suseela, 2020).

While primary metabolites are found ubiquitously across the plant domain, secondary metabolites exhibit a narrower dispersal and serve specific functions. Therefore, although they may not be indispensable for sustenance, they can play an important role in the survival and reproductive efficacy of plants (Santos, 2004). Secondary metabolites can be classified into three main categories: terpenes, phenolic compounds, and nitro-generated chemicals, based on their chemical composition (Taiz and Zeiger, 2002; Gupta et al., 2022, 2024).

Terpenes are typically characterized by their insolubility in water and are synthesized through the utilization of acetyl-CoA or intermediates derived from glycolysis. This course encompasses the study of plant hormones, carotenoids, essential oils, and triterpene saponins. These compounds collectively contribute to various biological processes such as photosynthesis, development, and growth. Plant essential oils, known for their insect-repellent properties, also play a crucial role in pollination. Additionally, they possess significant anti-inflammatory, antibacterial, analgesic, antifungal, and antioxidant properties. Triterpene saponins, on the other hand, exhibit detergent, emulsifying, foaming, antioxidant, and antimicrobial activities. Phenolic compounds encompass a variety of chemical substances, such as flavonoids, phenols, tannins, and lignans. These compounds are chiefly created and produced by the shikimic acid pathway, utilizing glycolysis forerunners such as phosphoenolpyruvate (PEP), as well as the pentose phosphate pathway (PPP), involving erythrose-4-phosphate. The flavonoids demonstrate the presence of a secondary aromatic ring, which is formed through the reaction of p-coumaroyl-CoA with three malonyl-CoA molecules obtained from the malonic acid route (Vermerris and Nicholson, 2008). Phenols are used as defenses against pests and pathogens, as pollinator attractants or agents for seed dispersal, as UV protection, and as signaling molecules in many interactions between plants and microorganisms (Steinkellner et al., 2007). Certain compounds in question are also of significance to human health due to their antibacterial, antioxidant, antiproliferative, antiviral, and anti-inflammatory attributes (Martens and Mithöfer, 2005).

The synthesis of nitrogenated compounds involves the utilization of amino acids, such as tryptophan, tyrosine, and lysine. This group of complexes involves alkaloids and cyanogenic glycosides. These chemical compounds exhibit protective characteristics against herbivores and feature medical attributes that are utilized for their pharmacological, stimulant, and sedative activities (Taiz and Zeiger, 2002).

Secondary plant metabolites are a significant reservoir of therapeutic agents that have been utilized for an extended period, particularly in developing nations, where a substantial portion of the population relies on these agents to fulfill their fundamental health needs (Agra et al., 2007). The production of secondary metabolites in plants and their subsequent medicinal properties can be influenced by various factors, such as seasonal fluctuations, daily cycles, plant maturity and growth, nutrient availability, temperature, precipitation levels, ultraviolet radiation, atmospheric composition, altitude, carbon dioxide levels, mechanical stimuli, and pathogenic attacks (Gobbo-Neto and Lopes, 2007).

13.2 ARBUSCULAR MYCORRHIZAL SYMBIOTIC ASSOCIATION

The potential to alter the synthesis of secondary metabolites exists when arbuscular mycorrhizal fungi (AMF) are introduced to plants (Maier et al., 1997). AMF, which fall within the taxonomic classification of the phylum Glomeromycota (Schüßler et al., 2001), form a mutualistic symbiotic association with the root systems of a majority of terrestrial plant species. This relationship encompasses the mutual exchange of photosynthetic products, nutrients, and water (Berbara et al., 2006).

The host plant typically undergoes augmented development in terms of heightened shoot and root height, as well as increased dry mass, due to the colonization of AMF. It has been observed that the concentrations of micronutrients exhibit a rise (Chu et al., 2001), and in macronutrients, for example, phosphorus (Matsubara et al., 2009), due to AMF. This increase in nutrient levels has been associated with elevated levels of photosynthetic pigments in leaves (Ratti et al., 2010) and also promote better production of primary metabolites (Manoharan et al., 2010).

The phenomenon by which the fungus can provide advantages to the host plant is referred to as AMF efficiency (Janos, 2007). Numerous parameters can be employed to make an estimation. The parameters under consideration encompass various aspects of plant characteristics, including leaf count, stem diameter, biomass, foliar area, fruit weight, seed count, and enzymatic activity (Chu et al., 2001), as well as physiological parameters, such as levels of photosynthetic pigments (Manoharan et al., 2010). Furthermore, the process of synthesizing primary metabolites is of significant importance (Ratti et al., 2010).

Several studies have demonstrated a positive association involving the biomass of plants that are inhabited by arbuscular mycorrhizal fungus (AMF) and the levels of secondary metabolites found in diverse medicinal plants. The plant species encompassed in this list are *Glycyrrhiza uralensis*, *Chlorophytum borivilianum*, *Satureja macrostema*, *Gymnema sylvestre*, *Dioscorea* spp., *Libidibia ferrea*, *Ocimum basilicum*, and *Salvia miltiorrhiza* (Dave et al., 2011; Zimare et al., 2013; Zolfaghari et al., 2013; Silva et al., 2014; Carreón-Abud et al., 2015; Lu et al., 2015; Ayeka et al., 2016; Chen et al., 2017). One advantageous characteristic of AMF is their potential to enhance the absorption of nutrients by plants (Bowles et al., 2016), As a result, this would have a direct or indirect influence on the concentrations of secondary metabolites (Yamawaki et al., 2013). The results demonstrated that the inclusion of *F. mosseae* had a beneficial effect on the development of shoot and root biomass, as well as the overall structure of the root system, and the accumulation of flavonoids in *Glycyrrhiza uralensis* plants under phosphorus-deficient nutritional conditions (Chen et al., 2017).

13.3 EFFECT OF AMF ON PLANTS UNDER ABIOTIC AND BIOTIC STRESS CONDITIONS

The growth of plants can be adversely impacted by various circumstances, including but not limited to drought, salinity, heavy metal contamination, pest infestation, and disease susceptibility. These conditions have been demonstrated to decrease the biomass of plants (Abeer and Didamony, 2014). The generation of secondary metabolites can be influenced by various factors. According to Hashem et al. (2018), the possible impact of AMF on the ability of plants to endure abiotic and biotic stressors may result in the enhancement of secondary metabolite synthesis. It is widely postulated that the activation of many defense response pathways is accountable for the heightened concentrations of various secondary metabolite categories, such as flavonoids and phenolics, in plants that are inhabited by AMF (Zeng et al., 2013). Elevated levels of terpenoids in the carotenoid pathway, flavonoids, phenolic compounds, and some alkaloids such as hyoscyamine and scopolamine in the phenylpropanoid pathway have been detected in plants that have undergone colonization by AMF (Kaur and Suseela, 2020). The various pathways involved in the plant–AMF symbiosis have distinct functions, including but not limited to stress tolerance, signaling, nutrition acquisition, in addition to defense against both biotic and abiotic encounters. Several classes of secondary metabolites whose synthesis was stimulated by AMF are given in Table 13.1.

TABLE 13.1

Plants Associated with Mycorrhiza with the Bioactive Compounds and Their Medicinal Value

Plant Name	Fungi	Bioactive Compound	Medicinal Value	References
Allium sativum	*Rhizophagus fasciculatus*	Alliin	Prevents the growth of bacteria, fungi, viruses, parasites, and cancer cells; protects against atherosclerosis and hypertension; has antioxidant and anticancerogenic characteristics	Borde et al., 2009
Achyranthes aspera	*Rhizophagus fasciculatus*	Flavonoids	Asthma, hypertension, and diabetes treatment, as well as of bronchitis, rheumatism, malaria fever, diarrhea, and cough	Tejavathi and Jayashree, 2013
Myracrodruon urundeuva	*Glomus* spp.	Imperatorin and total coumarins	Common cold, sinusitis, hay fever, bronchitis, acne, eczema, boils, canker sores, carbuncles, and rheumatism treatment	Zhao and He, 2011
Angelica archangelica	*Funneliformis mosseae*	Monoterpenoid and coumarin	Loss of appetite, anorexia, bronchitis, moderate gastrointestinal spasms, slow digestion, and flatulence	Zitterl-Eglseer et al., 2015

TABLE 13.1 (*Continued*)
Plants Associated with Mycorrhiza with the Bioactive Compounds and Their Medicinal Value

Plant Name	Fungi	Bioactive Compound	Medicinal Value	References
Anethum graveolens	*Glomus macrocarpum* and *Rhizophagus fasciculatus*	Anethole	Gastrointestinal distress, colic, and improved digestion	Kapoor et al., 2002
Coriandrum sativum	*Glomus hoi*	β-caryophyllene, p-cymene, geraniol	Effective against free radicals, diabetes, liver damage, bacteria, and fungi	Rydlová et al., 2016
Foeniculum vulgare	*Glomus macrocarpum* and *Rhizophagus fasciculatus*	Essential oil	Beneficial for the respiratory, reproductive, endocrine, and digestive systems	Kapoor et al., 2004
Trachyspermum ammi	*Rhizophagus fasciculatus*	Thymol	Useful for its antibacterial, antifungal, antioxidant, and antiseptic properties, as well as its analgesic, choleretic, cardiotonic, bronchodilatory, emetic, lithotriptic, and diuretic effects	Kapoor et al., 2002
Catharanthus roseus	*Glomus* spp.	Gymnemic acid	Coughs, lung infections, hyperglycemia, diuretic/hemorrhagic/wound healing effects	Andrade et al., 2013
Gymnema sylvestre	*Rhizophagus fasciculatus* and *Funneliformis mosseae*	Ephedrine and guanosine	Management of type 2 diabetes	Zimare et al., 2013
Pinellia ternate	*Rhizophagus intraradices* and *Funneliformis mosseae*	Atractylol	Relief of nausea and coughing	Guo et al., 2010
Atractylodes macrocephala	*Funneliformis mosseae*, *Claroideoglomus etunicatum*, *Glomus tortuosum*, and *Funneliformis mosseae*	Artemisinin content	Spleen strengthening, increased vitality, no more dampness or hydroschesis, and a calmer pregnancy	Yan-qi and Xue-li, 2005
Artemisia annua	*Rhizophagus irregularis*	Sesquiterpene lactones	Relief of symptoms of inflammation, fever, malaria, cough, and gastrointestinal distress	Domokos et al., 2018
Arnica montana	*Funneliformis geosporum* and *Funneliformis constrictum*	Essential oil, E-β-ocimene	Improvement in circulation, speeding of wound healing, and alleviation of arthritic discomfort	Kai et al., 2011
Artemisia umbelliformis	*Planticonsortium tenue*, *Rhizophagus intraradices*, *Claroideoglomus claroideum*, and *C. etunicatum*	Phenolics	Stopping coughs	Binet et al., 2011
Baccharis trimera	*Rhizophagus clarus*	Phenolics	Gastric and hepatic-protector; antioxidant; fights microbes, fungi, parasites, and inflammation	Freitas et al., 2004
Cynara cardunculus	*Rhizophagus intraradices* and *Funneliformis mosseae*	Phenolics and cichoric acid	Limit the risk of cancer and cardiovascular disease	Ceccarelli et al., 2010
Echinacea purpurea	*Rhizophagus intraradices*	Flavonoids	Pain relief from toothache to diarrhea, snake bite to psoriasis, epilepsy to rheumatoid arthritis, and cancer	Araim et al., 2009
Eclipta alba	*Glomus aggregatum*, *Funneliformis mosseae*, and *Rhizophagus fasciculatus*	Scopolamine and quercetin	Diseases of the digestive tract, the respiratory system, the skin, the liver, the spleen, and cuts and wounds all treatable	Tejavathi and Jayashree, 2013

(*Continued*)

TABLE 13.1 (*Continued*)
Plants Associated with Mycorrhiza with the Bioactive Compounds and Their Medicinal Value

Plant Name	Fungi	Bioactive Compound	Medicinal Value	References
Eclipta prostrata	*Rhizophagus irregularis, Funneliformis mosseae, Claroideoglomus etunicatum, Claroideoglomus microaggregatm,* and *Funneliformis geosporum*	Thymol derivatives	Care for patients with liver illness, type 2 diabetes, vertigo, and bleeding disorders	Vo et al., 2019
Inula ensifolia	*Rhizophagus clarus*	Stevioside, rebaudioside-A	Inhibition of the growth of human cancer	Zubek et al., 2010
Stevia rebaudiana	*Rhizophagus fasciculatus*	Total phenols, flavonoids, alkaloids, tannins, and saponins	Medicine for the improvement of cardiac function, blood flow, and blood pressure	Mandal et al., 2013
Wedelia chinensis	*Rhizophagus fasciculatus*	Total phenols and tannins	Remedies for stings, bites, fever, infections, kidney failure, colds, wounds, and amenorrhea	Nisha and Rajeshkumar, 2010
Commiphora leptophloeos	*Gigaspora albida* and *Claroideoglomus etunicatum*	Flavonoids	Relief of symptoms of asthma, colds, cough, kidney issues, inflammation, and stomach pain	Lima et al., 2017
Solanum viarum	*Glomus aggregatum* and *Trichoderma harzianum*	Withaferin-A	Treatment for malignancy, Addison's disease, and rheumatoid arthritis	Hemashenpagam and Selvaraj, 2011
Withania somnifera	*Rhizophagus irregularis*	Caffeic acid	Cancer therapy	Johny et al., 2021
Viola tricolor	*Rhizophagus irregularis*	Curcumin	Care for a wide range of respiratory and skin conditions	Zubek et al., 2015
Curcuma longa	*Glomus, Gigaspora,* and *Acaulospora* sp.	Total phenols content	Anti-inflammatory, antitumor, and arachidonic acid metabolism inhibition, with antioxidant properties	Dutta and Neog, 2016
Passiflora alata	*Claroideoglomus etunicatum* and *Rhizophagus intraradices*	Berberine, jatrorrhizine, palmatine	Multiple illnesses treatments, including sleep disorders, anxiety, and hysteria	Riter Netto et al., 2014
Phellodendron amurense	*Funneliformis mosseae, Claroideoglomus etunicatum, Diversispora epigaea,* and *Glomus diaphanum*	Total phenolic acids	Jaundice, diarrhea, high blood pressure, inflammation, and other liver-related conditions all treatable	Ji-hong and Guo-ting, 2006
Salvia miltiorrhiza	*Funneliformis geosporum or Acaulospora laevis*	Essential oils	Menstrual problems, heart disease, and stroke treatment	Wu et al., 2021
Thymus daenensis	*Funneliformis mosseae* and *Rhizophagus intraradices*	Thymol, p-cymene, and γ-terpinene	Digestive aid, carminative, expectorant, antitussive, and antispasmodic	Arpanahi et al., 2020
Cymbopogon citratus	*Funneliformis mosseae*	Essential oils, geranial, neral, and β-pinene	Remedy for sore throat, fever, bronchitis, rheumatism, irritable bowel syndrome, and toothache	Mirzaie et al., 2020
Valeriana jatamansi	*Rhizophagus intraradices*	Gallic acid, chlorogenic acid, catechin	Calming, antianxiety, antidepressant, antioxidant, and antibacterial properties	Jugran et al., 2015
Valeriana officinalis	*Rhizophagus intraradices*	Valerenic acid	Sleep-inducing, antispasmodic, and sedative properties	Nell et al., 2009
Gloriosa superba	*Funneliformis mossae, Rhizophagus fasciculatus, Gigaspora margarita,* and *Gigaspora gilmorei*	Colchicine content	Gout, rheumatoid arthritis, skin problems, and liver conditions	Pandey et al., 2014

TABLE 13.1 (*Continued*)
Plants Associated with Mycorrhiza with the Bioactive Compounds and Their Medicinal Value

Plant Name	Fungi	Bioactive Compound	Medicinal Value	References
Dioscorea spp. and yam	*Rhizophagus clarus*, *Claroideoglomus etunicatum*, and *Rhizophagus fasciculatus*	Polyphenols, flavonoids, and anthocyanin	Inhibition of lipid peroxidation, resistance to free radical attack, reduction of low-density lipoprotein (LDL), and prevention of cardiovascular disease	Lu et al., 2015
Euphorbia hirta	*Funneliformis mosseae*	Phenols, flavonoids, alkaloids, and terpenoids	Asthma, bronchitis, coryza, dysentery, jaundice, acne, gonorrhea, digestive issues, and tumors	Tejavathi and Jayashree, 2013
Anadenanthea colubrina	*Acaulospora longula* and *Gigaspora albida*	Catechin	Anti-inflammatory and respiratory disorder treatment	Pedone-Bonfim et al., 2013
Castanospermum austral	*Rhizophagus intraradices* and *Gigaspora margarita*	Castanospermine	Useful as HIV inhibitor and in AIDS therapy; also effective against cancer and inflammation	Abu-Zeyad et al., 1999
Glycine max	*Funneliformis mosseas*	Isoflavonoids	Cancer, heart disease, postmenopausal issues, diabetes, and several forms of neurodegenerative disease – all reduced	Morandi et al., 1984
Glycyrrhiza glabra	*Glomus hoi*, *Claroideoglomus etunicatum*, and *Claroideoglomus claroideum*	Glycyrrhizic acid	Antiviral properties and the ability to transport several drugs	Johny et al., 2021
Glycyrrhiza uralensis	*Funneliformis mosseae*	Glycyrrhizic acid, liquiritin, isoliquiritin, and isoliquiritigen	Promise of immunomodulation and tumor suppression	Chen et al., 2017
Libidibia ferrea	*Claroideoglomus etunicatum*	Total flavonoids	Prevention if ulcers, relief of inflammation, warding-off of cancer, killing of germs, cessation of bleeding, and repair of wounds	Silva et al., 2014
Medicago sativa	*Rhizophagus intraradics*	Formononetin	Powers against free radicals, inflammation, cancer, and immunosuppression	Volpin et al., 1994
Prosopis laevigata	*Gigaspora rosea*	Trigonelline	Heart disease prevention and therapy, as well as pharyngitis, dysentery, and ocular inflammations	Rojas-Andrade et al., 2003
Hypericum perforatum	*Rhizophagus intraradices*, *Funneliformis constrictum*, and *Funneliformis geosporum*	Naphthodianthrone-es, hypericin, pseudohypericin	Useful in treatment of excitability, neuralgia, anxiety, and depression (thanks to their calming and astringent characteristics)	Zubek et al., 2012
Coleus forskohlii	*Glomus bagyarajii* and *Scutellospora calospora*	Forskolin	Hypertension, heart disease, eczema, and asthma treatment	Sailo and Bagyaraj, 2005
Leucas aspera	*Funneliformis mosseae*	Alkaloids	Jaundice, dyspepsia, fever, helminthic manifestation, respiratory and skin illnesses – all aided by the carminative, antihistaminic, antipyretic, and antiseptic qualities of the herb	Tejavathi and Jayashree, 2013
Mentha arvensis	*Rhizophagus fasciculatus*	Terpenes content	Asthma, jaundice, liver disease, spleen disease, allergies, and jaundice	Gupta et al., 2002
Melissa officinalis	*Claroideoglomus etunicatum* and *Claoideoglomus claroideum*	Citronellal and neral	Sedative, spasmolytic, antibacterial, antioxidant, and antitumoral effects	Engel et al., 2016

(*Continued*)

TABLE 13.1 (Continued)

Plants Associated with Mycorrhiza with the Bioactive Compounds and Their Medicinal Value

Plant Name	Fungi	Bioactive Compound	Medicinal Value	References
Ocimum basilicum	*Gigarpora margarita*	Linalool and geraniol	Medicine for a wide variety of ailments, including migraines, colds, tummy troubles, skin tags, worm infestations, and renal problems	Rasouli-Sadaghiani et al., 2010
Origanum vulgare	*Claroideoglomus etunicatum*	Essential oil, p-cymene, γ-terpinene	Medications for stomachaches, toothaches, and inducing periods	Karagiannidis et al., 2012
Plectranthus amboinicus	*Rhizophagus clarus*	Carvacrol, α-Bergamotene, α-humulene	Activities that aid digestion, expectoration, muscle spasm relief, wound healing, and warding against infection	Merlin et al., 2020
Pogostemon cablin	*Claroideoglomus etunicatum*	Essential oils	Effective in relieving symptoms like motion sickness, diarrhea, the common cold, and headaches	Arpana et al., 2008
Satureja macrostema	*Rhizophagus irregularis*	β-Linalool, menthone, and pulegone	Inhibition of microbial growth	Carreón-Abud et al., 2015
Salvia officinalis	*Rhizophagus clarus*	Essential oil, camphor, α-humulene, and viridiflorol	Seizures, ulcers, gout, rheumatism, inflammation, dizziness, tremors, paralysis, diarrhea, and hyperglycemia	da Cruz et al., 2019

13.4 MODERN METHODS FOR THE PRODUCTION OF SECONDARY METABOLITES IN SYMBIOTIC THERAPEUTIC PLANTS

Plant secondary metabolites are frequently acquired by extracting specimens that are cultivated in their native environments. Approximately 95% of the medicinal plants utilized in the contemporary Indian market are acquired from their native habitats. The synthesis of secondary metabolites in plants grown in natural environments exhibits considerable variability and unpredictability owing to a range of abiotic and biotic influences. The conditions that can potentially disturb the metabolic pathways countable for the production of bioactive chemicals encompass severe alkalinity, temperatures, salinity, drought, and plant diseases (Dayani and Sabzalian, 2017). Furthermore, the overexploitation of medicinal plant species from their native ecosystems possesses the capacity to considerably heighten their susceptibility to extinction. The cultivation of medicinal plants in outdoor conditions might need a significant amount of time, especially when it comes to woody species, like *Taxus brevifolia*, as well as slow-growing perennial plants, such as *Panax ginseng*. These plants may necessitate multiple years to attain the intended level of metabolite production (Malik et al., 2011). Hence, it is imperative to explore alternate production strategies.

The practice of cultivating medicinal plants under controlled conditions can refine the purity, consistency, quality, biomass, and bioactivity output from botanical resources (Hayden, 2006). Numerous cultivation techniques take and boost the economic cultivation of secondary metabolites while also considering the feasibility of including AMF in the process.

13.5 GREENHOUSE CULTIVATION

Greenhouses are extensively utilized for year-round crop cultivation. The manipulation of ecological conditions, such as humidity and temperature, allows for the creation of ideal growth conditions for the specific crop or plant being cultivated. This practice promotes the development of the desired characteristics and ensures a reliable and abundant production of bioactive compounds of superior quality (Panwar et al., 2003). Several kinds of medicinal plants, including *Echinacea purpurea*, *Echinacea angustifolia*, *Withania somnifera*, *Ocimum basilicum*, and *Psoralea corylifolia*, have been cultivated in a greenhouse environment (Panwar et al., 2003; Zheng et al., 2006). Several plant species, including *Curcuma longa*, *Artemisia annua*, *Glycyrrhiza glabra*, *Coleus forskohlii*, and *Gloriosa superba*, have been deliberated about AMF in greenhouse conditions. Several investigations have documented a noteworthy augmentation in the synthesis of bioactive substances (Pandey et al., 2014). Therefore, the cultivation of medicinal plants in regulated greenhouse conditions, grouping with the utilization of AMF, presents a promising strategy for augmenting both the amount and quality of bioactive compounds on a substantial magnitude.

13.6 AEROPONICS

The aeroponics cultivation method entails the suspension of plant roots in a confined enclosure, resulting in light deprivation and exposure to a nutrient-dense water spray delivered via atomizers (Lakhiar et al., 2018). This system has been established for the cultivation of various plant species, including horticultural crops, such as *Cucumis sativus*, *Lactuca sativa*, and *Solanum lycopersicum* (Movahedi and Rostami, 2020). It has also been applied to the cultivation of medicinal herbs like *Crocus sativus*, *Anemopsis californica*, and *Valeriana officinalis* (Hayden, 2006), as well as medicinal crops like *Zingiber officinale* and *Arctium lappa*. The roots of plants are subjected to various techniques to gain secondary metabolites (Hayden et al., 2004). The cultivation of *Ocimum basilicum* utilizing aeroponic techniques resulted in a higher crop production, comparable amounts of flavonoid and phenolic components, and equivalent antioxidant properties in comparison to plants grown in a field. The cultivation of *Withania coagulans* and *Cichorium intybus* in an aeroponic arrangement resulted in significantly greater yields when compared to the cultivation of these plants in soil (Movahedi and Rostami, 2020). The utilization of this technique demonstrated high efficacy in the synthesis of bioactive compounds derived from the roots of medicinal plants. Notable examples include the generation of β-sitosterol in *Cannabis sativa* and chlorogenic acid in *Arctium lappa*. The primary locations for the manufacturing and storage of bioactive complexes in medicinal plants are the root apices (Watson et al., 2015).

Nevertheless, the extraction of these active compounds proves to be highly challenging using traditional techniques. The plant milking technology (Plant milking®) is used to extract and enhance compounds from *Morus alba*, a tree that holds significance in traditional Chinese medicine due to its high alkaloid and flavonoid content (Chajra et al., 2020).

Aeroponic culture systems have been established and applied for the formation of arbuscular mycorrhizal fungi inoculum. In these systems, both plant roots and AMF are exposed to a nutrient liquid mist.

To produce AMF, plants undergo pre-colonization before being introduced into the system. This is achieved by pre-cultivating plant seedlings and AMF propagules, both of which are first surface-sterilized in jars filled with a combination of sand and perlite. Subsequently, the plants that have not undergone colonization by external entities are relocated to the aeroponic container, facilitating the growth and development of their roots, as well as the creation of AMF. To inhibit the growth of algae, it is common practice to shield the container from light (Jarstfer and Sylvia, 1995). Various techniques can be employed to apply mist, which primarily diverge in the size of the dewdrops that are produced. These techniques include the use of an atomizing disk, hassled spray through a microirrigated outlet, the generation of an ultrasonically produced fog of nutrient solution with droplet dews ranging from 3 to 10 µm in diameter, and the utilization of ultrasonic nebulizer technology that results in microdroplets measuring 1 µm in diameter (Jarstfer and Sylvia, 1995; Mohammad et al., 2000). A substantial quantity of viable AMF propagules can be seen in aeroponic culture. This inoculant is subsequently utilized in a ground field experiment (Mohammad et al., 2004). Therefore, it is possible that aeroponic systems can be utilized in the cultivation of medicinal plants in conjunction with their AMF partners, with the aim of achieving significant biomass and the generation of secondary metabolites.

13.7 HYDROPONICS

Hydroponic systems encompass a variety of methods that facilitate the delivery of nutrients in the fluid form, either with or without a solid supportive medium to provide support for plant (Hayden, 2006). Various medicinal plants, including *Leonurus quinquelobatus*, *Mentha piperita*, *Salvia officinalis*, *Linum usitatissimum*, *Achillea millefolium*, *Echinacea angustifolia*, *Ocimum basilicum*, *Bidens tripartita*, *Hypericum perforatum*, *Tanacetum parthenium*, and *Leonurus sibiricus*, have been subjected to the application of such systems (Maggini et al., 2011). The utilization of these systems has facilitated the synthesis of bioactive substances, such as tropane alkaloids found in *Datura innoxia*, total phenols and rosmarinic acid present in *Ocimum basilicum*, and oil making in *Valeriana officinalis*.

Various hydroponic culture techniques are available for the significant production of AMF. The primary distinction between them is in the method of aeration and the manner in which the nutritional solution is administered. In the static hydroponic culture method, it is necessary to aerate the nutrient solution using an aeration pump to avoid oxygen deprivation in the mycorrhizal plants' roots (Ijdo et al., 2011). Substantial amounts of mycorrhizal *Linum usitatissimum* plant roots, extramatrical mycelium, and chlamydospores that were devoid of any remnants from solid substrate components, were obtained using this particular technique (Dugassa et al., 1995)

In an alternative hydroponic system known as the nutritional film technique (NFT), a narrow solution containing essential nutrients, referred to as a film, is directed through slanted channels, and sometimes referred to as gills. This setup facilitates the development of plant roots and the establishment of arbuscular mycorrhizal fungi (Ijdo et al., 2011). The cultivation of arbuscular mycorrhizal fungi has been employed as a technique later the late 1980s, resulting in the generation of several sporocarps by *F. mosseae*.

A novel, as well as cost-effective, *in vitro* bioreactor system utilizing plant-based approaches for the significant cultivation of arbuscular mycorrhizal fungi was introduced (Ijdo et al., 2011). In this experimental setup, *Medicago truncatula* roots and AMF of the *Glomus* type were cultivated within a sterilized tube. The tube was linked at both tops of a tank that contained decontaminated fluid culture medium. The nutrient solution is circulated throughout in the roots of the mycorrhizal system, providing nourishment to the plant and fungal partners. Meanwhile, the plant shoot undergoes development in a controlled growth chamber under open-air circumstances.

The hydroponic system has been utilized to examine the impact of the mycorrhizal symbiosis, specifically in terms of phosphorus uptake in maize plants (Garcés-Ruiz et al., 2017). Therefore, the incorporation of semi-hydroponic or hydroponic systems with therapeutic plants and AMF holds promise for enhancing the yield and production of secondary metabolites. Furthermore, it is possible to perform permeabilization procedures within these growing systems, to extract the substances released by roots in a non-destructive way, allowing several extractions from the same plants annually (Cartabia et al., 2021). *Datura innoxia* plants were grown under aquaculture circumstances without the presence of any AMF. Following cultivation, the roots of the plants were subsequently permeabilized using Tween 20. Consequently, a high concentration of tropane alkaloids, such as hyoscyamine and scopolamine, was detected in the liquid medium.

It is noteworthy that all the plants showed the ability to endure and persist following the process of rinsing and subsequent reintegration into the hydroponic system. The method described in this study enables the permeabilization of plant cells several times while maintaining their vitality. various methods of permeabilization, such as adjusting the dosage and period of Tween 20 and incorporating tropane alkaloids precursors, can be employed to enhance the formation of chemical compounds in the nutrient solution (Gontier et al., 2002).

13.8 *IN VITRO* PRODUCTION PROCESS

Micropropagation, also known as *in vitro* propagation, refers to the asexual reproduction of plants through the cultivation of tissues, cells, or organs in a controlled laboratory environment. The process entails the sterile cultivation of tissue or organ explants within sealed containers, utilizing specified culture media under carefully regulated conditions (Debnath and Arigundam, 2020).

13.9 PLANT *IN VITRO* CULTURE

The activities encompassed in this field of study include the propagation, preservation of germplasm, manufacture of bioactive chemicals, and hereditary enhancement of numerous species of medicinal plants (Nalawade and Tsay, 2004). Various protocols industrialized to facilitate the in vitro mass propagation of several plant species, including *Zingiber zerumbet*, *Pinellia ternate*, *Gentiana davidii*, *Anoectochilus formosanus*, *Limonium wrightii*, *Scrophularia yoshimura*, *Bupleurum falcatum*, *Dendrobium linawianum*, and *Fritillaria hupehensis*. These protocols utilize shoot morphogenesis as a means of propagation. Additionally, protocols have been established for *Corydalis yanhusuo* and *Angelica sinensis*, which employ somatic embryogenesis as the propagation method. Furthermore, cell suspension cultures have been successfully employed for the propagation of *Angelica sinensis*, *Angelica dahurica*, *Taxus mairei*, *Gentiana davidii*, and *Bupleurum falcatum* (Nalawade and Tsay, 2004).

The literature has documented the relationship between AMF and intact floras cultured *in vitro*. Rapid and consistent mycorrhization of *Medicago truncatula* seedlings by utilizing an *in vitro* cultivation approach that involved the use of a mycorrhizal donor plant. This involved placing the plantlets in a mycelial network that was actively growing from the mycorrhizal donor plant. Therefore, it is possible to expand their application to medicinal plants in order to rise the making of secondary metabolites (Voets et al., 2009).

13.10 CONCLUSION

AMFs have the potential to provide several advantages to medicinal plants, including the encouragement of growth and enhanced tolerance to stressful environmental circumstances. Considerable evidences suggest that AMF may increase the accumulation of bioactive chemicals in these plants. The use of mycorrhizal technology shows promises as a long-term strategy to boost medical plant development and secondary metabolite synthesis. The number of secondary metabolites produced by medicinal plants can be affected by a wide range of environmental and cultural conditions. Thus, it is crucial to consider these factors while honing the conditions for optimum yields of plants and the formation of associated secondary metabolites. In order to guarantee the efficacy of the bioactive compounds produced by mycorrhizal medicinal plants, a number of strategies have been proposed. Aeroponic, hydroponic, and semi-hydroponic systems are only a few examples of the substrate and substrate-free techniques at play here. It has also been reported how to use high-root colonization (HRC) techniques to grow micropropagated medicinal plants in vitro culture systems with arbuscular mycorrhizal plants or mycorrhizal donor plants. The utilization of these systems has the potential to create optimal environmental conditions for plants, leading to enhanced agricultural output and the synthesis of bioactive chemicals.

Secondary metabolite production is dependent on the type of AMF, the plant species involved, and the climatic conditions under which the plant is grown. AMF exhibit a lack of host specificity, yet they may display a preference for a particular host. Previous research has established that various species of AMF can elicit variations in the synthesis of bioactive secondary metabolites within the same host or genotype. Presence of *Glomus caledoniam* resulted in an upsurge in the production of rosmarinic acid and caffeic acid in *Ocimum basilicum*. On the other hand, the presence of *F. mosseae* only led to an enhancement in caffeic acid production. AFMs have been found to have a positive impact on the health of plants and the generation of secondary metabolites on a significant scale. This has proven advantageous in various sectors, such as the medical industry, and has the potential to contribute to the overall economic growth of a country.

REFERENCES

Abdel-Salam, E., A. Alatar, and M.A. El-Sheikh. 2018. Inoculation with arbuscular mycorrhizal fungi alleviates harmful effects of drought stress on damask rose. *Saudi Journal of Biological Sciences* 25: 1772–80. https://doi.org/10.1016/j.sjbs.2017.10.015.

Abeer, H., and E.L. Didamony. 2014. Alleviation of adverse impact of salinity on faba bean (*Vicia faba* L.) by arbuscular mycorrhizal fungi. *Pakistan Journal of Botany* 46: 2003–13. https://inis.iaea.org/search/search.aspx?orig_q=RN:46026328.

Abu-Zeyad, R., A.G. Khan, and C. Khoo. 1999. Occurrence of arbuscular mycorrhiza in *Castanospermum australe* A. Cunn. & C. Fraser and effects on growth and production of castanospermine. *Mycorrhiza* 9: 111–17. https://doi.org/10.1007/s005720050008.

Agra, M.F., P.F. Freitas, and J.M. Barbosa-Filho. 2007. Synopsis of the plants known as medicinal and poisonous in northeast of Brazil. *Brazilian Journal of Pharmacognosy* 17: 114–40. https://doi.org/10.1590/S0102-695X2007000100021.

Akiyama, K., and H. Hayashi. 2006. Strigolactones: Chemical signals for fungal symbionts and parasitic weeds in plant roots. *Annals of Botany* 97: 925–31. https://doi.org/10.1093/aob/mcl063.

Andrade, S.A.L., S. Malik, A.C.H.F. Sawaya, A. Bottcher, and P. Mazzafera. 2013. Association with arbuscular mycorrhizal fungi influences alkaloid synthesis and accumulation in *Catharanthus roseus* and *Nicotiana tabacum* plants. *Acta Physiologiae Plantarum* 35: 867–80. https://doi.org/10.1007/s11738-012-1130-8.

Araim, G., A. Saleem, J.T. Arnason, and C. Charest. 2009. Root colonization by an arbuscular mycorrhizal (AM) fungus increases growth and secondary metabolism of purple coneflower, *Echinacea purpurea* (L.) Moench. *Journal of Agricultural and Food Chemistry* 57: 2255–58. https://doi.org/10.1021/jf803173x.

Arpana, J., D.J. Bagyaraj, E.V.S. Prakasa Rao, T.N. Parameswaran, and B. Abdul Rahiman. 2008. Symbiotic response of patchouli [*Pogostemon cablin* (Blanco) Benth.] to different arbuscular mycorrhizal fungi. *Advances in Environmental Biology* 2: 20–24. https://www.aensiweb.com//old/aeb/2008/20-24.pdf.

Arpanahi, A.A., M. Feizian, G. Mehdipourian, and D.N. Khojasteh. 2020. Arbuscular mycorrhizal fungi inoculation improve essential oil and physiological parameters and nutritional values of *Thymus daenensis* Celak and *Thymus vulgaris* L. under normal and drought stress conditions. *European Journal of Soil Biology* 100 (September–October): 103217, 1–11. https://doi.org/10.1016/j.ejsobi.2020.103217.

Ayeka, P.A., Y. Bian, P.G. Mwitari, et al. 2016. Immunomodulatory and anticancer potential of Gan cao (*Glycyrrhiza uralensis* Fisch.) polysaccharides by CT-26 colon carcinoma cell growth inhibition and cytokine IL-7 upregulation *in vitro*. *BMC Complementary Medicine and Therapies* 16 (July): 206, 1–8. https://doi.org/10.1186/s12906-016-1171-4.

Beik, A., S. Joukar, and H. Najafipour. 2020. A review on plants and herbal components with antiarrhythmic activities and their interaction with current cardiac drugs. *Journal of Traditional and Complementary Medicine* 10: 275–87. https://doi.org/10.1016/j.jtcme.2020.03.002

Berbara, R.L.L., F.A. Souza, and H.M.A.C. Fonseca. 2006. III-Fungos micorrízicos arbusculares: muito além da nutrição. In *Nutrição mineral de plantas*, ed. M.S. Fernandes, 53–85. Viçosa, MG: SBCS. http://www.ufrrj.br/amfoods/arquivos/arq_publicacao/20_ARQ.pdf.

Binet, M.N., D. Van Tuinen, N. Deprêtre, N. Koszela, C. Chambon, and S. Gianinazzi. 2011. Arbuscular mycorrhizal fungi associated with *Artemisia umbelliformis* Lam, an endangered aromatic species in Southern French Alps, influence plant P and essential oil contents. *Mycorrhiza* 21: 523–35. https://doi.org/10.1007/s00572-010-0354-y.

Borde, M., M. Dudhane, and P.K. Jite. 2009. Role of bioinoculant (AM fungi) increasing in growth, flavor content and yield in *Allium sativum* L. under field condition. *Notulae Botanicae Horti Agrobotanici Cluj-Napoca* 37: 124–28. https://doi.org/10.15835/nbha3723289.

Bowles, T.M., F.H. Barrios-Masias, E.A. Carlisle, T.R. Cavagnaro, and L.E. Jackson. 2016. Effects of arbuscular mycorrhizae on tomato yield, nutrient uptake, water relations, and soil carbon dynamics under deficit irrigation in field conditions. *Science of the Total Environment* 566: 1223–34. https://doi.org/10.1016/j.scitotenv.2016.05.178.

Brundrett, M.C., and L. Tedersoo. 2018. Evolutionary history of mycorrhizal symbioses and global host plant diversity. *New Phytologist* 220: 1108–15. https://doi.org/10.1111/nph.14976.

Butler, M.S. 2004. The role of natural product chemistry in drug discovery. *Journal of Natural Products* 67: 2141–53. https://doi.org/10.1021/np040106y.

Carreón-Abud, Y., R. Torres-Martínez, B. Farfán-Soto, et al. 2015. Arbuscular mycorrhizal symbiosis increases the content of volatile terpenes and plant performance in *Satureja macrostema* (Benth.) Briq. *Boletín Latinoamericano y del Caribe de Plantas Medicinales y Aromáticas* 14: 273–79. https://www.redalyc.org/pdf/856/85641104002.pdf.

Cartabia, A., E. Tsiokanos, N. Tsafantakis, et al. 2021. The arbuscular mycorrhizal fungus *Rhizophagus irregularis* MUCL 41833 modulates metabolites production of *Anchusa officinalis* L. under semi-hydroponic cultivation. *Frontiers in Plant Science* 12 (September): 724352, 1–21. https://doi.org/10.3389/fpls.2021.724352.

Ceccarelli, N., M. Curadi, L. Martelloni, C. Sbrana, P. Picciarelli, and M. Giovannetti 2010. Mycorrhizal colonization impacts on phenolic content and antioxidant properties of artichoke leaves and flower heads two years after field transplant. *Plant and Soil* 335: 311–23. https://doi.org/10.1007/s11104-010-0417-z.

Chajra, H., A. Salwinski, A. Guillaumin, et al. 2020. Plant milking technology – an innovative and sustainable process to produce highly active extracts from plant roots. *Molecules* 25: 4162 (September): 1–22. https://doi.org/10.3390/molecules25184162.

Chen, M., M. Arato, L. Borghi, E. Nouri, and D. Reinhardt. 2018. Beneficial services of arbuscular mycorrhizal fungi–from ecology to application. *Frontiers in Plant Science* 9 (September): 1270, 1–14. https://doi.org/10.3389/fpls.2018.01270.

Chen, M., G. Yang, Y. Sheng, et al. 2017. *Glomus mosseae* inoculation improves the root system architecture, photosynthetic efficiency and flavonoids accumulation of liquorice under nutrient stress. *Frontiers in Plant Science* 8 (June): 931, 1–10. https://doi.org/10.3389/fpls.2017.00931.

Chen, S.L., H. Yu, H.M. Luo, Q. Wu, C.F. Li, and A. Steinmetz. 2016. Conservation and sustainable use of medicinal plants: Problems, progress, and prospects. 11 (July): 37, 1–10. https://doi.org/10.1186/s13020-016-0108-7.

Chu, E.Y., M. de Regina Freire Möller, and J.G. de Carvalho. 2001. Efeitos da inoculação micorrízica em mudas de gravioleira em solo fumigado e não fumigado. *Pesquisa Agropecuaria Brasileira* 36: 671–80. https://doi.org/10.1590/S0100-204X2001000400010.

Coleman, D.C., D.A. Crossley Jr, and P.F. Hendrix. 2004. 2-primary production processes in soils: Roots and rhizosphere associates. In *Fundamentals of soil ecology*, 2nd ed., eds. D.C. Coleman, D.A. Crossley Jr, and P.F. Hendrix, 23–46. Cambridge: Academic Press. https://doi.org/10.1016/B978-012179726-3/50003-4.

da Cruz, R.M.S., G.L.S. da Cruz, D.C. Dragunski, A.C.G. Junior, O. Alberton, and S.G.H. de Souza. 2019. Inoculation with arbuscular mycorrhizal fungi alters content and composition of essential oil of sage (*Salvia officinalis*) under different phosphorous levels. *Australian Journal of Crop Science* 13: 1617–24. https://www.cropj.com/cruz_13_10_2019_1617_1624.pdf.

Dambisya, Y.M., and G. Tindimwebwa. 2003. Traditional remedies in children around eastern cape, South Africa. *East African Medical Journal* 80: 402–05. https://www.ajol.info/index.php/eamj/article/view/8730.

Dave, S., J. Das, and J.C. Tarafdar. 2011. Effect of vesicular arbuscular mycorrhizae on growth and saponin accumulation in *Chlorophytum borivilianum*. *Science Asia* 37: 165–69. https://scienceasia.org/2011.37.n2/scias37_165.pdf.

Dayani, S., and M.R. Sabzalian. 2017. Production of secondary metabolites in medicinal plants through hydroponic system. In *Controlled environment agriculture-production of specialty crops providing human health benefits through hydroponics*, ed. M. Asaduzzaman, 33–53. New York: Nova Science Publishers.

de Vries, F.T., R.I. Griffiths, C.G. Knight, O. Nicolitch, and A. Williams. 2020. Harnessing rhizosphere microbiomes for drought-resilient crop production. *Science* 368: 270–74. https://www.science.org/doi/10.1126/science.aaz5192.

Debnath, S.C., and U. Arigundam. 2020. *In vitro* propagation strategies of medicinally important berry crop, lingonberry (*Vaccinium vitis-idaea* L.). *Agronomy* 10 (May): 744, 1–19. https://doi.org/10.3390/agronomy10050744.

Domokos, E., L. Jakab-Farkas, B. Darkó, et al. 2018. Increase in *Artemisia annua* plant biomass artemisinin content and guaiacol peroxidase activity using the arbuscular mycorrhizal fungus *Rhizophagus irregularis*. *Frontiers in Plant Science* 9 (April): 478, 1–9. https://doi.org/10.3389/fpls.2018.00478.

Dugassa, D.G., G. Grunewaldt-Stöcker, and F. Schönbeck. 1995. Growth of *Glomus intraradices* and its effect on linseed (*Linum usitatissimum* L.) in hydroponic culture. *Mycorrhiza* 5: 279–82. https://doi.org/10.1007/BF00204962.

Dutta, S.C., and B. Neog. 2016. Accumulation of secondary metabolites in response to antioxidant activity of turmeric rhizomes co-inoculated with native arbuscular mycorrhizal fungi and plant growth promoting rhizobacteria. *Scientia Horticulturae* 204: 179–84. https://doi.org/10.1016/j.scienta.2016.03.028.

Ekor, M. 2014. The growing use of herbal medicines: Issues relating to adverse reactions and challenges in monitoring safety. *Frontiers in Pharmacology* 4 (January): 177, 1–10. https://doi.org/10.3389/fphar.2013.00177.

Engel, R., K. Szabo, L. Abranko, K. Rendes, A. Füzy, and T. Takács. 2016. Effect of arbuscular mycorrhizal fungi on the growth and polyphenol profile of marjoram, lemon balm, and marigold. *Journal of Agricultural and Food Chemistry* 64: 3733–42. https://doi.org/10.1021/acs.jafc.6b00408.

Freitas, M.S.M., M.A. Martins, A.J.C. de Carvalho, and R.F.V. Carneiro. 2004. Growth and production of total phenols in carqueja [*Baccharis trimera* (less.)] in response to inoculation with arbuscular mycorrhizal fungi, in the presence and in the absence of mineral manuring. *Revista Brasileira de Plantas Medicinais* 6: 30–34.

Garcés-Ruiz, M., M. Calonne-Salmon, K. Plouznikoff, et al. 2017. Dynamics of short-term phosphorus uptake by intact mycorrhizal and non-mycorrhizal maize plants grown in a circulatory semi-hydroponic cultivation system. *Frontiers in Plant Science* 8 (August): 1471, 1–8. https://doi.org/10.3389/fpls.2017.01471.

Garg, A.K., M. Faheem, and S. Singh. 2021. Role of medicinal plant in human health disease. *Asian Journal of Plant Science & Research* 11: 19–21. https://www.imedpub.com/articles/role-of-medicinal-plant-in-human-health-disease.pdf.

Gobbo-Neto, L., and N.P. Lopes. 2007. Medicinal plants: Factors of influence on the content of secondary metabolites. *Quimica Nova* 30: 374–81. http://dx.doi.org/10.1590/S0100-40422007000200026.

Gontier, E., A. Clément, T.L.M. Tran, et al. 2002. Hydroponic combined with natural or forced root permeabilization: A promising technique for plant secondary metabolite production. *Plant Science* 163: 723–32. https://doi.org/10.1016/S0168-9452(02)00171-1.

Guo, Q., L. Cheng, and Z. Liu. 2010. Study on influence of arbuscular mycorrhizal fungi *Pinellia ternata* yield and chemical composition. *China Journal of Chinese Matera Medica* 35: 333–38. https://doi.org/10.4268/cjcmm20100317.

Gupta, A., S. Jain, and N. Verma. 2022. *Industrial applications of soil microbes*, vol. 1. Bentham Science Publishers Pte Ltd. https://doi.org/10.2174/97898150399551220101.

Gupta, A., S. Jain, and N. Verma. 2024. *Industrial applications of soil microbes*, vol. 3. Bentham Science Publishers Pte Ltd. https://doi.org/10.2174/97898150797531240301.

Gupta, M.L., A. Prasad, M. Ram, and S. Kumar. 2002. Effect of the vesicular–arbuscular mycorrhizal (VAM) fungus *Glomus fasciculatum* on the essential oil yield related characters and nutrient acquisition in the crops of different cultivars of menthol mint (*Mentha arvensis*) under field conditions. *Bioresource Technology* 81: 77–79. https://doi.org/10.1016/S0960-8524(01)00109-2.

Hamilton, A.C. 2004. Medicinal plants, conservation and livelihoods. *Biodiversity & Conservation* 13: 1477–17. https://doi.org/10.1023/B:BIOC.0000021333.23413.42.

Hashem, A., E.F. Abd_Allah, A.A. Alqarawi, and D. Egamberdieva. 2018. Arbuscular mycorrhizal fungi and plant stress tolerance. In *Plant microbiome: Stress response. Microorganisms for sustainability*, eds. D. Egamberdieva and P. Ahmad, 81–103. Singapore: Springer. https://doi.org/10.1007/978-981-10-5514-0_4.

Hayden, A.L. 2006. Aeroponic and hydroponic systems for medicinal herb, rhizome, and root crops. *Hort Science* 41: 536–38. https://doi.org/10.21273/HORTSCI.41.3.536

Hayden, A.L., L.A. Brigham, and G.A. Giacomelli. 2004. Aeroponic cultivation of ginger (*Zingiber officinale*) rhizomes. *Acta Horticulturae* 659: 397–402. https://doi.org/10.17660/ActaHortic.2004.659.52.

Hemashenpagam, N., and T. Selvaraj. 2011. Effect of arbuscular mycorrhizal (AM) fungus and plant growth promoting rhizomicroorganisms (PGPR's) on medicinal plant *Solanum viarum* seedlings. *Journal of Environmental Biology* 32: 579–83. https://pubmed.ncbi.nlm.nih.gov/22319872/.

Hussein, R.A., and A.A. El-Anssary. 2019. Plants secondary metabolites: The key drivers of the pharmacological actions of medicinal plants. In *Herbal medicine*, ed. P.F. Builders, 11–30. London: IntechOpen. http://dx.doi.org/10.5772/intechopen.76139.

Ijdo, M., S. Cranenbrouck, and S. Declerck. 2011. Methods for large-scale production of AM fungi: Past, present, and future. *Mycorrhiza* 21: 1–16. https://doi.org/10.1007/s00572-010-0337-z.

Jain, S., A. Gupta, and N. Verma. 2023. *Industrial applications of soil microbes*, vol. 2. Bentham Science Publishers Pte Ltd. https://doi.org/10.2174/97898150502641230201.

Jain, S., A. Gupta, and N. Verma. 2024. *Industrial applications of soil microbes*, vol. 4. Bentham Science Publishers Pte Ltd. https://doi.org/10.2174/97898151249961240401.

Janos, D.P. 2007. Plant responsiveness to mycorrhizas differs from dependence upon mycorrhizas. *Mycorrhiza* 17: 75–91. https://doi.org/10.1007/s00572-006-0094-1.

Jarstfer, A.G., and D.M. Sylvia. 1995. Aeroponic culture of VAM fungi. In *Mycorrhiza: Structure, function, molecular biology and biotechnology*, eds. A. Varma and B. Hock, 427–41. Berlin, Heidelberg: Springer. https://doi.org/10.1007/978-3-662-03779-9_18.

Ji-hong, F., and Y. Guo-ting. 2006. Effect of AM fungi on the content of berberine, jatrorrhizine and palmatine of *Phellodendron amurense* seedings. *Protection Forest Science and Technology* 5: 24–26.

Johny, L., D.M. Cahill, and A. Adholeya. 2021. AMF enhance secondary metabolite production in ashwagandha, licorice, and marigold in a fungi-host specific manner. *Rhizosphere* 17 (March): 100314. https://doi.org/10.1016/j.rhisph.2021.100314.

Jugran, A.K., A. Bahukhandi, P. Dhyani, et al. 2015. The effect of inoculation with mycorrhiza: AM on growth, phenolics, tannins, phenolic composition and antioxidant activity in *Valeriana jatamansi* Jones. *Journal of Soil Science and Plant Nutrition* 15: 1036–49. http://dx.doi.org/10.4067/S0718-95162015005000072.

Kai, G., H. Xu, C. Zhou, et al. 2011. Metabolic engineering tanshinone biosynthetic pathway in *Salvia miltiorrhiza* hairy root cultures. *Metabolic Engineering* 13: 319–27. https://doi.org/10.1016/j.ymben.2011.02.003.

Kapoor, R., B. Giri, and K.G. Mukerji. 2002. *Glomus macrocarpum*: A potential bioinoculant to improve essential oil quality and concentration in dill (*Anethum graveolens* L.) and carum [*Trachyspermum ammi* (Linn.) Sprague]. *World Journal of Microbiology and Biotechnology* 18: 459–63. https://doi.org/10.1023/A:1015522100497.

Kapoor, R., B. Giri, and K.G. Mukerji. 2004. Improved growth and essential oil yield and quality in *Foeniculum vulgare* mill on mycorrhizal inoculation supplemented with P-fertilizer. *Bioresource Technology* 93: 307–11. https://doi.org/10.1016/j.biortech.2003.10.028.

Karagiannidis, N., T. Thomidis, E. Panou-Filotheou, and C. Karagiannidou. 2012. Response of three mint and two oregano species to *Glomus etunicatum* inoculation. *Australian Journal of Crop Science* 6: 164–69. http://www.cropj.com/thomidis_6_1_2012_164_169.pdf.

Kaur, S., and V. Suseela. 2020. Unraveling arbuscular mycorrhiza-induced changes in plant primary and secondary metabolome. *Metabolites* 10 (August): 335, 1–30. https://doi.org/10.3390/metabo10080335.

Lakhiar, I.A., J. Gao, T.N. Syed, F.A. Chandio, and N.A. Buttar. 2018. Modern plant cultivation technologies in agriculture under controlled environment: A review on aeroponics. *Journal of Plant Interactions* 13: 338–52. https://doi.org/10.1080/17429145.2018.1472308.

Lima, C.S., H.R.S. Santos, U.P. de Albuquerque, and F.S.B. da Silva. 2017. Mycorrhizal symbiosis increase the level of total foliar phenols and tannins in *Commiphora leptophloeos* (Mart.) JB Gillett seedlings. *Industrial Crops and Products* 104: 28–32. https://doi.org/10.1016/j.indcrop.2017.04.020.

Lombardino, J.G., and J.A. Lowe III. 2004. The role of the medicinal chemist in drug discovery – Then and now. *Nature Reviews Drug Discovery* 3: 853–62. https://doi.org/10.1038/nrd1523.

Lu, F.C., C.Y. Lee, and C.L. Wang. 2015. The influence of arbuscular mycorrhizal fungi inoculation on yam (*Dioscorea* spp.) tuber weights and secondary metabolite content. *Peer J Publishing* 3 (September): e1266, 1–19. https://doi.org/10.7717/peerj.1266.

Maggini, R., C. Kiferle, L. Guidi, A. Pardossi, and A. Raffaelli. 2011. Growing medicinal plants in hydroponic culture. *Acta Horticulturae* 952: 697–04. https://doi.org/10.17660/ActaHortic.2012.952.88.

Maier, W., K. Hammer, U. Dammann, B. Schulz, and D. Strack. 1997. Accumulation of sesquiterpenoid cyclohexenone derivatives induced by an arbuscular mycorrhizal fungus in members of the Poaceae. *Planta* 202: 36–42. https://doi.org/10.1007/s004250050100.

Malik, S., R.M. Cusidó, M.H. Mirjalili, E. Moyano, J. Palazón, and M. Bonfill. 2011. Production of the anticancer drug taxol in *Taxus baccata* suspension cultures: A review. *Process Biochemistry* 46: 23–34. https://doi.org/10.1016/j.procbio.2010.09.004.

Mandal, S., H. Evelin, B. Giri, V.P. Singh, and R. Kapoor. 2013. Arbuscular mycorrhiza enhances the production of stevioside and rebaudioside-A in *Stevia rebaudiana* via nutritional and non-nutritional mechanisms. *Applied Soil Ecology* 72: 187–94. https://doi.org/10.1016/j.apsoil.2013.07.003.

Manoharan, P.T., V. Shanmugaiah, N. Balasubramanian, S. Gomathinayagam, M.P. Sharma, and K. Muthuchelian. 2010. Influence of AM fungi on the growth and physiological status of *Erythrina variegata* Linn. grown under different water stress conditions. *European Journal of Soil Biology* 46: 151–56. https://doi.org/10.1016/j.ejsobi.2010.01.001.

Martens, S., and A. Mithöfer. 2005. Flavones and flavone synthases. *Phytochemistry* 66: 2399–07. https://doi.org/10.1016/j.phytochem.2005.07.013.

Matsubara, Y.I., T. Ishigaki, and K. Koshikawa. 2009. Changes in free amino acid concentrations in mycorrhizal strawberry plants. *Scientia Horticulturae* 119: 392–96. https://doi.org/10.1016/j.scienta.2008.08.025.

Merlin, E., E. Melato, E.L.B. Lourenco, et al. 2020. Inoculation of arbuscular mycorrhizal fungi and phosphorus addition increase coarse mint (*Plectranthus amboinicus* Lour.) plant growth and essential oil content. *Rhizosphere* 15 (September): 100217. https://doi.org/10.1016/j.rhisph.2020.100217.

Mirzaie, M., A.R. Ladanmoghadam, L. Hakimi, and E. Danaee. 2020. Water stress modifies essential oil yield and composition, glandular trichomes and stomatal features of lemongrass (*Cymbopogon citratus*) inoculated with arbuscular mycorrhizal fungi. *Journal of Agricultural Science and Technology* 22: 1575–85. http://jast.modares.ac.ir/article-23-34457-en.html.

Mohammad, A., A.G. Khan, and C. Kuek. 2000. Improved aeroponic culture of inocula of arbuscular mycorrhizal fungi. *Mycorrhiza* 9: 337–39. https://doi.org/10.1007/s005720050278.

Mohammad, A., B. Mitra, and A.G. Khan. 2004. Effects of sheared-root inoculum of *Glomus intraradices* on wheat grown at different phosphorus levels in the field. *Agriculture, Ecosystems & Environment* 103: 245–49. https://doi.org/10.1016/j.agee.2003.09.017.

Morandi, D., J.A. Bailey, and V. Gianinazzi-Pearson. 1984. Isoflavonoid accumulation in soybean roots infected with vesicular-arbuscular mycorrhizal fungi. *Physiological Plant Pathology* 24: 357–64. https://doi.org/10.1016/0048-4059(84)90009–2.

Movahedi, Z., and M. Rostami. 2020. Production of some medicinal plants in aeroponic system. *Journal of Medicinal Plants and By-products* 9: 91–99. https://jmpb.areeo.ac.ir/article_122079_9583dadf45310c5a4772857e7433104e.pdf.

Nalawade, S.M., and H.S. Tsay. 2004. *In vitro* propagation of some important Chinese medicinal plants and their sustainable usage. *In Vitro Cellular & Developmental Biology – Plant* 40: 143–54. https://doi.org/10.1079/IVP2003504.

Nell, M., C. Wawrosch, S. Steinkellner, et al. 2009. Root colonization by symbiotic arbuscular mycorrhizal fungi increases sesquiterpenic acid concentrations in *Valeriana officinalis* L. *Planta Medica* 76: 393–98. https://www.thieme-connect.de/products/ejournals/abstract/10.1055/s-0029-1186180.

Nisha, M.C., and S. Rajeshkumar. 2010. Effect of arbuscular mycorrhizal fungi on growth and nutrition of *Wedilia chinensis* (Osbeck) Merril. *Indian Journal of Science and Technology* 3: 676–78. https://doi.org/10.17485/ijst/2010/v3i6.23.

Noceto, P.A., P. Bettenfeld, R. Boussageon, et al. 2021. Arbuscular mycorrhizal fungi, a key symbiosis in the development of quality traits in crop production, alone or combined with plant growth-promoting bacteria. *Mycorrhiza* 31: 655–69. https://doi.org/10.1007/s00572-021-01054-1.

Olsson, P.A., I. Thingstrup, I. Jakobsen, and E. Bååth. 1999. Estimation of the biomass of arbuscular mycorrhizal fungi in a linseed field. *Soil Biology and Biochemistry* 31: 1879–87. https://doi.org/10.1016/S0038-0717(99)00119–4.

Pandey, D.K., T. Malik, A. Dey, J. Singh, and R.M. Banik. 2014. Improved growth and colchicine concentration in *Gloriosa superba* on mycorrhizal inoculation supplemented with phosphorus-fertilizer. *African Journal of Traditional, Complementary and Alternative Medicines* 11: 439–46. http://dx.doi.org/10.4314/ajtcam.v11i2.30.

Panwar, N., S. Kothari, and N.S. Rathore. 2003. Suitability of greenhouse cultivation of medicinal plants. *Agricultural Engineering Today* 27: 52–61. https://www.indianjournals.com/ijor.aspx?target=ijor:aet&volume=27&issue=5and6&article=007.

Pedone-Bonfim, M.V.L., M.A. Lins, I.R. Coelho, A.S. Santana, F.S.B. Silva, and L.C. Maia. 2013. Mycorrhizal technology and phosphorus in the production of primary and secondary metabolites in cebil (*Anadenanthera colubrina* (Vell.) Brenan) seedlings. *Journal of the Science of Food and Agriculture* 93: 1479–84. https://doi.org/10.1002/jsfa.5919.

Rasouli-Sadaghiani, M.H., A. Hassani, M. Barin, Y.R. Danesh, and F. Sefidkon. 2010. Effects of arbuscular mycorrhizal (AM) fungi on growth, essential oil production and nutrients uptake in basil. *Journal of Medicinal Plants Research* 4: 2222–28. https://doi.org/10.5897/JMPR10.337.

Ratti, N., H.N. Verma, and S.P. Gautam. 2010. Effect of *Glomus* species on physiology and biochemistry of *Catharanthus roseus*. *Indian Journal of Microbiology* 50: 355–60. https://doi.org/10.1007/s12088-010-0012-2.

Riter Netto, A.F., M.S.M. Freitas, M.A. Martins, A.J.C. Carvalho, and J.A. Vitorazi Filho. 2014. Efeito de fungos micorrízicos arbusculares na bioprodução de fenóis totais e no crescimento de *Passiflora alata* Curtis. *Revista Brasileira de Plantas Medicinais* 16 (March): 1, 1–9. https://doi.org/10.1590/S1516-05722014000100001.

Rojas-Andrade, R., C. Cerda-García-Rojas, J. Frías-Hernández, L. Dendooven, V. Olalde-Portugal, and A. Ramos-Valdivia. 2003. Changes in the concentration of trigonelline in a semi-arid leguminous plant (*Prosopis laevigata*) induced by an arbuscular mycorrhizal fungus during the presymbiotic phase. *Mycorrhiza* 13: 49–52. https://doi.org/10.1007/s00572-002-0201-x.

Rydlová, J., M. Jelínková, K. Dušek, E. Dušková, M. Vosátka, and D. Püschel. 2016. Arbuscular mycorrhiza differentially affects synthesis of essential oils in coriander and dill. *Mycorrhiza* 26: 123–31. https://doi.org/10.1007/s00572-015-0652-5.

Sailo, G.L., and D.J. Bagyaraj. 2005. Influence of different AM-fungi on the growth, nutrition and forskolin content of *Coleus forskohlii*. *Mycological Research* 109: 795–98. https://doi.org/10.1017/S0953756205002832.

Santos, R.I. 2004. Metabolismo básico e origem dos metabólitos secundários. In *Farmacognosia: da planta ao medicamento*, eds. C.M.O. Simões, E.P. Schenkel, G. Gosmann, J.C.P. Mello, L.A. de Mentz, and P.R. Petrovick, 403–34. Porto Alegre: Florianópolis.

Schüßler, A., D. Schwarzott, and C. Walker. 2001. A new fungal phylum, the Glomeromycota: Phylogeny and evolution. *Mycological Research* 105: 1413–21. https://doi.org/10.1017/S0953756201005196.

Silva, F.A., F.S.B. Silva, and L.C. Maia. 2014. Biotechnical application of arbuscular mycorrhizal fungi used in the production of foliar biomolecules in ironwood seedlings [*Libidibia ferrea* (Mart. ex Tul.) LP Queiroz var. *ferrea*]. *Journal of Medicinal Plant Research* 8: 814–19. https://doi.org/10.5897/JMPR2014.5358.

Steinkellner, S., V. Lendzemo, I. Langer, et al. 2007. Flavonoids and strigolactones in root exudates as signals in symbiotic and pathogenic plant-fungus interactions. *Molecules* 12: 1290–06. https://doi.org/10.3390/12071290.

Taiz, L., and E. Zeiger. 2002. *Plant physiology*. 3rd ed. Sunderland: Sinauer Associates. https://exa.unne.edu.ar/biologia/fisiologia.vegetal/PlantPhysiologyTaiz2002.pdf.

Tejavathi, D.H., and D.R. Jayashree. 2013. Effect of AM fungal association on the growth performance of selected medicinal herbs. *Indian Journal of Applied Research* 3: 12–15. https://core.ac.uk/download/pdf/72802279.pdf.

Thirkell, T.J., M.D. Charters, A.J. Elliott, S.M. Sait, and K.J. Field. 2017. Are mycorrhizal fungi our sustainable saviours? Considerations for achieving food security. *Journal of Ecology* 105: 921–29. https://doi.org/10.1111/1365-2745.12788.

Vermerris, W., and R. Nicholson. 2008. Biosynthesis of phenolic compounds. In *Phenolic compound biochemistry*, eds. W. Vermerris and R. Nicholson, 63–149. Dordrecht: Springer. https://doi.org/10.1007/978-1-4020-5164-7_3.

Vo, A.T., I. Haddidi, H. Daood, Z. Mayer, and K. Posta. 2019. Impact of arbuscular mycorrhizal inoculation and growth substrate on biomass and content of polyphenols in *Eclipta prostrata*. *Hort Science* 54: 1976–83. https://doi.org/10.21273/HORTSCI14227–19.

Voets, L., I.E. de la Providencia, K. Fernandez, M. IJdo, S. Cranenbrouck, and S. Declerck. 2009. Extraradical mycelium network of arbuscular mycorrhizal fungi allows fast colonization of seedlings under *in vitro* conditions. *Mycorrhiza* 19: 347–56. https://doi.org/10.1007/s00572-009-0233-6.

Volpin, H., Y. Elkind, Y. Okon, and Y. Kapulnik. 1994. A vesicular arbuscular mycorrhizal fungus (*Glomus intraradix*) induces a defense response in alfalfa roots. *Plant Physiology* 104: 683–89. https://doi.org/10.1104/pp.104.2.683.

Watson, B.S., M.F. Bedair, E. Urbanczyk-Wochniak, et al. 2015. Integrated metabolomics and transcriptomics reveal enhanced specialized metabolism in *Medicago truncatula* root border cells. *Plant Physiology* 167: 1699–16. https://doi.org/10.1104/pp.114.253054.

Wu, Y.H., H. Wang, M. Liu, et al. 2021. Effects of native arbuscular mycorrhizae isolated on root biomass and secondary metabolites of *Salvia miltiorrhiza* Bge. *Frontiers in Plant Science* 12 (February): 617892, 1–13. https://doi.org/10.3389/fpls.2021.617892.

Yamawaki, K., A. Matsumura, R. Hattori, et al. 2013. Effect of inoculation with arbuscular mycorrhizal fungi on growth, nutrient uptake and curcumin production of turmeric (*Curcuma longa* L.). *Agricultural Sciences* 4: 66–71. http://dx.doi.org/10.4236/as.2013.42011.

Yan-qi, L.U., and H.E. Xue-li. 2005. Effects of AM fungi on the chemical composition and growth amount of *Atractylodes macrocephala* koidz seedling on different N levels. *Journal of Hebei University* 25: 650–53. https://xbzrb.hbu.edu.cn/EN/10.3969/j.issn.1000-1565.2005.06.019.

Zeng, Y., L.P. Guo, B.D. Chen, et al. 2013. Arbuscular mycorrhizal symbiosis and active ingredients of medicinal plants: Current research status and prospectives. *Mycorrhiza* 23: 253–65. https://doi.org/10.1007/s00572-013-0484-0.

Zhao, J.L., and X.L. He. 2011. Effects of AM fungi on drought resistance and content of chemical components in *Angelica dahurica*. *Acta Agriculturae Boreali Sinica* 19: 110–19. https://doi. 10.3724/SP.J.1011.2011.00110.

Zheng, Y., M. Dixon, and P. Saxena. 2006. Greenhouse production of *Echinacea purpurea* (L.) and *E. angustifolia* using different growing media, NO_3^-/NH^{4+} ratios and watering regimes. *Canadian Journal of Plant Science* 86: 809–15. https://doi.org/10.4141/P05-167.

Zimare, S.B., M.Y. Borde, P.K. Jite, and N.P. Malpathak. 2013. Effect of AM fungi (Gf, Gm) on biomass and gymnemic acid content of *Gymnema sylvestre* (Retz.) R. Br. ex Sm. *Proceedings of the National Academy of Sciences, India Section B: Biological Sciences* 83: 439–45. https://doi.org/10.1007/s40011-013-0159-9.

Zitterl-Eglseer, K., M. Nell, A. Lamien-Meda, et al. 2015. Effects of root colonization by symbiotic arbuscular mycorrhizal fungi on the yield of pharmacologically active compounds in *Angelica archangelica* L. *Acta Physiologiae Plantarum* 37 (January): 21, 1–11. https://doi.org/10.1007/s11738-014-1750-2.

Zolfaghari, M., V. Nazeri, F. Sefidkon, and F. Rejali. 2013. Effect of arbuscular mycorrhizal fungi on plant growth and essential oil content and composition of *Ocimum basilicum* L. *Iranian Journal of Plant Physiology* 3: 643–50. https://ijpp.saveh.iau.ir/article_540674_106e7cbd543d7f90dd7be07c661a7736.pdf.

Zubek, S., S. Mielcarek, and K. Turnau. 2012. Hypericin and pseudohypericin concentrations of a valuable medicinal plant *Hypericum perforatum* L. are enhanced by arbuscular mycorrhizal fungi. *Mycorrhiza* 22: 149–56. https://doi.org/10.1007/s00572-011-0391-1.

Zubek, S., K. Rola, A. Szewczyk, M.L. Majewska, and K. Turnau. 2015. Enhanced concentrations of elements and secondary metabolites in *Viola tricolor* L. induced by arbuscular mycorrhizal fungi. *Plant Soil* 390: 129–42. https://doi.org/10.1007/s11104-015-2388-6.

Zubek, S., A. Stojakowska, T. Anielska, and K. Turnau. 2010. Arbuscular mycorrhizal fungi alter thymol derivative contents of *Inula ensifolia* L. *Mycorrhiza* 20: 497–504. https://doi.org/10.1007/s00572-010-0306-6.

14 Arbuscular Mycorrhizal Fungi–Plant Interaction for Salinity Stress Alleviation

*Priyanka Chandra
(ICAR – Central Soil Salinity Research Institute, Karnal 132001, Haryana, India; priyanka.chandra921@gmail.com)

Arvind Kumar Rai
(ICAR – Central Soil Salinity Research Institute, Karnal 132001, Haryana, India; ak.rai@ icar.gov.in)

Parul Sundha
(ICAR – Central Soil Salinity Research Institute, Karnal 132001, Haryana, India; parul.sundha@icar.gov.in)

Nirmalendu Basak
(ICAR – Central Soil Salinity Research Institute, Karnal 132001, Haryana, India; nirmalendubasak@rediffmail.com)

Sandeep Bedwal
(ICAR – Central Soil Salinity Research Institute, Karnal 132001, Haryana, India; sbedwal519@gmail.com)

Rajender Kumar Yadav
(ICAR – Central Soil Salinity Research Institute, Karnal 132001, Haryana, India; rk.yadav@icar.gov.in)

14.1 INTRODUCTION

Due to its prevalence in all climates, an increase in the salinity of the soil is becoming a serious issue on a global scale. Growing salinization of arable land will cause a 50% loss of land, and up to 50% by 2050 (Basak et al., 2020). According to the Food and Agriculture Organization, over one billion hectares in more than 100 nations experience issues with soil salinity, and about 77 million hectares (5%) of the 1.5 billion ha of cultivated land around the globe are impacted by salinity (FAO, 2020). The severe effects of soil salinization on plant growth, productivity, and agricultural output are thoroughly described in several reports (Rai et al., 2021a). According to recent reports, soil salinization had a direct negative influence on the world economy and resulted in losses totaling $27.3 billion (Sharma et al., 2019; Kumar and Sharma, 2020). The excessive use of synthetic chemical fertilizers, the use of groundwater for irrigation, and flood irrigation practices are the main causes of the increasing soil salinization of arable land (Cuevas et al., 2019). These factors ultimately convert fertile soil into saline soil and reduce crop productivity by about 20%. High salt deposition has a negative impact on plant germination, development, and reproduction, which results in significant productivity losses (Kamran et al., 2019). Salinity is now reported as an important abiotic stress that has been linked to the following serious effects: (a) reduced plant growth, including seed germination, vegetative development, and reproductive development; (b) increased production of reactive oxygen species (ROS), which can cause damage to cell organelles, plasma membrane proteins, and lipids; (c) the accumulation of Na^+ and Cl^- ions in plant cells, which results in a severe ion imbalance and physiological dryness, is a negative effect of salinity stress (Eynard et al., 2005). Ion homeostasis disturbances affect several physiological and biochemical functions in plants (Guo et al., 2020).

A broad variety of microbial colonies plants in their natural habitat, affecting their morphological, physiological, biochemical, and molecular defenses against the shifting environment (Chandra et al., 2021). To deal with the soil salinity, these adaptive measures, however, lose their effectiveness. Arbuscular mycorrhizal (AM) fungi are

DOI: 10.1201/9781003408833-14

widespread among a variety of soil microorganisms colonizing the root and have been shown to enhance plant performance in stressful settings, hence increasing plant yield (Feng et al., 2002; Jain et al., 2023, 2024). Nearly 80% of all terrestrial plants, including the majority of crop species, are capable of developing a mutualistic interaction with the AM fungus (Treseder, 2013). These plants include both terrestrial and aquatic plants. In this regard, using biological processes like mycorrhizal application to lessen the effects of salt stress might be a better option. It has been demonstrated that the AM fungi use a variety of strategies to support plant development and salinity tolerance, including magnifying nutrient uptake and modifying the physiological and biochemical characteristics of the host plant (Tian et al., 2009). According to an extensive study, AM promotes improved plant development and yield under saline stress circumstances. Because of these benefits, AM has become a good replacement for bio-ameliorating saline soils (Chandra et al., 2022). It is possible to improve plant-fungi symbiosis in contemporary crop production systems by using these improved AM fungi and related microorganisms (Sheng et al., 2009; Gupta et al., 2022, 2024). This will raise crop yield and quality for a variety of commercially valuable plant species. The inoculation of AM fungi improves a variety of processes, such as photosynthesis, nutrient absorption, antioxidant defense, and osmolytes accumulation, which ultimately results in improved growth performance and increased plant salinity tolerance, according to numerous studies on a variety of plant species (Oliveira et al., 2005; Evelin et al., 2019). The several processes (nutrient absorption, biochemical, and physiological mechanisms) used by AM symbiosis in host plants to reduce salt stress are covered in the current chapter and are explained in the following sections.

14.2 IMPACT OF SALINITY STRESS ON CROP PHYSIOLOGY

One of the most detrimental abiotic stresses that impacts plant growth, development, and productivity is salinity stress. Due to their innate ability to adapt, plants have developed suitable coping mechanisms for such harsh situations with high salt concentrations (Rengasamy, 2016). The two main negative consequences that affect plants in high salinity are intracellular osmotic pressure and ion toxicity (Abhinandan et al., 2018; Waqas et al., 2019). Water availability is impacted by osmotic stresses, increased Na^+ and Cl^- absorption, which lowers the osmotic potential between the root and soil. Ion toxicity is caused by high levels of Na+, Cl^-, or SO_2^4, which prevents plants from absorbing nutrients. As a result of the high concentration of salt lowering the endogenous level of phytohormones, which prevents seedling establishment, seeds in saline soils do not germinate effectively (Ma et al., 2020).

Salinity impairs various physiological functions in plants; decreased stomatal conductivity limits C-fixation capacity, disrupting the catalytic activities of C-fixing enzymes, and degrades photosynthetic pigments. Plants growing in saline soil have shown a considerable loss in shoot and root biomass (Gupta and Huang, 2014). According to studies, salinity (10 mM NaCl) can reduce *Triticum aestivum* yield by up to 65%. Increased salt concentrations around the root decrease K, Ca, and Mg uptake, which disrupts stomatal conductance and transpiration rate (Muhu-Din Ahmed et al., 2022). Iron (Fe) is a crucial metal activator (cofactor) of many antioxidant enzymes that aid in controlling the activities that allow plants to survive, such as photosynthesis and chloroplast production. Salt-affected soils are deficient in iron content (Arif et al., 2020).

In such soil, plants exhibited interveinal chlorosis and decreased yield. In saline environments, the chloride ion (Cl^-) predominates, and its absorption is inversely linked to that of nitrogen. Increased Cl^- uptake brought on by salinity is frequently accompanied by a decrease in nitrogen uptake by plants (Hessini et al., 2019). The generation of ROS, including O_2, OH^-, and H_2O_2, is a result of oxidative stress brought on by salinity as a secondary stress. ROS are primarily found in the chloroplasts, mitochondria, and peroxisomes as a harmful consequence of aerobic metabolism. Their increased production results in lipid peroxidation, which disrupts membrane permeability and photosynthesis, destroys nucleic acids, blocks enzymes, and triggers programmed cell death (Bistgani et al., 2019). Due to the high concentrations of salts, *Brassica napus* and *Triticum aestivum* plants grown in saline circumstances had poor development and yield (EL Sabagh et al., 2021; Naheed et al., 2021; El-Badri et al., 2022).

14.3 SALT-AFFECTED SOILS

Salinity is the edaphic stresses that occur alone or in combination with other stresses (Eynard et al., 2005). The high electrical conductivity (EC) of the soil aqueous solutions, which is expressed as a significant amount of soluble salts, denotes salt-affected soils (SAS) (Chhabra, 2004). High osmotic pressure, restricted water intake, the presence of certain ions that affect nutrient availability, and an unbalanced equilibrium of nutrients in the SAS hinders the growth and vigour of crops in these soils. Saline and sodic soils are two types of salt-affected soils that differ from one another primarily in terms of their genesis, physico-chemical characteristics, and limitations on plant growth (Singh and Abrol, 1985; Abrol et al., 1988).

Their identification is based on the determination of EC, ESP, and SAR, and then further classified based on the USDA classification. White salt encrustation covers the surface of saline soils. They primarily contain sodium, calcium, and magnesium sulfates and chlorides. These soils' soil saturation paste extract has an electrical conductivity (ECe) of more than 4 dSm^{-1} at 25°C. Saline soil has a pH under 8.5 and an exchangeable sodium content of less than 15. Salts like sodium carbonate, sodium bicarbonate, and sodium silicate, which may be dissolved in alkaline solutions, are abundant in alkali soils/sodic soil. Additionally, they have an adequate amount of exchangeable salt, which leads to poor soil physical conditions that hinder plant growth. The exchangeable sodium percentage is larger than 15, the electrical conductivity of the saturation paste extract is less than $4dSm^{-1}$ at 25°C, and the pH ranges from 8.5 to 10. The pH of calcareous alkali soils and their ESP are closely related. While the pH range is higher than 8.5, saline-alkali soils have electrical conductivity of the saturation paste extract larger than $4dSm^{-1}$ and the exchangeable sodium percentage greater than 15. These soils develop as a result of a salinization and alkalization process occurring simultaneously. The characteristics of these soils will resemble those of alkali soils as long as there are too many soluble salts present. The soil becomes very severely alkaline (pH above 8.5) as a result of the leaching of extra soluble salts (Singh and Abrol, 1985; Chhabra, 2004; Abrol et al., 1988).

14.4 ARBUSCULAR MYCORRHIZAL FUNGI–PLANT INTERACTION

Arbuscular mycorrhizal fungi (AMF) are obligate biotrophs that must be grown with a living plant, and approximately 80% of terrestrial plant species can be colonized by AM fungi and establish symbiosis in nature. Therefore, the concept of rhizosphere has been extended to incorporate the soil surrounding both roots and mycorrhizal fungi, being referred to as the "mycorrhizosphere, which includes the more specific zone influenced by only (Zhang et al., 2022). Arbuscular mycorrhizal fungi (AMF) are obligatory biotrophs, meaning they can only thrive in the presence of living plants. Most terrestrial plant species have symbiotic relationships with AMF, about 80% of them. The term "rhizosphere" has been changed to "mycorrhizosphere" to define the soil next to both plant roots and AMF. Similarly, the term "hyphosphere" has been created to characterize the more precise zone influenced by the hyphae of AMF (Table 14.1).

In addition to the rhizosphere, the mycorrhizosphere and hyphosphere are also the main areas where plant-soil-microbe interactions occur (Zhang et al., 2022) (Figure 14.1). It is known that there are more than 255 AMF species, and several of them, including *Glomus etunicatum, Funneliformis mosseae,* and *Rhizophagus irregularis*, are employed as biofertilizer for numerous crops such as wheat, maize, and rice, and have either positive effects of AMF (Bharti et al., 2016; Emmanuel and Babalola, 2020; Deja-Sikora et al., 2023).

AMF can occasionally change the soil's composition and have an impact on the function of microorganisms (Wilde et al., 2009). AMF gives plants a healthier habitat and facilitates nutrient uptake under nutrient-deficient conditions as well (Rai et al., 2021b). The benefits of AMF as biofertilizers for crops, especially cereals, include increased nutrient uptake, improved physiological traits, and increased plant growth and productivity (Briat et al., 2020; Etesami et al., 2021). Since AMF rely on photosynthates from host plants, they are obligate biotrophs that cannot proliferate in the absence of carbon and a living plant. In exchange for access to host plants' carbon stores, the AM fungus improved plant growth and P uptake. Because P from the soil is exchanged for C from plant roots, AM symbioses are mutualistic (Walder et al., 2012). According to a report, AM fungi utilize 5–20% of the total carbon absorbed by the plants (Bonfante and Desirò, 2015; Luginbuehl et al., 2017). Recent research using an in vitro culture system showed that plants give AMF more carbohydrates than AM fungus-enhanced P (Wang et al., 2016). However, roots offer more carbohydrates, indicating that the relationship between plants and AM fungal partners maintains an equitable flow of carbon and phosphorus (Thirkell et al., 2020). The 30% carbon investment of the C3 plant flax yielded up to 94% of the phosphorus delivered by the mycorrhizal hyphal networks, however, the C4 plant sorghum invested 70% of carbon with 6% of phosphorus return (Walder et al., 2012). This shows that C-P transfer is not acceptable in the C3/C4 plant intercropping system (Keymer et al., 2017; Faghihinia and Jansa, 2022). Plant roots and extraradical hyphae of AM fungus are both capable of secreting protons, enzymes, and carbohydrates, which contribute to the variations in soil chemical and biological features between rhizosphere and bulk soil (Bonfante and Genre, 2010).

Numerous studies have demonstrated that the unbalanced uptake of anion/cation by AM fungal mycelium led to variance in the rhizosphere's soil acid-base properties (Aarle et al., 2002). When the soil was fertilized with NH_4^+, the pH was found to be reduced at the root-soil and hyphae-soil interfaces, but the hyphosphere soil became alkaline when treated with NO_3^+. By dissolving phosphate ions along with Ca_2^+ and phytate in calcareous soil, acidification caused by NH_4^+ in rhizosphere and hyphosphere increased the mobilization of sparingly soluble inorganic and organic phosphates (Li et al., 1991; Ortas et al., 2004). So, one of the key mechanisms in the mobilization of scarcely soluble phosphates by AM fungi in soil was acidification (Ezawa et al., 1995). Additionally, because phosphatase activity is strongly pH dependent, it is typically tested at pH 6.5 or pH 11.0 for acid and alkaline phosphatase, respectively

TABLE 14.1
Terms and Their Definitions

	Terms	Definition
1	Rhizosphere	The narrow zone of soil surrounding living roots
2	Mycorrhizosphere	The soil zone influenced by both roots and mycorrhiza
3	Hyphosphere	The zone of soil influenced by the hyphae of AMF

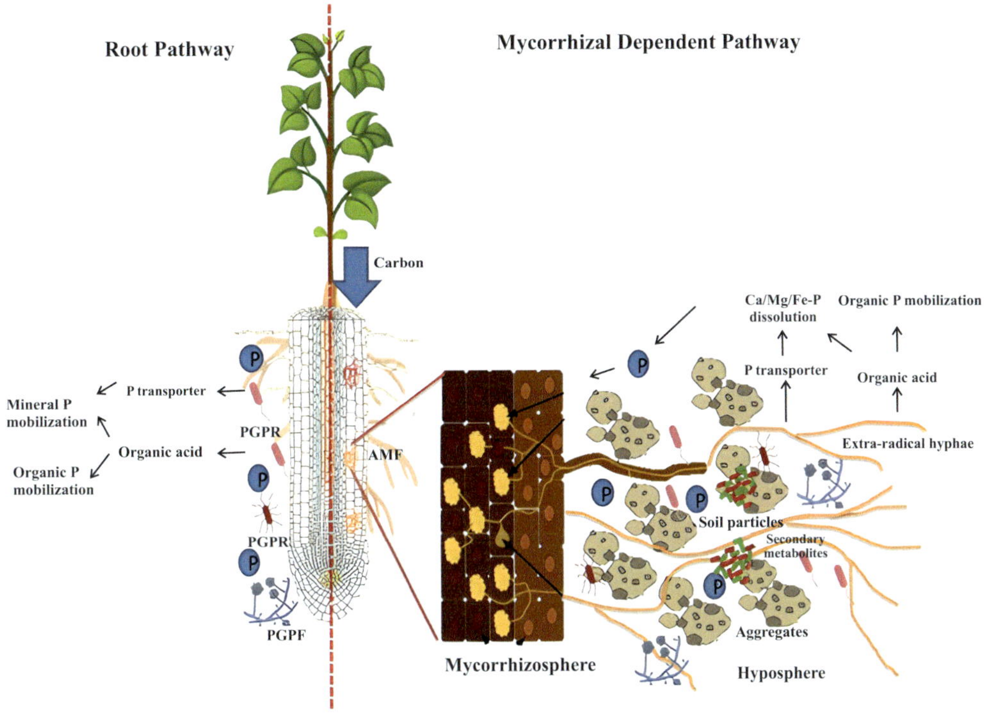

FIGURE 14.1 The description of the functioning of AMF in the rhizosphere.

(Johri et al., 2015). While alkaline phosphatase transcripts were reported in intraradical and extraradical hyphae in *Funneliformis mosseae*, the transcriptome of *Glomus intraradices* revealed that this AMF possesses phosphatase transcripts in its genome. The release of carbohydrates by AM fungal extraradical mycelia may be able to give hyphae-associated microorganisms with enough energy to function and allow for direct interaction (de Oliveira Mendes et al., 2014; Andrino et al., 2021; Etesami et al., 2021).

These exudates, which are released by AM extra-radical mycelia, affect the composition of the community and promote the proliferation of beneficial microorganisms (Jiang et al., 2021). The study was conducted to analyze the relationships between AMF and the bacteria they are associated (Chiu and Paszkowski, 2019). According to the findings, exudates contain carbohydrates, amino acids, and metabolites that can be used as a substrate or source of energy to promote the growth and metabolism of bacteria that are linked with the AM fungus (Tahat and Sijam, 2012; Shahzad et al., 2015). There is evidence that these compounds, which are present in root or hyphal exudates, promote the development of the bacteria that are related to them (Kobae, 2019).

14.5 ROLE OF AM FUNGI IN PLANT GROWTH PROMOTION

Plant and AMF have a symbiotic connection in which nutrients like organic carbon, lipids, and sugars are transferred (Luginbuehl et al., 2017). It is widely recognized that AMF colonization in plants can increase plants' intake of nutrients (Sheng et al., 2009; Lee et al., 2015). AMF application has become a fascinating tactic for bettering crop nutrition. About 90% of the plant's P, 60% of its N, and 20% of its carbon can be maintained by AMF. In different plants, AM fungi raise the levels of N, P, calcium (Ca), and other nutrients like sulfur (S), potassium (K), iron (Fe), copper (Cu), manganese (Mn), and zinc (Zn). Additionally, early plant growth and increased stress tolerance may be facilitated by AM fungi by stimulating physiological and biochemical pathways in cereal plants at different stages (Etesami et al., 2021; Rai et al., 2021b; Chandra et al., 2022).

Numerous studies show that AMF increases plant uptake of crucial macro- and micronutrients, which increases photosynthetic activity and, ultimately, biomass buildup. AMF can increase the uptake of inorganic nutrients, particularly phosphate, by nearly all plants (Saboor et al., 2021b). Additionally, AMF can encourage plants to absorb nutrients from nutrient-poor soils. AMF interaction has been shown to improve the phytoavailability of micronutrients like zinc and copper in addition to macronutrients (Saboor et al., 2021b). AMF enhances the host roots' capacity for surface absorption. In tomatoes treated with AMF, there was an increase in leaf area as well as nitrogen, potassium, calcium, and phosphorus contents, which indicated improved plant growth (Ingraffia et al., 2019).

AMF form symbiotic relationships with host plant roots to absorb vital nutrients from them and then exchange them for mineral nutrients, including N, P, K, Ca, Zn, and S (Saboor et al., 2021b). As a result, inside the root cells, AMF maintain the plants' nutritional needs even in unfavorable situations (Chandrasekaran, 2022). AMF create fungal structures like arbuscules that help in the interchange of inorganic minerals and carbon and phosphorus compounds, giving host plants a significant amount of vigour in the process (Wipf et al., 2019). As a result, they can greatly increase the phosphorus levels in both root and shoot systems. Mycorrhizal association enhances the phosphorus supply to the infected roots of host plants in phosphorus-limited environments. Increased photosynthetic activity and other leaf functions that are correlated with N, P, and carbon uptake have also been seen after AMF inoculation. In *Euonymus japonica*, during salinity stress, due to immediate fungal association, enhanced levels of P, Ca, and Kwas observed (Gómez-Bellot et al., 2015). AMF symbiosis also ensures nutrient uptake under drought stress. AMF inoculation in *Pistacia vera* showed high levels of P, K, Zn, and Mn under drought stress in a different study. Additionally, *Chrysanthemum morifolium* plant tissues had higher P and N contents following AMF inoculation (Wang et al., 2018), while *Leymus chinensis* had higher water and intercellular CO_2, P, and N contents following AMF inoculation, all of which enhanced seedling weight (Li et al., 2019; Khan et al., 2021).

It is reported that AMF stimulates growth under salinity stress by increasing the intake of nearly all necessary nutrients and, on the other hand, decreasing the uptake of Na and Cl (Paul Schreiner, 2007; Chandra et al., 2022; Chandrasekaran, 2022). By efficiently enhancing nutrient uptake, extra-radical mycelium (ERM) can enhance plant growth and development. It is well known that fungi can extract significant amounts of nitrogen (N) from dead and decomposing matter to survive, and because AMF have a larger biomass of N than other fungi, they are the largest contributor in the overall N pool, which is roughly equivalent to fine roots (Hodge and Storer, 2015). Due to the extra-radical hyphae of AMF that AMF absorb and assimilate inorganic N, AMF play a significant role in the N cycle. According to several reports, the AMF can transfer between 20 and 75% of the total N they absorb to their hosts (Sweeney et al., 2022). Improved chlorophyll contents result from better uptake of N in AMF-colonized plants (Balestrini et al., 2020). Several studies have also shown that plants treated with AMF have better nitrogen absorption (Thirkell et al., 2016, 2019; Fu et al., 2021). Olive plants inoculated with AMF demonstrated improved growth and macro- and micronutrient acquisition (Aganchich et al., 2022). Increased secondary metabolite levels in AMF-colonized strawberries led to better antioxidant properties (Robinson Boyer et al., 2016). Sugars, organic acids, vitamin C, flavonoids, and minerals in citrus fruit can all be improved by *Glomus versiforme* (Qiangsheng et al., 2007).

Additionally, increased accumulation of anthocyanins, chlorophyll, carotenoids, total soluble phenolics, tocopherols, and different mineral elements are transported by mycorrhizal symbiosis in several horticulture crops (Szczałba et al., 2019; Zhu et al., 2022). Crops like maize (Yun et al., 2018), yams (Lu et al., 2015), and potatoes (Ghobadi et al., 2020) demonstrated better mycorrhizal responsiveness in terms of their higher yields. The biosynthesis of beneficial phytochemicals in edible plants can also be improved by AMF (Rai et al., 2021b). *Trichoderma atroviride* and the AMF, *G. intraradices* and *G. mosseae,* increase the grain yield of wheat as a result of AMF's greater micronutrient availability and better absorption of N, P, and other micronutrients at grain filling stages, higher protein concentration. It has been reported that improved cereal crop quality and yield result from nutrient uptake that is selectively controlled according to crop stage (Cobb et al., 2016; Chandra et al., 2022).

According to the reports, AMF colonization boosts P uptake during flowering initiation, which influences early flowering and results in more flowers and consequently more grain. Under drought conditions, *G. intraradices* increases the output of foxtail millet (Gong et al., 2015) and the green and dry biomass in maize (Hagh et al., 2016). The plants' capacity to develop their resistance to biotic and abiotic stress conditions is strengthened by AMF. Similar to this, AMF and *Azotobacter chroococum* improved the cob weight, length, grain weight, and starch content of maize during the fruiting stage by facilitating N and P uptake (Song et al., 2020). Under salt stress, *F. mosseae*, *Acaulospora laevis*, and *G. margarita* increased rice productivity by 125% while also increasing spikelet fertility up to 125%, yield by 143%, and grain weight by 18.4% (Parvin et al., 2020).

Studies on barley revealed that AMF, such as *R. irregularis*, *Pacispora franciscana*, *F. mosseae*, *F. geosporum*, and *G. tenebrosum*, responded positively and boosted yields, kernel weight, and nutrient content (Khan et al., 2022; Pellegrino et al., 2022; Benaffari et al., 2023). The AM fungi *R. irregularis* and *G. mosseae* were found to enhance sorghum vegetative biomass and favor high accumulation of P, Zn, and Fe in the grain under drought stress conditions (Khan et al., 2022). Similarly, *Rhizophagus fasciculatus* and *Ambispora leptoticha* synergistically with *Pseudomonas,* possessing plant growth–promoting potential, boosting millet's growth and productivity (Mathimaran et al., 2020). *C. etunicatum* also increases millet yield and productivity (Schütz et al., 2022). Quinoa under drought stress treated with *G. mosseae* had increased plant shoot growth, dry weight, grain production, and nutritional quality (Benaffari et al., 2023). Because ammonium and nitrate ions in soil are easier to diffuse to the root than phosphate ions are, AMF plays a significant role in nitrogen uptake (Ingraffia et al., 2019). The extraradical hyphae of AM fungi can absorb inorganic N as both NO_3^- and NH_4^+ and organic N as amino acids, then transmit to plants and enhance plant nitrogen uptake, according to the use of ^{15}N-labeled technology. Additionally, mounting research suggests that extraradical mycelia of AM fungi can absorb significant N from a decomposing organic patch and boost host plants' ability to take nitrogen (Walder et al., 2012; Hodge and Storer, 2015).

It is generally agreed that AMF can mobilize the sparingly soluble P from soil using a variety of processes, which are enumerated in what follows: (1) Extraradical mycelia are extended, and (2) protons, organic anions, and phosphatase are released. The P depletion zone was caused by extraradical hyphae of the AM fungus *Glomus mosseae* that reached up to 11.7 cm from the surface of the root (Koide and Kabir, 2000). Phosphatase activity rises as a result of the hygrosphere's acidification, which is brought on by $NH4^+$ absorption through the extraradical mycelia of the AM fungus. This stimulates phytin-P mineralization and improves maize uptake of P from phytin-P (Yao et al., 2001; Etesami et al., 2021). Through their interactions, an AMF *Rhizophagus irregularis* and phosphate-solubilizing bacteria improved the mineralization of soil Na-phytate by raising phosphatase activity (Sharma et al., 2020). Additionally, the acquisition of microelements by plants is greatly influenced by the AM fungus. Using compartmented microcosms, it has been found that 53–62% of the total Cu uptake in *Trifolium repens* (Xiao et al., 2020) and the total of 25% of Zn uptake in maize (Saboor et al., 2021a) is accounted for by extraradical hyphae. Additionally, AM fungal inoculation had a substantial impact on the concentration of Fe in sorghum shoots, indicating that the exterior hyphae of AM fungi can mobilize and aid in Fe uptake from the hyphal compartment and transfer to host plants (Abdelhalim et al., 2022).

14.6 EFFECT OF AMF ON CROPS UNDER SALINITY STRESS

Salinity in the soil slows plant growth by impeding nutrient uptake and physiological development, which lowers yield productivity. Salt Stress also leads to the overproduction of ROS. The deleterious effects of salt stress on plants are also known to be ameliorated by AMF (Yilma, 2019). Under salinity stress, the AMF improved the growth rate, leaf water potential, and water use efficiency in the *Antirrhinum majus* plants (El-Nashar, 2017). According to the research, the AMF symbiosis has positive effects on physiological variables as photosynthetic rate, stomatal conductance, and leaf water relations in saline environments (Khan et al., 2021).

AMF greatly reduced the negative effects of salt stress on photosynthesis (Elhindi et al., 2016). Under salt stress, *Ocimum basilicum* L., mycorrhizal inoculation significantly increased photosynthetic rate, other gas exchange properties, chlorophyll content, and water usage efficiency (Elhindi et al., 2016). *Allium sativum* plants that have been inoculated with AMF demonstrated enhanced growth characteristics under salty circumstances, including leaf area index and fresh and dry biomass (Cantrell and Linderman, 2001). Recent investigations showed that mycorrhizal inoculation under saline conditions significantly increased fresh and dry weights as well as the N concentration of shoot and root (Emmanuel and Babalola, 2020). Additionally, plants with AMF exhibit improved synthesis of salicylic acid, jasmonic acid, and several critical inorganic nutrients (Evelin et al., 2019; Arif et al., 2020). For instance, under salt stress conditions, the AMF-treated *Cucumis sativus* plants had increased amounts of total P, Ca^{2+}, N, Mg^{2+}, and K^+ compared to the untreated plants (Agarwal et al., 2013). Under saline conditions, mycorrhizal inoculation of *Capsicum annuum* resulted in increased chlorophyll levels, Mg^{2+} and N absorption, and decreased Na+ transport (Yilma, 2019).

Furthermore, mycorrhizal lettuce plants produced more biomass, increased proline synthesis, increased N uptake, and showed changes in ionic relations, particularly decreased Na^+ accumulation, than non-mycorrhizal plants did under stress. Key growth regulator levels can be efficiently controlled by AMF inoculation (Azcón et al., 2003). For instance, under salt stress, an AMF-mediated increase in cytokinin concentration led to a noticeable photosynthate translocation was reported. Additionally, it was established that changes in the polyamine pool were the cause of the AMF-mediated growth enhancement under salt stress (Begum et al., 2019).

In addition, how salt impacts lettuce plants was significantly reduced by increased strigolactone in AMF-treated plants (Aroca et al., 2013). By inhibiting lipid membrane peroxidation under salinity stress, AMF-colonized plants can reduce oxidative stress. Additionally, it was found that the inoculation of AMF increased the accumulation of several organic acids, which in turn increased the osmoregulation process in plants grown under saline stress. For instance, the indirect involvement of AMF in plant osmoregulation under salt stress by seen in increased synthesis/accumulation of certain organic acids in maize plants growing in saline soil and inducing higher production of betaine (Sheng et al., 2011).

Due to the reduced rate of transpiration and oxidative stress brought on by a drought-like scenario caused by salinity stress, plant growth is negatively impacted. Through its negative effects on enzyme activity, ion absorption, and nutrient assimilation, drought stress has a detrimental impact on plant growth. However, there is compelling evidence that AMF reduces the effects of drought stress in a variety of crops, including wheat (Thirkell et al., 2020), barley (Sendek et al., 2019), maize (Begum et al., 2019), soybean (Oliveira et al., 2022), strawberry (Boyer et al., 2015), and onion (Cantrell and Linderman, 2001). The fact that roots have access to a lot of soil and that fungi's extraradical hyphae can withstand dryness may be the main reason for plant tolerance.

Through their exterior hyphae, which may directly absorb water and subsequently transmit it to host plants from the soil, AMF are well known for boosting water intake and enhancing drought tolerance in plants. These extraroot zone external hyphae serve as extension roots (Saboor et al., 2021b; Zhang et al., 2022). These hyphae enlarge the absorptive area by entering soil pores. Numerous investigations show a correlation between plant AMF connections and an increase in plant drought resistance. The quantity of water that the hyphae can absorb during drought stress may help the plant maintain turgor pressure, photosynthetic rates, and plant growth (Qiangsheng et al., 2007; Young et al., 2015; Aganchich et al., 2022).

In plants under drought stress, the plant-AMF interaction regulates several physio-biochemical activities, such as improved osmotic adjustment, stomatal regulation by regulating ABA metabolism, enhanced proline accumulation, or increased glutathione levels. Under drought stress, AMF increases root size and effectiveness, leaf area index, and biomass. Additionally, the AMF symbiosis improves stomatal conductance, transpiration rate, leaf water relations, and gas exchange (Mathur et al., 2019). The ABA responses that control stomatal conductance and other connected physiological processes can be aided by AMF. Through up-regulation of the antioxidant system, AMF-mediated growth and photosynthetic increase in C3 (*Leymus chinensis*) and C4 (*Hemarthria altissima*) plants (Li et al., 2019).

Two aquaporin genes, GintAQPF1 and GintAQPF2, were found in the extraradical mycelia of *Glomus intraradices* and cortical cells of plant roots under drought stress conditions, demonstrating the direct contribution of the AM fungus to plant drought tolerance (Li et al., 2013). The effects of salinity stress on photosynthesis, osmotic balance, enzymatic activity, and nutrient uptake prevent plants from growing and accumulating biomass. Through altering plant physiological processes, AMF regularly assist plants in minimizing the effects of salt stress (Giovannini et al., 2020).

14.7 SECONDARY METABOLITES OF MYCORRHIZA FOR PLANT GROWTH

Mycorrhizal fungi produce a variety of secondary metabolites, which play crucial roles in this mutualistic interaction. Secondary metabolites, unlike primary metabolites, are not directly involved in the basic metabolic processes of the organism but are essential for defense, communication, and ecological interactions (Sheng et al., 2011).

14.7.1 Types of Secondary Metabolites in AMF

AMF produces a diverse range of secondary metabolites, including terpenes, alkaloids, phenolics, and organic acids. These compounds are released into the soil or exchanged between the fungus and plant, where they influence plant health and growth. The main categories of secondary metabolites include:

14.7.1.1 Terpenes

Terpenes are a large group of secondary metabolites synthesized by mycorrhizal fungi. These compounds have various roles in plant-fungal interactions, including the enhancement of plant growth and defense. Some terpenes produced by AMF can stimulate plant root growth, increase plant resistance to pathogens, and improve plant tolerance to abiotic stresses like drought or extreme temperatures. Terpenes also contribute to the protection of mycorrhizal hyphae from microbial predation.

14.7.1.2 Phenolic Compounds

Phenolic compounds, such as flavonoids and lignans, are another class of secondary metabolites produced by AMF. These compounds are involved in the regulation of plant immune responses and the enhancement of nutrient uptake. Phenolics have been shown to affect plant cell wall formation and structural integrity, which aids in the formation of stronger root systems. Furthermore, these compounds can help plants resist oxidative stress, improve their water retention capacity, and boost their resistance to fungal pathogens.

14.7.1.3 Organic Acids

Organic acids, such as citric acid and oxalic acid, are commonly produced by AMF. These acids play an essential role in soil nutrient mobilization. Organic acids can chelate metal ions and make phosphorus and other nutrients more accessible to plants, particularly in soils with low nutrient availability. This enhances the uptake of essential nutrients, leading to better plant growth. The presence of these acids also influences the pH of the surrounding soil, which can impact microbial activity and nutrient cycling.

14.7.1.4 Alkaloids

Alkaloids, a diverse group of nitrogen-containing compounds, are produced by some AMF. These metabolites are known for their potential role in protecting the plant against herbivores, pathogens, and other environmental stresses. Alkaloids may help strengthen plant defense mechanisms by stimulating the production of protective proteins and antioxidants in plant cells. Additionally, alkaloids can influence the plant's signaling pathways, helping to regulate growth and stress responses.

14.8 FUNCTIONS OF SECONDARY METABOLITES IN PLANT GROWTH

14.8.1 Enhancement of Nutrient Uptake

One of the primary functions of secondary metabolites in AMF is the facilitation of nutrient acquisition. Organic acids released by the mycorrhizal hyphae improve the availability of essential nutrients, especially phosphorus, nitrogen, and trace elements. This is particularly important in nutrient-deficient soils, where plants rely heavily on the fungal symbionts to obtain adequate nutrition (Thirkell et al., 2020).

14.8.2 Plant Growth Promotion

AMF not only supply plants with nutrients but also produce secondary metabolites that enhance root development and overall plant growth. For example, terpenes and phenolic compounds can stimulate root branching and elongation, facilitating better anchorage and access to soil resources. These metabolites also promote the formation of a healthier root microbiome, which is critical for nutrient cycling and disease suppression.

14.8.3 Stress Tolerance

AMF contributes to plant resilience against various abiotic stresses such as drought, salinity, and temperature extremes. Secondary metabolites such as phenolic compounds and terpenes have been shown to improve the plant's ability to tolerate oxidative stress by reducing the production of reactive oxygen species (ROS) in response to stress. Additionally, organic acids can help regulate water uptake, ensuring that plants can maintain hydration during periods of water scarcity (Feng et al., 2002).

14.8.4 Disease Resistance

The secondary metabolites produced by AMF also play a role in plant defense mechanisms. These metabolites can stimulate the production of plant defense hormones such as jasmonic acid and salicylic acid, which activate plant immune responses. Moreover, the release of certain alkaloids and phenolic compounds can inhibit the growth of soilborne pathogens, further protecting the plant (Thirkell et al., 2020).

The secondary metabolites produced by mycorrhizal fungi play an indispensable role in the growth and health of plants. These compounds enhance nutrient uptake, promote growth, improve stress tolerance, and provide disease resistance, all of which are vital for plant survival, particularly in nutrient-poor or challenging environments. Understanding the biochemical interactions between mycorrhizal fungi and plants can lead to improved agricultural practices and sustainable crop production, as the application of mycorrhizal fungi can boost plant health and productivity. Further research into the specific secondary metabolites involved in these processes may reveal even more applications for these fascinating microorganisms in agriculture and environmental management.

14.9 CONCLUSION AND FUTURE PROSPECTS

The benefits of AMF symbiotic relationships with various plants in harsh environments, such as high salinity, and the secondary metabolites produced are summarized in this chapter. These relationships can encourage plant growth and productivity. It has been recently demonstrated that plants inoculated with AMF can effectively combat various environmental cues, like salinity, drought, nutrient stress, and extreme temperatures, and as a result, help increase the yield of several crops. The AMF have primarily been discussed as beneficial entities for nutrient uptake from soil. The consistent sustainability of contemporary global agricultural systems depends critically on encouraging the use of AMF. Unquestionably, using AMF to improve agriculture can drastically cut down on the usage of synthetic fertilizers and other chemicals, encouraging bio-healthy agriculture. Crop plants with increased growth and production due to AMF treatment may be able to meet the world's rising population's demand for food. Due to their extensive use, environmentally friendly technology will also be strongly promoted. The identification of genes and gene products mediating the AMF-mediated growth and development regulation in response to stressful stimuli should be the primary focus of future studies. Future studies in this area may focus on identifying the key physiological and metabolic processes under various environmental conditions, as well as the host and AMF-specific protein components governing symbiotic interaction. Crop productivity can be increased by comprehending how AMF-induced modulations in tolerance mechanisms and the crosstalk are activated to govern plant performance. To better understand their function in nature as a bio-fertilizer for sustainable agricultural production, AMF must be investigated at all levels. While years of research have increased our knowledge of the multi-complex mechanisms guiding plant-AMF symbiosis in reducing salt stress in plants, there are still several issues that need to be resolved. Future research into these problems will help us comprehend the procedure better.

REFERENCES

Aarle, I.M.V., P.A. Olsson, and B. Söderström. 2002. Arbuscular mycorrhizal fungi respond to the substrate pH of their extraradical mycelium by altered growth and root colonization. *New Phytologist* 155: 173–82. https://doi.org/10.1046/j.14698137.2002.00439.x.

Abdelhalim, T.S., N.A.J. Tia, K.A. Sir Elkhatim, et al. 2022. Exploring the potential of arbuscular mycorrhizal fungi (AMF) for improving health-promoting phytochemicals in sorghum. *Rhizospher* 24 (December): 100596. https://doi.org/10.1016/j.rhisph.2022.100596.

Abhinandan, K., L. Skori, M. Stanic, N.M.N. Hickerson, M. Jamshed, and M.A. Samuel. 2018. Abiotic stress signaling in wheat – an inclusive overview of hormonal interactions during abiotic stress responses in wheat. *Frontiers in Plant Science* 9 (June): 734, 1–25. https://doi.org/10.3389/fpls.2018.00734.

Abrol, I.P., J.S.P. Yadav, and F.I. Massoud. 1988. *Salt affected soils and their management*. FAO Soils Bulletin 39. https://www.fao.org/4/x5871e/x5871e00.htm.

Aganchich, B., S. Wahbi, A. Yaakoubi, H. El-Aououad, and J. Bota. 2022. Effect of arbuscular mycorrhizal fungi inoculation on growth and physiology performance of olive trees under regulated deficit irrigation and partial rootzone drying. *South African Journal of Botany* 148: 1–10. https://doi.org/10.1016/j.sajb.2022.03.051.

Agarwal, S., R. Lone, F. Farooq, R. Shuab, and K.K. Koul. 2013. *In vitro* establishment of arbuscular mycorrhizal fungi (AMF) in cucumber (*Cucumis sativus*) roots. *Advances in Bioresearch* 4: 155–59. https://soeagra.com/abr/abrdecember2013/24.pdf.

Andrino, A., G. Guggenberger, L. Sauheitl, S. Burkart, and J. Boy. 2021. Carbon investment into mobilization of mineral and organic phosphorus by arbuscular mycorrhiza. *Biology and Fertility of Soils* 57: 47–64. https://doi.org/10.1007/s00374-020-01505-5.

Arif, Y., P. Singh, H. Siddiqui, A. Bajguz, and S. Bajguz. 2020. Salinity induced physiological and biochemical changes in plants: An omic approach towards salt stress tolerance. *Plant Physiology and Biochemistry* 156: 64–77. https://doi.org/10.1016/j.plaphy.2020.08.042.

Aroca, R., J.M. Ruiz-Lozano, Á.M. Zamarreño, et al. 2013. Arbuscular mycorrhizal symbiosis influences strigolactone production under salinity and alleviates salt stress in lettuce plants. *Journal of Plant Physiology* 170: 47–55. https://doi.org/10.1016/j.jplph.2012.08.020.

Azcón, R., E. Ambrosano, and C. Charest. 2003. Nutrient acquisition in mycorrhizal lettuce plants under different phosphorus and nitrogen concentration. *Plant Science* 165: 1137–45. https://doi.org/10.1016/S0168-9452(03)00322–4.

Balestrini R., C. Brunetti, W. Chitarra, and L. Nerva. 2020. Photosynthetic traits and nitrogen uptake in crops: Which is the role of arbuscular mycorrhizal fungi? *Plants* 9 (August): 1105, 1–16. https://doi.org/10.3390/plants9091105.

Basak, N., A. Barman, P. Sundha, and A.K. Rai. 2020. Recent trends in soil salinity appraisal and management. In *Soil analysis: Recent trends and applications*, ed. A. Rakshit, S. Ghosh, S. Chakraborty, V. Philip, and A. Datta, 143–61 Singapore: Springer. https://doi.org/10.1007/978-981-15-2039-6_9.

Begum, N., M.A. Ahanger, Y. Su, et al. 2019 Improved drought tolerance by AMF inoculation in maize (*Zea mays*) involves physiological and biochemical implications. *Plants* 8 (December): 579, 1–20. https://doi.org/10.3390/plants8120579.

Benaffari, W., A. Boutasknit, M. Anli, N. Nasri, A. Meddich. 2023. Application of arbuscular mycorrhizal fungi alone or combined with different composts to improve physiological and biochemical attributes related to drought stress tolerance in quinoa. *Journal of Soil Science and Plant Nutrition* 23: 4250–66. https://doi.org/10.1007/s42729-023-01345-w.

Bharti, N., D. Barnawal, K. Wasnik, S.K. Tewari, and A. Kalra. 2016. Co-inoculation of *Dietzia natronolimnaea* and *Glomus intraradices* with vermicompost positively influences *Ocimum basilicum* growth and resident microbial community structure in salt affected low fertility soils. *Applied Soil* Ecology 100: 211–25. https://doi.org/10.1016/j.apsoil.2016.01.003.

Bistgani, Z.E., M. Hashemi, M. DaCosta, L. Craker, F. Maggi, and M.R. Morshedloo. 2019. Effect of salinity stress on the physiological characteristics, phenolic compounds and antioxidant activity of *Thymus vulgaris* L. and *Thymus daenensis* Celak. *Industrial Crops and Products* 135: 311–20. https://doi.org/10.1016/j.indcrop.2019.04.055.

Bonfante, P., and A. Desirò. 2015. Arbuscular mycorrhizas: The lives of beneficial fungi and their plant hosts. In *Principles of plant-microbe interactions*, ed. B. Lugtenberg, 235–45. Cham: Springer. https://doi.org/10.1007/978-3-319-08575-3_25.

Bonfante, P., and A. Genre. 2010. Mechanisms underlying beneficial plant–fungus interactions in mycorrhizal symbiosis. *Nature Communications* 1 (July): 48, 1–11. https://doi.org/10.1038/ncomms1046.

Boyer, L.B., W. Feng, N. Gulbis, et al. 2016. The use of arbuscular mycorrhizal fungi to improve strawberry production in coir substrate. *Frontiers in Plant Science* 7 (August): 1237, 1–9. https://doi.org/10.3389/fpls.2016.01237.

Boyer, L.R., P. Brain, X.M. Xu, and P. Jeffries. 2015 Inoculation of drought-stressed strawberry with a mixed inoculum of two arbuscular mycorrhizal fungi: Effects on population dynamics of fungal species in roots and consequential plant tolerance to water deficiency. *Mycorrhiza* 25: 215–27. https://doi.org/10.1007/s00572-014-0603-6.

Briat, J.F., A. Gojon, C. Plassard, et al. 2020. Reappraisal of the central role of soil nutrient availability in nutrient management in light of recent advances in plant nutrition at crop and molecular levels. *European Journal of Agronomy* 116 (May): 126069. https://doi.org/10.1016/j.eja.2020.126069.

Cantrell, I.C., and R.G. Linderman. 2001. Preinoculation of lettuce and onion with VA mycorrhizal fungi reduces deleterious effects of soil salinity. *Plant and Soil* 233: 269–81. https://doi.org/10.1023/A:1010564013601.

Chandra, P., A. Singh, K. Prajapat, A.K. Rai, and R.K. Yadav. 2022. Native arbuscular mycorrhizal fungi improve growth, biomass yield, and phosphorus nutrition of sorghum in saline and sodic soils of the semi–arid region. *Environmental and Experimental Botany* 201 (September): 104982. https://doi.org/10.1016/j.envexpbot.2022.104982.

Chandra, P., A. Wunnava, P. Verma, A. Chandra, and R.K. Sharma. 2021. Strategies to mitigate the adverse effect of drought stress on crop plants – influences of soil bacteria: A review. *Pedosphere* 31: 496–09. https://doi.org/10.1016/S1002-0160(20)60092–3.

Chandrasekaran, M. 2022. Arbuscular mycorrhizal fungi mediated enhanced biomass, root morphological traits and nutrient uptake under drought stress: A Meta-Analysis. *Journal of Fungi* 8 (June): 660, 1–12. https://doi.org/10.3390/jof8070660.

Chhabra, R. 2004. Classification of salt-affected soils. *Arid Land Research and Management* 19: 61–79. https://doi.org/10.1080/15324980590887344.

Chiu, C.H., and U. Paszkowski. 2019. Mechanisms and impact of symbiotic phosphate acquisition. *Cold Spring Harbor Perspectives in Biology* 11 (October): a034603, 1–33. https://doi.org/10.1101/cshperspect.a034603.

Cobb, A.B., G.W.T. Wilson, C.L. Goad, et al. 2016. The role of arbuscular mycorrhizal fungi in grain production and nutrition of sorghum genotypes: Enhancing sustainability through plant-microbial partnership. *Agriculture, Ecosystems & Environment* 233: 432–40. https://doi.org/10.1016/j.agee.2016.09.024.

Cuevas J., I.N. Daliakopoulos, F. del Moral, J.J. Hueso, and I.K.Tsanis. 2019. A review of soil-improving cropping systems for soil salinization. *Agronomy* 9 (June): 295, 1–22. https://doi.org/10.3390/agronomy9060295.

de Oliveira Mendes, G., A.L. Moreira de Freitas, O.L. Pereira, I.R. da Silva, N.B. Vassilev, and M.D. Costa. 2014. Mechanisms of phosphate solubilization by fungal isolates when exposed to different P sources. *Annals of Microbiology* 64: 239–49. https://doi.org/10.1007/s13213-013-0656-3.

Deja-Sikora, E., K. Werner, and K. Hrynkiewicz. 2023. AMF species do matter: *Rhizophagus irregularis* and *Funneliformis mosseae* affect healthy and PVY-infected *Solanum tuberosum* L. in a different way. *Frontiers in Microbiology* 14 (April): 1127278, 1–16. https://doi.org/10.3389/fmicb.2023.1127278.

El-Badri, A.M., M. Batool, I.A.A. Mohamed, et al. 2022. Mitigation of the salinity stress in rapeseed (*Brassica napus* L.) productivity by exogenous applications of bio-selenium nanoparticles during the early seedling stage. *Environmental Pollution* 310 (April): 119815, 1–16. https://doi.org/10.1016/j.envpol.2022.119815.

Elhindi, K., A.S.E. Din, E. Abdel-Salam, and A. Elgorban. 2016. Amelioration of salinity stress in different basil (*Ocimum basilicum* L.) varieties by vesicular-arbuscular mycorrhizal fungi. *Acta Agriculturae Scandinavica, Section B – Soil & Plant Science* 66: 583–92. https://doi.org/10.1080/09064710.2016.1204467.

El-Nashar, Y.I. 2017. Response of snapdragon (*Antirrhinum majus* L.) to blended water irrigation and arbuscular mycorrhizal fungi inoculation: Uptake of minerals and leaf water relations. *Photosynthetica* 55: 201–9. https://doi.org/10.1007/s11099-016-0650-7.

EL-Sabagh, A., M.S. Islam, M. Skalicky, et al. 2021. Salinity stress in wheat (*Triticum aestivum* L.) in the changing climate: Adaptation and management strategies. *Frontiers in Agronomy* 3 (July): 661932, 1–20. https://doi.org/10.3389/fagro.2021.661932.

Emmanuel, O.C., and O.O. Babalola. 2020. Productivity and quality of horticultural crops through co-inoculation of arbuscular mycorrhizal fungi and plant growth promoting bacteria. *Microbiological Research* 239 (July): 126569, 1–11. https://doi.org/10.1016/j.micres.2020.126569.

Etesami, H., B.R. Jeong, and B.R. Glick. 2021. Contribution of arbuscular mycorrhizal fungi, phosphate-solubilizing bacteria, and silicon to P uptake by plant. *Frontiers in Plant Science* 12 (July): 699618, 1–29. https://doi.org/10.3389/fpls.2021.699618.

Evelin, H., T.S. Devi, S. Gupta, and R. Kapoor. 2019. Mitigation of salinity stress in plants by arbuscular mycorrhizal symbiosis: Current understanding and new challenges. *Frontiers in Plant Science* 10 (April): 470, 1–21. https://doi.org/10.3389/fpls.2019.00470.

Eynard, A., R. Lal, and K. Wiebe. 2005. Crop response in salt-affected soils. *Journal of Sustainable Agriculture* 27: 5–50. https://doi.org/10.1300/J064v27n01_03.

Ezawa, T., M. Saito, and T. Yoshida. 1995. Comparison of phosphatase localization in the intraradical hyphae of arbuscular mycorrhizal fungi, *Glomus* spp. and *Gigaspora* spp. *Plant and Soil* 176: 57–63. https://doi.org/10.1007/BF00017675.

Faghihinia, M., and J. Jansa. 2022. Mycorrhiza governs plant-plant interactions through preferential allocation of shared nutritional resources: A triple ((13C, 15N and 33P) labeling study. *Frontiers in Plant Science* 13 (December): 1047270, 1–19. https://doi.org/10.3389/fpls.2022.1047270.

FAO. 2020. *The State of Food and Agriculture 2020. Overcoming Water Challenges in Agriculture*. Rome. https://doi.org/10.4060/cb1447en.

Feng, G., F. Zhang, X. Li, C.Tian, and Z. Rengel. 2002. Improved tolerance of maize plants to salt stress by arbuscular mycorrhiza is related to higher accumulation of soluble sugars in roots. *Mycorrhiza* 12: 185–90. https://doi.org/10.1007/s00572-002-0170-0.

Fu, D., Y. Rui, C. Zevenbergen, and R.P. Singh. 2021. Nitrogen absorption efficiency and mechanism in arbuscular mycorrhizal fungi – *Canna indica* symbiosis. *Chemosphere* 282 (November): 130708. https://doi.org/10.1016/j.chemosphere.2021.130708.

Ghobadi, M., M.M. Dehnavi, A.R. Yadavi, K. Parvizi, and D. Zafari. 2020. Reduced P fertilization improves Fe and Zn uptake in potato when inoculated with AMF in P, Fe and Zn deficient soil. *Rhizosphere* 15 (September): 100239. https://doi.org/10.1016/j.rhisph.2020.100239.

Giovannini, L., M. Palla, M. Agnolucci, et al. 2020. Arbuscular mycorrhizal fungi and associated microbiota as plant biostimulants: Research strategies for the selection of the best performing inocula. *Agronomy* 10 (January): 106. https://doi.org/10.3390/agronomy10010106.

Gómez-Bellot, M.J., M.F. Ortuño, P.A. Nortes, J. Vicente-Sánchez, S. Bañón, and M.J. Sánchez-Blanco. 2015. Mycorrhizal euonymus plants and reclaimed water: Biomass, water status and nutritional responses. *Scientia Horticulturae* 186: 61–69. https://doi.org/10.1016/j.scienta.2015.02.022.

Gong, M., X. You, and Q. Zhang. 2015. Effects of *Glomus intraradices* on the growth and reactive oxygen metabolism of foxtail millet under drought. *Annals of Microbiology* 65: 595–602. https://doi.org/10.1007/s13213-014-0895-y.

Guo, J., M. Du, H. Tian, and B. Wang. 2020. Exposure to high salinity during seed development markedly enhances seedling emergence and fitness of the progeny of the extreme halophyte *Suaeda salsa*. *Frontiers in Plant Science* 11 (August): 1291, 1–11. https://doi.org/10.3389/fpls.2020.01291.

Gupta, A., S. Jain, and N. Verma. 2022. *Industrial applications of soil microbes*, vol. 1. Bentham Science Publishers Pte Ltd. https://doi.org/10.2174/97898150399551220101.

Gupta, A., S. Jain, and N. Verma. 2024. *Industrial applications of soil microbes*, vol. 3. Bentham Science Publishers Pte Ltd. https://doi.org/10.2174/97898150797531240301.

Gupta, B., and B. Huang. 2014. Mechanism of salinity tolerance in plants: Physiological, biochemical, and molecular characterization. *International Journal of Genomics* 2014 (April): 701596, 1–19. https://doi.org/10.1155/2014/701596.

Hagh, E.D., B. Mirshekari, M.R. Ardakani, F. Farahvash, and F. Rejali. 2016. Optimizing phosphorus use in sustainable maize cropping via mycorrhizal inoculation. *Journal of Plant Nutrition* 39: 1348–56. https://doi.org/10.1080/01904167.2015.1086797.

Hessini, K., K. Issaoui, and S. Ferchichi. 2019. Interactive effects of salinity and nitrogen forms on plant growth, photosynthesis and osmotic adjustment in maize. *Plant Physiology and Biochemistry* 139: 171–78. https://doi.org/10.1016/j.plaphy.2019.03.005.

Hodge, A., and K. Storer. 2015. Arbuscular mycorrhiza and nitrogen: Implications for individual plants through to ecosystems. *Plant and Soil* 386: 1–19. https://doi.org/10.1007/s11104-014-2162-1.

Ingraffia, R., G. Amato, A.S. Frenda, and D. Giambalvo. 2019. Impacts of arbuscular mycorrhizal fungi on nutrient uptake, N2 fixation, N transfer, and growth in a wheat/faba bean intercropping system. *PLoS One* 14 (March): e0213672, 1–16. https://doi.org/10.1371/journal.pone.0213672.,

Jain, S., A. Gupta, and N. Verma. 2023. *Industrial applications of soil microbes*, vol. 2. Bentham Science Publishers Pte Ltd. https://doi.org/10.2174/97898150502641230201.

Jain, S., A. Gupta, and N. Verma. 2024. *Industrial applications of soil microbes*, vol. 4. Bentham Science Publishers Pte Ltd. https://doi.org/10.2174/97898151249961240401.

Jiang, F., L. Zhang, J. Zhou, T.S. George, and G. Feng. 2021. Arbuscular mycorrhizal fungi enhance mineralisation of organic phosphorus by carrying bacteria along their extraradical hyphae. *New Phytologist* 230: 304–15. https://doi.org/10.1111/nph.17081.

Johri, A.K., R. Oelmüller, M. Dua, et al. 2015. Fungal association and utilization of phosphate by plants: Success, limitations, and future prospects. *Frontiers in Microbiology* 6 (October): 984, 1–13. https://doi.org/10.3389/fmicb.2015.00984.

Kamran, M., A. Parveen, S. Ahmar, et al. 2019. An overview of hazardous impacts of soil salinity in crops, tolerance mechanisms, and amelioration through selenium supplementation. *International Journal of Molecular Sciences* 21 (December): 148, 1–27. https://doi.org/10.3390/ijms21010148.

Keymer, A., P. Pimprikar, V. Wewer, et al. 2017. Lipid transfer from plants to arbuscular mycorrhiza fungi. *Elife* 6 (July): e29107, 1–33. https://doi.org/10.7554/eLife.29107.

Khan, Y., S. Shah, and T. Hui. 2022. The roles of arbuscular mycorrhizal fungi in influencing plant nutrients, photosynthesis, and metabolites of cereal crops-A review. *Agronomy* 12 (September): 2191, 1–19. https://doi.org/10.3390/agronomy12092191.

Khan, Y., X. Yang, X. Zhang, T. Yaseen, L. Shi, and T. Zhang. 2021. Arbuscular mycorrhizal fungi promote plant growth of *Leymus chinensis* (Trin.) Tzvelev by increasing the metabolomics activity under nitrogen addition. *Grassland Science* 67: 128–38. https://doi.org/10.1111/grs.12299.

Kobae, Y. 2019. Dynamic phosphate uptake in arbuscular mycorrhizal roots under field conditions. *Frontiers in Environmental Science* 6 (January): 159, 1–12. https://doi.org/10.3389/fenvs.2018.00159.

Koide, R.T., and Z. Kabir. 2000. Extraradical hyphae of the mycorrhizal fungus *Glomus intraradices* can hydrolyse organic phosphate. *New Phytologist* 148: 511–17. https://doi.org/10.1046/j.1469-8137.2000.00776.x.

Kumar, P., and P.K. Sharma. 2020. Soil salinity and food security in India. *Frontiers in Sustainable Food Systems* 4 (October): 533781, 1–15. https://doi.org/10.3389/fsufs.2020.533781.

Lee, Y., R. Krishnamoorthy, G. Selvakumar, K. Kim, and T. Sa. 2015. Alleviation of salt stress in maize plant by co-inoculation of arbuscular mycorrhizal fungi and *Methylobacterium oryzae* CBMB20. *Journal of the Korean Society for Applied Biological Chemistry* 58: 533–40. https://doi.org/10.1007/s13765-015-0072-4.

Li, J., B. Meng, H. Chai, et al. 2019. Arbuscular mycorrhizal fungi alleviate drought stress in C3 (*Leymus chinensis*) and C4 (*Hemarthria altissima*) grasses via altering antioxidant enzyme activities and photosynthesis. *Frontiers in Plant Science* 10 (April): 499, 1–12. https://doi.org/10.3389/fpls.2019.00499.

Li, T., Y.J. Hu, Z.P. Hao, H. Li, and B.D. Chen. 2013. Aquaporin genes GintAQPF1 and GintAQPF2 from *Glomus intraradices* contribute to plant drought tolerance. *Plant Signaling & Behavior* 8 (February): e24030, 1–2. https://doi.org/10.4161/psb.24030.

Li, X-L., E. George, and H. Marschner. 1991. Phosphorus depletion and pH decrease at the root–soil and hyphae–soil interfaces of VA mycorrhizal white clover fertilized with ammonium. *New Phytologist* 119: 397–404. https://doi.org/10.1111/j.1469-8137.1991.tb00039.x.

Lu, F.C., C.Y. Lee, and C.L. Wang. 2015. The influence of arbuscular mycorrhizal fungi inoculation on yam (*Dioscorea* spp.) tuber weights and secondary metabolite content. *PeerJ* 3 (September): e1266, 1–19. https://doi.org/10.7717/peerj.1266.

Luginbuehl, L.H., G.N. Menard, S. Kurup, et al. 2017. Fatty acids in arbuscular mycorrhizal fungi are synthesized by the host plant. *Science* 356: 1175–78. https://doi.org/10.1126/science.aan0081.

Ma, Y., M.C. Dias, and H. Freitas. 2020. Drought and salinity stress responses and microbe-induced tolerance in plants. *Frontiers in Plant Science* 11 (November): 591911, 1–18. https://doi.org/10.3389/fpls.2020.591911.

Mathimaran, N., S. Jegan, M.N. Thimmegowda, et al. 2020. Intercropping transplanted pigeon pea with finger millet: Arbuscular mycorrhizal fungi and plant growth promoting rhizobacteria boost yield while reducing fertilizer input. *Frontiers in Sustainable Food Systems* 4 (June): 88, 1–12. https://doi.org/10.3389/fsufs.2020.00088.

Mathur, S., R.S. Tomar, and A. Jajoo. 2019. Arbuscular mycorrhizal fungi (AMF) protects photosynthetic apparatus of wheat under drought stress. *Photosynthesis Research* 139: 227–38. https://doi.org/10.1007/s11120-018-0538-4.

Muhu-Din Ahmed, H.G., Y. Zeng, H. Raza, et al. (2022) Characterization of wheat (*Triticum aestivum* L.) accessions using morpho-physiological traits under varying levels of salinity stress at seedling stage. *Frontiers in Plant Science* 13 (July): 953670, 1–13. https://doi.org/10.3389/fpls.2022.953670.

Naheed, R., H. Aslam, H. Kanwal, et al. 2021. Growth attributes, biochemical modulations, antioxidant enzymatic metabolism and yield in *Brassica napus* varieties for salinity tolerance. *Saudi Journal of Biological Sciences* 28: 5469–79. https://doi.org/10.1016/j.sjbs.2021.08.021.

Oliveira, R.S., M. Vosátka, J.C. Dodd, and P.M.L. Castro. 2005. Studies on the diversity of arbuscular mycorrhizal fungi and the efficacy of two native isolates in a highly alkaline anthropogenic sediment. *Mycorrhiza* 16: 23–31. https://doi.org/10.1007/s00572-005-0010-0.

Oliveira, T.C., J.S.R. Cabral, L.R. Santana, et al. 2022. The arbuscular mycorrhizal fungus *Rhizophagus clarus* improves physiological tolerance to drought stress in soybean plants. *Scientific Reports* 12 (May): 9044, 1–15. https://doi.org/10.1038/s41598-022-13059-7.

Ortas, I., D.L. Rowell, and P.J. Harris. 2004. Effect of mycorrhizae and pH change at the root-soil interface on phosphorus uptake by sorghum using a rhizocylinder technique. *Communications in Soil Science and Plant Analysis* 35: 1061–80. https://doi.org/10.1081/CSS-120030587.

Parvin, S., M.V. Geel, T. Yeasmin, E. Verbruggen, and O. Honnay. 2020. Effects of single and multiple species inocula of arbuscular mycorrhizal fungi on the salinity tolerance of a Bangladeshi rice (*Oryza sativa L.*) cultivar. *Mycorrhiza* 30: 431–44. https://doi.org/10.1007/s00572-020-00957-9.

Pellegrino, E., M. Nuti, and L. Ercoli. 2022. Multiple arbuscular mycorrhizal fungal consortia enhance yield and fatty acids of *Medicago sativa*: A two-year field study on agronomic traits and tracing of fungal persistence. *Frontiers in Plant Science* 13 (February): 814401, 1–19. https://doi.org/10.3389/fpls.2022.814401.

Qiangsheng, W., Z. Yingning, and X. Renxue. 2007. Effect of *Glomus versiforme* inoculation on reactive oxygen metabolism of *Citrus tangerine* leaves exposed to water stress. *Frontiers of Agriculture in China* 1: 438–43. https://doi.org/10.1007/s11703-007-0072-1.

Rai, A.K., N. Basak, and P. Sundha. 2021a. Saline and sodic ecosystems in the changing world. In *Soil science: Fundamentals to recent advances*, eds. A. Rakshit, S.K. Singh, P.C. Abhilash, and A. Biswas, 175–90. Singapore: Springer. https://doi.org/10.1007/978-981-16-0917-6_9.

Rai, A.K., A. Dinkar, N. Basak, et al. 2021b. Phosphorus nutrition of oats genotypes in acidic soils: Exploiting responsive plant-microbe partnership. *Applied Soil Ecology* 167 (November): 104094. https://doi.org/10.1016/j.apsoil.2021.104094.

Rengasamy, P. 2016. Soil chemistry factors confounding crop salinity tolerance – A review. *Agronomy* 6 (August): 53, 1–11. https://doi.org/10.3390/agronomy6040053.

Saboor, A., M.A. Ali, S. Danish, et al. 2021a. Effect of arbuscular mycorrhizal fungi on the physiological functioning of maize under zinc-deficient soils. *Scientific Reports* 11 (September): 18468, 1–11. https://doi.org/10.1038/s41598-021-97742-1.

Saboor, A., M.A. Ali, S. Hussain, et al. 2021b. Zinc nutrition and arbuscular mycorrhizal symbiosis effects on maize (*Zea mays L.*) growth and productivity. *Saudi Journal of Biological Sciences* 28: 6339–51. https://doi.org/10.1016/j.sjbs.2021.06.096.

Schreiner, R.P. 2007. Effects of native and nonnative arbuscular mycorrhizal fungi on growth and nutrient uptake of 'Pinot noir' (*Vitis vinifera* L.) in two soils with contrasting levels of phosphorus. *Applied Soil Ecology* 36: 205–15. https://doi.org/10.1016/j.apsoil.2007.03.002.

Schütz, L., K. Saharan, P. Mäder, T. Boller, and N. Mathimaran. 2022. Rate of hyphal spread of arbuscular mycorrhizal fungi from pigeon pea to finger millet and their contribution to plant growth and nutrient uptake in experimental microcosms. *Applied Soil Ecology* 169 (January): 104156, 1–10. https://doi.org/10.1016/j.apsoil.2021.104156.

Sendek, A., C. Arakoç, C. Wagg, et al. 2019. Drought modulates interactions between arbuscular mycorrhizal fungal diversity and barley genotype diversity. *Scientific Reports* 9 (July): 9650, 1–15. https://doi.org/10.1038/s41598-019-45702-1.

Shahzad, T., C. Chenu, P. Genet, et al. 2015. Contribution of exudates, arbuscular mycorrhizal fungi and litter depositions to the rhizosphere priming effect induced by grassland species. *Soil Biology & Biochemistry* 80: 146–55. https://doi.org/10.1016/j.soilbio.2014.09.023.

Sharma, P.C., A. Datta, A.K. Yadav, M. Choudhary, H.S. Jat, and A. McDonald. 2019. Effect of crop management practices on crop growth, productivity and profitability of rice–wheat system in western Indo-Gangetic plains. *Proceedings of the National Academy of Sciences, India Section B: Biological Sciences* 89: 715–27. https://doi.org/10.1007/s40011-018-0985-x.

Sharma, S., S. Compant, M.B. Ballhausen, S. Ruppel, and P. Franken. 2020. The interaction between *Rhizoglomus irregulare* and hyphae attached phosphate solubilizing bacteria increases plant biomass of *Solanum lycopersicum*. *Microbiological Research* 240 (November): 126556, 1–10. https://doi.org/10.1016/j.micres.2020.126556.

Sheng, M., M. Tang, H. Chen, B. Yang, F. Zhang, and Y. Huang. 2009. Influence of arbuscular mycorrhizae on the root system of maize plants under salt stress. *Canadian Journal of Microbiology* 55: 879–86. https://doi.org/10.1139/w09-031.

Sheng, M., M. Tang, F. Zhang, and Y. Huang. 2011. Influence of arbuscular mycorrhiza on organic solutes in maize leaves under salt stress. *Mycorrhiza* 21: 423–30. https://doi.org/10.1007/s00572-010-0353-z.

Singh, S.B., and I.P. Abrol. 1985. Effect of soil sodicity on the growth, yield and chemical composition of groundnut (*Arachis hypogaea* Linn.). *Plant and Soil* 84: 123–27. https://doi.org/10.1007/BF02197873.

Song, Y., J. Liu, and F. Chen. 2020. *Azotobacter chroococcum* inoculation can improve plant growth and resistance of maize to armyworm, *Mythimna separata* even under reduced nitrogen fertilizer application. *Pest Management Science* 76: 4131–40. https://doi.org/10.1002/ps.5969.

Sweeney, C.J., M. Bottoms, S. Ellis, et al. 2022. Arbuscular mycorrhizal fungi and the need for a meaningful regulatory plant protection product testing strategy. *Environmental Toxicology and Chemistry* 41: 1808–23. https://doi.org/10.1002/etc.5400.

Szczałba, M., T. Kopta, M. Gąstoł, and A. Sękara. 2019. Comprehensive insight into arbuscular mycorrhizal fungi, *Trichoderma* spp. and plant multilevel interactions with emphasis on biostimulation of horticultural crops. *Journal of Applied Microbiology* 127: 630–47. https://doi.org/10.1111/jam.14247.

Tahat, M.M., and K. Sijam. 2012. Arbuscular mycorrhizal fungi and plant root exudates bio-communications in the rhizosphere. *African Journal of Microbiology Research* 6: 7295–301. https://doi.org/10.5897/AJMR12.2250.

Thirkell, T., D. Cameron, and A. Hodge. 2019. Contrasting nitrogen fertilisation rates alter mycorrhizal contribution to barley nutrition in a field trial. *Frontiers in Plant Science* 10 (October): 1312, 1–9. https://doi.org/10.3389/fpls.2019.01312.

Thirkell, T.J., D.D. Cameron, and A. Hodge. 2016. Resolving the "nitrogen paradox" of arbuscular mycorrhizas: Fertilization with organic matter brings considerable benefits for plant nutrition and growth. *Plant, Cell & Environment* 39: 1683–90. https://doi.org/10.1111/pce.12667.

Thirkell, T.J., D. Pastok, and K.J. Field. 2020. Carbon for nutrient exchange between arbuscular mycorrhizal fungi and wheat varies according to cultivar and changes in atmospheric carbon dioxide concentration. *Global Change Biology* 26: 1725–38. https://doi.org/10.1111/gcb.14851.

Tian, H., J.P. Gai, J.L. Zhang, P. Christie, and X.L. Li. 2009. Arbuscular mycorrhizal fungi associated with wild forage plants in typical steppe of eastern inner Mongolia. *European Journal of Soil Biology* 45: 321–27. https://doi.org/10.1016/j.ejsobi.2009.05.003.

Treseder, K.K. 2013. The extent of mycorrhizal colonization of roots and its influence on plant growth and phosphorus content. *Plant and Soil* 371: 1–13. https://doi.org/10.1007/s11104-013-1681-5.

Walder, F., H. Niemann, M. Natarajan, M.F. Lehmann, T. Boller, and A. Wiemken. 2012. Mycorrhizal networks: Common goods of plants shared under unequal terms of trade. *Plant Physiology* 159: 789–97. https://doi.org/10.1104/pp.112.195727.

Wang, Y., M. Wang, Y. Li, A. Wu, and J. Huang. 2018. Effects of arbuscular mycorrhizal fungi on growth and nitrogen uptake of *Chrysanthemum morifolium* under salt stress. *PLoS One* 13 (April): e0196408, 1–14. https://doi.org/10.1371/journal.pone.0196408.

Wang, Z.G., Y.L. Bi, B. Jiang, et al. 2016. Arbuscular mycorrhizal fungi enhance soil carbon sequestration in the coalfields, northwest China. *Scientific Reports* 6 (October): 34336, 1–11. https://doi.org/10.1038/srep34336.

Waqas, M.A., C. Kaya, A. Riaz, et al. 2019. Potential mechanisms of abiotic stress tolerance in crop plants induced by thiourea. *Frontiers in Plant Science* 10 (October): 1336, 1–14. https://doi.org/10.3389/fpls.2019.01336.

Wilde, P., A. Manal, M. Stodden, E. Sieverding, U. Hildebrandt, and H. Bothe. 2009. Biodiversity of arbuscular mycorrhizal fungi in roots and soils of two salt marshes. *Environmental Microbiology* 11: 1548–61. https://doi.org/10.1111/j.1462-2920.2009.01882.x.

Wipf, D., F. Krajinski, D.V. Tuinen, G. Recorbet, and P.E. Courty. 2019. Trading on the arbuscular mycorrhiza market: From arbuscules to common mycorrhizal networks. *New Phytologist* 223: 1127–42. https://doi.org/10.1111/nph.15775.

Xiao, Y., Z. Zhao, L. Chen, and Y. Li. 2020. Arbuscular mycorrhizal fungi and organic manure have synergistic effects on *Trifolium repens* in Cd-contaminated sterilized soil but not in natural soil. *Applied Soil Ecology* 149 (May): 103485. https://doi.org/10.1016/j.apsoil.2019.103485.

Yao, Q., X. Li, G. Feng, and P. Christie. 2001. Mobilization of sparingly soluble inorganic phosphates by the external mycelium of an abuscular mycorrhizal fungus. *Plant and Soil* 230: 279–85. https://doi.org/10.1023/A:1010367501363.

Yilma, G. 2019. The role of mycorrhizal fungi in pepper (*Capsicum annuum*) production. *International Journal of Advanced Research in Biological Sciences* 6: 59–65. https://doi.org/10.22192/ijarbs.2019.06.12.008.

Young, T., D.D. Cameron, and G.K. Phoenix. 2015. Using AMF inoculum to improve the nutritional status of *Prunella vulgaris* plants in green roof substrate during establishment. *Urban Forestry and Urban Greening* 14: 959–67. https://doi.org/10.1016/j.ufug.2015.08.012.

Yun, P., L. Xu, S-S. Wang, L. Shabala, S. Shabala, and W-Y. Zhang. 2018. *Piriformospora indica* improves salinity stress tolerance in *Zea mays* L. plants by regulating Na^+ and K^+ loading in root and allocating K^+ in shoot. *Plant growth regulators* 86: 323–31. https://doi.org/10.1007/s10725-018-0431-3.

Zhang, L., J. Zhou, T.S. George, E. Limpens, and G. Feng. 2022. Arbuscular mycorrhizal fungi conducting the hyphosphere bacterial orchestra. *Trends in Plant Science* 27: 402–11. https://doi.org/10.1016/j.tplants.2021.10.008.

Zhu, B., T. Gao, D. Zhang, K. Ding, C. Li, and F. Ma. 2022. Functions of arbuscular mycorrhizal fungi in horticultural crops. *Scientia Horticulturae* 303 (September): 111219. https://doi.org/10.1016/j.scienta.2022.111219.

15 Endophytes
Enormous Source of Novel Bioactive Secondary Metabolites

Jyoti Bhagat
(Department of Microbiology, Guru Nanak Dev University, Amritsar 143005, India; jots_005@yahoo.com)

Amarjeet Kaur
(Department of Microbiology, Guru Nanak Dev University, Amritsar 143005, India; amarjeetmicro2@gmail.com)

**Priyanka Chandra*
(ICAR – Central Soil Salinity Research Institute, Karnal 132001, Haryana, India; priyanka.chandra921@gmail.com)

15.1 INTRODUCTION

Humanity is facing problems of the tremendous increase of microbial infections in the world's population. Drugs are being used as therapeutic agents for treating different diseases that are usually chemically synthesized and are associated with many side effects. Therefore, research efforts in the recent past have been directed to invent new, safer, and efficient natural metabolites that possess low toxicity and are easily obtainable with negligible environmental impact (Liu et al., 2001). The significance of natural products in the drug discovery and development processes has been reported extensively (Koehn and Carter, 2005). The majority of the natural compounds have been isolated from plants that constitute a large fraction of the current pharmacopoeias (Emiliani et al., 2014). Besides medicinal plants, various microorganisms have also been identified as a good source of bioactive metabolites (Sagar et al., 2010; Jain et al., 2023, 2024). Traditionally, the discoveries have come from microbial strains residing in soil but microorganisms from certain unusual ecological habitats or niches like plants, leaf litter, insects, and dung can also be expected to yield novel molecules. Therefore, there is a general call to bio-prospect new microbial sources, especially from less explored regions and niches, to maximize the isolation of novel bioactive metabolites.

A diverse group of microorganisms is reported to be associated with the plants (Ryan et al., 2008). Endophytes are the microorganisms that live inside the plant tissues, without eliciting the symptoms of disease, and are gaining importance as a source of unreported secondary metabolites (Azevedo and Araújo, 2007; Gupta et al., 2022, 2024). Bacon and White (2000) gave a widely accepted definition of endophytes as microorganisms that colonize living, internal tissues of plants without causing any immediate negative effects. Endophytes are ubiquitous and are associated with 300,000 plant species existing on earth that serve as hosts to one or more types of endophytes. It is estimated that as many as 1 million different microorganisms exist as endophytes (Strobel and Daisy, 2003).

Endophytes are present virtually in all parts (stem, leaves, roots, bark, and fruiting bodies) of the host plant, including seeds. Variation in endophytic microbes within the host plants is not only influenced by their region of location within the plant but also by the nature of the main chemical constituents present in the plant part (Da-Silva et al., 2013). Their distribution also varies with geographical region, age, and climatic conditions (Chareprasert et al., 2006). Bills et al. (2002) evaluated a metabolic distinction between temperate and tropical endophytic microflora. They demonstrated that endophytes isolated from tropical regions provide a larger number of highly active natural biomolecules than those isolated from temperate regions. The effect of the environment in which the host plant is present also influences the general metabolism of endophytic microbes that occur within it.

Endophytes interact with plants at the physiological and genetic level, and such interactions substantially influence plant fitness and vegetation dynamics (Nicoletti, 2014). Endophytes confer profound enhanced health benefits to the host either by producing certain functional biomolecules (Firakova et al., 2007) or by directly transferring the genes responsible for these metabolites to the host plant (Thongsandee et al., 2012), and in return, usually get nutrition and protection from their host plant. Endophytic community within plants or biomolecules produced by

DOI: 10.1201/9781003408833-15

them may enhance the host's tolerance to biotic and abiotic stresses, control plant pathogens (Krishnamurthy and Gnanamanickam, 1998), nematodes, and insects (Hallmann et al., 1997; Azevedo et al., 2000). They accelerate seedling emergence and plant growth, besides increasing agricultural production at decreased chemical doses (Adesemoye and Kloepper, 2009), resulting in more sustainable agricultural practices.

15.2 BIOMOLECULES OF ENDOPHYTIC ORIGIN

Endophytes are the chemical synthesizers inside the plants and produce a large number of biologically active metabolites with diverse chemical structures, including alkaloids, benzo pyranones, flavonoids, glycosides, phenolic acids, quinones, steroids, terpenoids, tetralones, xanthones, and others. The compounds produced by endophytes find wide-ranging application in medicine as antimicrobials, anticancers, antioxidants, insecticidal, immune suppressants, hypocholesterolemic agents, as well as therapeutically important enzyme inhibitors (Tan and Zou, 2001; Bérdy, 2005; Kingston, 2011; Kaul et al., 2012). Owing to genetic recombination between the endophytes and their host plants, endophytes share biosynthetic potential with the host plant. This is exemplified in the case of Taxol and Huperzine A. "Taxol," an anticancer compound produced by endophytic fungi *Taxomyces andreanae* and also by the host plants *Taxus* (Tan and Zou, 2001). Similarly, "Huperzine A," an alkaloid, showed acetylcholinesterase inhibitory (AChEI) potential, produced by endophytic fungi *Shiraia* sp. (Zhu et al., 2010) and *Cladosporium cladosporioides* (Zhang et al., 2011) isolated from the host plant *Huperzia serrata*, a known source of AChEI molecules. Only a few of the plants have been completely studied for their endophytic ecology, thus, only a handful of endophytes have been described yet, which means there is a great opportunity to find interesting and novel endophytic microorganisms capable of producing new and targeting natural products.

Endophytic actinomycetes *Streptomyces* sp. isolated from *Monstera* sp. produced 'Coronamycin', a novel peptide antibiotic active against human fungal pathogen *Cryptococcus neoformans*, pythiaceous fungi, and malarial parasite *Plasmodium falciparum* (Ezra et al., 2004). Waqas et al. (2014) reported endophytic fungi *Phoma glomerata* and *Paecilomyces formosus* isolated from *Glycine max* (L.) Merr and *Cucumis sativus* promoted plant growth by the production of gibberellins. Six spironaphthalenes produced by endophytic fungus *Berkleasmium* sp. isolated from *Dioscorea zingiberensis* showed antimicrobial, antileishmanial, and nematicidal activity (Shan et al., 2013).

15.3 ANTIMICROBIAL AGENTS

Drug resistance is a major problem for treating life-threatening diseases, which enhances the need to prospect new and safe bioactive compounds that are effective against resistant microbes (Strobel and Daisy, 2003). Endophytes isolated from different medicinal plants have been reported for their antimicrobial potential. Many endophytic fungi effective against bacterial and fungal pathogens have been described in the literature. Endophytic fungi *Lophodermium pinastri* and *Phoma herbarum* isolated from *Conifer needles* showed antibacterial activity against *E. coli* (Ravnikar et al., 2015). Pawle and Singh (2014) reported the broad-spectrum antimicrobial potential of endophytic fungus *Nigrospora* sp. isolated from *Ginkgo biloba*, effective against *Candida albicans*, *Geotrichum* sp., *Escherichia coli*, *Klebsiella* sp., and *Staphylococcus aureus*. *Xylaria* sp. isolated from the same plant species. *Ginkgo biloba* L. also showed broad-spectrum antimicrobial activity against *S. aureus*, *E. coli*, *Salmonella enteritidis*, *Aeromonas hydrophila*, *Yersinia* sp., *C. albicans*, and *Pseudomonas expansum* (Liu et al., 2008). Endophytic fungi *Aspergillus fumigatus*, *Phomopsis* sp., and *Botryosphaeria* sp. residing in different *Garcinia* sp. are very well known to produce metabolites that possess antimicrobial activity against *Bacillus subtilis*, *E. coli*, *Klebsiella pneumoniae*, *S. flexneri*, *S. aureus*, *Mycobacterium tuberculosis*, and *Microsporum gypseum* (Phongpaichit et al., 2006; Rukachaisirikul et al., 2007; Ruma et al., 2013). *Periconia* sp. isolated from *Piper longum* was reported to possess antibacterial activity against *M. tuberculosis* and *M. smegmatis* (Verma et al., 2011). El-Hady et al. (2014) demonstrated the antimicrobial potential of endophytic fungus *Aspergillus sydowii* strain W4–2 isolated from sponge *Agelas* sp. against *C. albicans*, *P. aeruginosa*, and *S. aureus*. Antimicrobial potential of endophytic fungi *Cumulospora marina* and *Pestalotiopsis* sp. isolated from *Acanthus ilicifolius* and *Acrostichum aureum* against *B. subtilis*, *S. aureus*, and *C. albicans* has been well documented (Maria et al., 2005). Endophytic *Alternaria tenuissima*, *Chaetomium truncatum*, and *Nigrospora oryzae* isolated from *Tylophora indica* showed antifungal agent production effective against *Fusarium oxysporum*, *Rhizoctonia solani*, and *Sclerotinia sclerotiorum* (Kumar et al., 2011).

Actinomycetes are the major source of antimicrobial compounds. Among approximately 10000 known drugs, 50–55% have been reported from actinomycetes, specifically from *Streptomycetes* (Subramani and Aalbersberg, 2012). In the recent past, endophytic actinomycetes and bacteria have also been documented for the production of various antimicrobial agents (Table 15.1).

15.4 ANTICANCER COMPOUNDS

Cancer is considered one of the major causes of death worldwide nowadays. Various secondary metabolites with anticancer activity have been investigated from endophytes. Taxol, for the treatment of advanced breast cancer, lung cancer, and refractory ovarian cancer, has been approved by the FDA (Cremasco and Starquit, 2010). Although Taxol is the most potent anticancer drug approved by the FDA, its supply has been limited due to the destructive collection of the Pacific Yew tree, which is the main source of Taxol. Production of taxol (paclitaxel) a diterpenoid, from endophytic fungus *Taxomyces andreanae* isolated from the Pacific Yew (*Taxus brevifolia*) has set the ball rolling for exploitation of endophytic fungi as store house of natural secondary metabolites and is fast emerging as alternative way to obtain a cheaper and more reliable anticancer product via microbial fermentation (Stierle et al., 1995). Production of Taxol from endophytic fungi is a viable and cheaper alternative that can protect ever-diminishing yew trees.

Many other Taxol-producing endophytic fungi have also been reported from different medicinal plants other than *Taxus* sp. Endophytic *Pestalotiopsis microspora* and *P. breviseta* isolated from *Theaceae* and *Adenocalymma alliaceum*, respectively, were also reported for the production of Taxol (Wei, 2004; Kathiravan and Raman, 2010). *Periconia* sp. isolated from *Torreya grandifolia* also showed potential to produce Taxol (Li et al., 1998). Gangadevi and Muthumary (2007) reported the production of Taxol by the endophytic fungus *Bartalinia robillardoides* Tassi, isolated from a medicinal plant, *Aegle marmelos* Correa ex Roxb. Besides Taxol, many other metabolites possessing anticancer activity have been isolated from endophytic microorganisms and listed in Table 15.2.

TABLE 15.1
Antimicrobial Compounds Produced by Endophytic Bacteria

S. No.	Endophytic Microorganism	Host Plant	Active Against	Reference
1	*Streptomycetes* sp.	*Rhododendron*	*Phytophthora cinnamomi* and *Pestalotiopsis sydowiana*	Shimizu et al., 2000
2	*Streptomyces*, *Streptosporangium*, *Microbispora*, *Streptoverticillium*, *Saccharomonospora* sp., and *Nocardia* sp.	*Azadirachta indica* A. Juss	*Pseudomonas fluorescens*, *Escherichia coli*, *Bacillus subtilis*, *Pythium*, and *Phytophthora* sp.	Verma et al., 2009
3	*Actinomycetes*	*Emblica officinalis*	*E. coli*, *S. typhi*, *K. pneumonia*, and *S. aureus*	Kumar et al., 2011
4	*S. antibioticus*	*Curcuma domestica*, *Isotoma longiflora*, and *Phaleria macrocarpa*	*B. subtilis*, *P. aeruginosa*, and *Staphylococcus aureus*	Sunaryanto and Mahsunah, 2013
5	*Streptomyces* sp. and *S. coelicolor*	*Rhizopora apiculata* and *Avicennia marina*	*P. aeruginosa*, *E. coli*, *P. vulgaris*, *S. aureus*, *S. typhi*, *Aspergillus niger*, *A. flavus*, *Penicillium* sp., and *A. fumigatus*	Gayathri and Muralikrishnan, 2013

TABLE 15.2
Anticancer Metabolites from Endophytic Microorganisms

S. No.	Endophyte	Host Plant	Metabolite	Cell Lines	References
1	*Fusarium* sp. G344	Milk thistle (*Silybum marianum*)	Beauvericin	PC-3	Raja et al., 2015
2	*Penicillium chrysogenum*	*Huperzia serrate*	C-25 steroid	Cytotoxic	Ying et al., 2014
3	*Aspergillus fumigatus* and *Fusarium* sp.	*Garcinia* spp.	Phloroglucinol	HeLa cells	Ruma et al., 2013
4	*Colletotrichum gloeosporioides*	*Barringtonia acutangula*	Biomolecule	Cytotoxic	Lakshmi and Selvi, 2013
5	*Cephalotheca faveolata*	*Eugenia jambolana* Lam.	Sclerotiorin	HCT-116	Giridharan et al., 2012
6	*Alternaria alternata*	*Coffea arabica* L.	-	HeLa	Fernandes et al., 2009
7	*A. fumigatus*	*Juniperus communis* L.	Podophyllotoxin	Cytotoxic	Kusari et al., 2009
8	*Pantoea agglomerans*	*Prunella vulgaris*	Cyclic peptide: cyclo-Pro-Val	Hep G2	Hsieh et al., 2009
9	*Ampelomyces* sp.	*Urospermum picroides*	Desmethyldiaportinoland altersolanol A	L5178Y cells	Aly et al., 2008
10	*F. oxysporum*	*Juniperus recurva*	Podophyllotoxin	Cytotoxic	Kour et al., 2008
11	*Phialocephala fortinii*	*Podophyllum peltatum*	Podophyllotoxin	Cytotoxic	Eyberger et al., 2006
12	*Trametes hirsuta*	*Podophyllum* sp.	Lignans	Cytotoxic	Puri et al., 2005
13	*Pestalotiopsis microspora*	*Theaceae* sp.	Taxol	Cytotoxic	Wei, 2004

TABLE 15.3
Different Enzyme Inhibitors Produced by Endophytic Microorganisms

S. No.	Endophyte	Host Plant	Inhibitor	References
1	*Streptomycete*, *Rhodococcus*, and *Arthrobacter*	*Cajanus lineata*, *Leucas ciliata*, *Rauwolfia* sp., and *Gomphostemma heyneanum*	α-amylase	Akshatha et al., 2014
2	*Heterodermia leucomela*	Macrolichen	Amylase	Karthik et al., 2011
3	*Streptomyces olivochromogenes*	*Tinospora crispa*	α-glucosidase	Pujiyanto et al., 2011
4	*Bipolaris sorokiniana*	*Rhazya stricta*	Acetylcholinesterase and urease	Khan et al., 2015
5	*Alternaria* sp.	*Ricinus communis*	Acetylcholinesterase	Singh et al., 2012
6	*Penicillium citreonigrum*	*Dryopteris setosa* (Thunb.)	α-glucosidase, acetylcholinesterase, and cytotoxicity	Yuan et al., 2014
7	*Aspergillus terreus*	*Bruguiera gymnoihiza* (Linn.)	Acetylcholinesterase and cytotoxicity	Deng et al., 2013
8	*Paecilomyces tenuis*	*Huperzia serrate*	Acetylcholinesterase	Su and Yang, 2014

15.5 ENZYME INHIBITORS

Inhibitors of enzymes are finding application in the management of various diseases and disorders therefore have therapeutic importance. Inhibitors also have application in the field of agriculture as a biocontrol agent. Endophytes have been associated with the production of a variety of enzyme inhibitors that are medically useful. The production of inhibitors of α-glucosidase and α-amylase has been reported from many endophytic actinomycetes and fungi (Pujiyanto et al., 2011). These inhibitors find application in the management of diabetes, obesity, and related disorders, and also help in pest management (Derosa and Maffioli, 2012; Jayaraj et al., 2013). Tyrosinase inhibitors find major application as therapeutics to treat hyperpigmentation-related disease called melasma (Kolbe et al., 2013). Another major inhibitor that has received considerable attention as a therapeutic is acetylcholinesterase inhibitors. AChE inhibitors have been used for the treatment of neurodegenerative diseases, including Alzheimer's disease, anticholinesterase poisoning, senile dementia, ataxia, Parkinson's disease, and myasthenia gravis. Table 15.3 summarizes the list of endophytes identified for production of inhibitors of different enzymes.

15.6 ACETYLCHOLINESTERASE INHIBITOR FROM ENDOPHYTES

Acetylcholinesterase (AChE, EC 3.1.1.7), is an enzyme involved in the metabolic hydrolysis of neurotransmitter acetylcholine (ACh) at cholinergic synapses in the central and peripheral nervous system (Yu and Dayan, 2005). The level of neurotransmitter ACh, in neurons and synaptic junctions, plays an important role in several neurodegenerative diseases, particularly Alzheimer's disease (AD), senile dementia, ataxia, Parkinson's disease, and myasthenia gravis (Soreq and Seidman, 2000). Hypothesis of the AD is the decline in cognitive function, which is linked to the loss of cholinergic neurotransmission in the hippocampus and cortex (Perry et al., 1978) due to the presence of ACh-hydrolyzing enzyme AChE.

One of the most accepted strategies in AD treatment is the use of acetyl cholinesterase inhibitors (AChEIs) (Darvesh et al., 2003) that enhance the signal transmission in nerve synapses, which is beneficial for curing central nervous system-related diseases (Coulthard et al., 2006). Another type of cholinesterase is butyrylcholinesterase (BuChE, EC 3.1.1.8). It has been suggested that the use of dual cholinesterase inhibitors targeting AChE and BuChE as targets may prove more effective in treating patients with AD (Martinez and Castro, 2006). These two forms of cholinesterase (AChE and BuChE) co-exist ubiquitously throughout the body. AChE and BuChE showed 65% amino acid sequence homology at the active center structure, but they are products of different genes on human chromosomes. AChE is produced specifically by 7q22, and BuChE is produced specifically by 3q26 genes. The exact physiological role of BuChE is still elusive, but it is generally viewed as a backup for the homologous AChE (Devkota et al., 2008). AChE is present in neurons and erythrocytes, whereas BuChE is mainly present in serum and glia.

15.6.1 Structure of AChE and Mechanism of Action

The first structure of AChE was discovered by Sussman et al. (1991) through X-ray analysis. AChE contains a long peptide monomer composed of 537 amino acids. AChE is a complex protein having a 20 Å deep groove, usually called a gorge, and it is made up of 14 aromatic residues that provide stabilization to the quaternary ammonium ions of ACh. Although the catalytic hydrolysis of substrate (ACh) is believed to take place in the base of this gorge, the initial binding of ACh occurs at the outer rim in a region called the "peripheral site". At the bottom of the gorge, there are

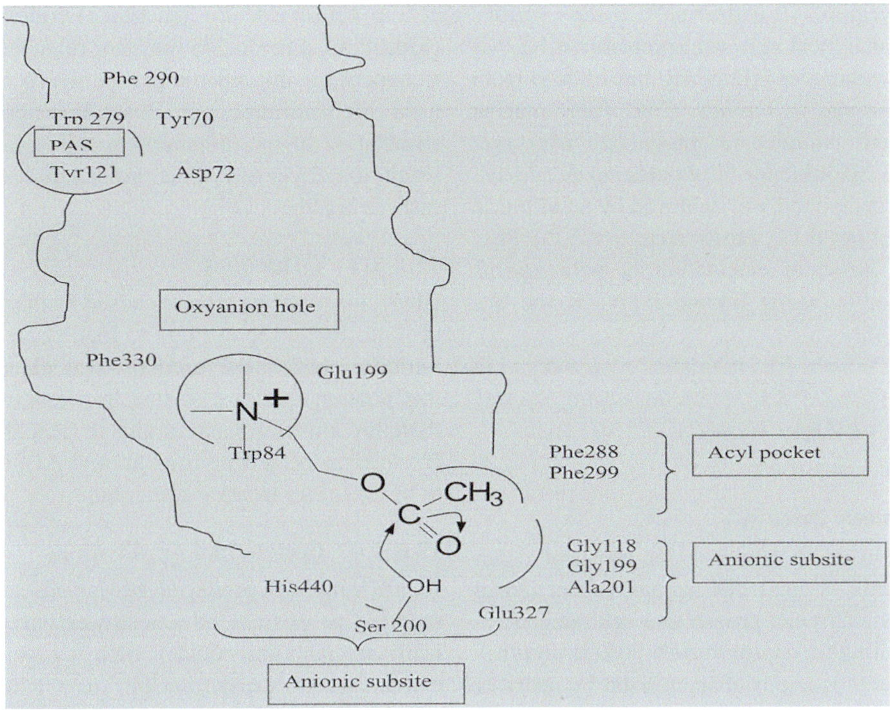

FIGURE 15.1 The binding regions of the AChE gorge.

Source: Houghton et al. (2006).

four main subsites "esteratic site", the "oxyanion hole", the "anionic subsite", and the "acyl pocket" (Figure 15.1). The hydrophobic site called "anionic subsite" is the choline-binding subsite and is largely comprised of aromatic residues, which are believed to bind to quaternary ammonium ligands by p-cation interactions (Houghton et al., 2006).

The substrate ACh enters the enzyme AChE gorge and choline, which is the hydrolysed product of AChE, then transferred back by choline acetyltransferase into the pre-synaptic nerve, where it is reformed again into acetylcholine by acetyl-CoA (Sussman et al., 1991). The binding site is composed of two subsites: the peripheral anionic site (PAS) corresponds to the entrance of the active site gorge, and the catalytic site (CAS) is situated close to the base of the active site gorge (Figure 15.1).

Galanthamine and huperzine, both alkaloidal inhibitors, are useful AD drugs that bind to the CAS region. Another inhibitor, propidium iodide, binds in the PAS region (Bourne et al., 2003), while inhibitors containing positively charged nitrogen specifically bind to the oxyanion hole (Houghton et al., 2006). However, certain tetracycline based inhibitory compounds bind to PAS and CAS simultaneously. Some AChE inhibitors bind not only to the enzyme's active site, but also to a "peripheral anionic site" that is near the protein surface. It is possible, therefore, that a dual-binding-site inhibitor might be able to alleviate cognitive deficits in AD patients and also act as a disease-modifying agent that delays the neurodegenerative cascade (Wang et al., 2008).

15.6.2 Purification of Acetylcholinesterase Inhibitor

The purification of AChEI has been reported from different sources using various techniques. Rhee et al. (2004) described the purification of "Ungeremine," an AChEI from the methanolic extract of *Nerine bowdenii* using a fluorometric flow assay an online preparative HPLC, centrifugal partition chromatography (CPC), 1H-NMR, 2D-NMR, and NOESY. An ethyl acetate extract of *Agrimonia pilosa* Ledeb yielded tiliroside, 3-methoxy quercetin, quercitrin, and quercetin that showed AChE inhibitory activity. Purification of these AChEIs was achieved by silica gel chromatography, Sephadex LH-20, and MIC gel column chromatography (Jung and Park, 2007). Purification and structure elucidation of novel AChEI "Hopeahainol A" using silica gel chromatography and single crystal X-ray spectroscopy was demonstrated by Ge et al. (2007). Arisugacin A and B, produced by *Penicillium* sp., a soil isolate known to possess AChE inhibitory activities, were purified and characterized using preparative HLPC and NMR, and MS (Kuno et al., 1996). Rao et al. (2001) described the purification of another AChEI, "4–(2,3,5-Trihydroxyphenyl) tetradecan-2-ol" from *Chrysosporium* sp. using solvent extraction and silica gel chromatography and structure elucidation by NMR, UV, and mass spectroscopy techniques. Gomaa and coworkers (2013) reported the purification and structure elucidation of novel AChE inhibitor "N,

N'-bis-(1-methyl-pyridin-2-yl)-hydrazine" using various chromatographic (analytical and semi-preparative HLPC) and spectroscopic techniques (1D NMR and EI-MS) from sea anemone *Gyrostoma helianthus*. Cold water marine organism *Synoicum pulmonaria* produced reversible, non-competitive AChE inhibitor "Pulmonarins A and B". Purification of the compound was achieved by solid phase extraction (SPE) followed by semi-preparative RP-HPLC with C_{18} column. Structure elucidation of pulmonarins A and B was carried out using 1D and 2D NMR and MS (Tadesse et al., 2014).

15.6.3 Therapeutic Applications of AChE Inhibitors

15.6.3.1 Alzheimer's Disease

AD is considered to be the third major cause of death worldwide after cardiovascular disease and cancer (Jiang et al., 2003). AD is a chronic, progressive neurodegenerative disease, accounting for approximately 50% of all types of dementia in elderly people is characterized by memory deterioration, abstract thinking & judgment, behavioral consequences, disability in daily living skills, impaired visuo-spatial skills, impairments in short-term and long-term memory loss, and language deterioration. Hypothesis of AD is the decline in cognitive function, which is linked to the loss of cholinergic neurotransmission in the hippocampus and cortex (Perry et al., 1978) due to the presence of ACh-hydrolyzing enzyme AChE. Therefore, in AD patients, the enhancement of brain cholinergic transmission remains a major goal for therapeutic agents. Inhibition of acetylcholinesterase might restore cholinergic balance and result in an increased level and duration of action of synaptic ACh (Jukic et al., 2012), serving as a treatment strategy for AD and related diseases.

Natural metabolite "Physostigmine", an indolic alkaloid, has been considered as the first principle AChE inhibitor isolated from the seeds of *Physostigma venenosum* Balfour (Calabar bean). Whereas, chemically synthesized tetrahydro aminacridine (tacrine), an amino acridine, is a noncompetitive, reversible inhibitor of AChE and BuChE, has been licensed in Canada as the first agent approved for treating the cognitive symptoms of AD (Tumiatti et al., 2010). Similarly, "Rivastigmine," a carbamate inhibitor of AChE and BuChE, was licensed in the UK (Holden and Kelly, 2002). Known AChEIs such as eserine, donepenzil, glanthamine, and rivastigmine are the only drugs approved for the treatment of AD as they promote an increase in concentration and duration of action of neurotransmitter ACh (Rollinger et al., 2004).

15.6.3.2 Myasthenia Gravis (MG)

Myasthenia gravis (MG) is an autoimmune disease, occurring due to the development of autoantibodies that destroy ACh in the neuromuscular junction. Main symptoms include muscle weakness and fatigue due to dysfunction of the neuromuscular junction. Pyridostigmine bromide (Mestinon), a reversible inhibitor of AChE, is the first-line treatment for myasthenia gravis, which is currently being used (Mehandiratta et al., 2014). It increases acetylcholine stimulation of the remaining acetylcholine receptors as a result, the effect of ACh is prolonged, leading to improvement in health.

15.6.3.3 Glaucoma

Glaucoma is an eye disease caused by the overaccumulation of fluid within the eye due to inadequate drainage. AChE inhibitors have been used to treat glaucoma, as vessels that contain fluid are dilated by overstimulation of ACh, therefore allowing faster drainage (Houghton et al., 2006). Physostigmine, a drug used to treat AD, is also considered to be useful for treating glaucoma.

15.6.3.4 Insecticidal Applications

AChE inhibitors produced by various sources have been reported to possess insecticidal activity. Many commercially available insecticides, such as organophosphates and carbamates (chlorpyrifos-CPF, methylchlorpyrifos-MCPF, and parathion-PTH) target the AChE of insects. These drugs are known to be specific, irreversible inhibitors of acetylcholinesterase. The toxicity of organophosphates and carbamates is based on their variable binding affinity to the serine hydroxyl group in the active site of AChE by phosphorylation or decarbamylation, which results in the inactivation of the enzyme (Ekström et al., 2006). The inhibition of AChE leads to an excess of acetylcholine accumulation at the synaptic clefts, resulting in over-stimulation of muscarinic and nicotinic receptors, causing lacrimation, muscular paralysis, salivation, convulsions, *etc*., and finally leading to death by asphyxiation. Ainge et al. (2002) investigated the extract of clubmoss, *Lycopodium varium*, for the isolation of the alkaloid huperzine A, as the major antifeedant and insecticidal component. Huperzine A showed insecticidal activity against the Australian carpet beetle (*Anthrenocerus australis*), the Australian sheep blowfly (*Lucilia cuprina*), and the webbing clothes moth (*Tineola bisselliella*). Endophytic microorganisms producing insecticidal compounds are summarized in Table 15.4.

15.6.3.5 Sources of Acetylcholinesterase Inhibitors

Acetylcholinesterase inhibitor molecules are produced by a variety of plants, fungi, bacteria, including endophytes. Traditionally, plants have been used as a source of AChEI for the treatment of AD. In this context, "Galanthamine," an alkaloid isolated from *Galanthus caucasicus*, is used for the treatment of mild to moderate AD. Metabolite "withanolide" isolated from *Withania somnifera* showed dual cholinesterase inhibitory activity against AChEI and BuChEI (Choudhary et al., 2004). *Bacopa monnieri* (Neer Brahmi) is considered to be a memory enhancer in Indian folklore and may act as a beneficial source for the treatment of AD (Ahirwar et al., 2012). Jazayeri and coworkers (2014) demonstrated the AChE inhibitory potential of extracts from

TABLE 15.4
Compounds Produced by Endophytic Microorganisms Exhibiting Insecticidal Activity

S. No	Endophyte	Host Plant	Insect	Reference
1	*Aspergillus oryzae*, *A. tamarii*, *A. versicolor*, and *Emericella nidulans*	*Rhizophora mucronata*	*Spodoptora litura* larvae	Abraham et al., 2015
2	Moroccan actinobacteria	Endemic aromatic and medicinal plants	*Ceratitis capitata* Wied.	Samri et al., 2015
3	*Alternaria* sp., *Aspergillus* sp., *Fusarium* sp., *Penicillin* sp., *Pesalotiopsis* sp., *Phoma* sp., and *Thizoctonia* sp.	*Tripeterygium* wilfordii	Mosquito and larvicidal activities	Han et al., 2013
4	*Alternaria* sp.	*Ricinus communis*	*Spodoptera litura*	Singh et al., 2012
5	*Heterodermia leucomela* (L)	Macrolichen	*Aedes aegypti*	Karthik et al., 2011
6	*Eupenicillium parvum*	*Azadirachta indica*	Insect repellent	Kusari et al., 2012
7	*Mycoleptodiscus indicus*	*Echinacea purpurea*	*Aedes aegypti*	Rosa et al., 2011
8	*Pencillium* sp.	*Derris elliptica* Benth	Aphids	Hu et al., 2005
9	*Neotyphodium typhnium*	*Poa ampla*	Mosquito and larvicidal activity	Ju et al., 1998

different medicinal plants, *viz. Brassica alba*, *B. nigra*, *Camellia sinensis*, *Cinchona officinalis*, *Citrus aurantifolia*, *Citrus aurantium*, *Ferula assafoetida*, *Humulus lupulus*, *Juglans regia*, *Juniperus sabina*, *Myristica fragrans*, *Pelargonium graveolens*, *Pistacia vera*, *Punica granatum*, *Rheum officinale*, *Rosa damascena*, *Salix alba*, and *Zizyphus vulgaris*. Studies conducted by Ahmed et al. (2003) reported that the methanolic extract of *Myricaria elegans* showed significant dual cholinesterase inhibitory (AChE and BuChE) activity. Methanolic extract of the traditional medicinal plant "Rangoon creeper" (*Quisqualis indica* Linn.) contains quisqualic acid that is known to possess strong antioxidant and AChE inhibitory activity (Wetwitayaklung et al., 2007). Extract of alkaloid-rich herb *Ammocharis coranica* showed potent AChE inhibition and is useful for treating mental illnesses or symptoms implicated in the pathophysiology of AD (Elisha et al., 2013).

Fungi are considered to be rich sources of bioactive metabolites. Many soil fungi have been reported to possess AChE inhibitory activity. *Chrysosporium lobatum* isolated from forest soil of Kaziranga National Park showed AChE inhibitory activity (Kumar et al., 2013). Teles and Takahashi (2013) reported the production of a new AChEI compound, "Paecilomide," that was isolated from the soil fungus *Paecilomyces lilacinus*. Dual cholinesterase (AChE and BuChE) inhibitory activity from soil isolates *Aspergillus terreus*, *Penicillium citrinum*, and *Penicillium* sp. has been demonstrated (Kuno et al., 1996; Kim et al., 2001, 2003).

Few reports on the production of AChEI by endophytic fungi have also been documented. Endophytic fungi *Cladosporium cladosporioides* (Zhang et al., 2011), *Shiraia* sp. (Zhu et al., 2010) and *Trichoderma* sp. (Dong et al., 2014) isolated from *Huperzia serrata* were found to produce AChE inhibitory molecule Hup A. *Sarcococca hookeriana* yielded two pregnane-type steriodal alkaloids hookerianamide H and I, along with three known alkaloids dictyophlebine, *N* alpha-methylepipachysamine D and sarcovagine C that showed good AChEI and BuChE inhibition (Devkota et al., 2007). Rodrigues and coworkers (2005) reported *Chaetomium* sp., *Guignardia mangiferae*, *Pestalotiopsis guepinii*, *Phomopsis* sp., *Physalospora* sp., *Xylaria* sp., and endophytes isolated from leaves of the plants belonging to families *Anacardiaceae*, *Apocynaceae*, *Leguminosae*, and *Palmae* as a source of dual cholinesterase inhibitory molecule. Endophytic fungus *Alternaria* sp. isolated from *Ricinus communis* was reported to inhibit acetylcholinesterase (Singh et al., 2012). Khan et al. (2015) reported the production of AChE inhibitory molecule "bipolarisenol" from the fungus *Bipolaris sorokiniana* from the leaves of alkaloid-rich ethno-medicinal plant *Rhazya stricta*. Marine endophytic fungus *Talaromyces* sp. produced three new bioactive compounds (talaromycesone A, isopentenyl xanthenone, talaroxanthenone) displaying potent AChE inhibitory activity (Wu et al., 2014). In addition, yeast *Yarrowia lipolytica* was also cited in literature as a source of AChE inhibitor compound (Lee et al., 2008).

Apart from fungi, diverse bacteria have also been reported for the production of AChEI. Wastewater lagoon cyanobacteria *Nostoc* produced alkaloid compound "nostocarboline", which showed cholinesterase inhibitory activity (Becher et al., 2005). "Geranylphenazinediol", a new phenazine natural biomolecule, was produced by the *Streptomyces* sp. isolated from a marine sediment sample with AChEI activity (Ohlendorf et al., 2012). Sangnoi et al. (2008) reported the production of a pyrroloquinolone molecule, "marinoquinoline", isolated from the marine gliding bacterium *Rapidithrix thailandica* with potent AChEI activity. Endophytic bacteria *Bacillus subtilis* isolated from marine sponge *Fasciospongia cavernosa* produced a metabolite that also exhibited appreciable AChE inhibitory activity (Pandey et al., 2014).

15.7 ASPARAGINASE FROM ENDOPHYTES

L-Asparaginase (L-asparagine amido hydrolase E.C. 3.5.1.1), which catalyzes the deamination of asparagine into aspartic acid and ammonia (Figure 15.2), is an enzyme of high therapeutic value that emerged as potent health care agent for the treatment of acute lymphocytic leukemia (Balasubraminium et al., 2012; Singhal and Swaroop, 2013). Asparaginase also has biotechnological applications in food industry to reduce acrylamide content in fried food/snacks.

L-asparaginase is produced by various plants and microorganisms (Dhanam and Kannan, 2014). Due to the difficulty in the extraction of asparaginase from plants, the search for novel microbial strains capable of producing asparaginase is a hot topic of ongoing research, although many microorganisms with the ability to produce asparaginase are known. The first microbial L-asparaginase was commercially produced from *Escherichia coli* (Mashburn and Wriston, 1964). Asparaginase from two strains, that is, *E. coli* and *Erwinia carotovora*, is currently in clinical use for the treatment of acute lymphoblastic leukemia (Senthil and Selvam, 2011). Clinical results have shown that asparaginase from *E. coli* causes toxicity and immunosuppression in addition to the development of resistance (Sahu et al., 2007). Therefore, the search for asparaginase from other microbial sources with low toxicity but with comparable therapeutic effects is important.

15.7.1 Purification of Asparaginase

Purification is an important step to analyze and characterize the functional aspects of the enzyme. Column chromatography, such as ion exchange and size exclusion, has been extensively used for the purification of asparaginases (Wei et al., 2008; Senthil and Selvam, 2011). There are several reports on the purification and characterization of asparaginase by a wide array of microorganisms using ammonium sulfate precipitation followed by ion exchange chromatography (DEAE) and gel filtration chromatography (Sephadex G 100, G 200) (Abraham, 2012; Borah et al., 2012; Jois et al., 2013; Kumar and Manonmani, 2013; Meghavarnam and Janakiraman, 2015). Loureiro et al. (2012) demonstrated the purification of asparaginase from *Aspergillus terreus* using DEAE-Sepharose followed by Sephacryl S-200 HR. Tropical soil fungus Fusarium Culmorum ASP-87 produced asparaginase that was purified using anion exchange DEAE cellulose, followed by gel filtration chromatography Sephadex G–100. The purified enzyme was stable at pH 8.0 and retained 100% activity for 24 h (Meghavarnam and Janakiraman, 2015). Asparaginase purified from *Fusarium* sp. has been identified as a promising source of antineoplastic agent, as it showed a potent cytotoxic effect against Ehrlich Ascites carcinoma cells, whereas no cytotoxic activity against normal human lymphocytes *in vitro*. The purified enzyme also showed optimal activity at alkaline pH and 37°C temperature, making it extremely valuable in chemotherapy (Abraham, 2012). Purification of asparaginase from sewage water isolate *E. coli* was achieved by DEAE cellulose, an anion exchange chromatography with its optimal activity at pH 6.0 and 55°C (Borah et al., 2012). The enzyme asparaginase from *Streptomyces* sp. was purified using ammonium sulfate precipitation followed by Sephadex G 100 and CM Sephadex C-50 gel filtration with an apparent molecular weight of 112 KDa (Narayana et al., 2008). The recombinant L-asparaginase (ECAR-LANS) from *E. carotovora* is a prospective therapeutic enzyme for leukaemia treatment. More than 90% enzyme purity, complemented with 72% recovery, was achieved with a single chromatographic purification step using SP sepharose (Krasotkina et al., 2004) whereas, Devi and Azmi (2012), demonstrated the single step purification of asparaginase from *E. caratovora* using SP sephadex with 24% yield of enzyme. The molecular weight of the purified enzyme was found to be 116 KDa and present in the form of dimer of dimers, revealed by SDS-PAGE. Purified enzyme was effective against Hep-2 cancerous cell lines.

15.7.2 Types of Asparaginase

Based on a study conducted by Borek and Jaskólski (2000) as well as on biochemical and crystallographic data, the known asparaginase sequences can be divided into three families (Figure 15.3). The first family corresponds to

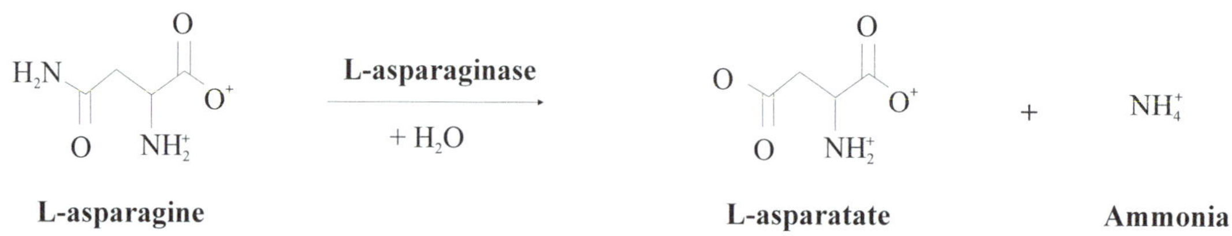

FIGURE 15.2 The asparaginase reaction.

Source: Borek and Jaskólski (2001).

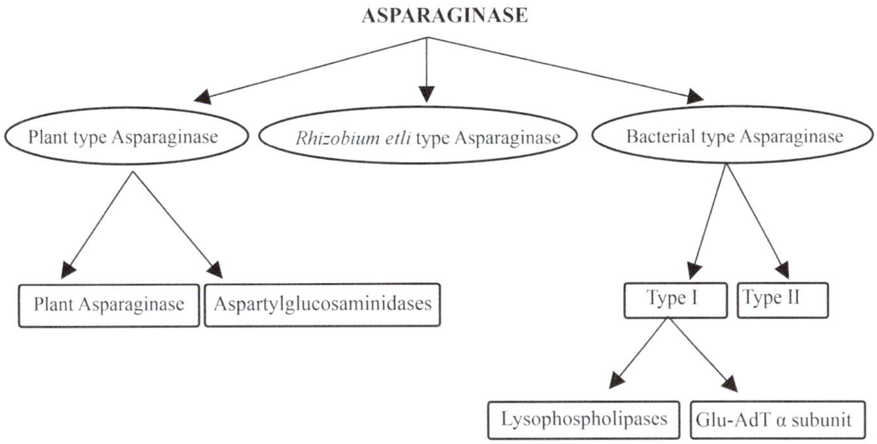

FIGURE 15.3 General classification of asparaginases.

Source: Borek and Jaskólski (2001).

plant-type asparaginases, the second, *Rhizobium etli* asparaginase, and the third, similar to bacterial-type asparaginases. Plant type asparaginase is further divided into plant asparaginase and aspartyl-glucosaminidases. Whereas, bacterial type asparaginase is categorized into type I and type II (Campbell et al., 1967). Type I asparaginases are constitutively expressed in the cytoplasm and characterized by enzymatic activity toward both L-asparagine and L-glutamine, and are further divided into lysophospholipases and Glu-AdT alpha subunit. In contrast, type II asparaginases are expressed in the periplasmic space of the bacterial membranes under anaerobic conditions and exhibit high specific activity against L-asparagine and are widely used as chemotherapeutic agents for acute lymphoblastic leukemia treatment (Duval et al., 2002).

15.7.3 Application of Asparaginase

15.7.3.1 Asparaginase as Chemotherapeutic Agent

Asparaginase is used for treating acute lymphoblastic leukemia in children, as neoplastic cells are unable to synthesize non-essential amino acid asparagine on their own due to the low expression of asparagine synthetase. Thus, these cells depend on high amounts of extracellular supply of asparagine for protein synthesis to keep up with the rapid proliferation rates, whereas normal cells can synthesize their asparagine. For this reason, the commonest therapeutic practice is to inject the asparaginase intravenously to decrease the blood concentration of L-asparagine, selectively affecting the neoplastic cells (Ebrahiminezhad et al., 2011). The World Health Organization (WHO) and Food and Drug Administration (FDA) have approved L-asparaginase for the treatment of acute lymphoblastic leukemia and lymphosarcoma. Asselin et al. (1989) studied the ability of L-asparaginase as a useful agent in killing tumor cells in patients undergoing treatment for acute lymphoblastic leukemia. Efficacy of asparaginase as an antitumor agent is related to its circulation, half-life in host serum, and its affinity for substrate asparagine (Schwartz et al., 1966; Boyse et al., 1967). Among bacterial sources, *E. coli* and *E. carotovora* asparaginase have sufficient circulation half-lives to be of clinical importance (Crowther, 1971; Howard and Carpenter, 1972) but elicit certain side effects such as hypersensitivity, anaphylaxis, and pancreatitis, and induce mild to severe immune responses. Acute lymphoblastic leukemia patient who has hypersensitivity to *E. coli* asparaginase could be treated with *Erwinia* asparaginase in combination with other chemotherapeutic drugs (Dhanam and Kannan, 2013). Various approaches have been used to overcome drawbacks. The half-life of asparaginase has been increased by conjugating it with polyethylene glycol (PEG). Pegasparaginase (PEG-asparaginase) is a modified version of the enzyme L-asparaginase and has a longer half-life than *E. coli* and *E. carotovora* asparaginase. PEG-asparaginase reduces the immunogenicity of the protein, increases its stability in the plasma of pretreated patients. A PEG conjugated or PEG *E. coli* asparaginase was found to be a useful alternative in patients with prior clinical hypersensitivity to native *E. coli* asparaginase. It can be administered at lower doses and for longer intervals. Elsapar, Erwinase, Kidrolase, and Oncaspar are the commercially available L-asparaginase (Theantana et al., 2009; Deokar et al., 2010).

15.7.3.2 Asparaginase in the Food Industry

Asparaginase also has biotechnological applications in the food industry as it is being used in the food industry to prevent acrylamide formation (Pedreschi et al., 2008; Kornbrust et al., 2012). Acrylamide is a toxic agent and causes neurotoxicity in humans. Formation of acrylamide is the result of the Maillard reaction between the free amino acid asparagine and the carbonyl group of reducing sugars like glucose that are present in large amounts in starchy foods. Asparagine acts as a precursor for acrylamide formation during the browning that occurs in frying, baking, and grilling carbohydrate-rich starchy food products like potatoes at temperatures exceeding 120°C (www.who.int/ipcs/publications/jecfa/reports/en/index.html). Therefore, lowering either the glucose or asparagine content in food can be expected to lower the acrylamide formation. By adding asparaginase before baking or frying the food, the formation of acrylamide is significantly reduced, as asparagine cannot take part in the Maillard reaction (Kornbrust et al., 2010).

Two commercial asparaginases used in the food industry are Acrylaway and Preventase. Acrylaway is manufactured by Novozymes company (Bagsvaerd, Denmark) and produced from *Aspergillus oryzae*, whereas Preventase is produced from *Aspergillus niger* by DSM Company (DSM Food specialities, Seclin Cédex, France). Treating French fries with Acrylaway before frying resulted in a 60% reduction in acrylamide formation (Pedreschi et al., 2008). Treatment of raw food stuff with asparaginase before thermal treatment at high temperature resulted in a 99 % lower acrylamide formation (Lindsay and Jang, 2005). Anese et al. (2011) concluded that acrylamide formation in biscuits was considerably reduced with no change in color and taste of the product when treated at intermediate concentrations of asparaginase. Glutaminase is another enzyme found with asparaginase. However, the presence of glutaminase along with asparaginase causes toxicity that leads to diabetes, leucopenia, liver dysfunction, pancreatitis, and neurological seizures (Duval et al., 2002). Therefore, the search for a new asparaginase enzyme with high therapeutic activity, different serological properties that should be free from or low glutaminase activity, is a desirable trait from food and chemical oncology perspectives.

15.7.4 Sources of Asparaginase

L-asparaginases are produced by various sources such as plants, animals, and microorganisms. Plants such as *Capsicum annum* L. and *Withania somnifera* have been reported for asparaginase production (Bano and Sivaramakrishnan, 1980; Oza et al., 2009). Several microorganisms, such as bacteria, fungi, yeast, actinomycetes, and algae, have been reported as potential sources of L-asparaginase (Verma et al., 2007; Dhanam and Kannan, 2013; Thakur et al., 2013). Enzymes produced by microbial sources are preferred over plant sources due to their economic production, consistency, stability, ease of process modification, purification, cost-effectiveness, and eco-friendly nature (Savitri et al., 2003). Different actinomycetes, *viz.*, *Streptomyces aureofasciculus*, *S. chattanoogenesis*, *S. hawaiiensis*, *S. orientalis*, *S. canus*, and *S. olivoviridis* isolated from estuarine fishes produced a significant amount of asparaginase activity (Sahu et al., 2007). Besides these, various endophytic microorganisms have also been reported for the production of asparaginase. Endophytic fungi *Aspergillus* sp. isolated from marine soft sponge *Aplysina fistularis* and salt marsh plant *Suaeda maritima* were reported to produce asparaginase (Ahmed et al., 2015; Kalyanasundaram et al., 2015). Asparaginase synthesis was also reported from the endophytic fungi *Colletotrichum* sp., *Eurotium* sp., *Fusarium* sp., *F. subglutinans*, *Phoma* sp., and *Penicillium* sp. isolated from *Cymbopogon citratus*, *Curcuma longa*, *Murraya koenigii*, *Thymus*, *Oldenlandia diffusa*, *Phomopsis* spp., and *Pereskia bleo*, respectively (Jalgaonwala and Mahajan, 2014; Manasa and Nalini, 2014; Masumi et al., 2014; Chow and Ting, 2015). In addition, endophytic bacteria *Bacillus methylotrophicus*, *B. siamensis*, *B. subtilis*, and *Serratia marcescens* isolated from *Centella asiatica*, *Houttuynia cordata*, *Litsea cubeba*, *Potentilla fulgens*, *Rubia cordifolia*, and *Spilanthes acmella* have been evaluated as a potent source of asparaginase (Joshi and Kulkarni, 2014). Table 15.5 summarizes the culture conditions and asparaginase activity production by different endophytic microbial sources.

15.8 CONCLUSION

Drugs being used as therapeutic agents for the treatment of various diseases and disorders are usually chemically synthesized and confer many side effects. Therefore, research efforts in the recent past have been directed to invent new, safer, and efficient natural metabolites that possess low toxicity and are easily obtainable with negligible environmental impact. Keeping this in mind, the present review emphasizes the importance of endophytic microorganisms as a potential source of novel bioactive compounds exhibiting different activities, such as antimicrobial, acetylcholine esterase inhibitory, and enzymatic activities. Different bioactive molecules produced by different endophytic microorganisms could prove to be novel sources of pharmaceuticals as well as agriculturally important.

TABLE 15.5
Comparative Analysis of Asparaginase Production from Different Microbial Sources

Microorganism		Source	Media Composition	Activity	Reference
Bacterial Source	*Citrobacter* sp.	Laboratory	Yeast autolysate and organic acids	451 U/mg	Bascomb et al., 1975
	Enterobacter aerogenes	National collection of microorganisms, Pune	$NaNO_3$, $MgSO_4$, $FeSO_4$, KCl, K_2HPO_4, and yeast extract	1.2 U/ml	Mukherjee et al., 2000
	Pseudomonas aeruginosa	Microbiological Resource Center, Egypt	Soya bean meal, and 0.01M phosphate buffer	17.9 U/mg	El-Bessoumy et al., 2004
	Thermoactinomyces thalpophilus	Procured from the lab	Starch, casein, NaCl, KNO_3, KH_2PO_4, $MgSO_4$, $CaCO_3$, and $FeSO_4$	1620 PU/g	Divakar et al., 2006
	Staphylococcus sp.	Soil isolate	M9 medium	55.6 IU/ml	Prakasham et al., 2007
	Amycolatopsis keratiniphila	Medicinal plant rhizosphere	Asparagine dextrose agar medium	3.05 um amm./ml/h	Khamna et al., 2009
	Streptomyces sp.	Marine sediment	M9 Medium	2.48 um amm/ml/h	Poorani et al., 2009
	Serratia marcescens SB08	Gut of sulfur butterfly (*Kricogonia lyside*)	Peptone, beef extract, yeast extract, and L-asparagine	256 IU/ml	Venil et al., 2009
	E. coli	Laboratory	M 9 medium	0.041 IU/mL	Abbas et al., 2010
	Pectobacterium carotovora	National Collection of Microorganisms, Pune	Carbon source, Na_2HPO_4, KH_2PO_4, NaCl, $MgSO_4$, $CaCl_2$, yeast extract, and peptone	14.56 U/ml	Kumar et al., 2010
	Nocardia levis MK-VL 113	Laterite soil	Asparagine, glycerol, 2HPO4, and trace salt solution	5.3 U/mg	Kavitha and Vijayalakshmi, 2012
	Bacillus sp.	Soil	Na_2HPO_4, KH_2PO_4, NaCl, L-asparagine, glycerol, $MgSO_4$, and $CaCl_2$	32 IU/ml	Mahajan et al., 2012
	S. capitis	*Mentha spicata* (mint) plant leaves	Na_2HPO_4, KH_2PO_4, NaCl, L-Asparagine, $MgSO_4$, $CaCl_2$, and Glucose	224 IU/ml	Udaya et al., 2013
	S. acrimycini	Marine actinomycetes	KH_2PO_4, L-asparagine, $MgSO_4.7H_2O$, $Cacl_2.2H_2O$, and glucose 3.0	18 U/ml	Selvam and Vishnupriya, 2013
	Streptomyces sp.	Soil isolate	Czapek Dox Agar Medium	Primary screening	Dhanam and Kannan, 2014
	Acinetobacter baumannii	Rhizosphere of cucumber and lettuce	M9 medium	2.78 U/mg	Muslim, 2014
	Myroides getamesis	Slaughterhouse	Glucose, asparagine, KCl, KH_2PO_4, $MgSO_4$, $CuNO_3$, $ZnSO_4$, and $FeSO_4$	Primary screening	Talluri et al., 2014
	Pseudomonas oryzihabitans	Plant *Hibiscus rosa sinensis*	Glucose, asparagine, KCl, K_2HPO_4, and $MgSO_4$	4.2 IU/ml	Bhagat et al., 2016
Fungal Source	*Aspergillus* sp.	Laboratory	Modified Czapek dox medium	41 IU/L	Sarquis et al., 2004
	A. niger	Culture collection, Banaras Hindu University	Modified Czapek dox medium	39.9 U/gds	Mishra, 2006
	Endophytic fungi H5BR7	Thailand medicinal plants, *Eupatorium odoratum*	Modified Czapek dox medium	1.53 IU/ml	Theantana et al., 2007
	Scopulariopsis strain FMG 133	Phyllosphere and rhizosphere of the mangrove ecosystem	Glucose asparagine medium	10.25 IU/ml	Gupta et al., 2009
	Colletotrichum sp. E5T9	Thai medicinal plant	Modified Czapek dox medium	0.4 IU/ml	Theantana et al., 2009

(Continued)

TABLE 15.5 (*Continued*)
Comparative Analysis of Asparaginase Production from Different Microbial Sources

Microorganism	Source	Media Composition	Activity	Reference
Bipolaris sp.	Brown rice	Modified Czapek dox medium	5.54 IU/ml	Lapmak et al., 2010
Emerciella nidulans	Soil isolate	Modified Czapek dox medium	1.1 IU/ml	Jayaramu et al., 2010
A. terreus	Soil isolate	Carob pod with 65% moisture	5.63 IU/ml	Siddalingeshwara et al., 2010
Flammulina velutipes	Lab	Glucose, asparagine, K_2PO_4, $MgSO_4$, yeast extract, and trace metal solution	16 IU/ml	Eisele et al., 2011
A. terreus MTCC 1782	MTCC, Chandigarh	Modified Czapek dox medium	36.64 IU/ml	Gurunathan and Sahadevan, 2011
Penicillium sp.	Soil isolate	Modified Czapek dox medium	15 U/ml	Rukachaisirikul et al., 2007
Fusarium sp.	Algae	Modified Czapek dox medium	70 U/ml	Thirunavukkarasu et al., 2011
A. terreus	Soil	Modified Czapek dox medium	9.3 IU/ml	Balasubraminium et al., 2012
P. brevicompactum	Culture collection, Egypt	Modified glucose Czapek Dox	3.80 U/mg	Elshafei et al., 2012
Penicillium sp.	Soil isolate	Modified Czapek dox medium	Primary screening	Mushtaq et al., 2012
Aspergillus sp.	Soil isolate	Modified Czapek dox medium and orange peel and sodium phosphate buffer	70.67 IU/g	Soniyamby et al., 2012
Cladosporium sp.	Lab	Wheat bran, moistened with 50 mM phosphate buffer (pH 6.3)	0.096 U/mg	Kumar et al., 2013
Alternaria sp.	Plant *Withania somnifera*	Modified Czapek dox medium	1.35 U/mg	Nagarajan et al., 2013
Mucor sp.	Soil isolate	Modified Czapek dox medium	69.43 U/mg	Thakur et al., 2013
Eutorium sp.	Plant *curcuma longa*	Modified Czapek dox medium	1.304 IU/ml	Jalgaonwala and Mahajan, 2014
Trichoderma viridae	Marine soil	Modified Czapek dox medium	78.2 U/mg	Lincoln and More, 2014
Fusarium sp.	Plant *Tabernaemontana heyneana*	Modified Czapek dox medium	1.006 U/ml	Manasa and Nalini, 2014
P. citrinum	Culture Collection Microbiology Lab, Odisha	Glucose, L asparagine, KH_2PO_4, and $MgSO_4$	899.3 U/mg	Patro and Gupta, 2014
Aspergillus sp.	Salt marsh plant (*Suaeda maritima*)	Modified Czapek dox medium	35.28 IU/ml	Kalyanasundaram et al., 2015
Talaromyces pinophilus	*Endophyte Curcuma amada*	Modified Czapek dox medium	108.95 IU/ml	Krishnapura and Belur, 2015
Rhodotorula rosa and *Candida utilis*	Procured from the lab	Glucose, asparagine, K_2HPO_4, KCl, $MgSO_4$, and yeast extract	53 U/ml, 20 U/ml	Arima et al., 1972
Cryptococcus, *Sporobolomyces*, *Trichosporon*, and *Torulopsis*	Institute for Fermentation Osaka	Sucrose, polypepton, beef extract, yeast extract, malt extract, KH_2PO_4, K_2HPO_4, and $MgSO_4$	Positive 0.3 U/ml	Imada et al., 1973
C. utilis	Laboratory	Glucose, polypepton, yeast extract, KH_2PO_4, and $MgSO_4$	3,900 U/mg protein	Sakamoto et al., 1977
C. utilis	NRRL USA	Glucose, peptone, and yeast nitrogen base	0.8 U/ml	Kil et al., 1995
Rhodosporidium toruloides	IMTECH Chandigarh	Mannitol, L-asparagine, K_2HPO_4, KH_2PO_4, KCl, $MgSO_4$, and yeast extract	$10{,}508 \times 10^3$ U/mg protein	Ramakrishnan and Joseph, 1996
Recombinant *Pichia pastoris*	*P. pastoris* GS115 (*his4*) (Invitrogen Corporation)	Phosphate buffer, yeast nitrogen base with ammonium sulfate and without amino acids, glycerol, and biotin	800 U/g	Ferrara et al., 2006

REFERENCES

Abbas, A.A., M.A. Sabbah, and O.A. Kathum. 2010. Partial purification and cytotoxic activity of L-asparaginase isolated from *Escherichia coli*. *Iraqi Journal of Science* 51: 290–94. https://www.iasj.net/iasj/download/eb7719e05c4619fe.

Abraham, A. 2012. Production, purification and characterization of extracellular L-asparaginase having anti neoplastic activity from *Fusarium* sp. *Journal of Advance Researches in Biological Sciences* 4: 293–301.

Abraham, S., A. Basukriadi, S. Pawiroharsono, and W. Sjamsuridzal. 2015. Insecticidal activity of ethyl acetate extracts from culture filtrates of mangrove fungal endophytes. *Mycobiology* 43: 137–49. https://doi.org/10.5941/MYCO.2015.43.2.137.

Adesemoye, A.O., and J.W. Kloepper. 2009. Plant–microbes interactions in enhanced fertilizer-use efficiency. *Applied Microbiology and Biotechnology* 85: 1–12. https://doi.org/10.1007/s00253-009-2196-0.

Ahirwar, S., M. Tembhre, S. Gour, and A. Namdeo. 2012. Anticholinesterase efficacy of *Bacopa monnieri* against the brain regions of rat-a novel approach to therapy for Alzheimer's disease. *Asian Journal of Experimental Science* 26: 65–70. http://www.ajesjournal.com/PDFs/12-1/12%20Anticholinesterase%20Efficacy%20of%20Bacopa%20monnieri%20against%20the%20Brain%20Regions%20of%20Rat.pdf.

Ahmed, M.M.A., F.N.A. Dahab, M.T. Taha, and S.M.F. Hassan. 2015. Production, purification and characterization of L-asparaginase from marine endophytic *Aspergillus* sp. ALAA-2000 under submerged and solid state fermentation. *Journal of Microbial and Biochemical Technology* 7: 165–72. http://dx.doi.org/10.4172/1948-5948.1000199.

Ahmed, W., B. Ahmad, M. Ahmad, Z. Iqbal, M. Nisar, and M. Ahmad. 2003. *In vitro* inhibition of acetylcholinesterase, butyrylcholinesterase and lipoxygenase by crude extract of *Myricaria elegans* Royle. *Journal of Biological Sciences* 3: 1046–49. https://scialert.net/abstract/?doi=jbs.2003.1046.1049.

Ainge, G.D., S.D. Lorimer, P.J. Gerard, and L.D. Ruf. 2002. Insecticidal activity of huperzine A from the New Zealand clubmoss, *Lycopodium varium*. *Journal of Agricultural and Food Chemistry* 50: 491–94. https://doi.org/10.1021/jf0106087.

Akshatha, V.J., M.S. Nalini, C. D'Souza, and H.S. Prakash. 2014. *Streptomycete* endophytes from anti-diabetic medicinal plants of the Western Ghats, inhibit alpha-amylase and promote glucose uptake. *Letters in Applied Microbiology* 58: 433–39. https://doi.org/10.1111/lam.12209.

Aly, A.H., R. Edrada-Ebel, I.D. Indriani, et al. 2008. Cytotoxic metabolites from the fungal endophyte *Alternaria* sp. and their subsequent detection in its host plant *Polygonum senegalense*. *Journal of Natural Products* 71: 972–80. https://doi.org/10.1021/np070447m.

Anese, M., B. Quarta, and J. Frias. 2011. Modelling the effect of asparaginase in reducing acrylamide formation in biscuits. *Food Chemistry* 126: 435–40. https://doi.org/10.1016/j.foodchem.2010.11.007.

Arima, K., T. Sakamoto, C. Araki, and G. Tamura. 1972. Production of extracellular L-asparaginases by microorganisms. *Agricultural and Biological Chemistry* 36: 356–61. https://www.tandfonline.com/doi/pdf/10.1080/00021369.1972.10860270.

Asselin, B.L., D. Ryan, C.N. Frantz, et al. 1989. *In vitro* and *in vivo* killing of acute lymphoblastic leukemia cells by L-asparaginase. *Cancer Research* 49: 4363–68. PMID: 2743326.

Azevedo, J.L., and W.L. Araújo. 2007. Diversity and applications of endophytic fungi isolated from tropical plants. In *Fungi: Multifaceted microbes*, eds. B.N. Ganguli and S.K. Deshmukh, 189–207. Boca Raton, SA: CRC Press.

Azevedo, J.L., W. Maccheroni, J.O. Pereira, and W.L. de Araújo. 2000. Endophytic microorganisms: A review on insect control and recent advances on tropical plants. *Electronic Journal of Biotechnology* 3: 40–65. https://tspace.library.utoronto.ca/bitstream/1807/1114/1/ej000005.pdf.

Bacon, C.W., and J. White. 2000. *Microbial endophytes*. New York, NY: Marcel Dekker Inc.

Balasubraminium, K., V. Ambikapathy, and A. Panneerselvam. 2012. Production, isolation and purification of L-asparaginase from *Aspergillus terreus* using submerged fermentation. *International Journal of Advances in Pharmaceutical Research* 3: 778–83.

Bano, M., and V.M. Sivaramakrishnan. 1980. Preparation and properties of L-asparaginase from green chillies (*Capsicum annum* L.). *Journal of Biosciences* 2: 291–97. https://doi.org/10.1007/BF02716861.

Bascomb, S., G.T. Banks, M.T. Skarstedt, A. Fleming, K.A. Bettelheim, and T.A. Connors. 1975. The properties and large-scale production of L-asparaginase from *Citrobacter*. *Microbiology* 91: 1–16. https://doi.org/10.1099/00221287-91-1-1.

Becher P.G., J. Beuchat, K. Gademann, and F. Jüttner. 2005. Nostocarboline: Isolation and synthesis of a new cholinesterase inhibitor from *Nostoc* 78–12A. *Journal of Natural Products* 68: 1793–95. https://doi.org/10.1021/np050312l.

Bérdy, J. 2005. Bioactive microbial metabolites. *The Journal of Antibiotics* 58: 1–26. https://doi.org/10.1038/ja.2005.1.

Bhagat, J., A. Kaur, and B.S. Chadha. 2016. Single step purification of asparaginase from endophytic bacteria *Pseudomonas oryzihabitans* exhibiting high potential to reduce acrylamide in processed potato chips. *Food and Bioproducts Processing* 99: 222–30. https://doi.org/10.1016/j.fbp.2016.05.010.

Bills, G., A. Dombrowski, F. Pelaez, J. Polishook, and Z. An. 2002. Recent and future discoveries of pharmacologically active metabolites from tropical fungi. In *Tropical mycology: Micromycetes*, eds. R. Watling, J.C. Frankland, A.M. Ainsworth, S. Isaac, and C.H. Robinson, 165–94. New York: CABI Publishing. https://www.cabidigitallibrary.org/doi/pdf/10.5555/20023069199.

Borah, D., R.N.S. Yadav, A. Sangra, L. Shahin, and A.K. Chaubey. 2012. Production, purification and process optimization of asparagines (an anticancer enzyme) from *E. coli*, isolated from sewage water. *Asian Journal of Pharmaceutical and Clinical Research* 5: 202–04. https://www.innovareacademics.in/journal/ajpcr/Vol5Suppl3/1226.pdf.

Borek, D., and M. Jaskólski. 2000. Crystallization and preliminary crystallographic studies of a new L-asparaginase encoded by the *Escherichia coli* genome. *Acta Crystallographica Section D* 56: 1505–07. https://doi.org/10.1107/S0907444900010076.

Borek, D., and M. Jaskólski. 2001. Sequence analysis of enzymes with asparaginase activity. *Acta Biochimia Polonica* 48: 893–902. https://pubmed.ncbi.nlm.nih.gov/11996000/.

Bourne, Y., P. Taylor, Z. Radić, and P. Marchot. 2003. Structural insights into ligand interactions at the acetylcholinesterase peripheral anionic site. *The EMBO Journal* 22: 1–12. https://doi.org/10.1093/emboj/cdg005.

Boyse, E.A., L.J. Old, H.A. Campbell, and L.T. Mashburn. 1967. Suppression of murine leukemias by L-asparaginase: Incidence of sensitivity among leukemias of various types:

Comparative inhibitory activities of guinea pig serum L-asparaginase and *Escherichia coli* L-asparaginase. *The Journal of Experimental Medicine* 125: 17–31. https://doi.org/10.1084/jem.125.1.17.

Campbell, H.A., L.T. Mashburn, E.A. Boyse, and L.J. Old. 1967. Two L-asparaginases from *Escherichia coli* B. Their separation, purification, and antitumor activity. *Biochemistry* 6: 721–30. https://doi.org/10.1021/bi00855a011.

Chareprasert, S., J. Piapukiew, S. Thienhirun, A.J.S. Whalley, and P. Sihanonth. 2006. Endophytic fungi of teak leaves *Tectona grandis* L. and rain tree leaves *Samanea saman* Merr. *World Journal of Microbiology and Biotechnology* 22: 481–86. https://doi.org/10.1007/s11274-005-9060-x.

Choudhary, M.I., S. Yousuf, S.A. Nawaz, and S. Ahmed. 2004. Cholinesterase inhibiting withanolides from *Withania somnifera*. *Chemical and Pharmaceutical Bulletin* 52: 1358–61. https://doi.org/10.1248/cpb.52.1358.

Chow, Y.Y., and A.S.Y. Ting. 2015. Endophytic L-asparaginase-producing fungi from plants associated with anticancer properties. *Journal of Advanced Research* 6: 869–76. https://doi.org/10.1016/j.jare.2014.07.005.

Coulthard, E., V. Singh-Curry, and M. Husain. 2006. Treatment of attention deficits in neurological disorders. *Current Opinion in Neurology* 19: 613–18. https://doi.org/10.1097/01.wco.0000247605.57567.9a.

Cremasco, M.A., and A.N. Starquit. 2010. Modeling for Taxol® separation in a simulated moving bed. *Brazilian Archives of Biology and Technology* 53: 1433–41. https://doi.org/10.1590/S1516-89132010000600020.

Crowther, D. 1971. L-asparaginase and human malignant disease. *Nature* 229: 168–71. https://doi.org/10.1038/229168a0.

Darvesh, S., R. Walsh, R. Kumar, et al. 2003. Inhibition of human cholinesterases by drugs used to treat Alzheimer disease. *Alzheimer Disease and Associated Disorders* 17: 117–26. https://doi.org/10.1097/00002093-200304000-00011.

Da-Silva, T.F., R.E. Vollu, D. Jurelevicius, et al. 2013. Does the essential oil of *Lippia sidoides* Cham. (pepper-rosmarin) affect its endophytic microbial community? *BMC Microbiology* 13 (February): 29, 1–14. https://doi.org/10.1186/1471-2180-13-29.

Deng, C.M., S.X. Liu, C.H. Huang, J.Y. Pang, and Y.C. Lin. 2013. Secondary metabolites of a mangrove endophytic fungus *Aspergillus terreus* (No. GX7-3B) from the South China sea. *Marine Drugs* 11: 2616–24. https://doi.org/10.3390/md11072616.

Deokar, V.D., M.D. Vetal, and L. Rodrigues. 2010. Production of intracellular L-asparaginase from *Erwinia carotovora* and its statistical optimization using response surface methodology (RSM). *International Journal of Chemical Science and Applications* 1: 25–36. https://bipublication.com/files/chv1I120104.pdf.

Derosa, G., and P. Maffioli. 2012. α-Glucosidase inhibitors and their use in clinical practice. *Archives of Medical Science* 8: 899–06. https://doi.org/10.5114/aoms.2012.31621.

Devi, S., and W. Azmi. 2012. One step purification of antileukemic of L-asparaginase from *Erwinia carotovora* MTCC 1428 and its application. *International Journal of Life Science and Pharma Research* 2: 36–45.

Devkota, K.P., B.N. Lenta, M.I. Choudhary, et al. 2007. Cholinesterase inhibiting and antiplasmodial steroidal alkaloids from *Sarcococca hookeriana*. *Chemical and Pharmaceutical Bulletin* 55: 1397–401. https://doi.org/10.1248/cpb.55.1397.

Devkota, K.P., B.N. Lenta, P.A. Fokou, and N. Sewald. 2008. Terpenoid alkaloids of the *Buxaceae* family with potential biological importance. *Natural Product Reports* 25: 612–30. https://doi.org/10.1039/B704958G.

Dhanam, J.G., and S. Kannan. 2013. L-asparaginase-types, perspectives and applications. *Advanced BioTech* 13: 1–5.

Dhanam, J.G., and S. Kannan. 2014. Screening and characterization of L-asparaginase producing *Streptomyces* isolated from soil samples of Periyar lake, Kumily. *Bioscience Discovery* 5: 50–54. https://jbsd.in/Vol%205%20No%201%20Jan%202014/Dhanam%2050–54.pdf.

Divakar, G., M. Sunitha, P. Vasu, P.U. Shanker, and P. Ellaiah. 2006. Optimization of process parameters for alkaline protease production under solid-state fermentation by *Thermoactinomyces thalpophilus* PEE 14. *Indian Journal of Biotechnology* 5: 80–83. https://nopr.niscpr.res.in/handle/123456789/5554.

Dong, L.H., S.W. Fan, Q.Z. Ling, B.B. Huang, and Z.J. Wei. 2014. Indentification of huperzine A-producing endophytic fungi isolated from *Huperzia serrata*. *World Journal of Microbiology and Biotechnology* 30: 1011–17. https://doi.org/10.1007/s11274-013-1519-6.

Duval, M., S.S. Ferster, X. Rialland, et al. 2002. Comparison of *Escherichia coli*-asparaginase with *Erwinia*-asparaginase in the treatment of childhood lymphoid malignancies: Results of a randomized European organization for research and treatment of cancer-children's leukemia group phase 3 trial. *Blood* 99: 2734–39. https://doi.org/10.1182/blood.V99.8.2734.

Ebrahiminezhad, A., S. Rasoul-Amini, and Y. Ghasemi. 2011. L-asparaginase production by moderate halophilic bacteria isolated from Maharloo Salt Lake. *Indian Journal of Microbiology* 51: 307–11. https://doi.org/10.1007/s12088-011-0158-6.

Eisele, N., D. Linke, K. Bitzer, S. Na'amnieh, M. Nimtz, and R.G. Berger. 2011. The first characterized asparaginase from a basidiomycete, *Flammulina velutipes*. *Bioresource technology* 102: 3316–21. https://doi.org/10.1016/j.biortech.2010.10.098.

Ekström, F., C. Akfur, A.K. Tunemalm, and S. Lundberg. 2006. Structural changes of phenylalanine 338 and histidine 447 revealed by the crystal structures of tabun-inhibited murine acetylcholinesterase. *Biochemistry* 45: 74–81. https://doi.org/10.1021/bi051286t.

El-Bessoumy, A.A., M. Sarhan, and J. Mansour. 2004. Production, isolation and purification of L-asparaginase from *Pseudomonas Aeruginosa* 50071 using solid-state fermentation. *Journal of Biochemistry and Molecular Biology* 37: 387–93. https://doi.org/10.5483/BMBRep.2004.37.4.387.

El-Hady, F.K.A., M.S. Abdel-Aziz, K.H. Shaker, Z.A. El-Shahid, and M.A. Ghani. 2014. Coral-derived fungi inhibit acetylcholinesterase, superoxide anion radical and microbial activities. *International Journal of Pharmaceutical Sciences Review and Research* 26: 301–08.

Elisha, I.L., E.E. Elgorashi, A.A. Hussein, G. Duncan, and J.N. Eloff. 2013. Acetylcholinesterase inhibitory effects of the bulb of *Ammocharis coranica* (Amaryllidaceae) and its active constituent lycorine. *South African Journal of Botany* 85: 44–47. https://doi.org/10.1016/j.sajb.2012.11.008.

Elshafei, A.M., M.M. Hassan, A.M. Abd-Elmontasr, D.A. Mahmoud, and D.H. Elghonemy. 2012. Purification, characterization and antitumor activity of L-asparaginase from *Penicillium brevicompactum* NRC 829. *Microbiology*

Research Journal International 2: 158–74. https://doi.org/10.9734/BMRJ/2012/1735.

Emiliani, G., A. Mengoni, I. Maida, et al. 2014. Linking bacterial endophytic communities to essential oils: Clues from *Lavandula angustifolia* Mill. *Evidence-Based Complementary and Alternative Medicine* 1 (May): 650905, 1–16. https://doi.org/10.1155/2014/650905.

Eyberger, A.L., R. Dondapati, and J.R. Porter. 2006. Endophyte fungal isolates from *Podophyllum peltatum* produce podophyllotoxin. *Journal of Natural Products* 69: 1121–24. https://doi.org/10.1021/np060174f.

Ezra, D., U.F. Castillo, G.A. Strobe, et al. 2004.Coronamycins, peptide antibiotics produced by a verticillate *Streptomyces* sp. (MSU-2110) endophytic on *Monstera* sp. *Microbiology* 150: 785–93. https://doi.org/10.1099/mic.0.26645-0.

Fernandes, M.D.R.V., T.A.C. e Silva, L.H. Pfenning, et al. 2009. Biological activities of the fermentation extract of the endophytic fungus *Alternaria alternata* isolated from *Coffea arabica* L. *Brazilian Journal of Pharmaceutical Sciences* 45: 677–85. https://doi.org/10.1590/S1984-82502009000400010.

Ferrara, M.A., N.M.B. Severino, J.J. Mansure, et al. 2006. Asparaginase production by a recombinant *Pichia pastoris* strain harbouring *Saccharomyces cerevisiae* ASP3 gene. *Enzyme and Microbial Technology* 39: 1457–63. https://doi.org/10.1016/j.enzmictec.2006.03.036.

Firakova, S., M. Sturdikova, and M. Muckova. 2007. Bioactive secondary metabolites produced by microorganisms associated with plants. *Biologia* 62: 251–57. https://doi.org/10.2478/s11756-007-0044-1.

Gangadevi, V., and J. Muthumary. 2007. Endophytic fungal diversity from young, mature and senescent leaves of *Ocimum basilicum* L. with special reference to Taxol production. *Indian Journal of Science and Technology* 1: 1–12. https://doi.org/10.17485/ijst/2008/v1i1/2.

Gayathri, P., and V. Muralikrishnan. 2013. Isolation and characterization of endophytic actinomycetes from mangrove plant for antimicrobial activity. *International Journal of Current Microbiology and Applied Science* 2: 78–89. https://www.ijcmas.com/vol-2–11/P.Gayathri%20and%20V.%20Muralikrishnan.pdf.

Ge, H.M., C.H. Zhu, D.H. Shi, et al. 2007. Hopeahainol A: An acetylcholinesterase inhibitor from *Hopea hainanensis*. *Chemistry – A European Journal* 14: 376–81. https://doi.org/10.1002/chem.200700960.

Giridharan, P., S.A. Verekar, A. Khanna, P.D. Mishra, and S.K. Deshmukh. 2012. Anticancer activity of sclerotiorin, isolated from an endophytic fungus *Cephalotheca faveolata* Yaguchi, Nishim. & Udagawa. *Indian Journal of Experimental Biololgy* 50: 464–68. https://pubmed.ncbi.nlm.nih.gov/22822525/.

Gomaa, M.N., A.M. Farag, A.M. Ayesh, and M.A. Embaby. 2013. Isolation and structure elucidation of acetyl cholinesterase inhibitor from *Gyrostoma helianthus* of the red sea, Egypt. *Journal of Cell and Animal Biology* 9: 16–25. https://doi.org/10.5897/JCAB2015.0431.

Gupta, A., S. Jain, and N. Verma. 2022. *Industrial applications of soil microbes*, vol. 1. Bentham Science Publishers Pte Ltd. https://doi.org/10.2174/97898150399551220101.

Gupta, A., S. Jain, and N. Verma. 2024. *Industrial applications of soil microbes*, vol. 3. Bentham Science Publishers Pte Ltd. https://doi.org/10.2174/97898150797531240301.

Gupta, N., S.J. Dash, and U.C. Basak. 2009. L-asparaginase from fungi of Bhitarkanika mangrove ecosystem. *Asia Pacific Journal of Molecular Biology and Biotechnology* 1: 27–30.

Gurunathan, B., and R. Sahadevan. 2011. Design of experiments and artificial neural network linked genetic algorithm for modeling and optimization of L-asparaginase production by *Aspergillus terreus* MTCC 1782. *Biotechnology and Bioprocess Engineering* 16: 50–58. https://doi.org/10.1007/s12257-010-0119-7.

Hallmann, J., A. Quadt-Hallmann, W.F. Mahaffe, and J.W. Kloepper. 1997. Bacterial endophytes in agricultural crops. *Canadian Journal of Microbiology* 43: 895–14. https://doi.org/10.1139/m97-131.

Han, L.R., Z.H. Wang, H.J. Zhang, L.S. Xue, J.T. Feng, and X. Zhang. 2013. Isolation of endophytic fungi from *Tripterygium wilfordii* and their insecticidal activities. *African Journal of Microbiology Research* 7: 771–76. https://doi.org/10.5897/AJMR12.1453.

Holden, M., and C. Kelly. 2002. Use of cholinesterase inhibitors in dementia. *Advances in Psychiatric Treatments* 8: 89–96. https://doi.org/10.1192/apt.8.2.89.

Houghton, P.J., Y. Ren, and M.J. Howes. 2006. Acetylcholinesterase inhibitors from plants and fungi. *Natural Product Reports* 23: 181–99. https://doi.org/10.1039/B508966M.

Howard, J.B., and F.H. Carpenter. 1972. L-asparaginase from *Erwinia carotovora* substrate specificity and enzymatic properties. *Journal of Biological Chemistry* 247: 1020–30. https://doi.org/10.1016/S0021-9258(19)45610-X.

Hsieh, P.W., L.C. Hsu, C.H. Lai, et al. 2009. Evaluation of the bioactivities of extracts of endophytes isolated from Taiwanese herbal plants. *World Journal of Microbiology and Biotechnology* 25: 1461–69. https://doi.org/10.1007/s11274-009-0036-0.

Hu, M.Y., G.H. Zhong, Z.T. Sun, G. Sh, H.M. Liu, and X.Q. Liu. 2005. Insecticidal activities of secondary metabolites of endophytic *Pencillium* sp. in *Derris elliptica* Benth. *Journal of Applied Entomology* 129: 413–17. https://doi.org/10.1111/j.1439-0418.2005.00988.x.

Imada, A., S. Igarasi, K. Nakahama, and M. Isono. 1973. Asparaginase and glutaminase activities of micro-organisms. *Microbiology* 76: 85–99. https://doi.org/10.1099/00221287-76-1-85.

Jain, S., A. Gupta, and N. Verma. 2023. *Industrial applications of soil microbes*, vol. 2. Bentham Science Publishers Pte Ltd. https://doi.org/10.2174/97898150502641230201.

Jain, S., A. Gupta, and N. Verma. 2024. *Industrial applications of soil microbes*, vol. 4. Bentham Science Publishers Pte Ltd. https://doi.org/10.2174/97898151249961240401.

Jalgaonwala R.E., and R.T. Mahajan. 2014. Production of anticancer enzyme asparaginase from endophytic *Eurotium* sp. isolated from rhizomes of *Curcuma longa*. *European Journal of Experimental Biology* 4: 36–43. https://www.primescholars.com/articles/production-of-anticancer-enzyme-asparaginase-from-endophyticeurotium-sp-isolated-from-rhizomes-of-curcuma-longa.pdf.

Jayaraj, S., S. Suresh, and R.K. Kadeppagari. 2013. Amylase inhibitors and their biomedical applications. *Starch* 65: 535–42. https://doi.org/10.1002/star.201200194.

Jayaramu, M., N.B. Hemalatha, C.P. Rajeshwari, K.G. Siddalingeswara, S.M. Mohsin, and N.S. Dutt. 2010. A novel approach for detection, confirmation and optimization of L-asparaginase from *Emericella Nidulans*. *Journal of Current Pharma Research; Karad* 1: 20–24. https://www.

proquest.com/docview/1287035762?sourcetype=Scholarly%20Journals#.

Jazayeri, S.B., A. Amanlou, N. Ghanadian, P. Pasalar, and M. Amanlou. 2014. A preliminary investigation of anticholinesterase activity of some Iranian medicinal plants commonly used in traditional medicine. *DARU Journal of Pharmaceutical Sciences* 22 (January): 17, 1–5. https://doi.org/10.1186/2008-2231-22-17.

Jiang, H., X. Luo, and D. Bai. 2003. Progress in clinical, pharmacological, chemical and structural biological studies of huperzine A: A drug of traditional Chinese medicine origin for the treatment of Alzheimer's disease. *Current Medicinal Chemistry* 10: 2231–52. https://doi.org/10.2174/0929867033456747.

Jois, S., S. Chandru, A. Chavan, R.Y. Lakshmikantha, and P.S. Rai. 2013. Study on production, purification and characterisation of L-asparaginase from *Escherichia coli* and *Pseudomonas aeruginosa*. *International Journal of Pharmaceutical, Chemical and Biological Sciences* 3: 565–70. https://www.ijpcbs.com/articles/study-on-productionpurification-and-characterisation-of-lasparaginase-from-escherichia-coliandpseudomonas-aeruginosa.pdf.

Joshi, R.D., and N.S. Kulkarni. 2014. Isolation of L-asparaginase producing endophytic bacteria from plants recommended for cancer therapy. *International Journal of Science and Research* 3: 506–09. https://www.ijsr.net/archive/v3i11/T0NUMTQ4ODc=.pdf.

Ju, Y., J.N. Sacalis, and C.C. Stil. 1998. Bioactive flavonoids from endophyte-infected blue grass (*Poa ampla*). *Journal of Agricultural and Food Chemistry* 46: 3785–88. https://doi.org/10.1021/jf980189m.

Jukic, M., F. Burcul, I. Carev, O. Politeo, and M. Milos. 2012. Screening for acetylcholinesterase inhibition and antioxidant activity of selected plants from *Croatia*. *Natural Products Research* 26: 1703–07. https://doi.org/10.1080/14786419.2011.602639.

Jung, M., and M. Park. 2007. Acetylcholinesterase inhibition by flavonoids from *Agrimonia pilosa*. *Molecules* 12: 2130–39. https://doi.org/10.3390/12092130.

Kalyanasundaram, I., J. Nagamuthu, B. Srinivasan, A. Pachayappan, and S. Muthukumarasamy. 2015. Production, purification and characterization of extracellular L-asparaginase from salt marsh fungal endophytes. *World Journal of Pharmacy and Pharmaceutical Sciences* 04: 663–77. https://storage.googleapis.com/innctech/wjpps/article_issue/1425125858.pdf.

Karthik, S., K.C. Nandini, T.R.P. Kekuda, K.S. Vinayaka, and S. Mukunda. 2011. Total phenol content, insecticidal and amylase inhibitory efficacy of *Heterodermia leucomela* (L). *Annals of Biological Research* 2: 38–43. https://www.cabidigitallibrary.org/doi/full/10.5555/20113309017.

Kathiravan, G., and V.S. Raman. 2010. *In vitro* Taxol production, by *Pestalotiopsis breviseta* – A first report. *Fitoterapia* 81: 557–64. https://doi.org/10.1016/j.fitote.2010.01.021.

Kaul, S., S. Gupta, M. Ahmed, and M.K. Dhar. 2012. Endophytic fungi from medicinal plants: A treasure hunt for bioactive metabolites. *Phytochemistry Reviews* 11: 487–505. https://doi.org/10.1007/s11101-012-9260-6.

Kavitha, A., and M. Vijayalakshmi. 2012. A study on L-asparaginase of *Nocardia levis* MK-VL 113. *The Scientific World Journal* 2012 (January): 160434, 1–5. https://doi.org/10.1100/2012/160434.

Khamna, S., A. Yokota, and S. Lumyong. 2009. L-asparaginase production by actinomycetes isolated from some Thai medicinal plant rhizosphere soils. *International Journal of Integrative Biology* 6: 22–26. http://cmuir.cmu.ac.th/jspui/handle/6653943832/59268.

Khan, A.L., L. Ali, J. Hussain, T.S. Rizvi, A. Al-Harrasi, and I.J. Lee. 2015. Enzyme inhibitory radicinol derivative from endophytic fungus *Bipolaris sorokiniana* LK12, associated with *Rhazya stricta*. *Molecules* 20: 12198–208. https://doi.org/10.3390/molecules200712198.

Kil, J.O., G.N. Kim, and I. Park. 1995. Extraction of extracellular L-asparaginase from *Candida utilis*. *Bioscience, Biotechnology and Biochemistry* 59: 749–50. https://doi.org/10.1271/bbb.59.749.

Kim, W.G., K.M. Cho, C.K. Lee, and I.D. Yoo. 2003. Terreulactones A, B, C, and D: Novel acetylcholinesterase inhibitors produced by *Aspergillus terreus* II. Physico-chemical properties and structure determination. *The Journal of Antibiotics* 56: 351–57. https://doi.org/10.7164/antibiotics.56.351.

Kim, W.G., N.K. Song, and I.D. Yoo. 2001. Quinolactacins Al and A2, new acetylcholinesterase inhibitors from *Penicillium citrinum*. *The Journal of Antibiotics* 54: 831–35. https://doi.org/10.7164/antibiotics.54.831.

Kingston, D.G.I. 2011. Modern natural products drug discovery and its relevance to biodiversity conservation. *Journal of Natural Products* 74: 496–11. https://doi.org/10.1021/np100550t.

Koehn F.E., and G.T. Carter. 2005. The evolving role of natural products in drug discovery. *Nature Reviews Drug Discovery* 4: 206–20. https://doi.org/10.1038/nrd1657.

Kolbe, L., T. Mann, W. Gerwat, et al. 2013. 4-n-butylresorcinol, a highly effective tyrosinase inhibitor for the topical treatment of hyperpigmentation. *Journal of the European Academy of Dermatology and Venereology* 27: 19–23. https://doi.org/10.1111/jdv.12051.

Kornbrust, B.A., T. Forman, and I. Matveeva. 2012. Applications of enzymes in breadmaking. In *Woodhead publishing series in food science, technology and nutrition, breadmaking*, 2nd ed., ed. S.P. Cauvain, 470–98. Cambridgeshire: Woodhead Publishing. https://doi.org/10.1533/9780857095695.2.470.

Kornbrust, B.A., M.A. Stringer, N.E.K. Lange, H.V. Hendriksen, R. Whitehurst, and M.V. Oort. 2010. Asparaginase – An enzyme for acrylamide reduction in food products. In *Enzymes in food technology,* eds. R.J. Whitehurst and M.V. Oort, 59–87. Oxford: John Wiley & Sons, Inc. https://doi.org/10.1002/9781444309935.ch4.

Kour, A., A.S. Shawl, S. Rehman, et al. 2008. Isolation and identification of an endophytic strain of *Fusarium oxysporum* producing podophyllotoxin from *Juniperus recurva*. *World Journal of Microbiology and Biotechnology* 24: 1115–21. https://doi.org/10.1007/s11274-007-9582-5.

Krasotkina, J., A.A. Borisova, Y.V. Gervaziev, and N.N. Sokolov. 2004. One-step purification and kinetic properties of the recombinant L-asparaginase from *Erwinia carotovora*. *Biotechnology and Applied Biochemistry* 39: 215–21. https://doi.org/10.1042/BA20030138.

Krishnamurthy, K., and S.S. Gnanamanickam. 1998. Short Communication: Induction of systemic resistance and salicylic acid accumulation in *Oryza sativa*, L. in the biological suppression of rice blast caused by treatments with *Pseudomonas* spp. *World Journal of Microbiology and Biotechnology* 14: 935–37. https://doi.org/10.1023/A:1008872925755.

Krishnapura, P.R., and P.D. Belur. 2015. Isolation and screening of endophytes from the rhizomes of some Zingiberaceae plants for L-asparaginase production. *Preparative Biochemistry and Biotechnology* 46: 281–87. https://doi.org/10.1080/10826068.2015.1031385.

Kumar, N.S.M., and H.K. Manonmani. 2013. Purification, characterization and kinetic properties of extracellular L-asparaginase produced by *Cladosporium* sp. *World Journal of Microbiology* and *Biotechnology* 29: 577–87. https://doi.org/10.1007/s11274-012-1213-0.

Kumar, N.S.M., R. Ramasamy, and H.K. Manonmani. 2013. Production and optimization of L-asparaginase from *Cladosporium* sp. using agricultural residues in solid state fermentation. *Industrial Crops and Product* 43: 150–58. https://doi.org/10.1016/j.indcrop.2012.07.023.

Kumar, S., V.V. Dasu, and K. Pakshirajan. 2010. Localization and production of novel L-asparaginase from *Pectobacterium carotovorum* MTCC 1428. *Process Biochem*istry 45: 223–29. https://doi.org/10.1016/j.procbio.2009.09.011.

Kumar, S., V.V. Dasu, and K. Pakshirajan. 2011. Purification and characterization of glutaminase-free L-asparaginase from *Pectobacterium carotovorum* MTCC 1428. *Bioresource Technology* 102: 2077–82. https://doi.org/10.1016/j.biortech.2010.07.114.

Kumar, S., N. Kaushik, R. Edrada-Ebel, R. Ebel, and P. Proksch. 2011. Isolation, characterization, and bioactivity of endophytic fungi of *Tylophora indica*. *World Journal of Microbiology and Biotechnology* 27: 571–77. https://doi.org/10.1007/s11274-010-0492-6.

Kuno, F., K. Otoguro, K. Shiomi, Y. Iwai, and S. Omura. 1996. Arisugacins A and B, novel and selective acetylcholinesterase inhibitors from *Penicillium* sp. FO–4259. I. Screening, taxonomy, fermentation, isolation and biological activity. *The Journal of Antibiotics* 49: 742–47. https://doi.org/10.7164/antibiotics.49.742.

Kusari, S., M. Lamshöft, and M. Spiteller. 2009. *Aspergillus fumigatus* Fresenius, an endophytic fungus from *Juniperus communis* L. Horstmann as a novel source of the anticancer pro-drug deoxypodophyllotoxin. *Journal of Applied Microbiology* 107: 1019–30. https://doi.org/10.1111/j.1365-2672.2009.04285.x.

Kusari, S., V.C. Verma, M. Lamshoeft, and M. Spiteller. 2012. An endophytic fungus from *Azadirachta indica* A. Juss. that produces azadirachtin. *World Journal of Microbiology and Biotechnology* 28: 1287–94. https://doi.org/10.1007/s11274-011-0876-2.

Lakshmi, P.J., and K.V. Selvi. 2013. Anticancer potentials of secondary metabolites from endophytes of *Barringtonia acutangula* and its molecular characterization. *International Journal of Current Microbiology and Applied Science* 2: 44–55. https://www.ijcmas.com/Archives/vol-2–2/Jayashree%20Lakshmi%20and%20Vanmathi%20Selvi.pdf.

Lapmak, K., S. Lumyong, S. Thongkuntha, P. Wongputtisin, and U. Sardsud. 2010. L-asparaginase production by *Bipolaris* sp. BR438 isolated from brown rice in Thailand. *Chiang Mai Journal of Science* 37: 160–64. https://www.thaiscience.info/journals/Article/CMJS/10905508.pdf.

Lee, D.H., J.S. Lee, S.H. Yi, and J.S. Lee. 2008. Production of the acetylcholinesterase inhibitor from *Yarrowia lipolytica* S–3. *Mycobiology* 36: 102–05. https://doi.org/10.4489/MYCO.2008.36.2.102.

Li, J.Y., R.S. Sidhu, E.J. Ford, D.M. Long, W.M. Hess, and G.A. Strobel. 1998. The induction of Taxol production in the endophytic fungus-*Periconia* sp. from *Torreya grandifolia*. *Journal of Industrial Microbiology and Biotechnology* 20: 259–64. https://doi.org/10.1038/sj.jim.2900521.

Lincoln, L., and S.S. More. 2014. Isolation and production of clinical and food grade L-asparaginase enzyme from fungi. *Journal of Pharmacognosy and Phytochemistry* 3: 177–83. https://www.phytojournal.com/vol3Issue3/Issue_sep_2014/48.1.pdf.

Lindsay, R.C., and S. Jang. 2005. Model systems for evaluating factors affecting acrylamide formation in deep fried foods. In *Chemistry and safety of acrylamide in food*, eds. M. Friedman and D. Mottram, 329–41. Boston, MA: Springer. https://doi.org/10.1007/0-387-24980-X_25.

Liu, C.H., W.X. Zou, H. Lu, and R.X. Tan. 2001. Antifungal activity of *Artemisia annua* endophyte cultures against phytopathogenic fungi. *Journal of Biotechnology* 88: 277–82. https://doi.org/10.1016/S0168-1656(01)00285–1.

Liu, X., M. Dong, X. Chen, M. Jiang, X. Lv, and J. Zhou. 2008. Antimicrobial activity of an endophytic *Xylaria* sp. YX-28 and identification of its antimicrobial compound 7-amino-4-methylcoumarin. *Applied Microbiology and Biotechnology* 78: 241–47. https://doi.org/10.1007/s00253-007-1305-1.

Loureiro, C.B., K.S. Borges, A.F. Andrade, L.G. Tone, and S. Said. 2012. Purification and biochemical characterization of native and pegylated form of L-asparaginase from *Aspergillus terreus* and evaluation of its antiproliferative activity. *Advances in Microbiology* 2: 138–45. http://dx.doi.org/10.4236/aim.2012.22019.

Mahajan, R.V., S. Saran, K. Kameswaran, V. Kumar, and R.K. Saxena. 2012. Efficient production of L-asparaginase from *Bacillus licheniformis* with low-glutaminse activity: Optimization, scale up and acrylamide degradation studies. *Bioresource Technology* 125: 11–16. https://doi.org/10.1016/j.biortech.2012.08.086.

Manasa, C., and M.S. Nalini. 2014. L-asparaginase activity of fungal endophytes from *Tabernaemontana heyneana* Wall. (Apocynaceae), endemic to the Western Ghats (India). *Hindawi Publishing Corporation International Scholarly Research Notices* 1 (November): 925131, 1–7. https://doi.org/10.1155/2014/925131.

Maria, G.L., K.R. Sridhar, and N.S. Raviraja. 2005. Antimicrobial and enzyme activity of mangrove endophytic fungi of southwest coast of India. *Journal of Agricultural Technology* 1: 67–80. http://ijat-aatsea.com/pdf/Maria_Sridhar%20page%2067–80.pdf.

Martinez, A., and A. Castro. 2006. Novel cholinesterase inhibitors as future effective drugs for the treatment of Alzheimer's disease. *Expert Opinion in Investigational Drugs* 15: 1–12. https://doi.org/10.1517/13543784.15.1.1.

Mashburn, L.T., and J.C. Wriston. 1964. Tumor inhibitory effect of L-asparaginase from *Escherichia coli*. *Archives in Biochemistry and Biophysics* 105: 450–53. https://doi.org/10.1016/0003-9861(64)90032–3.

Masumi, S., S. Mirzaei, R. Kalvandi, and D. Zafari. 2014. Asparaginase and amylase activity of thyme endophytic fungi. *Journal of Crop Protection* 3: 655–62. http://jcp.modares.ac.ir/article-3–9351-en.html.

Meghavarnam, A.K., and S. Janakiraman. 2015. Purification and characterization of therapeutic enzyme L-asparaginase from a tropical soil fungal isolate *Fusarium Culmorum* ASP–87. *MOJ Proteomics & Bioinformatics* 2: 171–75. https://doi.org/10.15406/mojpb.2015.02.00064.

Mehandiratta, M.M., S. Pandey, and T. Kuntzer. 2014. Acetylcholinesterase inhibitor treatment for myasthenia gravis. *Cochrane Library Cochrane Database of Systematic*

Reviews 10 (October): CD006986, 1–17. https://doi.org/10.1002/14651858.CD006986.pub3.

Mishra, A. 2006. Production of L-asparaginase, an anticancer agent, from *Aspergillus niger* using agricultural waste in solid state fermentation. *Applied Biochemistry and Biotechnology* 135: 33–42. https://doi.org/10.1385/ABAB:135:1:33.

Mukherjee, J., S. Majumdar, and T. Scheper. 2000. Studies on nutritional and oxygen requirements for production of L-asparaginase by *Enterobacter aerogenes*. *Applied Microbiology and Biotechnology* 53: 180–84. https://doi.org/10.1007/s002530050006.

Mushtaq, M.S., K.G. Siddalingeshwara, J. Karthic, N.M. Sunil, and K.S. Prathiba. 2012. Rapid screening and confirmation of L-asparaginase from *Penicillium* sp. *International Journal of Pharmacology and Pharmaceutics* 1: 147–50.

Muslim, S.N. 2014. Production, purification and characterization of a novel L-asparaginase from *Acinetobacter baumannii* with anticancerous activity. *International Journal of Current Engineering and Technology* 4 (February): 1, 1–6. https://inpressco.com/wp-content/uploads/2014/01/Paper26–112.pdf.

Nagarajan, A., N. Thirunavukkarasu, T.S. Surayanarayanan, and S.N. Gummadi. 2013. Screening and isolation of novel glutaminase free L-asparaginase from fungal endophytes. *Research Journal of Microbiology* 9: 163–76. https://scialert.net/abstract/?doi=jm.2014.163.176.

Narayana, K.J.P., K.G. Kumar, and M. Vijayalakshmi. 2008. L-asparaginase production by *Streptomyces albidoflavus*. *Indian Journal of Microbiology* 48: 331–36. https://doi.org/10.1007/s12088-008-0018-1.

Nicoletti, R., A. Fiorentino, and M. Scognamiglio. 2014. Endophytism of *Penicillium* species in woody plants. *The Open Mycology Journal* 8: 1–26. https://doi.org/10.2174/1874437001408010001.

Ohlendorf, B., D. Schulz, A. Erhard, K. Nagel, and J.F. Imhoff. 2012. Geranylphenazinediol, an acetylcholinesterase inhibitor produced by a *Streptomyces* species. *Journal of Natural Products* 75: 1400–04. https://doi.org/10.1021/np2009626.

Oza, V.P., S.D. Trivedi, P.P. Parmar, and R.B. Subramanian. 2009. *Withania somnifera* (ashwagandha): A novel source of L-asparaginase. *Journal of Integrative Plant Biology* 51: 201–06. https://doi.org/10.1111/j.1744-7909.2008.00779.x.

Pandey, S., A. Sree, D.P. Sethi, et al. 2014. A marine sponge associated strain of *Bacillus subtilis* and other marine bacteria can produce anticholinesterase compounds. *Microbial Cell Factories* 13 (February): 24, 1–9. https://doi.org/10.1186/1475-2859-13-24.

Patro, K.K.R., and N. Gupta. 2014. Impact of cultural and nutritional conditions on L-asparaginase production by *Penicillium citrinum* Thom. *International Journal of Pharma Medicine and Biological Sciences* 3 (July): 3, 1–7. https://www.ijpmbs.com/uploadfile/2015/0412/20150412043550260.pdf.

Pawle, G., and S.K. Singh. 2014. Antimicrobial, antioxidant activity and phytochemical analysis of an endophytic species of *Nigrospora* isolated from living fossil *Ginkgo biloba*. *Current Research in Environmental & Applied Mycology* 4: 1–9. https://www.creamjournal.org/pdf/Cream_4_1_1.pdf.

Pedreschi, F., K. Kaack, and K. Granby. 2008. The effect of asparaginase on acrylamide formation in French fries. *Food Chemistry* 109: 386–92. https://doi.org/10.1016/j.foodchem.2007.12.057.

Perry, E.K., R.H. Perry, G. Blessed, and B.E. Tomlinson. 1978. Changes in brain cholinesterases in senile dementia of Alzheimer type. *Neuropathology and Applied Neurobiology* 4: 273–77. https://doi.org/10.1111/j.1365-2990.1978.tb00545.x.

Phongpaichit, S., N. Rungjindamai, V. Rukachaisirikul, and J. Sakayaroj. 2006. Antimicrobial activity in cultures of endophytic fungi isolated from *Garcinia* species. *FEMS Immunology and Medical Microbiology* 48: 367–72. https://doi.org/10.1111/j.1574-695X.2006.00155.x.

Poorani, E., M.K. Saseetharan, and P. Dhevagi. 2009. L-asparaginase production by newly isolated marine *Streptomyces* sp. strain EPD 27: Molecular identification. *International Journal of Integrative Biology* 7: 150–55.

Prakasham, R.S., C.S. Rao, R.S. Rao, G.S. Lakshmi, and P.N. Sarma. 2007. L-asparaginase production by isolated *Staphylococcus* sp. – 6A: Design of experiment considering interaction effect for process parameter optimization. *Journal of Applied Microbiology* 102: 1382–91. https://doi.org/10.1111/j.1365-2672.2006.03173.x.

Pujiyanto, S., Y. Lestari, A. Suwanto, S. Budiarti, and L.K. Darusman. 2011. Alpha-glucosidase inhibitor activity and characterization of endophytic actinomycetes isolated from some Indonesian diabetic medicinal plants. *International Journal of Pharmacy and Pharmaceutical Sciences* 4: 327–33. http://repository.ipb.ac.id/handle/123456789/76309.

Puri, S.C, V. Verma, T. Amna, G.N. Qazi, and M. Spiteller. 2005. An endophytic fungus from *Nothapodytes foetida* that produces camptothecin. *Journal of Natural Products* 68: 1717–19. https://doi.org/10.1021/np0502802.

Raja, H.A., A. Kaur, T. El-Elimat, et al. 2015. Phylogenetic and chemical diversity of fungal endophytes isolated from *Silybum marianum* (L) Gaertn. (milk thistle). *Mycology* 6: 8–27. https://doi.org/10.1080/21501203.2015.1009186.

Ramakrishnan, M.S., and R. Joseph. 1996. Characterization of an extracellular asparaginase of *Rhodosporidium toruloides* CBS14 exhibiting unique physicochemical properties. *Canadian Journal of Microbiology* 42: 316–25. https://doi.org/10.1139/m96-047.

Rao, K.C.S., S. Divakar, N.G. Karanth, and A.P. Sattur. 2001. 14–(2', 3', 5'-Trihy droxyphenyl) tetradecan-2-ol, a novel acetylcholinesterase inhibitor from *Chrysosporium* sp. *The Journal of Antibiotics* 54: 848–49. https://doi.org/10.7164/antibiotics.54.848.

Ravnikar, M., M. Tercelj, D. Janes, B. Strukelj, and S. Kreft. 2015. Antibacterial activity of endophytic fungi isolated from *Conifer* needles. *African Journal of Biotechnology* 14: 867–71. https://doi.org/10.5897/AJB2014.14387.

Rhee, I.K., N. Appels, B. Hofte, et al. 2004. Isolation of the acetylcholinesterase inhibitor ungeremine from *Nerine bowdenii* by preparative HPLC coupled on-line to a flow assay system. *Biological and Pharmaceutical Bulletin* 27: 1804–09. https://doi.org/10.1248/bpb.27.1804.

Rodrigues, K.F., G.L. Costa, M.P. Carvalho, and R.A. Epifanio. 2005. Evaluation of extracts produced by some tropical fungi as potential cholinesterase inhibitors. *World Journal of Microbiology and Biotechnology* 21: 1617–21. https://doi.org/10.1007/s11274-005-8344-5.

Rollinger, J.M., A. Hornick, T. Langer, H. Stuppner, and H. Prast. 2004. Acetylcholinesterase inhibitory activity of scopolin and scopoletin discovered by virtual screening of natural products. *Journal of Medicinal Chemistry* 47: 6248–54. https://doi.org/10.1021/jm049655r.

Rosa, L.H., M.L.A. Vieira, B.B. Cota, et al. 2011. Endophytic fungi of tropical forests: A promising source of bioactive prototype molecules for the treatment of neglected diseases.

In *Drug development – A case study based insight into modern strategies*, ed. C. Rundfeldt, 469–86. InTech Open. https://doi.org/10.5772/27783.

Rukachaisirikul, V., J. Kaeobamrung, W. Panwiriyarat, et al. 2007. A new pyrone derivative from the entophytic fungus *Penicillium paxilli* PSU-A71. *Chemical and Pharmaceutical Bulletin* 55: 1383–84. https://doi.org/10.1248/cpb.55.1383.

Ruma, K., S. Kumar, and H.S. Prakash. 2013. Antioxidant, anti-inflammatory, antimicrobial and cytotoxic properties of fungal endophytes from *Garcinia* species. *International Journal of Pharmacy and Pharmaceutical Sciences* 5: 889–97.

Ryan, R.P., K. Germaine, A. Franks, D.J. Ryan, and D.N. Dowling. 2008. Bacterial endophytes: Recent developments and applications. *FEMS Microbiology Letters* 278: 1–9. https://doi.org/10.1111/j.1574-6968.2007.00918.x.

Sagar, S., M. Kaur, and K.P. Minneman. 2010. Antiviral lead compounds from marine sponges. *Marine drugs* 8: 2619–38. https://doi.org/10.3390/md8102619.

Sahu, M.K., K. Sivakumar, E. Poorani, T. Thangaradjou, and L. Kannan. 2007. Studies on L-asparaginase enzyme of actinomycetes isolated from estuarine fishes. *Journal of Environmental Biology* 28: 465–74. http://www.jeb.co.in/journal_issues/200704_apr07_supp/paper_19.pdf.

Sakamoto, T., C. Araki, T. Beppu, and K. Arima. 1977. Partial purification and some properties of extracellular asparaginase from *Candida utilis*. *Agricultural and Biological Chemistry* 41: 1359–64. https://www.jstage.jst.go.jp/article/bbb1961/41/8/41_8_1359/_pdf.

Samri, S.E., M. Baz, A. Jamjari, et al. 2015. Preliminary assessment of insecticidal activity of Moroccan actinobacteria isolates against mediterranean fruit fly (*Ceratitis capitata*). *African Journal of Biotechnology* 14: 859–66. https://academicjournals.org/journal/AJB/article-full-text-pdf/018346651259.

Sangnoi, Y., O. Sakulkeo, S. Yuenyongsawad, et al. 2008. Acetylcholinesterase-inhibiting activity of pyrrole derivatives from a novel marine gliding bacterium, *Rapidithrix thailandica*. *Marine Drugs* 6: 578–86. https://doi.org/10.3390/md6040578.

Sarquis, M.I.D.M., E.M.M. Olivelra, A.S. Santos, and G.L.D. Costa. 2004. Production of L-asparaginase by filamentous fungi. *Memorias do Instituto Oswaldo Cruz* 99: 489–92. https://doi.org/10.1590/S0074-02762004000500005.

Savitri, N. Astana, and W. Azmi. 2003. Microbial L-asparaginase: A potent antitumour enzyme. *Indian Journal of Biotechnology* 2: 184–94. https://nopr.niscpr.res.in/handle/123456789/11297.

Schwartz, J.H., J.Y. Reeves, and J.D. Broome. 1966. Two L-asparaginases from *E. coli* and their action against tumors. *Proceedings of the National Academy of Sciences (PNAS)* 56: 1516–19. https://doi.org/10.1073/pnas.56.5.1516.

Selvam, K., and B. Vishnupriya. 2013. Partial purification and cytotoxic activity of L-asparaginase from *Streptomyces acrimycini* NGP. *International Journal of Research and Pharmaceutical Biomedical Science* 4: 859–69.

Senthil, K.M., and K. Selvam. 2011. Isolation and purification of high efficiency L-asparaginase by quantitative preparative continuous elution SDS-PAGE electrophoresis. *Journal of Microbial and Biochemical Technology* 3: 73–83. https://doi.org/10.4172/1948-5948.1000055.

Shan, T., S. Lu, C. Luo, et al. 2013. Preparative separation of spirobisnaphthalenes from endophytic fungus *Berkleasmium* sp. Dzf12 by high-speed counter-current chromatography. *Molecules* 18: 12896–908. https://doi.org/10.3390/molecules181012896.

Shimizu, M., Y. Nakagawa, S.A.T.O. Yukio, et al. 2000. Studies on endophytic actinomycetes (I) *Streptomyces* sp. isolated from *Rhododendron* and its antifungal activity. *Journal of General Plant Pathology* 66: 360–66. https://doi.org/10.1007/PL00012978.

Siddalingeshwara, K.G., D.N. Dhatri, T. Pramoda, T. Vishwanatha, K.M. Sudipta, and S.M. Mohsin. 2010. Rapid screening and confirmation of L-glutaminase producing novel *Aspergillus wentii*. *International Journal of Chemtech Research* 2: 830–33. https://sphinxsai.com/s_v2_n2/CT_V.2No.2/ChemTech_Vol_2No.2_pdf/CT=11%20(830–833).pdf.

Singh, B., A. Thakur, S. Kaur, B.S. Chadha, and A. Kaur. 2012. Acetylcholinesterase inhibitory potential and insecticidal activity of an endophytic *Alternaria* sp. from *Ricinus communis*. *Applied Biochemistry and Biotechology* 168: 991–1002. https://doi.org/10.1007/s12010-012-9835-0.

Singhal, B., and K. Swaroop. 2013. Optimization of culture variables for the production of L-asparaginase from *Pectobacterium carotovorum*. *African Journal of Biotechnology* 12: 6959–67. https://www.ajol.info/index.php/ajb/article/view/139662.

Soniyamby, A.R., and B.V. Praveesh. 2012. Isolation and screening of L-asparaginase producing fungi from soil samples. *International Journal of Pharmacy and Pharmaceutical Sciences* 4: 279–82.

Soreq, H., and S. Seidman. 2000. Anti-sense approach to anticholinesterase therapeutics. *The Israel Medical Association Journal* 2: 81–85. https://pubmed.ncbi.nlm.nih.gov/10909423/.

Stierle, A., G. Strobel, D. Stierle, P. Grothaus, and G. Bignami. 1995. The search for a Taxol producing microorganisms among the endophytic fungi of the Pacific yew, *Taxus brevifolia*. *Journal of Natural Products* 58: 1315–24. https://doi.org/10.1021/np50123a002.

Strobel, G., and B. Daisy. 2003. Bioprospecting for microbial endophytes and their natural products. *Microbiology and Molecular Biology Review* 67: 491–02. https://doi.org/10.1128/mmbr.67.4.491-502.2003.

Su, J., and M. Yang. 2014. Huperzine A production by *Paecilomyces tenuis* YS-13, an endophytic fungus isolated from *Huperzia serrata*. *Natural Product Research* 29: 1035–41. https://doi.org/10.1080/14786419.2014.980245.

Subramani, R., and W. Aalbersberg. 2012. Marine actinomycetes: An ongoing source of novel bioactive metabolites. *Microbiological Research* 167: 571–80. https://doi.org/10.1016/j.micres.2012.06.005.

Sunaryanto, R., and A.H. Mahsunah. 2013. Isolation, purification, and characterization of antimicrobial substances from endophytic actinomycetes. *Makara Journal of Science* 17: 87–92. https://scholarhub.ui.ac.id/science/vol17/iss3/3.

Sussman, J.L., M. Harel, F. Frolow, et al. 1991. Atomic structure of acetylcholinesterase from *Torpedo californica*: A prototypic acetylcholine-binding protein. *Science* 253: 872–79. https://www.science.org/doi/10.1126/science.1678899.

Tadesse, M., J. Svenson, K. Sepcic, et al. 2014. Isolation and synthesis of pulmonarins A and B, acetylcholinesterase inhibitors from the colonial ascidian *Synoicum pulmonaria*. *Journal of Natural Products* 77: 364–69. https://doi.org/10.1021/np401002s.

Talluri, V.S.S.L.P., M. Bhavana, K.K. Siva, K.P. Anil, and S.V. Rajagopal. 2014. *Myroides gitamensis* sp. nov., L-asparaginase producing bacteria isolated from slaughter house soil sample in Visakhapatnam, India. *Journal of*

Microbial & Biochemical Technology 6: 144–47. https://doi.org/10.4172/1948-5948.1000135.

Tan, R.X., and W.X. Zou. 2001. Endophytes: A rich source of functional metabolites. *Natural Product Reports* 18: 448–59. https://doi.org/10.1039/B100918O.

Teles, A.P.C., and J.A. Takahashi. 2013. Paecilomide, a new acetylcholinesterase inhibitor from *Paecilomyces lilacinus*. *Microbiological Research* 168: 204–10. https://doi.org/10.1016/j.micres.2012.11.007.

Thakur, M., L. Lincoln, F.N. Niyonzima, and S.S. More. 2013. Isolation, purification and characterization of fungal extracellular L-asparaginase from *Mucor Hiemalis*. *Journal of Biocatalysis and Biotransformation* 2 (December): 2, 1–9. http://dx.doi.org/10.4172/2324-9099.1000108.

Theantana, T., K.D. Hyde, and S. Lumyong. 2007. Asparaginase production by endophytic fungi isolated from some Thai medicinal plants. *KMITL Science and Technology Journal* 7: 13–18. https://li01.tci-thaijo.org/index.php/cast/article/view/86787/68733.

Theantana, T., T. Hyde, and S. Lymyong. 2009. Asparginase production by endophytic fungi from Thai medicinal plants: Cytotoxicity properties. *International Journal of Integrative Biology* 7 (August): 1, 1–8.

Thirunavukkarasu, N., T.S. Suryanarayanan, T.S. Murali, J.P. Ravishankar, and S.N. Gummadi. 2011. L-asparaginase from marine derived fungal endophytes of seaweeds. *Mycosphere* 2: 147–55. https://www.mycosphere.org/pdf/MC2_2_No5.pdf.

Thongsandee, W., Y. Matsuda, and S. Ito. 2012. Temporal variations in endophytic fungal assemblages of *Ginkgo biloba* L. *Journal of Forest Research* 17: 213–18. https://doi.org/10.1007/s10310-011-0292-3.

Tumiatti, V., A. Minarini, M.L. Bolognesi, A. Milelli, M. Rosini, and C. Melchiorre. 2010. Tacrine derivatives and Alzheimer's disease. *Current Medicinal Chemistry* 17: 1825–38. https://doi.org/10.2174/092986710791111206.

Udaya, P., C.S. Rao, and B.R. Satish. 2013. Studies on L-asparaginase production by using *Staphylococcus capitis*. *Journal of Chemical, Biological and Physical Sciences* 3: 201–09. https://www.scirp.org/reference/referencespapers?referenceid=2162424.

Venil, C.K., K. Nanthakumar, K. Karthikeyan, and P. Lakshmanaperumalsamy. 2009. Production of L-asparaginase by *Serratia marcescens* SB08: Optimization by response surface methodology. *Iranian Journal of Biotechnology* 7: 10–18. https://www.ijbiotech.com/article_7091.html.

Verma, N., K. Kumar, G. Kaur, and S. Anand. 2007. L-asparaginase: A promising chemotherapeutic agent. *Critical Reviews in Biotechnology* 27: 45–62. https://doi.org/10.1080/07388550601173926.

Verma, V.C., S.K. Gond, A. Kumar, A. Mishra, R.N. Kharwar, and A.C. Gange. 2009. Endophytic actinomycetes from *Azadirachta indica* A. Juss.: Isolation, diversity and anti-microbial activity. *Microbial Ecology* 57: 749–56. https://doi.org/10.1007/s00248-008-9450-3.

Verma, V.C., E. Lobkovsky, A.C. Gange, S.K. Singh, and S. Prakash. 2011. Piperine production by endophytic fungus *Periconia* sp. isolated from *Piper longum* L. *Journal of Antibiotics* 64: 427–31. https://doi.org/10.1038/ja.2011.27.

Wang, Z.F., J. Yan, Y. Fu, et al. 2008. Pharmacodynamic study of FS-0311: A novel highly potent, selective acetylcholinesterase inhibitor. *Cellular and Molecular Neurobiology* 28: 245–61. https://doi.org/10.1007/s10571-007-9199-0.

Waqas, M., A.L. Khan, and I.J. Lee. 2014. Bioactive chemical constituents produced by endophytes and effects on rice plant growth. *Journal of Plant Interactions* 9: 478–87. https://doi.org/10.1080/17429145.2013.860562.

Wei, J.G. 2004. *Research about diversity of Pestalotiopsis clavispora in Theaceae and Taxus and molecular systematics research of Pestalotiopsis clavispora*. Thesis. Zhejiang University (In Chinese).

Wei, Y., J. Chen, R. Jia, M. Wang, and W. Wu. 2008. Purification of *Escherichia coli* L-asparaginase mutants by a native polyacrylamide gel electrophoresis. *Journal of Chromatographic Science* 46: 556–59. https://doi.org/10.1093/chromsci/46.6.556.

Wetwitayaklung, P., C. Limmatvapirat, T. Phaechamud, and S. Keokitichai. 2007. Kinetics of acetylcholinesterase inhibition of *Quisqualis indica* Linn. flower extract. *Science, Engineering and Health Studies* 1: 20–28. https://li01.tci-thaijo.org/index.php/sehs/article/view/7257.

Wu, B., B. Ohlendorf, V. Oesker, et al. 2014. Acetylcholinesterase inhibitors from a marine fungus *Talaromyces* sp. strain LF458. *Marine Biotechnology* 17: 110–19. https://doi.org/10.1007/s10126-014-9599-3.

Ying, Y.M., Z.Z. Zheng, L.W. Zhang, W.G. Shan, J.W. Wang, and Z.J. Zhan. 2014. Rare C_{25} steroids produced by *Penicillium chrysogenum* P1X, a fungal endophyte of *Huperzia serrata*. *Helvetica Chimica Acta* 97: 95–101. https://doi.org/10.1002/hlca.201300153.

Yu, A.J., and P. Dayan. 2005. Uncertainty, neuro-modulation and attention. *Neuron* 46: 681–92. https://doi.org/10.1016/j.neuron.2005.04.026.

Yuan, W.H., Z.W. Wei, P. Dai, et al. 2014. Halogenated metabolites isolated from *Penicillium citreonigrum*. *Chemistry and Biodiversity* 11: 1078–87. https://doi.org/10.1002/cbdv.201300349.

Zhang, Z.B., Q.G. Zeng, R.M. Yan, Y. Wang, Z.R. Zou, and D. Zhu. 2011. Endophytic fungus *Cladosporium cladosporioides* LF70 from *Huperzia serrata* produces huperzine A. *World Journal of Microbiology and Biotechnology* 27: 479–86. https://doi.org/10.1007/s11274-010-0476-6.

Zhu, D., J. Wang, Q. Zeng, Z. Zhang, and R. Yan. 2010. A novel endophytic huperzine A-producing fungus, *Shiraia* sp. Slf14, isolated from *Huperzia serrata*. *Journal of Applied Microbiology* 109: 1469–78. https://doi.org/10.1111/j.1365-2672.2010.04777.x.

16 Algal Secondary Metabolites

An Overview

Shreya Kapoor
(Department of Microbiology, Ram Lal Anand College, University of Delhi, and Department of Biotechnology, Delhi Technological University, New Delhi, India; shreyakapoor1103@gmail.com)

Sakshi Verma
(Department of Microbiology, Ram Lal Anand College, University of Delhi, India; vsakshi0103@gmail.com)

Sneh Priya
(Department of Microbiology, Ram Lal Anand College, University of Delhi, India; snehpriya2003@gmail.com)

Aishani Gupta
(Department of Psychology, Indraprastha College for Women, University of Delhi, India; aishanigupta565@gmail.com)

**Vandana Gupta*
(Department of Microbiology, Ram Lal Anand College, University of Delhi, India; vandanagupta@rla.du.ac.in)

16.1 INTRODUCTION

Algae are the primitive polyphyletic organisms that emerged about 2.5 billion years ago. Earlier, these were thought to be found exclusively in aquatic habitats, but recent studies have reported their ubiquitous presence and substantial involvement in preserving ecological balance and the overall health of Earth's ecosystem. Not only do they serve as the energy base of the food web in aquatic environments, but they also provide habitat to a variety of organisms. Additionally, microalgae have better photosynthetic efficiency and CO_2 sequestration than land plants, thereby significantly contributing to Earth's atmospheric composition. Over 800,000 algal species are known to exist in nature, exhibiting an enormous range of ecological adaptations (Arora et al., 2021). For the obstinate survival in varied habitats, algae produce a wide variety of bioactive compounds via complex metabolic pathways that provide them the ability to thrive in extreme conditions and competitive environments (Bhowmick et al., 2020). Protein, carbohydrates, and lipids are the primary metabolites essential for the survival and growth of organisms (Gupta et al., 2022, 2024). The secondary metabolites are secretory molecules, typically formed during the late growth phase, that is, idiophase. These substances do not take part in the main metabolism, however they provide additional survival benefits (Vidhyanandan et al., 2020; Abu-Ghosh et al., 2021; Babich et al., 2022; Fernando et al., 2022). The structural landscape of algal secondary metabolites is vast, ranging from low molecular weight compounds such as mycosporine-like amino acid (MAAs) extending to complex structures, for example, cyanotoxins. Additionally, these exhibit a high degree of chemical variation, encompassing alkaloids, carotenoids, phenolics, flavonoids, lignin, sterols, tannins, terpenoids, and others (Rico et al., 2017; Vidhyanandan et al., 2020). For the host organism, these metabolites perform a variety of functions, ranging from photo-protection to defense against predators and grazers, and facilitate communication with other organisms (Yang et al., 2023; Jain et al., 2023, 2024). Archaeological evidences suggest the utilization of algae as a part of human diet and traditional medicines for a thousand years, reflecting their rich nutritional profile and therapeutic potential. In this chapter, we explore the immense potential of algal secondary metabolites to be utilized in the production and development of new products for food, pharmaceuticals, agricultural, and cosmetic industries.

16.2 ALGAL SECONDARY METABOLITES: TYPES AND FUNCTIONS

Microorganisms develop secondary metabolites (SMs) when one or more nutrients in the culture media are low and during

DOI: 10.1201/9781003408833-16

biotic and abiotic environmental stresses. In nature, SMs typically perform a variety of supportive activities in addition to a significant ecological role (Purves et al., 2016). The synthesis of these SMs enables the microorganisms to retain their growth rates or to increase the likelihood of survival under unfavorable environmental conditions. Microalgae's potential for photosynthetic activity enables them to produce SMs with distinct chemical diversity and function that are created by ensuing biosynthetic processes (Kolackova et al., 2023). Algal SMs include toxins, gibberellins, alkaloids, antibiotics, biopolymers, *etc*. Various algal SMs and their potential uses are summarized in Table 16.1. Structures of some of the representative algal SMs are depicted in Figure 16.1. Environmental conditions have an impact on their production, and these are not synthesised until the microbe enters the stationary phase of growth, having almost completed its logarithmic growth phase. Complex biochemical mechanisms control its production (Abdel-Aziz et al., 2017). Similar to plants, algae need carbohydrates and other primary metabolites for cell growth and division. When the algal cells are actively expanding in the early stages of the growth, this is extremely crucial. Near the end of the growth cycle, as the organism approaches the end of its life cycle, secondary metabolites tend to be produced. This is because during this time, environmental factors, including nutrition nonavailability and limitation, and toxin exposure will cause stress to cells. These secondary metabolites assist in mitigating the effects of stressful conditions and trigger certain signalling pathways involved in the inflammatory responses (Jeong et al., 2020). Under stressful conditions, algae have an inherent tendency to synthesize and accumulate large amounts of secondary metabolites (Sreenikethanam et al., 2022). For instance, phlorotannin contents in brown macroalgae increase with increasing salinity in their habitats (Kamiya et al., 2010). Nutrient regime also influences phlorotannin and terpenoid contents in macroalgae (Vaidya & Stinchcombe, 2020), there are also many effects of irradiance on algal secondary metabolites. UV radiation can induce significant increase in phlorotannin concentration of *Ascophyllum nodosum*. Phlorotannins are known to absorb UV radiations thereby protecting the host from UV-induced damage. A decrease from the optimal temperature for growth to subliminal temperature can induce β-carotene synthesis and increased lipid content in *Dunaliella salina* cells, thereby promoting the formation of lipid-carotene globules in the chloroplast periphery (Li et al., 2010). Fucoxanthin is produced by the algal species under light stress as it offers photo-protection and plays a role as an accessory pigment in the light-harvesting complex (Din et al., 2022). Terpene synthases in algae provide chemical defenses against biotic stressors, including fungi, bacteria, *etc*. (Wei et al., 2019).

16.3 PRODUCTION AND EXTRACTION

16.3.1 Production Mechanism

Several compounds are produced via secondary metabolic pathways, including alkaloids, carotenoids, phenolic compounds, lignin, phytosterols, and others. Algal growth in extreme environments, such as nutrient deprivation and exposure to extremes of visible and UV light, as well as mechanisms for light harvesting, carbon acquisition, nutrient limitation, and assimilation of toxic compounds, all result in the production of unique secondary metabolites (Rico et al., 2017).

The addition of an organic carbon source, a change in light intensity, or a combination of the two can be used to influence microalgal metabolism to favor the accumulation of a particular secondary metabolite. Usually, open cultures are considered better for the production of pigments, however, such systems are prone to contamination. Additionally, it requires considerable surface for cultivation, and there is constant loss of water through evaporation, necessitating the need for closed systems. There are several parameters to be taken care of for optimal growth for the accumulation of desired metabolites. These include, selection of high-yielding strains, the thickness of the wall of the photobioreactor, mixing regime, optimal light, temperature, and oxygen availability (Márquez-Rocha et al., 2019). Furthermore, strategies like genetic engineering can be used to manipulate the algal metabolic pathways to augment the synthesis of the target compound. Subsequently, the resulting biomass is harvested, followed by cell lysis and extraction of the metabolite.

16.3.2 Extraction

A single solvent (usually methanol) or a combination of solvents may be used to extract metabolites from biomass, particularly when a large number of molecules with varying polarity are being sought. Both polar and nonpolar solvents use sequential extraction to create a succession of samples. Simple extraction techniques like decoction, maceration, liquid-liquid extraction, infusion, percolation, digestion, and hot continuous extraction (Soxhlet) are traditional extraction methods. However, they present certain limitations, for example, time-consuming and ecologically damaging. To address these problems, cutting-edge techniques like – pressurized liquid extraction (PLE), supercritical fluid extraction (SFE) can be utilised. SFE is highly beneficial as it provides better selectivity, less extraction time and doesn't utilise any toxic compounds (Alassali et al., 2016). Supercritical fluid allows for a quicker and more effective extraction process since it penetrates the biomass's structure more deeply. Nevertheless, this technology is being significantly impacted by high running costs, particularly for larger volumes (Singh et al., 2021).

16.3.3 Metabolic Profiling

Each secondary metabolite family has some specific chemical characteristics with different polarities, for instance, polyhydroxylated alkaloids being polar and terpenoids being non-polar. It is possible to profile an increasing

TABLE 16.1
Summary of Algal SMs, Their Mode of Action, and Current Applications

Class	SM	Source	Mode of Action	Application	Reference
Carotenoid	Lutein	*Dunaliella salina*, *Chlorella pyrenoidosa*, *Chlorella protothecoides*, *Chlorella sorokiniana*, and *Chlorella vulgaris*	Reduces the amount of blue light that reaches the photoreceptor, ROS scavenger	Prevention of age-related cataract and maculopathy; antitumor; food coloring agent; feed additive; antioxidant	Bule et al., 2018
	β-carotene	*D. salina*, *D. bardawil*, and *Spirulina maxima*	ROS scavenging, direct effect on the thymus, which increases the production of immune cells	Anti-inflammatory, antioxidant, food colorant, vitamin A supplement; reduction of macular degeneration and colorectal cancer risk; immunomodulatory; promotion of atherogenesis	Rico et al., 2017; Bule et al., 2018
	Astaxanthin	*Haematococcus pluvialis*, *C. Zofingiensis*, *C. vulgaris*, and *Chlorococcum* sp.	Scavenges ROS and other reactive species, favors the neutralization of reactive oxygen and nitrogen species, bonds with free radical to form a non-reactive product	Food colorant, poultry and aquaculture feed additive, antiaging, sun protection, antidiabetic; maintenance of membrane structure and decrease in membrane fluidity; useful in oral, colon, and liver cancers, cardiovascular diseases, and degenerative eye diseases	Rico et al., 2017; Bule et al., 2018; Pereira et al., 2021
	Violaxanthin	*D. tertiolecta* and *C. ellipsoidea*	Under high light intensity: converts to zeaxanthin through the action of the enzyme violaxanthin de-epoxidase	Anti-inflammatory, antitumor, photoprotection	Bule et al., 2018
	Zeaxanthin	*D. salina*, *Porphyridium cruentum*, *C. saccharophila*, and *Synechocystis* sp.	Functions as a blue light filter; anti-inflammation and antioxidation	Prevention of age-related cataract, reduction of macular degeneration	Rico et al., 2017; Murillo et al., 2019
	Lycopene	*Haematococcus pluvialis*, *C. vulgaris*, and *Porphyra umbilicalis*	Reduces mutagenesis, regulates cell cycle progression, and up-regulates the expression of connexin 43 (cell communication)	Antioxidant, dietary supplement, and anticancer	Heber and Lu, 2002; Rico et al., 2017
	Fucoxanthin	*Phaeodactylum tricornutum* and *Isochrysis* sp.	Induce G1 cell cycle arrest and apoptosis by modulating expression of various cellular molecules and cellular signal transduction pathways	Anticancer, antiobesity; prevention of cerebrovascular diseases	Rico et al., 2017; Bule et al., 2018
Polyunsaturated fatty acids (PUFAs)	Arachidonate and eicosapentaenoic acid	*C. vulgaris*, *Tetraselmis* sp., *Nannochloropsis oculate*, *Crypthecodinium cohnii*, *Nannochloropsis gaditana*, *Schizochytrium* sp., and *Ulkenia* sp.	Regulates signaling pathways that influence patterns of gene expression	Role in neurogenesis and neurotransmission; treatment of atherosclerosis, rheumatoid, arthritis, Alzheimer's, and psoriasis; nutritional supplements	Calder, 2012; Bule et al., 2018
Tocopherols	RRR-α-tocopherol and α-tocopherol	*Porphyridium* sp., *Spirulina platensis*, *Desmodesmus* sp., and *Tetraselmis suecica*	Radical scavenger; functions depending on the H-atom donating ability, location, and movement within the membrane	Vitamin E antioxidant activity, anti-inflammatory, antitumor, proper functioning of muscle and nerves, prevention of blood clots and boosting of immune system	Bule et al., 2018

(Continued)

TABLE 16.1 (*Continued*)
Summary of Algal SMs, Their Mode of Action, and Current Applications

Class	SM	Source	Mode of Action	Application	Reference
Phyco-biliproteins	Phycoerythrin	*Oscillatoria agardhi*, *Nostoc*, *Spirulina platensis*, and *Galdiera suphuraria*	Fluorescence of phycoerythrin protein quenched by the peroxyl radical generated by thermal decomposition of AAPH	Food and cosmetics coloring agent, fluorescence immunoassays reagent, antioxidant	Rico et al., 2017
	Phycocyanin	*Nostoc* UAM 206, *Spirulina platensis*, *Porphyridium aerugineum*, and *Porphyridium cruentum*	Activates the apoptotic pathway (the mitochondrial/cytochrome C pathway), activates caspase, and induces poly (ADP-ribose) polymerase-1 (PARP-1) cleavage, alters the Bcl-2/Bax ratio	Fluorescence immunoassays reagent, label for biological molecules, antioxidative function, anti-inflammatory activity, anticancer, immune enhancement, liver and kidney protection	Jiang et al., 2017
Phyto-steroids	Phytosterols	*Phormidium luridum*, *Phormidium autumnale*, *Scenedesmus*, *Ulva lactuca*, *Cystoseira adriatica*, and *Sargassum fusiforme*	Competitive exclusion of cholesterol from bile acid micelles inducing cholesterol precipitation in the intestinal tract, reducing its cholesterol reabsorption	Regulation of membrane fluidity and permeability; prevention of cardiovascular disorders	Rico et al., 2017; Fagundes and Wagner, 2021
Phenolics	Hydroxycinnamic acids	*Ch. vulgaris*, *Haematococcus pluvialis*, *Diacronema lutheri*, *Phaeodactylum* sp., *Tetraselmis suecica*, and *Porphyridium purpureum*	Radical-scavenging activity that is linked to their hydrogen- or electron-donating ability and to the stability of the resulting phenoxyl radicals	Anti-inflammatory, antioxidant	Teixeira et al., 2013
	4-hydroxy-benzaldehyde	*Spongiochloris spongiosa*, *Spirulina platensis*, *Anabaena doliolum*, *Nostoc* sp., and *Cylindrospermum* sp.	Potentiation of glucose uptake demonstrated in 3T3-L1 adipocytes, as well as the stimulation of glucose metabolism and the inhibition of hepatic glucose production	Flavorings and perfuming agents in the food and cosmetic industries; antidiabetic	Eddouks et al., 2014
	3,4-dihydroxybenzaldehyde	*Spongiochloris spongiosa* and *Spirulina platensis*	Inhibits enzymes involved in metabolism of lipid, protein, and carbohydrate	Managing several diseases, for example, antiepileptics, neuroleptics, antidepressants, cytostatic, immunosuppressants	Kim et al., 2022
MAAs	7-O-(β-arabinopyranosyl)-porphyra-334 and Shinorine	*Nostoc commune*, *Alexandrium tamarense*, *A. catenella*, *Synechocystis* sp., and *C. sorokiniana*	Blocks pyrimidine dimer formation; activates focal adhesion kinases (FAK) and mitogen activated protein kinases (MAPK)	Sunscreen, anti-inflammatory, antitumor, antioxidative, wound healing	Raj et al., 2021
Scytonemin	ScyA, ScyB, and SycC	*Fischerella muscicola*, *Nostoc commune*, *Chlorogloeopsis* sp., *Scytonema* sp., and *Nostoc punctiforme*	Accumulates within the extracellular sheath or slime and absorbs and screens UV rays	Anti-inflammatory, antiproliferation, sun protection; an extra layer of defense	Balskus et al., 2011
Polyphenolics	Phlorotannins	*Eisenia arborea*, *Eisenia bicyclis*, *Ecklonia kurome*, *Ecklonia cava*, and *Laminaria digitata*	Inhibits viral fusion and mRNA entry, improves the phagocytic activity of macrophages NK cells, activates T and B lymphocytes, inhibits the production of metalloproteinase	Antidiabetic, and antiobesity, antioxidant, anti-inflammatory, immunomodulatory, antiviral, and anticancer	Khan et al., 2022
Terpenoid	Mono-, di-, tri-, tetra-, mero-, and sesquiterpenoids	*Porphyridium purpureum*, *Erythrolobus australicus*, *Chondrococcus hornemannii*, and *Portieria hornemannii*	Repairs the damaged DNA and biomolecules	Anticancer and antioxidant	Khan et al., 2022

number of compounds by using different analytical systems such as GC-MS (gas chromatography-mass spectrometry), LC-MS (liquid chromatography-mass spectrometry), and HPLC-DAD (high-performance liquid chromatography with diode-array detection) coupled to a sequential extraction.

To profile secondary metabolites, polar and non-polar molecules are sequentially extracted using 50% ethanol (v/v) and dichloromethane, respectively. The goal of gas-liquid chromatography is to remove volatile substances from a complicated combination by passing the sample through a heated column partitioned between an inert gas under pressure and a thin layer of non-volatile liquid coated on an inert support inside the column. The technique uses the temperature of vaporization particular to each compound to separate them from a solution.

A mixture of substances in solution can be separated into individual compounds using high-pressure liquid chromatography. Use of a solvent mixture that is pumped under pressure through the column in this method separates each component based on their solubility and size (Rispail et al., 2005).

16.4 PROPERTIES OF ALGAL SECONDARY METABOLITES

16.4.1 Chemical Properties

Alkaloids are compounds derived from amino acids and contain heterocyclic nitrogen particles (El-Beltagi et al., 2019; Mohiuddin, 2019). Marine algae mostly contain alkaloids belonging to 2-phenylethylamine and indole groups, while halogenated alkaloids are specific to particular algae (El-Beltagi et al., 2019). The chemical characteristics of alkaloids, like ionization profile and their capacity to form stable salts, indicate their potential to be used as drug candidates (Souza et al., 2020).

Carotenoids are a group of naturally occurring lipophilic pigments that are chemically isoprenoid polyenes composed of pure carbon skeletons (Gupta et al., 2021; Mekinić et al., 2023). Carotenoids have two roles; firstly, they act as accessory light-harvesting pigments in photosystems, and secondly, they are photo-protective molecules (Gupta et al., 2021). Due to their chemical diversity, carotenoids have various biochemical and physiological functions, like anticancer, antioxidant, anti-inflammatory, chemo-protective, *etc.* (Mekinić et al., 2023).

Phenolic compounds are made of aromatic benzene rings containing hydroxyl groups and functional derivatives bonded to the ring (Mekinić et al., 2019). Algae produce various phenolic compounds ranging from simple molecules like phenolic acids to more complex polyphenols like phlorotannins, terpenoids, *etc.* They play an important role as cell wall components to cope with environmental conditions and also provide chemical defense (Mekinić et al., 2019; Del Mondo et al., 2021).

Sterols belong to a class of isoprenoid lipids and play an important role in maintaining membrane fluidity and permeability. Oils synthesized by algae possess a greater number of sterols compared to plants, and some microalgae even synthesize certain sterols like 7-dehydroporiferasterol that are not present in plants and display outstanding bioactivities (Potijun et al., 2021).

16.4.2 Antibacterial Potential

Antibiotic resistance is increasing at an alarming rate, requiring the development of fast and efficient alternatives to conventional antibiotics, and algae can be considered a boon for novel drug discovery due to their ability to synthesize highly active SMs possessing potential antibacterial properties (Bhowmick et al., 2020). Microalgae can survive in all types of environments, and the stress they face plays an important role in the production of antibacterial compounds, including alkaloids, fatty acids, indoles, macrolides, phenols, terpenes, and glycolipids, among others (Rojas et al., 2020; Little et al., 2021).

A study reported that methanol extracts from three local microalgae, *Chlorella* sp. UKM8, *Scenedesmus* sp. UKM9 and *Chlorella sorokiniana* (UKM2), when tested against Gram-positive and Gram-negative bacteria, showed antibacterial activities. *Chlorella* sp. (UKM8) methanol extracts displayed the highest antibacterial activity, and the major compounds providing this property were phenol, hexadecenoic acid, phytol, octadecadienoic acid, and bicycloheptane (Shaima et al., 2022).

Marine algae produce polyphenolic SMs with high antimicrobial activity, which can be utilized as supplements to antibiotic therapy (Besednova et al., 2020). The methicillin-resistant *Staphylococcus aureus* (MRSA) is a multidrug-resistant pathogen that can lead to diseases such as sepsis, pneumonia, endocarditis, erysipelas, osteomyelitis, and foodborne diseases. Various algal seaweed extracts have been tested against MRSA. The extracts obtained from algae *Padina tetrastromatica* and *Padina gymnosporia* have shown significant antibacterial activity against clinical isolates of *Staphylococcus* and the ethyl acetate extracts from *P. tetrastromatica* for the treatment of MRSA infections (Besednova et al., 2020).

Cyanobacteria are another useful algal source to obtain antibacterial metabolites. For example, noscomin isolated from *Nostoc commune*, malyngamides, and amides of the fatty acid (8)–7(S)-methoxytetradec-4(E)-enoate isolated from *Lyngbya majuscule*, alkaloids ambiguine H isonitrile and ambiguine I isonitrile isolated from *Fischerella* sp., and kawaguchipeptin B isolated from toxin-producing *Microcystis aeruginosa* are all antibacterial compounds with promising applications against antibiotic-resistant bacteria (Barzkar et al., 2019).

16.4.3 Anticarcinogenic Potential

Cancer is a major threat to public health due to the increasing cancer-related death rates, and there is an urgent need for the development of new anticancer drugs (Sanjeewa

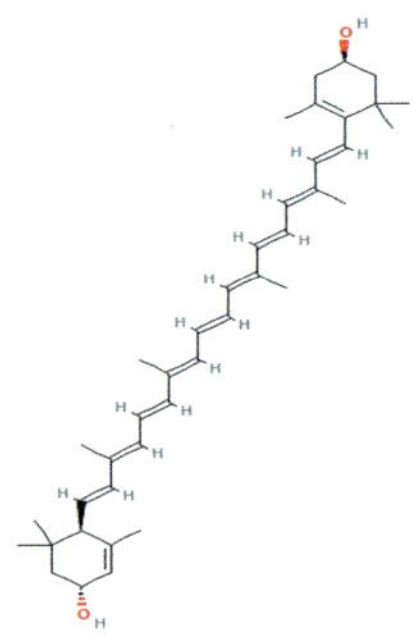
Lutein ($C_{40}H_{56}O_2$)

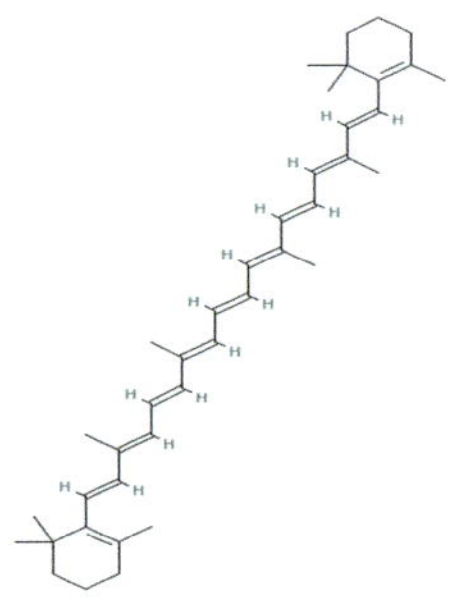
Beta-carotene ($C_{40}H_{56}$)

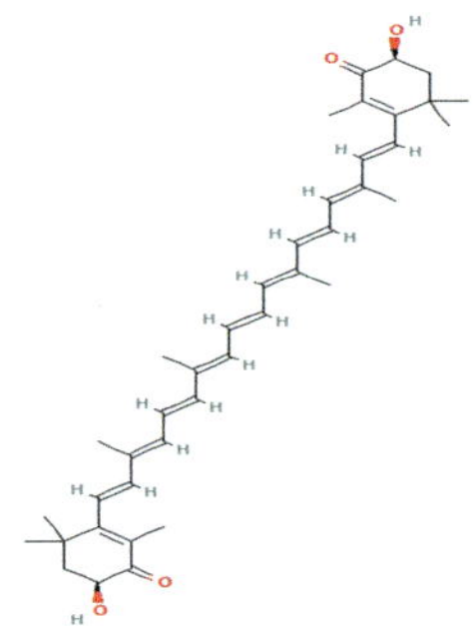
Astaxanthin ($C_{40}H_{52}O_4$)

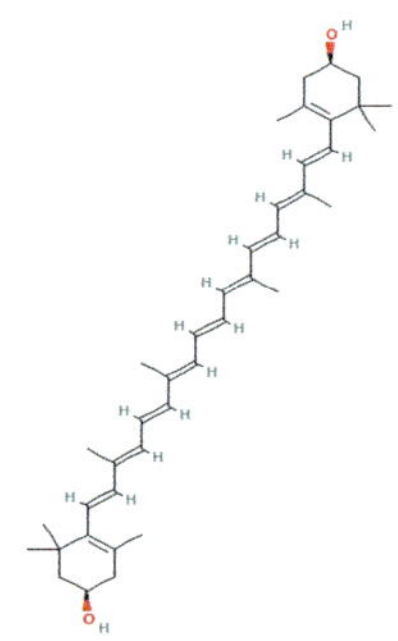
Zeaxanthin ($C_{40}H_{56}O_2$)

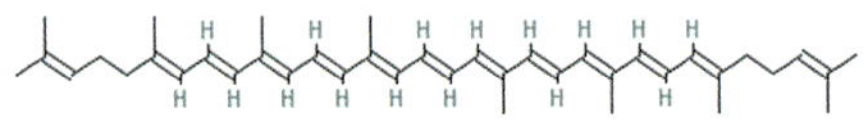
Lycopene ($C_{40}H_{56}$)

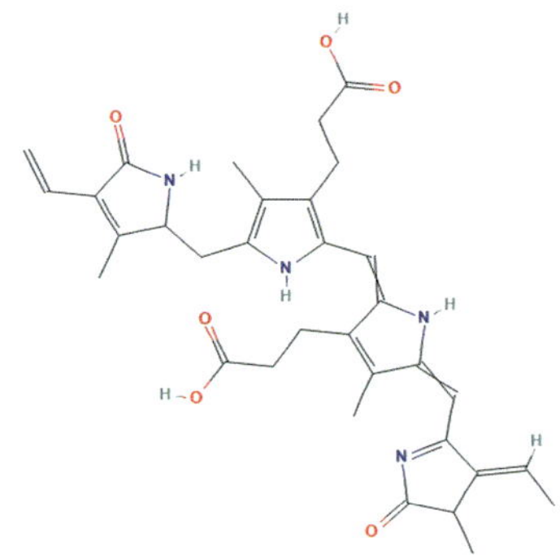
Phycoerythrobilin ($C_{33}H_{38}N_4O_6$)

FIGURE 16.1 Structures of some of the representative secondary metabolites obtained from algae.

et al., 2017; Alves et al., 2018). The current chemotherapeutic strategies and synthetic drugs used for treatment have side effects on patients' health and are not completely effective due to which compounds obtained from natural resources such as algae are gaining attention (Sanjeewa et al., 2017; Mansur et al., 2020).

The structural diversity and uniqueness of algal secondary metabolites make them promising candidates for cancer therapy, and in a study, Halomon, a pentahalogenated monoterpene obtained from the red alga, *Portieria hornemannii*, exhibited highly differential cytotoxicity in the NCI-60-cell-line panel, with brain, renal, and colon cancer cell lines being the most sensitive (Lefranc et al., 2019). Extract fractions from *Chlorella sorokiniana* and *Chaetoceros calcitrans* exhibited significant anticancer activity at 1–3 µg/mL, suggesting that these metabolites may be further investigated for the development of anticancer drugs (Andrade et al., 2018).

Carotenoids, tetraterpenoid organic pigments, have been studied for their anti-inflammatory, antitumor, and anticancer activities (Nuzzo et al., 2022; Sharma et al., 2023). β-carotene derived from *Dunaliella salina*, *Chlamydomonas reinhardtii*, *Tetraselmis suecica*, and *Isochrysis galbana* showed anticancer potential against breast, liver, prostate, pancreatic, gastric, colorectal cancers, neuroblastoma, and non-Hodgkin lymphoma (Sharma et al., 2023). The compound inhibits various signalling pathways in cancer cells and is also found to display immune-activating properties (Nuzzo et al., 2022; Sharma et al., 2023). Flavonoids also possess the ability to kill cancer cells by autophagy and induce apoptosis (Ferdous and Yusof, 2021). Some flavonoids obtained from algae, including rutin (3′,4′,5,7-tetrahydroxy-flavone-3-rutinoside), quercetin (pentahydroxyflavone), morin (3,5,7,2',4'-pentahydroxyflavone), catechin (flavan-3-ol), apigenin (4′,5,7-trihydroxyflavone), *etc*., displayed potential anticancer and antitumor activities (Ferdous and Yusof, 2021).

16.4.4 Anti-Inflammatory Potential

Inflammation is a nonspecific response generated by the body against any harmful stimuli and can be induced due to mechanical injuries, chemical, biological, and physical agents, and immunological disorders (Thiyagarasaiyar et al., 2020). The current treatment available for inflammatory skin diseases and respiratory problems can cause side effects. Algal metabolites can be used as an alternative for the development of more effective and less toxic anti-inflammatory agents due to their potential biofunctional properties (Tan et al., 2021; Leitner et al., 2022). Various secondary metabolites derived from seaweeds, including alginic acid, eckol, dieckol, fucosterol, gallic acid, diphlorethohydroxycarmalol, phlorotannins, and flavonoids, exhibit anti-inflammatory activities against air pollutant-induced respiratory problems (Tan et al., 2021).

Green alga *Chromochloris zofingiensis* was found to exhibit anti-inflammatory properties, and it was revealed that secondary metabolites such as Lyso-DGTS (LDGTS), 5-phenylvaleric acid, and oleamide were responsible for anti-inflammatory activity (Leitner et al., 2022). Eicosapentaenoic acid (EPA) obtained from ethanol extract of *Nannochloropsis oceanica* was found to impart anti-inflammatory, antioxidative, and antiamyloidogenesis activities in an LPS-induced AD mouse model, suggesting its possible neuroprotective effects (Barbalace et al., 2019). Studies have reported that extracts obtained from *Ecklonia cava* displayed potential anti-inflammatory activities, and phlorotannins were responsible for their protective effects in PC12 cells (Barbalace et al., 2019; Thiyagarasaiyar et al., 2020). Marine algae produce fucoxanthin, a natural product of carotenoids, and possess a variety of biological activities, including anti-inflammatory potential, which may be attributed to its strong antioxidant capacity, making it a promising compound for the treatment of human inflammation-related diseases (Liu et al., 2020).

16.5 APPLICATIONS AND USAGE OF ALGAL SECONDARY METABOLITES IN DAILY LIFE

16.5.1 Cosmeceutical Industries

The cosmeceutical market is flourishing with time, driven by a shift toward an ultramodern life. Given the side effects of the current cosmetics, there has been a surge in interest in natural bioactive ingredients. The global market demand for organic cosmetics is anticipated to reach approximately $54.5 billion by 2027. Across generations, the application of algal compounds has been instrumental in mitigating skin diseases as these compounds exhibit a wide range of pharmacological activities with minimal cytotoxicity in human cells. Algal SMs like scytonemin, sporopollenin, carotenoids, and MAAs have shown potential in mitigating the effects of highly damaging, unfavorable UV radiation, thereby serving as effective sun-protecting agents (Table 16.2).

16.5.2 Agricultural Sector

Algal secondary metabolites have properties that indicate their potential use in better crop production and enhanced crop protection as they function as a nutritional source, increase production of biomass, and improve plant tolerance to stress factors (*e.g.*, drought, salt, pollution, pathogens, *etc.*) (González-Pérez et al., 2022). Bio stimulants using microalgae are a sustainable option for agriculture, replacing chemical pesticides and fertilizers since microalgae can produce metabolites which can also benefit industrial purposes as shown in Table 16.3 (González-Pérez et al., 2022).

Algae can be used to develop biopesticides since their extracts possess effective antimicrobial, herbicidal, nematicidal, and insecticidal properties against crop pathogens (Asimakis et al., 2022). Few pathogens against which they have been proven to be effective include the bacterial genera *Agrobacterium*, *Pseudomonas*, *Xanthomonas*, and *Erwinia*, fungal genera such as *Fusarium*, *Verticillium*, *Rhizoctonia*, *Phytophthora*, and *Phoma* and viral pathogens including tobacco mosaic virus (TMV) and potato virus X (PVX) (Asimakis et al., 2022). In comparison to biopesticides developed from other microorganisms, algae-based biopesticides are easier to maintain, requiring fewer inputs. Industries in developing countries are becoming more focused on algal biofertilizers, biostimulants, algal sensors, and biopesticides, which is a great advancement toward sustainable agriculture (Pattanayak et al., 2021).

Phytohormones such as auxins, gibberellins, cytokinins, abscisic acid, and ethylene produced by green algae could be extracted and widely employed in agriculture (Pan et al., 2019). For instance, auxins display herbicidal properties and promote the growth and development of plants, and algal cytokinins act as antimicrobial agents against diseases in tomato (Pan et al., 2019) (Table 16.3).

TABLE 16.2
Cosmeceutical Potential of Algal Metabolites

Compound	Microalgal Source	Cosmeceutical Application	Reference
Lutein	*Leptolyngbya foveolarum*	Antioxidants	Lopes et al., 2020
Sporopollenin	*Chroococcidiopsis* sp.	Water loss prevention	He et al., 2016; Tampucci et al., 2022
MAAs	*Porphyra umbilicalis*	UV protection	Lawrence et al., 2019
Beta-carotene	*Dunaliella, Haematococcus*	Antiaging effects	Wang et al., 2022
Fucoxanthin	*Halamphora coffeaeformis* and *Phaeodactylum tricornutum*	Protection of skin from photoaging	Zasada and Budzisz, 2019
Glycoin	*Spirulina subsalsa*	Moisturizing agent; cheek skin elasticity	Haskins et al., 2018
Halogenated terpenoids	*Plocamium costatum*	Skin whitening agents	Mago et al., 2023
Phlorotannins	*Ecklonia cava*	Curtailing skin pigmentation	Chakdar and Pabbi, 2017
Cantaxanthin	*Nannochloropsis*	Tanning goods	Stoyneva-Gärtner et al., 2020
Phycocyanin	*Porphyridium purpureum*	Used in eye shadows	Mourelle et al., 2017

TABLE 16.3
Summary of the Role of Algal Secondary Metabolites in the Agricultural Sector

Algal Compound	Source Algae	Application	Reference
Abscisic acid	*Chlorella minutissima*	Reduces severity of early blight disease by targeting *Solanum lycopersicum*	Mrid et al., 2021
Auxins	*Chlorella pyrenoidosa*	Promotes plant growth and development; exhibits herbicidal activities	Munira et al., 2018
Cytokinins	*Gracilaria*	Elevates plant growth and cell division, tolerance to abiotic stress like drought, antifungal and antibiotic properties	Dmytryk and Chojnacka, 2018; Win et al., 2018
Ethylene	*Chlorella pyrenoidosa*	Contributes to senescence of xylem cells, elevates auxin transport, and advances root hair growth for efficient mineral and water absorption	Woodson, 2018
Gibberellins	*Chlamydomonas reinhardtii*	Activates amylase genes in aleurone cells, thereby stimulating seed germination, and enhances biomass production	Nabti et al., 2017
Chorellins	*Chlorella sp.*	Inhibits pathogenic bacteria and biopesticides	Abu-Ghosh et al., 2021
Cryptophycin 1	*Nostoc sp.*	Blocks the cell cycle of *Cryptococcus*	Fidor et al., 2019

Algae (microalgae and cyanobacteria) are capable of removing heavy metals (*e.g.*, cadmium, chromium, lead, *etc.*), trace elements (*e.g.*, copper, iron, manganese, copper, *etc.*), hydrocarbons and others, from soil and are great bioremediation agents which can play major role in improving soil quality and fertility, greatly benefiting agriculture and society at large (Ferreira et al., 2023). Recently, due to increased contamination of land, the accumulation of these heavy metals in food such as rice grains was reported, which leads to a negative impact on human health, and the capacity of algae to degrade pollutants enzymatically could be utilized to fight this problem (Ferreira et al., 2023).

16.5.3 Commercial and Valuable Bioproducts

Algae, in general, produce an extensive collection of bioactive compounds that can be directly or indirectly used for bioplastic production. Lipids, cellulose, and lactic acids serve as the key ingredients for the generation of microalga-based bioplastics (Chong et al., 2022; Sreenikethanam and Bajhaiya, 2022). Microalgae-based thermoplastics have also been produced by thermally and mechanically induced polymerization of the algal proteins, followed by their blending with polyethylene. Under environmental or nutritional stress, different microalgal species have been reported to accumulate polyhydroxyalkanoates (PHA) that can be harnessed as substrates for bioplastic production. PHAs offer promising benefits, including, but not confined to, complete biodegradation, high biocompatibility, higher resistance to UV rays and hydrolytic enzymes (Chong et al., 2022). Polyhydroxybutyrate is among the well-recognized polyhydroxyalkanoates to be utilized as a bioplastic (Sreenikethanam and Bajhaiya, 2022). These have high tensile strength (40 MPa), high potential mechanical performance, and a favorable melting point of 179°C. Its inherent thermoplastic, brittle, and hydrophobic characteristics are attributed to the presence of methyl and ester functional groups in its structure. The promising benefits of PHB-based bioplastics render them a feasible replacement for conventional plastics (Arora et al., 2023). These find diverse applications across different sectors, including food packaging, veterinary, protection nets, and greenhouse films. These can additionally be used for wound dressing, sutures, *etc.* The inherent porous nature of PHB facilitates the release of the drug to the wound. By its high molecular weight and non-adhesive properties, PHB can be used as a drug carrier and an antimicrobial agent, respectively. Additionally, it can also be used for the encapsulation of fertilizers (Sreenikethanam and Bajhaiya, 2022).

16.5.4 Therapeutic Uses

Macro and microalgae have a wide range of specialised metabolic pathways for producing bioactive compounds exhibiting noteworthy anti-inflammatory, antioxidant, antimicrobial, antidiabetic, and anticancer properties (Uma et al., 2023).

16.5.5 Importance of Halide Metabolites

Halogenation of metabolites elevates the inherent biological activity, rendering them valuable assets in therapeutic research. It further improves the pharmacokinetic and pharmacodynamic properties of the compounds, thereby increasing oral bioavailability. Recently, halides have gained attention because of their promising roles in optimising drugs from the hit to lead molecules due to their role in improving the drug binding affinity (Kumar et al., 2023). Both micro and macroalgae produce a diverse array of organohalides displaying substantial potential for therapeutic applications (Table 16.4).

TABLE 16.4
Therapeutic Potential of Halide Metabolites Obtained from Algae

Therapeutic Role	Halide Metabolites	Algal Source	Reference
Antibacterial	Bromophycolides	*Callophycus serratus*	Lane et al., 2009; Karpiński, 2019
	Bromoditerpenes	*Sphaerococcus coronopifolius*	Rodrigues et al., 2015; Carpena et al., 2022
	Fallachromenoic acid	*Sargassum fallax*	Rushdi et al., 2020
Antimalarial	Bromophycolides	*Callophycus serratus*	Prashar et al., 2022
	C15 acetogeninen-ynes	*Laurencia dendroidea*	Bawakid et al., 2017
Antitumor	Halomon 17	*Portieria hornemanii*	Rosa et al., 2019
	Mertensene	*Pterocladiella capillacea*	Tarhouni-Jabberi et al., 2017
Immunomodulatory	Neorogioldiol	*Laurencia glandulifera*	Pradhan et al., 2020
	Bis-benzyl ether	*Polysiphonia morrowii*	Choi et al., 2018
	Halogenated monoterpene and mertensene	*Pterocladiella capillacea*	Pradhan et al., 2020
Antidiabetic	Bromophenol	*Rhodomela confervoides*	Carpena et al., 2022
Antiviral	Acetogenins	*Laurencia mariannensis*	Palaniveloo et al., 2020
	Sesquiterpenes	*Laurencia*	Cikoš et al., 2021

16.5.6 Carotenoids and Their Bioavailability

Several studies, both *in vivo* and *in vitro*, have unveiled the therapeutic potential of carotenoids and their derivatives, encompassing cardio-protective, antioxidant, anticancer, anti-inflammatory, neuro-protective, ocular protective, hepato-protective, and immune-modulatory activities (Table 16.1). The compounds must be delivered to target tissues following ingestion to mediate these activities. The fraction of ingested carotenoids reaching the systemic circulation is known as bioavailability, which significantly affects the design and activity of functional foods and therapeutic agents. Bioaccessibility is a part of bioavailability, elucidating the fraction of ingested carotenoids released from the food matrix to the gastrointestinal lumen, followed by its solubilization in the micelles, facilitating its absorption. Numerous factors, including physical entrapment of the compounds in the matrix, chemical structure, and physicochemical properties of the carotenoid, the digestive enzymes, and the physiological conditions, affect the size and composition of the micelles, thereby affecting the extent of absorption and thus bioaccessibility. Besides, the species of the source algae also affects the bioaccessibility of the carotenoids, for example, low accessibility is exhibited by the ones derived from *Chlorella vulgaris* and *Nannochloropsis* sp., while the ones obtained from *Phaeodactylum tricornutum* show higher bioaccessibility (27–52%) (Bernaerts et al., 2020; Molteni et al., 2022). The variation in cell wall composition might be responsible for the distinct bioaccessibility of the carotenoids. Following this, the configuration of the compound significantly affects its micellarization during digestion and thus bioaccessibility. Cis-isomers appear to have a bent shape and better incorporation into the micelles, however, trans-isomers have a rigid acyclic structure, limiting their incorporation into micelles (Demarco et al., 2022). Micellarization is further affected by the lipophilicity of the compounds; the higher the lipophilicity, more is the micellarization and bioaccessibility. Owing to their less hydrophobic characteristic, xanthophylls have more bioaccessibility. Over that, several studies have taken into account the impact of cooking style on the bioaccessibility of carotenoids, for example, boiled and fried eggs having more zeaxanthin and lutein bioaccessibility than the scrambled ones.

There are several strategies to increase the bioavailability of the compounds, which involve the fabrication of efficient delivery systems, including liposome-based formulations, biopolymer-based nanoparticles. For instance, zein-propylene glycol alginate composite nanoparticles protect β-carotene against the acidic environment of the gastrointestinal tract. Emulsion-based delivery systems are suitable as vehicles to deliver carotenoids, as these enhance solubility, stability, and bioavailability of encapsulated compounds. These allow the generation of mixed micelle particles with a hydrophobic core that are easy to solubilize and are efficient in transporting lipophilic compounds (Molteni et al., 2022).

16.5.7 Beneficial Health Impact in Aquaculture

Aquaculture is one of the rapidly burgeoning sectors of the food industry, expected to witness 7.1% expansion by 2027. The production of commercial fish and other seafood predominantly relies upon fish feed. However, increasing the fish feed to meet the ever-expanding demand for aquaculture would lead to an ecosystem imbalance in the ocean. Thus, it is highly imperative to uncover environment-friendly and sustainable alternatives for feed. Algal metabolites, owing to their ability to support the growth and reproduction of certain aquatic organisms, are being exploited as a source of feedstock in aquaculture. This is both economically and ecologically beneficial. There is a growing consensus in the research community regarding the suitability of algal metabolites as beneficial dietary supplements (Chen et al., 2020). Astaxanthin has recently gained attention as a sustainable nutritional resource for the rearing of economic and ornamental fish as it promotes weight gain, boosts immunity, and disease resistance. A substantial increase in the specific growth rate of rainbow trout (a highly nutritious, widely consumed fish) was observed upon increasing the dietary supplementation of astaxanthin. Additionally, it has been reported to ameliorate phagocytic and lysozyme activity in fish and promote the synthesis of antioxidases, for example, superoxide dismutase, thereby increasing their resistance to diseases and improving the survival efficiency (Lu et al., 2021). Notably, in Albino Oscar fish infected with *Aeromonas hydrophila*, supplementation of astaxanthin significantly decreased the mortality rates. A cell wall component of *Euglena* sp., *viz*., paramylon, has displayed an immunostimulatory role in different aquatic organisms like Atlantic salmon, mussels, *etc*. (Nagappan et al., 2021). Besides, algal metabolites are utilised for imparting pigmentation to the commercial aquaculture products, for example, astaxanthin supplementation for providing pink coloured fillet in the farmed salmonid fish, and carotenoids are utilised to enhance orange colour in gonads, contributing to its attractive appearance and thus increasing consumers' demand (Lu et al., 2021)

16.5.8 Nutraceuticals

Nutraceuticals, commonly known as functional foods or medicinal foods, are a group of bioactive compounds that amalgamate nutritional value with physiological benefits. These are utilised to improve the overall health of individuals and reduce the incidence of chronic diseases. The regular involvement of nutraceuticals in the diet provides significant long-term health benefits. For millennia, macro/microalgae, renowned for their substantial nutritional worth, have been incorporated into foods in Asian countries like Japan and Korea (Tanna and Mishra, 2018; Kumari et al., 2023). Besides, the burgeoning global demand for sustainable resources for producing essential bioactive compounds with nutraceutical potential has sparked a resurgence of interest in algal-derived metabolites. High-value potential dietary supplements from algae predominantly include carotenoids, such as β-carotene, astaxanthin, fucoxanthin, and lutein, having substantial antioxidant, anti-inflammatory, and anticancer effects (Abu-Ghosh et al., 2021). While carotenoids can be derived from various resources, including plants and synthetic materials, algae stand out as the most advantageous choice due to their significantly reduced side effects and toxicity, better production efficiency, and enhanced bioactivity. As a point of case, marigold flowers, a traditional source of lutein, typically contain around 0.3mg/g of compound. In contrast, algal cells have been reported to possess concentrations as high as 4mg/g (Patel et al., 2022). Lutein supplements are recognised as effective functional foods demonstrating the ability to ameliorate cardiovascular diseases, age-related macular degeneration, and several cancers. Fucoxanthin has also been shown to have substantial anti-inflammatory and antioxidant properties in conjunction with hepato-protective, antidiabetic, and antiobesity potential. Its role in managing obesity is attributed to its ability to induce uncoupling protein 1 (UCP1) in abdominal white adipose tissue. Several obesity management products comprising of fucoxanthin are prominently marketed under the trade names fücoTHIN® and Solaray® (Wang et al., 2020). Besides, phycobiliproteins are essential algal metabolites emerging as a potential replacement for toxic and carcinogenic synthetic food colorants, coupled with an added perk of having therapeutic value. At present, the market features more than a hundred companies operating at a commercial scale for algal metabolite production, highlighting a notable expansion within this field (Figure 16.2). Cyanotech Corporation, USA (for the production of Bioastin and astaxanthin), BlueBioTech Int., Germany (phycocyanin and astaxanthin production), Algatechnologies Ltd, USA (AstaPure production) are some of the examples of such companies (Kumari et al., 2023).

16.6 CHALLENGES AND FUTURE PROSPECTS OF ALGAL SECONDARY METABOLITES

Algal compounds have shown potential antibiotic, antifungal, and antitoxic properties, but the development of useful products such as biopesticides, nutraceuticals, and pharmaceuticals using these compounds is challenging. The main constraints limiting large-scale production of the algal metabolites are high operational costs, non-standardization, and poorly developed production techniques, which are attributed to the lack of strain robustness, complex metabolic profiles, relatively slow growth rate, and low productivity. The pathways involved in metabolite production and their mechanism of action are still undiscovered (Costa et al., 2019). Genetic engineering for strain improvement gives us hope in this aspect, but gene manipulation of microalgae is still not well-studied, even though advanced gene editing tools are available (Sreenikethanam et al., 2022). Transformation is another method being explored for higher metabolite production, and the efficiency for this depends on the selection of an appropriate gene transfer method since algal cell wall composition varies significantly from species to species, and rigid cell wall makes gene deliveries difficult leading to less efficient product recovery (Sreenikethanam et al., 2022). Furthermore, the algal metabolites are prone to oxidation upon storage, which significantly impacts the quality of the derived commercial products.

Given the aforementioned challenges, it is of utmost importance to increase the biomass and metabolite

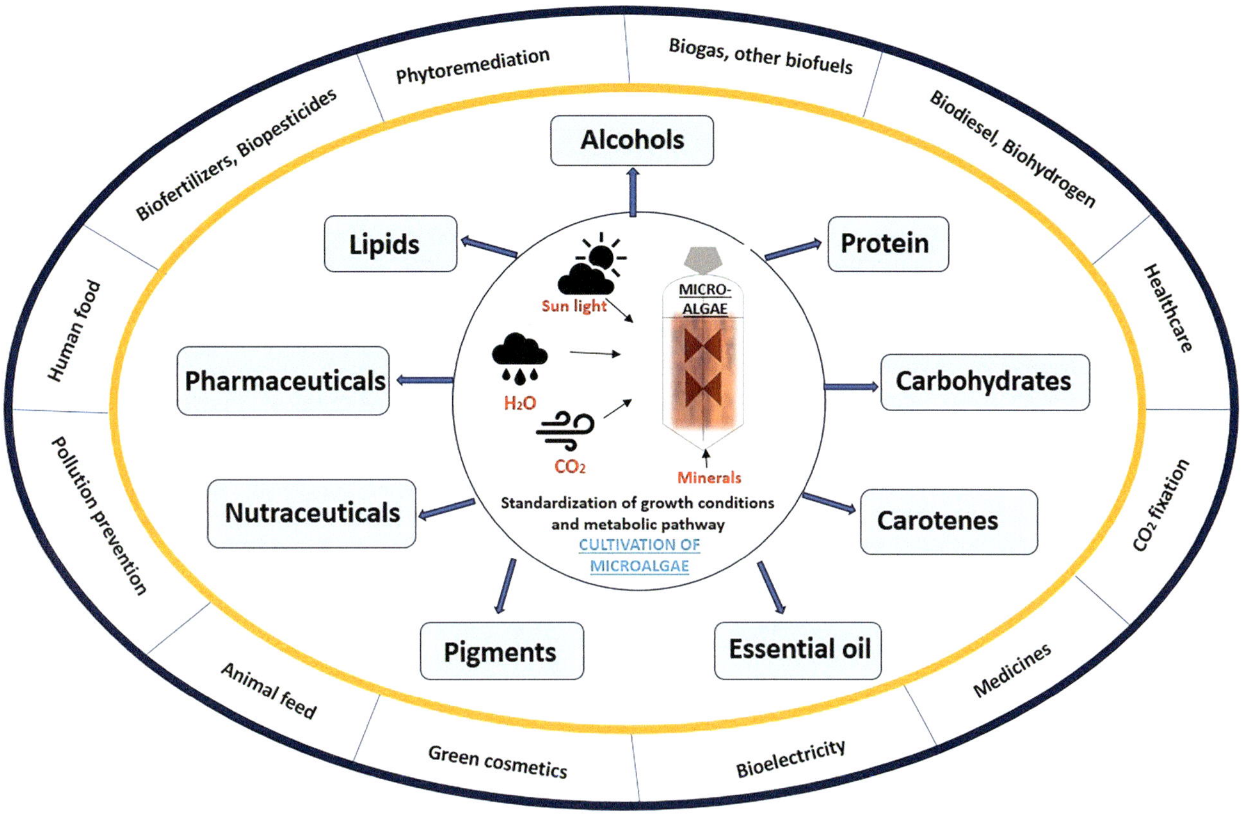

FIGURE 16.2 Large-scale cultivation of microalgae, different products obtained from them, and use of these products in various industries.

production while simultaneously reducing the production cost. Former can be achieved by optimizing the algal metabolic pathways and augmenting photosynthetic efficiency, while recycling the waste products will assist in reducing the overall cost (Lu et al., 2021). In the recent past, the importance of qualitative and quantitative evaluation of the algal metabolites has been recognised. The integrative and multidisciplinary approaches combining omics data have emerged as promising avenues for this purpose, which enable functional characterization of the metabolites and provide insights into the underpinning mechanisms (Hughes et al., 2021). The knowledge of physiological effects in conjunction with the mechanism of action will serve as the groundwork for precisely choosing the algal compounds for formulating effective pharmaceutical, agricultural, and nutraceutical products. The increase in production of metabolites by altering growth conditions and improving genetic potential is being explored as one of the promising techniques (Sreenikethanam et al., 2022). Besides, advancements in screening and extraction technologies will assist in the identification and isolation of novel metabolites with higher nutraceutical, pharmaceutical, agricultural, and cosmetic value. Following this, there should be constant efforts to expand clinical research for validating and affirming the established *in vitro* discoveries (Nowruzi et al., 2020).

16.7 CONCLUSION

Algae are a vast and diverse group of organisms, encompassing an astonishing diversity, each uniquely tailored to its environment, having a capacious range of biochemical and physiological characteristics. These are considered an important raw material for the production of vitamins, bioactive compounds, amino acids, and pigments. Algal SMs, owing to their significant benefits in the aspects of sustainability, safety control, productivity, and production area, are emerging into the limelight (Rico et al., 2017; Vidhyanandan et al., 2020).

Algal SMs exhibit a rich chemical diversity, including lactam rings, cyclic peptides, massive macrolide rings, unsaturated polyacetylene, and polyene bonds. The versatility of their applications in sectors as diverse as nutraceutical and aquaculture feed to agriculture, therapeutics, and pharmaceuticals to cosmetics is underpinned by their remarkable attributes spanning from potent anti-inflammatory, antiaging, and antioxidant roles to a rich nutritional profile. Numerous algal metabolites have been reported to be efficient against multidrug-resistant microorganisms and to mitigate the effects of chronic diseases like diabetes. Beyond this, they also demonstrate remarkable potential to absorb UV radiation, maintain plant health, and promote their growth. There are the challenges faced in establishing an algal metabolites-based market, including but not limited to complexity in assessing and understanding the ability of algae to produce target metabolites in sufficient quantities to be utilised under standard and repeatable conditions. Over that, there is an imperative demand to formulate optimized strategies for enhancing algal growth, pre-treating biomass, and isolating the target metabolite. The amalgamation of technological advancements and an interdisciplinary approach involving metabolic engineering and an omics approach will provide a better understanding of metabolic pathways of existing algal species and identify optimum growth conditions, simultaneously allowing for the systematic screening of unexamined algal species for active metabolites with better nutritional, cosmeceutical, and pharmaceutical potential.

REFERENCES

Abdel-Aziz, S.M., M.M.A. Elsoud, and A.A.H. Anise. 2017. Microbial biosynthesis: A repertory of vital natural products. In *Food biosynthesis*, eds. A.M. Grumezescu and A.M. Holban, 25–54. Elsevier Inc and Academic Press. https://doi.org/10.1016/B978-0-12-811372-1.00003-8.

Abu-Ghosh, S., Z. Dubinsky, V. Verdelho, and D. Iluz. 2021. Unconventional high-value products from microalgae: A review. *Bioresource Technology* 329 (June): 124895. https://doi.org/10.1016/j.biortech.2021.124895.

Alassali, A., I. Cybulska, G.P. Brudecki, R. Farzanah, and M.H. Thomsen. 2016. Methods for upstream extraction and chemical characterization of secondary metabolites from algae biomass. *Advanced Techniques in Biology & Medicine* 4 (February): 1000163, 1–16. https://doi.org/10.4172/2379-1764.1000163.

Alves, C., J. Silva, S. Pinteus, et al. 2018. From marine origin to therapeutics: The antitumor potential of marine algae-derived compounds. *Frontiers in Pharmacology* 9 (August): 777, 1–24. https://doi.org/10.3389/fphar.2018.00777.

Andrade, K.A.M., C. Lauritano, G. Romano, and A. Ianora. 2018. Marine microalgae with anti-cancer properties. *Marine Drugs* 16 (May): 165, 1–17. https://doi.org/10.3390/md16050165.

Arora, K., P. Kumar, D. Bose, X. Li, and S. Kulshrestha. 2021. Potential applications of algae in biochemical and bioenergy sector. *3 Biotech* 11 (May): 296. https://doi.org/10.1007/s13205-021-02825-5.

Arora, Y., S. Sharma, and V. Sharma. 2023. Microalgae in bioplastic production: A comprehensive review. *Arabian Journal for Science and Engineering* 48: 7225–41. https://doi.org/10.1007/s13369-023-07871-0.

Asimakis, E., A.A. Shehata, W. Eisenreich, et al. 2022. Algae and their metabolites as potential bio-pesticides. *Microorganisms* 10 (January): 307, 1–32. https://doi.org/10.3390/microorganisms10020307.

Babich, O., S. Sukhikh, V. Larina, et al. 2022. Algae: Study of edible and biologically active fractions, their properties and applications. *Plants* 11 (March): 780, 1–30. https://doi.org/10.3390/plants11060780.

Balskus, E.P., R.J. Case, and C.T. Walsh. 2011. The biosynthesis of cyanobacterial sunscreen scytonemin in intertidal microbial mat communities. *FEMS Microbiology Ecology* 77: 322–32. https://doi.org/10.1111/j.1574-6941.2011.01113.x.

Barbalace, M.C., M. Malaguti, L. Giusti, A. Lucacchini, S. Hrelia, and C. Angeloni. 2019. Anti-inflammatory activities of marine algae in neurodegenerative diseases. *International Journal of Molecular Sciences* 20 (June): 3061, 1–20. https://doi.org/10.3390/ijms20123061.

Barzkar, N., S.T. Jahromi, H.B. Poorsaheli, and F. Vianello. 2019. Metabolites from marine microorganisms, micro, and macroalgae: Immense scope for pharmacology. *Marine Drugs* 17 (August): 464, 1–29. https://doi.org/10.3390/md17080464.

Bawakid, N.O., W.M. Alarif, N.A. Alburae, et al. 2017. Isolaurenidificin and bromlaurenidificin, two new C15-acetogenins from the red alga *Laurencia obtusa*. *Molecules* 22 (May): 807, 1–8. https://doi.org/10.3390/molecules22050807.

Bernaerts, T.M.M., H. Verstreken, C. Dejonghe, et al. 2020. Cell disruption of *Nannochloropsis* sp. improves *in vitro* bioaccessibility of carotenoids and ω3-LC-PUFA. *Journal of Functional Foods* 65 (January): 103770, 1–10. https://doi.org/10.1016/j.jff.2019.103770.

Besednova, N.N., B.G. Andryukov, T.S. Zaporozhets, et al. 2020. Algae polyphenolic compounds and modern antibacterial strategies: Current achievements and immediate prospects. *Biomedicines* 8 (September): 342, 1–19. https://doi.org/10.3390/biomedicines8090342.

Bhowmick, S., A. Mazumdar, A. Moulick, and V. Adam. 2020. Algal metabolites: An inevitable substitute for antibiotics. *Biotechnology Advances* 43 (November): 107571. https://doi.org/10.1016/j.biotechadv.2020.107571.

Bule, M.H., I. Ahmed, F. Maqbool, M. Bilal, and H.M.N. Iqbal. 2018. Microalgae as a source of high-value bioactive compounds. *Frontiers in Bioscience-Scholar* 10: 197–216. https://doi.org/10.2741/s509.

Calder, P.C. 2012. Mechanisms of action of (n-3) fatty acids. *The Journal of Nutrition* 142: 592S–9S. https://doi.org/10.3945/jn.111.155259.

Carpena, M., P. Garcia-Perez, P. Garcia-Oliveira, et al. 2022. Biological properties and potential of compounds extracted from red seaweeds. *Phytochemistry Reviews* 22: 1509–40. https://doi.org/10.1007/s11101-022-09826-z.

Chakdar, H., and S. Pabbi. 2017. Algal pigments for human health and cosmeceuticals. In *Algal green chemistry*, eds. R.P. Rastogi, D. Madamwar, and A. Pandey, 171–88. Amsterdam: Elsevier. https://doi.org/10.1016/B978-0-444-63784-0.00009-6.

Chen, F., Y. Xiao, X. Wu, Y. Zhong, Q. Lu, and W. Zhou. 2020. Replacement of feed by fresh microalgae as a novel technology to alleviate water deterioration in aquaculture. *RSC Advances* 10: 20794–800. https://doi.org/10.1039/D0RA03090B.

Choi Y.K., B-R. Ye, E-A. Kim, et al. 2018. Bis (3-bromo-4,5-dihydroxybenzyl) ether, a novel bromophenol from the marine red alga *Polysiphonia morrowii* that suppresses LPS-induced inflammatory response by inhibiting ROS-mediated ERK signaling pathway in RAW 264.7 macrophages. *Biomedicine & Pharmacotherapy* 103: 1170–77. https://doi.org/10.1016/j.biopha.2018.04.121.

Chong, J.W.R., X. Tan, K.S. Khoo, et al. 2022. Microalgae-based bioplastics: Future solution towards mitigation of plastic wastes. *Environmental Research* 206 (April): 112620. https://doi.org/10.1016/j.envres.2021.112620.

Cikoš, A.-M., M. Jurin, R. Čož-Rakovac, D. Gašo-Sokač, S. Jokić, and I. Jerković. 2021. Update on sesquiterpenes from red macroalgae of the *Laurencia* genus and their biological activities (2015–2020). *Algal Research* 56 (June): 102330. https://doi.org/10.1016/j.algal.2021.102330.

Costa, J.A.V., B.C.B. Freitas, C.G. Cruz, J. Silveira, and M.G. Morais. 2019. Potential of microalgae as biopesticides to contribute to sustainable agriculture and environmental development. *Journal of Environmental Science and Health, Part B* 54: 366–75. https://doi.org/10.1080/03601234.2019.1571366.

Del Mondo, A., A. Smerilli, L. Ambrosino, et al. 2021. Insights into phenolic compounds from microalgae: Structural variety and complex beneficial activities from health to nutraceutics. *Critical Reviews in Biotechnology* 41: 155–71. https://doi.org/10.1080/07388551.2021.1874284.

Demarco, M., J.O. de Moraes, Â.P. Matos, R.B. Derner, F.D.F. Neves, and G. Tribuzi. 2022. Digestibility, bioaccessibility and bioactivity of compounds from algae. *Trends in Food Science & Technology* 121: 114–28. https://doi.org/10.1016/j.tifs.2022.02.004.

Din, N.A.S., A.S.M. Alayudin, N.S. Sofian-Seng, et al. 2022. Brown algae as functional food source of fucoxanthin: A review. *Foods* 11 (July): 2235, 1–35. https://doi.org/10.3390/foods11152235.

Dmytryk, A., and K. Chojnacka. 2018. Algae as fertilizers, biostimulants, and regulators of plant growth. In *Algae biomass: Characteristics and applications*, eds. K. Chojnacka, P. Wieczorek, G. Schroeder, and I. Michalak, 115–22. Cham: Springer. https://doi.org/10.1007/978-3-319-74703-3_10.

Eddouks, M., A. Bidi, B.E.L. Bouhali, and N.A. Zeggwagh. 2014. Insulin resistance as a target of some plant-derived phytocompounds. In *Studies in natural products chemistry*, ed. Atta-ur-Rahman, 351–73. Elsevier. https://doi.org/10.1016/B978-0-444-63430-6.00011-4.

El-Beltagi, H.S., H.I. Mohamed, and M.M.A. El-Enain. 2019. Role of secondary metabolites from seaweeds in the context of plant development and crop production. In *Seaweeds as plant fertilizer, agricultural biostimulants and animal fodder*, eds. L. Pereira, K. Bahcevandziev, and N.H. Joshi, 64–79. Boca Raton: CRC Press. https://doi.org/10.1201/9780429487156.

Fagundes, M.B., and R. Wagner. 2021. Sterols biosynthesis in algae. In *Bioactive compounds – Biosynthesis, characterization and applications*, eds. L.Q. Zepka, T.C. do Nascimento, and E. Jacob-Lopes, 1–12. IntechOpen. https://doi.org/10.5772/intechopen.96719.

Ferdous, U.T., and Z.N.B. Yusof. 2021. Insight into potential anticancer activity of algal flavonoids: Current status and challenges. *Molecules* 26 (November): 6844, 1–12. https://doi.org/10.3390/molecules26226844.

Fernando, I.P.S., W. Lee, and G. Ahn. 2022. Marine algal flavonoids and phlorotannins; An intriguing frontier of biofunctional secondary metabolites. *Critical Reviews in Biotechnology* 42: 23–45. https://doi.org/10.1080/07388551.2021.1922351.

Ferreira, A., C.R. Bastos, C. Marques-dos-Santos, F.G. Acién-Fernandez, and L. Gouveia. 2023. Algae culture for agriculture: From past to future. *Frontiers in Agronomy* 5 (June): 1064041, 1–26. https://doi.org/10.3389/fagro.2023.1064041.

Fidor, A., R. Konkel, and H. Mazur-Marzec. 2019. Bioactive peptides produced by Cyanobacteria of the genus *Nostoc*: A review. *Marine Drugs* 17 (September): 561, 1–16. https://doi.org/10.3390/md17100561.

González-Pérez, B.K., A.M. Rivas-Castillo, A. Valdez-Calderón, and M.A. Gayosso-Morales. 2022. Microalgae as biostimulants: A new approach in agriculture. *World Journal of Microbiology and Biotechnology* 38 (November): 4. https://doi.org/10.1007/s11274-021-03192-2.

Gupta, A., S. Jain, and N. Verma. 2022. *Industrial applications of soil microbes*, vol. 1. Bentham Science Publishers Pte Ltd. https://doi.org/10.2174/97898150399551220101.

Gupta, A., S. Jain, and N. Verma. 2024. *Industrial applications of soil microbes*, vol. 3. Bentham Science Publishers Pte Ltd. https://doi.org/10.2174/97898150797531240301.

Gupta, A.K., K. Seth, K. Maheshwari, et al. 2021. Biosynthesis and extraction of high-value carotenoid from algae. *Frontiers in Bioscience (Landmark edition)* 26: 171–90. https://doi.org/10.52586/4932.

Haskins, A.H., D.J. Buglewicz, H. Hirakawa, A. Fujimori, Y. Aizawa, and T.A. Kato. 2018. Palmitoyl ascorbic acid 2-glucoside has the potential to protect mammalian cells from high-LET carbon-ion radiation. *Scientific Reports* 8 (September): 13822, 1–7. https://doi.org/10.1038/s41598-018-31747-1.

He, X., J. Dai, and Q. Wu. 2016. Identification of sporopollenin as the outer layer of cell wall in microalga *Chlorella protothecoides*. *Frontiers in Microbiology* 7 (June): 1047, 1–11. https://doi.org/10.3389/fmicb.2016.01047.

Heber, D., and Q.-Y. Lu. 2002. Overview of mechanisms of action of lycopene. *Experimental Biology and Medicine* 227: 920–23. https://doi.org/10.1177/153537020222701013.

Hughes, A.H., F. Magot, A.F. Tawfike, et al. 2021. Exploring the chemical space of macro- and micro-algae using comparative metabolomics. *Microorganisms* 9 (February): 311, 1–16. https://doi.org/10.3390/microorganisms9020311.

Jain, S., A. Gupta, and N. Verma. 2023. *Industrial applications of soil microbes*, vol. 2. Bentham Science Publishers Pte Ltd. https://doi.org/10.2174/97898150502641230201.

Jain, S., A. Gupta, and N. Verma. 2024. *Industrial applications of soil microbes*, vol. 4. Bentham Science Publishers Pte Ltd. https://doi.org/10.2174/97898151249961240401.

Jeong, Y., S.-H. Cho, H. Lee, et al. 2020. Current status and future strategies to increase secondary metabolite production from Cyanobacteria. *Microorganisms* 8 (November): 1849, 1–24. https://doi.org/10.3390/microorganisms8121849.

Jiang, L., Y. Wang, Q. Yin, et al. 2017. Phycocyanin: A potential drug for cancer treatment. *Journal of Cancer* 8: 3416–29. https://doi.org/10.7150/jca.21058.

Kamiya, M., T. Nishio, A. Yokoyama, et al. 2010. Seasonal variation of phlorotannin in sargassacean species from the coast of the sea of Japan. *Phycological Research* 58: 53–61. https://doi.org/10.1111/j.1440-1835.2009.00558.x.

Karpiński, T.M. 2019. Marine macrolides with antibacterial and/or antifungal activity. *Marine Drugs* 17 (April): 241, 1–25. https://doi.org/10.3390/md17040241.

Khan, F., G.-J. Jeong, M.S.A. Khan, N. Tabassum, and Y.-M. Kim. 2022. Seaweed-derived phlorotannins: A review of multiple biological roles and action mechanisms. *Marine Drugs* 20 (June): 384, 1–31. https://doi.org/10.3390/md20060384.

Kim, E.-A., E.-J. Han, J. Kim, et al. 2022. Anti-allergic effect of 3,4-dihydroxybenzaldehyde isolated from *Polysiphonia morrowii* in IgE/BSA-stimulated mast cells and a passive cutaneous anaphylaxis mouse model. *Marine Drugs* 20 (February): 133, 1–13. https://doi.org/10.3390/md20020133.

Kolackova, M., A. Janova, M. Dobesova, et al. 2023. Role of secondary metabolites in distressed microalgae. *Environmental Research* 224 (May): 115392. https://doi.org/10.1016/j.envres.2023.115392.

Kumar, S., G. Choudhir, and A. Singh. 2023. Halide metabolites from the algal biomass: A promising therapeutic for targeting rare diseases. In *Algal metabolites*, eds. J. Sangeetha, D. Thangadurai, S. Islam, and R. Hospet, 30. New York: Apple Academic Press. https://doi.org/10.1201/9781003346081.

Kumari, A., Garima, and N. Bharadvaja. 2023. A comprehensive review on algal nutraceuticals as prospective therapeutic agent for different diseases. *3 Biotech* 13 (January): 44. https://doi.org/10.1007/s13205-022-03454-2.

Lane, A.L., E.P. Stout, A.-S. Lin, et al. 2009. Antimalarial bromophycolides J–Q from the Fijian red alga *Callophycus serratus*. *The Journal of Organic Chemistry* 74: 2736–42. https://doi.org/10.1021/jo900008w.

Lawrence, K.P., P.F. Long, and A.R. Young. 2019. Mycosporine-like amino acids for skin photoprotection. *Current Medicinal Chemistry* 25: 5512–27. https://doi.org/10.2174/0929867324666170529124237.

Lefranc, F., A. Koutsaviti, E. Ioannou, et al. 2019. Algae metabolites: From *in vitro* growth inhibitory effects to promising anticancer activity. *Natural Product Reports* 36: 810–41. https://doi.org/10.1039/c8np00057c.

Leitner, P.D., T. Jakschitz, R. Gstir, et al. 2022. Anti-inflammatory extract from soil algae *Chromochloris zofingiensis* targeting TNFR/NF-κB signaling at different levels. *Cells* 11 (April): 1407, 1–21. https://doi.org/10.3390/cells11091407.

Li, X., H.-Y. Hu, and Y. Zhang. 2010. Growth and lipid accumulation properties of a freshwater microalga *Scenedesmus* sp. under different cultivation temperature. *Bioresource Technology* 102: 3098–102. https://doi.org/10.1016/j.biortech.2010.10.055.

Little, S.M., G.N. Senhorinho, M. Saleh, et al. 2021. Antibacterial compounds in green microalgae from extreme environments: A review. *Algae* 36: 61–72. https://doi.org/10.4490/algae.2021.36.3.6.

Liu, M., W. Li, Y. Chen, X. Wan, and J. Wang. 2020. Fucoxanthin: A promising compound for human inflammation-related diseases. *Life sciences* 255 (August): 117850. https://doi.org/10.1016/j.lfs.2020.117850.

Lopes, G., D. Clarinha, and V. Vasconcelos. 2020. Carotenoids from cyanobacteria: A biotechnological approach for the topical treatment of psoriasis. *Microorganisms* 8 (February): 302, 1–19. https://doi.org/10.3390/microorganisms8020302.

Lu, Q., H. Li, Y. Zou, H. Liu, and L. Yang. 2021. Astaxanthin as a microalgal metabolite for aquaculture: A review on the synthetic mechanisms, production techniques, and practical application. *Algal Research* 54 (April): 102178. https://doi.org/10.1016/j.algal.2020.102178.

Mago, Y., Y. Sharma, Y. Thakran, A. Mishra, S. Tewari, and N. Kataria. 2023. Next-generation organic beauty products obtained from algal secondary metabolites: A sustainable development in cosmeceutical industries. *Molecular Biotechnology* (August). https://doi.org/10.1007/s12033-023-00841-9.

Mansur, A.A., M.T. Brown, and R.A. Billington. 2020. The cytotoxic activity of extracts of the brown alga *Cystoseira tamariscifolia* (Hudson) Papenfuss, against cancer cell lines changes seasonally. *Journal of Applied Phycology* 32: 2419–29. https://doi.org/10.1007/s10811-019-02016-z.

Márquez-Rocha, F.J., D. Palma-Ramírez, P. García-Alamilla, J.F. López-Hernández, I.S. Santiago-Morales, and A.I. Flores-Vela. 2019. Microalgae cultivation for secondary metabolite production. In *Microalgae – From physiology to application*, ed. M. Vítová, 1–17. IntechOpen. https://doi.org/10.5772/intechopen.88531.

Mekinić, I.G., D. Skroza, V. Šimat, I. Hamed, M. Čagalj, and Z.P. Perković. 2019. Phenolic content of brown algae (Pheophyceae) species: Extraction, identification, and quantification. *Biomolecules* 9 (June): 244, 1–25. https://doi.org/10.3390/biom9060244.

Mekinić, I.G., V. Šimat, N.B. Rathod, I. Hamed, and M. Čagalj. 2023. Algal carotenoids: Chemistry, sources, and application. *Foods* 12 (July): 2768, 1–30. https://doi.org/10.3390/foods12142768.

Mohiuddin, A.K. 2019. Chemistry of secondary metabolites. *Annals of Clinical Toxicology* 2 (May): 1014, 1–22. https://www.remedypublications.com/open-access/chemistry-of-secondary-metabolites–1124.pdf.

Molteni, C., C.L. Motta, and F. Valoppi. 2022. Improving the bioaccessibility and bioavailability of carotenoids by means of nanostructured delivery systems: A comprehensive review. *Antioxidants* 11 (September): 1931, 1–31. https://doi.org/10.3390/antiox11101931.

Mourelle, M.L., C.P. Gómez, and J.L. Legido. 2017. The potential use of marine microalgae and cyanobacteria in cosmetics and thalassotherapy. *Cosmetics* 4 (November): 46, 1–14. https://doi.org/10.3390/cosmetics4040046.

Mrid, R.B., B. Benmrid, J. Hafsa, H. Boukcim, M. Sobeh, and A. Yasri. 2021. Secondary metabolites as biostimulant and bioprotectant agents: A review. *Science of The Total Environment* 777 (July): 146204. https://doi.org/10.1016/j.scitotenv.2021.146204.

Munira, S., A. Farenhorst, K. Sapkota, D. Nilsson, and C. Sheedy. 2018. Auxin herbicides and pesticide mixtures in groundwater of a Canadian Prairie province. *Journal of Environmental Quality* 47: 1462–67. https://doi.org/10.2134/jeq2018.05.0202.

Murillo, A.G., S. Hu, and M.L. Fernandez. 2019. Zeaxanthin: Metabolism, properties, and antioxidant protection of eyes, heart, liver, and skin. *Antioxidants* 8 (September): 390, 1–18. https://doi.org/10.3390/antiox8090390.

Nabti, E., B. Jha, and A. Hartmann. 2017. Impact of seaweeds on agricultural crop production as biofertilizer. *International Journal of Environmental Science and Technology* 14: 1119–34. https://doi.org/10.1007/s13762-016-1202-1.

Nagappan, S., P. Das, M. Abdul Quadir, et al. 2021. Potential of microalgae as a sustainable feed ingredient for aquaculture. *Journal of Biotechnology* 341: 1–20. https://doi.org/10.1016/j.jbiotec.2021.09.003.

Nowruzi, B., G. Sarvari, and S. Blanco. 2020. The cosmetic application of cyanobacterial secondary metabolites. *Algal Research* 49 (August): 101959. https://doi.org/10.1016/j.algal.2020.101959.

Nuzzo, G., G. Senese, C. Gallo, et al. 2022. Antitumor potential of immunomodulatory natural products. *Marine Drugs* 20 (June): 386, 1–12. https://doi.org/10.3390/md20060386.

Palaniveloo, K., M. Rizman-Idid, T. Nagappan, and S.A. Razak. 2020. Halogenated metabolites from the diet of *Aplysia dactylomela* Rang. *Molecules* 25 (February): 815, 1–18. https://doi.org/10.3390/molecules25040815.

Pan, S., J. Jeevanandam, and M.K. Danquah. 2019. Benefits of algal extracts in sustainable agriculture. In *Grand challenges in algae biotechnology*, eds. A. Hallmann and P.H. Rampelotto, 501–34. Cham: Springer. https://doi.org/10.1007/978-3-030-25233-5_14.

Park, Y.H., S.-I. Han, B. Oh, et al. 2022. Microalgal secondary metabolite productions as a component of biorefinery: A review. *Bioresource Technology* 344 (January): 126206. https://doi.org/10.1016/j.biortech.2021.126206.

Patel, A., U. Rova, P. Christakopoulos, and L. Matsakas. 2022. Microalgal lutein biosynthesis: Recent trends and challenges to enhance the lutein content in microalgal cell factories. *Frontiers in Marine Science* 9 (October): 1015419, 1–9. https://doi.org/10.3389/fmars.2022.1015419.

Pattanayak, S., S. Das, and G. Biswal. 2021. Crop disease control through application of algal biopesticides: Alternate route map in organic farming. In *Biopesticides in organic farming*, ed. L.P. Awasthi, 123–26. Boca Raton: CRC Press. https://doi.org/10.1201/9781003027690.

Pereira, C.P.M., A.C.R. Souza, A.R. Vasconcelos, P.S. Prado, and J.J. Name. 2021. Antioxidant and anti-inflammatory mechanisms of action of astaxanthin in cardiovascular diseases (review). *International Journal of Molecular Medicine* 47: 37–48. https://doi.org/10.3892/ijmm.2020.4783.

Potijun, S., S. Jaingam, N. Sanevas, S. Vajrodaya, and A. Sirikhachornkit. 2021. Green microalgae strain improvement for the production of sterols and squalene. *Plants* 10 (August): 1673, 1–13. https://doi.org/10.3390/plants10081673.

Pradhan, B., R. Nayak, S. Patra, B.P. Jit, A. Ragusa, and M. Jena. 2020. Bioactive metabolites from marine algae as potent pharmacophores against oxidative stress-associated human diseases: A comprehensive review. *Molecules* 26 (December): 37, 1–25. https://doi.org/10.3390/molecules26010037.

Prashar, C., N. Thakur, S. Chakraborti, S.S.A. Hussain, K. Vashisht, and K.C. Pandey. 2022. The landscape of nature-derived antimalarials-potential of marine natural products in countering the evolving *Plasmodium*. *Frontiers in Drug Discovery* 2 (December): 1065231, 1–15. https://doi.org/10.3389/fddsv.2022.1065231.

Purves, K., L. Macintyre, D. Brennan, et al. 2016. Using molecular networking for microbial secondary metabolite bioprospecting. *Metabolites* 6 (January): 2, 1–18. https://doi.org/10.3390/metabo6010002.

Raj, S., A.M. Kuniyil, A. Sreenikethanam, P. Gugulothu, R.B. Jeyakumar, and A.K. Bajhaiya. 2021. Microalgae as a source of mycosporine-like amino acids (MAAs); Advances and future prospects. *International Journal of Environmental Research and Public Health* 18 (November): 12402, 1–15. https://doi.org/10.3390/ijerph182312402.

Rico, M., A.G. González, J. Santana-Casiano, M. Gonzalez-Davila, N. Pérez, and M.S. de Tangil. 2017. Production of primary and secondary metabolites using algae. In *Prospects and challenges in algal biotechnology*, eds. B.N. Tripathi and D. Kumar, 311–26. Singapore: Springer. https://doi.org/10.1007/978-981-10-1950-0_12.

Rispail, N., R. Nash, and K.J. Webb. 2005. Secondary metabolite profiling. In *Lotus japonicus handbook*, ed. A.J. Márquez, 341–48. Dordrecht: Springer. https://doi.org/10.1007/1-4020-3735-X_33.

Rodrigues, D., C. Alves, A. Horta, et al. 2015. Antitumor and antimicrobial potential of bromoditerpenes isolated from the red alga, *Sphaerococcus Coronopifolius*. *Marine Drugs* 13: 713–26. https://doi.org/10.3390/md13020713.

Rojas, V., L. Rivas, C. Cárdenas, and F. Guzmán. 2020. Cyanobacteria and eukaryotic microalgae as emerging sources of antibacterial peptides. *Molecules* 25 (December): 5804, 1–27. https://doi.org/10.3390/molecules25245804.

Rosa, G.P., W.R. Tavares, P.M.C. Sousa, A.K. Pagès, A.M.L. Seca, and D.C.G.A. Pinto. 2019. Seaweed secondary metabolites with beneficial health effects: An overview of successes in *in vivo* studies and clinical trials. *Marine Drugs* 18 (December): 8, 1–35. https://doi.org/10.3390/md18010008.

Rushdi, M.I., I.A.M. Abdel-Rahman, H. Saber, et al. 2020. Pharmacological and natural products diversity of the brown algae genus *Sargassum*. *RSC Advances* 10: 24951–72. https://doi.org/10.1039/D0RA03576A.

Sanjeewa, K.K.A., J-S. Lee, W-S. Kim, and Y-J. Jeon. 2017. The potential of brown-algae polysaccharides for the development of anticancer agents: An update on anticancer effects reported for fucoidan and laminaran. *Carbohydrate Polymers* 177: 451–59. https://doi.org/10.1016/j.carbpol.2017.09.005.

Shaima, A.F., N.H.M. Yasin, N. Ibrahim, M.S. Takriff, D. Gunasekaran, and M.Y.Y. Ismaeel. 2022. Unveiling antimicrobial activity of microalgae *Chlorella sorokiniana* (UKM2), *Chlorella* sp. (UKM8) and *Scenedesmus* sp. (UKM9). *Saudi Journal of Biological Sciences* 29: 1043–52. https://doi.org/10.1016/j.sjbs.2021.09.069.

Sharma, R., A.S. Mondal, and N. Trivedi. 2023. Anticancer potential of algae-derived metabolites: Recent updates and breakthroughs. *Future Journal of Pharmaceutical Sciences* 9 (May): 44, 1–44. https://doi.org/10.1186/s43094-023-00492-2.

Singh, S., D.K. Verma, M. Thakur, et al. 2021. Supercritical fluid extraction (SCFE) as green extraction technology for high-value metabolites of algae, its potential trends in food and human health. *Food Research International* 150 (December): 110746. https://doi.org/10.1016/j.foodres.2021.110746.

Souza, C.R.M., Bezerra, W.P., and Souto, J.T. 2020. Marine alkaloids with anti-inflammatory activity: Current knowledge and future perspectives. *Marine drugs* 18 (March): 147, 1–17. https://doi.org/10.3390/md18030147.

Sreenikethanam, A., and A. Bajhaiya. 2022. Algae based bio-plastics: Future of green economy. In *Biorefineries – selected processes*, ed. K. Biernat, 1–19. IntechOpen. https://doi.org/10.5772/intechopen.100981.

Sreenikethanam, A., S. Raj, R.J. Banu, P. Gugulothu, and A.K. Bajhaiya. 2022. Genetic engineering of microalgae for secondary metabolite production: Recent developments, challenges, and future prospects. *Frontiers in Bioengineering and Biotechnology* 10 (March): 836056, 1–14. https://doi.org/10.3389/fbioe.2022.836056.

Stoyneva-Gärtner, M., B. Uzunov, and G. Gärtner. 2020. Enigmatic microalgae from aeroterrestrial and extreme habitats in cosmetics: The potential of the untapped natural sources. *Cosmetics* 7 (April): 27, 1–22. https://doi.org/10.3390/cosmetics7020027.

Tampucci, S., G. Tofani, P. Chetoni, et al. 2022. Sporopollenin microcapsule: Sunscreen delivery system with photoprotective properties. *Pharmaceutics* 14 (September): 2041, 1–19. https://doi.org/10.3390/pharmaceutics14102041.

Tan, P.X., K. Thiyagarasaiyar, C-Y. Tan, et al. 2021. Algae-derived anti-inflammatory compounds against particulate matters-induced respiratory diseases: A systematic review. *Marine Drugs* 19 (May): 317, 1–15. https://doi.org/10.3390/md19060317.

Tanna, B., and A. Mishra. 2018. Metabolites unravel nutraceutical potential of edible seaweeds: An emerging source of functional food. *Comprehensive Reviews in Food Science and Food Safety* 17: 1613–24. https://doi.org/10.1111/1541-4337.12396.

Tarhouni-Jabberi, S., O. Zakraoui, E. Ioannou, et al. 2017. Mertensene, a halogenated monoterpene, induces G2/M cell cycle arrest and caspase dependent apoptosis of human colon adenocarcinoma HT29 cell line through the modulation of ERK-1/-2, AKT and NF-κB signaling. *Marine Drugs* 15 (July): 221, 1–14. https://doi.org/10.3390/md15070221.

Teixeira, J., A. Gaspar, E.M. Garrido, J. Garrido, and F. Borges. 2013. Hydroxycinnamic acid antioxidants: An electrochemical overview. *BioMed Research International* 2013 (June): 251754, 1–11. https://doi.org/10.1155/2013/251754.

Thiyagarasaiyar, K., B.H. Goh, Y.J. Jeon, and Y.Y. Yow. 2020. Algae metabolites in cosmeceutical: An overview of current applications and challenges. *Marine Drugs* 18 (June): 323, 1–33. https://doi.org/10.3390/md18060323.

Uma, V.S., Z. Usmani, M. Sharma, et al. 2023. Valorisation of algal biomass to value-added metabolites: Emerging trends and opportunities. *Phytochemistry Reviews* 22: 1015–40. https://doi.org/10.1007/s11101-022-09805-4.

Vaidya, P., and J.R. Stinchcombe. 2020. The potential for genotype-by-environment interactions to maintain genetic variation in a model legume-rhizobia mutualism. *Plant Communications* 1 (November): 100114, 1–12. https://doi.org/10.1016/j.xplc.2020.100114.

Vidhyanandan, L.M., S.M. Kumar, and S.T. Sukumaran. 2020. Algal metabolites and phyco-medicine. In *Plant metabolites: Methods, applications and prospects*, eds. S.T. Sukumaran, S. Sugathan, and S. Abdulhameed, 291–316. Singapore: Springer. https://doi.org/10.1007/978-981-15-5136-9_13.

Wang, C., X. Chen, Y. Nakamura, C. Yu, and H. Qi. 2020. Fucoxanthin activities motivate its nano/micro-encapsulation for food or nutraceutical application: A review. *Food & Function* 11: 9338–58. https://doi.org/10.1039/D0FO02176H.

Wang, J., X. Hu, J. Chen, T. Wang, X. Huang, and G. Chen. 2022. The extraction of β-carotene from microalgae for testing their health benefits. *Foods* 11 (February): 502, 1–19. https://doi.org/10.3390/foods11040502.

Wei, G., Q. Jia, X. Chen, et al. 2019. Terpene biosynthesis in red algae is catalyzed by microbial type but not typical plant terpene synthases. *Plant Physiology* 179: 382–90. https://doi.org/10.1104/pp.18.01413.

Win, T.T., G.D. Barone, F. Secundo, and P. Fu. 2018. Algal biofertilizers and plant growth stimulants for sustainable agriculture. *Industrial Biotechnology* 14: 203–11. https://doi.org/10.1089/ind.2018.0010.

Woodson, W.R. 2018. Pollination signals and flower senescence. In *Annual plant reviews online*, eds. S.D. O'Neill and J.A. Roberts, 294–11. John Wiley & Sons, Inc. https://doi.org/10.1002/9781119312994.apr0055.

Yang, Y., S.H.A. Hassan, M.K. Awasthi, et al. 2023. The recent progress on the bioactive compounds from algal biomass for human health applications. *Food Bioscience* 51 (February): 102267. https://doi.org/10.1016/j.fbio.2022.102267.

Zasada, M., and E. Budzisz. 2019. Retinoids: Active molecules influencing skin structure formation in cosmetic and dermatological treatments. *Advances in Dermatology and Allergology* 36: 392–97. https://doi.org/10.5114/ada.2019.87443.

17 Unlocking the Potential of Algal Secondary Metabolites

Renu Sharma
(Department of Botany, University of Rajasthan, Jaipur, India; renu.chotia88@gmail.com)

**Tara Maheshwari*
(Department of Botany, University of Rajasthan, Jaipur, India; taramaheshwari98@gmail.com)

Apexa Pareek
(Department of Botany, University of Rajasthan, Jaipur, India; pareekapeksha34@gmail.com)

Payal Lodha
(Department of Botany, University of Rajasthan, Jaipur, India; payallodha10@yahooo.in)

17.1 INTRODUCTION

The term "algae" refers to a large diversity of photosynthetic microbial organisms that are unrelated phylogenetic entities and have a range of variation from picoplankton cells to macroalgal kelps. Predefined taxonomic and chemical diversity between and within algal groups and its implications for commercial application are mentioned in detail by Stengel et al. (2011).

As a result of advancements in the chemical characterization and analytical techniques of new compounds as well as the capability to screen for various bioactivities in high-throughput systems, recent research has been focused on a variety of novel uses of algae in a wide range of industrial sectors such as fuel, agriculture, health care, cosmetics, and pharmaceuticals. Additionally, their distinct traits reflect similar metabolic processes that might present chances for acquiring rare compounds (Wilkie et al., 2011).

The unique metabolic reactions that occur in algae and produce diverse secondary metabolites such as caretenoids, polyphenols, polysaccharides, peptides, proteins, alkaloids, terpenoids, polyunsaturated fatty acids, mycosporine-like amino acids, and halogenated derivatives of polyphenols, *etc.* include light harvesting mechanisms, acquisition of carbon and the role of nitrogen (N) and sulfur (S) assimilation, growth in extreme environmental conditions such as nutrient limitation and exposure to extremes of visible and UV light (Beardall and Raven, 2012), and stress conditions that elevate some metabolites for growth and recovery. Their major ecological role is to scavenge reactive oxygen species (ROS), including free radicals generated during exposure to UV radiation. Additionally, their toxicological properties (toxins) protect the algae from predators (Connan et al., 2006). These secondary metabolites give a spectrum of functionality and a wide range of bioactivity, such as antioxidant, anti-inflammatory, anticoagulant, immunomodulatory, anticancer, antidiabetic, antihypertensive, *etc.* (Wijesinghe and Jeon, 2011). There is a need for understanding the determining concentration of bioactive compounds in wild and cultivated algae since bioactivity is particularly extract-specific and may vary based on location, growth conditions like temperature, pH, salinity, and seasonality, even within species range.

With a focus on bioactive secondary metabolites from micro and macroalgae and special attention to the polyphenolic compounds with their methods of extraction and factors affecting the growth and production of secondary metabolites, we will discuss here the involvement of these compounds in various industrial applications and their untapped research potential.

17.2 TYPES OF ALGAL SECONDARY METABOLITES

Microalgae are photon-dependent cell factories that convert carbon-containing organic compounds and CO_2 to a wide range of substrates such as lipids, carbohydrates, proteins, vitamins, pigments, phycocolloids, sterols, waxes, ketones, hydrocarbons, carotenoids, lectins, mycosporine, polyketides, *etc.* Secondary metabolites are diverse chemical molecules that are not essential to growth and development but act as defense molecules and protect the organism in unfavorable conditions (Jain et al., 2023, 2024). Biosynthesis of secondary metabolites initiates from basic pathways, such as glycolysis or shikimic acid pathways, and subsequently emerges mostly depending on the type of cell, developmental stages, and environmental circumstances (Patra et al., 2013; Gupta et al., 2022, 2024).

DOI: 10.1201/9781003408833-17

17.2.1 Pigments

In nature, all photosynthetic organisms like cyanobacteria, algae, and higher plants, as well as some non-photosynthetic organisms such as fungi and bacteria, have the beauty of pigmentation derived from the carotenoids.

All photoautotrophic algae involved in the light harvesting process comprise chlorophyll (mainly a) and various other accessory pigments depending on the particular algal group (Table 17.1). Chl a and b, along with phycobilins and carotenoids, are present in green algae; Chl a and Chl d are found in brown algae. Carotenoids are a group of naturally occurring, fat-soluble (lipophilic) pigments. Carbon atoms with variable numbers of double bonds (polyene backbone) are interlinked to form the basic structure of the carotenoid molecule, obtained from the isoprenoid pathway. The number of conjugated double bonds in their skeleton determines the colour of the carotenoid. Based on their chemical structure, carotenoids with a pure carbon backbone are referred to as carotenes (cyclized or uncyclized, α-carotene, βcarotene, γ-carotene, lycopene, phytoene) and oxygenated carotenes are known as xanthophylls (lutein, zeaxanthin, astaxanthin, violaxanthin, canthaxanthin, echineone, β-cryptoxanthin, fucoxanthin, peridinin, neoxanthin are known xanthophylls) (Edge et al., 1997; Ambati et al., 2019; Cheng et al., 2020). Under stress conditions, secondary carotenoids like astaxanthin and canthaxanthin are stored in high amounts in cytoplasmic lipid globules. The presence of a characteristic pink or red color in some stressed algae is due to the accumulation of a carotenoid protective layer (Begum et al., 2016).

Carotenoids primarily act as accessory light-harvesting pigments in the photosystem, resulting in extending the range of solar radiation (wavelength) that is not trapped by chlorophyll and, hence, supporting the process of photosynthesis to a greater extent. Secondly, it acts as a photoprotector by dispersing extra energy and reactive toxic oxygen molecules. Thus, carotenoids maintain pigment-protein complexes and stabilize the solidity of membranes required by cells for survival and growth (Gong and Bassi, 2016). Astaxanthin is called 'Super Vit-E' because of its high antioxidant activity. Canthaxanthin and peridinin showed anticarcinogenic activities, and neoxanthin lessened lung cancer (Buono et al., 2014). Zeaxanthin and lutein are referred to as eye vitamins as they protect the eyes from damaging UV and blue light (Ventura et al., 2017).

TABLE 17.1
Main Accessory Pigments in Algae

Group	Main Accessory Pigments	Algal Species	Reference
Cyanobacteria (Chl a)	β-carotene	*Anabaena* sp., *Nostoc punctiforme*, and *Gloeobacter violaceus*	Takaichi and Mochimaru, 2007
	β-carotene, nostoxanthin, and zeaxanthin	*Thermosynechococcus elongatus*, *Synechococcus*, and *Spirulina* sp.	
	β-carotene and zeaxanthin	*Synechocystis* sp. and *Prochlorococcus marinus*	
Rhodophyta (Chl a and d)	β-carotene and zeaxanthin	*Cyanidium caldarium*, *Porphyridium*, and *Rhodosorus marinus*	Takaichi and Mochimaru, 2007
	β-carotene, zeaxanthin, α-carotene, and lutein	*Chondracanthus volans*, *Gelidium johnstonii*, *Plocamium violaceum*, *Lithothrix aspergillum*, *Batrachospermum* sp., and *Porphyra perforata*	
	Antheraxanthin	*Bossiella californica* and *Corallina pilulifera*	
Chlorophyta (Chl a and b)	Astaxanthin and β-carotene	*Chlorococcum* sp.	Vuppaladadiyam et al., 2018
	β-carotene, zeaxanthin, lutein, and trans β-carotene	*Dunaliella salina*	
	Astaxanthin and β-carotene	*H. pluvialis*	
Phaeophyta (Chl a, c_1, and c_2)	β-carotene, luteoxanthin, and fucoxanthin	*Fucus serratus* and *F. vasiculosus*	Haugan and Liaaen-Jensen, 1994
	β-carotene and fucoxanthin	*Laminaria digitata* and *L. saccharina*	
	β-carotene, fucoxanthin, and violaxanthin	*Pelvitia* sp. and *Ascophyllum nodosum*	
Bacillariophyta (Chl a, c_1, and c_2)	β-carotene and fucoxanthin	*Cyclotella meneghiniana*, *Phaeodactylum tricornutum*, and *Odontella aurita*	Kuczynska et al., 2015
Dinophyta (Chl a and c_2)	β-carotene, peridinin, and diadinoxanthin	*T. sanguine* and *Baldinia anauniensis*	Frassanito et al., 2006

17.2.2 Polyphenolic Secondary Metabolites

Polyphenolic compounds, or phenols, comprise a large and diverse group of compounds consisting of hydroxyl groups (-OH) directly attached to aromatic hydrocarbon groups that are found in a variety of terrestrial and marine algae. Due to their antioxidant properties and wide range of biological effects, they are used as an important component of human and animal nutrition. Polyphenols have a variety of roles in algal development from early embryonic stages to mature cells, reproduction, growth, cell wall construction, and polyspermy (Schoenwaelder, 2002). Kaur and Das (2011) mentioned that phenolic compounds obtained from marine algae have potential as antioxidants, antiaging, anti-inflammatory, antibacterial, antifungal, cytotoxic, antimalarial, antiproliferative, anticancer, *etc.*

17.2.2.1 Flavonoid

In algae, flavonoids mainly occur as glycosides and sometimes as aglycones. Flavonoids have a skeleton of 15 carbons arranged as two phenyl rings (A and B) linked with a three-carbon unit forming a heterocyclic ring. Based on the position of the benzenoid substituent, it can be divided into two groups: firstly, flavonoids that contain flavanones, flavonols, flavones, flavan-3-ols, and anthocyanidins; and secondly, isoflavonoids that contain isoflavones and isoflavans (Fernando et al., 2022).

Kaempferol, a type of flavonol, was recently reported in *Nannochloris* sp. and *Tetraselmis suecica* (Haoujar et al., 2019). The highest phenol and flavonoid content was found in the ethyl acetate fraction of *Gracilaria edulis*, which is also a potential source of several potent antidiabetic compounds (Gunathilaka et al., 2019). The important groups of flavonoids and their derivatives have been listed in Table 17.2.

17.2.2.2 Lipids

Microalgae work as mini factories powered by solar energy that convert carbon dioxide to a variety of lipids (fats, phospholipids, steroids, and waxes) and provide the food to support the whole aquatic ecosystem. The function of lipids in algal cells is vital as components of the membrane, storage products, metabolites, and a source of energy supplies. Stress situations, such as nitrogen unavailability, high temperatures, pH changes, and high salt concentrations, can enhance lipid productivity (Kwak et al., 2016).

Microalgae are the principal producers of polyunsaturated fatty acids (hereafter, PUFAs), and they contain the enzymes desaturases and elongases, which are essential for PUFAs biosynthesis (Jeon et al., 2017). PUFA, with numerous double bonds, offers great nutritional and medicinal properties. PUFA productions were reported in three marine microalgae: *Nannochloropsis oculata* (Eustigmatophyceae), *Thalassiosira pseudonana* (Bacillariophyceae), and Haptophyte *Pavlova lutheri* (Tonon et al., 2002). Algae contain a vast variety of fatty acids in their lipid storage (Guschina and Harwood, 2006). As the need for pure PUFAs grows, certain alternative sources are being examined. The high lipid yield was obtained from *Tetraselmis* sp. (11%) and *Scenedesmus almeriensis* (10%) (Hernández et al., 2014).

Biodiesel production from Microalgae has the potential to completely displace the demand for fossil diesel. Microalgae that show oil content of 20–70% are *Botryococcus braunii*, *Chlorella* sp., *Crypthecodinium cohnii*, *Cylindrotheca* sp., *Dunaliella primolecta*, *Isochrysis* sp., *Monallanthus salina*, *Nannochloris* sp., *Nannochloropsis* sp., *Neochloris oleoabundans*, *Nitzschia* sp., *Phaeodactylum tricornutum*, *Schizochytrium* sp., and *Tetraselmis suecica* (Chisti, 2007).

17.2.2.3 Terpenoids

Terpenoids are a vast family of chemical substances found in many different species, and they account for over 60% of all marine naturally occurring substances recorded from algae, with Rhodophytes containing the highest quantity (nearly 50%), whereas Phaeophytes include roughly 36% of all documented algal terpenoids (Leal et al., 2013). The sesquiterpene class is dominated by red algae, while the

TABLE 17.2
Types of Flavonoids Derived from Different Algal Species

Group	Example	Algal Species	Reference
Flavonols	Catechin	*A. spicifera* and *Nannochloris* sp.	Zeng et al., 2001
	Quercetin and myricetin	*A. spicifera* and *Himanthalia elongate*	Rajauria, 2018
	Kaempferol	*H. pluvialis* and *Nannochloris* sp.	Klejdus et al., 2010
Flavanones	Naringenin	*H. pluvialis*	Klejdus et al., 2010
Isoflavones	Daidzin, genistein, sissotrin, formononetin, and biochanin A	*Sargassum vulgare*, *S. muticum*, *S. pallidum*, *Hypnea spinella*, *Porphyra* sp., *Undaria pinnatifida*, *Chondrus crispus*, *Halopytis incurvus*, *Scenedesmus* sp., and *Nostoc* sp.	Klejdus et al., 2010
Flavones	Luteolin, liquiritigenin, and apigenin	*D. lutheri*, *H. pluvialis*, and *Sargassum pallidum*	Klejdus et al. (2010); Liu et al. (2009)

diterpene class is dominated by brown algae, which have the most macroalgal natural products, with almost 500 representatives (Young et al., 2015). Phenolic terpenoids have also been characterized in brown and red macroalgae. Reddy and Urban (2009) reported meroditerpenoids (plastoquinones, chromanols, and chromenes) found almost exclusively in the brown algae Sargassaceae. The Rhodomelaceae (especially *Laurencia* spp.) family of red algae contains diterpenes and sesquiterpenes (Masuda et al., 2002).

17.2.2.4 Steroids

Steroids and steroid-like triterpenes are found in three groups of macroalgae: red (Rhodophyta), green (Chlorophyta), and brown (Phaeophyta). Red algae produce a variety of ecdysteroid-like molecules, such as pinnasterol (Viano et al., 2011). Extraction from red algae, *Eucheuma cottonii*, also shows the presence of bioactive steroids (Fasya et al., 2019).

Phytoecdysteroids are oxidized sterols produced by brown and green algae, for example, the cathasterone from the brown alga *Polycladia* (*Cystoseira*) *myrica* (Jiang et al., 2008) and the green alga *Ulva lactuca* (Awad, 2000). Five new sterols were examined from the brown alga *Sargassum carpophyllum* as active compounds (Tang et al., 2002). Steroids from the brown alga, *Turbinaria conoides*, were evaluated for antiviral and antihistaminic activity (Kumar et al., 2011). Ten stigmastane-type steroids were isolated from *Dictyopteris undulata* (Feng et al., 2018).

17.2.2.5 Phlorotannins

Phlorotannins are polymerized derivatives of phloroglucinol units. The biosynthesis pathway of bioactive phlorotannin has not been studied in detail and remains poorly understood. Two proposed pathways for the biosynthesis of phlorotannins are (A) the biosynthesis of phloroglucinol by malonyl-CoA catalyzed via PKS1 (Meslet-Cladière et al., 2013) and (B) the oxidative coupling of phloroglucinol.

The intrinsic factors (life history, age, size, and reproductive status) as well as extrinsic factors like seasonal variations, environmental conditions, nutrient content, salinity, availability of light, the incidence of ultraviolet and cosmic radiation, and herbivory affect the biosynthesis of phlorotannins in algae (Singh and Sidana, 2013).

Phlorotannin traps light in the middle and far regions of the UV spectrum and thus plays a vital role a photoprotective from ultraviolet radiation (Gómez and Huovinen, 2010). They have potential biofunctional activities that include antioxidant, anti-inflammatory, anticancer, immunomodulatory, antidiabetes, inhibition of enzymes such as tyrosinase, angiotensin-I-converting enzyme (ACE-I), and acetylcholinesterase (AChE), antiviral, antibacterial, *etc.*, but dominantly most attention has been given to antioxidant activity (Zolotariova et al., 2019). The most common source of polyphenolic compounds is the brown alga, *Ecklonia cava* includes various phlorotannins, that is, phloroglucinol, eckol, fucodiphloroethol G, phlorofucofuroeckol A, dieckol, 7-phloroeckol, and 6, 60-bieckol (Li et al., 2009; Stengel et al., 2011), and other sources are listed in Table 17.3.

17.2.2.6 Other Polyphenolic Compounds

Mycosporine-like amino acids (MAAs) are soluble in water and are also considered one of the phenolic compounds produced and characterized mainly from Cyanobacteria and Rhodophyta, but also found in the green alga *Prasiola* spp., but results are in dilemma. MAAs had also been extracted from a red alga, *Gracilaria cornea*, dinoflagellates, *A. tamarense*, and *H. triquetra* (Laurion and Roy, 2009).

Other phenolic compounds that have been isolated, like colpol, tichocarpol, coumarins, and some vanillic acid derivatives from different algal species, are tabulated in Table 17.4.

TABLE 17.3
Classes of Phlorotannin with Associated Algal Species

Class of Phlorotannin	Other Derivatives	Algal Species	Reference
Fucols (phlorotannins with aryl linkages)	Difucol, trifucol, tetrafucol A, and tetrafucol B	*Himanthalia elongate*, *Analipus japonicus*, *F. vesiculosus*, *B. bifurcate*, and *Scytothamnus australis*	Glombitza et al., 1975, 1977
Phlorethols (phlorotannins with ether linkages)	Fuhalols, phlorethols, bifuhalol, and diphlorethol	*F. vesiculosus*, *C. mascalocarpum*, *C. angustifolium*, *Cystoseira tamariscifoli*, *Laminaria ochroleuca*, *Sargassum spinuligerum*, and *S. muticum*	Ragan and Jamieson, 1982; Li et al., 2017
Fucophlorethols (with both aryl and ether linkages)	Fucophlorethol A, fucophlorethol B, and fucodiphlorethol	*F. vesiculosus*, *S. spinuligerum*, *Cystoseira baccata*, *H. elongate*, *L. ochroleuca*, and *E. cava*	Glombitza et al., 1997; Tsao, 2010
Eckols and carmalols (with dibenzo [1,4] dioxin linkages)	Eckstolonol, bieckols, 3-phloroeckol, and fucofuroeckols A	*Ecklonia*, *Ishige okamurae*, *Eisenia arborea*, and *E. kurome*	Moon et al., 2011; Kim et al., 2015
	Diphlorethohydroxycarmalol (C146) and triphlorethohydroxycarmalol (C147)	*C. maschalocarpum* and *I. okamurae*	Nagayama et al., 2002

TABLE 17.4
Other Polyphenols Reported in Algae

Polyphenols	Class	Name of Algal Species	Reference
Bromophenols	Rhodophyceae	*Pterocladia capillacea, Odonthalia corymbifera, Rhodomela confervoides, Jania rubense*, and *Asparagopsis armata*	Whitfield et al., 1999
	Phaeophyceae	*Padina arborescens, Sargassum siliquastrum, Lobophora variegata*, and *Sargassum fallax*	Ishii et al., 2004
	Chlorophyceae	*Codium fragile, Avrainvillea longicaulis, Avrainvillea nigricans*, and *Avrainvillea rawsonii*	Whitfield et al., 1999
MAAs	Rhodophyceae, chlorophyceae, and dinoflagellates	*Porphyra* sp., *Gracilaria cornea, Prasiola* spp., *Dunaliella tertiolecta, A. tamarense*, and *H. triqueta*	Carreto and Carignan, 2011; Laurion and Roy, 2009
Tichocarpol	Rhodophyceae	*Tichocarpus crinitus*	Ishii et al., 2004
Colpol	Phaeophyceae	*Colpomenia sinuosa*	Green et al., 1993
Coumarins	Phaeophyceae	*Dasycladus vermicularis*	Perez-Rodriguez et al., 2003
Vanillic acid	Phaeophyceae	*Cladophora socialis*	Feng et al., 2007
Lignins	Rhodophyta	*Calliarthron cheilosporioides*	Freile-Pelegrin and Robledo, 2013

17.2.3 Polysaccharides

Polysaccharides are biomacromolecules found in the cell walls of marine algae. Polysaccharides derived from marine algae are frequently associated with pharmacological activities such as antioxidant, anticancer, and immunomodulatory properties; they are also anticoagulant and have high demand in the food and cosmetic industries (de Jesus Raposo et al., 2015; Hamed et al., 2015). Their bioactivities are often tightly connected to their chemical factors, such as kinds, ratios of monosaccharide components, molecular sizes, and glycosidic linkage (Ho et al., 2014). The most extensively exploited algal polysaccharides are agar, carrageenan, alginate, and galactan, which are used as food additives like gelling and emulsifying agents, flocculants and hydrating agents, stabilizers, thickening agents, *etc* (Bernal and Llamas, 2012). *Kappaphycus alvarezii* is used on an industrial scale to provide carrageenan to the food industries (Rupert et al., 2022). Other rhodophyceae members, for example, *Mastocarpus stellatus, Porphyra haitanensis, Gracilaria fisheri*, and *Gelidium crinale* are the main sources of agar, carrageenan, and sulfated galactan (Wang and Weller, 2006; Kalitnik et al., 2013; Imjongjairak et al., 2016).

Phaeophyceae or brown algae, such as *Cystoseira sedoides, Laminaria japonica* (Cheng et al., 2020), *Sargassum fusiforme* (Jin et al., 2014), *Scytothamnus australis, etc.*, produce mainly laminaran, which has anticancerous and antioxidant activities. The green algae *Ulva armoricana* and *U. lactuca*, (Anisha et al., 2023) have antiviral and antiradiation properties. Alginates (hydrocolloid) were also successfully isolated from *Ascophyllum nodosum, Durvillaea potatorum, Ecklonia maxima, Laminaria/Saccharina* sp., *Lessonia nigrescens*, and *Macrocystis pyrifera* (Bixler and Porse, 2011; FAO, 2014).

17.2.4 Vitamins

Vitamins are one of the important organic micronutrients that are not synthesized directly in organisms and therefore must be obtained externally from the diet. The biomass of microalgae is a valuable source of nearly all essential vitamins (e.g., A, B1, B2, B6, B12, C, E, nicotinate, biotin, folic acid, and pantothenic acid) and a balanced mineral content (e.g., Na, K, Ca, Mg, Fe, Zn, and trace minerals) (Becker, 2004). Kelp (*Macrocystis pyrifera*) contains the active form of vitamin E (α-tocopherol). Table 17.5 shows a list of vitamins extracted from different algal species. *Palmaria palmata* and *Porphyra tenera* have large quantities of vitamins A, B1, B2, and B12 (Kolanjinathan et al., 2014). The vitamin C content of *Gracilaria changii, Himanthalia elongata, Porphyra umbilicalis*, and *Eisenia arborea* is approximately the same as that found in tomatoes and lettuce (34.4 mg/100 g dry weight) matches the value as reported in various vitamin C sources (Wells et al., 2017).

Brown algae (like *Ascophyllum* and *Fucus* sp.) have a higher composition of vitamin E than green and red seaweeds (Burtın, 2003). Bioproducts, including proteins, omega-3, -6, and -9 fatty acids, chlorophylls, antioxidants, carotenoids, and polysaccharides, obtained from microalgae *Dunaliella, Haematococcus, Chlorella*, and *Parachlorella* spp., have a variety of applications (Borowitzka, 2013; Abu-Ghosh et al., 2021).

17.2.5 Toxins

The harmful algal bloom (HAB) occurred due to toxic phytoplankton in either marine or freshwater systems. More than 250 species of microalgae were examined to produce so-called algal blooms. Approximately one-fourth of algal bloom species are characterized as producing toxins (Hallegraeff, 2003). Several potent toxins are responsible

TABLE 17.5
Vitamins from Algal Species

Vitamin	Algae Species	Reference
Vitamin A	*Palmaria palmata* and *Porphyra tenera*	Kolanjinathan et al., 2014
Vitamin B1 and B2	*T. suecica*, *Isochrysis galbana*, *D. teriolecta*, *Chlorella stigmatophora*, *Tetraselmis* sp., *Nannochloropsis* sp.	Fabregas et al., 1990; Brown et al., 1999
Vitamin B3	*Nannochloropsis gaditana* and *Chlorella* sp.	Edelmann et al., 2019
Vitamin B5, B7	*T. suecica*, *Isochrysis galbana*, *D. teriolecta*, and *Chlorella stigmatophora*	Fabregas et al., 1990
Vitamin C	*Nannochloropsis* sp., *Chaetoceros* spp., *Chroomonas salina*, *Nannochloropsis oculata*, *Pavlova lutheri*, *Gracilaria changii*, *Himanthalia elongata*, and *Porphyra umbilicalis*	Wells et al., 2017; Chan and Matanjun, 2017; Ilyas et al., 2023
Vitamin E	*Haematococcus pluvialis*, *Microchloropsis salina*, *Chlamydomonas nivalis*	Mudimu et al., 2017

TABLE 17.6
Major Toxins-Producing Algal Species

Toxins	Producers/Algal Species	References
Cyanotoxins		
Neurotoxins and anatoxin-a	*Aphanizomenon*, *Planktothrix*, and *Anabaena*	Falconer et al. (2005)
Saxitoxins	*Alexandrium*, *Pyrodinium*, *Gymnodinium*, *Anabaena circinalis*, *Aphanizomenon gracile*, *Cylindrospermopsis raciborskii*, and *Lyngbya wollei*	Pearson et al. (2010)
Cyanopeptolin	*Microcystis* and *Planktothrix*	Faltermann et al. (2014)
Hepatotoxins		
Microcystins	*Anabaena*, *Planktothrix*, *Nostoc*, *Anabaenopsis*, *Fischerella*, *Hapalosiphon*, *Limnothrix*, *Lyngbya*, *Microcystis*, *Rivularia*, *Synechocystis*, and *Synechococcus*	Pearson et al. (2010)
Nodularins	*Nodularia*	Mulvenna et al. (2012)
Cylindrospermopsins	*Cylindrospermopsis raciborskii*, *Aphanizomenon ovalisporum*, and *Aphanizomenon flos-aquae*	Kehr et al. (2011)
Dermatotoxins		
Lyngbyatoxin-a	*Lyngbya*	Funari and Testai, 2008
Aplysiatoxin	Cyanobacteria	Stewart et al. (2006)

for a wide array of human illnesses, aquatic mammal and bird mortality, and an extensive reduction in the number of fish. Major toxins with their algal species are listed in Table 17.6.

Paralytic shellfish (PSP) toxins, especially Saxitoxin, were obtained by several members of gonyaulacoid or gymnodinioid dinoflagellates, including *Alexandrium*, *Gymnodinium*, and *Pyrodinium* (Pereira et al., 2004).

Cylindrospermopsins (cyanotoxins that are produced by cyanobacteria) are produced by *Aphanizomenon ovalisporum*, *Cylindrospermopsis raciborskii*, *Raphidiopsis curvata*, and *Umezakia natans* (Briand et al., 2003).

The highest production of toxin is generally found in optimal conditions for cell growth (Kaebernick and Neilan, 2001). However, the production of cylindrospermopsin by *C. raciborskii* is negatively correlated with growth rate (Griffiths and Saker, 2003), as is the case for several allelopathic compounds. *Ascophyllum nodosum* produces defenses in the presence of grazing cues (Toth and Pavia, 2000). Waterborne cues induce chemical defense in a marine alga.

The toxins' production may be a defense mechanism against predators. Some PUFAs and their derivatives have the ability to be antipredators (Watson and Cruz-Rivera, 2003). Such fatty acids are produced by the hydrolysis of lipids and result in the disruption of the grazer's cell wall, which makes it an efficient defense against grazers (Watson and Cruz-Rivera, 2003; Leflaive and Ten-Hage, 2007).

17.3 EXTRACTION OF SECONDARY METABOLITES

A single solvent (usually methanol) or a combination of solvents may be used to extract biomass for the production of natural products, particularly when a large number of substances with varying polarities are sought after. Both polar and non-polar solvents are being used for sequential extraction (Buss and Butler, 2004). Different valuable chemical extraction techniques are discussed further in this chapter. These techniques range from simple to sophisticated. In addition, a summary of the analysis methods and the conditions used is provided in Table 17.7.

17.3.1 Traditional Procedures of Extraction

Compounds from plants are extracted using various mechanical and chemical procedures, like solvent extraction and steam distillation (Shirsath et al., 2012). Soxhlet, hydrodistillation, and alcohol-based maceration are some of the methods employed for bioactive component extraction (Wang and Weller, 2006). The best procedure is chosen based on the characteristics of the target compound to obtain the highest yield and purest product. The utilization of traditional Soxhlet extraction procedures is frequently restricted by the mass transfer resistances brought on by the presence of more than one phase in the system (Marathe et al., 2019). As a result of these constraints and the dramatic rise in demand for marine algal bioactives, there is a pressing need to create efficient, selective, cost-effective, and environmentally friendly extraction technologies that can rapidly boost yields without negatively impacting the marine ecosystem (Ibañez et al., 2012).

17.3.2 Novel Methods of Extraction

A few recently developed extraction methods with the potential to fulfil a significant number of the requirements, such as microwave-assisted extraction (MAE), ultrasound-assisted extraction (UAE), supercritical fluid extraction (SFE), and pressurised liquid extraction (PLE), are included in this category. The principles and mechanisms of action, advantages and disadvantages over more traditional approaches, and possible appropriateness for the extraction of marine algae bioactives will be explored in these technologies.

17.3.2.1 Lipid Extraction-

There are many techniques used for lipid extraction from microalgae, like solvent extraction (Kochert method and Soxhlet method), microwave-aided extraction, Ultrasonic extraction, and electroporation (Biller et al., 2013; Hernández et al., 2014). Some solvent-free, easy, and eco-friendly techniques can also be used for lipid extraction, including isotonic extraction and the osmotic pressure method. Additionally, the leftover microalgal waste from these extraction methods can be used as animal feed or transformed into liquid fuels (Ho et al., 2014).

17.3.2.2 Protein Extraction

Solvent extraction, which solubilizes the proteins in organic solvents with surfactants while keeping their functional qualities, has been successfully used to recover proteins. The type of salt employed in the bulk aqueous phase, pH, ionic strength, and other variables all affect the extraction process (Vanthoor-Koopmans et al., 2013). Proteins can be extracted by the SC-CO_2 extraction method without the use of flammable organic solvents (Bjornsson et al., 2012).

TABLE 17.7
Different Types of Extraction Methods

Components	Methods	Solvent	Reference
Carbohydrates	Chemical hydrolysis	Without solvent	Karemore and Sen, 2016
Lipids	Folch method	Chloroform–methanol	Grima et al., 2013
	Soxhlet extraction	Hexane	Grima et al., 2013
	Bligh and Dyer	Chloroform and methanol	Kumar et al., 2015
Proteins	SC-CO_2 extraction	Without the use of flammable organic solvents	Bjornsson et al., 2012
Polyphenol	Solid–liquid extraction	Ethanol, methanol, acetone	Machu et al., 2015
			Imbs et al., 2018
Fatty acids	Bligh and Dyer and direct saponification	Chloroform–methanol mixture; potassium hydroxide and ethanol	Li et al., 2014
Phycobiliprotein	Aqueous two-phase extraction	Organic solvent needed	Patil et al., 2008
Carotenoids	Conventional solvent extraction method	n-hexane, dichloromethane, dimethyl ether	Khoo et al., 2020
	Supercritical fluid extraction	Diethyl ether, acetone, methanol, ethanol	Guedes et al., 2013
	Subcritical fluid extraction	Ethane and ethylene	Marzorati et al., 2020
	Pressurized liquid extraction	1,1,1,2-tetrafluoroethane and dimethyl ether	Gallego et al., 2021

17.3.2.3 Polyunsaturated Fatty Acids Extraction

Supercritical fluid extraction (SFE), direct saponification, solvent extraction and sonication, and Bligh and Dyer (B & D) extraction are some of the techniques used to analyse the fatty acids in marine microalgae (Li et al., 2014). Direct saponification uses ethanol solvent with potassium hydroxide as a catalyst; the Bligh and Dyer technique uses a chloroform-methanol mixture. According to Li et al. (2014), the SC-CO_2 approach produced the highest fatty acid levels. To extract EPA and DHA from microalgae, the SFE process was shown to be the most efficient because it produced the maximum yield and used less energy (Yen et al., 2015).

17.3.2.4 Phenol Extraction

In recent years, numerous techniques have been developed for the isolation of polyphenols from algae, like heat reflux extraction and Soxhlet extraction, but these conventional methods are of low efficiency as they require high energy and a long processing time. Ibañez et al. (2012) and Luo et al. (2010) studied several high-efficiency methods, including microwave-assisted extraction (MAE) and ultrasound-assisted extraction (UAE) (Herrera, 2005). The algal matrix is extremely vulnerable to microwave irradiation because of its high natural moisture content. Other alternatives to traditional extraction techniques include pressurised liquid extraction (PLE), which uses solvents kept in a liquid state by high temperatures and pressures throughout the extraction process, and supercritical fluid extraction (SFE), which uses fluids at temperatures and pressures above their critical points (Klejdus et al., 2009). HPLC can be integrated with contemporary detection techniques like mass spectrometry (MS) and nuclear magnetic resonance (NMR) to provide quick structural analysis and compound identification with little sample processing (Steevensz et al., 2012). Liquid chromatography also allows the large-scale separation and purification of specific molecules.

17.4 FACTORS AFFECTING ALGAL SECONDARY METABOLITES

Any organism removed from its natural environment will experience species or strain-specific growth conditions, but in extremely stressful situations, it will manufacture particular metabolites to adapt to and get over the biotic or abiotic stress conditions. Microalgae have excellent tolerance to abiotic stress conditions and produce useful metabolites. As a tool integrated with microalgal biorefinery for its sustainable growth, this particular trait of microalgae may be used to produce desired metabolites under abiotic stress (temperature, light, pH, nutrient availability, concentrations of CO_2 and O_2, salinity, cell density, and fragility). Microalgae have developed from being a basic food source to high-end goods and an energy feedstock in the present day. Most algae show abundant plasticity concerning the composition of chemicals, whether single or in combinations of multiple environmental factors.

Influence of abiotic stress conditions: The most common alteration that is found in algal secondary metabolites is the production of bioactive compounds (mainly phenolic and PUFAs) to adapt to the extreme environmental conditions for survival.

17.4.1 Temperature

Each species has its optimal range of growth temperature, for which it is possible to classify the algae as psychrophiles (<15°C), mesophiles (<50°C), and thermophiles (>50°C). No photosynthetic organisms have been reported to be able to grow beyond 75°C, due to chlorophyll instability. The indirect effect of temperature stress may enhance other stresses and antioxidant response mechanisms (Sampath-Wiley et al., 2008).

17.4.2 UV Radiation

It is also a determining factor for secondary metabolites like phlorotannin; UV radiation stimulates phlorotannin production but negatively suppresses the inhibition of photosynthesis, along with DNA damage. Swanson and Druehl (2002) reported the protective role of UV in *Ascophyllum nodosum*, *Fucus vesiculosus* (Koivikko et al., 2005), and *Macrocystis integrifolia*.

17.4.3 pH

pH plays an important role in the regulation of enzymatic activity, ammonia toxicity, and the availability of phosphorus and inorganic carbon (Khalil et al., 2010).

17.4.4 Heavy Metal Contaminants

They can induce changes in the chemical constituents of micro- and macroalgae, with particular effects on pigment composition (Pinto et al., 2003). Some algal species can selectively accumulate metals and, hence, may be researched to purify contaminated wastewater through bioabsorption. Brown algae (*Sargassum* sp.) and red algae (*Gracilaria* sp.) levels of protein were positively correlated with nitrogen levels in water (Marinho-Soriano et al., 2006), as well as the composition and concentration of other phenolic compounds such as MAAs of the red alga *Asparagopsis armata* (Figueroa et al., 2008).

Concentration and antioxidant activity of phenolic compounds were varied with variation in thallus parts of two *Laminaria* species, that is, *A. nodosum* and *S. muticum*, as reported by Connan et al. (2006), and accumulation of secondary metabolites was related to different herbivore defense levels exhibited by different thallus regions in brown *A. nodosum* (Pavia et al., 2002). Microalgal growth and their lipid and fatty acid composition are closely related to the nutrient content of their cultivation medium.

17.4.5 Salinity

It is another factor that can alter the biochemical composition as studied in *Gracilaria* sp. grown under different salinity conditions, which shows several oxidative stress responses and wide variation in phenolics, antioxidative enzymes, phycobiliproteins, and fatty acids (Kumar et al., 2010). To cope with salinity stress, microalgae react by absorbing compatible solutes, up-regulating carotenoid production, changing membrane composition by up-regulating ion transporters, and producing aquaporins. High-salinity-tolerant algae such as *Dunaliella salina* produce sucrose, glycerol, proline, glycine, betaine, *etc*., as compatible solutes (osmo-protectants) under salinity stress (Paliwal et al., 2017).

Under biotic factors, algal chemical ecology and the functional roles of secondary metabolites actively participate in antifouling, antiherbivore, or antimicrobial defenses. The most well-known biotic interactions involving algae are lichen symbioses, which have a high potential value for bioactive lichen substances. Allelopathical activity of algal chemicals has toxic effects, for example, on macroalgal chemicals active toward microalgae, plants, bacteria, or other marine organisms like invertebrates and vertebrates (Leflaive and Ten-Hage, 2007). Chemodiversity, which is further compounded by the effects of the local environment, is implied by genetic variability. When examining linked bioactivities, especially in cases where substances have numerous biological roles, it is important to take into account the physiological processes, factors, and environmental interactions, regulate biochemical composition.

17.5 APPLICATIONS OF ALGAL SECONDARY METABOLITES

Marine algae are utilised for a variety of purposes across a wide range of industries, including agriculture, cosmetics, food, pharmaceutical industries, and biofuel production. It is interesting to note that the Food and Drug Administration (FDA) allows the use of phycocyanin, which is typically found in red algae and cyanobacteria, as a natural colour addition in food and cosmetics because it is non-toxic, naturally occurring, and biodegradable (Pagels et al., 2019). Human hepatocarcinoma and human breast carcinoma have been demonstrated to be inhibited by polyunsaturated fatty acids, phenolics, and flavonoids from *Taonia atomaria* and *Laurencia papillose. Table 17.8 shows the applications of different compounds obtained from algae.*

Since a carotenoid like astaxanthin is regarded as a powerful antioxidant and plays a significant part in the process of scavenging free radicals in the human body, it has become increasingly popular as a dietary supplement for people (Ambati et al., 2019). Commercial applications for microalgal pigment include use as a natural food colourant and as a cosmetic ingredient. Beta-carotene has been shown to inhibit the growth of cancer cells, and it has also been linked to a lower risk of high cholesterol and cardiovascular disease.

TABLE 17.8
Applications of Different Types of Secondary Metabolites

Compounds	Type	Application	Reference
Carbohydrate	Polysaccharides	Immune system, inflammatory reactions, and anti-HIV activity	Witvrouw and De Clercq, 1997
Lipids		Antitumor, antioxidant, antimicrobial, antivirus (with various degrees), and nutritional supplements	Abd El et al., 2014 El Baz et al., 2014 Ghazala et al., 2010
Protein	Phycobilliproteins	Antioxidant, anticancer, antihypertensive, antiatherosclerotic, and immunomodulatory effects	Fan et al., 2014
Carotenoid	Astaxanthin	Anti-inflammatory, hepatoprotector (retinal, brain, and spinal cord neurons), and prevents cardiovascular diseases	Monroy-Ruiz et al., 2011
	β-carotene	Anti-inflammatory, antioxidant, immunoprotector, dermoprotector, hepatoprotector; breast, colon, lung, liver, and skin cancers	Murthy et al., 2005
	Fucoxanthin	Antioxidant, antiangiogenic, anti-inflammatory, photo- and neuroprotective; inhibits DNA damage and H_2O_2-induced apoptosis; reduces nitric oxide, prostaglandin, TNF-α, and histamine levels in *in vivo* models	Peng et al., 2011 Heo et al., 2008
Phenol	Phlorotannin	Anti-HIV, anticancer, bactericidal, radioprotective, antiallergic, and other health-beneficial	Kim et al., 2011
	Dieckols	Antihyperlipidemic effect and hydroxyl-methyl glutaryl coenzyme A reductase inhibition in high-fat-diet-induced obese mice	Oben et al., 2007
	Polyphenol	Improved fasting blood glucose level and decreased blood total cholesterol and glycated serum protein levels in streptozotocin-induced diabetic mice	Sharifuddin et al., 2015

17.5.1 In Agricultural Industries

Organic fertilizers (source microalgae) are emerging as new, environmentally friendly pesticides to control pests and environment from the harmful effects of synthetic fertilizers.

Phycocolloids, such as alginates, agars, and carrageenans, are commonly extracted from macroalgae and used as thickeners, stabilisers, and in the encapsulation of probiotics, as well as in the formation of gels and water-soluble films (Milledge et al., 2015; Gomez et al., 2020).

Because macroalgae flourish in water, the production of algal biofuels, sometimes known as "third-generation biofuels," has been contemplated as a potential strategy for lowering reliance on fuels derived from petroleum and avoiding a struggle for arable land between food and energy production. Microalgae are a more viable option for biodiesel production because of the simplicity of their life cycle, which can be managed, and their reproduction rate can be increased (Culaba et al., 2020). Once the lipids have been extracted, the microalgal biomass can be used to make renewable energy or fuel. It has been proven to contain the same amount of energy as 80% of petroleum. Choosing the right microalgae strain is essential because the lipid content varies greatly depending on the species and the growth circumstances (Milano et al., 2016). Biofuel is made from a variety of different microalgae, including *Chlorella* spp., *Dunaliella salina*, *Haematococcus pluvialis*, *Spirulina platensis*, *Porphyridium cruentum*, *Microcystis aeruginosa*, and *Scenedesmus obliquus* (Shuba and Kifle, 2018).

17.6 CONCLUSION

In the last few years, many industries such as cosmetics, fuel, pharmaceuticals, and food have developed their interest in secondary metabolites, which are obtained from various algal species of different classes of algae. Rhodophyceae, Chlorophyceae, and Phaeophyceae show chemical diversity related to primary and secondary metabolites with several bioactivities such as antitumor, anti-inflammatory, antidiabetic, antivirus, neuroprotective, *etc.* Micro and macro algae produce secondary metabolites as excretory products under stressed conditions such as exposure to UV light, changes in environmental conditions, that is, temperature, salinity, *etc.* The mostly studies on increasing the production of secondary metabolites are based on *in vitro.* There is a need for advancement in technology and coupling with genetic engineering, optimal culture conditions for the development of knowledge of secondary metabolites efficiency, and their use in various industries.

REFERENCES

Abd El Baky, H.H., F.K. El Baz, G.S. El Baroty, M.M.S. Asker, and E.A. Ibrahim. 2014. Phospholipids of some marine macroalgae: Identification, antivirus, anticancer and antimicrobial bioactivities. *Der Pharma Chemica* 6: 370–82. https://www.derpharmachemica.com/pharma-chemica/phospholipids-of-some-marine-macroalgae-identification-antivirus-anticancer-and-antimicrobial-bioactivities.pdf.

Abu-Ghosh, S., Z. Dubinsky, V. Verdelho, and D. Iluz. 2021. Unconventional high-value products from microalgae: A review. *Bioresource Technology* 329 (June): 124895, 1–12. https://doi.org/10.1016/j.biortech.2021.124895.

Ambati, R.R., D. Gogisetty, R.G. Aswathanarayana, et al. 2019. Industrial potential of carotenoid pigments from microalgae: Current trends and future prospects. *Critical Reviews in Food Science and Nutrition* 59: 1880–902. https://doi.org/10.1080/10408398.2018.1432561.

Anisha, G.S., T. Augustianath, S. Padmakumari, R.R. Singhania, A. Pandey, and A.K. Patel. 2023. Ulvan from green macroalgae: Bioactive properties advancing tissue engineering, drug delivery systems, food industry, agriculture and water treatment. *Bioresource Technology Reports* 22 (June): 101457. https://doi.org/10.1016/j.biteb.2023.101457.

Awad, N.E. 2000. Biologically active steroid from the green alga *Ulva lactuca*. *Phytotherapy Research* 14: 641–43. https://doi.org/10.1002/1099-1573(200012)14:8<641::AID-PTR668>3.0.CO;2-R.

Beardall, J., and J.A. Raven. 2012. Algal metabolism. *Encyclopedia of Life Science*. https://doi.org/10.1002/9780470015902.a0000321.pub2.

Becker, W. 2004. Microalgae in human and animal nutrition. In: *Handbook of microalgal culture: Biotechnology and applied phycology*, ed., A. Richmond, 312–51. Oxford: Blackwell, John Wiley & Sons, Inc. https://doi.org/10.1002/9780470995280.ch18.

Begum, H., F.M. Yusoff, S. Banerjee, H. Khatoon, and M Shariff. 2016. Availability and utilization of pigments from microalgae. *Critical Reviews in Food Science and Nutrition* 56: 2209–22. https://doi.org/10.1080/10408398.2013.764841.

Bernal, P., and M.A. Llamas. 2012. Promising biotechnological applications of antibiofilm exopolysaccharides. *Microbial Biotechnology* 5: 670–73. https://doi.org/10.1111/j.1751-7915.2012.00359.x.

Biller, P., C. Friedman, and A.B. Ross. 2013. Hydrothermal microwave processing of microalgae as a pre-treatment and extraction technique for bio-fuels and bio-products. *Bioresource Technology* 136: 188–95. https://doi.org/10.1016/j.biortech.2013.02.088.

Bixler, H.J., and H. Porse. 2011. A decade of change in the seaweed hydrocolloids industry. *Journal of Applied Phycology* 23: 321–35. https://doi.org/10.1007/s10811-010-9529-3.

Bjornsson, W.J., K.M. MacDougall, J.E. Melanson, S.J.B. O'Leary, and P.J. McGinn. 2012. Pilot-scale supercritical carbon dioxide extractions for the recovery of triacylglycerols from microalgae: A practical tool for algal biofuels research. *Journal of Applied Phycology* 24: 547–55. https://doi.org/10.1007/s10811-011-9756-2.

Borowitzka, M.A. 2013. High-value products from microalgae – their development and commercialisation. *Journal of Applied Phycology* 25: 743–56. https://doi.org/10.1007/s10811-013-9983-9.

Briand, J.F., S. Jacquet, C. Bernard, and J.F. Humbert. 2003. Health hazards for terrestrial vertebrates from toxic cyanobacteria in surface water ecosystems. *Veterinary Research* 34: 361–77. https://doi.org/10.1051/vetres:2003019.

Brown, M.R., M. Mular, I. Miller, C. Farmer, and C. Trenerry. 1999. The vitamin content of microalgae used in aquaculture. *Journal of Applied Phycology* 11: 247–55. https://doi.org/10.1023/A:1008075903578.

Buono, S., A.L. Langellotti, A. Martello, F. Rinna, and V. Fogliano. 2014. Functional ingredients from microalgae.

Food & Function 5: 1669–85. https://doi.org/10.1039/C4FO00125G.

Burtın, P. 2003. Nutritional value of seaweeds. *Electronic Journal of Environmental, Agricultural and Food Chemistry* 2: 498–503. http://marineagronomy.org/sites/default/files/NUTRITIONAL%20VALUE%20OF%20SEAWEEDS.pdf.

Buss, A.D., and M.S. Butler. 2004. A new model for utilising chemical diversity from natural sources. *Drug Development Research* 62: 362–70. https://doi.org/10.1002/ddr.10389.

Carreto, J.I., and M.O. Carignan. 2011. Mycosporine-like amino acids: Relevant secondary metabolites. Chemical and ecological aspects. *Marine Drugs* 9: 387–446. https://doi.org/10.3390/md9030387.

Chan, P.T., and P. Matanjun. 2017. Chemical composition and physicochemical properties of tropical red seaweed, *Gracilaria changii*. *Food Chemistry* 221: 302–10. https://doi.org/10.1016/j.foodchem.2016.10.066.

Cheng, S.H., H.E. Khoo, K.W. Kong, K.N. Prasad, and C.M. Galanakis. 2020. Extraction of carotenoids and applications. In *Carotenoids: Properties, processing and applications*, ed. C.M. Galanakis, 259–88. Academic Press. https://doi.org/10.1016/B978-0-12-817067-0.00008-7.

Cheng, Y., L. Wu, C. Fang, et al. 2020. Synthesis of porous carbon materials derived from *laminaria japonica* via simple carbonization and activation for supercapacitors. *Journal of Materials Research and Technology* 9: 3261–71. https://doi.org/10.1016/j.jmrt.2020.01.022.

Chisti, Y. 2007. Biodiesel from microalgae. *Biotechnology Advances* 25: 294–306. https://doi.org/10.1016/j.biotechadv.2007.02.001.

Connan, S., F. Delisle, E. Deslandes, and E. Ar Gall. 2006. Intra-thallus phlorotannin content and antioxidant activity in Phaeophyceae of temperate waters. *Botanica Marina* 49: 39–46. https://doi.org/10.1515/BOT.2006.005.

Culaba, A.B., A.T. Ubando, P.M.L. Ching, W.H. Chen, and J.S. Chang. 2020. Biofuel from microalgae: Sustainable pathways. *Sustainability* 12 (September): 8009, 1–19. https://doi.org/10.3390/su12198009.

de Jesus Raposo, M.F., A.M.B. De Morais, and R.M.S.C. De Morais. 2015. Marine polysaccharides from algae with potential biomedical applications. *Marine Drugs* 13: 2967–3028. https://doi.org/10.3390/md13052967.

Edelmann, M., S. Aalto, B. Chamlagain, S. Kariluoto, and V. Piironen. 2019. Riboflavin, niacin, folate and vitamin B12 in commercial microalgae powders. *Journal of Food Composition and Analysis* 82 (May): 103226, 1–10. https://doi.org/10.1016/j.jfca.2019.05.009.

Edge, R., D.J. McGarvey, and T.G. Truscott. 1997. The carotenoids as anti-oxidants – a review. *Journal of Photochemistry and Photobiology B: Biology* 41: 189–200. https://doi.org/10.1016/S1011-1344(97)00092–4.

El Baz F.K., G.S. El-Baroty, A.E. Ibrahim and H.H. Abd El Baky. 2014. Cytotoxicity, antioxidants and antimicrobial activities of lipids extracted from some marine algae. *Journal of Aquaculture Research & Development* 5 (November): 1000284, 1–5. https://doi.org/10.4172/2155-9546.1000284.

Fabregas, J., and C. Herrero. 1990. Vitamin content of four marine microalgae. Potential use as source of vitamins in nutrition. *Journal of Industrial Microbiology and Biotechnology* 5: 259–63. https://doi.org/10.1007/BF01569683.

Falconer, I.R., and A.R. Humpage. 2005. Health risk assessment of cyanobacterial (blue-green algal) toxins in drinking water. *International Journal of Environmental Research and Public Health* 2: 43–50. https://doi.org/10.3390/ijerph2005010043.

Faltermann, S., S. Zucchi, E. Kohler, J.F. Blom, J. Pernthaler, and K. Fent. 2014. Molecular effects of the cyanobacterial toxin cyanopeptolin (CP1020) occurring in algal blooms: Global transcriptome analysis in zebrafish embryos. *Aquatic Toxicology* 149: 33–39. https://doi.org/10.1016/j.aquatox.2014.01.018.

Fan, X., L. Bai, L. Zhu, L. Yang, and X. Zhang. 2014. Marine algae-derived bioactive peptides for human nutrition and health. *Journal of Agricultural and Food Chemistry* 62: 9211–22. https://doi.org/10.1016/j.aquatox.2014.01.018.

FAO. 2014. *The state of world fisheries and aquaculture: Opportunities and challenges 2014*. Food and Agriculture Organization of the United Nations Rome. https://openknowledge.fao.org/server/api/core/bitstreams/b673bef5-f7a3-43eb-baf9-05221a9c34ef/content.

Fasya, A.G., A. Baderos, A.D.R. Madjid, S. Amalia, and D.S. Megawati. 2019. Isolation, identification and bioactivity of steroids compounds from red algae *Eucheuma cottonii* petroleum ether fraction. *AIP Conference Proceedings* 2120 (July): 030025, 1–7. https://doi.org/10.1063/1.5115629.

Feng, M.T., T. Wang, A.H. Liu, et al. 2018. PTP1B inhibitory and cytotoxic C-24 epimers of Δ^{28}-24-hydroxy stigmastane-type steroids from the brown alga *Dictyopteris undulata* Holmes. *Phytochemistry* 146: 25–35. https://doi.org/10.1016/j.phytochem.2017.11.013.

Feng, Y., A.R. Carroll, R. Addepalli, G.A. Fechner, V.M. Avery, and R.J. Quinn. 2007. Vanillic acid derivatives from the green algae *Cladophora socialis* as potent protein tyrosine phosphatase 1B inhibitors. *Journal of Natural Products* 70: 1790–92. https://doi.org/10.1021/np070225o.

Fernando, I.P.S., W. Lee, and G. Ahn. 2022. Marine algal flavonoids and phlorotannins; an intriguing frontier of biofunctional secondary metabolites. *Critical Reviews in Biotechnology* 42: 23–45. https://doi.org/10.1080/07388551.2021.1922351.

Figueroa, F.L., A. Bueno, N. Korbee, R. Santos, L. Mata, and A. Schuenhoff. 2008. Accumulation of mycosporine-like amino acids in *Asparagopsis armata* grown in tanks with fishpond effluents of gilthead sea bream, *Sparus aurata*. *Journal of the World Aquaculture Society* 39: 692–99. https://doi.org/10.1111/j.1749-7345.2008.00199.x.

Frassanito, R., G. Flaim, I. Mancini, and G. Guella. 2006. High production of unexpected carotenoids in Dinophyceae. Astaxanthin esters from the freshwater dinoflagellate *Tovellia sanguinea*. *Biochemical Systematics and Ecology* 34: 843–53. https://doi.org/10.1016/j.bse.2006.05.013.

Freile-Pelegrin, Y., and D. Robledo. 2013. Bioactive phenolic compounds from algae. In *Bioactive compounds from marine foods: Plant and animal sources*, eds. B. Hernandez-Ledesma and M. Herrero, 113–29. Chichester: John Wiley & Sons Ltd. https://doi.org/10.1002/9781118412893.ch6.

Funari, E., and E. Testai. 2008. Human health risk assessment related to cyanotoxins exposure. *Critical Reviews in Toxicology* 38: 97–125. https://doi.org/10.1080/10408440701749454.

Gallego, R., M. Bueno, A.M. Chourio, E. Ibáñez, M.D.A. Saldaña, and M. Herrero. 2021. Use of high and ultra-high pressure based-processes for the effective recovery of bioactive compounds from *Nannochloropsis oceanica* microalgae. *The Journal of Supercritical Fluids* 167 (January): 105039. https://doi.org/10.1016/j.supflu.2020.105039.

Ghazala, B., B. Naila, and M. Shameel. 2010. Fatty acids and biological activities of crude extracts of freshwater algae from Sindh. *Pakistan Journal of Botany* 42: 1201–12. http://mail.pakbs.org/pjbot/PDFs/42(2)/PJB42(2)1201.pdf.

Glombitza, K.W., M. Keusgen, and S. Hauperich. 1997. Fucophlorethols from the brown algae *Sargassum spinuligerum* and *Cystophora torulosa*. *Phytochemistry* 46: 1417–22. https://doi.org/10.1016/S0031-9422(97)00499-8.

Glombitza, K.W., H.W. Rauwald, and G. Eckhardt. 1975. Fucole, polyhydroxyoligophenyle aus *Fucus vesiculosus*. *Phytochemistry* 14: 1403–05. https://doi.org/10.1016/S0031-9422(00)98637-0.

Glombitza, K.W., H.W. Rauwald, and G. Eckhardt. 1977. Phlorotannine aus *Himanthalia elongata*. *Phytochemistry* 16: 1614. https://doi.org/10.1016/0031-9422(77)84042-9.

Gómez, I., and P. Huovinen. 2010. Induction of phlorotannins during UV exposure mitigates inhibition of photosynthesis and DNA damage in the kelp *Lessonia nigrescens*. *Photochemistry and Photobiology* 86: 1056–63. https://doi.org/10.1111/j.1751-1097.2010.00786.x.

Gomez, L.P., C. Alvarez, M. Zhao, et al. 2020. Innovative processing strategies and technologies to obtain hydrocolloids from macroalgae for food applications. *Carbohydrate Polymers* 248 (November): 116784. https://doi.org/10.1016/j.carbpol.2020.116784.

Gong, M, and A. Bassi. 2016. Carotenoids from microalgae: A review of recent developments. *Biotechnology Advances* 34: 1396–412. https://doi.org/10.1016/j.biotechadv.2016.10.005.

Green, D., Y. Kashman, and A. Miroz. 1993. Colpol, a new cytotoxic C6-C4-C6 metabolite from the alga *Colpomenia sinuosa*. *Journal of Natural Products* 56: 1201–02. https://doi.org/10.1021/np50097a033.

Griffiths, D.J., and M.L. Saker. 2003. The Palm Island mystery disease 20 years on: A review of research on the cyanotoxin cylindrospermopsin. *Environmental Toxicology* 18: 78–93. https://doi.org/10.1002/tox.10103.

Grima, E.M., F.G.A. Fernández, and A.R. Medina. 2013. Downstream processing of cell mass and products. In *Handbook of microalgal culture: Applied phycology and biotechnology*, eds. A. Richmond and Q. Hu, 267–309. John Wiley & Sons, Ltd. https://doi.org/10.1002/9781118567166.ch14

Gupta, A., S. Jain, and N. Verma. 2022. *Industrial applications of soil microbes*, vol. 1. Bentham Science Publishers Pte Ltd. https://doi.org/10.2174/97898150399551220101.

Gupta, A., S. Jain, and N. Verma. 2024. *Industrial applications of soil microbes*, vol. 3. Bentham Science Publishers Pte Ltd. https://doi.org/10.2174/97898150797531240301.

Guedes, A.C., M.S. Gião, A.A. Matias, et al. 2013. Supercritical fluid extraction of carotenoids and chlorophylls a, b and c, from a wild strain of *Scenedesmus obliquus* for use in food processing. *Journal of Food Engineering* 116: 478–82. https://doi.org/10.1016/j.jfoodeng.2012.12.015.

Gunathilaka, T.L., K.W. Samarakoon, P. Ranasinghe, and L.D.C. Peiris. 2019. *In-vitro* antioxidant, hypoglycemic activity, and identification of bioactive compounds in phenol-rich extract from the marine red algae *Gracilaria edulis* (Gmelin) Silva. *Molecules* 24 (October): 3708, 1–17. https://doi.org/10.3390/molecules24203708.

Guschina I.A., and J.L. Harwood. 2006. Lipids and lipid metabolism in eukaryotic algae. *Progress in Lipid Research* 45: 160–86. https://doi.org/10.1016/j.plipres.2006.01.001.

Hallegraeff, G.M. 2003. Harmful algal blooms: A global review. In *Manual on harmful marine microalgae*, eds. G.M. Hallegraeff, D.M. Anderson, and A.D. Cembella, 25–49. Place de Fontenoy, Paris: UNESCO Publishing. https://unesdoc.unesco.org/ark:/48223/pf0000131711.

Hamed, I., F. Özogul, Y. Özogul, and J.M. Regenstein. 2015. Marine bioactive compounds and their health benefits: A review. *Comprehensive Reviews in Food Science and Food Safety* 14: 446–65. https://doi.org/10.1111/1541-4337.12136.

Haoujar, I., F. Cacciola, J. Abrini, et al. 2019. The contribution of carotenoids, phenolic compounds, and flavonoids to the antioxidative properties of marine microalgae isolated from Mediterranean Morocco. *Molecules* 24 (November): 4037, 1–17. https://doi.org/10.3390/molecules24224037.

Haugan, J.A., and S. Liaaen-Jensen. 1994. Algal carotenoids 54. Carotenoids of brown algae (Phaeophyceae). *Biochemical Systematics and Ecology* 22: 31–41. https://doi.org/10.1016/0305-1978(94)90112-0.

Heo, S.J., S.C. Ko, S.M. Kang, et al. 2008. Cytoprotective effect of fucoxanthin isolated from brown algae *Sargassum siliquastrum* against H2O2-induced cell damage. *European Food Research and Technology* 228: 145–51. https://doi.org/10.1007/s00217-008-0918-7.

Hernández, D., M. Solana, B. Riaño, M.C. García-González, and A. Bertucco. 2014. Biofuels from microalgae: Lipid extraction and methane production from the residual biomass in a biorefinery approach. *Bioresource Technology* 170: 370–78. https://doi.org/10.1016/j.biortech.2014.07.109.

Herrera, M.C., and M.L. De Castro. 2005. Ultrasound-assisted extraction of phenolic compounds from strawberries prior to liquid chromatographic separation and photodiode array ultraviolet detection. *Journal of Chromatography A* 1100: 1–7. https://doi.org/10.1016/j.chroma.2005.09.021.

Ho, S.H., J.S. Chang, Y.Y. Lai, and C.N.N. Chen. 2014. Achieving high lipid productivity of a thermotolerant microalga *Desmodesmus* sp. F2 by optimizing environmental factors and nutrient conditions. *Bioresource Technology* 156: 108–16. https://doi.org/10.1016/j.biortech.2014.01.017.

Ibañez, E., M. Herrero, J.A. Mendiola, and M. Castro-Puyana. 2012. Extraction and characterization of bioactive compounds with health benefits from marine resources: Macro and micro algae, cyanobacteria, and invertebrates. In *Marine bioactive compounds: Sources, characterization and applications*, ed. M. Hayes, 55–98. Boston, MA: Springer. https://doi.org/10.1007/978-1-4614-1247-2_2.

Ilyas, Z., A.A. Redha, Y.S. Wu, F.Z. Ozeer, and R.E. Aluko. 2023. Nutritional and health benefits of the brown seaweed *Himanthalia elongata*. *Plant Foods for Human Nutrition* 78: 233–42. https://doi.org/10.1007/s11130-023-01056-8.

Imbs, T.I., and T.N. Zvyagintseva. 2018. Phlorotannins are polyphenolic metabolites of brown algae. *Russian Journal of Marine Biology* 44: 263–73. https://doi.org/10.1134/S106307401804003X.

Imjongjairak, S., K. Ratanakhanokchai, N. Laohakunjit, C. Tachaapaikoon, P. Pason, and R. Waeonukul. 2016. Biochemical characteristics and antioxidant activity of crude and purified sulfated polysaccharides from *Gracilaria fisheri*. *Bioscience, Biotechnology, and Biochemistry* 80: 524–32. https://doi.org/10.1080/09168451.2015.1101334.

Ishii, A., T. Okino, M. Suzuki, and Y. Machiguchi. 2004. Tichocarpols A and B, two novel phenylpropanoids with feeding-deterrent activity from the red alga *Tichocarpus crinitus*. *Journal of Natural Products* 67: 1764–66. https://doi.org/10.1021/np0498509.

Jain, S., A. Gupta, and N. Verma. 2023. *Industrial applications of soil microbes*, vol. 2. Bentham Science Publishers Pte Ltd. https://doi.org/10.2174/97898150502641230201.

Jain, S., A. Gupta, and N. Verma. 2024. *Industrial applications of soil microbes*, vol. 4. Bentham Science Publishers Pte Ltd. https://doi.org/10.2174/97898151249961240401.

Jeon, S., B.R. Jeong, and Y.K. Chang. 2017. Chemicals and fuels from microalgae. In *Consequences of microbial interactions with hydrocarbons, oils, and lipids: Production of fuels and chemicals. Handbook of hydrocarbon and lipid microbiology*, ed. S. Lee, 1–22. Cham: Springer. https://doi.org/10.1007/978-3-319-31421-1_384-2.

Jiang R.W., A.L. Lane, L. Mylacraine, et al. 2008. Structures and absolute configurations of sulfate-conjugated triterpenoids including an antifungal chemical defense of the green macroalga *Tydemania expeditionis*. *Journal of Natural Products* 71: 1616–19. https://doi.org/10.1021/np800307h.

Jin, W., W. Zhang, J. Wang, et al. 2014. Characterization of laminaran and a highly sulfated polysaccharide from *Sargassum fusiforme*. *Carbohydrate Research* 385: 58–64. https://doi.org/10.1016/j.carres.2013.12.009.

Kaebernick, M., and B.A. Neilan. 2001. Ecological and molecular investigations of cyanotoxin production. *FEMS Microbiology Ecology* 35: 1–9. https://doi.org/10.1111/j.1574-6941.2001.tb00782.x.

Kalitnik, A.A., A.O.B. Barabanova, V.P. Nagorskaya, et al. 2013. Low molecular weight derivatives of different carrageenan types and their antiviral activity. *Journal of Applied Phycology* 25: 65–72. https://doi.org/10.1007/s10811-012-9839-8.

Karemore, A., and R. Sen. 2016. Downstream processing of microalgal feedstock for lipid and carbohydrate in a biorefinery concept: A holistic approach for biofuel applications. *RSC Advances* 6: 29486–96. https://doi.org/10.1039/C6RA01477A.

Kaur, S., and M. Das. 2011. Functional foods: An overview. *Food Science and Biotechnology* 20 (August): 861. https://doi.org/10.1007/s10068-011-0121-7.

Kehr, J.C., D.G. Picchi, and E. Dittmann. 2011. Natural product biosyntheses in cyanobacteria: A treasure trove of unique enzymes. *Beilstein Journal of Organic Chemistry* 7: 1622–35. https://doi.org/10.3762/bjoc.7.191.

Khalil, Z.I., M.M.S. Asker, S. El-Sayed, and I.A. Kobbia. 2010. Effect of pH on growth and biochemical responses of *Dunaliella bardawil* and *Chlorella ellipsoidea*. *World Journal of Microbiology and Biotechnology* 26: 1225–31. https://doi.org/10.1007/s11274-009-0292-z.

Khoo, K.S., K.W. Chew, G.Y. Yew, S. Manickam, C.W. Ooi, and P.L. Show. 2020. Integrated ultrasound-assisted liquid biphasic flotation for efficient extraction of astaxanthin from *Haematococcus pluvialis*. *Ultrasonics Sonochemistry* 67 (October): 105052. https://doi.org/10.1016/j.ultsonch.2020.105052.

Kim, E.K., Y. Tang, Y.S. Kim, et al. 2015. First evidence that *Ecklonia cava*-derived dieckol attenuates MCF-7 human breast carcinoma cell migration. *Marine Drugs* 13: 1785–97. https://doi.org/10.3390/md13041785.

Kim, S.K., N.V. Thomas, and X. Li. 2011. Anticancer compounds from marine macroalgae and their application as medicinal foods. *Advances in Food and Nutrition Research* 64: 213–24. https://doi.org/10.1016/B978-0-12-387669-0.00016-8.

Klejdus, B., J. Kopecký, L. Benešová, and J. Vacek. 2009. Solid-phase/supercritical-fluid extraction for liquid chromatography of phenolic compounds in freshwater microalgae and selected cyanobacterial species. *Journal of Chromatography A* 1216: 763–71. https://doi.org/10.1016/j.chroma.2008.11.096.

Klejdus, B., L. Lojková, M. Plaza, M. Šnóblová, and D. Štěrbová. 2010. Hyphenated technique for the extraction and determination of isoflavones in algae: Ultrasound-assisted supercritical fluid extraction followed by fast chromatography with tandem mass spectrometry. *Journal of Chromatography A* 1217: 7956–65. https://doi.org/10.1016/j.chroma.2010.07.020.

Koivikko, R., J. Loponen, T. Honkanen, and V. Jormalainen. 2005. Contents of soluble, cell-wall-bound and exuded phlorotannins in the brown alga *Fucus vesiculosus*, with implications on their ecological functions. *Journal of Chemical Ecology* 31: 195–212. https://doi.org/10.1007/s10886-005-0984-2.

Kolanjinathan, K., P. Ganesh, and P. Saranraj. 2014. Pharmacological importance of seaweeds: A review. *World Journal of Fish and Marine Sciences* 6: 1–15. http://idosi.org/wjfms/wjfms6(1)14/1.pdf.

Kuczynska, P., M. Jemiola-Rzeminska, and K. Strzalka. 2015. Photosynthetic pigments in diatoms. *Marine Drugs* 13: 5847–81. https://doi.org/10.3390/md13095847.

Kumar, M., P. Kumari, V. Gupta, C.R.K. Reddy, and B. Jha. 2010. Biochemical responses of red alga *Gracilaria corticata* (Gracilariales, Rhodophyta) to salinity induced oxidative stress. *Journal of Experimental Marine Biology and Ecology* 391: 27–34. https://doi.org/10.1016/j.jembe.2010.06.001.

Kumar, R.R., P.H. Rao, and M. Arumugam. 2015. Lipid extraction methods from microalgae: A comprehensive review. *Frontiers in Energy Research* 2 (January): 61, 1–9. https://doi.org/10.3389/fenrg.2014.00061.

Kumar, S.S., Y. Kumar, M.S.Y. Khan, J. Anbu, and E. De Clercq. 2011. Antihistaminic and antiviral activities of steroids of *Turbinaria conoides*. *Natural Product Research* 25: 723–29. https://doi.org/10.1080/14786411003781515.

Kwak, H.S., J.Y.H. Kim, H.M. Woo, E. Jin, B.K. Min, and S.J. Sim. 2016. Synergistic effect of multiple stress conditions for improving microalgal lipid production. *Algal Research* 19: 215–24. https://doi.org/10.1016/j.algal.2016.09.003.

Laurion, I., and S. Roy. 2009. Growth and photoprotection in three dinoflagellates (including two strains of *Alexandrium tamarense*) and one diatom exposed to four weeks of natural and enhanced ultraviolet-b radiation[1]. *Journal of Phycology* 45: 16–33. https://doi.org/10.1111/j.1529-8817.2008.00618.x.

Leal, M.C., M.H.G. Munro, J.W. Blunt, et al. 2013. Biogeography and biodiscovery hotspots of macroalgal marine natural products. *Natural Product Reports* 30: 1380–90. https://doi.org/10.1039/C3NP70057G.

Leflaive, J.P., and L. Ten-Hage. 2007. Algal and cyanobacterial secondary metabolites in freshwaters: A comparison of allelopathic compounds and toxins. *Freshwater Biology* 52: 199–214. https://doi.org/10.1111/j.1365-2427.2006.01689.x.

Li, Y., X. Fu, D. Duan, X. Liu, J. Xu, and X. Gao. 2017. Extraction and identification of phlorotannins from the brown alga, *Sargassum fusiforme* (Harvey) Setchell. *Marine Drugs* 15 (February): 49, 1–15. https://doi.org/10.3390/md15020049.

Li, Y., F.G. Naghdi, S. Garg, et al. 2014. A comparative study: The impact of different lipid extraction methods on current microalgal lipid research. *Microbial Cell Factories* 13 (January): 14, 1–9. https://doi.org/10.1186/1475-2859-13-14.

Li, Y., Z.J. Qian, B. Ryu, S.H. Lee, M.M. Kim, and S.K. Kim. 2009. Chemical components and its antioxidant properties *in vitro*: An edible marine brown alga, *Ecklonia cava*. *Bioorganic & Medicinal Chemistry* 17: 1963–73. https://doi.org/10.1016/j.bmc.2009.01.031.

Liu, X., C.Y. Wang, C.L. Shao, et al. 2009. Chemical constituents from *Sargassum pallidum* (Turn.) C. Agardh.

Biochemical Systematics and Ecology 37: 127–29. https://doi.org/10.1016/j.bse.2009.01.009.

Luo, H., B. Wang, C. Yu, and Y. Xu. 2010. Optimization of microwave-assisted extraction of polyphenols from *Enteromorpha prolifra* by orthogonal test. *Chinese Herbal Medicines* 2: 321–25. https://doi.org/10.3969/j.issn.1674-6384.2010.04.011.

Machu, L., L. Misurcova, J.V. Ambrozova, et al. 2015. Phenolic content and antioxidant capacity in algal food products. *Molecules* 20: 1118–33. https://doi.org/10.3390/molecules20011118.

Marathe, S.J., S.B. Jadhav, S.B. Bankar, K.K. Dubey, and R.S. Singhal. 2019. Improvements in the extraction of bioactive compounds by enzymes. *Current Opinion in Food Science* 25: 62–72. https://doi.org/10.1016/j.cofs.2019.02.009.

Marinho-Soriano, E., P.C. Fonseca, M.A.A. Carneiro, and W.S.C. Moreira. 2006. Seasonal variation in the chemical composition of two tropical seaweeds. *Bioresource Technology* 97: 2402–06. https://doi.org/10.1016/j.biortech.2005.10.014.

Marzorati, S., A. Schievano, A. Idàb, and L, Verotta. 2020. Carotenoids, chlorophylls and phycocyanin from Spirulina: Supercritical CO2 and water extraction methods for added value products cascade. *Green Chemistry* 22: 187–96. https://doi.org/10.1039/C9GC03292D.

Masuda, M., S. Kawaguchi, T. Abe, T. Kawamoto, and M. Suzuki. 2002. Additional analysis of chemical diversity of the red algal genus *Laurencia* (Rhodomelaceae) from Japan. *Phycological Research* 50: 135–44. https://doi.org/10.1046/j.1440-1835.2002.00267.x.

Meslet-Cladière, L., L. Delage, C.J.J. Leroux, et al. 2013. Structure/function analysis of a type III polyketide synthase in the brown alga *Ectocarpus siliculosus* reveals a biochemical pathway in phlorotannin monomer biosynthesis. *The Plant Cell* 25: 3089–103. https://doi.org/10.1105/tpc.113.111336.

Milano, J., H.C. Ong, H.H. Masjuki, et al. 2016. Microalgae biofuels as an alternative to fossil fuel for power generation. *Renewable and Sustainable Energy Reviews* 58: 180–97. https://doi.org/10.1016/j.rser.2015.12.150.

Milledge, J.J., A. Staple, and P.J. Harvey. 2015. Slow pyrolysis as a method for the destruction of Japanese wireweed, *Sargassum muticum*. *Environment and Natural Resources Research* 5: 28–37. http://dx.doi.org/10.5539/enrr.v5n1p28.

Monroy-Ruiz, J., M.Á. Sevilla, R. Carrón, and M.J. Montero. 2011. Astaxanthin-enriched-diet reduces blood pressure and improves cardiovascular parameters in spontaneously hypertensive rats. *Pharmacological Research* 63: 44–50. https://doi.org/10.1016/j.phrs.2010.09.003.

Moon, H.E., M.N. Islam, B.R. Ahn, et al. 2011. Protein tyrosine phosphatase 1B and α-glucosidase inhibitory phlorotannins from edible brown algae, *Ecklonia stolonifera* and *Eisenia bicyclis*. *Bioscience, Biotechnology, and Biochemistry* 75: 1472–80. https://doi.org/10.1271/bbb.110137.

Mudimu, O., I.K. Koopmann, N. Rybalka, T. Friedl, R. Schulz, and W. Bilger. 2017. Screening of microalgae and cyanobacteria strains for α-tocopherol content at different growth phases and the influence of nitrate reduction on α-tocopherol production. *Journal of Applied Phycology* 29: 2867–75. https://doi.org/10.1007/s10811-017-1188-1.

Mulvenna, V., K. Dale, B. Priestly, et al. 2012. Health risk assessment for cyanobacterial toxins in seafood. *International Journal of Environmental Research and Public Health* 9: 807–20. https://doi.org/10.3390/ijerph9030807.

Murthy, K.N.C., A. Vanitha, J. Rajesha, M.M. Swamy, P.R. Sowmya, and G.A. Ravishankar. 2005. *In vivo* antioxidant activity of carotenoids from *Dunaliella salina* – a green microalga. *Life Sciences* 76: 1381–90. https://doi.org/10.1016/j.lfs.2004.10.015.

Nagayama, K., Y. Iwamura, T. Shibata, I. Hirayama, and T. Nakamura. 2002. Bactericidal activity of phlorotannins from the brown alga *Ecklonia kurome*. *Journal of Antimicrobial Chemotherapy* 50: 889–93. https://doi.org/10.1093/jac/dkf222.

Oben, J., E. Enonchong, D. Kuate, et al. 2007. The effects of ProAlgaZyme novel algae infusion on metabolic syndrome and markers of cardiovascular health. *Lipids in Health and Disease* 6 (September): 20, 1–9. https://doi.org/10.1186/1476-511X-6-20.

Pagels, F., A.C. Guedes, H.M. Amaro, A. Kijjoa, and V. Vasconcelos. 2019 Phycobiliproteins from cyanobacteria: Chemistry and biotechnological applications. *Biotechnology Advances* 37: 422–43. https://doi.org/10.1016/j.biotechadv.2019.02.010.

Paliwal, C., M. Mitra, K. Bhayani, et al. 2017. Abiotic stresses as tools for metabolites in microalgae. *Bioresource Technology* 244: 1216–26. https://doi.org/10.1016/j.biortech.2017.05.058.

Patil, G., S. Chethana, M.C. Madhusudhan, and K.S.M.S. Raghavarao. 2008. Fractionation and purification of the phycobiliproteins from *Spirulina platensis*. *Bioresource Technology* 99: 7393–96. https://doi.org/10.1016/j.biortech.2008.01.028.

Patra, B., C. Schluttenhofer, Y. Wu, S. Pattanaik, and L. Yuan. 2013. Transcriptional regulation of secondary metabolite biosynthesis in plants. *Biochimica et Biophysica Acta (BBA)-Gene Regulatory Mechanisms* 1829: 1236–47. https://doi.org/10.1016/j.bbagrm.2013.09.006.

Pavia, H., G.B. Toth, and P. Åberg. 2002. Optimal defense theory: Elasticity analysis as a tool to predict intraplant variation in defenses. *Ecology* 83: 891–97. https://doi.org/10.1890/0012-9658(2002)083[0891:ODTEAA]2.0.CO;2.

Pearson, L., T. Mihali, M. Moffitt, R. Kellmann, and B. Neilan. 2010. On the chemistry, toxicology and genetics of the cyanobacterial toxins, microcystin, nodularin, saxitoxin and cylindrospermopsin. *Marine Drugs* 8: 1650–80. https://doi.org/10.3390/md8051650.

Peng, J., J.P. Yuan, C.F. Wu, and J.H. Wang. 2011. Fucoxanthin, a marine carotenoid present in brown seaweeds and diatoms: Metabolism and bioactivities relevant to human health. *Marine Drugs* 9: 1806–28. https://doi.org/10.3390/md9101806.

Pereira, P., R. Li, W. Carmichael, E. Dias, and S. Franca. 2004. Taxonomy and production of paralytic shellfish toxins by the freshwater cyanobacterium *Aphanizomenon gracile* LMECYA40. *European Journal of Phycology* 39: 361–68. https://doi.org/10.1080/09670260410001714723.

Perez-Rodriguez, E., J. Aguilera, and F.L. Figueroa. 2003. Tissular localization of coumarins in the green alga *Dasycladus vermicularis* (Scopoli) Krasser: A photoprotective role? *Journal of Experimental Botany* 54: 1093–100. https://doi.org/10.1093/jxb/erg111.

Pinto, E., T.C.S. Sigaud-kutner, M.A.S. Leitão, O.K. Okamoto, D. Morse, and P. Colepicolo. 2003. Heavy metal–induced oxidativestress in algae. *Journal of Phycology* 39: 1008–18. https://doi.org/10.1111/j.0022-3646.2003.02–193.x.

Ragan, M.A., and W.D. Jamieson. 1982. Oligomeric polyphloroglucinols from *Fucus vesiculosus*: Photoplate mass spectrometric investigation. *Phytochemistry* 21: 2709–11. https://doi.org/10.1016/0031-9422(82)83103–8.

Rajauria, G. 2018. Optimization and validation of reverse phase HPLC method for qualitative and quantitative assessment of polyphenols in seaweed. *Journal of Pharmaceutical and Biomedical Analysis* 148: 230–37.

Reddy, P., and S. Urban. 2009. Meroditerpenoids from the southern Australian marine brown alga *Sargassum fallax*. *Phytochemistry* 70: 250–55. https://doi.org/10.1016/j.phytochem.2008.12.007.

Rupert, R., K.F. Rodrigues, V.Y. Thien, and W.T.L. Yong. 2022. Carrageenan from *Kappaphycus alvarezii* (Rhodophyta, Solieriaceae): Metabolism, structure, production, and application. *Frontiers in Plant Science* 13 (May): 859635, 1–16. https://doi.org/10.3389/fpls.2022.859635.

Sampath-Wiley, P., C.D. Neefus, and L.S. Jahnke. 2008. Seasonal effects of sun exposure and emersion on intertidal seaweed physiology: Fluctuations in antioxidant contents, photosynthetic pigments and photosynthetic efficiency in the red alga *Porphyra umbilicalis* Kützing (Rhodophyta, Bangiales). *Journal of Experimental Marine Biology and Ecology* 361: 83–91. https://doi.org/10.1016/j.jembe.2008.05.001.

Schoenwaelder, M.E.A. 2002. The occurrence and cellular significance of physodes in brown algae. *Phycologia* 41: 125–39. https://doi.org/10.2216/i0031-8884-41-2-125.1.

Sharifuddin, Y., Y.X. Chin, P.E. Lim, and S.M. Phang. 2015. Potential bioactive compounds from seaweed for diabetes management. *Marine Drugs* 13: 5447–91. https://doi.org/10.3390/md13085447.

Shirsath, S.R., S.H. Sonawane, and P.R. Gogate. 2012. Intensification of extraction of natural products using ultrasonic irradiations – A review of current status. *Chemical Engineering and Processing: Process Intensification* 53: 10–23 https://doi.org/10.1016/j.cep.2012.01.003.

Shuba, E.S., and D. Kifle. 2018. Microalgae to biofuels: 'Promising'alternative and renewable energy, review. *Renewable and Sustainable Energy Reviews* 81: 743–55. https://doi.org/10.1016/j.rser.2017.08.042.

Singh, I.P., and J. Sidana. 2013. Phlorotannins. In *Functional ingredients from algae for foods and nutraceuticals*, ed. H. Domínguez, 181–204. Elsevier: Woodhead Publishing. https://doi.org/10.1533/9780857098689.1.181.

Steevensz, A.J., S.L. MacKinnon, R. Hankinson, et al. 2012. Profiling phlorotannins in brown macroalgae by liquid chromatography–high resolution mass spectrometry. *Phytochemical Analysis* 23: 547–53. https://doi.org/10.1002/pca.2354.

Stengel, D.B., S. Connan, and Z.A. Popper. 2011. Algal chemodiversity and bioactivity: Sources of natural variability and implications for commercial application. *Biotechnology Advances* 29: 483–501. https://doi.org/10.1016/j.biotechadv.2011.05.016.

Stewart, I., P.J. Schluter, and G.R. Shaw. 2006. Cyanobacterial lipopolysaccharides and human health–a review. *Environmental Health* 5 (March): 7, 1–23. https://doi.org/10.1186/1476-069X-5–7.

Swanson, A.K., and L.D. Druehl. 2002. Induction, exudation and the UV protective role of kelp phlorotannins. *Aquatic Botany* 73: 241–53. https://doi.org/10.1016/S0304-3770(02)00035–9.

Takaichi, S., and M. Mochimaru. 2007. Carotenoids and carotenogenesis in cyanobacteria: Unique ketocarotenoids and carotenoid glycosides. *Cellular and Molecular Life Sciences* 64 (July): 2607. https://doi.org/10.1007/s00018-007-7190-z.

Tang, H.F., Y.H. Yi, X.S. Yao, Q.Z. Xu, S.Y. Zhang, and H.W. Lin. 2002. Bioactive steroids from the brown alga *Sargassum carpophyllum*. *Journal of Asian Natural Products Research* 4: 95–101. https://doi.org/10.1080/10286020290027362.

Tonon, T., D. Harvey, T.R. Larson, and I.A. Graham. 2002. Long chain polyunsaturated fatty acid production and partitioning to triacylglycerols in four microalgae. *Phytochemistry* 61: 15–24. https://doi.org/10.1016/S0031-9422(02)00201–7.

Toth, G.B., and H. Pavia. 2000. Water-borne cues induce chemical defense in a marine alga (*Ascophyllum nodosum*). *Proceedings of the National Academy of Sciences* 97: 14418–20. https://doi.org/10.1073/pnas.250226997.

Tsao, R. 2010. Chemistry and biochemistry of dietary polyphenols. *Nutrients* 2: 1231–46. https://doi.org/10.3390/nu2121231.

Vanthoor-Koopmans, M., R.H. Wijffels, M.J. Barbosa, and M.H.M. Eppink. 2013. Biorefinery of microalgae for food and fuel. *Bioresource Technology* 135: 142–49. https://doi.org/10.1016/j.biortech.2012.10.135.

Ventura, S.P.M., B.P. Nobre, F. Ertekin, et al. 2017. Extraction of value added compounds from microalgae. In *Microalgae-based biofuels and bioproducts*, eds. C. Gonzalez-Fernandez and R. Muñoz, 461–83. Elsevier and Woodhead Publishing. https://doi.org/10.1016/B978-0-08-101023-5.00019-4.

Viano, Y., D. Bonhomme, A. Ortalo-Magné, et al. 2011. Dictyotadimer A, a new dissymmetric bis-diterpene from a brown alga of the genus *Dictyota*. *Tetrahedron Letters* 52: 1031–35. https://doi.org/10.1016/j.tetlet.2010.12.095.

Vuppaladadiyam, A.K., P. Prinsen, A. Raheem, R. Luque, and M. Zhao. 2018. Microalgae cultivation and metabolites production: A comprehensive review. *Biofuels, Bioproducts and Biorefining* 12: 304–24. https://doi.org/10.1002/bbb.1864.

Wang, L., and C.L. Weller. 2006. Recent advances in extraction of nutraceuticals from plants. *Trends in Food Science & Technology* 17: 300–12. https://doi.org/10.1016/j.tifs.2005.12.004.

Watson, S.B., and E. Cruz-Rivera. 2003. Algal chemical ecology: An introduction to the special issue. *Phycologia* 42: 319–23. https://doi.org/10.2216/i0031-8884-42-4-319.1.

Wells, M.L., P. Potin, J.S. Craigie, et al. 2017. Algae as nutritional and functional food sources: Revisiting our understanding. *Journal of Applied Phycology* 29: 949–82. https://doi.org/10.1007/s10811-016-0974-5.

Whitfield, F.B., F. Helidoniotis, K.J. Shaw, and D. Svoronos. 1999. Distribution of bromophenols in species of marine algae from eastern Australia. *Journal of Agricultural and Food Chemistry* 47: 2367–73. https://doi.org/10.1021/jf981080h.

Wijesinghe, W.A.J.P., and Y.J. Jeon. 2011. Biological activities and potential cosmeceutical applications of bioactive components from brown seaweeds: A review. *Phytochemistry Reviews* 10: 431–43. https://doi.org/10.1007/s11101-011-9214-4.

Wilkie A.C., S.J. Edmundson, and J.G. Duncan. 2011. Indigenous algae for local bioresource production: Phycoprospecting. *Energy for Sustainable Development* 15: 365–71. https://doi.org/10.1016/j.esd.2011.07.010.

Witvrouw, M., and E. De Clercq. 1997. Sulfated polysaccharides extracted from sea algae as potential antiviral drugs. *General Pharmacology: The Vascular System* 29: 497–511. https://doi.org/10.1016/S0306-3623(96)00563–0.

Yen, H.W., S.C. Yang, C.H. Chen, Jesisca, and J.S. Chang. 2015. Supercritical fluid extraction of valuable compounds from microalgal biomass. *Bioresource Technology* 184: 291–96. https://doi.org/10.1016/j.biortech.2014.10.030.

Young, R.M., K.M. Schoenrock, J.L. von Salm, C.D. Amsler, and B.J. Baker. 2015. Structure and function of macroalgal natural products. In *Natural products from marine algae. Methods in molecular biology*, eds. D. Stengel and S. Connan, 39–73. New York: Humana Press. https://doi.org/10.1007/978-1-4939-2684-8_2.

Zeng, L.M., C.J. Wang, J.Y. Su, et al. 2001. Flavonoids from the red alga *Acanthophora spicifera*. *Chinese Journal of Chemistry* 19: 1097–100. https://doi.org/10.1002/cjoc.20010191116.

Zolotariova, Y.K., V.M. Mokrosnop, and S.S. Stepanov. 2019. Polyphenol compounds of macroscopic and microscopic algae. *International Journal on Algae* 21: 5–24. https://doi.org/10.1615/InterJAlgae.v21.i1.10.

18 Microalgae as a Sustainable Source of Value-Added Secondary Metabolites

Manisha Bisht
(Department of Chemistry, M.B. Govt. Post Graduate College, Haldwani 263139, India; manishatanuja@gmail.com)

Pramod Rawat
(Department of Biotechnology, Graphic Era (Deemed-to-Be University), Dehradun 248002, India; pramodrawat.bt@geu.ac.in)

Yashaswi Singh
(Department of Biotechnology, Graphic Era (Deemed-to-Be University), Dehradun 248002, India; yashaswi.singh@gmail.com)

**Manoj Pal*
(Department of Microbiology, Graphic Era (Deemed-to-Be University), Dehradun 248002, India; manoj19823@gmail.com)

18.1 INTRODUCTION

In today's world, we are facing many environmental problems, and the demand for resources is increasing rapidly worldwide (Asif and Muneer, 2007). Because of this, finding sustainable alternatives has become incredibly important. Among the many possibilities being explored, microalgae have emerged as a captivating and promising solution, offering a multitude of benefits that span beyond their minuscule size (Ali et al., 2022). Microalgae, often overlooked in comparison to their larger botanical counterparts, are microscopic photosynthetic organisms that inhabit diverse aquatic environments worldwide (Metting, 1996). Despite their small size, these tiny powerhouses play an essential role in balancing our planet's ecosystems. As primary producers, they harness solar energy to convert carbon dioxide into oxygen, shaping the foundation of aquatic food webs and influencing global biogeochemical cycles. Beyond their ecological significance, microalgae exhibit a prodigious repertoire of biochemical diversity, producing an impressive array of secondary metabolites such as carotenoids, chlorophylls, phycobiliproteins, fatty acids, polyunsaturated fatty acids, antioxidants, bioactive lipids, phenolic compounds, vitamins, bioactive pigments, and sterols (Russell et al., 2022; Gupta et al., 2022, 2024).

These specialized compounds are distinct from primary metabolites required for basic cellular functions. For decades, researchers have unraveled the remarkable therapeutic, pharmaceutical, and industrial potential that lies within these bioactive molecules (Khavari et al., 2021). The importance of secondary metabolites cannot be overstated, as they serve as a source of pharmaceutical drugs, nutraceuticals, antioxidants, and other valuable products that cater to human health, wellness, and industrial needs (Kiran and Mohan, 2021; Jain et al., 2023, 2024). As society transitions toward more sustainable practices, the demand for eco-friendly alternatives to conventional resources has never been more pressing. In this context, microalgae's remarkable ability to synthesize and accumulate diverse secondary metabolites has drawn considerable attention from the scientific community and industries alike. The exploration of microalgae as a sustainable source of value-added secondary metabolites is underpinned by several compelling factors. Their rapid growth rates, high biomass productivity, and minimal land and water requirements make them attractive candidates for scalable cultivation with a reduced environmental footprint (Sarwer et al., 2022; Singh et al., 2011). Furthermore, their ability to thrive in diverse environmental conditions, including wastewater and saline environments, positions them as an adaptable and renewable resource for metabolite production (Suganya et al., 2016). This chapter aims to shed light on the diverse groups of microalgae and their potential to produce various types of secondary metabolites with their potential applications across various sectors. Moreover, the challenges and opportunities associated with the commercialization of microalgae-derived secondary metabolites are also discussed. In summary, this chapter aims to showcase the exceptional potential of microalgae as a sustainable source of value-added secondary metabolites, paving the way for greener industries, enhanced human well-being, and a more

DOI: 10.1201/9781003408833-18

harmonious coexistence with our precious planet. As the boundaries of biotechnology and sustainable practices continue to expand, microalgae stand poised to play a pivotal role in shaping a brighter and more sustainable future for generations to come.

18.2 COMMON HIGH-VALUE MICROALGAE SPECIES

Microalgae encompass a vast and diverse group of microscopic organisms, commonly found in aquatic environments such as oceans, lakes, rivers, and even in damp terrestrial habitats. They belong to various taxonomic groups, including cyanobacteria (blue-green algae), green algae, diatoms, dinoflagellates, and euglenoids, among others. The sheer abundance and adaptability of microalgae species make them an intriguing subject of study for researchers exploring their potential applications. Each microalgae species possesses unique characteristics and biochemical compositions, leading to significant variations in their ecological roles and capabilities. Some species are adept at thriving in extreme conditions, such as high salinity or temperature, while others are more resilient to fluctuations in nutrient availability. This diverse range of species contributes to the potential discovery of novel microalgae with valuable secondary metabolites that could be harnessed for various applications. Some common microalgae and their derived secondary metabolites are shown in Table 18.1.

Within the vast pool of microalgae, certain species have garnered particular attention due to their production of high-value secondary metabolites. For instance:

18.2.1 *HAEMATOCOCCUS PLUVIALIS*

It is a single-celled freshwater microalgae belonging to the Chlorophyceae family. It is found in various aquatic habitats worldwide and has gained significant recognition as the richest and most promising source for natural astaxanthin production on a commercial scale (Cheong et al., 2020; Lim et al., 2018). Astaxanthin, also known as (3,3′-dihydroxy-β, β-1-carotene-4,4′-dione), is a bright blood-red secondary carotenoid. *H. pluvialis* can directly synthesize astaxanthin by subjecting its cells to specific stresses (Shah et al., 2016).

18.2.2 *CHLORELLA VULGARIS*

It belongs to the Oocytaceae family and measures between 2 to 10 μm in size. It is known for its high protein content, accounting for 42–58% of the dry weight of its biomass. Numerous secondary metabolites have been extracted from *C. vulgaris*, including carotenoids, lutein, astaxanthin, mycosporine-like amino acids (MAAs), omega-3 and omega-6 fatty acids, and phenolic compounds.

TABLE 18.1
Microalgae and Their Derived Value-Added Secondary Metabolites

S. No.	Microalgae	Secondary metabolites	Reference
1	*Halamphora coffeaeformis*, *Phaeodactylum tricornutum*, and *Odontella auritacan*	Fucoxanthin	Abu-Ghosh et al., 2021
2	*Olisthodiscus luteus*	Fucosterol	Sañé et al., 2023
3	*Oscillatoria* sp.	Hoshinolactam	Verma et al., 2022
4	*Nostoc commune*	Mycosporine-like amino acids (MAAs)	Matsui et al., 2011
5	*Lyngbya majuscula*	Curacin A	Gerwick et al., 1994
6	*Symploca* sp.	Dolastatin 10	Luesch et al., 2001
7	*Nostoc flagelliforme*	Nostoflan	Kanekiyo et al., 2005
8	*Microcystis ichthyoblabe*	Cyclic depsipeptides and ichthyopeptins A and B	Zainuddin et al., 2007
9	*Moorea producens*	Microcolin A and malyngamide S	Gerwick and Moore, 2012
10	*Haematococcus lacustris*	Astaxanthin	Oslan et al., 2022
11	*Schizochytrium* sp.	DHA	Perez-Velazquez et al., 2019
12	*Tetraselmis* sp.	PUFA (Omega 3 (ω3), omega 6 (ω6), and omega 9 (ω9))	Dammak et al., 2016
13	*Amphidinium carterae*	Amphidinols (APDNs)	Molina-Miras et al., 2018
14	*Oscillatoria* sp.	Hoshinolactam	Ogawa et al., 2017
15	*Symploca hydnoides* and *Lyngbya majuscule*	Dolastatins	Mondal et al., 2020
16	*Scytonema* sp.	Scytonemin	Rastogi et al., 2014

18.2.3 *Spirulina* spp.

It is a group of cyanobacteria, specifically a type of blue-green algae, that belongs to the family Cyanophyceae. It has gained considerable attention for its nutritional and potential health benefits, especially due to its high content of essential nutrients, proteins, vitamins, and minerals. *Spirulina* spp. is abundant in producing a range of valuable secondary metabolites, including phycocyanin, carotenoids, chlorophyll, gamma-linolenic acid (GLA), tocopherols (vitamin E), *etc.*

18.2.4 *Dunaliella salina*

Dunaliella salina is a type of halophilic (salt-loving), unicellular, non-motile, and biflagellate green alga that is commonly found in saline environments, such as salt flats, salt pans, and hypersaline lakes. It is named after the Italian botanist Michelangelo Merisi da Caravaggio, who used the pseudonym "*Dunaliella*" to refer to a type of seaweed. *Dunaliella salina* synthesizes various secondary metabolites such as beta-carotene, glyceryldiesters, linoleic acid and alpha-linolenic acid, mannitol, and tocopherols.

18.2.5 *Nannochloropsis* spp.

Nannochloropsis spp. are a genus of small (2 to 5 μm), unicellular, marine microalgae belonging to the class Eustigmatophyceae. They are widely distributed in marine environments and have gained significant attention in recent years due to their potential applications in various fields, including aquaculture, biofuel production, and biotechnology. Some common secondary metabolites isolated from *Nannochloropsis* spp. are eicosapentaenoic acid (EPA), docosahexaenoic acid (DHA), fucoxanthin, and phycobiliproteins.

18.2.6 *Chlorella* spp.

Chlorella spp. are a group of green microalgae belonging to the genus *Chlorella*. These single-celled organisms range in size from 2 to 10 μm and are widely distributed in freshwater environments, and some species can also be found in marine habitats. *Chlorella* is one of the most extensively studied microalgae due to its nutritional profile and potential applications in various industries. It is used to isolate secondary metabolites: chlorophylls, carotenoids, C-phycocyanin, gamma-linolenic acid (GLA), alpha-linolenic acid (ALA), vitamin B12, and chlorella growth factor (CGF).

18.2.7 *Isochrysis* spp.

Isochrysis spp. are a group of marine microalgae belonging to the class Eustigmatophyceae. These golden-brown microalgae are commonly found in marine environments and are an essential part of the marine food web. *Isochrysis* cells are typically oval or ellipsoidal and are small, measuring about 5 to 10 micrometers in diameter. They have a characteristic golden-brown colour due to the presence of pigments like fucoxanthin. *Isochrysis* spp. have gained considerable attention due to their unique biochemical composition and potential applications in various industries. *Isochrysis* spp. are used to isolate fucoxanthin, omega-3 fatty acids, mycosporine-like amino acids (MAAs), cyclic imides, and betaine lipids.

18.2.8 *Tetraselmis* spp.

Tetraselmis spp. are a group of marine microalgae belonging to the class Prasinophyceae. These green microalgae are commonly found in marine environments and are known for their ecological importance and potential applications in various industries. *Tetraselmis* cells are typically small and pear-shaped, with two anterior flagella and two posterior flagella, giving them a characteristic tetrahedral shape. The size of *Tetraselmis* cells can vary depending on the species, ranging from about 5 to 20 micrometers in diameter. Various secondary metabolites are isolated from *Tetraselmis* spp. such as fatty acids, carotenoids, phytosterols, tocopherols (vitamin E), and cyclic imides.

18.2.9 *Botryococcus braunii*

Botryococcus braunii is a species of green microalgae that belongs to the family Botryococcaceae. It is a unique and intriguing microorganism known for its ability to produce and secrete large quantities of hydrocarbons, making it of particular interest in the field of biofuel research and biotechnology. *Botryococcus braunii* is found in various aquatic environments, including freshwater, brackish water, and marine environments. It is a colonial microalga, meaning it forms colonies or clusters of cells. The individual cells are spherical to ovoid in shape and range in size from 10 to 40 micrometers in diameter. The colonies can vary in size, and they may be visible to the naked eye, appearing as greenish-brown globules on the water surface. Secondary metabolites isolated from *Botryococcus braunii* are triterpenoids and polysaccharides.

18.2.10 *Scenedesmus* spp.

Scenedesmus spp. are a group of green microalgae belonging to the class Chlorophyceae. They are widely distributed in freshwater environments, and some species can also be found in brackish water and marine habitats. *Scenedesmus* is one of the most common and well-studied genera of green algae. *Scenedesmus* cells are typically small and have a distinct colonial or filamentous organization. Each cell is single-celled and is usually oval or ellipsoidal, with two or more cells arranged in a row or side by side in a flat plane. The arrangement of cells in a colony gives *Scenedesmus* its characteristic star-like appearance. Various species are used to isolate the secondary metabolites such as carotenoids, fatty acids, tocopherols (vitamin E), phycobiliproteins, flavonoids, and glycolipids.

18.3 FACTORS INFLUENCING SECONDARY METABOLITE PRODUCTION IN MICROALGAE

The production of secondary metabolites in microalgae is influenced by several factors, including intrinsic and extrinsic factors. Understanding of these factors is critical for optimizing metabolite yield and developing cost-effective production strategies. Some key factors include:

18.3.1 Nutrient Availability

The availability of essential nutrients, such as nitrogen, phosphorus, and trace elements, can significantly impact metabolite synthesis (Leflaive and Ten-Hage, 2007). Nutrient limitation or excess can induce stress responses, leading to altered metabolite profiles (Gauthier et al., 2020).

18.3.2 Light Intensity and Quality

Light is an essential energy source for photosynthetic microalgae. The intensity and spectral quality of light can influence the rate of photosynthesis and, consequently, the production of secondary metabolites (Darko et al., 2014).

18.3.3 Temperature

Microalgae species have specific temperature ranges in which they thrive optimally (Shohael et al., 2006). Temperature variations can affect enzyme activities and biosynthetic pathways, thus influencing secondary metabolite production (Kumari et al., 2022).

18.3.4 Salinity

For marine microalgae, salinity plays a crucial role in regulating osmotic balance and physiological processes, which can, in turn, affect the secondary metabolite synthesis (BenMoussa-Dahmen et al., 2016).

18.3.5 Environmental Stress

Stressors such as UV radiation, high salinity, nutrient deprivation, and temperature extremes can induce microalgae to produce protective secondary metabolites (Chen et al., 2017).

18.3.6 Genetic and Strain Variability

Different strains and genetic variants within a species may exhibit varying metabolite production capabilities, highlighting the importance of strain selection and genetic manipulation for targeted metabolite yield (Park et al., 2022b).

By following the intricate interplay of these factors, researchers can devise cultivation strategies and biotechnological interventions to enhance the synthesis of desired secondary metabolites from microalgae. The discussion of microalgae diversity and potential delves into the vast array of microalgae species and their unique biochemical compositions, paving the way for exploring novel high-value microalgae with promising secondary metabolites. Moreover, understanding the factors influencing metabolite production is vital for harnessing the full potential of microalgae as a sustainable and renewable resource for value-added secondary metabolites across various industries. As researchers continue to unveil the untapped potential of these microscopic wonders, the future holds great promise for advancements in biotechnology and sustainable practices.

18.4 EXTRACTION AND ANALYTICAL TECHNIQUES

18.4.1 Methods for Extracting Secondary Metabolites from Microalgae

The extraction of secondary metabolites from microalgae presents a crucial step in harnessing their potential for various applications. A plethora of extraction methods have been employed to obtain these valuable bioactive compounds. Among the commonly used techniques are solvent extraction, supercritical fluid extraction, and microwave-assisted extraction. Solvent extraction involves the use of organic solvents to selectively dissolve the target metabolites from the microalgal biomass. Supercritical fluid extraction utilizes carbon dioxide in a supercritical state as a solvent to achieve efficient extraction at moderate temperatures, reducing thermal degradation of sensitive compounds. Microwave-assisted extraction employs microwave irradiation to enhance the extraction process, resulting in higher yields and reduced extraction times. Each method has its advantages and limitations, and researchers continuously explore innovative approaches to optimize extraction efficiency while considering factors like environmental impact and cost-effectiveness.

18.4.1.1 Solvent Extraction

Different organic solvents have been employed to extract valuable secondary metabolites, primarily based on the solubility of biomolecules. In a similar context, Kwang et al. (2008) utilized 100% ethanol supplemented with 1% butylated hydroxytoluene (BHT) to isolate mainly violaxanthin from *C. ellipsoidea* and lutein from *C. vulgaris*. Similarly, Fajardo et al. (2007) used 96% vol/vol ethanol followed by a two-phase system having hexane and water to extract oil-rich substrate such as eicosapentaenoic acid (EPA) from *Phaeodactylum tricornutum*. Fatty acid content from the dry mass of *Haematococcus pluvialis* was isolated by (Cerón et al., 2007) using 100% acetone followed by characterization using TLC. Tokuşoglu and Ünal, 2003, used water, chloroform, and methanol mixture to extract fatty acids from microalgae (*Spirulina platensis*, *Chlorella vulgaris*, and *Isochrysis galbana*).

Rizwanul Fattah et al. (2020) used hexane solvent in a Soxhlet condenser to isolate lipid extracts of *Chlorella* sp. and *Spirulina* sp. Patil and Kaliwal (2019) used a variety of solvents (chloroform, acetone, ethanol, methanol, and aqueous solvents) for the extraction of value-added products from *Scenedesmus bajacalifornicus* BBKLP–07. Group has employed Soxhlet extraction for the isolation of various phytochemical constituents, including alkaloids, flavonoids, glycosides, phenols, lignins, saponins, sterols, tannins, anthraquinone, and reducing sugar using the standard procedures. Ferreira et al. (2013) have shown the potential of Soxhlet using n-hexane as a solvent to extract oil compounds from *Nannochloropsis* sp. The obtained results have shown 40.7% (dry weight biomass) recovery of oil products from the alga mass. Although solvent extraction methods have been applied in the extraction of various biomolecules but nowadays, these methods are less popular due to their low recovery (Michalak and Chojnacka, 2014).

18.4.1.2 Supercritical Fluid Extraction (SFE)

Supercritical fluid extraction (SFE) is an advanced technique that uses a supercritical fluid, like carbon dioxide (CO_2), to extract compounds from solids or liquids (Anklam et al., 1998). The supercritical fluid's unique properties allow it to efficiently dissolve the desired compounds. SFE is eco-friendly, avoids organic solvents, and is ideal for heat-sensitive compounds due to its low-temperature operation. It finds applications in various industries, including food, pharmaceuticals, and herbal products (Herrero et al., 2010). SFE has been widely employed in the past for the extraction of several secondary metabolites (Michalak et al., 2016). The extraction of Carotenoids and fat-soluble vitamins from *T. obliquus* biomass was studied by Chronopoulou et al. (2019) using SFE and has demonstrated the effect of varying the SFE polarity on the efficiency of recovery. The potential of SFE in the recovery of β-carotenoid from *Botryococcus braunii*, *Chlorella vulgaris*, *Dunaliella salina*, and *Arthrospira maxima* was exhibited by (Mendes et al., 2003). The extraction of Astaxanthin and β-carotene from two major microalgal strains, *viz.*, *Haematococcus pluvialis* and *Dunaliella salina*, was comparatively optimized by Rammuni et al. (2019) using various extraction techniques, including SFE. Fujii (2012) used SFE methods to optimize the extraction of astaxanthin and chlorophyll from *Haematococcus pluvialis*. Macías-Sánchez et al. (2010) extracted lutein and β-carotene from *S. almeriensis* using SFE under different experimental conditions. SFE provides a rapid extraction rate and high yield while being environmentally friendly due to its reduced reliance on organic solvents. However, its drawbacks, such as the substantial initial investment required and the limited polarity of supercritical CO_2, hinder its widespread adoption.

18.4.1.3 Microwave-Assisted Extraction (MAE)

This method involves the use of microwave energy (300 MHz to 300 GHz) to facilitate the release of bioactive compounds from microalgae cells. The process works by generating rapid internal heating, which causes the cell walls to rupture and allows for the easy extraction of valuable secondary metabolites (Kubrakova and Toropchenova, 2008). One of the key advantages of MAE is its ability to significantly reduce extraction time compared to traditional methods, making it a time-efficient and productive approach. Additionally, MAE is considered environmentally friendly since it requires minimal or no use of harmful organic solvents, reducing the overall environmental impact (Mwaurah et al., 2020). Moreover, the controlled and mild heating conditions of MAE help in preserving the integrity and bioactivity of the extracted metabolites (Wen et al., 2020). There are two main types of MAE systems, namely, closed vessel and open vessel. Closed vessels are used for the extraction of target compounds at higher temperature and pressure conditions, whereas open vessel systems are used for extractions carried out at atmospheric pressure conditions (Kaufmann and Christen, 2002). In terms of cost-effectiveness, MAE proves to be a more viable choice than SFE. Nevertheless, it necessitates an additional separation step to eliminate solid residues, which is not required in SFE (Wang and Weller, 2006). MAE has been employed for the extraction of various valuable secondary metabolites from microalgae (Salinas-Salazar et al., 2019). A study conducted by Dang et al. (2018) has shown the potential of MAE for the extraction of total phenolic content of brown alga, *Sargassum vestitum,* and found more efficient than ultrasound-assisted extraction (UAE). Juin et al. (2015) employed MAE for the extraction of phycoerythrin (PE), phycocyanin (PC), and allophycocyanin (APC). They deduced that MAE proves to be a rapid and productive method for efficiently extracting and pre-purifying phycobiliproteins, even from microalgae with a substantial exopolysaccharide-rich cell wall. Toan et al. (2021) optimized the extraction of phlorotannin (PhE) and exhibited the highest yield of phlorotannin (5.59 ± 0.11 mg PhE/g). However, we can improve the recovery or yield of bioactive compounds by implementing changes, such as variation in operating conditions, repetitive extraction, combined approaches (such as UAE-assisted), and by exploring novel microwave system designs.

18.4.2 Analytical Techniques for Identification and Quantification of Secondary Metabolites

Accurate identification and quantification of extracted secondary metabolites are imperative for understanding their biological activities and potential applications (Beatrice-Lindner et al., 2018). A wide array of sophisticated analytical techniques has emerged to facilitate comprehensive analysis. Gas chromatography-mass spectrometry (GC-MS) and liquid chromatography-mass spectrometry

(LC-MS) are extensively used for their high sensitivity and ability to separate complex mixtures (Bhowmick et al., 2020). Nuclear magnetic resonance (NMR) spectroscopy offers valuable structural information, aiding in the elucidation of unknown compounds (Bisht et al., 2021). Additionally, high-performance liquid chromatography (HPLC) (Fernando et al., 2022) and spectrophotometric methods provide valuable insights into metabolite composition and concentration (Mehariya et al., 2021). Furthermore, advanced techniques like metabolomics, which involve the global profiling of metabolites, have become instrumental in understanding the holistic metabolic landscape of microalgae (Bisht et al., 2021). The integration of multiple analytical tools enables a more comprehensive characterization of secondary metabolites, empowering researchers to unveil the full potential of microalgae-derived compounds.

18.4.2.1 Chromatographic and Mass Spectroscopic Methods

The isolation and purification of these valuable compounds from microalgae require sophisticated techniques to separate complex mixtures into individual components. Chromatography, a powerful separation method, plays a crucial role in achieving the high levels of purity necessary for further study and application.

18.4.2.1.1 Thin-Layer Chromatography (TLC)

TLC involves the separation of compounds on a thin layer of adsorbent material, typically silica gel or aluminium oxide, coated onto a solid support such as a glass plate. A mixture of compounds is applied to the TLC plate, and as the solvent travels up the plate by capillary action, the individual compounds migrate at different rates based on their interactions with the stationary phase. Once separated, these compounds can be scraped off the plate, eluted, and collected for further analysis. Jaime et al. (2005) used preparative thin layer chromatography (TLC) with other sophisticated techniques, such as HPLC, to purify and determine different chlorophylls and carotenoids in *S. platensis* extract. The group has optimized the recovery of different carotenoids: β- and R-carotene, β-cryptoxanthin, asthaxanthin, zeaxanthin, lutein, echinenone, oscillaxanthin, and myxoxanthophyll. Rodríguez-Meizoso et al. (2008) explored the novel *Phormidium* sp. for different antioxidant compounds and used TLC with other methods, including HPLC and GC-MS. Bhuvana et al. (2019) analyzed phytochemical constituents: phenolic, flavonoids, pigments, protein, and carbohydrates from two microalgae species, *viz. Nannochloropsis oculata* and *Chlorella vulgaris*. They initially used TLC, followed by HPLC, and GC-MS for the characterization of bioactive compounds. Although TLC is a simple and traditional chromatographic method for the initial characterization of a mixed population of bioactive molecules, it is still not enough for a high level of purification.

18.4.2.1.2 High Performance Liquid Chromatography (HPLC)

High performance liquid chromatography (HPLC) is a powerful analytical technique used for the separation, identification, and quantification of various compounds in a mixture (Ren et al., 2008). HPLC has enabled researchers to study a wide range of secondary metabolites from algae, including pigments (such as chlorophylls and carotenoids), lipids, phenolic compounds, alkaloids, and other bioactive molecules (Mantoura and Llewellyn, 1983). It provides valuable information about the chemical composition of algae, which can have implications for various industries, including pharmaceuticals, cosmetics, and biotechnology. Inbaraj et al. (2006) have developed an HPLC-based improved method for the extraction of carotenoids from *Chlorella pyrenoidosa*. The group has resolved 32 carotenoids within 49 min by using a C30 column and internal standard β-apo-8′-carotenal with a flow rate of 1 mL/min and detection at 450 nm. Jin et al. (2019) used an HPLC method based on matrix solid-phase dispersion (MSPD) for the simultaneous determination of fucoxanthin, lutein, and astaxanthin in different microalgae. They used solid phase extraction (SPE) method using C8 adsorbent, the SPE column was made up of microalgal matrix/sorbent (1:2). The column first was first eluted with 4 mL methanol (containing 0.1% BHT) followed by the analysis of eluent by HPLC using C8 column and mobile phase; acetonitrile (ACN)/H_2O. In a similar approach, Manfellotto et al. (2020) used a gene overexpression method for the higher yield of carotenoids in the unicellular Alga, *Phaeodactylum tricornutum*. Castro-Puyana et al. (2013) developed an optimized method for the extraction of carotenoids from *Neochloris oleoabundans* using pressurized liquid extraction with food-grade solvents such as ethanol and limonene. The obtained extract was analyzed by high-performance liquid chromatography–tandem mass spectrometry. Todorović et al. (2021) demonstrated the novel astaxanthin extraction approach from unicellular green microalga, *Haematococcus pluvialis*. The group has used three different solvents: methyl tert-butyl ether (MTBE), hexane isopropanol (HEX-IPA), and acetone (ACE) in HPLC coupled with diode array detection and the quadrupole time-of-flight mass spectrometry (HPLC-DAD and LC-QTOF-MS) technique. They have isolated eleven astaxanthin monoesters, which accounted for 78.8% of the total astaxanthin pool, six astaxanthin diesters (20.5% of total), while free astaxanthin represented the smallest fraction (0.7%). Grujić et al. (2022) isolated 16 different carotenoids from *H. pluvialis* and characterized them through HPLC-DAD and LC-QTOF-MS. Please refer to Table 18.2 for chromatographic methods employed for the determination of secondary metabolites from microalgae.

With all these advantages, there are some challenges with HPLC, depending on the minimum concentration of metabolite to be detected and the complexity of the pre-treated sample. Nowadays, the trend of secondary metabolite detection

TABLE 18.2
Common Chromatographic Methods Employed for the Determination of Secondary Metabolites from Microalgae

S. No.	Microalgae	Compound	Method	Reference
1	*Saccharopolyspora* sp. NFXS83	Fucoxanthin	HPLC and C18-column	Bertrand et al., 2023
2	*Chlamydomona sreinhardtii* and *Scenedesmus quadricauda*	Phenols, flavonoids, carotenoids, SAM, S-adenosylhomocysteine, 5'-methylthioadenosine, and 5-methylcytosine (5-mC)	HPLC-MS/MS	Bacova et al., 2019
3	*Scenedesmus* sp.	Carotenoids	Supercritical fluid chromatography	Abrahamsson et al., 2012
4	*Nannochloropsis gaditana*, *Chlorella* sp., *Haematococcus pluvialis*, *Scenedesmus almeriensis*, *Isochrysis galbana*, *Tetraselmis suecica*, and *Karlodinium veneficum*	Carotenoid	HPLC	Cerón-García et al., 2018
5	*Desmodesmus protuberans*, *Desmodesmus denticulatus* var. *linearis*, *Chlamydomonas planctogloea*, *Coelastrum sphaericum*, *Parachlorella kessleri*, *Chlorella zofingiensis*, and *Coelastrum sphaericum*	Lutein, astaxanthin, g-linolenic acid, eicosapentaenoic acid (EPA), and docosahexaenoic acid (DHA)	HPLC/APCI-MS/MS and GC-FID	Soares et al., 2019
6	*Spirulina platensis*	C-phycocyanin and allophycocyanin	Anion exchanger streamline-DEAE	Bermejo et al., 2006
7	*Ettlia* sp., *Monoraphidium* sp., *Chlorella vulgaris*, and *Nannochloropsis oceanica*	Lipid content	Gas chromatography	Mishra et al., 2014
8	*Porphyridium cruentum*	B-phycoerythrin	Ion-exchange chromatography with a DEAE-cellulose column	Bermejo et al., 2003
9	*Spongiochloris spongiosa* and cyanobacterial strains (*Spirulina platensis*, *Anabaena doliolum*, *Nostoc* sp., and *Cylindrospermum* sp.)	Benzoic acid derivatives (p-hydroxybenzoic, protocatechuic, gallic, vanillic, and syringic acid), hydroxybenzaldehydes (4-hydroxybenzaldehyde and 3,4-dihydroxybenzaldehyde), and cinnamic acid derivatives (o-coumaric, p-coumaric, caffeic, ferulic, sinapic, and chlorogenic acid)	Solid-phase/supercritical-fluid extraction (SPE/SFE) with subsequent reversed-phase HPLC	Klejdus et al., 2009
10	*Chlorella vulgaris*	Lutein	High-performance countercurrent chromatography (HPCCC)	Fábryová et al., 2019
11	*Haematococcus pluvialis*	Astaxanthin	GC–MS and LC–MS	Su et al., 2014
12	*Paracoccus pantotrophus* FMR19	Long-chain alkanes, fatty alcohols, fatty acid methyl esters, and aromatic hydrocarbons	GC–MS	Begum et al., 2016
13	*Prorocentrum lima*, *Protoceratium reticulatum*, and *Alexandrium ostenfeldii*	Toxins [okadaic acid (OA), dinophysistoxin-1 (DTX-1), pectenotoxin-2 (PTX-2), azaspiracid-1 (AZA-1), yessotoxin (YTX), and 13-desmethyl spirolide C (SPX-1)]	Ultra-high-performance liquid chromatography coupled to high-resolution Orbitrap mass spectrometry (UHPLC-HR-Orbitrap MS)	Orellana et al., 2015
14	*Galdieria sulphuraria*	Oleamide, palmitamide, and pheophorbide A	Ultra-high-performance liquid chromatography coupled to high-resolution mass spectrometry (UHPLC-HRMS)	Ambrosino et al., 2023
15	*Tetradesmus obliquus* (*Scenedesmus acutus*)	Lectin	Anion exchange chromatography using a DEAE Sephadex A-50 column, and FPLC-Superdex G-75 HR10/300GL	Silva et al., 2020

is based on the conjunction of multiple techniques such as HPLC-MS, HPLC-UV-Visible spectroscopy, LC-NMR, *etc.*

18.4.2.2 Structural Characterization of Secondary Metabolites

Structural characterization of secondary metabolites is an important aspect for understanding the therapeutic potential and mechanism of action (Yang et al., 2010). Various advanced spectroscopic methods, such as nuclear magnetic resonance (NMR) spectroscopy and mass spectroscopy, are available for the structural elucidation of secondary metabolites (Park et al., 2022a).

Nuclear magnetic resonance (NMR) spectroscopy furnishes valuable insights into the molecular structure, aiding in the identification of a compound's stereochemistry. In contrast, Mass Spectrometry offers the means to determine a compound's molecular weight and fragmentation patterns, thus facilitating compound identification and offering glimpses into its structural attributes. Other methodologies, such as X-ray crystallography alongside high-resolution transmission electron microscopy (HR-TEM), contribute to comprehending the crystal geometry of biomolecules, leading to a holistic grasp of their shape and spatial configuration. NMR spectroscopy employs diverse pulse sequences to amass multiple NMR spectra, enabling the correlation of nuclear spins and the establishment of atom connectivity (Gomes et al., 2017). Two-dimensional NMR techniques like correlation spectroscopy (COSY) (Numan et al., 2022), heteronuclear multiple-quantum coherence (HMQC) and heteronuclear multiple-bond correlation (HMBC) (Shaala and Youssef, 2015) unravel interatomic relationships and bond connectivity within the compound. Additionally, advanced NMR methods such as nuclear overhauser effect spectroscopy (NOESY) provide valuable data concerning atom spatial arrangement, facilitating the determination of molecular conformation and stereochemistry (Betancur et al., 2020). Employing specific NMR methodologies like focused TOCSY (Total Correlation Spectroscopy) and distinct HSQC (heteronuclear single-quantum coherence) (Gupta et al., 2013) permits the recognition and allocation of particular functional groups within the substance. On the whole, sophisticated NMR approaches assume a pivotal role in unravelling the structural attributes of bioactive compounds originating from microorganisms, contributing to their characterization and enhancing comprehension of their biological functionalities. Some commonly used spectroscopic methods employed for the characterization of secondary metabolites are mentioned in Table 18.3.

TABLE 18.3
Spectroscopic Methods Used for the Structural Characterization of Secondary Metabolites Derived from Microalgae

S. No.	Microalgae	Spectroscopic Method	Bioactive Compound	Reference
1	*Chlorella sorokiniana* and *Desmodesmus* sp.	NMR	Lipid content	Srivastava et al., 2017
2	*Dunaliella* sp., *Dunaliella salina*, *Chaetoceros calcitrans*, *Chaetoceros gracilis*, and *Tisochrysis lutea*	NMR	1,4/2,5-cyclohexanetetrol (CHT) and dimethylsulfoniopropionate (DMSP), 2,3-dihydroxypropane-1-sulfonate (DHSP), (digalactosyldiacyglycerols (DGDGs), sulphoquinovosyl diacylglycerols (SQDGs), amino acids, carbohydrates, scyllo-inositol, mannitol, lactic acid, and homarine	Iglesias et al., 2019
3	*Chaetoceros calcitrans*	NMR	Fucoxanthin, astaxanthin, violaxanthin, zeaxanthin, canthaxanthin, and lutein	Azizan et al., 2018
4	*Nannochloropsis oceanica*, *Chaetoceros muelleri*, and *Conticribra weissflogii*	^{1}H HR-MAS NMR	EPA and PUFAS	Caprara et al., 2023
5	*Vulcanodinium rugosum*	^{1}H NMR, ^{13}C NMR, HSQC-NOESY, and HSQC-ROESY	Portimine	Selwood et al., 2013
6	*Lyngbya majuscula*	HSQC, COSY, and HSQMBC	Quinoline alkaloid, malyngamide T, and a tryptophan derivative	Nogle and Gerwick, 2003
7	*Isochrysis galbana*	1H NMR, COSY, TOCSY, HSQC, and HMBC	Chlorophyll-a and fucoxanthin, polar fatty acids (PLs, GLs, DGDGs, SGDGs), and PUFAs	Aguilera-Sáez et al., 2019
8	*Amphidinium carterae*	HRESIMS, COSY, TOCSY, HSQC, HSQC-TOCSY, and H2BC	Amphidinol 24, Amphidinol 25, and Amphidinol 26	Morales-Amador et al., 2021
9	*Chlorella vulgaris*	COSY, TOCSY, HSQC, HMBC, H2BC, 1D ROESY	Exopolysaccharides (EPSs), and oligomers	Capek et al., 2020
10	*Karenia brevis*	^{1}H NMR, ^{13}C NMR, COSY, HMBC, and TOCSY	Brevisamide	Satake et al., 2008
11	*Microcystis*, *Anabaena*, *Oscillatoria*, and *Nostoc*	MALDI-TOF	Microcystins, micropeptin, and anabaenopeptolin	Erhard et al., 1997

Another cutting-edge approach utilized for characterizing bioactive molecules isolated from microorganisms is X-ray crystallography. Much like 2D-NMR techniques, this method enables researchers to ascertain the three-dimensional configuration of these compounds on an atomic scale, offering vital insights into their chemical constitution and spatial alignment. By cultivating single crystals of the bioactive substance and exposing them to X-ray diffraction, scientists can gauge the angles and intensities of diffracted X-rays, which are subsequently employed to compute the electron density dispersion within the crystal lattice (Wang and Seyedsayamdost, 2017). This data facilitates the creation of a precise molecular model, unveiling the placements of individual atoms and their interconnections within the compound. X-ray crystallography assists in comprehending the stereochemistry, molecular interactions, and overall structure of the bioactive entity, thereby supporting the design of more potent drugs and therapeutic interventions. Moreover, this technique contributes to unravelling structure-activity relationships (Tojo et al., 2014), thereby aiding in the refinement and innovation of novel pharmaceuticals derived from microorganisms.

18.4.3 Challenges and Advancements in Extraction and Analysis

Despite substantial progress in extraction and analytical techniques, challenges persist in the pursuit of exploiting microalgae-derived secondary metabolites to their fullest. The complexity of microalgal matrices demands further optimization of extraction methods to ensure higher yields and purity of target metabolites while minimizing the extraction of unwanted components. Analytical techniques must also address issues related to sensitivity, selectivity, and reproducibility to accurately quantify and identify metabolites at trace levels. Moreover, the diversity of microalgae species and their metabolite profiles necessitate tailored approaches for each specific case. Advancements in technology, such as miniaturization of extraction systems, automation, and high-throughput analytical platforms, hold promise in overcoming these challenges. Furthermore, interdisciplinary collaboration between biologists, chemists, and engineers is crucial to leverage collective expertise and develop innovative strategies for efficient extraction, analysis, and utilization of microalgae-derived secondary metabolites in various industries, including pharmaceuticals, nutraceuticals, and biofuels.

18.5 Future Scope of Microalgae-Derived Secondary Metabolites

The future scope of microalgae-derived secondary metabolites holds immense promise across various industries, including pharmaceuticals, nutraceuticals, cosmetics, agriculture, and biofuels. These metabolites exhibit diverse and valuable properties such as antioxidant, antimicrobial, anti-inflammatory, and anticancer activities, making them prime candidates for novel drug development and functional ingredient formulation. Additionally, microalgae-derived metabolites can contribute to sustainable agriculture by enhancing crop yields and protecting plants against pests and diseases. Despite this potential, significant challenges persist.

The optimization of cultivation conditions for increased secondary metabolite production, efficient extraction techniques to maximize yields, and the development of scalable, cost-effective processes remain key hurdles. Furthermore, ensuring the consistency, purity, and stability of extracted metabolites, along with navigating complex regulatory frameworks, presents ongoing challenges. Addressing these obstacles through interdisciplinary collaboration, technological innovation, and sustainable practices will be essential in realizing the full potential of microalgae-derived secondary metabolites in shaping the future of various industries.

18.6 Conclusion

The exploration of microalgae as a sustainable source of value-added secondary metabolites has illuminated a remarkable avenue of scientific interest with far-reaching implications. This book chapter has delved into the intricate world of microalgae, uncovering their astonishing ability to synthesize a diverse array of secondary metabolites with multifaceted benefits for industries spanning pharmaceuticals, nutraceuticals, cosmetics, *etc.* In the current scenario, we are at the edge of a paradigm shift toward greener and more sustainable practices, where microalgae have emerged as nature's intricate biofactories, capable of harnessing sunlight and simple nutrients to produce compounds of immense value. However, the journey toward harnessing this potential is not without its challenges. The quest for optimal cultivation strategies, innovative extraction methodologies, and scalable production processes persists, underscoring the need for relentless research and development. Moreover, the integration of ethical and regulatory considerations, in harmony with ecological balance, adds layers of complexity to the path forward. Yet, as we navigate these challenges, we remain inspired by the prospect of a future where microalgae-derived secondary metabolites revolutionize industries, improve human health, and contribute to a more sustainable and harmonious coexistence with the natural world. Through collaborative efforts, technological advancements, and a deepening understanding of microalgal biology, we embark on a transformative journey toward realizing the boundless potential of microalgae as a sustainable source of invaluable secondary metabolites.

References

Abrahamsson, V., I. Rodriguez-Meizoso, and C. Turner. 2012. Determination of carotenoids in microalgae using supercritical fluid extraction and chromatography. *Journal of Chromatography A* 1250: 63–68. https://doi.org/10.1016/j.chroma.2012.05.069.

Abu-Ghosh, S., Z. Dubinsky, V. Verdelho, and D. Iluz. 2021. Unconventional high-value products from microalgae:

A review. *Bioresource Technology* 329 (June): 124895. https://doi.org/10.1016/j.biortech.2021.124895.
Aguilera-Sáez, L.M., A.C. Abreu, J. Camacho-Rodríguez, C.V. González-López, M. Del Carmen Cerón-García, and I. Fernández. 2019. NMR metabolomics as an effective tool to unravel the effect of light intensity and temperature on the composition of the marine microalgae *Isochrysis galbana*. *Journal of Agricultural and Food Chemistry* 67: 3879–89. https://doi.org/10.1021/acs.jafc.8b06840.
Ali, S.S., S.G. Mastropetros, M. Schagerl, et al. 2022. Recent advances in wastewater microalgae-based biofuels production: A state-of-the-art review. *Energy Reports* 8: 13253–80. https://doi.org/10.1016/j.egyr.2022.09.143.
Ambrosino, A., A. Chianese, C. Zannella, et al. 2023. *Galdieria sulphuraria*: An extremophilic alga as a source of antiviral bioactive compounds. *Marine Drugs* 21 (June): 383, 1–18. https://doi.org/10.3390/md21070383.
Anklam, E., H. Berg, L. Mathiasson, M. Sharman, and F. Ulberth. 1998. Supercritical fluid extraction (SFE) in food analysis: A review. *Food Additives and Contaminants*, 15: 729–50. https://doi.org/10.1080/02652039809374703.
Asif, M., and T. Muneer. 2007. Energy supply, its demand and security issues for developed and emerging economies. *Renewable and Sustainable Energy Reviews* 11: 1388–13. https://doi.org/10.1016/j.rser.2005.12.004.
Azizan, A., M.S.A. Bustamam, M. Maulidiani, et al. 2018. Metabolite profiling of the microalgal diatom chaetoceros calcitrans and correlation with antioxidant and nitric oxide inhibitory activities via 1H NMR-based metabolomics. *Marine Drugs* 16 (May): 154, 1–19. https://doi.org/10.3390/md16050154.
Bacova, R., B. Klejdus, P. Ryant, N.Cernei, V. Adam, and D. Huska. 2019. The effects of 5-azacytidine and cadmium on global 5-methylcytosine content and secondary metabolites in the freshwater microalgae *Chlamydomonas reinhardtii* and *Scenedesmus quadricauda*. *Journal of Phycology* 55: 329–42. https://doi.org/10.1111/jpy.12819.
Beatrice-Lindner, P., J.A. Garrido-Cardenas, C. Sepulveda, and F.G. Acien-Fernandez. 2018. A new approach for detection and quantification of microalgae in industrial-scale microalgal cultures. *Applied Microbiology and Biotechnology* 102: 8429–36. https://doi.org/10.1007/s00253-018-9268-y.
Begum, I.F., R. Mohankumar, M. Jeevan, and K. Ramani. 2016. GC–MS analysis of bio-active molecules derived from *Paracoccus pantotrophus* FMR19 and the antimicrobial activity against bacterial pathogens and MDROs. *Indian Journal of Microbiology* 56: 426–32. https://doi.org/10.1007/s12088-016-0609-1.
BenMoussa-Dahmen, I., H. Chtourou, F. Rezgui, S. Sayadi, and A. Dhouib. 2016. Salinity stress increases lipid, secondary metabolites and enzyme activity in *Amphora subtropica* and *Dunaliella* sp. for biodiesel production. *Bioresource Technology* 218: 816–25. https://doi.org/10.1016/j.biortech.2016.07.022
Bermejo, R., F.G Acién, M.J. Ibáñez, J.M. Fernández, E. Molina, and J.M. Alvarez-Pez. 2003. Preparative purification of B-phycoerythrin from the microalga *Porphyridium cruentum* by expanded-bed adsorption chromatography. *Journal of Chromatography B* 790: 317–25. https://doi.org/10.1016/S1570-0232(03)00168–5.
Bermejo, R., M.A. Felipe, E.M. Talavera, and J.M. Alvarez-Pez. 2006. Expanded bed adsorption chromatography for recovery of phycocyanins from the microalga *Spirulina platensis*. *Chromatographia* 63: 59–66. https://doi.org/10.1365/s10337-005-0702-9.
Bertrand, C.D.F., R. Martins, F. Quintas-Nunes, P. Reynolds-Brandão, M.T.B. Crespo, and F.X. Nascimento. 2023. *Saccharopolyspora* sp. NFXS83 in marine biotechnological applications: From microalgae growth promotion to the production of secondary metabolites. *Microorganisms* 11 (March): 902, 1–17. https://doi.org/10.3390/microorganisms11040902.
Betancur, L.A., A.M. Forero, D.M. Vinchira-Villarraga, et al. 2020. NMR-based metabolic profiling to follow the production of anti-phytopathogenic compounds in the culture of the marine strain *Streptomyces* sp. PNM–9. *Microbiological Research* 239 (May): 126507, 1–8. https://doi.org/10.1016/j.micres.2020.126507.
Bhowmick, S., A. Mazumdar, A. Moulick, and V. Adam. 2020. Algal metabolites: An inevitable substitute for antibiotics. *Biotechnology Advances* 43 (November): 107571. https://doi.org/10.1016/j.biotechadv.2020.107571.
Bhuvana, P., P. Sangeetha, V. Anuradha, and M.S. Ali. 2019. Spectral characterization of bioactive compounds from microalgae: *N. Oculata* and *C. Vulgaris*. *Biocatalysis and Agricultural Biotechnology* 19 (May): 101094. https://doi.org/10.1016/j.bcab.2019.101094.
Bisht, B., V. Kumar, P. Gururani, et al. 2021. The potential of nuclear magnetic resonance (NMR) in metabolomics and lipidomics of microalgae – A review. *Archives of Biochemistry and Biophysics* 710 (October): 108987. https://doi.org/10.1016/j.abb.2021.108987.
Capek, P., M. Matulová, M Šutovská, J. Barboríková, M. Molitorisová, and I. Kazimierová. 2020. *Chlorella vulgaris* α-L-arabino-α-L-rhamno-α,β-D-galactan structure and mechanisms of its anti-inflammatory and anti-remodelling effects. *International Journal of Biological Macromolecules* 162: 188–98. https://doi.org/10.1016/j.ijbiomac.2020.06.151.
Caprara, C.S.C., T.K. Mathias, M.de F.C. Santos, et al. 2023. Application of 1H HR-MAS NMR-based metabolite fingerprinting of marine microalgae. *Metabolites* 13 (January): 202, 1–11. https://doi.org/10.3390/metabo13020202.
Castro-Puyana, M., M. Herrero, I. Urreta, et al. 2013. Optimization of clean extraction methods to isolate carotenoids from the microalga *Neochloris oleoabundans* and subsequent chemical characterization using liquid chromatography tandem mass spectrometry. *Analytical and Bioanalytical Chemistry* 405: 4607–16. https://doi.org/10.1007/s00216-012-6687-y.
Cerón, M.C., M.C. García-Malea, J. Rivas, et al. 2007. Antioxidant activity of *Haematococcus pluvialis* cells grown in continuous culture as a function of their carotenoid and fatty acid content. *Applied Microbiology and Biotechnology* 74: 1112–19. https://doi.org/10.1007/s00253-006-0743-5.
Cerón-García, M.C., C.V. González-López, J. Camacho-Rodríguez, L. López-Rosales, F. García-Camacho, and E. Molina-Grima. 2018. Maximizing carotenoid extraction from microalgae used as food additives and determined by liquid chromatography (HPLC). *Food Chemistry* 257: 316–24. https://doi.org/10.1016/j.foodchem.2018.02.154.
Chen, B., C. Wan, M.A. Mehmood, J.S. Chang, F. Bai, and X. Zhao. 2017. Manipulating environmental stresses and stress tolerance of microalgae for enhanced production of lipids and value-added products–A review. *Bioresource Technology* 244: 1198–206. https://doi.org/10.1016/j.biortech.2017.05.170.
Cheong, J.Y., M. Muskhazli, A.A.N. Azwady, S.A. Ahmad, and A.A. Adli. 2020. Three dimensional optimisation for the enhancement of astaxanthin recovery from shrimp shell wastes by *Aeromonas hydrophila*. *Biocatalysis and*

Agricultural Biotechnology 27 (August): 101649. https://doi.org/10.1016/j.bcab.2020.101649.

Chronopoulou, L., C.D. Bosco, F.D. Caprio, et al. 2019. Extraction of carotenoids and fat-soluble vitamins from *Tetradesmus obliquus* microalgae: An optimized approach by using supercritical CO2. *Molecules* 24 (July): 2581, 1–14. https://doi.org/10.3390/molecules24142581.

Dammak, M., S.M. Haase, R. Miladi, et al. 2016. Enhanced lipid and biomass production by a newly isolated and identified marine microalga. *Lipids in Health and Disease* 15 (December): 209, 1–13. https://doi.org/10.1186/s12944-016-0375-4.

Dang, T.T., M.C. Bowyer, I.A.V. Altena, and C.J. Scarlett. 2018. Optimum conditions of microwave-assisted extraction for phenolic compounds and antioxidant capacity of the brown alga *Sargassum vestitum*. *Separation Science and Technology* 53: 1711–23. https://doi.org/10.1080/01496395.2017.1414845.

Darko, E., P. Heydarizadeh, B. Schoefs, and M.R. Sabzalian. 2014. Photosynthesis under artificial light: The shift in primary and secondary metabolism. *Philosophical Transactions of the Royal Society B: Biological Sciences* 369 (March): 20130243, 1–7. https://doi.org/10.1098/rstb.2013.0243.

Erhard, M., H. von Döhren, and P. Jungblut. 1997. Rapid typing and elucidation of new secondary metabolites of intact cyanobacteria using MALDI-TOF mass spectrometry. *Nature Biotechnology* 15: 906–09. https://doi.org/10.1038/nbt0997-906.

Fábryová, T., J. Cheel, D. Kubáč, et al. 2019. Purification of lutein from the green microalgae *Chlorella vulgaris* by integrated use of a new extraction protocol and a multi-injection high performance counter-current chromatography (HPCCC). *Algal Research* 41 (August): 101574. https://doi.org/10.1016/j.algal.2019.101574.

Fajardo, A.R., L.E. Cerdán, A.R. Medina, F.G.A. Fernández, P.A.G. Moreno, and E.M. Grima. 2007. Lipid extraction from the microalga *Phaeodactylum tricornutum*. *European Journal of Lipid Science and Technology* 109: 120–26. https://doi.org/10.1002/ejlt.200600216.

Fattah, I.M.R., M.Y. Noraini, M Mofijur, et al. 2020. Lipid extraction maximization and enzymatic synthesis of biodiesel from microalgae. *Applied Sciences* 10 (September): 6103, 1–18. https://doi.org/10.3390/app10176103.

Fernando, I.P.S., W.W. Lee, and G. Ahn. 2022. Marine algal flavonoids and phlorotannins; An intriguing frontier of biofunctional secondary metabolites. *Critical Reviews in Biotechnology* 42: 23–45. https://doi.org/10.1080/07388551.2021.1922351.

Ferreira, A.F., L.A Ribeiro, A.P. Batista, et al. 2013. A biorefinery from *Nannochloropsis* sp. microalga – energy and CO2 emission and economic analyses. *Bioresource Technology* 138: 235–44. https://doi.org/10.1016/j.biortech.2013.03.168.

Fujii, K. 2012. Process integration of supercritical carbon dioxide extraction and acid treatment for astaxanthin extraction from a vegetative microalga. *Food and Bioproducts Processing* 90: 762–66. https://doi.org/10.1016/j.fbp.2012.01.006.

Gauthier, M.R., G.N.A. Senhorinho, and J.A. Scott. 2020. Microalgae under environmental stress as a source of antioxidants. *Algal Research* 52 (December): 102104. https://doi.org/10.1016/j.algal.2020.102104.

Gerwick, W.H., and B.S. Moore. 2012. Lessons from the past and charting the future of marine natural products drug discovery and chemical biology. *Chemistry and Biology* 19: 85–98. https://doi.org/10.1016/j.chembiol.2011.12.014.

Gerwick, W.H., P.J. Proteau, D.G. Nagle, E. Hamel, A. Blokhin, and D.L. Slate. 1994. Structure of curacin A, a novel antimitotic, antiproliferative and brine shrimp toxic natural product from the marine cyanobacterium *Lyngbya majuscula*. *Journal of Organic Chemistry* 59: 1243–45. https://doi.org/10.1021/jo00085a006.

Gomes, A.R., A.C. Duarte, and T.A.P. Rocha-Santos. 2017. Analytical techniques for discovery of bioactive compounds from marine fungi. In *Fungal metabolite*, eds. J.M. Mérillon and K. Ramawat, 415–34. Cham: Springer. https://doi.org/10.1007/978-3-319-25001-4_9.

Grujić, V.J., B. Todorović, J. Ambrožič-Dolinšek, R. Kranvogl, and T. Ciringer. 2022. Diversity and content of carotenoids and other pigments in the transition from the green to the red stage of *Haematococcus pluvialis* microalgae identified by HPLC-DAD and LC-QTOF-MS. *Plants* 11 (April): 1026, 1–14. https://doi.org/10.3390/plants11081026.

Gupta, A., S. Jain, and N. Verma. 2022. *Industrial applications of soil microbes*, vol. 1. Bentham Science Publishers Pte Ltd. https://doi.org/10.2174/97898150399551220101.

Gupta, A., S. Jain, and N. Verma. 2024. *Industrial applications of soil microbes*, vol. 3. Bentham Science Publishers Pte Ltd. https://doi.org/10.2174/97898150797531240301.

Gupta, V., Thakur, R.S., C.R.K. Reddy, and B. Jha. 2013. Central metabolic processes of marine macrophytic algae revealed from NMR based metabolome analysis. *RSC Advances* 3: 7037–47. https://doi.org/10.1039/c3ra23017a.

Herrero, M., J.A. Mendiola, A. Cifuentes, and E. Ibáñez. 2010. Supercritical fluid extraction: Recent advances and applications. *Journal of Chromatography A* 1217: 2495–511. https://doi.org/10.1016/j.chroma.2009.12.019.

Iglesias, M.J., R. Soengas, I. Probert, et al. 2019. NMR characterization and evaluation of antibacterial and antiobiofilm activity of organic extracts from stationary phase batch cultures of five marine microalgae (*Dunaliella* sp., *D. salina*, *Chaetoceros calcitrans*, *C. gracilis and Tisochrysis lutea*). *Phytochemistry* 164: 192–205. https://doi.org/10.1016/j.phytochem.2019.05.001.

Inbaraj, B.S., J.T. Chien, and B.H. Chen. 2006. Improved high performance liquid chromatographic method for determination of carotenoids in the microalga *Chlorella pyrenoidosa*. *Journal of Chromatography A* 1102: 193–99. https://doi.org/10.1016/j.chroma.2005.10.055.

Jaime, L., J.A. Mendiola, M. Herrero, et al. 2005. Separation and characterization of antioxidants from *Spirulina platensis* microalga combining pressurized liquid extraction, TLC, and HPLC-DAD. *Journal of Separation Science* 28: 2111–19. https://doi.org/10.1002/jssc.200500185.

Jain, S., A. Gupta, and N. Verma. 2023. *Industrial applications of soil microbes*, vol. 2. Bentham Science Publishers Pte Ltd. https://doi.org/10.2174/97898150502641230201.

Jain, S., A. Gupta, and N. Verma. 2024. *Industrial applications of soil microbes*, vol. 4. Bentham Science Publishers Pte Ltd. https://doi.org/10.2174/97898151249961240401.

Jin, G., Y. Liu, S, Xue, et al. 2019. Determination of three carotenoids in microalgae by matrix solid-phase dispersion extraction and high-performance liquid chromatography. *Chromatographia* 82: 1593–601. https://doi.org/10.1007/s10337-019-03795-w.

Juin, C., J.R. Chérouvrier, V. Thiéry, et al. 2015. Microwave-assisted extraction of phycobiliproteins from *Porphyridium purpureum*. *Applied Biochemistry and Biotechnology* 175: 1–15. https://doi.org/10.1007/s12010-014-1250-2.

Kanekiyo, K., J.B. Lee, K. Hayashi, et al. 2005. Isolation of an antiviral polysaccharide, nostoflan, from a terrestrial cyanobacterium, *Nostoc flagelliforme*. *Journal of Natural Products* 68: 1037–41. https://doi.org/10.1021/np050056c.

Kaufmann, B., and P. Christen. 2002. Recent extraction techniques for natural products: Microwave-assisted extraction and pressurised solvent extraction. *Phytochemical Analysis* 13: 105–13. https://doi.org/10.1002/pca.631.

Khavari, F., M. Saidijam, M. Taheri, and F. Nouri. 2021. Microalgae: Therapeutic potentials and applications. *Molecular Biology Report* 48: 4757–65. https://doi.org/10.1007/s11033-021-06422-w.

Kiran, B.R., and S.V. Mohan. 2021. Microalgal cell biofactory – therapeutic, nutraceutical and functional food applications. *Plants* 10 (April): 836, 1–31. https://doi.org/10.3390/plants10050836.

Klejdus, B., J. Kopecký, L. Benešová, and J. Vacek. 2009. Solid-phase/supercritical-fluid extraction for liquid chromatography of phenolic compounds in freshwater microalgae and selected cyanobacterial species. *Journal of Chromatography A* 1216: 763–71. https://doi.org/10.1016/j.chroma.2008.11.096.

Kubrakova, I.V., and E.S. Toropchenova. 2008. Microwave heating for enhancing efficiency of analytical operations (review). *Inorganic Materials* 44: 1509–19. https://doi.org/10.1134/S0020168508140069.

Kumari, S., S. Satapathy, M. Datta, and S. Kumar. 2022. Adaptation of microalgae to temperature and light stress. In *Plant stress: Challenges and management in the new decade. Advances in science, technology & innovation*, eds. S. Roy, P. Mathur, A.P. Chakraborty, and S.P. Saha, 123–34. Chem: Springer. https://doi.org/10.1007/978-3-030-95365-2_8.

Kwang, H.C., Y.K. Song, D.U. Lee. 2008. Antiproliferative effects of carotenoids extracted from *Chlorella ellipsoidea* and *Chlorella vulgaris* on human colon cancer cells. *Journal of Agricultural and Food Chemistry* 56: 10521–26. https://doi.org/10.1021/jf802111x.

Leflaive, J., and L. Ten-Hage. 2007. Algal and cyanobacterial secondary metabolites in freshwaters: A comparison of allelopathic compounds and toxins. *Freshwater Biology* 52: 199–214. https://doi.org/10.1111/j.1365-2427.2006.01689.x.

Lim, K.C., F.M. Yusoff, M. Shariff, and M.S. Kamarudin. 2018. Astaxanthin as feed supplement in aquatic animals. *Reviews in Aquaculture* 10: 738–73. https://doi.org/10.1111/raq.12200.

Luesch, H., R.E. Moore, V.J. Paul, S.L. Mooberry, and T.H. Corbett. 2001. Isolation of dolastatin 10 from the marine cyanobacterium *Symploca* species VP642 and total stereochemistry and biological evaluation of its analogue symplostatin 1. *Journal of Natural Products* 64: 907–10. https://doi.org/10.1021/np010049y.

Macías-Sánchez, M.D., J.M. Fernandez-Sevilla, F.G.A. Fernández, M.C.C. García, and E.M. Grima. 2010. Supercritical fluid extraction of carotenoids from *Scenedesmus almeriensis*. *Food Chemistry* 123: 928–35. https://doi.org/10.1016/j.foodchem.2010.04.076.

Manfellotto, F., G.R. Stella, M.I. Ferrante, A. Falciatore, and C. Brunet. 2020. Engineering the unicellular alga *Phaeodactylum tricornutum* for enhancing carotenoid production. *Antioxidants* 9 (August): 757, 1–13. https://doi.org/10.3390/antiox9080757.

Mantoura, R.F.C., and C.A. Llewellyn. 1983. The rapid determination of algal chlorophyll and carotenoid pigments and their breakdown products in natural waters by reverse-phase high-performance liquid chromatography. *Analytica Chimica Acta* 151: 297–14. https://doi.org/10.1016/S0003-2670(00)80092-6.

Matsui, K., E. Nazifi, S. Kunita, N. Wada, S. Matsugo, and T. Sakamoto. 2011. Novel glycosylated mycosporine-like amino acids with radical scavenging activity from the cyanobacterium *Nostoc commune*. *Journal of Photochemistry and Photobiology* 105: 81–89. https://doi.org/10.1016/j.jphotobiol.2011.07.003.

Mehariya, S., R.K. Goswami, O.P. Karthikeysan, P. Verma. 2021. Microalgae for high-value products: A way towards green nutraceutical and pharmaceutical compounds. *Chemosphere* 280 (October): 130553. https://doi.org/10.1016/j.chemosphere.2021.130553.

Mendes, R.L., B.P. Nobre, M.T. Cardoso, A.P. Pereira, A.F. Palavra. 2003. Supercritical carbon dioxide extraction of compounds with pharmaceutical importance from microalgae. *Inorganica Chimica Acta* 356: 328–34. https://doi.org/10.1016/S0020-1693(03)00363-3.

Metting, F.B. 1996. Biodiversity and application of microalgae. *Journal of Industrial Microbiology and Biotechnology* 17: 477–89. https://doi.org/10.1007/bf01574779.

Michalak, I., and K. Chojnacka. 2014. Algal extracts: Technology and advances. *Engineering in Life Sciences* 14: 581–91. https://doi.org/10.1002/elsc.201400139.

Michalak, I., B. Górka, P.P. Wieczorek, et al. 2016. Supercritical fluid extraction of algae enhances levels of biologically active compounds promoting plant growth. *European Journal of Phycology* 51: 243–52. https://doi.org/10.1080/09670262.2015.1134813.

Mishra, S.K., W.I. Suh, W. Farooq, et al. 2014. Rapid quantification of microalgal lipids in aqueous medium by a simple colorimetric method. *Bioresource Technology* 155: 330–33. https://doi.org/10.1016/j.biortech.2013.12.077.

Molina-Miras, A., A. Morales-Amador, C.R. de Vera, et al. 2018. A pilot-scale bioprocess to produce amphidinols from the marine microalga *Amphidinium carterae*: Isolation of a novel analogue. *Algal Research* 31: 87–98. https://doi.org/10.1016/j.algal.2018.01.010.

Mondal, A., S. Bose, S. Banerjee, et al. 2020. Marine cyanobacteria and microalgae metabolites – A rich source of potential anticancer drugs. *Marine Drugs* 18 (September): 476, 1–59. https://doi.org/10.3390/md18090476.

Morales-Amador, A., A. Molina-Miras, and L. López-Rosales. 2021. Isolation and structural elucidation of new amphidinol analogues from *Amphidinium carterae* cultivated in a pilot-scale photobioreactor. *Marine Drugs* 19 (July): 432, 1–13. https://doi.org/10.3390/md19080432.

Mwaurah, P.W., S. Kumar, N. Kumaret et al. 2020. Novel oil extraction technologies: Process conditions, quality parameters, and optimization. *Comprehensive Reviews in Food Science and Food Safety* 19: 3–20. https://doi.org/10.1111/1541-4337.12507.

Nogle, L.M., and W.H. Gerwick. 2003. Diverse secondary metabolites from a Puerto Rican collection of *Lyngbya majuscula*. *Journal of Natural Products* 66: 217–20. https://doi.org/10.1021/np020332c.

Numan, M., M. Shah, S. Asaf, N. Ur Rehman, and A. Al-Harrasi. 2022. Bioactive compounds from endophytic bacteria *Bacillus subtilis* strain EP1 with their antibacterial activities. *Metabolites* 12 (December): 1228, 1–10. https://doi.org/10.3390/metabo12121228.

Ogawa, H., A. Iwasaki, S. Sumimoto, et al. 2017. Isolation and total synthesis of hoshinolactam, an antitrypanosomal lactam from a marine cyanobacterium. *Organic Letters* 19: 890–93. https://doi.org/10.1021/acs.orglett.7b00047.

Orellana, G., L. Van Meulebroek, S. Van Vooren, et al. 2015. Quantification and profiling of lipophilic marine toxins in microalgae by UHPLC coupled to high-resolution orbitrap mass spectrometry. *Analytical and Bioanalytical Chemistry* 407: 6345–56. https://doi.org/10.1007/s00216-015-8637-y.

Oslan, S.N.H., S.N. Oslan, R. Mohamad, et al. 2022. Bioprocess strategy of *Haematococcus lacustris* for biomass and astaxanthin production keys to commercialization: Perspective and future direction. *Fermentation* 8 (April): 179, 1–14. https://doi.org/10.3390/fermentation8040179.

Park, J.D., Y. Li, K. Moon, E.J. Han, S.R. Lee, and M.R. Seyedsayamdost. 2022a. Structural elucidation of cryptic algaecides in marine algal-bacterial symbioses by NMR spectroscopy and microED. *Angewandte Chemie – International Edition* 61, e202114022 (December): 1–12. https://doi.org/10.1002/anie.202114022.

Park, Y.H., S.H. Han, B. Oh, et al. 2022b. Microalgal secondary metabolite productions as a component of biorefinery: A review. *Bioresource Technology* 344 (January): 126206. https://doi.org/10.1016/j.biortech.2021.126206.

Patil, L., and B.B. Kaliwal. 2019. Microalga *Scenedesmus bajacalifornicus* BBKLP-07, a new source of bioactive compounds with *in vitro* pharmacological applications. *Bioprocess and Biosystems Engineering* 42: 979–94. https://doi.org/10.1007/s00449-019-02099-5.

Perez-Velazquez, M., D.M. Gatlin, M.L. González-Félix, et al. 2019. Effect of fishmeal and fish oil replacement by algal meals on biological performance and fatty acid profile of hybrid striped bass (*Morone crhysops* ♀ × *M. saxatilis* ♂). *Aquaculture* 507: 83–90. https://doi.org/10.1016/j.aquaculture.2019.04.011.

Rammuni, M.N., T.U. Ariyadasa, P.H.V. Nimarshana, and R.A. Attalage. 2019. Comparative assessment on the extraction of carotenoids from microalgal sources: Astaxanthin from *H. pluvialis* and β-carotene from *D. salina*. *Food Chemistry* 277: 128–34. https://doi.org/10.1016/j.foodchem.2018.10.066.

Rastogi, R.P., R.R. Sonani, and D. Madamwar. 2014. The high-energy radiation protectant extracellular sheath pigment scytonemin and its reduced counterpart in the cyanobacterium *Scytonema* sp. R77DM. *Bioresource Technology* 171: 396–400. https://doi.org/10.1016/j.biortech.2014.08.106.

Ren, M.T., J. Chen, Y. Song, L.S. Sheng, P. Li, and L.W. Qi. 2008. Identification and quantification of 32 bioactive compounds in *Lonicera* species by high performance liquid chromatography coupled with time-of-flight mass spectrometry. *Journal of Pharmaceutical and Biomedical Analysis* 48: 1351–60. https://doi.org/10.1016/j.jpba.2008.09.037.

Rodríguez-Meizoso, I., L. Jaime, S. Santoyo, et al. 2008. Pressurized fluid extraction of bioactive compounds from *Phormidium species*. *Journal of Agricultural and Food Chemistry* 56: 3517–23. https://doi.org/10.1021/jf703719p.

Russell, C., C. Rodriguez, and M. Yaseen. 2022. High-value biochemical products and applications of freshwater eukaryotic microalgae. *The Science of the Total Environment* 809 (October): 151111, 1–49. https://doi.org/10.1016/j.scitotenv.2021.151111.

Salinas-Salazar, C., J.S. Garcia-Perez, R. Chandra, C. Castillo-Zacarias, H.M.N. Iqbal, and R. Parra-Saldívar. 2019. Methods for extraction of valuable products from microalgae biomass. In *Microalgae biotechnology for development of biofuel and wastewater treatment*, eds. M.A. Alam and Z. Wang, 245–63. Singapore: Springer. https://doi.org/10.1007/978-981-13-2264-8_11.

Sañé, E., A. Del Mondo, L. Ambrosino, A. Smerilli, C. Sansone, and C. Brunet. 2023. The Recent advanced in microalgal phytosterols: Bioactive ingredients along with human-health driven potential applications. *Food Reviews International* 39: 1859–78. https://doi.org/10.1080/87559129.2021.1938115.

Sarwer, A., S.M. Hamed, A.I. Osman, et al. 2022. Algal biomass valorization for biofuel production and carbon sequestration: A review. *Environmental Chemistry Letters* 20: 2797–51. https://doi.org/10.1007/s10311-022-01458-1.

Satake, M., A.J. Bourdelais, R.M. Van Wagoner, D.G. Baden, and J.L.C. Wright. 2008. Brevisamide: An unprecedented monocyclic ether alkaloid from the dinoflagellate *Karenia brevis* that provides a potential model for ladder-frame initiation. *Organic Letters* 10: 3465–68. https://doi.org/10.1021/ol801243n.

Selwood, A.I., A.L. Wilkins, R. Munday, F. Shi, L.L. Rhodes, and P.T. Holland. 2013. Portimine: A bioactive metabolite from the benthic dinoflagellate *Vulcanodinium rugosum*. *Tetrahedron Letters* 54: 4705–07. https://doi.org/10.1016/j.tetlet.2013.06.098.

Shaala, L.A., and D.T.A. Youssef. 2015. Identification and bioactivity of compounds from the fungus *Penicillium* sp. CYE-87 isolated from a marine tunicate. *Marine Drugs* 13: 1698–09. https://doi.org/10.3390/md13041698.

Shah, M.M.R., Y. Liang, J.J. Cheng, and M. Daroch. 2016. Astaxanthin-producing green microalga *Haematococcus pluvialis*: From single cell to high value commercial products. *Frontiers in Plant Science* 7 (April): 531, 1–28. https://doi.org/10.3389/fpls.2016.00531.

Shohael, A.M., M.B. Ali, K.W. Yu, E.J. Hahn, and K.Y. Paek. 2006. Effect of temperature on secondary metabolites production and antioxidant enzyme activities in *Eleutherococcus senticosus* somatic embryos. *Plant Cell, Tissue and Organ Culture* 85: 219–28. https://doi.org/10.1007/s11240-005-9075-x.

Silva, A.J., V.L.R. Cavalcanti, A.L.F. Porto, W.A. Gama, R.M.P. Brandão-Costa, and R.P. Bezerra. 2020. The green microalgae *Tetradesmus obliquus* (*Scenedesmus acutus*) as lectin source in the recognition of ABO blood type: Purification and characterization. *Journal of Applied Phycology* 32: 103–10. https://doi.org/10.1007/s10811-019-01923-5.

Singh, A., P.S. Nigam, and J.D. Murphy. 2011. Renewable fuels from algae: An answer to debatable land based fuels. *Bioresource Technology* 102: 10–16. https://doi.org/10.1016/j.biortech.2010.06.032.

Soares, A.T., D.C. da Costa, A.A.H. Vieira, and N.R.A. Filho. 2019. Analysis of major carotenoids and fatty acid composition of freshwater microalgae. *Heliyon* 5: e01529 (April): 1–20. https://doi.org/10.1016/j.heliyon.2019.e01529.

Srivastava, G. Nishchal, and V.V. Goud. 2017. Salinity induced lipid production in microalgae and cluster analysis (ICCB 16-BR_047). *Bioresource Technology* 242: 244–52. https://doi.org/10.1016/j.biortech.2017.03.175.

Su, Y., J. Wang, M. Shi, et al. 2014. Metabolomic and network analysis of astaxanthin-producing *Haematococcus pluvialis* under various stress conditions. *Bioresource Technology* 170: 522–29. https://doi.org/10.1016/j.biortech.2014.08.018.

Suganya, T., M. Varman, H.H. Masjuki, and S. Renganathan. 2016. Macroalgae and microalgae as a potential source for commercial applications along with biofuels production: A biorefinery approach. *Renewable and Sustainable*

Energy Reviews 55: 909–41. https://doi.org/10.1016/j.rser.2015.11.026.

Toan, T.Q., T.D. Phong, D.D. Tien, et al. 2021. Optimization of microwave-assisted extraction of phlorotannin from *Sargassum swartzii* (Turn.) C. Ag. with ethanol/water. *Natural Product Communications* 16: 1–11. https://doi.org/10.1177/1934578X21996184.

Todorović, B., V.J. Grujić, A.U. Krajnc, R. Kranvogl, and J. Ambrožič-Dolinšek. 2021. Identification and content of astaxanthin and its esters from microalgae *Haematococcus pluvialis* by HPLC-DAD and LC-QTOF-MS after extraction with various solvents. *Plants* 10 (October): 2413I, 1–14. https://doi.org/10.3390/plants10112413.

Tojo, S., T. Kohno, T. Tanaka, et al. 2014. Crystal structures and structure-activity relationships of imidazothiazole derivatives as IDO1 inhibitors. *ACS Medicinal Chemistry Letters* 5: 1119–23. https://doi.org/10.1021/ml500247w.

Tokuşoglu, Ö., and M.K. Ünal. 2003. Biomass nutrient profiles of three microalgae: *Spirulina platensis*, *Chlorella vulgaris*, and *Isochrisis galbana*. *Journal of Food Science*, 68: 1144–48. https://doi.org/10.1111/j.1365-2621.2003.tb09615.x.

Verma, S., S. Thapa, N. Siddiqui, and H. Chakdar. 2022. Cyanobacterial secondary metabolites towards improved commercial significance through multiomics approaches. *World Journal of Microbiology and Biotechnology* 38 (April): 100. https://doi.org/10.1007/s11274-022-03285-6.

Wang, L., and C.L Weller. 2006. Recent advances in extraction of nutraceuticals from plants. *Trends in Food Science and Technology* 17: 300–12. https://doi.org/10.1016/j.tifs.2005.12.004.

Wang, R., and M.R. Seyedsayamdost. 2017. Roseochelin B, an algaecidal natural product synthesized by the roseobacter *Phaeobacter inhibens* in response to algal sinapic acid. *Organic Letters* 19: 5138–41. https://doi.org/10.1021/acs.orglett.7b02424.

Wen, L., Z. Zhang, D.W. Sun, S.P. Sivagnanam, and B.K. Tiwari. 2020. Combination of emerging technologies for the extraction of bioactive compounds. *Critical Reviews in Food Science and Nutrition* 60: 1826–41. https://doi.org/10.1080/10408398.2019.1602823.

Yang, P.W., M.G. Li, J.Y. Zhao, et al. 2010. Oligomycins A and C, major secondary metabolites isolated from the newly isolated strain *Streptomyces diastaticus*. *Folia Microbiologica* 55: 10–16. https://doi.org/10.1007/s12223-010-0002-0.

Zainuddin, E.N., R. Mentel, V. Wray, et al. 2007. Cyclic depsipeptides, ichthyopeptins A and B, from *Microcystis ichthyoblabe*. *Journal of Natural Products* 70: 1084–88. https://doi.org/10.1021/np060303s.

19 Functional and Nutritional Potential of Freshwater Microalgae

An Eco-Friendly Solution to Global Food

*Neetu Mishra
(Department of Family and Community Sciences (Home Science), Science Faculty Campus, University of Allahabad, Allahabad 211002, UP, India; neetum1976@gmail.com;)

Neha Mishra
(Department of Food Nutrition and Public Health, NAI, SHUATS, Prayagraj 211007, UP, India; nehamishra@shiats.edu.in)

19.1 INTRODUCTION

Continuous rising population and decreasing resources created an urgent need for the search of nutrient-rich novel food sources that require limited resources (Matos, 2019). Microalgae represent an interesting possibility to meet the nutritional demand of the population. These simple organisms can grow in non-potable water and non-arable land and are also widely utilized in wastewater treatment and CO_2 mitigation. Microalgae, often referred to as "microscopic powerhouses," hold immense potential as a source to meet human nutrition needs to combat malnutrition, hidden hunger, and food shortages (Nwoba, 2017). These tiny aquatic organisms are packed with a wide array of essential nutrients and bioactive compounds, making them an attractive and sustainable option for addressing these nutritional challenges and improving overall human health. Microalgae comprise a diverse group of photosynthetic microorganisms that live in diverse habitats, including both freshwater and marine water (Zhang et al., 2018).

Freshwater microalgae are abundant and diverse. These microorganisms have evolved over billions of years to adapt to diverse ecological niches, giving rise to an impressive reservoir of biomolecules that can improve the nutritional status of humans and have the potential to be developed as food sources to address the global food crisis. They also hold significant importance for various ecological, industrial, and research purposes. Freshwater habitats host a wide range of microalgal species, while often overlooked compared to marine microalgae. In terms of nutritional properties, compared to oceanic/seawater microalgae, freshwater microalgae contain nearly the same protein and amino acids, lipids and fatty acids, carbohydrates, and vitamins. Moreover, freshwater microalgae display an interesting nutritional composition, with a balanced protein concentration and amino acid profile, which is comparable or superior to the conventional protein sources (Coelho et al., 2021). Certain species of microalgae, such as *Spirulina* and *Chlorella*, are exceptional sources of protein. *Spirulina*, for instance, contains all nine essential amino acids, making it a complete protein source comparable to traditional protein-rich foods like meat and eggs. Microalgae also offer a range of other essential nutrients, including n-3 and n-6 polyunsaturated fatty acids, carbohydrates, vitamins, minerals, carotenoids, and antioxidants. These nutrients have various health benefits and play crucial roles in maintaining overall well-being. Microalgae can supplement and enhance the nutritive value of traditional food and animal feed, making them a valuable addition to address nutritional challenges (Conde et al., 2021). In addition to their nutritional value, microalgae have been found to possess various bioactive compounds that contribute to their health benefits, such as antioxidant, antibacterial, antiviral, and anti-inflammatory properties, which can support immune function, reduce inflammation, and protect against oxidative stress. They have been used as food for centuries because of their good nutritional value. However, there are limited studies on microgreens and still rarely investigated for use as an alternative food. Research into the nutritional values and health benefits is in its infancy. Only a small handful of microalgal species are being produced at a commercial scale for use as human food or protein supplements

Recently, it received attention in the food and pharmaceutical industries because of its high nutritional value, array of secondary metabolites, and certain pigments such as phycocyanin, carotenoids, and chlorophyll (Jain et al., 2023, 2024). Microalgae-derived supplements are increasingly popular, offering a convenient way to fill nutritional gaps and support overall health (Gupta et al., 2022, 2024). In this exploration, we will delve into the potential of freshwater microalgae for a sustainable solution to address global nutritional challenges. The nutritional attributes and vital role of freshwater microalgae as functional food are also reviewed in the chapter.

DOI: 10.1201/9781003408833-19

19.2 MICROALGAE AS A SUSTAINABLE SOLUTION TO FOOD INSECURITY

Microalgae cultivation is highly sustainable and eco-friendly compared to traditional agriculture and livestock farming. They are the oldest form of life, generally having high photosynthetic efficiency and biomass productivity (Hussain et al., 2021). In addition, they can survive in extreme adverse environmental conditions that include a wide range of temperatures, salinity, and pH values, and different light intensities. Furthermore, they accumulate diverse secondary metabolites under mild stress or adverse growth conditions that have therapeutic importance. The cultivation of microalgae produces fewer greenhouse gases and requires less water compared to traditional food production. Microalgae can be cultivated without competing with traditional crops for land or fresh water, making them an environmentally friendly solution (Kusmayadi et al., 2021). Moreover, microalgae cultivation requires minimal inputs such as fertilizers or pesticides, reducing the environmental impact compared to conventional agriculture practices (Quintana et al., 2011). It is worth mentioning that microalgae have a high growth rate and can be harvested throughout the year, providing a consistent and reliable source of nutrition. Microalgae can thrive in various environments, including saltwater and wastewater, potentially mitigating water pollution issues and utilizing otherwise underutilized resources. In conclusion, microalgae hold great potential as a sustainable and nutritious source of food to address global nutritional challenges.

19.3 NUTRITIONAL COMPOSITION OF MICROALGAE

Microalgae, often referred to as the "superfood of the future," have gained significant attention in recent years due to their rich nutritional composition. These microscopic organisms contain a wide range of essential nutrients, including proteins, omega-3 fatty acids, vitamins, and minerals, making them a potential sustainable solution for combating malnutrition and meeting the growing global demand for food (Nagi et al., 2021). The nutritional composition of freshwater microalgae can vary significantly depending on the species, environmental conditions, and growth stages (Prihanto et al., 2022). The macronutrient compositions of common microalgae are presented in Table 19.1. Green microalgae, such as *Chlorella*, *Spirulina*, *Dunaliella*, and *Haematococcus,* are the most explored green microalgae because of their remarkable nutritional composition. They are packed with vitamins, minerals, essential amino acids, and antioxidants (Lopes et al., 2022). These nutrients can help to support overall health and may be particularly beneficial for individuals with nutrient deficiencies.

TABLE 19.1
Macronutrient Composition of Common Microalgae

Microalgae	Protein	Carbohydrate	Lipid	Reference
Arthrospira platensis	53.3–70.3	11.57–23.9	5.8–19.55	Mostafa and El-Gendy, 2017
Chlorella vulgaris	50.2–54.7	19.2–41.92	3–25.7	Almeida et al., 2017; Belotti et al., 2014; Zainan et al., 2018
Chlorella pyrenoidosa	44	6	27	Yadavalli et al., 2014
Scenedesmus obliquus	50–56	10–17	12–14	Becker, 2007
Scenedesmus quadricauda	34.69	0.63	15.99	Arguelles, 2018
Dunaliella salina	57	32	6	Becker, 2007
Haematococcus pluvialis	48	27	15	Gouveia et al., 2008
E. gracilis	39–61	14–18	14–20	Becker, 2007
C. reinhardtii	32.7–48	17.8–21	15.5–17	Becker, 2007; Nakanishi, 2021
Spirulina platensis	13.30	30.21	48.36	Jena et al., 2011
Chlorella sorokiniana	9.90	35.67	18.81	Chen et al., 2014
Nannochloropsis oceanic	24.80	22.70	19.10	Cheng et al., 2014
Dunaliella tertiolecta	2.87	21.69	61.32	Shuping et al., 2010
Scenedesmus dimorphus	8–18	21–52	16–40	Castrillón et al., 2013
Chlorococum humicola	–	32.50	–	Harun and Danquah, 2011
Chlamydomonas reinhardtii	64.70	22.60	12.60	Mahdy et al., 2014
Spirogyra sp.	6–20	33–64	11–21	Milano et al., 2016
Porphyridium cruentum	28–39	40–57	9–14	Milano et al., 2016

19.3.1 Protein

Freshwater microalgae are an abundant source of high-quality protein, often containing all essential amino acids, making them a valuable protein source for vegetarians and vegans (Bleakley and Hayes, 2017). It's important to note that the protein content varies with species, nutrient availability, light intensity, and cultivation methods.

On average, freshwater microalgae are known to contain protein levels ranging from approximately 30% to 60% of their dry weight. Some species have relatively low levels of protein, while most species have over 40% crude protein content, based on dry mass (Chacón-Lee and González-Mariño, 2010). The protein-rich microalgae species for example, *Arthrospira platensis* (*Spirulina*), *Chlorella vulgaris*, *Dunaliella salina*, *Haematococcus pluvialis*, and *Scenedesmus obliquus*, are considered valuable sources of high-quality protein. They show a very high content of protein with values higher than soybean, corn, and wheat. For example, *H. pluvialis* has a higher arginine content than eggs and soybeans. *C. vulgaris* has a lysine content of 8.4%, which is higher than eggs (5.3%) and soybeans (6.4%). *Spirulina*, in particular, has protein levels that can exceed 60% of its dry weight. *Spirulina* and *Chlorella* are the most commonly exploited industrial species due to their high protein contents (51–58% of dry matter) and favorable essential amino acid profiles (Becker, 2006; Ismail and Hwang, 2020). Furthermore, previous studies have shown that they contain all eight essential amino acids in equal or higher amounts than recommended by the WHO/FAO.

In recent years, there has been a surge of interest in the utilization of microalgae as a sustainable protein source to address the growing global demand for protein-rich foods. Traditional protein sources, such as livestock farming, are associated with significant environmental impacts, including deforestation, greenhouse gas emissions, and excessive water usage. In contrast, microalgae cultivation is exceptionally eco-friendly, requiring minimal land and freshwater resources. Microalgae-based proteins have low land requirements (<2.5 m^2/kg) compared to animal and plant-based proteins (42–252 m^2). Furthermore, microalgae can be cultivated in a controlled environment, allowing for year-round production and reducing the risk of contamination.

19.3.2 Lipid

Microalgal lipids are an important source of polysaccharide fatty acids, including linoleic acid, γ-linolenic acid (GLA), eicosapentaenoic acid (EPA), docosahexaenoic acid (DHA), and arachidonic acid. The polyunsaturated fatty acids (PUFAs), particularly omega-3 fatty acids, including eicosapentaenoic acid (EPA), docosahexaenoic acid (DHA), are crucial for human health, supporting brain function, reducing inflammation, and promoting cardiovascular well-being. Polyunsaturated fatty acid has low oxidative stability and produces rancidity in food products. Although microalgal oils are more stable and less prone to oxidation compared to fish oil, reducing the risk of rancidity. Microalgal lipids are a more sustainable source as they can be cultivated in controlled environments, reducing the pressure on wild fish populations and the risk of overfishing. It is free from contaminants such as heavy metals, PCBs, and dioxins, as they are cultivated in a closed system. Freshwater microalgae have a higher lipid content than seaweed (Ambrozova et al., 2014). Dolganyuk et al. (2022) studied the lipid complex of different microalgae and found a huge variation in the fatty acid composition, especially in the n-3 to n-6 ratio, with *Dunaliella salina* containing the highest amount of n-3 fatty acids (ALA and DHA). Previous studies reported that several freshwater microalgae, including *Chlorella vulgaris*, *Dunaliella salina*, and *Arthrospira platensis* accumulate high amounts of lipids, varying from 30%–70% lipids (Torres-Tiji et al., 2020).

19.3.3 Pigments

Microalgae contain various pigments, including carotenoids, chlorophyll a, b, and c. They play essential roles in photosynthesis and offer a range of health benefits for humans. Microalgae-derived pigments, especially chlorophylls, carotenoids, astaxanthin, lutein, lycopene, and canthaxanthin, phycobiliproteins, as shown in Table 19.2. They have unique colors and molecular structures, respectively, and show different physiological activities and health effects in the human body (Mimouni et al., 2012; Ngamwonglumlert et al., 2017). These pigments have gained attention due to their health-associated benefits and their potential applications in nutrition, food coloring, and pharmaceuticals (Novoveská et al., 2019). The freshwater green algae *Haematococcus pluvialis* is commercially important as a source of astaxanthin (yellow-to-red pigment), which constitutes 4–5% of dry biomass. It is one of nature's most potent antioxidants, known for its ability to protect cells from oxidative stress and inflammation. Astaxanthin shows antioxidant properties that are 10 times that of β-carotene and above 500 times that of α-tocopherol (Dufosse, 2008). Another microalga, *Dunaliella salina*, is renowned for its high beta-carotene content (yellow pigment). *Dunaliella salina* produces 10–14% beta-carotene content of its dry mass, which is used as a natural colorant and contributes reddish coloration (Sathasivam and Juntawong, 2013). Beta-carotene content has the potential to enhance vision and strengthen the immune system. Furthermore, *Spirulina* contains a blue-green pigment, phycocyanin, which has been the focus of scientific research for its antioxidant and anti-inflammatory properties. Due to their potent antioxidant properties, these pigments have gained commercial prominence in the fields of food, nutraceuticals, and cosmetics. Additionally, they have demonstrated beneficial effects in addressing various health concerns, including inflammatory diseases, cardiovascular issues, liver disorders, nervous system conditions, cancer, metabolic syndrome, diabetes, and gastrointestinal ailments.

TABLE 19.2
Prominent Pigments in Common Freshwater Microalgae

Species	Pigments	References
Dunaliella salina	β-carotene, astaxanthin, zeaxanthin, lutein, and cryptoxanthin	Dufossé et al., 2005; Rabbani et al., 1998
H. pluvialis	Astaxanthin, canthaxanthin, and lutein	Buono et al., 2014
C. vulgaris	Canthaxanthin, astaxanthin, β-carotene, and echinenone	Patias et al., 2017
C. sorokiniana	Chlorophyll and carotenoids	Miazek et al., 2017; Ogbonna et al., 2021
S. obliquus	Astaxanthin, lutein, β-carotene, and echinenone	Patias et al., 2017; Qin et al., 2008
Scenedesmus dimorphus	Astaxanthin, chlorophyll, lutein, and zeaxanthin	Sallehudin et al., 2018; Ahmad et al., 2022
Ankistrodesmus	Lutein, chlorophyll, and ß-carotene	Sallehudin et al., 2018; Ogbonna et al., 2021
B. braunii	Neoxanthin, loroxanthin, violaxanthin, lutein, α-carotene, β-carotene, and *echinenone*	Tonegawa et al., 1998; Ambati et al., 2019
Spirulina	β-carotene, astaxanthin, β-carotene, astaxanthin, β-carotene and astaxanthin	Sun et al., 2012

19.3.4 Vitamins

Microalgae are also rich sources of vitamins, including B vitamins (such as B12 in *Chlorella*), vitamin K, and provitamin A (beta-carotene), contributing to overall health and vitality. *Haslea ostrearia* is a rich source of vitamin E (tocopherols). *P. cruentum* produces high quantities of vitamins E and C, as well as β-carotene (vitamin A). The microalga *D. salina* readily produces vitamins A and E, pyridoxine, nicotinic acid, thiamine, riboflavin, and biotin (Koyande et al., 2019).

19.4 MICROALGAE UTILIZATION FOR FOOD FORMULATIONS

The potential applications of microalgae in functional foods are vast and encompass a wide range of health benefits. Microalgae are considered a novel food source due to their nutritional contents, such as lipids, proteins, vitamins, and trace minerals, as shown in Table 19.3. Currently, microalgae are commercially used in the food industry for the formulation of food products such as pasta (Fradique et al., 2010), biscuits, candy, bread, noodles, and desserts (Batista et al., 2008) to improve techno-functional and nutritional properties of traditional food products. Microalgae, especially *Spirulina,* have been identified as potential sources of high-protein foods needed by malnourished people. *A. platensis* (at 2.6%, *w/w*) was added to enhance levels of protein, minerals, and lipids by 22.6%, 46.4%, and 28.1%, respectively (Lucas et al., 2018). The developed snack has scored 82% of overall acceptability with the positive effect of colouration and no effect on texture and flavor. The addition of *A. platensis* (*Spirulina*) powder in bread dough produces vibrant green or blue-green bread. The green hue can be visually appealing, and the mild *spirulina* flavor complements the sweet bread's taste. Hafsa et al. (2014) formulated bread by substituting 1% and 3% (w/w) with *A. platensis* and found an increase in the bread's mineral, protein, and fat content. The study revealed that the 3% formulation has scored the highest nutritional value, although it encountered decreased consumer approval due to the green color imparted to the bread by the microalgae. In support, studies reported that the addition of microalgae would not affect texture parameters and produce a significant effect on colour only (García-Segovia et al., 2017).

Microalgae-derived powders or extracts can be added to energy bars, granola bars, and snack bites to enhance their nutritional profile. They provide vitamins, minerals, and antioxidants, making the snacks more nutritious. Microalgae's bioactive compounds can be harnessed to create functional snacks. For example, astaxanthin-rich microalgae can be used to make antioxidant-rich snacks, promoting health benefits. Functional foods are not only intended to provide essential nutrients but also to offer additional physiological benefits that can promote overall wellness and reduce the risk of chronic diseases. Pigments derived from microalgae are gaining recognition as environmentally friendly dyes, valuable nutraceuticals, and potent antioxidants. They hold significant commercial potential across a range of industries, including food and dairy, alternative food products, dietary supplements, pharmaceuticals, and cosmetics. Microalgae pigments, such as *Spirulina's* blue-green color, can be used as natural food colorants. They can impart attractive colors to snack products, such as blue or green hues in gummies, candies, or coatings.

Dried microalgae biomass fortified snacks were enriched with valuable amino acids and polyunsaturated fatty acids. Microalgae are characterized by their high polyunsaturated fatty acid, especially omega-3 fatty acid offers various health benefits. Microalgae provide a plant-based source of essential omega-3s, making them suitable for vegetarians and vegans. The neutral taste and odor of microalgal omega-3s make them versatile for use in various food products without impacting flavor; however, fish oil has a fishy taste and odour, which some people find unpleasant. Previous studies have incorporated microalgae oil into

TABLE 19.3
Utilization of Microalgae in Food Formulation

Microalgae	Food Product	Nutritional Content	Sensory Evaluation	References
Arthrospira platensis and *Chlorella vulgaris*	Bread	Increase in polyphenols and high antioxidant	Decrease in color in bread; decrease in the volume of the loaf	Sukhikh et al., 2022
Dunaliella salina	Pasta	Increase in chlorophyll A, chlorophyll B, and carotene contents	Weight and volume of pasta increased; color of pasta darker	El-Baz et al., 2017
A. platensis and *C. vulgaris*	Cookies	Higher protein and phenolic content	High color and texture	Batista et al., 2017
Spirulina platensis	Dark chocolate	Increase in carotenoid content	Dark in color and bitter taste	Ekantari et al., 2019
S. platensis	Milk chocolate	Increase in carotenoid content	Bitter taste	Ekantari et al., 2019

functional foods, such as fortified beverages, dairy alternatives, and nutritional supplements. For instance, microalgae omega-3, especially DHA, are added to juices, milk, and other beverages, especially for infants and children (Araujo and Peteiro, 2021). Previous studies have shown that the addition of *A. platensis* biomass can promote significant impacts on a food product's microstructure, nutritional and functional properties, namely in bakery products such as biscuits (Batista et al., 2017) and pasta (De Marco et al., 2014). *Chlorella vulgaris* and *Arthrospira platensis*, in combination added to wheat flour to produce noodles that can improve the nutritional quality. *Chlorella* biomass additives in pasta recipes to enrich the product with biologically active phytonutrients, as well as to achieve the desired color range without the use of synthetic dyes. Tohamy et al. (2018) formulated processed cheese fortified with *Chlorella vulgaris* at levels of 2, 4, and 6% (*w*/*w*) showed enhanced magnesium, potassium, selenium, zinc, iron, and antioxidant potential. Lafarga et al. (2019) developed microalgal-added broccoli soup, which added *Arthrospira*, *Chlorella*, and *Tetraselmis* species at 0.5%-2% and observed enhanced antioxidant capacity with high levels of bioaccessible polyphenols (i.e., between 32.9 and 45.6 mg/100 mL) for each microalgae/broccoli-formulated soup. Furthermore, Los et al. (2018) developed a dehydrated soup containing *A. platensis* at a level of 15% (w/w). This formulation led to enhancements in the soup's nutritional content, including increased levels of protein, fiber, chlorophyll, lipids, and antioxidant capacity. Importantly, when evaluated through sensory assessment, the soup's acceptability remained uncompromised.

19.5 CONCLUSION

In an era where health-conscious consumers are increasingly seeking natural and sustainable sources of functional foods, microalgae have emerged as a promising and versatile solution. Their remarkable adaptability, nutrient density, and sustainability make them a valuable resource in the quest for healthier and more eco-friendly food choices. Additionally, microalgae offer numerous health benefits due to their antioxidant, antibacterial, antiviral, and anti-inflammatory properties. Furthermore, nutrient-rich microalgae can be effectively utilized in the formulation of bakery and snack products to enhance their nutritional value, color, and flavor.

In conclusion, microalgae offer an array of opportunities for enhancing the nutritional content and sustainability of food formulations. Their diverse applications, from protein-rich ingredients to natural food colorants and functional additives, make them a versatile choice for food manufacturers looking to meet the evolving demands of health-conscious and environmentally aware consumers. As research and technology continue to advance, the integration of microalgae into various food products is likely to expand, contributing to healthier and more sustainable food choices.

REFERENCES

Ahmad, N., J.R. Mounsef, and R. Lteif. 2022. Pigment production by *Scenedesmus dimorphus* using different low-cost and alternative culture media. *Journal of Chemical Technology & Biotechnology* 97: 287–94. https://doi.org/10.1002/jctb.6940.

Almeida, H.N., G.Q. Calixto, B.M.E. Chagas, et al. 2017. Characterization and pyrolysis of *Chlorella vulgaris* and *Arthrospira platensis*: Potential of bio-oil and chemical production by Py-GC/MS analysis. *Environmental Science and Pollution Research* 24: 14142–50. https://doi.org/10.1007/s11356-017-9009-2.

Ambati, R.R., D. Gogisetty, R.G. Aswathanarayana, et al. 2019. Industrial potential of carotenoid pigments from microalgae: Current trends and future prospects. *Critical Reviews in Food Science and Nutrition* 59: 1880–902. https://doi.org/10.1080/10408398.2018.1432561.

Ambrozova, J.V., L. Misurcova, R. Vicha, et al. 2014. Influence of extractive solvents on lipid and fatty acids content of edible freshwater algal and seaweed products, the green microalga *Chlorella kessleri* and the cyanobacterium *Spirulina platensis*. *Molecules* 19: 2344–60. https://doi.org/10.3390/molecules19022344.

Araujo, R., and C. Peteiro. 2021. *Algae as food and food supplements in Europe, EUR 30779 EN*. Luxembourg: Publications Office of the European Union. https://doi.org/10.2760/049515, JRC125913.

Arguelles, E.D.L.R. 2018. Proximate analysis, antibacterial activity, total phenolic content and antioxidant capacity of a green microalgae *Scenedesmus quadricauda* (Turpin) Brébisson. *Asian Journal of Microbiology Biotechnology and Environmental Sciences* 20: 150–58. https://www.envirobiotechjournals.com/issues/article_abstract.php?aid=8495&iid=244&jid=1.

Batista, A.P., L. Gouveia, M.C. Nunes, J.M. Franco, and A. Raymundo. 2008. Microalgae biomass as a novel functional ingredient in mixed gel systems. In *Gums and stabilisers for the food industry*, Vol. 14, eds. P.A. Williams and G.O. Phillips, 487–94. Cambridge: RSC Publishing. https://doi.org/10.1039/9781847558312-00487.

Batista, A.P., A. Niccolai, P. Fradinho, et al. 2017. Microalgae biomass as an alternative ingredient in cookies: Sensory, physical and chemical properties, antioxidant activity and in vitro digestibility. *Algal Research* 26: 161–71. https://doi.org/10.1016/j.algal.2017.07.017.

Becker, E.W. 2007. Micro-algae as a source of protein. *Biotechnology Advances* 25: 207–10. https://doi.org/10.1016/j.biotechadv.2006.11.002.

Belotti, G., B. De Caprariis, P. De Filippis, M. Scarsella, and N. Verdone. 2014. Effect of *Chlorella vulgaris* growing conditions on bio-oil production via fast pyrolysis. *Biomass and Bioenergy* 61: 187–95. https://doi.org/10.1016/j.biombioe.2013.12.011.

Bleakley, S., and M. Hayes. 2017. Algal proteins: Extraction, application, and challenges concerning production. *Foods* 6 (April): 1–34. https://doi.org/10.3390/foods6050033.

Buono, S., A.L. Langellotti, A. Martello, F. Rinna, and V. Fogliano. 2014. Functional ingredients from microalgae. *Food & Function* 5: 1669–85. https://doi.org/10.1039/c4fo00125g.

Castrillón, L.J.R., M.E.R. Carmona, and Y.V. Salazar. 2015. *Microalgas para la industria alimenticia*. Bolivariana: Universidad Pontificia Bolivariana.

Chacón-Lee, T.L., and G.E. González-Mariño. 2010. Microalgae for "healthy" foods-possibilities and challenges. *Comprehensive Reviews in Food Science and Food Safety* 9: 655–75. https://doi.org/10.1111/j.1541-4337.2010.00132.x.

Chen, W.H., Z.Y. Wu, and J.S. Chang. 2014. Isothermal and non-isothermal torrefaction characteristics and kinetics of microalga *Scenedesmus obliquus* CNW-N. *Bioresource Technology* 155: 245–51. https://doi.org/10.1016/j.biortech.2013.12.116.

Cheng, J., R. Huang, T. Yu, T. Li, J. Zhou, and K. Cen. 2014. Biodiesel production from lipids in wet microalgae with microwave irradiation and bio-crude production from algal residue through hydrothermal liquefaction. *Bioresource Technology*. 151: 415–18. https://doi.org/10.1016/j.biortech.2013.10.033.

Coelho, D.F.M., C.M.R.P.M. Alfaia, J.M.P. Assunção, et al. 2021. Impact of dietary *Chlorella vulgaris* and carbohydrate-active enzymes incorporation on plasma metabolites and liver lipid composition of broilers. *BMC Veterinary Research* 17 (June): 229, 1–13. https://scite.ai/reports/10.1186/s12917-021-02932-8.

Conde, T.A., I. Zabetakis, A. Tsoupras, et al. 2021. Microalgal lipid extracts have potential to modulate the inflammatory response: A critical review. *International Journal of Molecular Sciences* 22 (September): 9825, 1–15. https://scite.ai/reports/10.3390/ijms22189825.

De Marco, E.R., M.E. Steffolani, C.S. Martínez, and A.E. León. 2014. Effects of *Spirulina* biomass on the technological and nutritional quality of bread wheat pasta. *LWT – Food Science and Technology* 58: 102–08. https://doi.org/10.1016/j.lwt.2014.02.054.

Dolganyuk, V., A. Andreeva, S. Sukhikh et al. 2022. Study of the physicochemical and biological properties of the lipid complex of marine microalgae isolated from the coastal areas of the eastern water area of the Baltic Sea. *Molecules* 27 (September): 5871, 1–19. https://doi.org/10.3390/molecules27185871.

Dufosse, L. 2008. Pigments from microalgae and microorganisms: Sources of food colorants. *In Food colorants: Chemical and functional properties*, ed. C. Socaciu, 399–26. Boca Raton: CRC Press. https://doi.org/10.1201/9781420009286.

Dufossé, L., P. Galaup, A. Yaron, et al. 2005. Microorganisms and microalgae as sources of pigments for food use: A scientific oddity or an industrial reality? *Trends in Food Science & Technology* 16: 389–06. https://doi.org/10.1016/j.tifs.2005.02.006.

Ekantari, N., S.A. Budhiyanti, W. Fitriya, A.B. Hamdan, and C. Riaty. 2019. Stability of chocolate bars fortified with nanocapsules carotenoid of *Spirulina platensis IOP Conference Series: Earth and Environmental Science* 370 (June): 012079, 1–14. https://doi.org/10.1088/1755-1315/370/1/012079.

El-Baz, F.K., S.M. Abdo, and A.M.S. Hussein. 2017. Microalgae *Dunaliella salina* for use as food supplement to improve pasta quality. *International Journal of Pharmaceutical Sciences Review and Research* 46: 45–51. https://globalresearchonline.net/journalcontents/v46–2/10.pdf.

FAO/WHO. 1973. *Joint FAO/WHO Ad Hoc expert committee on energy and protein requirements*. World Health Organization & Food and Agriculture Organization of the United Nations. https://iris.who.int/handle/10665/41042.

Fradique, M., A.P. Batista, M.C. Nunes, L. Gouveia, N.M. Bandarra, and A. Raymundo. 2010. Incorporation of *Chlorella vulgaris* and *Spirulina maxima* biomass in pasta products. Part 1: Preparation and evaluation. *Journal of the Science of Food and Agriculture* 90: 1656–64. https://doi.org/10.1002/jsfa.3999.

García-Segovia, P., M.J.P. Moreno, I.F. Lara, and J.M. Monzó. 2017. Effect of microalgae incorporation on physicochemical and textural properties in wheat bread formulation. *Food Science and Technology International* 23: 437–47. https://doi.org/10.1177/1082013217700259.

Gouveia, L., A.P. Batista, A. Raymundo, and N.M. Bandarra. 2008. *Spirulina maxima* and *Diacronema vlkianum* microalgae in vegetable gelled desserts. *Nutrition and Food Science* 38: 492–01. https://doi.org/10.1108/00346650810907010.

Gupta, A., S. Jain, and N. Verma. 2022. *Industrial applications of soil microbes*, vol. 1. Bentham Science Publishers Pte Ltd. https://doi.org/10.2174/97898150399551220101.

Gupta, A., S. Jain, and N. Verma. 2024. *Industrial applications of soil microbes*, vol. 3. Bentham Science Publishers Pte Ltd. https://doi.org/10.2174/97898150797531240301.

Hafsa, Y.A., D. Amel, S. Samia, and S. Sidahmed. 2014. Evaluation of nutritional and sensory properties of bread enriched with spirulina. *Annals Food Science and Technology* 15: 270–75.

Harun, R., and M.K. Danquah. 2011. Enzymatic hydrolysis of microalgal biomass for bioethanol production. *Chemical Engineering Journal* 168: 1079–84. https://doi.org/10.1016/j.cej.2011.01.088.

Hussain, F., S.Z. Shah, H. Ahmad, et al. 2021. Microalgae an ecofriendly and sustainable wastewater treatment option: Biomass application in biofuel and bio-fertilizer production. A review. *Renewable and Sustainable Energy Reviews* 137 (March): 110603. https://doi.org/10.1016/j.rser.2020.110603.

Ismail, I., Y.H. Hwang, and S.T. Joo. 2020. Meat analog as future food: A review. *Journal of Animal Science and Technology* 62: 111–20. https://doi.org/10.5187/jast.2020.62.2.111.

Jain, S., A. Gupta, and N. Verma. 2023. *Industrial applications of soil microbes*, vol. 2. Bentham Science Publishers Pte Ltd. https://doi.org/10.2174/97898150502641230201.

Jain, S., A. Gupta, and N. Verma. 2024. *Industrial applications of soil microbes*, vol. 4. Bentham Science Publishers Pte Ltd. https://doi.org/10.2174/97898151249961240401.

Jena, U., K.C. Das, and J.R. Kastner. 2011. Effect of operating conditions of thermochemical liquefaction on biocrude production from *Spirulina platensis*. *Bioresource Technology* 102: 6221–29. https://doi.org/10.1016/j.biortech.2011.02.057.

Koyande, A.K., K.W. Chew, K. Rambabu, Y. Tao, D.T. Chu, and P.L. Show. 2019. Microalgae: A potential alternative to health supplementation for humans. *Food Science and Human Wellness* 80: 16–24. https://doi.org/10.1016/j.fshw.2019.03.001.

Kusmayadi, A., Y.K. Leong, H.W. Yen, C.Y. Huang, and J.S. Chang. 2021. Microalgae as sustainable food and feed sources for animals and humans–biotechnological and environmental aspects. *Chemosphere* 271 (January): 129800. https://doi.org/10.1016/j.chemosphere.2021.129800.

Lafarga, T., F.G. Acién-Fernández, M. Castellari, S. Villaró, G. Bobo, and I. Aguiló-Aguayo. 2019. Effect of microalgae incorporation on the physicochemical, nutritional, and sensorial properties of an innovative broccoli Soup. *LWT – Food Science and Technology* 111: 167–74. https://doi.org/10.1016/j.lwt.2019.05.037.

Lopes, P.A., D. Coelho, and J.A.M. Prates. 2022. Testimony on a successful lab protocol to disrupt *Chlorella vulgaris* microalga cell wall. *PLoS One* 17 (May): e0268565, 1–9. https://doi.org/10.1371/journal.pone.0268565.

Los, P.R., D.R.S. Simões, R.S. Leone, B.C. Bolanho, T. Cardoso, and E.D.G. Danesi. 2018. Viability of peach palm by-product, *Spirulina platensis*, and spinach for the enrichment of dehydrated soup. *Pesquisa Agropecuária Brasileira* 53: 1259–67. https://doi.org/10.1590/S0100-204X2018001100008.

Lucas, B.F., M.G. de Morais, T.D. Santos, and J.A.V. Costa. 2018. *Spirulina* for snack enrichment: Nutritional, physical and sensory evaluations. *LWT-Food Science and Technology* 90: 270–76. https://doi.org/10.1016/j.lwt.2017.12.032.

Mahdy, A., L. Mendez, M. Ballesteros, and C. González-fernández. 2014. Enhanced methane production of *Chlorella vulgaris* and *Chlamydomonas reinhardtii* by hydrolytic enzymes addition. *Energy Conversion and Management* 85: 551–57. https://doi.org/10.1016/j.enconman.2014.04.097.

Matos, Â.P. 2019. Microalgae as a potential source of proteins. In *Proteins: Sustainable source, processing and applications*, ed. C.M. Galanakis, 63–96. Academic Press, Elsevier Inc. https://doi.org/10.1016/B978-0-12-816695-6.00003-9.

Miazek, K., C. Remacle, A. Richel, and D. Goffin. 2017. Beech wood *Fagus sylvatica* dilute-acid hydrolysate as a feedstock to support *Chlorella sorokiniana* biomass, fatty acid and pigment production. *Bioresource Technology* 230: 122–31. https://doi.org/10.1016/j.biortech.2017.01.034.

Milano, A., H.C. Ong, H.H. Masjuki, et al. 2016. Microalgae biofuels as an alternative to fossil fuel for power generation. *Renewable and Sustainable Energy Reviews* 58: 180–97. https://doi.org/10.1016/j.rser.2015.12.150.

Mimouni, V., L. Ulmann, V. Pasquet, et al. 2012. The potential of microalgae for the production of bioactive molecules of pharmaceutical interest. *Current Pharmaceutical Biotechnology* 13: 2733–50. https://doi.org/10.2174/138920112804724828.

Mostafa, S.S.M., and N.S. El-Gendy. 2017. Evaluation of fuel properties for microalgae *Spirulina platensis* bio-diesel and its blends with Egyptian petro-diesel. *Arabian Journal of Chemistry* 10: S2040–50. https://doi.org/10.1016/j.arabjc.2013.07.034.

Nagi, G.K., A.K. Minhas, S. Gaur, P. Jain, and S. Mandal. 2021. Integration of algal biofuels with bioremediation coupled industrial commodities towards cost-effectiveness. *Frontiers in Energy Research* 9 (August): 735141, 1–16. https://scite.ai/reports/10.3389/fenrg.2021.735141.

Nakanishi, A., Y. Sakihma, and N. Ozawa. 2021. Improvement of growth of *Chlamydomonas reinhardtii* in CO_2 – stepwisely aerating condition. *Journal of Applied Biotechnology Reports* 8: 37–40. https://doi.org/10.30491/jabr.2020.237342.1246.

Ngamwonglumlert, L., S. Devahastin, and N. Chiewchan. 2017. Molecular structure, stability and cytotoxicity of natural green colorants produced from *Centella asiatica* L. leaves treated by steaming and metal complexations. *Food Chemistry*. 232: 387–94. https://doi.org/10.1016/j.foodchem.2017.04.034.

Novoveská, L., M.E. Ross, M.S. Stanley, R. Pradelles, V. Wasiolek, and J-F. Sassi. 2019. Microalgal carotenoids: A review of production, current markets, regulations, and future direction. *Marine Drugs* 17 (November): 640, 1–21. https://doi.org/10.3390/md17110640.

Nwoba, E.G. 2017. Effect of selected light spectra on the growth of *Chlorella* spp. (Chlorophyta). *Nigerian Journal of Biotechnology* 32: 69–76. http://dx.doi.org/10.4314/njb.v32i1.10.

Ogbonna, J.C., N.O. Nweze, and C.N. Ogbonna. 2021. Effects of light on cell growth, chlorophyll, and carotenoid contents of *Chlorella sorokiniana* and *Ankistrodesmus falcatus* in poultry dropping medium. *Journal of Applied Biology & Biotechnology* 92: 157–63. http://dx.doi.org/10.7324/JABB.2021.9215.

Patias, L.D., A.S. Fernandes, F.C. Petry, A.Z. Mercadante, E. Jacob-Lopes, and L.Q. Zepka. 2017. Carotenoid profile of three microalgae/cyanobacteria species with peroxyl radical scavenger capacity. *Food Research International* 100: 260–66. https://doi.org/10.1016/j.foodres.2017.06.069.

Prihanto, A.A., Y.D. Jatmiko, R. Nurdiani, A. Miftachurrochmah, and M. Wakayama. 2022. Freshwater microalgae as promising food sources: Nutritional and functional properties. *The Open Microbiology Journal* 16 (Mach): e187428582206200, 1–14. https://doi.org/10.2174/18742858-v16-e2206200.

Qin, S., G.X. Liu, and Z.Y. Hu. 2008. The accumulation and metabolism of astaxanthin in *Scenedesmus obliquus* (Chlorophyceae). *Process Biochemistry* 43: 795–02. https://doi.org/10.1016/j.procbio.2008.03.010.

Quintana, N., F. Van der Kooy, M.D. Van de Rhee, G.P. Voshol, and R. Verpoorte. 2011. Renewable energy from Cyanobacteria: Energy production optimization by metabolic pathway engineering. *Applied Microbiology and Biotechnology* 91: 471–90. https://doi.org/10.1007/s00253-011-3394-0.

Rabbani, S., P. Beyer, J. Lintig, P. Hugueney, and H. Kleinig. 1998. Induced β-carotene synthesis driven by triacylglycerol deposition in the unicellular alga *Dunaliella bardawil*. *Plant Physiology* 116: 1239–48. https://doi.org/10.1104/pp.116.4.1239.

Sallehudin, N.J., R.A. Raus, M. Mustapa, R. Othman, and M. Mel. 2018. Screening of lutein content in several fresh-water microalgae. *International Food Research Journal* 25: 2307–12. http://www.ifrj.upm.edu.my/25%20(06)%202018/(11).pdf.

Sathasivam, R., and N. Juntawong. 2013. Modified medium for enhanced growth of *Dunaliella* strains. *International Journal of Current Science* 5: 67–73. https://rjpn.org/ijcspub/papers/IJCSP10A1374.pdf.

Shuping, Z., W. Yulong, Y. Mingde, L. Chun, and T. Junmao. 2010. Pyrolysis characteristics and kinetics of the marine microalgae *Dunaliella tertiolecta* using thermogravimetric analyzer. *Bioresource Technology* 101: 359–65. https://doi.org/10.1016/j.biortech.2009.08.020.

Sukhikh, S., S. Ivanova, V. Dolganyuk, et al. 2022. Evaluation of the prospects for the use of microalgae in functional bread production. *Applied Sciences* 12 (December): 12563, 1–15. https://doi.org/10.3390/app122412563.

Sun, X., Y. Chang, Y. Ye, et al. 2012. The effect of dietary pigments on the coloration of Japanese ornamental carp (koi, *Cyprinus carpio* L.). *Aquaculture* 342: 62–68. https://doi.org/10.1016/j.aquaculture.2012.02.019.

Tohamy, M.M., M.A. Ali, H.A-G. Shaaban, A.G. Mohammad, and A.M. Hasanain. 2018. Production of functional spreadable processed cheese using *Chlorella vulgaris*. *Acta Scientiarum Polonorum Technologia Alimentaria* 17: 347–58. http://dx.doi.org/10.17306/J.AFS.2018.0589.

Tonegawa, I., S. Okada, M. Murakami, and K. Yamaguchi. 1998. Pigment composition of the green microalga *Botryococcus braunii* Kawagushi–1. *Fisheries Science* 64: 305–08. https://doi.org/10.2331/fishsci.64.305.

Torres-Tiji, Y., F.J. Fields, and S.P. Mayfield. 2020. Microalgae as a future food source. *Biotechnology Advances* 41 (July–August): 107536. https://doi.org/10.1016/j.biotechadv.2020.107536.

Yadavalli, R., C.S. Rao, R.S. Rao, and R. Potumarthi. 2014. Dairy effluent treatment and lipids production by *Chlorella pyrenoidosa* and *Euglena gracilis*: Study on open and closed systems. *Asia-Pacific Journal of Chemical Engineering* 9: 368–73. https://doi.org/10.1002/apj.1805.

Zainan, N.H., S.C. Srivatsa, F. Li, and S. Bhattacharya. 2018. Quality of bio-oil from catalytic pyrolysis of microalgae *Chlorella vulgaris*. *Fuel* 223: 12–19. https://doi.org/10.1016/j.fuel.2018.02.166.

Zhang, S., Y. He, B. Sen, et al. 2018. Alleviation of reactive oxygen species enhances PUFA accumulation in *Schizochytrium* sp. through regulating genes involved in lipid metabolism. *Metabolic Engineering Communications* 6: 39–48. https://doi.org/10.1016/j.meteno.2018.03.002.

20 Marine Microbial Secondary Metabolites

An Overview and Importance

P. Trapti
(ITM University, Gwalior, MP, India; Email: trapti.pathak@itmuniversity.ac.in)

20.1 INTRODUCTION

Marine environments have a vast range and largely unexplored source of biodiversity, contributing a wealth of biological resources. Among these resources, marine microorganisms have gained considerable attention due to their ability to produce a wide range of secondary metabolites.

Marine microbial secondary metabolites are diverse compounds produced by microorganisms inhabiting marine environments. They possess a wide range of biological activities and have significant applications in pharmaceutical drug discovery, biotechnology, and ecological research. Their exploration and study hold great promise for addressing various challenges in human health, industry, and environmental sustainability.

Marine microbial secondary metabolites are simple and small organic molecules with unique chemical structure, produced by marine microbes, that is, bacteria, fungi, and algae that belong exclusively marine ecosystem (Blunt et al., 2012, 2014, 2016). These organic metabolites are not essential or required for the existence or survivability of the microbes that produce them, but they can provide a competitive advantage in their environment using serving as organic chemical weapons, antibiotics, or signaling molecules. These metabolites play a crucial role in the ecological interactions and survival strategies of marine organisms.

These small, simple organic molecules also have potential as a good source of new drug discovery, agricultural products, and industrial materials because of their chemical nature and structure (Kim, 2017). They also have shown many recent advances promising properties related to biological activities, that is, anticancerous, antiviral, antibacterial, antifungal, *etc.*

As the study of marine microbial secondary metabolites acts as an interdisciplinary field, it involves marine microbiology, aquatic biology, chemistry, ecology, biotechnology, pharmacology, agriculture, and many other branches of applied sciences.

The research and development of marine microbial secondary metabolites has great potential to revolutionize new research fields like pharmaceutical drug discovery, biotechnology, and environmental research due to their diverse chemical structures and potent biological activities, and they give new solutions to some of the world's most promising health and environmental challenges. The organic molecules have gained significant attention in recent research areas like pharmaceutical drug discovery, biotechnology, and environmental research due to their diverse properties, characteristics, molecular structures and potent biological activities.

20.2 KEY ROLES OF MARINE MICROBIAL SECONDARY METABOLITES IN APPLIED SCIENCES

Some important aspects of marine microbial secondary metabolites are why scientists are focusing on them. They are as follows:

20.2.1 Production and Function

Microbes produce secondary metabolites as a part of their metabolic pathways. While primary metabolites are essential for basic cellular functions, secondary metabolites are not directly involved in growth or reproduction. Instead, they often serve ecological functions, such as defense against predators, competition for resources, or communication within microbial communities (Gupta et al., 2022, 2024).

20.2.2 Chemical Diversity

Marine microbial secondary metabolites exhibit an extraordinary chemical diversity, ranging from simple organic compounds to complex structures. They can include alkaloids, peptides, polyketides, terpenoids, and many other classes of compounds (Kim, 2017). The marine environment offers unique conditions, such as high salinity, temperature variations, and competition, which influence the synthesis of these compounds.

20.2.3 Biological Activities

One of the most remarkable aspects of marine microbial secondary metabolites is their broad spectrum of biological activities. Many of these compounds exhibit antimicrobial,

DOI: 10.1201/9781003408833-20

antifungal, antiviral, antitumor, anti-inflammatory, and antioxidant properties. These activities make them valuable in drug discovery efforts, as they can serve as lead compounds or provide structural scaffolds for the development of new pharmaceuticals (Jain et al., 2023, 2024).

20.2.4 Uses in Pharmaceutical

Marine microbial secondary metabolites have shown great potential in the development of drugs for various diseases. For example, some compounds derived from marine microorganisms have demonstrated antitumor effects, leading to the development of anticancer drugs like trabectedin. Other compounds have exhibited antiviral properties and may contribute to the development of treatments for viral infections.

20.2.5 Uses in Biotechnological Science

Besides their pharmaceutical potential, marine microbial secondary metabolites have other biotechnological applications. They can be used in the development of new antibiotics, enzyme inhibitors, bioactive coatings, and agricultural products. Additionally, they play a role in the production of pigments, flavorings, and natural products used in the food and cosmetic industries.

20.2.6 Role in Ecology and Environment

Marine microbial secondary metabolites have ecological implications beyond their roles in defense and competition. They can influence the structure and dynamics of marine ecosystems by affecting the interactions between organisms, including symbiotic relationships, predator-prey interactions, and the establishment of biofilms. They also play a role in chemical signaling and communication within microbial communities.

20.2.7 Role in Discovery and Research & Development Area

Exploring marine microbial secondary metabolites involves various research methods. Scientists often isolate and culture marine microorganisms in the laboratory to study their metabolite production. They employ techniques such as chemical extraction, purification, and structural elucidation to identify and characterize these compounds. Advanced analytical methods, including mass spectrometry and nuclear magnetic resonance, aid in the analysis of complex chemical structures.

20.2.8 Future Perspectives & Challenges

The discovery and exploitation of marine microbial secondary metabolites face challenges, including the need for sustainable and environmentally friendly collection methods, as well as the identification of novel compounds from unexplored marine environments. Advances in genomics, metagenomics, and synthetic biology offer promising avenues for unlocking the full potential of marine microbial secondary metabolites.

20.3 MARINE MICROBES

20.3.1 Introduction and Definition

Marine microbes, also known as marine microorganisms or marine microbiota, refer to the diverse group of microscopic organisms that inhabit marine environments. These organisms are too small to be seen with the naked eye and play vital roles in their marine ecosystems, where they are present and contributing to various biogeochemical processes and forming the foundation of marine food webs.

20.3.2 Classification of Marine Microbes

Marine microorganisms that produce secondary metabolites have vast diversity as shown in Table 20.1. They comprise a wide range of microorganisms, each with its unique characteristics and classification (Rateb and Ebel, 2011; Rédou et al., 2015). The major groups and classifications of marine microbes are as follows:

20.3.2.1 Phytoplankton

Phytoplankton are photosynthetic microorganisms that form the base of the marine food web. They are responsible for serving to the Earth's oxygen production and serve as the primary producers in marine ecosystems, for example, Diatoms, Dinoflagellates, and Coccolithophores.

20.3.2.2 Bacteria

Bacteria are prokaryotic microorganisms, and they are found richly in marine habitats. They perform roles, that is, nutrient cycling, decomposition, and symbiotic relationships with other organisms. They are classified based on their various characteristics, like their shape, metabolic activities, and genetic variability. Some common groups of marine bacteria that produce secondary metabolites are, that is, Cyanobacteria, Proteobacteria, and Bacteroidetes (Bull et al., 2005; Fenical and Jensen, 2006).

20.3.2.3 Archaea

Archaea are primitive, single-celled microbes and are different from bacteria and eukaryotes. They inhabit different marine habitats, like as extreme environments such as hydrothermal vents. They play vital roles in the nitrogen and carbon cycles. Some examples are Euryarchaeota, Thaumarchaeota, methanogens, *etc*.

20.3.2.4 Viruses

Marine viruses, also known as marine bacteriophages, are minute infective agents that infect and replicate in marine bacteria and archaea. They are the most abundant biological entities in the marine environment that influence microbial diversity and ecosystem dynamics. They are classified into their sub-classes based on their genetic material, shape, morphology, and other topographies. Some examples are Myoviridae, Podoviridae, and Siphoviridae.

20.3.2.5 Protists

Protists are eukaryotic microorganisms that include a diversified range of microorganisms. They were classified into to following different groups, that is, Foraminifera, Radiolaria, Ciliates, and Dinoflagellates.

20.3.2.6 Fungi

Many species of fungi belong to marine habitats. Among them, some species have been discovered that have a great potential to produce bioactive compounds known as secondary metabolites. For example, *Aspergillus* sp., *Penicillium* sp., *etc.*

Note: This classification of marine microbes is based on current scientific knowledge and is subject to ongoing research. However, marine microbes have great diversity, and their classifications may also evolve.

20.4 ROLE AND IMPORTANCE OF MARINE MICROBES

Marine microbes play significant roles in the marine environment and are important for various ecological and global processes. The following are some important features that highlight the significance of marine microbes:

20.4.1 Primary Production

Marine microorganisms, especially phytoplankton, are responsible for providing the Earth's primary production. Because using photosynthesis, they can convert sunlight, carbon dioxide, and nutrients into organic matter, as a result, oxygen is produced as a by-product. This procedure makes the basic platform of marine food webs and supports the growth and existence of higher trophic levels.

20.4.2 Nutrient Cycling

Marine microorganisms play a crucial role in nutrient cycling processes also which is necessary for the functioning of an ecosystem. Their main contribution is to the breaking down and recycling of organic matter. So that the nutrients can be released, like nitrogen, phosphorus, and iron, back into the water column of the marine habitat, and these recycled nutrients are accessible for uptake by other marine organisms, facilitating the continuous flow of energy and nutrients in the marine environment.

20.4.3 Carbon Cycle and Climate Regulation

Marine microorganisms, mainly phytoplankton and bacteria, play an important role in the global carbon cycle. They can absorb CO_2 from the atmosphere via photosynthesis and integrate it into their cells. When these organisms die or are used by other organisms, the carbon can sink to the ocean floor, which contributes to long-term carbon sequestration. Moreover, some marine microorganisms, like coccolithophores, with their shells, which are made of $CaCO_3$, can accumulate as deposition in the form of limestone over millions of years, which promotes sequestering carbon from the atmosphere.

20.4.4 Biogeochemical Cycling

Marine microbes are also important for many biogeochemical cycles, like nitrogen, sulfur, iron cycles, *etc.* They complete main procedures like nitrogen fixation, denitrification, ammonification, and nitrification, which regulate the availability and transformation of these important elements in the marine environment. These cycles have implications for nutrient availability, water quality, and ecosystem health globally.

20.4.4.1 As Secondary Metabolites Producer

The marine ecosystem is incredibly diverse, hosting a wide array of microorganisms, like bacteria, fungi, and algae. Some of these microbial groups can produce

TABLE 20.1
Examples of Secondary Metabolites Produced by Marine Microbes

S. No.	Group of Marine Microorganisms	Examples
1	Phytoplankton	Diatoms, dinoflagellates, and coccolithophores
2	Bacteria	SAR11, *Prochlorococcus*, and *Vibrio*
3	Archaea	Methanogens, *etc.*
4	Viruses	Myoviridae and Podoviridae
5	Protista	Foraminifera, Radiolaria, and *Paramecium*
6	Fungi	*Aspergillus* sp., *Penicillium* sp., *Dichotomomyces* sp., *etc.*

Note: These examples show only a segment of the vast group of marine microbes present in the marine habitats. They are mutually serving in the marine environmental functioning.

secondary metabolites, which are small organic compounds and are important for the survivability of the organisms (Scheuer, 1990; Blunt et al., 2012, 2014, 2016; Kim, 2017).

20.4.4.2 Symbiotic Relationships

Marine microorganisms also form symbiotic associations with other organisms (Rosenberg and Koren, 2006), that is, corals, sponges, and certain marine animals. Like coral reefs depend on a symbiotic relationship between corals and photosynthetic dinoflagellates (Taylor et al., 2007). The dinoflagellates provide nutrients to corals and give the vibrant colours and productivity of coral reef ecosystems.

20.4.4.3 Disease Regulation

Some marine microorganisms, especially bacteria and viruses, play a crucial role in regulating disease dynamics in organisms that are present in the marine environment. They can act as pathogens, which cause diseases in marine animals. Though they also play their role in controlling the population and preventing the spread of diseases via some mechanisms.

So, the understanding about the role of marine microorganisms is complex for effective ecosystem management, conservation efforts, and predicting the impacts of environmental changes, like climate change and pollution, on marine habitats.

20.5 SECONDARY METABOLITES

20.5.1 Introduction and Definition

Secondary metabolites are organic compounds that are produced by various organisms, including plants, animals, fungi, and bacteria. Unlike primary metabolites, which are involved in many necessary life mechanisms, that is, growth and development, life sustainability *etc.* These compounds are not required directly for the organism's survivability. Although they frequently play significant roles in ecological interactions and have important uses in many advanced fields, that is, medicine, agriculture, and industry.

Secondary metabolites are structurally and functionally diversified, and they assist several ecological drives. For example, alkaloids, terpenoids, phenolic, flavonoids, and antibiotics. These organic compounds are usually synthesized through complex biochemical pathways and are frequently produced in response to environmental stimuli, like predation, competition, or stress.

The secondary metabolites also play their crucial roles in defense against herbivores, pathogens, or competing organisms. Some secondary metabolites also have antimicrobial properties. These help organisms defend against pathogens or inhibit the growth of microorganisms that are competitive with them.

Besides these functions, secondary metabolites have important uses in various fields, like in medical science, where secondary metabolites serve as a good source of therapeutic compounds. There are many drugs, including antibiotics, anticancerous drugs, and pain killers, that are derived from secondary metabolites produced by marine microbes, that is, fungi, bacteria, *etc.* These compounds often have unique and complex structures that can interact with specific molecular targets in the body, which leads to therapeutic effects. They also have their agricultural applications, industrial applications, *etc.*

We can define them as *"Secondary metabolites are the small organic molecules produced by microorganisms, and they are important for their life sustainability and development. However, they are not directly required for the organism's survivability, which is the producer for them."*

Studying secondary metabolites and their biosynthesis pathways has become an important area of research in fields like natural product chemistry, pharmacology, and biotechnology. Scientists are continuously exploring the vast potential of secondary metabolites, both for understanding ecological interactions and for developing novel compounds with therapeutic, agricultural, or industrial applications.

So, the secondary metabolites have a vast range of beneficial effects on the living organisms that produce these chemical compounds, as well as on other individual that exist with them in their environment. Such as: Secondary metabolites produced by marine microbes act as a defense mechanism against predators, parasites, or pathogens, as signal molecules in between organisms, which allowing them to communicate to each other and coordinate behaviors, can also help organisms to adapt their environment by allowing them to tolerate adverse conditions, can used to develop new drugs and therapies. These metabolites can be toxic, repellent, or otherwise harmful to the attacker, also helping to deter or eliminate threats, and also play their role in biofilm formation, bioremediation, *etc.*

Therefore, the secondary metabolites are a good source of chemical diversity that offers several opportunities for scientific research and the development of novel organic compounds with their significant role and potential in therapeutic, medical, agricultural, industrial, and other commercial uses and applications.

Thus, secondary metabolites play a key role in the survival and success for their existence of organisms present in their habitats or environment.

20.5.2 Marine Ecosystem as a Rich Source of Secondary Metabolites

The marine ecosystem is made up of biotic and abiotic components and has a vast range of unmapped sources of secondary metabolites. Marine organisms, including microorganisms, algae, invertebrate animals, and fish, may produce a vast range of secondary metabolites with chemically diverse structures and bioactivities. Besides other marine organisms, marine micro-organisms are a good source of secondary metabolites mainly because of their vast range of diversified world, richness, and their unique adaptations according to their surrounding environment (Rateb and Ebel, 2011; Rédou et al., 2015). They produce these small organic compounds with unique chemical structures and biological properties that have great potential for many new research areas, that is, drug discovery, pesticides, and many other biotechnological approaches. There are many marine bacteria, have been shown to produce these compounds with anticancerous, antimicrobial, antifungal, antiviral, and many other biological properties. These compounds can develop a defense mechanism against predators, competitors, and pathogens in the exciting and vibrant marine environment.

So, the marine ecosystem is indeed a rich source of secondary metabolites, and its exploration has revealed a plethora of unique and bioactive compounds with diverse chemical structures and valuable applications (Scheuer, 1990). The investigation of the marine environment for secondary metabolites has been facilitated by recent advances in new technologies, including genomics, metabolomics, and synthetic biology. These techniques allow the determination, identification, and characterization of secondary metabolites and the genetic pathways involved in their production. Thus, the marine environment is a good and untouched source of secondary metabolites that have a great potential to produce different new compounds with their many important uses. Further investigation and utilization of these resources are compulsory to untapped the full potential of especially marine microbial secondary metabolites. The following are the factual explanations that the marine ecosystem is considered a treasure trove of secondary metabolites:

20.5.2.1 Biodiversity

The marine habitats are amazingly diversified, which host a wide range of organisms, such as bacteria, fungi, algae, sponges, corals, mollusks, and many others. This biodiversity rises the likelihood of discovering novel secondary metabolites. Different types of marine organisms have developed unique chemical strategies to adaptation to their specific ecological niches and make them as potential sources of diverse and bioactive compounds.

20.5.2.2 Complexity in Chemical Structure

Marine microbes produce secondary metabolites that regularly exhibit a complex and intricate chemical structure. The marine ecosystem shows challenges like competition for resources, predation, and tough physical circumstances, which may lead to the evolution of chemically diverse compounds with great potential and biological activities. The complexity in structure in marine secondary metabolites improves their potential for interactions with specific molecular targets and eligible them as a valuable material in drug development and discovery.

20.5.2.3 Defense and Adaptation

Marine microbes also encounter various ecological pressures, including predation, competition, and colonization by pathogens. So, for survivability and to thrive in such environments, they must have developed classy chemical defense mechanisms. Secondary metabolites produced by marine organisms act as deterrents, toxins, antifouling agents, and antimicrobial compounds, and protect themselves from predators, competitors, and pathogens. Thus, these compounds frequently exhibit powerful bioactivity and can be used to develop new drugs or eco-friendly pesticides.

20.5.3 Extreme and Unexplored Environments

The marine environment includes wide and unexplored habitats, like deep-sea trenches, hydrothermal vents, and cold-water coral reefs. These marine habitats are categorized by extreme situations, containing high pressures, low temperatures, and limited sunlight, which pose challenges to the survivability of organisms in the marine environment. The organisms blooming in these habitats show adaptation to unique biochemical pathways for coping with such extreme situations, which results in the production of novel secondary metabolites with potent and valuable properties.

20.5.3.1 Symbiotic Relationships

Many marine organisms, including marine microbes also involved in symbiotic relationships, and this symbiotic interaction frequently comprises the exchange of chemical signals and metabolites. Some chemical compounds formed by these symbionts are essential for the health and survivability of the coral reef marine habitat. Discovering these symbiotic relationships can uncover unique secondary metabolites with interesting bioactivities (Rosenberg and Koren, 2006).

The exploration of the marine environment for secondary metabolites has yielded many findings, including potent anticancerous drugs, antibiotics, anti-inflammatory compounds, and enzymes that have industrial uses (Scheuer, 1990). Although it is important to note that marine organisms are habitually slow-growing and sensitive to environmental changes, and require sustainable and responsible methodologies in the collection and extraction of natural products from marine habitats.

Thus, the vast biodiversity, chemical complexity, adaptation approaches, unexplored habitats, and symbiotic relationships in the marine environment make it an exceptionally promising and valuable source of secondary metabolites with diversified uses in medicine, agriculture, industry, and many other areas.

20.6 MARINE MICROBES AS PRODUCERS OF SECONDARY METABOLITES

Many marine microorganisms act as a good source of secondary metabolites, which are organic bioactive compounds that are not directly involved in the growth, development, or reproduction of an organism but have a vast range of functions, such as defense against predators, communication, and competition for resources. Such as: antibiotics, antifouling agents, anticancer agents, pigments, toxicants, *etc*., produced by different marine microbes as secondary metabolites, and they have a great potential to be used in medical and clinical, agricultural, pharmacology, and many other fields, as well as in research and development areas.

The marine microbial world, including bacteria, archaea, fungi, and unicellular algae or microalgae, is well known as to produce a vast range of secondary metabolites with their potential in many biological activities (Karthikeyan et al., 2022). Some of the major kinds of marine microorganisms that are known to produce secondary metabolites are marine bacteria, that is, Actinomycetes, as producers of secondary metabolites, including many antibiotics, marine fungi, that is, *Aspergillus*, *Penicillium*, *Cladosporium*, marine microalgae, *etc*. These microorganisms have adapted to bloom in variable marine habitats and have evolved complex biochemical pathways to produce a vast range of bioactive compounds.

20.6.1 Why Marine Microbes Act as Producers of Secondary Metabolites

As we know, some microbes that belong to marine habitats can produce secondary metabolites because of some important abilities. These are as follows:

20.6.1.1 Adaptation to Environmental Encounters

Marine microorganisms face different ecological encounters, like nutrient limitations, temperature variation and salinity, and competition for resources. For survivability and competition in these marine habitats, they have evolved biochemical mechanisms, such as the production of secondary metabolites. These organic compounds enable them to adapt to changes in the environment, ward off competitors, and defend against pathogens.

20.6.1.2 Antipredatory Strategies and Chemical Defense

Marine microorganisms that produce secondary metabolites as chemical defenses against predators or competing organisms. Those compounds can inhibit or depress the growth of other microbes and serve as a competitive benefit. Some marine bacteria produce compounds that deter the growth of other bacteria present in the same habitat and protect themselves in their ecological niche and resources.

20.6.1.3 Communication and Quorum Sensing

Secondary metabolites produced by marine microorganisms play a key role in intercellular communication and quorum sensing. Quorum sensing is a method by which microbes can coordinate their behavior, which is based on population density. Secondary metabolites also perform as signaling molecules, which allow microorganisms to sense the presence of neighboring individuals, control gene expression, and coordinate with some activities like biofilm formation or the release of virulence factors.

20.6.1.4 Nutrient Cycling and Ecological Interactions

Marine microorganisms, that is, bacteria and microalgae, contribute to necessary biogeochemical cycles, like nitrogen, sulfur, and carbon cycles. They produce secondary metabolites, involved in these cycles and facilitating to acquisition, transformation, and recycling of nutrients. These organic compounds can also affect the availability of nutrients for other organisms, shaping the dynamics of marine environments and contributing to nutrient cycling and the productivity of the ecosystem.

20.6.1.5 As Symbiont

Marine microorganisms are also engaged in symbiotic relationships with other organisms. These symbiotic associations frequently involve the exchange of nutrients and chemical signals. Marine microbes associated with symbiotic hosts can produce secondary metabolites that give advantage to the host organism by providing protection against pathogens, enhancing nutrient availability, or facilitating other mutually profitable interactions.

20.7 APPLICABILITY IN APPLIED SCIENCES

These compounds (secondary metabolites) produced by marine microorganisms have incredible prospective for many new and advanced applied science branches, such as secondary metabolites have biotechnological applications. Many of these compounds have novel chemical structures and bioactive characters that make them important for drug development and discovery, pharmaceutical development, and biotechnological processes. For example, in antibiotics, antifungal agents, antiviral compounds, and enzymes with industrial applications.

Now, future research is needed to fully explore the diversity of the marine microbial world and the use of these organisms to produce new and utilizable bioactive organic secondary metabolites. Exploring the diversified microbial communities present in the marine ecosystems and their produced secondary metabolites offers a rich source of bioactive compounds with great potential uses in medicine, agriculture, and industry. Furthermore, ongoing research and advances in biotechnological techniques enable the discovery, production, and optimization of these secondary metabolites for various practical uses.

20.8 FACTORS INFLUENCING THE PRODUCTION OF SECONDARY METABOLITES THROUGH MARINE MICROORGANISMS

The production of secondary metabolites by marine microorganisms is a multifaceted and active process and can be influenced by different environmental and biological factors. To understand these factors and how they have their impact to better predict and help to optimize the production of these important bioactive compounds in the form of secondary metabolites. The production of secondary metabolites by marine microorganisms is influenced by various factors that influence their growth, physiology, and biochemical pathways. A few important factors, influence the formation of secondary metabolites via marine microbes are as follows:

20.8.1 Nutritional Availability

Nutritional availability, such as carbon, nitrogen, phosphorus, trace elements, *etc*., plays a vital role in secondary metabolite production by microbes. Limits as well as excesses of particular nutrients have an impact on metabolic activities and biosynthesis of secondary metabolites. For example, the limitation of nitrogen can influence the production of antibiotics through marine bacteria.

20.8.2 Environmental Factors

Environmental factors like temperature, salinity, light intensity, and pH can also influence the formation of secondary metabolites by microbes present in the marine environment. Different marine microbes have particular environmental requirements for optimum growth and secondary metabolite formation by microbes. If the changes occur in these environmental circumstances, then they lead to changes in the metabolic activities and the types of secondary metabolites produced by microbes.

20.8.3 Diversity and Substrate Availability

The diversity and availability of substrates in the marine habitat can influence the biosynthesis of secondary metabolites. Marine microbes can consume different organic compounds, which are derived from plants, algae, or other microbes present in the same marine environment as substrates for biosynthesis of secondary metabolites. The availability of specific substrates can also induce the appearance of certain biosynthetic genes and pathways.

20.8.4 Interactions with Other Organisms

The marine microbial fauna and flora interact with other organisms also in their ecosystem, that is, symbiotic relationships or competition for resources, which can influence secondary metabolite synthesis in marine microbes. Symbiotic associations, where marine microbes associate with other organisms, that is, corals or sponges, can lead to the formation of exclusive secondary metabolites that are not synthesized under a free-living environment. Also, competition with other microbes can encourage the synthesis of secondary metabolites with antimicrobial properties.

20.8.5 Genetic Regulation and Signaling

The regulation of the production of secondary metabolites is performed at the genetic level. Genes responsible for this regulation can control the expression of biosynthetic genes, which are involved in secondary metabolite synthesis. Environmental signals and intercellular communication, that is, quorum sensing, can moderate the expression of these regulatory genes and influence the production of secondary metabolites. Quorum sensing allows microbes to sense population density and coordinate their behavior, including the secondary metabolite production by microorganisms.

20.8.6 Stress and Environmental Stimuli

Environmental stress, like variation in temperature, salinity, and nutritional availability, can elicit the formation of secondary metabolites. Marine microbes may produce secondary metabolites as a defensive mechanism against stressors or as a part of their survivability management.

20.8.7 Ecological and Evolutionary Factors

The ecological role and evolutionary history of marine microbes can have an impact on the production of secondary metabolites. Certain marine microbes have developed to synthesize unique secondary metabolites as adaptations to their ecological niche or as a protective mechanism against predators or competitors.

So, the understanding of these influencing factors and their reactions is important for manipulating and optimizing secondary metabolites production by marine microbial fauna and flora.

By means of influencing culture conditions, nutritional availability, and genetic regulation and signaling, scientists can improve the desired secondary metabolites production and explore their possible uses in various advanced research fields, such as medicine, agriculture, marine microbiology, pharmacology, biotechnology, *etc*.

20.9 CHARACTERISTICS AND PROPERTIES OF MARINE MICROBIAL SECONDARY METABOLITES

As secondary metabolites are small organic molecules produced by marine living microorganisms and are non-essential for their primary metabolic functions, like growth, reproduction, *etc*. But these compounds often have unique functions so that they can help the organism to survive and exist in its surroundings or environment because marine microbial secondary metabolites possess a wide range of specific characteristics and properties in them, which makes them novel, valuable, and unique from other compounds. The following are some important characteristics and properties of these compounds:

20.9.1 Chemical and Structural Array

Marine microbial secondary metabolites exhibit a remarkable structural range, with a wide diversity of chemical groups and typical molecular architectures. This diversification arises from the complex biosynthesis and metabolic pathways and enzymatic changes in marine microbes. The structural complexity of these compounds frequently contributes to their bioactivities and prospects for their uses in new scientific and technological research fields.

When we understand the structural diversity of marine microbial secondary metabolites, these include a vast range of chemical structures, like polyketides, peptides, alkaloids, terpenes, and many others. These small organic compounds possess complex and novel frameworks that are commonly not found in terrestrial organisms. The marine habitats provide different ecological niches, which lead to the evolution of diversified metabolic pathways within microorganisms belonging to the marine environment, resulting in the production of structurally diversified secondary metabolites.

If we look into the chemical diversity of secondary metabolites produced by marine microorganisms, then these compounds also show chemical diversification in their chemical functionalities and modifications. Some common chemical modifications are oxidation, reduction, glycosylation, acetylation, methylation, and halogenation. These modifications can significantly influence the bioactivities and pharmacokinetic properties of these compounds.

The chemical divergence of marine microbial secondary metabolites is believed to arise from different factors, that is, the special physiological and ecological characteristics of marine microbes, and the challenges they face in the marine ecosystem as well. These challenges, like competition for resources and defense against predators, have driven the evolution of different metabolic pathways and the production of bioactive compounds.

Thus, the chemical and structural diversity of marine microbial secondary metabolites is evidence of the huge potential of the marine ecosystem as a good source of unique and bioactive compounds. Understanding and exploring this diversification possibly provide valuable insights in the research area of drug discovery biotechnology, and many other applications in applied sciences.

20.9.2 As Bioactive Compounds

Several marine microbial secondary metabolites show effective bioactivity and display numerous properties. Marine microbial secondary metabolites exhibit a vast area of bioactivities, which gives attraction with significant attention in many fields, like pharmaceuticals, biotechnology, drug discovery, marine microbiology, genetic engineering, agriculture, and many more. These activities have been identified extensively, and some of them are also remarkable. These compounds may play their role as antimicrobial agents, anticancerous compounds, anti-inflammatory agents, antiviral substances, enzymatic inhibitors, or neuroactive compounds. As bioactive compounds, marine microbial secondary metabolites arise from their interactions with definite molecular targets in living organisms, which makes them attractive for their therapeutic and biomedical and clinical approaches. Because of these properties, it is important and remarkable that the different bioactivities of marine microbial secondary metabolites are very much diversified, and have great potential via their applications of these compounds, which is why they are still being explored. Constant research and discovery efforts in this field hold the potential of undiscovered new bioactivities and growing the scope of their uses in medicinal, agricultural, biotechnological, and many other industries.

20.9.3 As Antimicrobial Agents

There are various marine microbial secondary metabolites that hold the power to act as antimicrobial agents with their potential antimicrobial properties against other microbes, that is, bacteria, fungi, and viruses. These organic compounds possess the possibilities for combating drug-resistant pathogens and have also been observed as a leader to the development of new antibiotics and antifungal agents (Thawabteh et al., 2023).

20.9.4 As Anticancerous Agents

Many marine microbial secondary metabolites possess established and significant anticancerous properties (Karthikeyan et al., 2022; Mayer et al., 2010). These compounds act as inhibitors for the growth of cancerous or tumor cells and also induce apoptosis (the process of programmed cell death), and interfere with many different cancer-related procedures, like angiogenesis, metastasis, *etc.* A few secondary metabolites have even advanced into clinical trials for cancer treatment.

20.9.5 As Anti-Inflammatory and Immunomodulatory Agents

Some Marine microbial secondary metabolites have also been established to have anti-inflammatory and immunomodulatory mechanisms. These compounds can moderate the immune response, reduce inflammation, and alleviate symptoms related to inflammatory diseases, that is, rheumatoid arthritis, inflammatory bowel disease, asthma, and others.

20.9.6 As Antioxidants

Oxidative stress is associated with several diseases and aging processes as well. The marine microbial secondary metabolites also display antioxidant activity, which helps in neutralizing harmful free radicals and gives protection to cells from oxidative harm. So, these compounds also have their possible uses in combating oxidative stress-related sicknesses.

20.9.7 As Neuroprotective Compounds

A few marine microbial secondary metabolites have shown neuroprotective properties using promoting neuronal cell survivability, enhancing cognitive function, and protection against neurodegenerative diseases, that is, Alzheimer's, Parkinson's diseases, *etc.*

20.9.8 As Antiviral Agents

There are some marine microbial secondary metabolites that have been observed as antiviral agents in their environment because they possess antiviral properties also against an array of viruses, like human immunodeficiency virus (HIV), herpes simplex virus (HSV), influenza virus, hepatitis viruses, *etc.* These compounds can act as inhibitors for viral replication, entry, or other critical phases in the life cycle of viruses.

20.9.9 Possess Antifouling Property

Marine microbes frequently face many challenges from biofouling, unwanted colonization at the surfaces by marine organisms like barnacles, algae, *etc.* The marine microbial secondary metabolites have also been recognized as effective antifouling agents, because they help to prevent the attachment and growth of these organisms on submerged surfaces of marine habitats. Antifouling properties make them valuable for developing environmentally friendly antifouling coatings and treatments. Some compounds, that is, barettin and onnamide, have been demonstrated to have effective antifouling activities, inhibiting the settlement and growth of fouling organisms.

20.9.10 As Antiparasitic Agents

Marine microbial secondary metabolites have also been investigated as Antiparasitic agents against several parasites present in the marine environment, like protozoa and helminths. These organic compounds have potent therapeutic agents for the treatment of parasitic infections.

20.9.11 As Enzyme Inhibitors and Biocatalysts

There are several marine microbial secondary metabolites, act as inhibitors to specific enzymes, for example, proteases, kinases, and phosphatases, which are involved in different disease processes. Moderating enzyme activity can have therapeutic uses in some circumstances, that is, cancer, inflammatory disorders, and cardiovascular diseases. In addition to that, marine microbial secondary metabolites have displayed potential as biocatalysts, also, they are facilitating chemical reactions in many industrial processes and biosynthesis of complex molecules.

20.9.12 Unique and Novel Compounds

The marine environment is a rich source of novel and unique secondary metabolites. The novelty and uniqueness of marine microbial secondary metabolites make them valuable resources for drug discovery, biotechnology, and other applicable fields. Exploring and harnessing the novelty of these compounds can lead to the development of new therapeutics, agrochemicals, and other innovative products. Marine microorganisms have evolved distinct biosynthetic pathways and biochemical strategies to adapt to marine habitats. Consequently, the secondary metabolites produced by marine microbes often possess novel chemical structures and functional groups not commonly found in terrestrial organisms. These novel compounds have the potential to serve as lead compounds for drug development and offer opportunities for innovation. With their novelty and uniqueness, these compounds lie in their different chemical structures and bioactivities, which differ from those found in terrestrial organisms. Several factors make them unique and novel compounds. They are like:

- Their structural and chemical diversity (as mentioned earlier) makes them different from other compounds. The marine metabolites possess specific complex and unusual frameworks, incorporating their diversified chemical modifications and functionalities.
- Marine microbes have contact with a wide pool of chemical precursors, derived from the marine ecosystem. For example, exclusive amino acids, fatty acids, sugars, and other building blocks that can contribute to the synthesis of unique secondary metabolites act as uncommon building blocks in the production of chemical compounds.
- They also have evolutionary adaptations. Because marine microbes have evolved via specialized mechanisms to adapt to the encounters modelled in marine habitats, like competition for resources, predation, and dangerous circumstances. These adaptations are involved in the formation of secondary metabolites with their uniqueness, that is, certain marine bacteria produce chemical compounds that serve as antimicrobial agents, as inhibitors to the growth of competitors, or as antifouling agents to prevent biofilm formation (Qian et al., 2022).
- They show ecological interactions because the marine habitats are teeming with diversified

complex organisms, for example, macroscopic organisms, that is, sponges, corals, and mollusks, as well as microbes like bacteria and fungi. Marine microbes are frequently involved in multifaceted ecological interactions with other organisms, like symbiosis relationships and competition. These interactions may influence the production of unique and novel secondary metabolites that serve special ecological functions, that is, defense against predators, communication, and nutrient acquisition.

- It has already been established that marine microbes deal with unique challenges in defending themselves against their predators, pathogens, and competitive organisms in the tough marine habitat. Thus, they have developed sophisticated chemical defense mechanisms, which include the synthesis of secondary metabolites with effective bioactivities. These compounds can deter predators, inhibit the growth of pathogens, compete for limited resources, and are involved in chemical welfare and defense with others in their environment.
- They also show untapped biodiversity. The ocean environment is largely unexplored, with huge areas yet to be studied. This untapped biodiversity provides immense potential for the discovery of novel and unique marine microbial secondary metabolites. Thus, the discovery of new compounds with unique structures and bioactivities is expected to accelerate.

20.9.13 Ecological Significance

Marine microbial secondary metabolites play important ecological roles for the biodiversity of marine ecosystems (Rosenberg and Koren, 2006). They can serve as chemical defenses against predation, competition, and microbial infections, as already discussed as mentioned earlier. A few compounds contribute to the communication and signaling between microbes, and influence their behavior and interactions. Marine microbial secondary metabolites also play a significant role in nutritional cycling and can shape the structure and dynamics of marine populations. The ecological significance of these compounds can be understood by the following aspects:

- Marine fauna encounters various challenges in their environment, like the risk of predation and intense competition for resources. Microbial secondary metabolites play a crucial role in providing chemical defense mechanisms that safeguard the producing organism from consumption or being outcompeted. These compounds can act as a deterrent to predators, hinder the growth of competing microorganisms, or discourage herbivores from grazing on them.
- Microorganisms' residence in marine environments establishes complicated biofilms on different surfaces like rocks, shells, and other organisms. Biofilms are organized communities of microorganisms enclosed within a matrix of extracellular polymeric substances. The involvement of marine microbial secondary metabolites is essential in the formation and nourishment of biofilms, facilitating the colonization of surfaces and the establishment of thriving microbial communities (Qian et al., 2022).
- Microbes inhabiting marine environments partake in intricate interactions and communication with other organisms. Secondary metabolites actively participate in these interactions, facilitating communication among microbial species and between microbes and other organisms, including higher marine organisms. These compounds serve as signaling molecules, enabling the coordination of behaviors and ecological functions within microbial communities.
- Microbial secondary metabolites can contribute to nutrient cycling in marine ecosystems. Some compounds are involved in the degradation and recycling of organic matter, enabling the breakdown of complex molecules into simpler forms that can be utilized by other organisms. This process is essential for the recycling of nutrients and the maintenance of ecosystem productivity.
- Microbial secondary metabolites play a vital role in the nutrient cycling of marine ecosystems. Certain compounds are actively involved in the decomposition and recycling of organic matter, facilitating the breakdown of complex molecules into simpler forms that can be readily utilized by other organisms. This essential process ensures the efficient recycling of nutrients, promoting the overall productivity and balance of the ecosystem.
- A multitude of marine organisms establish symbiotic associations with microbes. These relationships are frequently influenced by microbial secondary metabolites. For instance, specific metabolites produced by marine bacteria or fungi can foster the growth and well-being of host organisms. This intricate interaction between organisms and microbial secondary metabolites forms the dynamics and outcomes of the symbiotic relationship in the marine environment.
- The biosynthesis of secondary metabolites improves the chemical diversity in marine environments, thus providing a foundation for biodiversity. These compounds exert influence on species interactions and community dynamics by modulating the distribution and abundance of organisms. They can function as allelochemicals, impacting the growth and survivability of other

living creatures, thus influencing the complete structure and composition of biocommunities. This complicated interaction between secondary metabolites and marine ecosystems emphasizes their essential role in shaping the ecological fabric of marine environments.

Therefore, the ecological significance of marine microbial secondary metabolites is pivotal for understanding the functioning and flexibility of marine environments. It exposes the interconnectedness and complexity of microbial communities and their importance in maintaining the health and sustainability of marine ecosystems.

20.9.14 Adaptation to Extreme Conditions

Marine microorganisms inhabit diverse and often extreme marine environments, such as deep-sea trenches, hydrothermal vents, and Polar regions. Secondary metabolites contribute to their adaptation and survival in these harsh conditions. These compounds display remarkable adaptations to thrive in extreme circumstances present in marine habitats. A few examples are noticeable:

20.9.14.1 Temperature Adaptations

For instance, metabolites produced by cold-adapted microbes, like psychrophilic, possess special structures that ensure stability and activity even at low temperatures. On another hand, metabolites from thermophilic microbes are designed to maintain functionality under high-temperature conditions, allowing them to thrive in environments with elevated temperatures. These adaptations enable marine microbes to efficiently cope with the temperature variations created in the marine environment.

20.9.14.2 Pressure Adaptations

The deep-sea marine habitat faces immense pressure, reaching a hundred times atmospheric pressure. Deep-sea microbial secondary metabolites have undergone adaptations to thrive under these extreme conditions. These compounds may possess a novel structural character that makes them resistant to pressure-induced modifications. Additionally, some metabolites exhibit altered activity patterns, enabling them to maintain functionality even at high pressures. These adaptations allow microbes to live in the deep-sea to efficiently navigate and thrive in the extreme pressure systems of the deep-sea environment.

20.9.14.3 Salinity Adaptations

Marine environments show diverse salinity levels, ranging from hypersaline habitats like salt pans to brackish estuaries. Microbial secondary metabolites have evolved adaptations to cope with and exploit these varying salinity conditions. For instance, halophilic microorganisms produce metabolites that are specifically tailored to be stable and active in high-salt environments. These metabolites enable microorganisms to thrive and fulfill their ecological roles in marine habitats with different salinity levels.

20.9.14.4 Nutrient Limitations

Marine ecosystems, especially open ocean areas far from coastlines, often face a nutrient shortage. In this view, microbial secondary metabolites have changed strategies to competently scavenge and utilize limited nutrients. Certain metabolites act as chelators, supporting microbes in acquiring necessary metals that may be rare. Others show a role in the capture and utilization of rare organic nutrients, ensuring the effective use of available resources. These adaptations enable marine microbes to thrive in nutrient-limited environments by efficiently obtaining and using vital nutrients for their growth and existence.

20.9.14.5 UV Radiation Protection

Marine microbes in sunny surface waters are exposed to high levels of UV radiation, which can lead to DNA damage and disruption of cellular processes. To counter these harmful effects, certain microbial secondary metabolites serve as sunscreens or photo-protective compounds. These metabolites act as shields, safeguarding microbes from the harmful effects of UV radiation and preventing damage to their cellular components. This adaptation enables marine microbes to flourish and maintain their biological activities even under intense UV radiation exposure in surface waters.

20.9.14.6 Osmotic Stress Tolerance

Variations in water salinity and osmotic conditions cause challenges to marine microbes. Microbial secondary metabolites play a vital role in osmoregulation and protection against osmotic stress. These compounds can help maintain cellular hydration, balance internal osmolarity, and protect cellular structures from osmotic damage.

Thus, understanding these adaptations can give insights into the molecular mechanisms underlying their survival strategies and may also have uses in numerous fields, including biotechnology, pharmaceuticals, and environmental research.

20.9.15 Biotechnological Potential

Marine microbial secondary metabolites have significant biotechnological potential. Their bioactivity, structural diversity, and novelty make them attractive for drug discovery, natural product-based therapies, and the development of new pharmaceuticals. They also find applications in biotechnology, a few of which are as follows:

20.9.15.1 Drug Discovery and Drug Design

Marine microbial secondary metabolites are an emerging and promising source in the field of drug discovery. Those compounds have various bioactive properties, encompassing antimicrobial, anticancer, antiviral, and anti-inflammatory activities. They offer a wealth of unique chemical scaffolds that can be screened and optimized for their therapeutic potential. Particularly, many marine-derived compounds, that is, cytarabine (an anticancer drug) and ziconotide (used for pain management), have been successfully developed from marine microbial metabolites. Therefore, the marine microbial secondary metabolites can contribute to the development of innovative therapeutics.

20.9.15.2 Antibiotic and Antimicrobial Agents

In sunlit of the mounting antibiotic resistance crisis, the quest for unique antimicrobial agents has become supreme. Marine microbial secondary metabolites have been confirmed to retain strong antimicrobial activities against a wide range of pathogenic bacteria, including strains resistant to multiple drugs. These compounds serve as a valuable resource for the discovery and development of new antibiotics and antimicrobial drugs to efficiently combat infectious diseases. Their novel chemical structures and mechanisms of action provide promising avenues for addressing the crucial need for unique therapeutics to combat drug-resistant pathogens.

20.9.15.3 Enzyme Production and Biocatalysis

Marine microbial secondary metabolites provide a valued resource for the production of enzymes with excellent properties and healthiness, even in adverse conditions. These enzymes find uses in varied biotechnological methods, including biocatalysis, bioremediation, and biofuel production. For instance, enzymes derived from marine microbes have been coupled for the synthesis of pharmaceuticals, degradation of pollutants through biodegradation processes, and conversion of renewable biomass into biofuels. The novel character of these enzymes enables them to catalyze complex reactions significantly, opening up avenues for sustainable and environment-friendly biotechnological applications through various industries.

20.9.16 Agricultural Applications

A few marine microbial secondary metabolites have verified possibilities in agricultural fields. They can act as natural pesticides, promoting plant growth or enhancing crop resistance against pathogens and environmental stressors. These metabolites give environment-friendly substitutes to synthetic agrochemicals, thus contributing to sustainable agriculture practices.

20.9.17 Industrial Applications

Marine microbial secondary metabolites have their uses in many industries. They can be used as natural pigments for food and cosmetic industries, as bioactive compounds in nutraceuticals, and as bio surfactants for bioremediation and industrial mechanisms. These novel properties and diversified chemical structures of these metabolites enable them to be valuable resources for industrial applications.

Thus, to understand exploring and coupling the biotechnological perspective of marine microbial secondary metabolites not only provides opportunities for the development of unique therapeutics and solutions to social challenges but also aids in the conservation and sustainable use of marine resources.

20.9.17.1 Chemical Interactions

Marine microbial secondary metabolites can also participate in chemical interactions and ecological relationships. They may act as allopathic compounds, influencing the growth and survival of other marine organisms. Secondary metabolites can also be involved in symbiotic relationships, where they contribute to the communication and exchange of nutrients between different organisms, such as corals and their symbiotic microorganisms. We can understand their chemical interactions in their environment as follows:

20.9.17.2 Allopathy

Marine microbial secondary metabolites can show allopathic effects, where they release chemical compounds which give an impact on the growth, development, and survival ability of other living creatures in their locality. These compounds can also impede the growth of competing microbes or deter grazing by herbivores, providing a competitive benefit to the producing organism. Allopathic interactions can influence the composition and structure of microbial communities in marine ecosystems.

20.9.17.3 Mutualistic Interactions

Marine microbial secondary metabolites can also facilitate mutualistic interactions between various living creatures within the environment. Like, metabolites produced by certain bacteria or fungi can encourage the growth and health of host organisms, that is, marine plants, sponges, corals, or other marine invertebrates. These metabolites may deliver vital nutrients, increase immune responses, or protect the host from pathogens, developing a mutually advantageous association.

20.9.17.4 Defense Mechanisms

Several marine microbial secondary metabolites perform as chemical defenses, protecting the producing organism from predation or competition. These compounds can also prevent predators, either via being toxic or indigestible, or through inducing behavioral or physiological actions that deter predation. Additionally, some metabolites have antimicrobial properties so that they prevent the growth of potential pathogens or competing microbes.

20.9.17.5 Trophic Interactions

Chemical interactions of marine microbial secondary metabolites also influence trophic interactions in marine food webs. These compounds can disturb the feeding behavior and preferences of predators, influence the flow of energy and nutrients through the ecosystem as well. They may attract or prevent specific organisms, leading to selective feeding or escape responses.

20.9.17.6 Signaling and Communication

These microbial secondary metabolites in marine habitats play a critical role in communication and signaling among living creatures. These compounds can perform as signaling molecules, facilitating communication between various microbial species and other organisms, that is, higher marine organisms. By chemical signaling, microbes can coordinate their behaviors, form cooperative relationships, or respond to environmental signals.

Understanding about chemical interactions of marine microbial secondary metabolites is important for unscrambling the complex dynamics of marine habitats. These interactions provide shape to species distributions, community structures, and ecological functions as well, and finally, they contribute to the complete health and functioning of marine ecosystems. Moreover, exploring the chemical diversification and bioactivities of marine microbial secondary metabolites can deliver insights to their prospective utilization in advanced fields like medicinal, biotechnological, environmental management, *etc.*

Thus, the characteristics and properties of marine microbial secondary metabolites make them valuable resources for scientific research, drug discovery, and biotechnological applications. Discovering their chemical diversity, bioactivity, and ecological roles contributes to our understanding of natural product chemistry and the potential for harnessing these compounds for various beneficial purposes.

20.10 METHODS & PROCESS FOR DISCOVERY AND CHARACTERIZATION OF MARINE MICROBIAL SECONDARY METABOLITES

The discovery and characterization of marine microbial secondary metabolites involve a systematic process that combines various methods and techniques. The following are the step-by-step methods usually employed in the discovery and characterization process of secondary metabolites (Figure 20.1):

20.10.1 Sample Collection

The process begins with the collection of marine microbial samples from different marine environments, such as seawater, sediments, or marine organisms like sponges or corals. Sampling can be done using techniques such as water filtration, sediment coring, or extraction from marine organisms. The collection of samples includes many steps to ensure the attainment of diversified and representative samples from various marine habitats. These steps are as follows:

20.10.2 Site Selection

20.10.2.1 To Determine Target Marine Environment

Marine microbial secondary metabolites can be present in different marine habitats, that is, coastal waters, coral reefs, deep-sea habitats, marine sediments, *etc.* So, it is required for sample collection to choose the target habitat from which we can collect the sample, either with specific environmental conditions or based on the research objective.

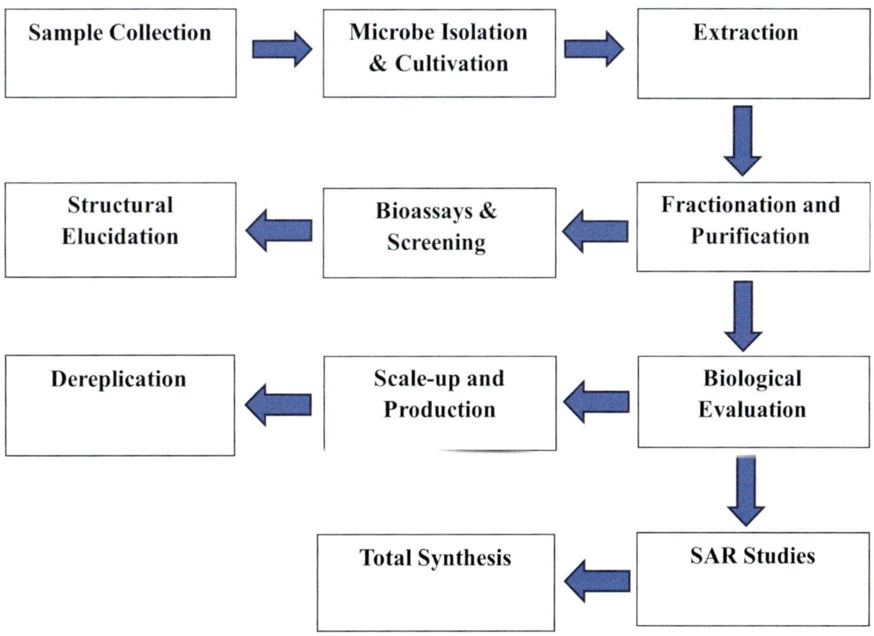

FIGURE 20.1 Ray chart of discovery and characterization of marine microbial secondary metabolites.

20.10.2.2 To Consider Biodiversity Hotspots

For that, select areas known for rich species diversity or regions with specific ecological characteristics, as they are more likely to host diverse microorganisms' communities.

20.10.2.3 To Take into Account Anthropogenic Influences

Examine the potent sources of pollution or anthropogenic activities that may affect the microorganism's communities and select sample sites away from such troubles.

20.10.3 Methods for Sample Collection

The following methods are applicable to collect the sample of marine microbial secondary metabolites according to their presence in various marine environments:

20.10.3.1 Water Sampling

First of all, select the marine habitat for the collection of the water sample. For that, always use sterilized containers, for example, Niskin bottles or sterile plastic bags, to prevent the collected sample from contamination.

20.10.3.2 Sediment Sampling

By using cores or grabs, we can collect samples from marine sediments. Get samples from different depths to capture various microbial communities with variations in them.

20.10.3.3 Sponge or Coral Sampling

By using sterile tools, that is, scalpels, first harvest sponges, or coral, to avoid contamination. Then ensure minimum damage to the host organisms as well.

20.10.3.4 Biofilm Sampling

To collect biofilm samples from submerged surfaces, scrape or swab the surfaces by using sterilized instruments.

20.10.3.5 *In situ* Collection

To collect the sample without environmental disturbance, we can utilize in situ techniques, that is, diffusion chambers, *in situ* incubations, or filtration methods. By using these methods, the capture of metabolites can be done directly from their sources.

20.10.4 Preservation Techniques

After the collection of samples, it is necessary to preserve them appropriately. For that we can apply the following:

20.10.4.1 Cold preservation

Keep samples at 2–8°C temperature (chilled) during collection to slow down microbial metabolism and minimize changes in metabolite composition.

20.10.4.2 Freezing

For long-term storage, at −80°C (which is an ultra-low temperature), freeze the collected samples to preserve microorganisms' diversity and metabolite integrity. Use cryoprotants, that is, glycerol, to protect the samples during freezing.

20.10.4.3 Fixatives

In collected samples, add appropriate fixatives (like formaldehyde) to preserve microorganisms' community structures for downstream analysis, like DNA sequencing or microscopy.

20.10.5 Transportation of Collected Samples

This can be done by the following:

20.10.5.1 Maintain Low Temperature

During transportation of collected samples, utilize insulated containers or coolers with ice packs to keep the samples at a low temperature and prevent them from degradation.

20.10.5.2 Minimize Exposure to Light

Protect the samples from direct sunlight or artificial light, because some secondary metabolites are also sensitive to light.

20.10.5.3 Minimize Transportation Time

Make sure to arrange for accelerated shipping or transport the collected samples to the laboratory as quickly as possible to avoid extended exposure to suboptimal conditions.

Although it is typical to follow proper sterile methods during the entire sample collection procedure to avoid contamination and ensure the accuracy of subsequent analysis.

Moreover, researchers should observe relevant ethical guidelines and obtain essential permits when collecting samples from protected marine habitats or indigenous territories.

20.10.6 Isolation and Cultivation

After sample collection, the microorganisms need to be isolated and cultivated in the laboratory. There, several cultivation techniques, including enrichment cultures, dilution-to-extinction, or specialized growth media, may be used to encourage the growth of specific microbial species or groups (Scheuer, 1990; Hayes and Britton, 2017). So, the isolation and cultivation of marine microbial secondary metabolites includes the following steps to get pure cultures of microbes and stimulate the production of secondary metabolites:

20.10.6.1 Sample Preparation

We can prepare the samples as follows:

20.10.6.1.1 Defrost Frozen Samples

If we collected frozen samples, then allow them to defrost slowly at a low temperature to prevent damage from the microbes.

20.10.6.1.2 Homogenize the Sample

For homogenization of solid samples such as sediment, sponge, utilize sterile tools or mix them in a sterile blender to ensure even distribution of microbes.

20.10.6.1.3 Serial Dilution

To perform serial dilution on the sample to reduce the microbial density and facilitate the isolation of individual microbial strains. It involves diluting the sample in sterile saline or an appropriate diluent. For the preparation of serial dilutions of the sample in sterile seawater or suitable growth medium to reduce microbial density and facilitate the isolation of individual strains.

20.10.6.2 Isolation of Microbes

There are several techniques for isolation microbes from the samples. Some of them are as follows:

20.10.6.2.1 Plating Techniques

In this technique, first spread a known volume of the diluted sample on solid nutrient agar media, and utilize a sterile spreader or pour the sample on the agar surface and spread it evenly. Then incubate the plates under suitable conditions (like temperature, salinity) for the growth of microbes.

20.10.6.2.2 Streak Plate Method

Utilize a sterile loop to streak individual colonies on agar plates to obtain isolated microbial colonies.

20.10.6.2.3 Single-Cell Isolation

For that, utilize micro-pipettes, micro-capillary tubes, or micromanipulators under a microscope to isolate a single microbial cell into isolated culture media or wells of microliter plates.

20.10.6.3 Cultivation and Optimization

For the cultivation and optimization of microbes, the following are applicable:

20.10.6.3.1 Selection of Media

Prepare and use specific culture media intended to stimulate the growth and secondary metabolite production of target microbes. It may include adjusting the salinity, pH, and nutrient composition of the medium to imitate the natural marine habitat.

20.10.6.3.2 Incubation Conditions

For this, set suitable incubation parameters like temperature, agitation, and oxygen levels according to the requirements of the isolated microbe or known conditions for secondary metabolite production.

20.10.6.3.3 Sub-Culturing

Transfer actively growing colonies or cultures to fresh media periodically to maintain their viability and prevent contamination.

20.10.6.4 Screening for Secondary Metabolite Production

It includes the following steps:

20.10.6.4.1 Bioassays

Various bioassays are available for the detection of the production of bioactive compounds by the cultivated microbes. These assays include testing the culture extracts against target organisms or disease models to assess their potential pharmaceutical, agricultural, or industrial applications.

20.10.6.4.2 Analytical Techniques

Different analytical techniques, such as high-performance liquid chromatography (HPLC), gas chromatography-mass spectrometry (GC-MS), nuclear magnetic resonance (NMR), or mass spectrometry (MS), can be used to identify and quantify specific secondary metabolites.

20.11 SCALING UP AND FERMENTATION OPTIMIZATION

These are as follows:

20.11.1 Scaling up Cultures

Once a promising microbe is identified, scale up the cultivation to high volumes, like shake flasks or bioreactors, to increase secondary metabolite production.

20.11.2 Fermentation Optimization

Optimize fermentation conditions, including aeration, agitation, nutrient supplementation, and culture duration, to enhance the quality and yield of secondary metabolites.

Through this process, it is typical to maintain sterile techniques to prevent contamination from other microorganisms and ensure the purity of the isolated strains. Additionally, researchers should keep detailed records of each isolated strain and its associated secondary metabolites for further studies or potential applications.

20.12 EXTRACTION AND FRACTIONATION

Once the microbial cultures are established, secondary metabolites are extracted from the microbial biomass. Various extraction techniques, such as solvent extraction, solid-phase extraction, or liquid-liquid extraction, are employed to recover the metabolites from the microbial cells or culture media.

Fractionation and purification are necessary steps in the isolation and purification of marine microbial secondary metabolites (Hayes and Britton, 2017). These processes help separate and concentrate the desired compounds from the complex mixture obtained during extraction. The following are the techniques used for marine microbial secondary metabolites, fractionation, and purification:

20.12.1 Liquid-Liquid Extraction (Partitioning)

Liquid-liquid extraction is frequently used as an initial step in fractionation. It involves the partitioning of the crude extract between two immiscible solvents with different polarities. The extract is dissolved in a suitable solvent (such as ethyl acetate or dichloromethane) and mixed with another immiscible solvent (such as water or a buffered solution). The target compounds distribute themselves between the two phases based on their solubility, allowing for the separation of different classes of compounds.

20.12.2 Solid-Phase Extraction (SPE)

Solid-phase extraction is an extensively applied method for fractionation and purification. It uses solid sorbents packed in cartridges or disks to selectively adsorb and retain compounds of interest while removing unwanted impurities. The crude extract is passed through the solid-phase extraction column, and different fractions are obtained by eluting the compounds with solvents of varying polarities. SPE is effective in concentrating target compounds and removing interfering substances.

20.12.3 Chromatographic Techniques

Chromatography is widely used in the fractionation and purification of marine microbial secondary metabolites because of its high resolving power. Many chromatographic techniques are applied. Some are as follows:

20.12.3.1 Column Chromatography

It is the separation of compounds based on their differential adsorption and elution from a column packed with an adsorbent material (such as silica gel or reversed-phase C18). The sample is loaded onto the column, followed by elution with solvents of increasing polarity to obtain different fractions.

20.12.3.2 Thin-Layer Chromatography (TLC)

TLC is a simple and cost-effective method for analyzing and separating compounds. It utilizes a thin layer of adsorbent material (usually silica gel or cellulose) coated on a flat plate. The sample is applied as spots on the plate, and the plate is then developed in a solvent system. Compounds migrate on the plate at different rates based on their affinity for the stationary phase, allowing for separation and visualization under UV light or using appropriate staining techniques.

20.12.3.3 High-Performance Liquid Chromatography (HPLC)

HPLC is a powerful and widely used chromatographic technique for fractionation and purification. It employs a high-pressure pump to force the sample through a packed column containing a stationary phase. The separation is achieved by the differential interaction of compounds with the stationary phase and the mobile phase. Various modes of HPLC, such as normal phase, reversed phase, ion exchange, or size exclusion chromatography, can be employed based on the chemical properties of the target compounds.

20.12.3.4 Preparative HPLC

Preparative HPLC is an extension of analytical HPLC and is specifically designed for the large-scale purification of target compounds. It applies a similar principle but utilizes larger columns and higher flow rates to handle larger sample volumes. Preparative HPLC is used for the isolation of larger quantities of pure compounds for further analysis and biological testing.

20.12.4 Crystallization

Crystallization is a technique that exploits the differences in solubility and crystal formation of compounds. Under controlled solvent composition, temperature, and concentration, the target compounds can be induced to form crystals while impurities remain in solution. Crystallization is often used as a final purification step to obtain highly pure compounds.

Thus, these fractionation and purification techniques can be applied separately or in combination to isolate and purify marine microbial secondary metabolites on the basis of the nature of the target compounds and their solubility

20.13 BIOASSAYS AND SCREENING

The fractions and purified compounds are subjected to bioassays to assess their biological activities. Bioassays involve testing the samples against various target organisms or biological systems to evaluate their antimicrobial, anticancer, or other relevant activities. Screening methods can include microbial growth inhibition assays, enzyme assays, or cell-based assays. A few are as follows:

20.13.1 Antimicrobial Assays

It will determine the ability of the compounds to inhibit the growth of microorganisms, including bacteria, fungi, and viruses. The most common antimicrobial assays include the disk diffusion method (Kirby-Bauer test), broth dilution method (MIC – Minimum Inhibitory Concentration determination), and time-kill kinetics. These assays help to identify potential antibiotics or antifungal agents.

20.13.2 Cytotoxicity and Anticancer Assays

It evaluates the ability of the compounds to inhibit the growth of cancer cells or induce cytotoxic effects. These are performed using various cancer cell lines derived from different tissues. Common assays include the MTT (3–(4,5-dimethylthiazol-2-yl)-2,5-diphenyltetrazolium bromide) assay, cell viability assays, and apoptosis assays.

20.13.3 Anti-Inflammatory Assays

It is suitable to measure the ability of compounds to inhibit inflammation or modulate inflammatory markers. Common assays include the inhibition of pro-inflammatory cytokine production (*e.g.*, IL-1β, TNF-α), measurement of reactive oxygen species (ROS), or evaluation of enzyme activity (*e.g.*, cyclooxygenase, lipoxygenase).

20.13.4 Antioxidant Assays

These assays determine the ability of compounds to scavenge free radicals and protect against oxidative damage. Common assays include the DPPH (2,2-diphenyl-1-picrylhydrazyl) assay, ABTS [2,2'-azino-bis (3-ethylbenzothiazoline-6-sulfonic acid)] assay, and oxygen radical absorbance capacity (ORAC) assay.

20.13.5 Enzyme Inhibition Assays

It is suitable to assess the ability of compounds to inhibit specific enzymes involved in disease processes. For example, compounds can be tested for their ability to inhibit enzymes such as acetylcholinesterase (related to Alzheimer's disease), tyrosinase (related to melanogenesis), or proteases (related to cancer metastasis).

20.13.6 Agricultural Bioassays

This will evaluate the potential of compounds for agricultural applications, such as insecticidal, herbicidal, or plant growth–promoting activities. Assays can include insect mortality assays, seed germination assays, or plant growth promotion assays.

20.13.7 Other Specific Bioassays

Depending on the targeted application, many specific bioassays can be performed. They may include tests for antiviral activity, antiparasitic activity, antidiabetic activity, neuroprotective activity, or wound healing activity, among others.

Note: Bioassays can be performed using different experimental models, including in vitro (cell-based assays), ex vivo (tissue or organ culture), or in vivo (animal models). The choice of bioassay depends on the specific biological target, intended application, and available resources.

Furthermore, bioassays and screening methods help to identify the most promising compounds with potential therapeutic, agricultural, or industrial value. Positive results from these assays can guide further investigations into the mechanism of action, pharmacokinetics, and toxicity profiles of the compounds, ultimately leading to the development of new drugs, agrochemicals, or other valuable products derived from marine microbial secondary metabolites.

20.14 STRUCTURAL ELUCIDATION

The characterized bioactive fractions or compounds undergo structural elucidation to determine their chemical structures (Scheuer, 1990). Advanced analytical techniques like NMR, MS, IR, and X-ray crystallography are utilized to analyze the compounds' molecular structures and configurations. The details of the techniques are as follows:

20.14.1 Nuclear Magnetic Resonance (NMR) Spectroscopy

NMR spectroscopy is one of the most powerful techniques for structural elucidation. It provides valuable information about the connectivity, stereochemistry, and functional groups present in the compound. Various NMR experiments, including 1D (proton, carbon-13) and 2D (COSY, HSQC, HMBC) NMR experiments, are done to analyze the NMR spectra and deduce the compound's structure.

20.14.2 Mass Spectrometry (MS)

Mass spectrometry is widely applied to determine the molecular weight and element composition of marine microbial secondary metabolites. Techniques such as electrospray ionization mass spectrometry (ESI-MS) or matrix-assisted laser desorption/ionization mass spectrometry (MALDI-MS) provide information about the molecular ions, fragment ions, and isotopic patterns, aiding in structural identification and confirmation.

20.14.3 Infrared (IR) Spectroscopy

IR spectroscopy is applied to identify the functional groups of compounds. It helps to determine the presence of specific bonds such as carbonyl (C=O), hydroxyl (OH), or amino (NH) groups, providing valuable information for structural characterization.

20.14.4 Ultraviolet-Visible (UV-Vis) Spectroscopy

UV-vis spectroscopy is utilized to analyze the absorption and electronic transitions of marine microbial secondary metabolites. It helps to identify chromophores or conjugated systems present in the molecule, providing insights into its structural features.

20.14.5 X-ray Crystallography

X-ray crystallography is a prevailing technique for the determination of the three-dimensional structure of marine microbial secondary metabolites. It requires X-ray diffraction data of the compound to be crystallized. The data is then analyzed to generate an electron density map, allowing for the precise determination of the compound's atomic positions and molecular conformation.

20.14.6 Mass Spectrometry Imaging (MSI)

MSI combines mass spectrometry with spatial information, which allows the visualization of the distribution of marine microbial secondary metabolites within a sample. It delivers insights into the compound's localization within microbial cells or marine organisms, aiding in understanding their ecological roles or interactions.

20.14.7 Computer-Aided Structure Elucidation (CASE) Tools

CASE tools apply computational algorithms to guess the structure of marine microbial secondary metabolites based on NMR, MS, and IR data. These tools can assist in generating structural proposals and narrowing down the possible structures of the compounds.

The combination of these techniques and methods, make enables the scientists to elucidate the chemical structures of marine microbial secondary metabolites. To understand the compounds' structures, researchers can determine their possible biological activities, mechanisms of action, and applications in different fields such as medicinal, biotechnological, and natural products and drug discovery.

20.15 CHARACTERIZATION OF BIOLOGICAL ACTIVITIES

The isolated compounds are verified for their biological activities in more detail, including the mechanism of action, toxicity, and potential uses. This involves further experiments, such as cell-based assays, enzymatic assays, and *in vivo* studies applying animal models or *in vitro* studies on human cells. Its objectives are to understand the compounds' interactions with living creatures and their probable therapeutic, agricultural, or industrial worth.

20.16 GENOME MINING AND BIOINFORMATICS

With the invention of genomic sequencing technologies, genome mining has become a vital approach for discovering unique secondary metabolite biosynthetic gene clusters in marine microbes. Bioinformatics tools and algorithms are applied to identify and analyze the genes involved in secondary metabolite biosynthesis, which aids in forecasting the chemical structures and possible bioactivities of the compounds.

20.17 BIOLOGICAL AND PHARMACOLOGICAL EVALUATION

The isolated and characterized compounds undergo biological and pharmacological evaluations to evaluate their potential as therapeutic agents, drug leads, or biotechnological products. It involves a wide range of *in vitro* and *in vivo* studies to examine their mechanisms of action, pharmacokinetics, toxicity, and efficacy.

20.18 SCALE-UP AND PRODUCTION

Once promising compounds are identified, they can be further studied for scale-up production. It involves optimizing fermentation conditions, downstream processing, and yield improvement.

20.19 STRUCTURE-ACTIVITY RELATIONSHIP (SAR) STUDIES

Examine the structure-activity relationship of the identified bioactive compounds to understand the main characteristics responsible for their bioactivities. It can be monitoring additional alterations or synthesis to enhance potential or selectivity.

20.20 NATURAL PRODUCT TOTAL SYNTHESIS

If the compound is challenging to isolate in sufficient quantities from natural sources, then total synthesis can be followed to produce the compound in the laboratory.

20.21 STRUCTURE-BASED DRUG DESIGN

By utilizing the information gained from the SAR studies and structural analysis to design and synthesize analogs or derivatives with improved properties or targeted activities.

The discovery and characterization of marine microbial secondary metabolites involve an iterative and interdisciplinary approach, combining microbiology, chemistry, bioassays, genomics, and bioinformatics. By employing these methods and techniques, researchers can unlock the vast potential of marine microbial secondary metabolites and explore their applications in various fields, including medicine, biotechnology, and environmental science.

20.22 APPLICATIONS OF MARINE MICROBIAL SECONDARY METABOLITES

Marine microbial secondary metabolites have diverse applications across various fields due to their bioactive properties and unique chemical structures. Here are some of the key applications of marine microbial secondary metabolites:

20.22.1 Pharmaceutical and Drug Discovery

Marine microbial secondary metabolites have been a rich source of compounds for drug discovery and pharmaceutical development. Many of these compounds exhibit potent bioactivities, including antimicrobial, anticancer, antiviral, anti-inflammatory, and neuroactive properties (König and

Wright, 1997; Mayer et al., 2010). Examples include the antiviral drug acyclovir derived from a marine sponge, the anticancer agent trabectedin from a marine tunicate, and the antibiotic erythromycin produced by marine bacteria. Marine microbial secondary metabolites continue to be explored for their potential in treating various diseases and disorders.

20.22.2 Biotechnology and Industrial Applications

Marine microbial secondary metabolites find applications in biotechnology and industrial sectors (Taylor et al., 2007). Certain compounds have shown enzymatic activities that are valuable for biocatalysis and industrial processes. For instance, marine microbial enzymes can be used in the production of biofuels, bioplastics, and specialty chemicals. Additionally, marine secondary metabolites with antifouling properties have potential applications in developing environmentally friendly coatings to prevent biofouling on ships, offshore structures, and aquaculture facilities.

20.22.3 Agricultural and Crop Protection

Some marine microbial secondary metabolites possess bioactivity against plant pathogens and pests, making them potential candidates for agricultural applications. These compounds can be used in the development of environmentally friendly biopesticides and biocontrol agents to protect crops from diseases and pests. They offer alternatives to conventional chemical pesticides, contributing to sustainable and eco-friendly agriculture practices.

20.22.4 Environmental and Bioremediation Applications

Marine microbial secondary metabolites have the potential to address environmental challenges. Some compounds exhibit properties that can be utilized in bioremediation efforts to mitigate pollution and remediate contaminated environments. For example, certain marine microbial secondary metabolites have shown capabilities in degrading or detoxifying pollutants, such as petroleum hydrocarbons or heavy metals, thereby aiding in environmental cleanup.

20.22.5 Cosmetics and Personal Care Products

Marine microbial secondary metabolites have found applications in the cosmetics and personal care industry. Many compounds possess properties such as antioxidant, antiaging, skin brightening, or moisturizing effects, making them valuable ingredients in skincare products. Marine-derived compounds like collagen, hyaluronic acid, and various peptides are used in cosmetics for their skin rejuvenating and hydrating properties.

20.22.6 Food and Nutraceuticals

Some marine microbial secondary metabolites have potential applications in the food and nutraceutical industries. Compounds with antimicrobial properties can be used as natural preservatives to extend the shelf life of food products. Additionally, certain compounds derived from marine microorganisms have shown antioxidant, anti-inflammatory, or cholesterol-lowering properties, making them attractive for the development of functional foods and nutraceutical supplements.

20.22.7 Chemical and Natural Product Research

Marine microbial secondary metabolites serve as valuable research tools for chemical and natural product studies (Fenical and Jensen, 2006; Blunt et al., 2016; Newman and Cragg, 2016). Their unique structures, chemical properties, and bioactivities provide opportunities for understanding structure-activity relationships, developing new synthetic methodologies, and exploring novel chemical reactions. Additionally, studying marine microbial secondary metabolites contributes to our understanding of natural product chemistry and the discovery of new chemical scaffolds.

The applications of marine microbial secondary metabolites are continually expanding as research progresses and new compounds are discovered. These compounds hold promise for addressing various societal needs, from developing new drugs to protecting the environment and improving industrial processes.

20.23 IMPORTANCE AND SIGNIFICANCE OF MARINE MICROBIAL SECONDARY METABOLITES FOR THE ENVIRONMENT AND TO MARINE MICROBES

Marine microbial secondary metabolites play a crucial role in both the marine environment and the microorganisms that produce them. Here are some key points of their importance and significance:

20.23.1 Defense for Chemicals

Marine microbial secondary metabolites serve as a chemical defense mechanism for both the producing microorganisms and the surrounding marine environment. They help protect microorganisms from predation, competition, and microbial infections by inhibiting the growth or survival of potential threats. These compounds can deter predators, prevent the colonization of surfaces by other microorganisms, and control the population dynamics within marine ecosystems.

20.23.2 Ecological Interactions

Marine microbial secondary metabolites participate in complex ecological interactions within marine ecosystems. They contribute to species-species and species-environment interactions, influencing the composition and structure of microbial communities (Rosenberg and Koren, 2006). Some compounds exhibit allopathic effects, inhibiting the growth of competing organisms and shaping community dynamics. Secondary metabolites can also act as signaling molecules, facilitating communication and interactions between microorganisms.

20.23.3 Nutrient Cycling

Marine microbial secondary metabolites play a role in nutrient cycling within marine ecosystems. Some compounds are involved in the breakdown of complex organic matter, facilitating the recycling of nutrients and the release of essential elements into the environment. Through enzymatic activities and decomposition processes, secondary metabolites contribute to the overall nutrient availability and cycling in marine habitats.

20.23.4 Biofilm Formation and Communication

Biofilms, which are complex microbial communities attached to surfaces, are prevalent in marine environments. Marine microbial secondary metabolites contribute to biofilm formation and maintenance. They can regulate biofilm formation, inhibit or promote the attachment of microorganisms to surfaces, and facilitate the communication and cooperation among biofilm-associated microbes (Qian et al., 2022). These compounds influence biofilm architecture, stability, and resilience.

20.23.5 Adaptation to Marine Conditions

Marine microbial secondary metabolites enable microorganisms to adapt to the challenging conditions of the marine environment. The production of specialized metabolites allows microorganisms to thrive in extreme conditions such as high pressures, temperature variations, salinity fluctuations, or exposure to UV radiation. These compounds protect against environmental stressors and support the survival and persistence of marine microorganisms.

20.23.6 Biotic Interactions

Marine microbial secondary metabolites can interact with other organisms in the marine ecosystem. Some compounds establish symbiotic relationships with marine organisms, such as corals, sponges, or algae, contributing to their health and survival. These interactions can involve the exchange of nutrients, signaling molecules, or the production of compounds that support the symbiotic partner's growth and function.

20.23.7 Biotechnological and Medical Applications

Marine microbial secondary metabolites have significant biotechnological and medical applications. Many compounds exhibit bioactive properties, such as antimicrobial, anticancer, antiviral, or anti-inflammatory activities, making them valuable sources for drug discovery and pharmaceutical development. The chemical diversity and novelty of marine microbial secondary metabolites offer potential for the development of new therapeutic agents, diagnostic tools, and biotechnological products (Taylor et al., 2007).

Understanding the importance and significance of marine microbial secondary metabolites contributes to our knowledge of marine ecosystems, their functioning, and the adaptation strategies of marine microorganisms. Exploring and harnessing these compounds have implications for environmental conservation, human health, and sustainable biotechnological advancements.

20.24 FUTURE OUTCOMES

The future outcomes of marine microbial secondary metabolites hold great promise and potential. These marine microbial secondary metabolites are likely to involve continued drug discoveries, alternative approaches to combat antibiotic resistance, advancements in industrial applications, therapeutic innovations, sustainable sourcing practices, and technological advancements. The potential of these compounds to address current and emerging challenges in health care, agriculture, and industry makes them a promising field of research and exploration.

So, the future perspectives of these compounds can be understood as follows:

- These compounds could lead to the development of new treatments for various diseases, including antibiotic-resistant infections, cancer, neurodegenerative disorders, and other unmet medical needs.
- Marine microbial secondary metabolites may also attract therapeutic innovations beyond traditional drug development. For example, they could serve as leads for the development of diagnostic tools, imaging agents, or targeted drug delivery systems.
- Marine microbial secondary metabolites offer a potential source of alternative antimicrobial agents. Future research may uncover compounds that possess novel modes of action, targeting resistant pathogens and providing effective alternatives to conventional antibiotics.
- Marine microbial secondary metabolites have shown promise in various industrial applications beyond pharmaceuticals.
- They have the potential to be utilized as biocontrol agents in agriculture, leading to more sustainable and environmentally friendly practices.
- Marine microbial communities and their secondary metabolite production improves, bioengineering techniques can be employed to enhance compound production or modify existing structures for improved efficacy. This could involve

genetic manipulation of microorganisms to optimize production yields or create novel compounds with desired properties.

- This could involve genetic manipulation of microorganisms to optimize production yields or create novel compounds with desired properties.
- As the current demand for marine microbial secondary metabolites grows, it is essential to ensure sustainable sourcing and conservation practices.
- Future outcomes may involve the development of responsible harvesting techniques, cultivation methods, and conservation strategies to protect marine ecosystems and biodiversity while accessing these valuable resources.
- The future outcomes of marine microbial secondary metabolites will depend on collaborative efforts between scientists, researchers, industries, and policymakers. Sharing knowledge, resources, and expertise across disciplines and regions can lead to accelerated discoveries and advancements.
- Moreover, the development of advanced technologies, such as high-throughput screening methods, metagenomics, and data analysis tools, will facilitate more efficient exploration and characterization of marine microbial secondary metabolites.

20.25 CONCLUSION

Marine microbial secondary metabolites are a rich source of bioactive compounds with applications in pharmaceuticals, agriculture, and biotechnology. Marine bacteria, fungi, and algae produce structurally diverse metabolites with antibiotic, anticancer, and antiviral properties. The discovery process includes sampling, microbial isolation, fermentation, extraction, purification, and bioactivity screening. These compounds often surpass terrestrial counterparts in potency and novelty.

Beyond medicine, they serve as biocontrol agents, nutraceuticals, and industrial enzymes. Challenges such as low yields, rediscovery of known molecules, and sustainable sourcing require advanced technologies and interdisciplinary approaches. Understanding their ecological roles aids in conservation and enhances their responsible utilization in various industries.

The discovery process involves sample collection, microbial isolation, fermentation, metabolite extraction, purification, and bioactivity screening. Marine-derived compounds have led to significant drug developments, such as antibiotics, anticancer, and antiviral agents, often exhibiting advantages over terrestrial alternatives. Beyond medicine, they are used in agriculture, nutraceuticals, enzymes, and bio surfactants.

Challenges in this field include remote sample collection, low compound yields, and frequent rediscovery of known molecules. Overcoming these requires interdisciplinary collaboration, advanced technologies, and sustainable practices. Marine microbes like *Aspergillus* and *Penicillium*, along with red algae, have yielded notable bioactive compounds. Analytical advancements such as mass spectrometry and NMR spectroscopy aid in characterizing these metabolites, while ecological studies help understand their roles in microbial interactions and chemical signaling, supporting sustainable resource utilization.

REFERENCES

Blunt, J.W., B.R. Copp, R.A. Keyzers, M.H.G. Munro, and M.R. Prinsep. 2012. Marine natural products. *Natural Product Reports* 29: 144–222. https://doi.org/10.1039/C2NP00090C.

Blunt, J.W., B.R. Copp, R.A. Keyzers, M.H.G. Munro, and M.R. Prinsep. 2014. Marine natural products. *Natural Product Reports* 31: 160–258. https://doi.org/10.1039/C3NP70117D.

Blunt, J.W., B.R. Copp, R.A. Keyzers, M.H. Munro, and M.R. Prinsep. 2016. Marine natural products. *Natural Product Reports* 33: 382–431. https://doi.org/10.1039/C5NP00156K.

Bull, A.T., J.E.M. Stach, A.C. Ward, and M. Goodfellow. 2005. Marine actinobacteria: Perspectives, challenges, future directions. *Antonie van Leeuwenhoek* 87: 65–79. https://doi.org/10.1007/s10482-004-6562-8.

Fenical W., and P.R. Jensen. 2006. Developing a new resource for drug discovery: Marine actinomycete bacteria. *Nature Chemical Biology* 2: 666–73. https://doi.org/10.1038/nchembio841.

Gupta, A., S. Jain, and N. Verma. 2022. *Industrial applications of soil microbes*, vol. 1. Bentham Science Publishers Pte Ltd. https://doi.org/10.2174/97898150399551220101.

Gupta, A., S. Jain, and N. Verma. 2024. *Industrial applications of soil microbes*, vol. 3. Bentham Science Publishers Pte Ltd. https://doi.org/10.2174/97898150797531240301.

Jain, S., A. Gupta, and N. Verma. 2023. *Industrial applications of soil microbes*, vol. 2. Bentham Science Publishers Pte Ltd. https://doi.org/10.2174/97898150502641230201.

Jain, S., A. Gupta, and N. Verma. 2024. *Industrial applications of soil microbes*, vol. 4. Bentham Science Publishers Pte Ltd. https://doi.org/10.2174/97898151249961240401.

Kim, S.K. 2017. *Marine OMICS: Principles and applications*. Boca Raton, FL: CRC Press.

Karthikeyan, A., A. Joseph, and B.G. Nair. 2022. Promising bioactive compounds from the marine environment and their potential effects on various diseases. *Journal of Genetic Engineering and Biotechnology* 20 (December): 14. https://doi.org/10.1186/s43141-021-00290-4.

König, G.M., and A.D. Wright. 1997. Sesquiterpene content of the antibacterial dichloromethane extract of the marine red alga *Laurencia obtusa*. *Planta Medica* 63: 186–87. https://doi.org/10.1055/s-2006-957643.

Mayer, A.M.S., A.D. Rodríguez, R.G.S. Berlinck, and N. Fusetani. 2010. Marine pharmacology in 2007–8: Marine compounds with antibacterial, anticoagulant, antifungal, anti-inflammatory, antimalarial, antiprotozoal, antituberculosis, and antiviral activities; affecting the immune and nervous system, and other miscellaneous mechanisms of action. *Comparative Biochemistry and Physiology Part C: Toxicology & Pharmacology* 153: 191–222. https://doi.org/10.1016/j.cbpc.2010.08.008.

Newman, D.J., and G.M. Cragg. 2016. Natural products as sources of new drugs from 1981 to 2014. *Journal of Natural Products* 79: 629–61. https://doi.org/10.1021/acs.jnatprod.5b01055.

Qian, P.Y., A. Cheng, R. Wang, and R. Zhang. 2022. Marine biofilms: Diversity, interactions and biofouling. *Nature Reviews Microbiology* 20: 671–84. https://doi.org/10.1038/s41579-022-00744-7

Rateb, M.E., and R. Ebel. 2011. Secondary metabolites of fungi from marine habitats. *Natural Product Reports* 28: 290–344. https://doi.org/10.1039/C0NP00061B.

Rédou V., M. Navarri, L. Meslet-Cladière, G. Barbier, and G. Burgaud. 2015. Species richness and adaptation of marine fungi from deep-subseafloor. *Applied and Environmental Microbiology* 81: 3571–83. https://doi.org/10.1128/AEM.04064-14.

Rosenberg, E., and O. Koren. 2006. Vibrios in coral health and disease. In *The biology of vibrios*, eds. F.L. Thompson, B. Austin, and J. Swings, 231–38. Washington, DC: ASM Press. https://doi.org/10.1128/9781555815714.

Scheuer, P.J. 1990. Some marine ecological phenomena: Chemical basis and biomedical potential. *Science* 248: 173–77. https://www.science.org/doi/10.1126/science.2183350.

Taylor, M.W., R. Radax, D. Steger, and M. Wagner. 2007. Sponge-associated microorganisms: Evolution, ecology, and biotechnological potential. *Microbiology and Molecular Biology Reviews* 71: 295–347. https://doi.org/10.1128/mmbr.00040-06.

Thawabteh, A.M., Z. Swaileh, M. Ammar, et al. 2023. Antifungal and antibacterial activities of isolated marine compounds. *Toxins* 15 (January): 93, 1–21. https://doi.org/10.3390/toxins15020093.

21 The Promise of Marine Microbiome

Exploring the Functional Dynamics of Secondary Metabolites from Marine Origin

Prabha Susan Philip
(Radiotracer Laboratory, College of Agriculture, Kerala Agricultural University, Vellanikkara, Kerala, India; prabha0825@gmail.com)

Sneha S. Nair
(Department of Food Safety, Government of Kerala, Changanacherry Circle, Kottayam, Kerala, India; snehasasidharan05@gmail.com)

21.1 INTRODUCTION

Microorganisms defined by a marine habitat, including the sea, ocean, brackish waters, *etc*., are referred to as *marine microorganisms*. They are diverse in nature and share space and food while building unique associations within the environment. The microbiome includes bacteria, fungi, viruses, protozoans, and algae. Marine microorganisms are estimated to make up about 70% to 90% of the ocean's biomass. The importance of these microbes lies in the effective roles they play in these habitats. They act as catalysts in all biogeochemical cycles, including carbon, nitrogen, and oxygen. They are the key players for the degradation of organic matter in the marine ecosystem, thereby maintaining the carbon cycle and also acting as agents for bioremediation (Gupta et al., 2022, 2024). The phototrophic microorganisms further outdo their role as oxygen producers by carrying out photosynthesis, releasing 50% of oxygen to the ecosystem. Marine microorganisms also play a vital role in marine food webs by occupying the critical bottom trophic level and ensuring a sustainable supply of food. The ability of these microbes to survive in environments such as high shear stress, high salinity, high light intensity, and low temperatures has resulted in the production of a large array of fascinating and structurally complex molecules through novel metabolic pathways present in them. These pathways can be exploited for biotechnology applications, including the manufacture of industrial products and novel bioactive compounds that may have application as pharmaceuticals (Jain et al., 2023, 2024).

21.2 MARINE MICROBIAL SPATIAL DIVERSITY

The marine environment is the largest part of the biosphere, with varying degrees of light, temperature, salinity, pressure, and available nutrients throughout its depth. Based on the proximity to the ocean surface, the marine ecosystem can be divided into pelagic and benthic regions. The pelagic region is further divided into neritic, epipelagic, meso, and bathypelagic regions (Hunter-Cevera et al., 2005).

The neritic zone is characterized by abundant nutrients and biological activity as it is close to land. The primary producers of this zone include passively drifting phytoplankton. The epipelagic/euphotic zone is the most important zone for photosynthesis. They are extremely diverse, varying from photosynthetic cyanobacteria to plant-like diatoms, but marine photosynthesis is dominated by microalgae, which, together with cyanobacteria, are collectively called phytoplankton (Parker et al., 2008).

The mesopelagic or "twilight" zone is the zone of increased hydrostatic pressure, diminished light, high inorganic nutrient concentrations, and episodic food supply. The insufficiency of light makes this zone a highly diverse and active community of viruses, bacteria, archaea, protists, zooplankton, and nekton (Koppelmann and Frost, 2008; Arístegui et al., 2009) and devoid of any phytoplankton. The zooplanktons and nektons are known to dominate this zone. The archaeal subclass Crenarchaeota has contributed up to 40% of picoplankton in the mesopelagic realm (Karner et al., 2001; Varela et al., 2008). For example, SAR11 ecotype II is the dominant contributor to the SAR 11 clade in the mesopelagic zone (Morris et al., 2005; Carlson et al., 2009). Viral abundance in mesopelagic waters is of the order 10^{10}–10^{12} viruses m^{-3} (Weinbauer et al., 2003) and does not always decrease exponentially with depth. Whereas prokaryotic, zooplankton, and micronekton biomass have shown an exponential decline with depth, from the euphotic to bathypelagic zones, by 2 to 4 orders of magnitude (Arístegui et al., 2009). However, microbial diversity has been shown in some cases to increase with depth (DeLong et al., 2006; Treusch et al., 2009). Hydrothermal vents are also a common feature in some areas of the bathypelagic zone. Organic carbon produced in the photic zone is transported

DOI: 10.1201/9781003408833-21

to the deeper layers, known as particulate organic carbon (POC) and dissolved organic carbon (DOC) determine the prokaryotic biomass in this zone, as the absence of light prevents any phototrophic life here. Archaea make up a larger portion of the total prokaryote cell abundance in this zone. Research has shown an abundance of heterotrophic nanoflagellates (HNF) and ciliates, which feed on the prokaryotic population in the bathypelagic zone (Nagata et al., 2010). Microbes occur as aggregates in these deep oceans with Actinobacteria, Alphaproteobacteria, Gammaproteobacteria, and Flavobacteriia (Bacteroidetes) dominating surface-related taxa in the bathypelagic zone, and Thaumarchaeota, Deltaproteobacteria, OM190 (Planctomycetes), and Planctomycetacia (Planctomycetes) dominating the endemic bathypelagic communities (Ruiz-González et al., 2020).

In the benthic zone, biomass is dominated by domains of bacteria, archaea, and unicellular eukaryotes, which play an important role in carbon cycling and nutrient regeneration (Middelburg et al., 2007) mainly through chemotrophy. These deep-sea microorganisms are adapted to extremes of temperature, pressure, and oxygen conditions and are relatively difficult to isolate. Recent metagenomics reveals that at least 2.9×10^{29} prokaryotes inhabit the first few meters of the benthic zone (Kallmeyer et al., 2021), suggesting their abundance in the deeper layers. Viruses are also a prominent population in the benthic region, as many studies have shown a positive correlation of viral abundance with an increase in prokaryotic population.

21.3 MARINE MICROBIAL PHYLOGENETIC DIVERSITY

The ocean is a home to many microbes, but the habitat attributes important to the microbes include concentrations of nutrients, light levels, pH, ultraviolet exposure and solar flux, temperature, oxygen saturation, and redox (Hunter-Cevera et al., 2005). Studies showed that the marine ecosystem provides a microhabitat to all three domains of life (archaea, bacteria, and eukarya) as well as viruses. The composition changes based on the abiotic conditions prevailing in the microhabitat and the ability of the microorganism to adapt to the same. In this section, we will discuss briefly about the phylogenetic diversity of microbes across the ocean.

21.3.1 Bacteria

Bacteria are presumably the most abundant microorganisms in the marine ecosystem. They play a key role in nutrient recycling and are major contributors to the food web. They are either free-living or symbiotic. The most abundant of them is the Proteobacteria.

21.3.1.1 Proteobacteria

Some 16S rRNA surveys have revealed the phylum Proteobacteria to be the most abundant phylum in all major habitats. They are Gram-negative bacteria with diverse metabolic activity. They are further divided into alpha-, beta-, gamma-, delta-, epsilon-, and zetaproteobacteria.

21.3.1.1.1 Alphaproteobacteria

Alpha proteobacteria mostly predominates the euphotic and mesopelagic zones and display a wide range of metabolic pathways. Members of the SAR clade dominate this group. Members of the order Pelagibacterales; *Pelagibacter ubique*, the first cultivated member of the SAR clade 11, is considered to be one of the most abundant and smallest organisms in the ocean. Characterized by a complete biosynthetic pathway for all 20 amino acids and all but a few cofactors, this alphaproteobacteria act as heterotrophic bacterioplanktons and break down dissolved organic matter, leading to carbon and nitrogen recycling (Rappé et al., 2002). Members of the order Caulobacterales, *Caulobacter crescentus*, *Algimonas*, *Hellea*, *Woodsholea*, and *Oceanicaulis*, are the other common inhabitants of the upper zones of the marine ecosystem. Their ability to live in oligotrophic habitats and the presence of distinctive stalks by which they attach firmly to algae, stones, thereby forming biofilm, make them the first to colonize any water body (Munn, 2019). *Spingomonas* is yet another alphaproteobacterium that can survive under oligotrophic conditions as well as highly polluted marshes, which makes them excellent candidates for bioremediation. They can also be exploited biotechnologically for the production of extracellular polymers such as sphingans (*e.g.*, gellan, welan, and rhamsan), which are used extensively in food and other industries (Glaeser and Kämpfer, 2014). Marine microbes are also known to fix nitrogen in this ecosystem. Plant-associated nitrogen fixer, *Rhodopseudomonas,* and chemoautotrophic nitrogen fixer, *Nitrobacter*, belonging to the alpha proteobacteria, regulate the nitrogen cycles in the marine ecosystem.

Pondering through the depths of the marine ecosystem, members of the group Rhizobiales, the pink-pigmented *Methylobacterium*, the type II methanotrophs *Methylocystis* and *Methylosinus*, and the facultative methylotrophs *Methylobrevis* are seen as common inhabitants of the marine sediments. They play an important role in the global cycling of the greenhouse gas methane (Knief, 2015). Alphaproteobacteria are also known to inhabit anaerobic niches of microbial mats, for example, the purple nonsulfur bacteria *Rhodospirillum* and the magnetotactic species *Magnetospirillum* are a few examples (Baldani et al., 2014).

21.3.1.1.2 Betaproteobacteria

β-proteobacteria are not widely found in the oligotrophic ocean but are abundant in freshwater habitats (Hiorns et al., 1997). In the marine ecosystem, they account for only ~5–10% of retrieved 16S rRNA gene sequences, many of which occur in very low abundance, as part of the biosphere (Munn, 2019). They either live as copiotrophic heterotrophs or as chemolithotrophs.

21.3.1.1.3 *Gammaproteobacteria*

In marine environmental gene surveys, gammaproteobacteria account for 25–30% of retrieved sequences, and bacteria in many of the clades have not yet been cultured (Munn, 2019). The order class includes Alteromonadales, which are mostly marine and comprise *Agarivorans*, *Alteromonas*, *Glaciecola*, *Marinobacter*, *Colwellia*, *Pseudoalteromonas*, *etc.* Most of them are ecologically relevant copiotrophs that inhabit almost every marine environment from surface waters to bathypelagic depths (Mikhailov et al., 2006). The bioluminescence-producing Vibrionaceae, including genus *Photobacterium phosphoreum*, *P. leiognathi*, *Aliivibrio fischeri*, *Vibrio campbellii*, and *V. harveyi*, can be used as biosensors in many biotechnological applications, including genetic engineering (Shaw et al., 1992). Other marine members of this class include *Oceanospirillum*, which can survive halophilic conditions and are used in bioremediation of oil spills; the purple sulfur bacteria (*Chromatium*, *Nitrosococcus*, *Phaeobacterium*, *Thiococcus*, and *Thiospirillum*); and the most diverse Thiotrichaceae, containing the sulfur oxidisers (*Beggiatoa*, *Thioploca*, *Thiothrix*, and *Thiomargarita*).

21.3.1.1.4 *Delta- and Epsilonproteobacteria*

The deltaproteobacteria are primarily sulfate and sulfur-reducing bacteria (*Desulfovibrio*), and bacterial predators (*Bdellovibrio*). Epsilonproteobacteria, by contrast, contain many species that oxidize the H_2S produced by the sulfate and sulfur reducers, contributing to productivity at hydrothermal vents (Brock et al., 2003).

21.3.1.2 Phylum Cyanobacteria

Cyanobacteria, commonly referred to as blue-green algae, are bacteria with the ability to carry out photosynthesis due to the presence of pigments chlorophyll a and phycobiliproteins. They are known to occupy several marine niches, including benthic and planktonic environments, including sea ice, shallow sediments, and microbial mats on the surface of inanimate objects, algae, or animal tissue. The majority of cyanobacteria include *Prochlorococcus*, which is the dominant phototroph in oligotrophic ocean waters, followed by *Synechococcus*, *Calothrix*, *Trichodesmium*, *Oscillatoria*, *etc.* In the marine environment, cyanobacteria are the major primary producers as members of the phytoplankton (Table 21.1).

21.3.2 Firmicutes

Firmicutes include Gram-positive bacteria, mostly which are characterized by the production of extremely resistant endospores, which allow them to resist high temperatures, irradiation, and desiccation, making them excellent candidates to survive under thermophilic and halophilic conditions such as hydrothermal vents and salterns. Bacteria such as *Bacillus* spp. and *Paenibacillus* spp. are known to produce a range of extracellular hydrolytic enzymes that degrade polysaccharides, proteins, lipids, and nucleic acids, which play an important role in organic decomposition in sediments. *Lactobacillus* and *Pediococcus* of this group contribute to probiotic applications in aquaculture. This group also hosts the pathogen *Clostridium* and the uncultivated giant bacteria *Epulopiscium fishelsoni*, which shares a symbiotic relationship with herbivorous surgeonfish (Munn, 2019).

21.3.3 Actinomycetales

Actinomycetes are a group of bacteria that predominates in terrestrial ecosystems. Due to their ability to decompose a variety of substrates and their antagonistic activity, they have become an active component of the marine microbial community. They play a major role in decomposition and heterotrophic nutrient cycling, production of antibiotics, melanins, enzymes, and anticancerous compounds (Grossart et al., 2004). Common marine members of the genera include *Dietzia*, *Rhodococcus*, *Streptomyces*, *Salinispora*, *Marinophilus*, *Solwaraspora*, *Salinibacterium*, *Aeromicrobium marinum*, *Williamsia maris*, and *Verrucosispora* (Magarvey et al., 2004; Bull et al., 2005; Jensen et al., 2005).

21.3.4 Other Common Genera

Other common marine microbial genera include *Aquifex*, *Hydrogenothermus*, and *Persephonella*, which are

TABLE 21.1
Characteristic Features of Common Marine Cyanobacteria

Genus	Habitat	Characteristic Features
Synechococcus and *Prochlorococcus*	All parts of the photic zone in the oligotrophic oceans with very low nutrient levels	The two most abundant photosynthetic organisms on Earth
Richelia intracellularis and *Calothrix rhizosoleniae*	Symbionts inside the cells of diatom	Live epiphytically on diatom
Crocosphaera watsonii	Association with phytoplankton	Bioluminescence
Trichodesmium	Surface of oceans	Massive blooms
Phormidium, *Oscillatoria*, and *Lyngbya*	Develop at interfaces between sediments and the overlying water	Biofilm

TABLE 21.2
Phylogenetic Diversity of Marine Archaebacteria

Class	Genus	Microhabitat
Euryarchaeota	*Methanococcus*, *Methanothermococcus*, *Methanoculleus*, *Methanospirillum*, *Methanosarcina*, and *Methanopyrus*	Marine sediments and hydrothermal vents
	Thermococcus, *Pyrococcus*, *Paleococcus*, *Archaeglobus*, and *Ferroglobus*	Marine sediments and hydrothermal vents
	Halorubrum, *Halobacterium*, *Halococcus*, *Haloarcula*, and *Haloquadratum*	Deep subterranean salt mines
Nanoarchaeota	*Nanoarchaeum equitans*	Hot vent of Iceland
Crenarchaeota	*Pyrodictium* and *Staphylothermus marinus*	Hydrothermal vents
Thaumarchaeota	*Nitrosopumilus*	Arctic sediments

Source: Goldin and Gorbach (1992).

extremely thermophilic and primary producers of marine hydrothermal vents. Bacteria in the phylum Chloroflexota are also found in a wide range of habitats, and members show diverse metabolism, including anoxygenic photosynthesis and aerobic or anaerobic heterotrophy (Munn, 2019). Members of the classes Sphingobacteria, Flavobacteriia, and Cytophaga are also an essential group of bacteria as they produce a wide range of hydrolytic extracellular enzymes, which are responsible for the degradation of carbohydrate polymers, chitin, and proteins (Munn, 2019).

21.3.5 Marine Archaebacteria

Archaebacteria are a group of bacteria (Table 21.2) that resemble eubacteria, but despite the morphological similarities, archaea possess many unique genes that code for certain enzymes, which help them survive extremes of habitat. These enzymes can be used in biotechnological industries through improved modern technologies due to their stability at extremes of pH, organic solvents, and high ionic strength.

21.3.6 Marine Fungi

Fungi are ubiquitous organisms present in every ecosystem, including marine habitats. They exhibit a tremendous potential in nutrient cycling and in the production of novel bioactive secondary metabolites such as steroids, polyketides, peptides, alkaloids, and polysaccharides, which are antimicrobial, anticancer, antiviral, antioxidant, and anti-inflammatory (Shin, 2020). Marine fungi can be obligate or facultative. Scientists for many years have pondered over the question of whether facultative fungal genes can be considered marine. Many facultative fungi have been identified to survive in their terrestrial ecosystem too. To be recognized as truly marine, a fungus must rely on the marine habitat for part or all of its life cycle. To be a part of the marine ecosystem, fungi need morphological and molecular adaptations. High levels of salinity and pressure can pose a threat to their survival. Marine fungi combat these conditions with the production of gels and slimes and through modified pathways, such as the high-osmolarity-glycerol (HOG) signaling pathway (Gladfelter et al., 2019).

Fungal diversity and abundance in the marine ecosystem are the most debated topics. Studies show that marine fungi represent less than 1% of the known species (Jones et al., 2019). They live as saprobes, mutualists, endobionts, and parasites in the marine ecosystem. Recent studies reveal marine fungi to be distributed in 769 genera, 226 families, 88 orders, 22 classes, and 7 phyla into *Aphelidiomycota*, *Ascomycota*, *Basidiomycota*, *Blastocladiomycota*, *Chytridiomycota*, *Mortierellomycota*, and *Mucoromycota* (Deshmukh et al., 2018).

Marine fungi have been isolated from different marine habitats such as deep sea (*Aspergillus*, *Cladosporium*, *Penicillium*, *Cadophora*, *Candida*, and *Pichia*), subsea floor (*Exophiala*, *Candida*, *Cryptococcus*, *Trichosporon*, *Rhodotorula*, *Rhodosporidium*, and *Malassezia*), hydrothermal vents (*Aureobasidium*, *Rhodotorula*, and *Exophiala*), coastal and ocean waters (*Candida*, *Debaryomyces*, *Cryptococcus*, *Trichosporon*, *Rhodotorula*, *Rhodosporidium*, *Malassezia*, *Phaeosphaeria*, *Aspergillus*, *Cladosporium*, *Chytridium*, *Rhizophydium*, *Lobulomyces*, and *Spizellomyces*), and polar systems (*Rhizophydium*, *Lobulomyces*, *Podochytrium*, *Rhizoclosmatium*, *Cladochytrium*, *Cyclopsomyces*, *Mesochytrium*, *Polychytrium*, *Cladosporium*, and *Penicillium)* (Munn, 2019; Grossart et al., 2004)

The need to explore the marine fungi lies in their ability to degrade complex hydrocarbons, and the production of a variety of antibacterial, antiviral, and anticancer agents (Deshmukh et al., 2918). Multiple studies reveal that fungi are a significant component of the marine microbiome; it is crucial to broaden this field of investigation into this underexplored marine microbiota.

21.4 FUNCTIONAL DYNAMICS OF MARINE MICROBIOME

The marine ecosystem provides an arsenal of functional dynamics (Figure 21.1). The vast marine habitat with varying degrees of environmental stress holds house for

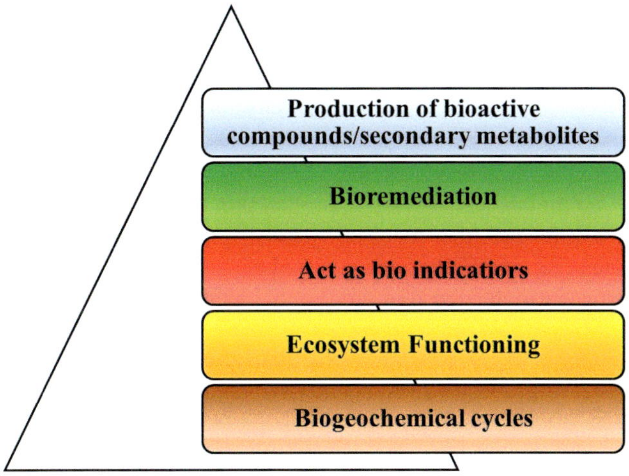

FIGURE 21.1 Illustration depicting the dynamic functions performed by marine microbiome.

an innumerable population of diverse microbes. These microbes have created a microhabitat or niche for themselves to survive the abiotic stress through unique metabolic pathways that can, in a way, lead to bioremediation. These unique pathways can become key steps in many biogeochemical cycles, such as carbon, nitrogen, and sulfur cycles, recycling the essential nutrients in the ocean and keeping its balance in the atmosphere. They act as microbial symbionts of many marine animals, such as sponges, corals, and many other marine invertebrates, forming obligate symbiotic associations with them (Taylor et al., 2007). In this chapter, we focus mainly on the secondary metabolites produced by the microbiota and their biological use.

21.5 FUNCTIONALLY DIVERSE SECONDARY METABOLITES

Metabolites are biomolecules produced by plants, fungi, and bacteria to perform physiological functions, and they can be categorized into primary and secondary metabolites based on their role and requirements. Primary metabolites are universally distributed organic compounds and are vital for the normal functioning of cells, for example, carbohydrates, lipids, and proteins. Secondary metabolites produced by plants, fungi, and bacteria are not directly involved in the growth, development, or reproduction of the organism. Secondary metabolites are heterogeneous low-molecular-weight organic compounds that are not directly involved in the primary metabolic activities of an organism but instead serve various specialized functions, such as defense against predators, attraction of pollinators, communication, competition, toxicity against other microbes, and pathogenicity, for example, alkaloids, steroids, phenolics, flavonoids. Unlike primary metabolites, which are essential for basic life processes, like respiration and photosynthesis, secondary metabolites serve various specialized functions, such as defense against predators, attraction of pollinators, and competition with other organisms.

The marine habitat is known for its immense diversity, offering an untapped reservoir of underexploited metabolite producers. This attracts researchers worldwide to discover more about the marine resources. Different classes of marine microbes are bacteria, fungi, and microalgae.

Bioactive compounds identified from marine bacteria have antibacterial, antiviral, antifungal, anti-quorum-sensing, and anticancer properties (Valliappan et al., 2014; Carroll et al., 2021).

21.5.1 Antimicrobial Products

The rich ecological diversity offered by marine habitats is a promising source of numerous novel bioactive compounds with antimicrobial properties. Owing to the untapped resources it possesses, the marine environment has been referred to as a goldmine of natural products. The complex marine environment is identified with different entities of microorganisms associated with marine plants and animals, along with those derived from marine sediment and water (Leal et al., 2013). The first antibiotic, penicillin, was discovered in the 1920s; since then, secondary metabolites from microorganisms are assumed to be a primary constituent of novel drug compounds with antimicrobial activities (Bhatnagar et al., 2010).

The six phyla of marine bacteria (Actinobacteria, Bacteroidetes, Cyanobacteria, Firmicutes, Planctomycetes, and Proteobacteria) seem to be the frequent primary sources of antimicrobial compounds among the various phyla investigated so far (Graça et al., 2015; Núñez-Montero and Bar rientos, 2018; Yasir, 2018). Actinomycetes, *Streptomyces* sp., produced a secondary metabolite, chlorocatechelin A, which can inhibit the growth of a wide range of bacterial and fungal pathogens. The secondary metabolites ambiguine-K and M isonitrile, isolated from the marine cyanobacterium *Fischerella ambigua* strain UTEX 1903, have exhibited intense antibacterial activity against *Mycobacterium tuberculosis*, with MIC values of 6.6 and 7.5 μM, respectively (Mo et al., 2009).

21.5.2 Anticancer Drugs

The prime reason leading to cancer is DNA damage caused by numerous oxidative lesions. Accumulation of oxidative lesions by the production of free radicals and other reactive oxygen species (ROS) over the ages risks the body to cancer. The majority of anticancer drugs available for ages are reported to be of natural origin. Marine microfloras, like bacteria, actinobacteria, cyanobacteria, fungi, and microalgae, altogether offer a great scope of discovery of novel drugs. Lack of a parallel ethnomedical history as compared to terrestrial habitats, together with the relative technical difficulties in collecting the marine floral samples, has led to underutilization of marine drugs as therapeutic drugs and henceforth remaining underrepresented in current pharmacopoeia (Jimeno et al, 2004). Most of the marine bioactive compounds with anticancer properties have demonstrated possible cytotoxic actions by inhibition of cell proliferation, their antimitotic behavior (antitubulin impacts), induced apoptosis, and inhibition of movement, invasion, or metastatic potential of cancer cells.

The common bioactive compounds having anticancer activity from these marine sources are alkaloids (staurosporine, ambigols, amycolactam, and marinoquinoline A), polyketides (chromone, engyodontiumones H, pestalpolyol I, hytidchromone A, B, C, and E), terpenes (meroterpenes, diterpene, and scopararane I), peptides (beauvericin, polymyxin B, and other non-ribosomal peptides), nucleosides (cytarabine, gemcitabine, and other nucleoside analogues), and carbohydrates (laminaran, alginic acid, and other sulfated polysaccharides). Marine bacteria are reported to be a source of numerous anticancer agents like bryostatins, discodermolide, eleutherobin, and sarcodictyin, *etc.* Several Research findings (Goldin and Gorbach, 1992) have reported the role of *Lactobacilli* in decreasing the induction of experimental colon cancer. Dietary supplements of *Lactobacilli* stimulate and modulate the mucosal immune system by reducing the production of proinflammatory cytokines through actions on NFκB pathways, resulting in increased production of anti-inflammatory cytokines IL-10 and host defense peptides such as β-defensin 2, enhancing IgA defenses and influencing dendritic cell maturation, as well as modulation of cell proliferation and apoptosis, through cell responses to short-chain fatty acids (Devine and Marsh, 2009).

Kahalalide F (KF), a depsipeptide compound believed to be synthesized by microbes associated with a mollusc, induces cytotoxicity and blocks the cell cycle in G1 phase in a p53-independent manner. *In vitro* studies showed that KF works against solid tumors with an interesting pattern of selectivity in prostate cancer cell lines. *In vivo* experiments demonstrate the activity of KF in breast and colon cancers.

Marine actinomycetes *Streptomyces* species also yield trioxacarcins D–F, in addition to the known trioxacarcins A–C. Gutingimycin, a highly polar trioxacarcin derivative from *Streptomyces* species, was isolated from the sediment of the Laguna de Terminos, Gulf of Mexico. Micromonosporaceae is a promising family of actinomycetes producing anticancer agents targeting proteasome function (Maskey et al., 2004). Thiocoraline, a novel bioactive depsipeptide isolated from *Micromonospora marina*, is selectively cytotoxic against lung and colon cancer cell lines and melanoma. It potentially inhibits RNA synthesis (Erba et al., 1999).

Studies have reported that the novel anticancer antibiotics chandrananimycins A (2), B (3), and C (4) with phenoxazin-3-one moiety have been isolated from the marine *Actinomadura* sp. (Maskey et al., 2004). These compounds inhibited the proliferation of cancer cell lines CCL HT29 (colon cancer cell), MEXF 514L (melanoma cells), LXFA 526L, LXFL 529L (lung cancer cells), CNCL SF268, LCL H460, MACL MCF-7 (breast cancer cells), and PRCL PC3M, RXF 631L (kidney tumor).

21.5.3 Antidiabetic Drugs

Diabetes is characterized by a lack of insulin secretion and insulin resistance, leading to two types of diabetes: type 1 diabetes, wherein beta cells of pancreatic islets, which produce the peptide hormone insulin, are destroyed by an autoimmune disorder, resulting in decreased insulin production, while type 2 diabetes is a situation that leads to reduced production of insulin in the body for proper functioning, or cells are not responding to the insulin, producing insulin resistance (Liu et al., 2014). The enzyme glucosidase produces catalytic action on glycosidic bonds in sugars. Particularly, β-glucosidase plays a key role in the degradation of polysaccharides and polysaccharides, and processing of glycoproteins and glycolipids. Marine bacteria like *Aka coralliphora* associated with marine sponge produce bioactive compounds having glucosidase inhibitor potential. Cyanobacteria species are also reported to have glycosidase inhibitors. Actinomycetes, *Streptomyces* sp., produce pyrostatins A and B with specific inhibitory activity against N-acetyl glucosaminidase.

21.5.4 Pigments

Microbial pigments derived from marine bacteria can be a promising approach to tackle the detrimental effects of these synthetic compounds. The marine bacteria produce various pigments like carotene, melanin, phenazine, pyrrole, violacein, and quinones. A range of pigmented compounds, including prodigiosin, astaxanthin, pyocyanin, melanin, and beta carotene, were found to be produced by bacterial species isolated from marine sediments or seawater, such as *Streptomyces* sp., *Pontibacter korlensis* sp., *Pseudomonas* sp., *Bacillus* sp., and *Vibrio* sp., respectively.

These pigments belong to a range of compound classes. For instance, carotenes are a subclass of carotenoids that have unsaturated polyhydrocarbon structures, prodiginines have a pyrrolyldipyrromethene core structure, tambjamines are alkaloid molecules, while violacein compounds are indole derivatives derived from tryptophan metabolism. Most of the pigments so produced by these microorganisms are reported to have industrial as well as therapeutic applications. The most valuable pigments in the international market, owing to their numerous industrial applications as nutritional supplements, antibiotics, and skin care, are beta-carotene, lutein, and astaxanthin. It is also being utilized in aquaculture industries to carry out the pigmentation of shrimps, trout, and salmon. At the industrial scale, astaxanthin production is accomplished using *Paracoccus* sp. (Venil et al., 2013; Ramesh et al., 2019).

The anticancer activity of many biopigments isolated from marine microbiota has been reported in scientific research studies, and a single pigment has been found to express anticancer activity against multiple cancerous cell lines. The efficiency of marine-derived BPs against several cancerous cell lines astaxanthin and 2-(p-hydroxybenzyl) prodigiosin (HBPG), with cytotoxicity against human breast cancer cell line (MCF-7), and human ovarian adenocarcinoma cell line, respectively; violacein, with antiproliferative activity in HeLa cells and also its efficacy as an anticancer drug in cervix cell carcinoma; tambjamines, showing apoptosis-inducing effects against glioblastoma cell line (SF-295), M14 melanoma cell line (MDA-MB-435), ileocecal colorectal adenocarcinoma cell line (HCT-8), and promyelocyticleukemia cells (HL-60); carotenoid pigments, exhibiting antitumor activity on esophageal squamous cancerous cells and many others, are already cited by researchers as an evidence for the role of biopigments in anticancer action (Fehér et al., 2008; Alem et al., 2020; Pinkerton et al., 2010; Afra et al., 2017).

21.5.5 Cosmetic Industry

The marine environment is under continuous exposure to sun rays, including UV A and UV B. As ultraviolet rays have high mutagenic power, many microbes residing in the upper marine habitat produce photo-protective compounds such as mycosporines and mycosporine-like amino acids (MAAs), which are now being exploited widely in the cosmetic industry. Common examples of microorganisms include *Pseudonocardia* sp. strain P1 (Actinomycetales) and *Micrococcus p. AK-334, which have been shown to produce MAA [58].* Microalgal-derived fucoxanthin, astaxanthin are other secondary metabolites that act as antioxidants and play a vital role in cosmetic creams and sunscreens. Squalene, yet another most commonly used antioxidant in the cosmetic industry, has been isolated from *Aurantiochytrium acetophilum* (Ganuza et al., 2019). *Fatty acids such as DHA isolated from Schizochytrium, Aurantiochytrium,* and *Ulkenia are a common ingredient of antiaging creams* (Cohen and Ratledge, 2015). Skin whitening agents such as kojic acid and N-acyl dehydrotyrosine derivative have been successfully isolated from *Alternaria* sp. and *Thalassotalea* sp., respectively, which have shown high potential in inhibiting tyrosinase activity (Deering et al., 2016). Marine organisms are also major producers of moisturizing molecules such as fatty acids and polysaccharides, which are commonly used in cosmetics. Algae-derived omega-6 polyunsaturated fatty acids, especially C-18 linoleic acid and gamma-linolenic acid, such as those found in marine microalgae, are common emulsifiers in moisturizing skin (Guillerme et al., 2017). Marine microbes are also a source of natural pigments, such as anthocyanins, carotenoids, and chlorophyll, which can be used as a natural source of color in nail polish, lipstick, and other cosmetics, thereby exposing yourself to less synthetic color and pigments.

21.6 CONCLUSION

Marine microbes play an important role in nature. In the marine ecosystem, they act as catalysts for major biogeochemical cycles, providing a continuous source of food and sustaining marine life. In addition, they act as an untapped source of secondary metabolites, vividly used for industrial applications. Despite their importance, very little is known about marine microbial diversity due to their unculturability. Traditionally, microbiology had relied on time-consuming techniques, such as isolation of pure culture and enrichment. With the advent of new microbial research fields, such as omics, PCR-based screening assays, combination biosynthesis, and ribosome engineering, used alone or in combination with classical methods, we can get a better understanding of the complex role microbes play in the ecosystem. Metabolic engineering methods that can create hybrid peptides are promising methods in the field of microbial biotechnology for synthesizing novel pharmaceutical products of marine origin.

REFERENCES

Afra, S., A. Makhdoumi, M.M. Matin, and J. Feizy. 2017. A novel red pigment from marine *Arthrobacter* sp. G20 with specific anticancer activity. *Journal of Applied Microbiology* 123: 1228–36. https://doi.org/10.1111/jam.13576.

Alem, D., J.J. Marizcurrena, V. Saravia, et al. 2020. Production and antiproliferative effect of violacein, a purple pigment produced by an Antarctic bacterial isolate. *World Journal Microbiology and Biotechnology* 36: 120. https://doi.org/10.1007/s11274-020-02893-4.

Arístegui, J., J.M. Gasol, C.M. Duarte, and G.J. Herndld. 2009. Microbial oceanography of the dark ocean's pelagic realm. *Limnology and Oceanography* 54: 1501–29. https://doi.org/10.4319/lo.2009.54.5.1501.

Baldani, J.I., S.S. Videira, K.R. dos Santos Teixeira, et al. 2014. The family Rhodospirillaceae. In *The prokaryotes*, eds. E. Rosenberg, E.F. DeLong, S. Lory, E. Stackebrandt, and F. Thompson, 533–618. Berlin, Heidelberg: Springer. https://doi.org/10.1007/978-3-642-30197-1_300.

Bhatnagar, I., and S.K. Kim. 2010. Immense essence of excellence: Marine microbial bioactive compounds. *Marine Drugs* 8: 2673–701. https://doi.org/10.3390/md8102673.

Brock, T.D., M.T. Madigan, J.M. Martinko, and J. Parker. 2003. *Brock biology of microorganisms*. Upper Saddle River, NJ: Prentice-Hall.

Bull, A.T., J.E.M. Stach, A.C. Ward, and M. Goodfellow. 2005. Marine actinobacteria: Perspectives, challenges, future directions. *Antonie van Leeuwenhoek* 87: 65–79. https://doi.org/10.1007/s10482-004-6562-8.

Carlson, C.A., R. Morris, R. Parsons, et al. 2009. Seasonal dynamics of SAR11 populations in the euphotic and mesopelagic zones of the northwestern Sargasso Sea. *The ISME Journal* 3: 283–95. https://doi.org/10.1038/ismej.2008.117.

Carroll, A.R., B.R. Copp, R.A. Davis, et al. 2021. Marine natural products. *Natural Product Report* 38: 362–413. https://doi.org/10.1039/D2NP00083K.

Cohen, Z., and C. Ratledge. 2015. *Single cell oils: Microbial and algal oils*. New York: AOCS Publishing. https://doi.org/10.4324/9781003040101.

Deering, R.W., J. Chen, J. Sun, et al. 2016. N-acyl dehydrotyrosines, tyrosinase inhibitors from the marine bacterium *Thalassotalea* sp. PP2–459. *Journal of Natural Products* 79: 447–50. https://doi.org/10.1021/acs.jnatprod.5b00972.

DeLong, E.F., C.M. Preston, T. Mincer, et al. 2006. Community genomics among stratified microbial assemblages in the ocean's interior. *Science* 311: 496–503. https://doi.org/10.1126/science.1120250.

Deshmukh, S.K., V. Prakash, and N. Ranjan. 2018. Marine fungi: A source of potential anticancer compounds. *Frontiers in Microbiology* 8 (January): 2536, 1–24. https://doi.org/10.3389/fmicb.2017.02536.

Devine, D.A., and P.D. Marsh. 2009. Prospects for the development of probiotics and prebiotics for oral applications. *Journal of Oral Microbiology* 1 (May): 1949, 1–11. https://doi.org/10.3402/jom.v1i0.1949.

Erba, E., D. Bergamaschi, S. Ronzoni, et al. 1999. Mode of action of thiocoraline, a natural marine compound with anti-tumour activity. *British Journal of Cancer* 80: 971–80. https://doi.org/10.1038/sj.bjc.6690451.

Fehér, D., R.S. Barlow, P.S. Lorenzo, and T.K. Hemscheidt. 2008. A 2-substituted prodiginine, 2-(p-hydroxybenzyl) prodigiosin, from *Pseudoalteromonas rubra*. *Journal of Natural Products* 71: 1970–72. https://doi.org/10.1021/np800493p.

Ganuza, E., S. Yang, M. Amezquita, et al. 2019. Genomics, biology and phylogeny *Aurantio chytriumacetophilum* sp. nov. (*Thraustro chytriaceae*), including first evidence of sexual reproduction. *Protist* 170: 209–32. https://doi.org/10.1016/j.protis.2019.02.004.

Gladfelter, A.S., T.Y. James, and A.S. Amend. 2019. Marine fungi. *Current Biology* 29: 191–95. https://doi.org/10.1016/j.cub.2019.02.009.

Glaeser, S.P., P. Kämpfer. 2014. The family *Sphingomonadaceae*. In *The prokaryotes*, eds. E. Rosenberg, E.F. DeLong, S. Lory, E. Stackebrandt, and F. Thompson, 641–707. Berlin, Heidelberg: Springer. https://doi.org/10.1007/978-3-642-30197-1_302.

Goldin, B.R., and S.L. Gorbach. 1992. Probiotics for humans. In *Probiotics*, ed. R. Fuller, 355–76. London, UK: Chapman and Hall. http://dx.doi.org/10.1007/978-94-011-2364-8.

Graça, A.P., F. Viana, J. Bondoso, et al. 2015. The antimicrobial activity of heterotrophic bacteria isolated from the marine sponge *Erylusdeficiens* (Astrophorida, Geodiidae). *Frontiers in Microbiology* 6 (May): 389, 1–14. https://doi.org/10.3389/fmicb.2015.00389.

Grossart, H.P., A. Schlingloff, M. Bernhard, et al. 2004. Antagonistic activity of bacteria isolated from organic aggregates of the German Wadden Sea. *FEMS Microbiology Ecology* 47: 387–96. https://doi.org/10.1016/S0168-6496(03)00305–2.

Guillerme, J.B., C. Couteau, and L. Coiffard. 2017. Applications for marine resources in cosmetics. *Cosmetics* 4 (September): 35, 1–15. https://doi.org/10.3390/cosmetics4030035.

Gupta, A., S. Jain, and N. Verma. 2022. *Industrial applications of soil microbes*, vol. 1. Bentham Science Publishers Pte Ltd. https://doi.org/10.2174/97898150399551220101.

Gupta, A., S. Jain, and N. Verma. 2024. *Industrial applications of soil microbes*, vol. 3. Bentham Science Publishers Pte Ltd. https://doi.org/10.2174/97898150797531240301.

Hiorns, W.D., B.A. Methé, S.A. Nierzwicki-Bauer, and J.P. Zehr. 1997. Bacterial diversity in Adirondack mountain lakes as revealed by 16S rRNA gene sequences. *Applied and Environmental Microbiology* 63: 2957–60. https://doi.org/10.1128/aem.63.7.2957-2960.1997.

Hunter-Cevera, J., D. Karl, and M. Buckley. 2005. *Marine microbial diversity: The key to earth's habitability*. Washington: American Academy of Microbiology. https://www.ncbi.nlm.nih.gov/books/NBK559439/pdf/Bookshelf_NBK559439.pdf.

Jain, S., A. Gupta, and N. Verma. 2023. *Industrial applications of soil microbes*, vol. 2. Bentham Science Publishers Pte Ltd. https://doi.org/10.2174/97898150502641230201.

Jain, S., A. Gupta, and N. Verma. 2024. *Industrial applications of soil microbes*, vol. 4. Switzerland: Bentham Science Publishers Pte Ltd. https://doi.org/10.2174/9789815124996124 0401.

Jensen, P.R., E. Gontang, C. Mafnas, et al. 2005. Marine actinomycete diversity from tropical Pacific Ocean sediments. *Environmental Microbiology* 7: 1039–48. https://doi.org/10.1111/j.1462-2920.2005.00785.x.

Jimeno, J., G. Faircloth, J.F. Sousa-Faro, et al. 2004. New marine derived anticancer therapeutics—a journey from the sea to clinical trials. *Marine Drugs* 2: 14–29. https://doi.org/10.3390/MD201014.

Jones, E.B.G., K.L. Pang, M.A. Abdel-Wahab, et al. 2019. An online resource for marine fungi. *Fungal Diversity* 96: 347–433. https://doi.org/10.1007/s13225-019-00426-5.

Kallmeyer, J., R. Pockalny, R.R. Adhikari, et al. 2021. Global distribution of microbial abundance and biomass in subseafloor sediment. *Proceedings of the National Academy of Sciences* 109: 16213–16. https://doi.org/10.1073/pnas. 1203849109.

Karner, M.B., E.F. DeLong, and D.M. Karl. 2001. Archaeal dominance in the mesopelagic zone of the Pacific Ocean. *Nature* 409: 507–10. https://doi.org/10.1038/35054051.

Knief, C. 2015. Diversity and habitat preferences of cultivated and uncultivated aerobic methanotrophic bacteria evaluated based on pmoA as molecular marker. *Frontiers in Microbiology* 6 (December): 1346, 1–38. https://doi.org/10.3389/fmicb.2015.01346.

Koppelmann, R., and J. Frost. 2008. The ecological role of zooplankton in the twilight and dark zones of the ocean. In *Biological oceanography research trends*, ed. L.P. Mertens, 67–130. Nova Science Publishers, Inc.

Leal, M.C., M.H. Munro, J.W. Blunt, et al. 2013. Biogeography and biodiscovery hotspots of macroalgal marine natural products. *Natural Product Reports* 30: 1380–90. https://doi.org/10.1039/C3NP70057G.

Liu, Y., Z. Gao, Q. Guo, et al. 2014. Anti-diabetic effects of CTB-APSL fusion protein in type 2 diabetic mice. *Marine Drugs* 12: 1512–29. https://doi.org/10.3390/md12031512.

Magarvey, N.A., J.M. Keller, V. Bernan, et al. 2004. Isolation and characterization of novel marine-derived actinomycete taxa rich in bioactive metabolites. *Applied Environmental Microbiology* 70: 7520–29. https://doi.org/10.1128/AEM.70.12.7520-7529.2004.

Maskey, R.P., M. Sevvana, I. Usón, E. Helmeke, H. Laatsch. 2004. Gutingimycin: A highly complex metabolite from a marine streptomycete. *Angewandte Chemie International Edition* 43: 1281–83. https://doi.org/10.1002/anie.200352312.

Middelburg, J.J., and F.J. Meysman. 2007. Burial at sea. Science 316: 1294–95. https://doi.org/10.1126/science.1144001.

Mikhailov, V.V., L.A. Romanenko, and E.P. Ivanova. 2006. The genus *Alteromonas* and related proteobacteria. In *The prokaryotes*, Vol. 6, eds. M. Dworkin, S. Falkow, E. Rosenberg, K.H. Schleifer, and E. Stackebrandt, 597–645. New York, NY: Springer. https://doi.org/10.1007/0-387-30746-X_20.

Mo, S., A. Krunic, G. Chlipala, and J. Orjala. 2009. Antimicrobial ambiguine isonitriles from the cyanobacterium *Fischerella ambigua*. *Journal of Natural Products* 72: 894–99. https://doi.org/10.1021/np800751j.

Morris, R.M., K.L. Vergin, J.-C. Cho, M.S. Rappé, C.A. Carlson, S.J. Giovannoni. 2005 Temporal and spatial response of bacterioplankton lineages to annual convective overturn at the Bermuda Atlantic time-series study site. *Limnology and Oceanography* 50: 1687–96. https://doi.org/10.4319/lo.2005.50.5.1687.

Munn, C.B. 2019. Marine microbiology: Ecology & applications. Boca Raton: CRC Press. https://doi.org/10.1201/9780429061042.

Nagata, T., C. Tamburini, J. Arístegui, et al. 2010. Emerging concepts on microbial processes in the bathypelagic ocean–ecology, biogeochemistry, and genomics. *Deep Sea Research Part II: Topical Studies in Oceanography* 57: 1519–36. https://doi.org/10.1016/j.dsr2.2010.02.019.

Núñez-Montero, K., and L. Barrientos. 2018. Advances in Antarctic research for antimicrobial discovery: A comprehensive narrative review of bacteria from Antarctic environments as potential sources of novel antibiotic compounds against human pathogens and microorganisms of industrial importance. *Antibiotics* 7 (October): 90, 1–23. https://doi.org/10.3390/antibiotics7040090.

Parker, M.S., T. Mock, and E.V. Armbrus. 2008. Genomic insights into marine microalgae. *Annual Review of Genetics* 42: 619–45. https://doi.org/10.1146/annurev.genet.42.110807.091417.

Pinkerton, D.M., M.G. Banwell, M.J. Garson, et al. 2010. Antimicrobial and cytotoxic activities of synthetically derived tambjamines C and E-J, BE-18591, and a related alkaloid from the marine bacterium *Pseudoalteromonas tunicata*. *Chemistry & Biodiversity* 7: 1311–24. https://doi.org/10.1002/cbdv.201000030.

Ramesh, C.H., N.V. Vinithkumar, and R. Kirubagaran. 2019 Marine pigmented bacteria: A prospective source of antibacterial compounds. *Journal of Natural Science, Biology and Medicine* 10: 104–13. https://doi.org/10.4103/jnsbm.JNSBM_201_18.

Rappé, M.S., S.A. Connon, K.L. Vergin, and S.J. Giovannoni. 2002 Cultivation of the ubiquitous SAR11 marine bacterioplankton clade. *Nature* 418: 630–33. https://doi.org/10.1038/nature00917.

Ruiz-González, C., M. Mestre, M. Estrada, et al. 2020 Major imprint of surface 573 plankton on deep ocean prokaryotic structure and activity. *Molecular Ecology* 29: 1820–38. https://doi.org/10.1111/mec.15454.

Shaw, J.J., F. Dane, D. Geiger, and J.W. Kloepper. 1992. Use of bioluminescence for detection of genetically engineered microorganisms released into the environment. *Applied and Environmental Microbiology* 58: 267–73. https://doi.org/10.1128/aem.58.1.267-273.1992.

Shin, H.J. 2020. Natural products from marine fungi. *Marine drugs* 18 (April): 230, 1–4. https://doi.org/10.3390/md18050230.

Taylor, M.W., R. Radax, D. Steger, and M. Wagner. 2007. Sponge-associated microorganisms: Evolution, ecology, and biotechnological potential. *Microbiology and Molecular Biology Reviews* 2: 295–347. https://doi.org/10.1128/MMBR.00040-06.

Treusch, A.H., K.L. Vergin, L.A. Finlay, et al. 2009. Seasonality and vertical structure of microbial communities in an ocean gyre. *The ISME Journal* 3: 1148–63. https://doi.org/10.1038/ismej.2009.60.

Valliappan, K., W. Sun, and Z. Li. 2014. Marine actinobacteria associated with marine organisms and their potentials in producing pharmaceutical natural products. *Applied Microbiology and Biotechnology* 98: 7365–77. https://doi.org/10.1007/s00253-014-5954-6.

Varela, M.M., H.M. Van Aken, E. Sintes, and G.J. Herndl. 2008. Latitudinal trends of *Crenarchaeota* and bacteria in the meso-and bathypelagic water masses of the Eastern North Atlantic. *Environmental Microbiology* 10: 110–24. https://doi.org/10.1111/j.1462-2920.2007.01437.x.

Venil, C.K., Z.A. Zakaria, and W.A. Ahmad. 2013. Bacterial pigments and their applications. *Process Biochemistry* 48: 1065–79. https://doi.org/10.1016/j.procbio.2013.06.006.

Weinbauer, M.G., I. Brettar, and M.G. Höfle. 2003. Lysogeny and virus-induced mortality of bacterioplankton in surface, deep, and anoxic marine waters. *Limnology and Oceanography* 48: 1457–65. https://doi.org/10.4319/lo.2003.48.4.1457.

Yasir, M. 2018. Analysis of bacterial communities and characterization of antimicrobial strains from cave microbiota. *Brazilian Journal of Microbiology* 49: 248–57. https://doi.org/10.1016/j.bjm.2017.08.005.

Index

V

X

Y

Z

For Product Safety Concerns and Information please contact our
EU representative GPSR@taylorandfrancis.com Taylor & Francis
Verlag GmbH, Kaufingerstraße 24, 80331 München, Germany